Applied Calculus

Sixth Edition, Hybrid

Stefan Waner
Hofstra University

Steven R. Costenoble
Hofstra University

Australia • Brazil • Japan • Korea • Mexico • Singapore • Spain • United Kingdom • United States

Applied Calculus, Sixth Edition, Hybrid
Stefan Waner, Steven R. Costenoble

Publisher: Richard Stratton

Development Editor: Jay Campbell

Editorial Assistant: Danielle Hallock

Media Editor: Andrew Coppola

Brand Manager: Gordon Lee

Marketing Communications Manager: Linda Yip

Content Project Manager: Alison Eigel Zade

Senior Art Director: Linda May

Print Buyer: Doug Bertke

Rights Acquisition Specialist: Shalice Shah-Caldwell

Production Service: MPS Limited

Text Designer: RHDG Design

Cover Designer: Chris Miller

Cover Image: William Dudziak (dudziak.com)

Compositor: MPS Limited

Library of Congress Control Number: 2012947818

Student Edition:
ISBN-13: 978-1-285-05640-1
ISBN-10: 1-285-05640-X

Brooks/Cole
20 Channel Center Street
Boston, MA 02210
USA

Cengage Learning is a leading provider of customized learning solutions with office locations around the globe, including Singapore, the United Kingdom, Australia, Mexico, Brazil and Japan. Locate your local office at **international.cengage.com/region**

Cengage Learning products are represented in Canada by Nelson Education, Ltd.

For your course and learning solutions, visit **www.cengage.com.**

Purchase any of our products at your local college store or at our preferred online store **www.cengagebrain.com.**

Instructors: Please visit **login.cengage.com** and log in to access instructor-specific resources.

Printed in Canada
1 2 3 4 5 6 7 16 15 14 13 12

Brief Contents

Contents

CHAPTER 3 Introduction to the Derivative 151

CHAPTER 4 Techniques of Differentiation with Applications 219

CHAPTER 5 Further Applications of the Derivative 277

CHAPTER 9 Trigonometric Models 483

Preface

Applied Calculus, sixth edition, hybrid, is intended for a one- or two-term course for students majoring in business, the social sciences, or the liberal arts. Like the earlier editions, the sixth edition, hybrid of *Applied Calculus* is designed to address the challenge of generating enthusiasm and mathematical sophistication in an audience that is often underprepared and lacks motivation for traditional mathematics courses. We meet this challenge by focusing on real-life applications and topics of current interest that students can relate to, by presenting mathematical concepts intuitively and thoroughly, and by employing a writing style that is informal, engaging, and occasionally even humorous.

The sixth edition, hybrid goes further than earlier editions in implementing support for a wide range of instructional paradigms: from traditional face-to-face courses to online distance learning courses, from settings incorporating little or no technology to courses taught in computerized classrooms, and from classes in which a single form of technology is used exclusively to those incorporating several technologies. We fully support three forms of technology in this text: TI-83/84 Plus graphing calculators, spreadsheets, and powerful online utilities we have created for the book. In particular, our comprehensive support for spreadsheet technology, both in the text and online, is highly relevant for students who are studying business and economics, where skill with spreadsheets may be vital to their future careers.

Key Features of the Hybrid Edition

Many mathematics courses are evolving into lecture-lab courses or into courses in which all assignments—homework and even tests—are delivered online. Furthermore, distance learning is growing rapidly. Instructors are increasingly faced with the challenge of integrating technology and teaching the concepts and skills of statistics.

In this hybrid text, as with the hardcover text, the goal is to provide instructors and students with the tools they can use to meet this challenge.

What does hybrid mean? A hybrid text involves an integration of several products and can be used best in a course with a strong blend of lecture time and online course work.

For this hybrid edition, the end-of-section exercises have been removed from the text and are available exclusively in Enhanced WebAssign, an easy-to-use online homework system. This slightly smaller book will be more manageable for the student who is spending his or her homework time on a computer.

Although this text has been tailored for instructors and students who use online homework, the focus of the expositions remains unchanged.

Our Approach to Pedagogy

Real-World Orientation We are confident that you will appreciate the diversity, breadth, and abundance of examples and exercises included in this edition. A large number of these are based on real, referenced data from business, economics, the life sciences, and the social sciences. Examples and exercises based on dated information have generally been replaced by more current versions; applications based on unique or historically interesting data have been kept.

Adapting real data for pedagogical use can be tricky; available data can be numerically complex, intimidating for students, or incomplete. We have modified and streamlined many of the real-world applications, rendering them as tractable as any "made-up" application. At the same time, we have been careful to strike a pedagogically sound balance between applications based on real data and more traditional "generic" applications. Thus, the density and selection of real data–based applications has been tailored to the pedagogical goals and appropriate difficulty level for each section.

Readability We would like students to read this book. We would like students to *enjoy* reading this book. Thus, we have written the book in a conversational and student-oriented style, and have made frequent use of question-and-answer dialogues to encourage the development of the student's mathematical curiosity and intuition. We hope that this text will give the student insight into how a mathematician develops and thinks about mathematical ideas and their applications.

Rigor We feel that mathematical rigor need not be antithetical to the kind of applied focus and conceptual approach that are earmarks of this book. We have worked hard to ensure that we are always mathematically honest without being unnecessarily formal. Sometimes we do this through the question-and-answer dialogs and sometimes through the "Before we go on..." discussions that follow examples, but always in manner designed to provoke the interest of the student.

Five Elements of Mathematical Pedagogy to Address Different Learning Styles The "Rule of Four" is a common theme in many texts. Implementing this approach, we discuss many of the central concepts **numerically, graphically,** and **algebraically** and clearly delineate these distinctions. The fourth element, **verbal communication** of mathematical concepts, is emphasized through our discussions on translating English sentences into mathematical statements and in our extensive Communication and Reasoning exercises at the end of each section. A fifth element, **interactivity,** is implemented through expanded use of question-and-answer dialogues but is seen most dramatically in the student Website. Using this resource, students can interact with the material in several ways: through interactive tutorials in the form of games, chapter summaries, and chapter review exercises, all in reference to concepts and examples covered in sections and with online utilities that automate a variety of tasks, from graphing to regression and matrix algebra.

Exercise Sets Our comprehensive collection of exercises provides a wealth of material that can be used to challenge students at almost every level of preparation and includes everything from straightforward drill exercises to interesting and rather challenging applications. The exercise sets have been carefully graded to move from straightforward basic exercises and exercises that are similar to examples in the text to more interesting and advanced ones, marked as "more advanced" for easy reference. There are also several much more difficult exercises, designated as "challenging." We have also included, in virtually every section of every chapter, interesting applications based on real data, Communication and Reasoning exercises that help students articulate mathematical concepts and recognize common errors, and exercises ideal for the use of technology.

Many of the scenarios used in application examples and exercises are revisited several times throughout the book. Thus, for instance, students will find themselves using a variety of techniques, from graphing through the use of derivatives and elasticity, to analyze the same application. Reusing scenarios and important functions provides unifying threads and shows students the complex texture of real-life problems.

New to This Edition

Content

- Chapter 1 (page 35): We now include, in Section 1.1, careful discussion of the common practice of representing functions as equations and vice versa; for instance, a cost equation like $C = 10x + 50$ can be thought of as defining a cost *function* $C(x) = 10x + 50$. Instead of rejecting this practice, we encourage the student to see this connection between functions and equations and to be able to switch from one interpretation to the other.

  Our discussion of functions and models in Section 1.2 now includes a careful discussion of the algebra of functions presented through the context of important applications rather than as an abstract concept. Thus, the student will see from the outset *why* we want to talk about sums, products, etc. of functions rather than simply *how* to manipulate them.
- Chapter 3 (page 151): Our discussion of limits now discusses extensively when, and why, substitution can be used to obtain a limit. We now also follow the usual convention of allowing only one-sided limits at endpoints of domains. This approach also applies to derivatives, where we now disallow derivatives at endpoints of domains, as is the normal convention.
- Chapter 4 (page 219): The closed-form formula for the derivative of $|x|$, introduced in Section 4.1, is now more fully integrated into the text, as is that for its antiderivative (in Chapter 6). (It is puzzling that it is not standard fare in other calculus books.)
- Chapter 6 (page 333): The sections on antiderivatives and substitution have been reorganized and streamlined and now include discussion of the closed-form antiderivative of $|x|$ and well as new exercises featuring absolute values.

  The definite integral is now introduced in the realistic context of the volume of oil released in an oil spill comparable in size to the *BP* 2011 Gulf oil spill.
- Chapter 8 (page 433): The discussion of level curves in Section 8.1 is now more extensive and includes added examples and exercises.
- **Case Studies:** A number of the Case Studies at the ends of the chapters have been extensively revised, using updated real data, and continue to reflect topics of current interest, such as spending on housing construction, modeling tax revenues, and controlling pollution.

Current Topics in the Applications

- We have added and updated numerous real data exercises and examples based on topics that are either of intense current interest or of general interest to contemporary students, including Facebook, XBoxes, iPhones, Androids, iPads, foreclosure rates, the housing crisis, subprime mortgages, the *BP* 2011 Gulf oil spill, and the U.S. stock market "flash crash" of May 6, 2010. (Also see the list, in the inside back cover, of the corporations we reference in the applications.)

Exercises

- We have expanded the chapter review exercise sets to be more representative of the material within the chapter. Note that all the applications in the chapter review exercises revolve around the fictitious online bookseller, *OHaganBooks.com,* and the various—often amusing—travails of *OHaganBooks.com* CEO John O'Hagan and his business associate Marjory Duffin.
- We have added many new conceptual Communication and Reasoning exercises, including many dealing with common student errors and misconceptions.

End-of-Chapter Technology Guides

- Our end-of-chapter detailed Technology Guides now discuss the use of spreadsheets in general rather than focusing exclusively on Microsoft® Excel, thus enabling readers to use any of the several alternatives now available, such as Google's online Google Sheets®, Open Office®, and Apple's Numbers®.

Continuing Features

Case Study **Checking up on Malthus**

Robert Nickelsberg/Getty Images

In 1798 Thomas R. Malthus (1766–1834) published an influential pamphlet, later expanded into a book, titled *An Essay on the Principle of Population As It Affects the Future Improvement of Society.* One of his main contentions was that population grows geometrically (exponentially) while the supply of resources such as food grows only arithmetically (linearly). This led him to the pessimistic conclusion that population would always reach the limits of subsistence and precipitate famine, war, and ill health, unless population could be checked by other means. He advocated "moral restraint," which includes the pattern of late marriage common in Western Europe at the time and now common in most developed countries and which leads to a lower reproduction rate.

- **Case Studies** Each chapter ends with a section entitled "Case Study," an extended application that uses and illustrates the central ideas of the chapter, focusing on the development of mathematical models appropriate to the topics. These applications are ideal for assignment as projects, and to this end we have included groups of exercises at the end of each.
- **Before We Go On** Most examples are followed by supplementary discussions, which may include a check on the answer, a discussion of the feasibility and significance of a solution, or an in-depth look at what the solution means.
- **Quick Examples** Most definition boxes include quick, straightforward examples that a student can use to solidify each new concept.
- **Question-and-Answer Dialogue** We frequently use informal question-and-answer dialogues that anticipate the kinds of questions that may occur to the student and also guide the student through the development of new concepts.

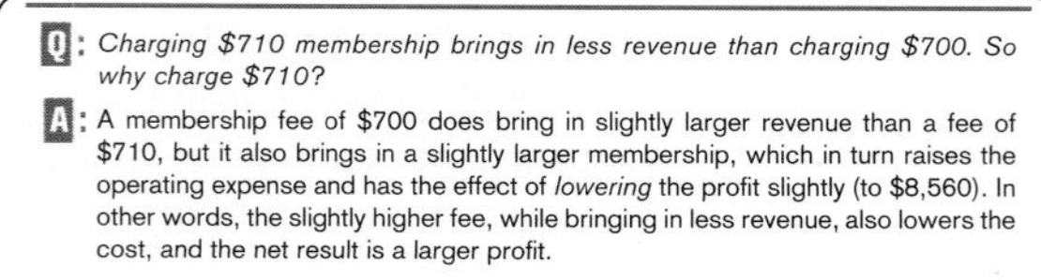

Q: *Charging \$710 membership brings in less revenue than charging \$700. So why charge \$710?*

A: A membership fee of \$700 does bring in slightly larger revenue than a fee of \$710, but it also brings in a slightly larger membership, which in turn raises the operating expense and has the effect of *lowering* the profit slightly (to \$8,560). In other words, the slightly higher fee, while bringing in less revenue, also lowers the cost, and the net result is a larger profit.

- **Marginal Technology Notes** We give brief marginal technology notes to outline the use of graphing calculator, spreadsheet, and Website technology in appropriate examples. When necessary, the reader is referred to more detailed discussion in the end-of-chapter Technology Guides.
- **End-of-Chapter Technology Guides** We continue to include detailed TI-83/84 Plus and Spreadsheet Guides at the end of each chapter. These Guides are referenced liberally in marginal technology notes at appropriate points in the chapter, so instructors and students can easily use this material or not, as they

prefer. Groups of exercises for which the use of technology is suggested or required appear throughout the exercise sets.

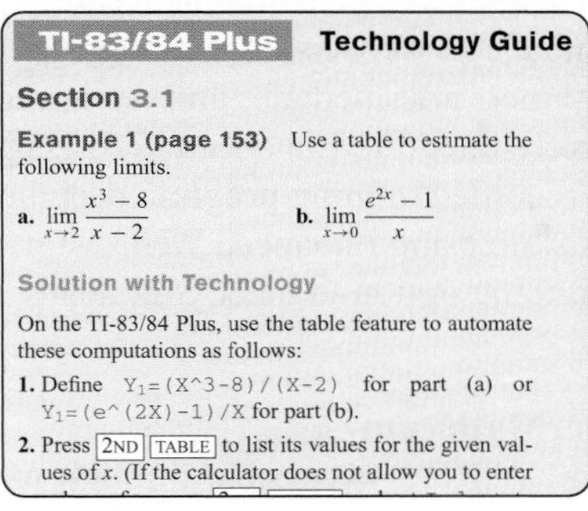

TI-83/84 Plus **Technology Guide**

Section 3.1

Example 1 (page 153) Use a table to estimate the following limits.

a. $\lim_{x \to 2} \frac{x^3 - 8}{x - 2}$ **b.** $\lim_{x \to 0} \frac{e^{2x} - 1}{x}$

Solution with Technology

On the TI-83/84 Plus, use the table feature to automate these computations as follows:

1. Define `Y1=(X^3-8)/(X-2)` for part (a) or `Y1=(e^(2X)-1)/X` for part (b).
2. Press [2ND] [TABLE] to list its values for the given values of x. (If the calculator does not allow you to enter

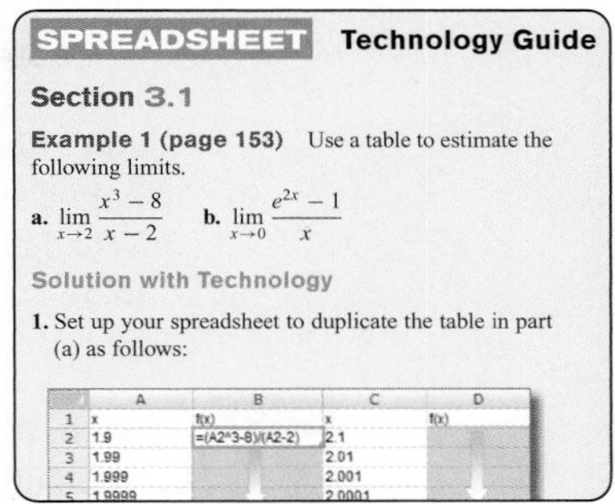

SPREADSHEET **Technology Guide**

Section 3.1

Example 1 (page 153) Use a table to estimate the following limits.

a. $\lim_{x \to 2} \frac{x^3 - 8}{x - 2}$ **b.** $\lim_{x \to 0} \frac{e^{2x} - 1}{x}$

Solution with Technology

1. Set up your spreadsheet to duplicate the table in part (a) as follows:

	A	B	C	D
1	x	f(x)	x	f(x)
2	1.9	=(A2^3-8)/(A2-2)	2.1	
3	1.99		2.01	
4	1.999		2.001	
5	1.9999		2.0001	

- **Communication and Reasoning Exercises for Writing and Discussion** These are exercises designed to broaden the student's grasp of the mathematical concepts and develop modeling skills. They include exercises in which the student is asked to provide his or her own examples to illustrate a point or design an application with a given solution. They also include "fill in the blank" type exercises, exercises that invite discussion and debate, and exercises in which the student must identify common errors. These exercises often have no single correct answer.

Supplemental Material

For Instructors and Students

Enhanced WebAssign®

Content

Exclusively from Cengage Learning, Enhanced WebAssign® combines the exceptional mathematics content in Waner and Costenoble's text with the most powerful online homework solution, WebAssign. Enhanced WebAssign engages students with immediate feedback, rich tutorial content, videos, animations, and an interactive eBook, helping students to develop a deeper conceptual understanding of the subject matter. The interactive eBook contains helpful search, highlighting, and note-taking features.

Instructors can build online assignments by selecting from thousands of text-specific problems, supplemented if desired with problems from any Cengage Learning textbook. Flexible assignment options give instructors the ability to choose how feedback and tutorial content is released to students as well as the ability to release assignments conditionally based on students' prerequisite assignment scores. Increase student engagement, improve course outcomes, and experience the superior service offered through CourseCare. Visit us at http://webassign.net/cengage or www.cengage.com/ewa to learn more.

Service

Your adoption of Enhanced WebAssign® includes CourseCare, Cengage Learning's industry leading service and training program designed to ensure that you have everything that you need to make the most of your use of Enhanced WebAssign. CourseCare provides one-on-one service, from finding the right solutions for your course to training and support. A team of Cengage representatives, including Digital

Solutions Managers and Coordinators as well as Service and Training Consultants, assists you every step of the way. For additional information about CourseCare, please visit www.cengage.com/coursecare.

Our Enhanced WebAssign training program provides a comprehensive curriculum of beginner, intermediate, and advanced sessions, designed to get you started and effectively integrate Enhanced WebAssign into your course. We offer a flexible online and recorded training program designed to accommodate your busy schedule. Whether you are using Enhanced WebAssign for the first time or an experienced user, there is a training option to meet your needs.

www.WanerMath.com
The authors' Website, accessible through www.WanerMath.com and linked within Enhanced WebAssign, has been evolving for more than a decade and has been receiving increasingly more recognition. Students, raised in an environment in which computers permeate both work and play, now use the Internet to engage with the material in an active way. The following features of the authors' Website are fully integrated with the text and can be used as a personalized study resource as well as a valuable teaching aid for instructors:

- *Interactive tutorials* on almost all topics, with guided exercises, which also can be used in classroom instruction or in distance learning courses
- *More challenging game versions of tutorials* with randomized questions that complement the traditional interactive tutorials and can be used as in-class quizzes
- *Detailed interactive chapter summaries* that review basic definitions and problem-solving techniques and can act as pre-test study tools
- *Downloadable Excel tutorials* keyed to examples given in the text
- *Online utilities for use in solving many of the technology-based application exercises.* The utilities, for instructor use in class and student use out of class, include a function grapher and evaluator that now also graphs derivatives and offers curve-fitting, regression tools, an interactive Riemann sum grapher with an improved numerical integrator, and a line entry calculator on the main page that does derivatives, expands polynomial expressions, does numerical integration and gives Taylor series.
- *Chapter true-false quizzes* with feedback for many incorrect answers
- *Supplemental topics* including interactive text and exercise sets for selected topics not found in the printed texts
- *Spanish versions* of chapter summaries, tutorials, game tutorials, and utilities

For Students

Student Solutions Manual *by Waner and Costenoble*
ISBN: 9781285085524
The student solutions manual provides worked-out solutions to the odd-numbered exercises in the text, plus problem-solving strategies and additional algebra steps and review for selected problems.

To access this and other course materials and companion resources, please visit **www.cengagebrain.com**. At the CengageBrain.com home page, search for the ISBN of your title (from the back cover of your book) using the search box at the top of the page. This will take you to the product page where free companion resources can be found.

Microsoft Excel Computer Laboratory Manual *by Anne D. Henriksen*
This laboratory manual uses Microsoft Excel to solve real-world problems in a variety of scientific, technical, and business disciplines. It provides hands-on experience to demonstrate for students that calculus is a valuable tool for solving practical, real-world problems while helping students increase their knowledge of Microsoft Excel. The manual is a set of self-contained computer exercises that are meant to be used over the course of a 15-week semester in a separate, 75-minute computer laboratory period. The weekly labs parallel the material in the text. Available at www.cengagebrain.com.

For Instructors

Complete Solution Manual *by Waner and Costenoble*
ISBN: 9781285085531
The instructor's solutions manual provides worked-out solutions to all of the exercises in the text.

Solution Builder *by Waner and Costenoble*
ISBN: 9781285085548
This time-saving resource offers fully worked instructor solutions to all exercises in the text in customizable online format. Adopting instructors can sign up for access at www.cengage.com/solutionbuilder.

PowerLecture™ with ExamView® computerized testing *by Waner and Costenoble*
ISBN: 9781285085579
This CD-ROM provides the instructor with dynamic media tools for teaching, including Microsoft® PowerPoint® lecture slides, figures from the book, and the Test Bank. You can create, deliver, and customize tests (both print and online) in minutes with ExamView® computerized testing, which includes Test Bank items in electronic format. In addition, you can easily build solution sets for homework or exams by linking to Solution Builder's online solutions manual.

Instructor's Edition
ISBN: 9781285056418

www.WanerMath.com
The Instructor's Resource Page at www.WanerMath.com features an expanded collection of instructor resources, including an updated corrections page, an expanding set of author-created teaching videos for use in distance learning courses, and a utility that automatically updates homework exercise sets from the fifth edition to the sixth.

Acknowledgments

This project would not have been possible without the contributions and suggestions of numerous colleagues, students, and friends. We are particularly grateful to our colleagues at Hofstra and elsewhere who used and gave us useful feedback on previous editions. We are also grateful to everyone at Cengage for their encouragement and guidance throughout the project. Specifically, we would like to thank Richard Stratton and Jay Campbell for their unflagging enthusiasm and Alison Eigel Zade for whipping the book into shape.

We would also like to thank our accuracy checker, Jerrold Grossman, and the numerous reviewers who provided many helpful suggestions that have shaped the development of this book.

Christopher Brown, *California Lutheran University*
Nathan Carlson, *California Lutheran University*
Scott Fallstrom, *University of Oregon*
Latrice Laughlin, *University of Alaska Fairbanks*
Michael Price, *University of Oregon*
Christopher Quarles, *Everett Community College*
Leela Rakesh, *Central Michigan University*
Tom Rosenwinkel, *Concordia University Texas*
Larry Taylor, *North Dakota State University*

Stefan Waner
Steven R. Costenoble

O

Precalculus Review

DreamPictures/Taxi/Getty Images

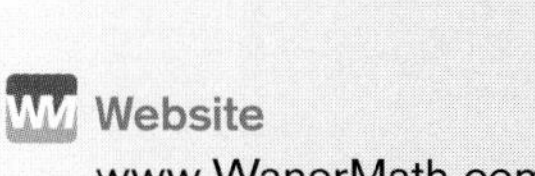

Website
www.WanerMath.com

- At the Website you will find section-by-section interactive tutorials for further study and practice.

Introduction

In this chapter we review some topics from algebra that you need to know to get the most out of this book. This chapter can be used either as a refresher course or as a reference.

There is one crucial fact you must always keep in mind: The letters used in algebraic expressions stand for numbers. All the rules of algebra are just facts about the arithmetic of numbers. If you are not sure whether some algebraic manipulation you are about to do is legitimate, try it first with numbers. If it doesn't work with numbers, it doesn't work.

0.1 Real Numbers

The **real numbers** are the numbers that can be written in decimal notation, including those that require an infinite decimal expansion. The set of real numbers includes all integers, positive and negative; all fractions; and the irrational numbers, those with decimal expansions that never repeat. Examples of irrational numbers are

$$\sqrt{2} = 1.414213562373\ldots$$

and

$$\pi = 3.141592653589\ldots$$

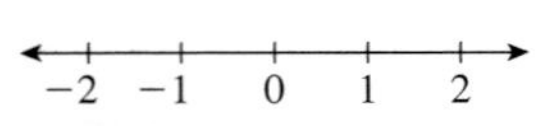

Figure 1

It is very useful to picture the real numbers as points on a line. As shown in Figure 1, larger numbers appear to the right, in the sense that if $a < b$ then the point corresponding to b is to the right of the one corresponding to a.

Intervals

Some subsets of the set of real numbers, called **intervals**, show up quite often and so we have a compact notation for them.

Interval Notation

Here is a list of types of intervals along with examples.

	Interval	*Description*	*Picture*	*Example*
Closed	$[a, b]$	Set of numbers x with $a \le x \le b$	a b (includes end points)	$[0, 10]$
Open	(a, b)	Set of numbers x with $a < x < b$	a b (excludes end points)	$(-1, 5)$
Half-Open	$(a, b]$	Set of numbers x with $a < x \le b$	a b	$(-3, 1]$
	$[a, b)$	Set of numbers x with $a \le x < b$	a b	$[0, 5)$

Infinite	$[a, +\infty)$	Set of numbers x with $a \le x$	a	$[10, +\infty)$
	$(a, +\infty)$	Set of numbers x with $a < x$	a	$(-3, +\infty)$
	$(-\infty, b]$	Set of numbers x with $x \le b$	b	$(-\infty, -3]$
	$(-\infty, b)$	Set of numbers x with $x < b$	b	$(-\infty, 10)$
	$(-\infty, +\infty)$	Set of all real numbers		$(-\infty, +\infty)$

Operations

There are five important operations on real numbers: addition, subtraction, multiplication, division, and exponentiation. "Exponentiation" means raising a real number to a power; for instance, $3^2 = 3 \cdot 3 = 9$; $2^3 = 2 \cdot 2 \cdot 2 = 8$.

A note on technology: Most graphing calculators and spreadsheets use an asterisk * for multiplication and a caret sign ^ for exponentiation. Thus, for instance, 3×5 is entered as `3*5`, $3x$ as `3*x`, and 3^2 as `3^2`.

When we write an expression involving two or more operations, like

$$2 \cdot 3 + 4$$

or

$$\frac{2 \cdot 3^2 - 5}{4 - (-1)}$$

we need to agree on the order in which to do the operations. Does $2 \cdot 3 + 4$ mean $(2 \cdot 3) + 4 = 10$ or $2 \cdot (3 + 4) = 14$? We all agree to use the following rules for the order in which we do the operations.

Standard Order of Operations

Parentheses and Fraction Bars First, calculate the values of all expressions inside parentheses or brackets, working from the innermost parentheses out, before using them in other operations. In a fraction, calculate the numerator and denominator separately before doing the division.

Quick Examples

1. $6(2 + [3 - 5] - 4) = 6(2 + (-2) - 4) = 6(-4) = -24$
2. $\dfrac{(4 - 2)}{3(-2 + 1)} = \dfrac{2}{3(-1)} = \dfrac{2}{-3} = -\dfrac{2}{3}$
3. $3/(2 + 4) = \dfrac{3}{2 + 4} = \dfrac{3}{6} = \dfrac{1}{2}$
4. $(x + 4x)/(y + 3y) = 5x/(4y)$

Exponents Next, perform exponentiation.

Quick Examples

1. $2 + 4^2 = 2 + 16 = 18$
2. $(2 + 4)^2 = 6^2 = 36$

Note the difference.

3. $2\left(\frac{3}{4-5}\right)^2 = 2\left(\frac{3}{-1}\right)^2 = 2(-3)^2 = 2 \times 9 = 18$
4. $2(1 + 1/10)^2 = 2(1.1)^2 = 2 \times 1.21 = 2.42$

Multiplication and Division Next, do all multiplications and divisions, from left to right.

Quick Examples

1. $2(3 - 5)/4 \cdot 2 = 2(-2)/4 \cdot 2$ — Parentheses first
   $= -4/4 \cdot 2$ — Left-most product
   $= -1 \cdot 2 = -2$ — Multiplications and divisions, left to right
2. $2(1 + 1/10)^2 \times 2/10 = 2(1.1)^2 \times 2/10$ — Parentheses first
   $= 2 \times 1.21 \times 2/10$ — Exponent
   $= 4.84/10 = 0.484$ — Multiplications and divisions, left to right
3. $4\frac{2(4-2)}{3(-2 \cdot 5)} = 4\frac{2(2)}{3(-10)} = 4\frac{4}{-30} = \frac{16}{-30} = -\frac{8}{15}$

Addition and Subtraction Last, do all additions and subtractions, from left to right.

Quick Examples

1. $2(3 - 5)^2 + 6 - 1 = 2(-2)^2 + 6 - 1 = 2(4) + 6 - 1$
   $= 8 + 6 - 1 = 13$
2. $\left(\frac{1}{2}\right)^2 - (-1)^2 + 4 = \frac{1}{4} - 1 + 4 = -\frac{3}{4} + 4 = \frac{13}{4}$
3. $3/2 + 4 = 1.5 + 4 = 5.5$
4. $3/(2 + 4) = 3/6 = 1/2 = 0.5$

Note the difference.

5. $4/2^2 + (4/2)^2 = 4/2^2 + 2^2 = 4/4 + 4 = 1 + 4 = 5$

Entering Formulas

Any good calculator or spreadsheet will respect the standard order of operations. However, we must be careful with division and exponentiation and use parentheses as necessary. The following table gives some examples of simple mathematical expressions and their equivalents in the functional format used in most graphing calculators, spreadsheets, and computer programs.

Mathematical Expression	*Formula*	*Comments*
$\frac{2}{3-x}$	`2/(3-x)`	Note the use of parentheses instead of the fraction bar. If we omit the parentheses, we get the expression shown next.
$\frac{2}{3}-x$	`2/3-x`	The calculator follows the usual order of operations.
$\frac{2}{3\times 5}$	`2/(3*5)`	Putting the denominator in parentheses ensures that the multiplication is carried out first. The asterisk is usually used for multiplication in graphing calculators and computers.
$\frac{2}{x}\times 5$	`(2/x)*5`	Putting the fraction in parentheses ensures that it is calculated first. Some calculators will interpret `2/3*5` as $\frac{2}{3\times 5}$, but `2/3(5)` as $\frac{2}{3}\times 5$.
$\frac{2-3}{4+5}$	`(2-3)/(4+5)`	Note once again the use of parentheses in place of the fraction bar.
2^3	`2^3`	The caret ^ is commonly used to denote exponentiation.
2^{3-x}	`2^(3-x)`	Be careful to use parentheses to tell the calculator where the exponent ends. Enclose the *entire exponent* in parentheses.
2^3-x	`2^3-x`	Without parentheses, the calculator will follow the usual order of operations: exponentiation and then subtraction.
3×2^{-4}	`3*2^(-4)`	On some calculators, the negation key is separate from the minus key.
$2^{-4\times 3}\times 5$	`2^(-4*3)*5`	Note once again how parentheses enclose the entire exponent.
$100\left(1+\frac{0.05}{12}\right)^{60}$	`100*(1+0.05/12)^60`	This is a typical calculation for compound interest.
$PV\left(1+\frac{r}{m}\right)^{mt}$	`PV*(1+r/m)^(m*t)`	This is the compound interest formula. PV is understood to be a single number (present value) and not the product of P and V (or else we would have used `P*V`).
$\frac{2^{3-2}\times 5}{y-x}$	`2^(3-2)*5/(y-x)` or `(2^(3-2)*5)/(y -x)`	Notice again the use of parentheses to hold the denominator together. We could also have enclosed the numerator in parentheses, although this is optional. (Why?)
$\frac{2^y+1}{2-4^{3x}}$	`(2^y+1)/(2-4^(3*x))`	Here, it is necessary to enclose both the numerator and the denominator in parentheses.
$2^y+\frac{1}{2}-4^{3x}$	`2^y+1/2-4^(3*x)`	This is the effect of leaving out the parentheses around the numerator and denominator in the previous expression.

Accuracy and Rounding

When we use a calculator or computer, the results of our calculations are often given to far more decimal places than are useful. For example, suppose we are told that a square has an area of 2.0 square feet and we are asked how long its sides are. Each side is the square root of the area, which the calculator tells us is

$$\sqrt{2} \approx 1.414213562$$

However, the measurement of 2.0 square feet is probably accurate to only two digits, so our estimate of the lengths of the sides can be no more accurate than that. Therefore, we round the answer to two digits:

$$\text{Length of one side} \approx 1.4 \text{ feet}$$

The digits that follow 1.4 are meaningless. The following guide makes these ideas more precise.

Significant Digits, Decimal Places, and Rounding

The number of **significant digits** in a decimal representation of a number is the number of digits that are not leading zeros after the decimal point (as in .0005) or trailing zeros before the decimal point (as in 5,400,000). We say that a value is **accurate to *n* significant digits** if only the first n significant digits are meaningful.

When to Round

After doing a computation in which all the quantities are accurate to no more than n significant digits, round the final result to n significant digits.

Quick Examples

1. 0.00067 has two significant digits. — The 000 before 67 are leading zeros.
2. 0.000670 has three significant digits. — The 0 after 67 is significant.
3. 5,400,000 has two or more significant digits. — We can't say how many of the zeros are trailing.[1]
4. 5,400,001 has 7 significant digits. — The string of zeros is not trailing.
5. Rounding 63,918 to three significant digits gives 63,900.
6. Rounding 63,958 to three significant digits gives 64,000.
7. $\pi = 3.141592653...$ $\frac{22}{7} = 3.142857142...$ Therefore, $\frac{22}{7}$ is an approximation of π that is accurate to only three significant digits (3.14).
8. $4.02(1 + 0.02)^{1.4} \approx 4.13$ — We rounded to three significant digits.

[1]If we obtained 5,400,000 by rounding 5,401,011, then it has three significant digits because the zero after the 4 is significant. On the other hand, if we obtained it by rounding 5,411,234, then it has only two significant digits. The use of scientific notation avoids this ambiguity: 5.40×10^6 (or 5.40 E6 on a calculator or computer) is accurate to three digits and 5.4×10^6 is accurate to two.

One more point, though: If, in a long calculation, you round the intermediate results, your final answer may be even less accurate than you think. As a general rule,

When calculating, don't round intermediate results. Rather, use the most accurate results obtainable or have your calculator or computer store them for you.

When you are done with the calculation, *then* round your answer to the appropriate number of digits of accuracy.

0.1 EXERCISES

Access end-of-section exercises online at **www.webassign.net**

0.2 Exponents and Radicals

In Section 1 we discussed exponentiation, or "raising to a power"; for example, $2^3 = 2 \cdot 2 \cdot 2$. In this section we discuss the algebra of exponentials more fully. First, we look at *integer* exponents: cases in which the powers are positive or negative whole numbers.

Integer Exponents

Positive Integer Exponents

If a is any real number and n is any positive integer, then by a^n we mean the quantity $a \cdot a \cdot \cdots \cdot a$ (n times); thus, $a^1 = a$, $a^2 = a \cdot a$, $a^5 = a \cdot a \cdot a \cdot a \cdot a$. In the expression a^n the number n is called the **exponent**, and the number a is called the **base**.

Quick Examples

$3^2 = 9$ $\qquad 2^3 = 8$

$0^{34} = 0$ $\qquad (-1)^5 = -1$

$10^3 = 1{,}000$ $\qquad 10^5 = 100{,}000$

Negative Integer Exponents

If a is any real number *other than zero* and n is any positive integer, then we define

$$a^{-n} = \frac{1}{a^n} = \frac{1}{a \cdot a \cdot \cdots \cdot a} \quad (n \text{ times})$$

Quick Examples

$2^{-3} = \frac{1}{2^3} = \frac{1}{8}$ $\quad$ $1^{-27} = \frac{1}{1^{27}} = 1$

$x^{-1} = \frac{1}{x^1} = \frac{1}{x}$ $\quad$ $(-3)^{-2} = \frac{1}{(-3)^2} = \frac{1}{9}$

$y^7y^{-2} = y^7\frac{1}{y^2} = y^5$ $\quad$ 0^{-2} is not defined

Zero Exponent

If a is any real number other than zero, then we define

$$a^0 = 1$$

Quick Examples

$3^0 = 1$ $\quad$ $1{,}000{,}000^0 = 1$

0^0 is not defined

When combining exponential expressions, we use the following identities.

Exponent Identity	Quick Examples
1. $a^m a^n = a^{m+n}$	$2^3 2^2 = 2^{3+2} = 2^5 = 32$ $x^3x^{-4} = x^{3-4} = x^{-1} = \frac{1}{x}$ $\frac{x^3}{x^{-2}} = x^3\frac{1}{x^{-2}} = x^3x^2 = x^5$
2. $\frac{a^m}{a^n} = a^{m-n}$ if $a \neq 0$	$\frac{4^3}{4^2} = 4^{3-2} = 4^1 = 4$ $\frac{x^3}{x^{-2}} = x^{3-(-2)} = x^5$ $\frac{3^2}{3^4} = 3^{2-4} = 3^{-2} = \frac{1}{9}$
3. $(a^n)^m = a^{nm}$	$(3^2)^2 = 3^4 = 81$ $(2^x)^2 = 2^{2x}$
4. $(ab)^n = a^nb^n$	$(4 \cdot 2)^2 = 4^2 2^2 = 64$ $(-2y)^4 = (-2)^4y^4 = 16y^4$
5. $\left(\frac{a}{b}\right)^n = \frac{a^n}{b^n}$ if $b \neq 0$	$\left(\frac{4}{3}\right)^2 = \frac{4^2}{3^2} = \frac{16}{9}$ $\left(\frac{x}{-y}\right)^3 = \frac{x^3}{(-y)^3} = -\frac{x^3}{y^3}$

Caution

- In the first two identities, the bases of the expressions must be the same. For example, the first gives $3^2 3^4 = 3^6$, but does *not* apply to $3^2 4^2$.
- People sometimes invent their own identities, such as $a^m + a^n = a^{m+n}$, which is wrong! (Try it with $a = m = n = 1$.) If you wind up with something like $2^3 + 2^4$, you are stuck with it; there are no identities around to simplify it further. (You can factor out 2^3, but whether or not that is a simplification depends on what you are going to do with the expression next.)

EXAMPLE 1 Combining the Identities

$$\begin{aligned}\frac{(x^2)^3}{x^3} &= \frac{x^6}{x^3} && \text{By (3)}\\ &= x^{6-3} && \text{By (2)}\\ &= x^3\end{aligned}$$

$$\begin{aligned}\frac{(x^4y)^3}{y} &= \frac{(x^4)^3y^3}{y} && \text{By (4)}\\ &= \frac{x^{12}y^3}{y} && \text{By (3)}\\ &= x^{12}y^{3-1} && \text{By (2)}\\ &= x^{12}y^2\end{aligned}$$

EXAMPLE 2 Eliminating Negative Exponents

Simplify the following and express the answer using no negative exponents.

a. $\dfrac{x^4y^{-3}}{x^5y^2}$ **b.** $\left(\dfrac{x^{-1}}{x^2y}\right)^5$

Solution

a. $\dfrac{x^4y^{-3}}{x^5y^2} = x^{4-5}y^{-3-2} = x^{-1}y^{-5} = \dfrac{1}{xy^5}$

b. $\left(\dfrac{x^{-1}}{x^2y}\right)^5 = \dfrac{(x^{-1})^5}{(x^2y)^5} = \dfrac{x^{-5}}{x^{10}y^5} = \dfrac{1}{x^{15}y^5}$

Radicals

If a is any non-negative real number, then its **square root** is the non-negative number whose square is a. For example, the square root of 16 is 4, because $4^2 = 16$. We write the square root of n as $\sqrt{n}$. (Roots are also referred to as **radicals**.) It is important to remember that $\sqrt{n}$ is never negative. Thus, for instance, $\sqrt{9}$ is 3, and not -3, even though $(-3)^2 = 9$. If we want to speak of the "negative square root" of 9, we write it as $-\sqrt{9} = -3$. If we want to write both square roots at once, we write $\pm\sqrt{9} = \pm 3$.

The **cube root** of a real number a is the number whose cube is a. The cube root of a is written as $\sqrt[3]{a}$ so that, for example, $\sqrt[3]{8} = 2$ (because $2^3 = 8$). Note that we can take the cube root of any number, positive, negative, or zero. For instance, the cube root of -8 is $\sqrt[3]{-8} = -2$ because $(-2)^3 = -8$. Unlike square roots, the cube root of a number may be negative. In fact, the cube root of a always has the same sign as a.

Higher roots are defined similarly. The **fourth root** of the *non-negative* number a is defined as the non-negative number whose fourth power is a, and written $\sqrt[4]{a}$. The **fifth root** of any number a is the number whose fifth power is a, and so on.

Note We cannot take an even-numbered root of a negative number, but we can take an odd-numbered root of any number. Even roots are always positive, whereas odd roots have the same sign as the number we start with. ■

EXAMPLE 3 ***n*th Roots**

$\sqrt{4} = 2$	Because $2^2 = 4$
$\sqrt{16} = 4$	Because $4^2 = 16$
$\sqrt{1} = 1$	Because $1^2 = 1$
If $x \geq 0$, then $\sqrt{x^2} = x$	Because $x^2 = x^2$
$\sqrt{2} \approx 1.414213562$	$\sqrt{2}$ is not a whole number.
$\sqrt{1+1} = \sqrt{2} \approx 1.414213562$	First add, then take the square root.[2]
$\sqrt{9+16} = \sqrt{25} = 5$	Contrast with $\sqrt{9} + \sqrt{16} = 3 + 4 = 7$.
$\dfrac{1}{\sqrt{2}} = \dfrac{\sqrt{2}}{2}$	Multiply top and bottom by $\sqrt{2}$.
$\sqrt[3]{27} = 3$	Because $3^3 = 27$
$\sqrt[3]{-64} = -4$	Because $(-4)^3 = -64$
$\sqrt[4]{16} = 2$	Because $2^4 = 16$
$\sqrt[4]{-16}$ is not defined	Even-numbered root of a negative number
$\sqrt[5]{-1} = -1$, since $(-1)^5 = -1$	Odd-numbered root of a negative number
$\sqrt[n]{-1} = -1$ if n is any odd number	

Q: *In the example we saw that $\sqrt{x^2} = x$ if x is non-negative. What happens if x is negative?*

A: If x is negative, then x^2 is positive, and so $\sqrt{x^2}$ is still defined as the non-negative number whose square is x^2. This number must be $|x|$, the **absolute value of *x***, which is the non-negative number with the same size as x. For instance, $|-3| = 3$, while $|3| = 3$, and $|0| = 0$. It follows that

$$\sqrt{x^2} = |x|$$

for every real number x, positive or negative. For instance,

$$\sqrt{(-3)^2} = \sqrt{9} = 3 = |-3|$$

and $$\sqrt{3^2} = \sqrt{9} = 3 = |3|.$$

In general, we find that

$$\sqrt[n]{x^n} = x \text{ if } n \text{ is odd, and } \sqrt[n]{x^n} = |x| \text{ if } n \text{ is even.}$$

[2]In general, $\sqrt{a+b}$ means the square root of the *quantity* $(a + b)$. The radical sign acts as a pair of parentheses or a fraction bar, telling us to evaluate what is inside before taking the root. (See the Caution on the next page.)

We use the following identities to evaluate radicals of products and quotients.

Radicals of Products and Quotients

If a and b are any real numbers (non-negative in the case of even-numbered roots), then

$$\sqrt[n]{ab} = \sqrt[n]{a}\sqrt[n]{b} \qquad \text{Radical of a product = Product of radicals}$$

$$\sqrt[n]{\frac{a}{b}} = \frac{\sqrt[n]{a}}{\sqrt[n]{b}} \quad \text{if } b \neq 0 \qquad \text{Radical of a quotient = Quotient of radicals}$$

Notes

- The first rule is similar to the rule $(a \cdot b)^2 = a^2b^2$ for the square of a product, and the second rule is similar to the rule $\left(\frac{a}{b}\right)^2 = \frac{a^2}{b^2}$ for the square of a quotient.
- ***Caution*** There is no corresponding identity for addition:

  $\sqrt{a+b}$ is *not* equal to $\sqrt{a} + \sqrt{b}$

(Consider $a = b = 1$, for example.) Equating these expressions is a common error, so be careful! ■

Quick Examples

1. $\sqrt{9 \cdot 4} = \sqrt{9}\sqrt{4} = 3 \times 2 = 6$ Alternatively, $\sqrt{9 \cdot 4} = \sqrt{36} = 6$
2. $\sqrt{\frac{9}{4}} = \frac{\sqrt{9}}{\sqrt{4}} = \frac{3}{2}$
3. $\frac{\sqrt{2}}{\sqrt{5}} = \frac{\sqrt{2}\sqrt{5}}{\sqrt{5}\sqrt{5}} = \frac{\sqrt{10}}{5}$
4. $\sqrt{4(3+13)} = \sqrt{4(16)} = \sqrt{4}\sqrt{16} = 2 \times 4 = 8$
5. $\sqrt[3]{-216} = \sqrt[3]{(-27)8} = \sqrt[3]{-27}\sqrt[3]{8} = (-3)2 = -6$
6. $\sqrt{x^3} = \sqrt{x^2 \cdot x} = \sqrt{x^2}\sqrt{x} = x\sqrt{x} \quad \text{if } x \geq 0$
7. $\sqrt{\frac{x^2+y^2}{z^2}} = \frac{\sqrt{x^2+y^2}}{\sqrt{z^2}} = \frac{\sqrt{x^2+y^2}}{|z|}$ We can't simplify the numerator any further.

Rational Exponents

We already know what we mean by expressions such as x^4 and a^{-6}. The next step is to make sense of *rational* exponents: exponents of the form p/q with p and q integers as in $a^{1/2}$ and $3^{-2/3}$.

Q: *What should we mean by $a^{1/2}$?*

A: The overriding concern here is that all the exponent identities should remain true. In this case the identity to look at is the one that says that $(a^m)^n = a^{mn}$. This identity tells us that

$$(a^{1/2})^2 = a^1 = a.$$

That is, $a^{1/2}$, when squared, gives us a. But that must mean that $a^{1/2}$ is the *square root* of a, or

$$a^{1/2} = \sqrt{a}.$$

A similar argument tells us that, if q is any positive whole number, then

$$a^{1/q} = \sqrt[q]{a}, \text{ the } q\text{th root of } a.$$

Notice that if a is negative, this makes sense only for q odd. To avoid this problem, we usually stick to positive a.

Q: *If p and q are integers (q positive), what should we mean by $a^{p/q}$?*

A: By the exponent identities, $a^{p/q}$ should equal both $(a^p)^{1/q}$ and $(a^{1/q})^p$. The first is the qth root of a^p, and the second is the pth power of $a^{1/q}$, which gives us the following.

Conversion Between Rational Exponents and Radicals

If a is any non-negative number, then

$$a^{p/q} = \sqrt[q]{a^p} = \left(\sqrt[q]{a}\right)^p.$$

↑ Using exponents ↑ ↑ Using radicals

In particular,

$$a^{1/q} = \sqrt[q]{a}, \text{ the } q\text{th root of } a.$$

Notes

- If a is negative, all of this makes sense only if q is odd.
- All of the exponent identities continue to work when we allow rational exponents p/q. In other words, we are free to use all the exponent identities even though the exponents are not integers. ■

Quick Examples

1. $4^{3/2} = (\sqrt{4})^3 = 2^3 = 8$
2. $8^{2/3} = (\sqrt[3]{8})^2 = 2^2 = 4$
3. $9^{-3/2} = \dfrac{1}{9^{3/2}} = \dfrac{1}{(\sqrt{9})^3} = \dfrac{1}{3^3} = \dfrac{1}{27}$
4. $\dfrac{\sqrt{3}}{\sqrt[3]{3}} = \dfrac{3^{1/2}}{3^{1/3}} = 3^{1/2-1/3} = 3^{1/6} = \sqrt[6]{3}$
5. $2^2 2^{7/2} = 2^2 2^{3+1/2} = 2^2 2^3 2^{1/2} = 2^5 2^{1/2} = 2^5\sqrt{2}$

EXAMPLE 4 Simplifying Algebraic Expressions

Simplify the following.

a. $\dfrac{(x^3)^{5/3}}{x^3}$ **b.** $\sqrt[4]{a^6}$ **c.** $\dfrac{(xy)^{-3}y^{-3/2}}{x^{-2}\sqrt{y}}$

Solution

a. $\dfrac{(x^3)^{5/3}}{x^3} = \dfrac{x^5}{x^3} = x^2$

b. $\sqrt[4]{a^6} = a^{6/4} = a^{3/2} = a \cdot a^{1/2} = a\sqrt{a}$

c. $\dfrac{(xy)^{-3}y^{-3/2}}{x^{-2}\sqrt{y}} = \dfrac{x^{-3}y^{-3}y^{-3/2}}{x^{-2}y^{1/2}} = \dfrac{1}{x^{-2+3}y^{1/2+3+3/2}} = \dfrac{1}{xy^5}$

Converting Between Rational, Radical, and Exponent Form

In calculus we must often convert algebraic expressions involving powers of x, such as $\dfrac{3}{2x^2}$, into expressions in which x does not appear in the denominator, such as $\dfrac{3}{2}x^{-2}$. Also, we must often convert expressions with radicals, such as $\dfrac{1}{\sqrt{1+x^2}}$, into expressions with no radicals and all powers in the numerator, such as $(1+x^2)^{-1/2}$. In these cases, we are converting from **rational form** or **radical form** to **exponent form.**

Rational Form

An expression is in **rational form** if it is written with positive exponents only.

Quick Examples

1. $\dfrac{2}{3x^2}$ is in rational form.
2. $\dfrac{2x^{-1}}{3}$ is not in rational form because the exponent of x is negative.
3. $\dfrac{x}{6} + \dfrac{6}{x}$ is in rational form.

Radical Form

An expression is in **radical form** if it is written with integer powers and roots only.

Quick Examples

1. $\dfrac{2}{5\sqrt[3]{x}} + \dfrac{2}{x}$ is in radical form.
2. $\dfrac{2x^{-1/3}}{5} + 2x^{-1}$ is not in radical form because $x^{-1/3}$ appears.
3. $\dfrac{1}{\sqrt{1+x^2}}$ is in radical form, but $(1+x^2)^{-1/2}$ is not.

Exponent Form

An expression is in **exponent form** if there are no radicals and all powers of unknowns occur in the numerator. We write such expressions as sums or differences of terms of the form

Constant × (Expression with $x)^p$ As in $\frac{1}{3}x^{-3/2}$

Quick Examples

1. $\frac{2}{3}x^4 - 3x^{-1/3}$ is in exponent form.
2. $\frac{x}{6} + \frac{6}{x}$ is not in exponent form because the second expression has x in the denominator.
3. $\sqrt[3]{x}$ is not in exponent form because it has a radical.
4. $(1+x^2)^{-1/2}$ is in exponent form, but $\frac{1}{\sqrt{1+x^2}}$ is not.

EXAMPLE 5 Converting from One Form to Another

Convert the following to rational form:

a. $\frac{1}{2}x^{-2} + \frac{4}{3}x^{-5}$ **b.** $\frac{2}{\sqrt{x}} - \frac{2}{x^{-4}}$

Convert the following to radical form:

c. $\frac{1}{2}x^{-1/2} + \frac{4}{3}x^{-5/4}$ **d.** $\frac{(3+x)^{-1/3}}{5}$

Convert the following to exponent form:

e. $\frac{3}{4x^2} - \frac{x}{6} + \frac{6}{x} + \frac{4}{3\sqrt{x}}$ **f.** $\frac{2}{(x+1)^2} - \frac{3}{4\sqrt[5]{2x-1}}$

Solution For (a) and (b), we eliminate negative exponents as we did in Example 2:

a. $\frac{1}{2}x^{-2} + \frac{4}{3}x^{-5} = \frac{1}{2}\cdot\frac{1}{x^2} + \frac{4}{3}\cdot\frac{1}{x^5} = \frac{1}{2x^2} + \frac{4}{3x^5}$

b. $\frac{2}{\sqrt{x}} - \frac{2}{x^{-4}} = \frac{2}{\sqrt{x}} - 2x^4$

For (c) and (d), we rewrite all terms with fractional exponents as radicals:

c. $\frac{1}{2}x^{-1/2} + \frac{4}{3}x^{-5/4} = \frac{1}{2}\cdot\frac{1}{x^{1/2}} + \frac{4}{3}\cdot\frac{1}{x^{5/4}}$

$= \frac{1}{2}\cdot\frac{1}{\sqrt{x}} + \frac{4}{3}\cdot\frac{1}{\sqrt[4]{x^5}} = \frac{1}{2\sqrt{x}} + \frac{4}{3\sqrt[4]{x^5}}$

d. $\frac{(3+x)^{-1/3}}{5} = \frac{1}{5(3+x)^{1/3}} = \frac{1}{5\sqrt[3]{3+x}}$

For (e) and (f), we eliminate any radicals and move all expressions involving x to the numerator:

e. $\frac{3}{4x^2} - \frac{x}{6} + \frac{6}{x} + \frac{4}{3\sqrt{x}} = \frac{3}{4}x^{-2} - \frac{1}{6}x + 6x^{-1} + \frac{4}{3x^{1/2}}$

$= \frac{3}{4}x^{-2} - \frac{1}{6}x + 6x^{-1} + \frac{4}{3}x^{-1/2}$

f. $\dfrac{2}{(x+1)^2} - \dfrac{3}{4\sqrt[5]{2x-1}} = 2(x+1)^{-2} - \dfrac{3}{4(2x-1)^{1/5}}$

$= 2(x+1)^{-2} - \dfrac{3}{4}(2x-1)^{-1/5}$

Solving Equations with Exponents

EXAMPLE 6 Solving Equations

Solve the following equations:

a. $x^3 + 8 = 0$ **b.** $x^2 - \dfrac{1}{2} = 0$ **c.** $x^{3/2} - 64 = 0$

Solution

a. Subtracting 8 from both sides gives $x^3 = -8$. Taking the cube root of both sides gives $x = -2$.

b. Adding $\frac{1}{2}$ to both sides gives $x^2 = \frac{1}{2}$. Thus, $x = \pm\sqrt{\frac{1}{2}} = \pm\frac{1}{\sqrt{2}}$.

c. Adding 64 to both sides gives $x^{3/2} = 64$. Taking the reciprocal (2/3) power of both sides gives

$$(x^{3/2})^{2/3} = 64^{2/3}$$

$$x^1 = \left(\sqrt[3]{64}\right)^2 = 4^2 = 16$$

so $x = 16$.

0.2 EXERCISES

Access end-of-section exercises online at **www.webassign.net** ENHANCED WebAssign

0.3 Multiplying and Factoring Algebraic Expressions

Multiplying Algebraic Expressions

Distributive Law

The **distributive law** for real numbers states that

$$a(b \pm c) = ab \pm ac$$

$$(a \pm b)c = ac \pm bc$$

for any real numbers a, b, and c.

Quick Examples

1. $2(x-3)$ is *not* equal to $2x-3$ but is equal to $2x-2(3)=2x-6$.
2. $x(x+1)=x^2+x$
3. $2x(3x-4)=6x^2-8x$
4. $(x-4)x^2=x^3-4x^2$
5. $(x+2)(x+3)=(x+2)x+(x+2)3$
   $=(x^2+2x)+(3x+6)=x^2+5x+6$
6. $(x+2)(x-3)=(x+2)x-(x+2)3$
   $=(x^2+2x)-(3x+6)=x^2-x-6$

There is a quicker way of expanding expressions like the last two, called the "FOIL" method (First, Outer, Inner, Last). Consider, for instance, the expression $(x+1)(x-2)$. The FOIL method says: Take the product of the first terms: $x \cdot x = x^2$, the product of the outer terms: $x \cdot (-2) = -2x$, the product of the inner terms: $1 \cdot x = x$, and the product of the last terms: $1 \cdot (-2) = -2$, and then add them all up, getting $x^2-2x+x-2=x^2-x-2$.

EXAMPLE 1 FOIL

a. $(x-2)(2x+5) = 2x^2 + 5x - 4x - 10 = 2x^2+x-10$

↑ First ↑ Outer ↑ Inner ↑ Last

b. $(x^2+1)(x-4)=x^3-4x^2+x-4$

c. $(a-b)(a+b)=a^2+ab-ab-b^2=a^2-b^2$

d. $(a+b)^2=(a+b)(a+b)=a^2+ab+ab+b^2=a^2+2ab+b^2$

e. $(a-b)^2=(a-b)(a-b)=a^2-ab-ab+b^2=a^2-2ab+b^2$

The last three are particularly important and are worth memorizing.

Special Formulas

$(a-b)(a+b)=a^2-b^2$	Difference of two squares
$(a+b)^2=a^2+2ab+b^2$	Square of a sum
$(a-b)^2=a^2-2ab+b^2$	Square of a difference

Quick Examples

1. $(2-x)(2+x)=4-x^2$
2. $(1+a)(1-a)=1-a^2$
3. $(x+3)^2=x^2+6x+9$
4. $(4-x)^2=16-8x+x^2$

Here are some longer examples that require the distributive law.

EXAMPLE 2 Multiplying Algebraic Expressions

a. $(x+1)(x^2+3x-4) = (x+1)x^2 + (x+1)3x - (x+1)4$
$= (x^3+x^2) + (3x^2+3x) - (4x+4)$
$= x^3 + 4x^2 - x - 4$

b. $\left(x^2 - \frac{1}{x} + 1\right)(2x+5) = \left(x^2 - \frac{1}{x} + 1\right)2x + \left(x^2 - \frac{1}{x} + 1\right)5$
$= (2x^3 - 2 + 2x) + \left(5x^2 - \frac{5}{x} + 5\right)$
$= 2x^3 + 5x^2 + 2x + 3 - \frac{5}{x}$

c. $(x-y)(x-y)(x-y) = (x^2 - 2xy + y^2)(x-y)$
$= (x^2 - 2xy + y^2)x - (x^2 - 2xy + y^2)y$
$= (x^3 - 2x^2y + xy^2) - (x^2y - 2xy^2 + y^3)$
$= x^3 - 3x^2y + 3xy^2 - y^3$

Factoring Algebraic Expressions

We can think of factoring as applying the distributive law in reverse—for example,

$$2x^2 + x = x(2x+1),$$

which can be checked by using the distributive law. Factoring is an art that you will learn with experience and the help of a few useful techniques.

Factoring Using a Common Factor

To use this technique, locate a **common factor**—a term that occurs as a factor in each of the expressions being added or subtracted (for example, x is a common factor in $2x^2 + x$, because it is a factor of both $2x^2$ and x). Once you have located a common factor, "factor it out" by applying the distributive law.

Quick Examples

1. $2x^3 - x^2 + x$ has x as a common factor, so
$2x^3 - x^2 + x = x(2x^2 - x + 1)$

2. $2x^2 + 4x$ has $2x$ as a common factor, so
$2x^2 + 4x = 2x(x+2)$

3. $2x^2y + xy^2 - x^2y^2$ has xy as a common factor, so
$2x^2y + xy^2 - x^2y^2 = xy(2x + y - xy)$

4. $(x^2+1)(x+2) - (x^2+1)(x+3)$ has x^2+1 as a common factor, so
$(x^2+1)(x+2) - (x^2+1)(x+3) = (x^2+1)[(x+2) - (x+3)]$
$= (x^2+1)(x + 2 - x - 3)$
$= (x^2+1)(-1) = -(x^2+1)$

5. $12x(x^2-1)^5(x^3+1)^6 + 18x^2(x^2-1)^6(x^3+1)^5$ has $6x(x^2-1)^5(x^3+1)^5$ as a common factor, so

$$\begin{aligned} &12x(x^2-1)^5(x^3+1)^6 + 18x^2(x^2-1)^6(x^3+1)^5 \\ &= 6x(x^2-1)^5(x^3+1)^5[2(x^3+1)+3x(x^2-1)] \\ &= 6x(x^2-1)^5(x^3+1)^5(2x^3+2+3x^3-3x) \\ &= 6x(x^2-1)^5(x^3+1)^5(5x^3-3x+2) \end{aligned}$$

We would also like to be able to reverse calculations such as $(x+2)(2x-5) = 2x^2-x-10$. That is, starting with the expression $2x^2-x-10$, we would like to **factor** it to get the expression $(x+2)(2x-5)$. An expression of the form ax^2+bx+c, where a, b, and c are real numbers, is called a **quadratic** expression in x. Thus, given a quadratic expression ax^2+bx+c, we would like to write it in the form $(dx+e)(fx+g)$ for some real numbers d, e, f, and g. There are some quadratics, such as x^2+x+1, that cannot be factored in this form at all. Here, we consider only quadratics that do factor, and in such a way that the numbers d, e, f, and g are integers (whole numbers; other cases are discussed in Section 5). The usual technique of factoring such quadratics is a "trial and error" approach.

Factoring Quadratics by Trial and Error

To factor the quadratic ax^2+bx+c, factor ax^2 as $(a_1x)(a_2x)$ (with a_1 positive) and c as c_1c_2, and then check whether or not $ax^2+bx+c = (a_1x \pm c_1)(a_2x \pm c_2)$. If not, try other factorizations of ax^2 and c.

Quick Examples

1. To factor x^2-6x+5, first factor x^2 as $(x)(x)$, and 5 as (5)(1):

$(x+5)(x+1) = x^2+6x+5.$ No good

$(x-5)(x-1) = x^2-6x+5.$ Desired factorization

2. To factor $x^2-4x-12$, first factor x^2 as $(x)(x)$, and -12 as $(1)(-12)$, $(2)(-6)$, or $(3)(-4)$. Trying them one by one gives

$(x+1)(x-12) = x^2-11x-12.$ No good

$(x-1)(x+12) = x^2+11x-12.$ No good

$(x+2)(x-6) = x^2-4x-12.$ Desired factorization

3. To factor $4x^2-25$, we can follow the above procedure, or recognize $4x^2-25$ as the difference of two squares:

$$4x^2-25 = (2x)^2-5^2 = (2x-5)(2x+5).$$

Note: Not all quadratic expressions factor. In Section 5 we look at a test that tells us whether or not a given quadratic factors.

Here are examples requiring either a little more work or a little more thought.

EXAMPLE 3 Factoring Quadratics

Factor the following: **a.** $4x^2 - 5x - 6$ **b.** $x^4 - 5x^2 + 6$

Solution

a. Possible factorizations of $4x^2$ are $(2x)(2x)$ or $(x)(4x)$. Possible factorizations of -6 are $(1)(-6)$, $(2)(-3)$. We now systematically try out all the possibilities until we come up with the correct one.

$(2x)(2x)$ and $(1)(-6)$:	$(2x+1)(2x-6) = 4x^2 - 10x - 6$	No good
$(2x)(2x)$ and $(2)(-3)$:	$(2x+2)(2x-3) = 4x^2 - 2x - 6$	No good
$(x)(4x)$ and $(1)(-6)$:	$(x+1)(4x-6) = 4x^2 - 2x - 6$	No good
$(x)(4x)$ and $(2)(-3)$:	$(x+2)(4x-3) = 4x^2 + 5x - 6$	Almost!
Change signs:	$(x-2)(4x+3) = 4x^2 - 5x - 6$	Correct

b. The expression $x^4 - 5x^2 + 6$ is not a quadratic, you say? Correct. It's a quartic (a fourth degree expression). However, it looks rather like a quadratic. In fact, it is quadratic *in* x^2, meaning that it is

$$(x^2)^2 - 5(x^2) + 6 = y^2 - 5y + 6$$

where $y = x^2$. The quadratic $y^2 - 5y + 6$ factors as

$$y^2 - 5y + 6 = (y-3)(y-2)$$

so

$$x^4 - 5x^2 + 6 = (x^2 - 3)(x^2 - 2)$$

This is a sometimes useful technique.

Our last example is here to remind you why we should want to factor polynomials in the first place. We shall return to this in Section 5.

EXAMPLE 4 Solving a Quadratic Equation by Factoring

Solve the equation $3x^2 + 4x - 4 = 0$.

Solution We first factor the left-hand side to get

$$(3x - 2)(x + 2) = 0.$$

Thus, the product of the two quantities $(3x - 2)$ and $(x + 2)$ is zero. Now, if a product of two numbers is zero, one of the two must be zero. In other words, either $3x - 2 = 0$, giving $x = \frac{2}{3}$, or $x + 2 = 0$, giving $x = -2$. Thus, there are two solutions: $x = \frac{2}{3}$ and $x = -2$.

0.3 EXERCISES

Access end-of-section exercises online at **www.webassign.net** ENHANCED WebAssign

0.4 Rational Expressions

Rational Expression

A **rational expression** is an algebraic expression of the form $\frac{P}{Q}$, where P and Q are simpler expressions (usually polynomials) and the denominator Q is not zero.

Quick Examples

1. $\dfrac{x^2-3x}{x}$ $\quad P = x^2-3x,\ Q = x$
2. $\dfrac{x+\frac{1}{x}+1}{2x^2y+1}$ $\quad P = x+\dfrac{1}{x}+1,\ Q = 2x^2y+1$
3. $3xy - x^2$ $\quad P = 3xy - x^2,\ Q = 1$

Algebra of Rational Expressions

We manipulate rational expressions in the same way that we manipulate fractions, using the following rules:

	Algebraic Rule	**Quick Example**
Product:	$\dfrac{P}{Q}\cdot\dfrac{R}{S}=\dfrac{PR}{QS}$	$\dfrac{x+1}{x}\cdot\dfrac{x-1}{2x+1}=\dfrac{(x+1)(x-1)}{x(2x+1)}=\dfrac{x^2-1}{2x^2+x}$
Sum:	$\dfrac{P}{Q}+\dfrac{R}{S}=\dfrac{PS+RQ}{QS}$	$\dfrac{2x-1}{3x+2}+\dfrac{1}{x}=\dfrac{(2x-1)x+1(3x+2)}{x(3x+2)}=\dfrac{2x^2+2x+2}{3x^2+2x}$
Difference:	$\dfrac{P}{Q}-\dfrac{R}{S}=\dfrac{PS-RQ}{QS}$	$\dfrac{x}{3x+2}-\dfrac{x-4}{x}=\dfrac{x^2-(x-4)(3x+2)}{x(3x+2)}=\dfrac{-2x^2+10x+8}{3x^2+2x}$
Reciprocal:	$\dfrac{1}{\left(\frac{P}{Q}\right)}=\dfrac{Q}{P}$	$\dfrac{1}{\left(\frac{2xy}{3x-1}\right)}=\dfrac{3x-1}{2xy}$
Quotient:	$\dfrac{\left(\frac{P}{Q}\right)}{\left(\frac{R}{S}\right)}=\dfrac{P}{Q}\cdot\dfrac{S}{R}=\dfrac{PS}{QR}$	$\dfrac{\left(\frac{x}{x-1}\right)}{\left(\frac{y-1}{y}\right)}=\dfrac{xy}{(x-1)(y-1)}=\dfrac{xy}{xy-x-y+1}$
Cancellation:	$\dfrac{P\cancel{R}}{Q\cancel{R}}=\dfrac{P}{Q}$	$\dfrac{(x-1)(xy+4)}{(x^2y-8)(x-1)}=\dfrac{xy+4}{x^2y-8}$

Caution Cancellation of summands is *invalid*. For instance,

$$\frac{\not{x} + (2xy^2 - y)}{\not{x} + 4y} = \frac{(2xy^2 - y)}{4y} \quad \text{✗ WRONG!} \quad \text{Do } \textit{not} \text{ cancel a summand.}$$

$$\frac{\not{x}(2xy^2 - y)}{4\not{x}y} = \frac{(2xy^2 - y)}{4y} \quad \text{✔ CORRECT} \quad \text{Do cancel a factor.}$$

Here are some examples that require several algebraic operations.

EXAMPLE 1 Simplifying Rational Expressions

a. $$\frac{\left(\frac{1}{x+y} - \frac{1}{x}\right)}{y} = \frac{\left(\frac{x-(x+y)}{x(x+y)}\right)}{y} = \frac{\left(\frac{-y}{x(x+y)}\right)}{y} = \frac{-y}{xy(x+y)} = -\frac{1}{x(x+y)}$$

b. $$\frac{(x+1)(x+2)^2 - (x+1)^2(x+2)}{(x+2)^4} = \frac{(x+1)(x+2)[(x+2)-(x+1)]}{(x+2)^4}$$

$$= \frac{(x+1)(x+2)(x+2-x-1)}{(x+2)^4} = \frac{(x+1)(x+2)}{(x+2)^4} = \frac{x+1}{(x+2)^3}$$

c. $$\frac{2x\sqrt{x+1} - \frac{x^2}{\sqrt{x+1}}}{x+1} = \frac{\left(\frac{2x\left(\sqrt{x+1}\right)^2 - x^2}{\sqrt{x+1}}\right)}{x+1} = \frac{2x(x+1) - x^2}{(x+1)\sqrt{x+1}}$$

$$= \frac{2x^2 + 2x - x^2}{(x+1)\sqrt{x+1}} = \frac{x^2 + 2x}{\sqrt{(x+1)^3}} = \frac{x(x+2)}{\sqrt{(x+1)^3}}$$

0.4 EXERCISES

Access end-of-section exercises online at **www.webassign.net** ENHANCED WebAssign

0.5 Solving Polynomial Equations

Polynomial Equation

A **polynomial equation** in one unknown is an equation that can be written in the form

$$ax^n + bx^{n-1} + \cdots + rx + s = 0$$

where $a, b, \ldots, r$, and s are constants.

We call the largest exponent of x appearing in a nonzero term of a polynomial the **degree** of that polynomial.

Quick Examples

1. $3x + 1 = 0$ has degree 1 because the largest power of x that occurs is $x = x^1$. Degree 1 equations are called **linear** equations.
2. $x^2 - x - 1 = 0$ has degree 2 because the largest power of x that occurs is x^2. Degree 2 equations are also called **quadratic equations**, or just **quadratics**.
3. $x^3 = 2x^2 + 1$ is a degree 3 polynomial (or **cubic**) in disguise. It can be rewritten as $x^3 - 2x^2 - 1 = 0$, which is in the standard form for a degree 3 equation.
4. $x^4 - x = 0$ has degree 4. It is called a **quartic**.

Now comes the question: How do we solve these equations for x? This question was asked by mathematicians as early as 1600 BCE. Let's look at these equations one degree at a time.

Solution of Linear Equations

By definition, a linear equation can be written in the form

$$ax + b = 0. \qquad \text{a and b are fixed numbers with $a \neq 0$.}$$

Solving this is a nice mental exercise: Subtract b from both sides and then divide by a, getting $x = -b/a$. Don't bother memorizing this formula; just go ahead and solve linear equations as they arise. If you feel you need practice, complete the exercises for this section at www.webassign.net. (You can access WebAssign with the access code that came with this book.)

Solution of Quadratic Equations

By definition, a quadratic equation has the form

$$ax^2 + bx + c = 0. \qquad \text{a, b, and c are fixed numbers and $a \neq 0$.}$$ [3]

The solutions of this equation are also called the **roots** of $ax^2 + bx + c$. We're assuming that you saw quadratic equations somewhere in high school but may be a little hazy about the details of their solution. There are two ways of solving these equations—one works sometimes, and the other works every time.

Solving Quadratic Equations by Factoring (works sometimes)

If we can factor[4] a quadratic equation $ax^2 + bx + c = 0$, we can solve the equation by setting each factor equal to zero.

[3] What happens if $a = 0$?

[4] See the section on factoring for a review of how to factor quadratics.

Quick Examples

1. $x^2 + 7x + 10 = 0$
   $(x + 5)(x + 2) = 0$ — Factor the left-hand side.
   $x + 5 = 0$ or $x + 2 = 0$ — If a product is zero, one or both factors is zero.
   Solutions: $x = -5$ and $x = -2$
2. $2x^2 - 5x - 12 = 0$
   $(2x + 3)(x - 4) = 0$ — Factor the left-hand side.
   $2x + 3 = 0$ or $x - 4 = 0$
   Solutions: $x = -3/2$ and $x = 4$

Test for Factoring

The quadratic $ax^2 + bx + c$, with a, b, and c being integers (whole numbers), factors into an expression of the form $(rx + s)(tx + u)$ with r, s, t, and u integers precisely when the quantity $b^2 - 4ac$ is a perfect square. (That is, it is the square of an integer.) If this happens, we say that the quadratic **factors over the integers**.

Quick Examples

1. $x^2 + x + 1$ has $a = 1$, $b = 1$, and $c = 1$, so $b^2 - 4ac = -3$, which is not a perfect square. Therefore, this quadratic does not factor over the integers.
2. $2x^2 - 5x - 12$ has $a = 2$, $b = -5$, and $c = -12$, so $b^2 - 4ac = 121$. Because $121 = 11^2$, this quadratic does factor over the integers. (We factored it above.)

Solving Quadratic Equations with the Quadratic Formula (works every time)

The solutions of the general quadratic $ax^2 + bx + c = 0$ $(a \neq 0)$ are given by

$$x = \frac{-b \pm \sqrt{b^2 - 4ac}}{2a}.$$

We call the quantity $\Delta = b^2 - 4ac$ the **discriminant** of the quadratic (Δ is the Greek letter delta), and we have the following general rules:

- If Δ is positive, there are two distinct real solutions.
- If Δ is zero, there is only one real solution: $x = -\dfrac{b}{2a}$. (Why?)
- If Δ is negative, there are no real solutions.

Quick Examples

1. $2x^2 - 5x - 12 = 0$ has $a = 2$, $b = -5$, and $c = -12$.

$$x = \frac{-b \pm \sqrt{b^2 - 4ac}}{2a} = \frac{5 \pm \sqrt{25 + 96}}{4} = \frac{5 \pm \sqrt{121}}{4} = \frac{5 \pm 11}{4}$$

$$= \frac{16}{4} \text{ or } -\frac{6}{4} = 4 \text{ or } -3/2$$

Δ is positive in this example.

2. $4x^2 = 12x - 9$ can be rewritten as $4x^2 - 12x + 9 = 0$, which has $a = 4$, $b = -12$, and $c = 9$.

$$x = \frac{-b \pm \sqrt{b^2 - 4ac}}{2a} = \frac{12 \pm \sqrt{144 - 144}}{8} = \frac{12 \pm 0}{8} = \frac{12}{8} = \frac{3}{2}$$

Δ is zero in this example.

3. $x^2 + 2x - 1 = 0$ has $a = 1$, $b = 2$, and $c = -1$.

$$x = \frac{-b \pm \sqrt{b^2 - 4ac}}{2a} = \frac{-2 \pm \sqrt{8}}{2} = \frac{-2 \pm 2\sqrt{2}}{2} = -1 \pm \sqrt{2}$$

The two solutions are $x = -1 + \sqrt{2} = 0.414\ldots$ and $x = -1 - \sqrt{2} = -2.414\ldots$

Δ is positive in this example.

4. $x^2 + x + 1 = 0$ has $a = 1$, $b = 1$, and $c = 1$. Because $\Delta = -3$ is negative, there are no real solutions.

Δ is negative in this example.

Q: *This is all very useful, but where does the quadratic formula come from?*

A: To see where it comes from, we will solve a general quadratic equation using "brute force." Start with the general quadratic equation.

$$ax^2 + bx + c = 0.$$

First, divide out the nonzero number a to get

$$x^2 + \frac{bx}{a} + \frac{c}{a} = 0.$$

Now we **complete the square:** Add and subtract the quantity $\frac{b^2}{4a^2}$ to get

$$x^2 + \frac{bx}{a} + \frac{b^2}{4a^2} - \frac{b^2}{4a^2} + \frac{c}{a} = 0.$$

We do this to get the first three terms to factor as a perfect square:

$$\left(x + \frac{b}{2a}\right)^2 - \frac{b^2}{4a^2} + \frac{c}{a} = 0.$$

(Check this by multiplying out.) Adding $\frac{b^2}{4a^2} - \frac{c}{a}$ to both sides gives:

$$\left(x + \frac{b}{2a}\right)^2 = \frac{b^2}{4a^2} - \frac{c}{a} = \frac{b^2 - 4ac}{4a^2}.$$

Taking square roots gives

$$x + \frac{b}{2a} = \frac{\pm\sqrt{b^2 - 4ac}}{2a}.$$

Finally, adding $-\frac{b}{2a}$ to both sides yields the result:

$$x = -\frac{b}{2a} + \frac{\pm\sqrt{b^2 - 4ac}}{2a}$$

or

$$x = \frac{-b \pm \sqrt{b^2 - 4ac}}{2a}.$$

Solution of Cubic Equations

By definition, a cubic equation can be written in the form

$$ax^3 + bx^2 + cx + d = 0. \qquad a, b, c, \text{ and } d \text{ are fixed numbers and } a \neq 0.$$

Now we get into something of a bind. Although there is a perfectly respectable formula for the solutions, it is very complicated and involves the use of complex numbers rather heavily.[5] So we discuss instead a much simpler method that *sometimes* works nicely. Here is the method in a nutshell.

Solving Cubics by Finding One Factor

Start with a given cubic equation $ax^3 + bx^2 + cx + d = 0$.

Step 1 By trial and error, find one solution $x = s$. If a, b, c, and d are integers, the only possible *rational* solutions[6] are those of the form $s = \pm$(factor of d)/(factor of a).

Step 2 It will now be possible to factor the cubic as

$$ax^3 + bx^2 + cx + d = (x - s)(ax^2 + ex + f) = 0$$

To find $ax^2 + ex + f$, divide the cubic by $x - s$, using long division.[7]

Step 3 The factored equation says that either $x - s = 0$ or $ax^2 + ex + f = 0$. We already know that s is a solution, and now we see that the other solutions are the roots of the quadratic. Note that this quadratic may or may not have any real solutions, as usual.

Quick Example

To solve the cubic $x^3 - x^2 + x - 1 = 0$, we first find a single solution. Here, $a = 1$ and $d = -1$. Because the only factors of ± 1 are ± 1, the only possible rational solutions are $x = \pm 1$. By substitution, we see that $x = 1$ is a solution. Thus, $(x - 1)$ is a factor. Dividing by $(x - 1)$ yields the quotient $(x^2 + 1)$. Thus,

$$x^3 - x^2 + x - 1 = (x - 1)(x^2 + 1) = 0$$

so that either $x - 1 = 0$ or $x^2 + 1 = 0$.

Because the discriminant of the quadratic $x^2 + 1$ is negative, we don't get any real solutions from $x^2 + 1 = 0$, so the only real solution is $x = 1$.

[5] It was when this formula was discovered in the 16th century that complex numbers were first taken seriously. Although we would like to show you the formula, it is too large to fit in this footnote.

[6] There may be *irrational* solutions, however; for example, $x^3 - 2 = 0$ has the single solution $x = \sqrt[3]{2}$.

[7] Alternatively, use "synthetic division," a shortcut that would take us too far afield to describe.

Possible Outcomes When Solving a Cubic Equation

If you consider all the cases, there are three possible outcomes when solving a cubic equation:

1. One real solution (as in the Quick Example on page 25)
2. Two real solutions (try, for example, $x^3 + x^2 - x - 1 = 0$)
3. Three real solutions (see the next example)

EXAMPLE 1 Solving a Cubic

Solve the cubic $2x^3 - 3x^2 - 17x + 30 = 0$.

Solution First we look for a single solution. Here, $a = 2$ and $d = 30$. The factors of a are ± 1 and ± 2, and the factors of d are $\pm 1, \pm 2, \pm 3, \pm 5, \pm 6, \pm 10, \pm 15$, and ± 30. This gives us a large number of possible ratios: $\pm 1, \pm 2, \pm 3, \pm 5, \pm 6, \pm 10, \pm 15, \pm 30, \pm 1/2$, $\pm 3/2, \pm 5/2, \pm 15/2$. Undaunted, we first try $x = 1$ and $x = -1$, getting nowhere. So we move on to $x = 2$, and we hit the jackpot, because substituting $x = 2$ gives $16 - 12 - 34 + 30 = 0$. Thus, $(x - 2)$ is a factor. Dividing yields the quotient $2x^2 + x - 15$. Here is the calculation:

$$
\begin{array}{r|l}
 & 2x^2 + x - 15 \\
\hline
x - 2 & 2x^3 - 3x^2 - 17x + 30 \\
 & \underline{2x^3 - 4x^2} \\
 & \quad x^2 - 17x \\
 & \quad \underline{x^2 - \ \ 2x} \\
 & \qquad -15x + 30 \\
 & \qquad \underline{-15x + 30} \\
 & \qquad\qquad 0.
\end{array}
$$

Thus,

$$2x^3 - 3x^2 - 17x + 30 = (x - 2)(2x^2 + x - 15) = 0.$$

Setting the factors equal to zero gives either $x - 2 = 0$ or $2x^2 + x - 15 = 0$. We could solve the quadratic using the quadratic formula, but, luckily, we notice that it factors as

$$2x^2 + x - 15 = (x + 3)(2x - 5).$$

Thus, the solutions are $x = 2$, $x = -3$ and $x = 5/2$.

Solution of Higher-Order Polynomial Equations

Logically speaking, our next step should be a discussion of quartics, then quintics (fifth degree equations), and so on forever. Well, we've got to stop somewhere, and cubics may be as good a place as any. On the other hand, since we've gotten so far, we ought to at least tell you what is known about higher order polynomials.

Quartics Just as in the case of cubics, there is a formula to find the solutions of quartics.[8]

[8]See, for example, *First Course in the Theory of Equations* by L. E. Dickson (New York: Wiley, 1922), or *Modern Algebra* by B. L. van der Waerden (New York: Frederick Ungar, 1953).

Quintics and Beyond All good things must come to an end, we're afraid. It turns out that there is no "quintic formula." In other words, there is no single algebraic formula or collection of algebraic formulas that gives the solutions to all quintics. This question was settled by the Norwegian mathematician Niels Henrik Abel in 1824 after almost 300 years of controversy about this question. (In fact, several notable mathematicians had previously claimed to have devised formulas for solving the quintic, but these were all shot down by other mathematicians—this being one of the favorite pastimes of practitioners of our art.) The same negative answer applies to polynomial equations of degree 6 and higher. It's not that these equations don't have solutions; it's just that they can't be found using algebraic formulas.[9] However, there are certain special classes of polynomial equations that can be solved with algebraic methods. The way of identifying such equations was discovered around 1829 by the French mathematician Évariste Galois.[10]

---

[9]What we mean by an "algebraic formula" is a formula in the coefficients using the operations of addition, subtraction, multiplication, division, and the taking of radicals. Mathematicians call the use of such formulas in solving polynomial equations "solution by radicals." If you were a math major, you would eventually go on to study this under the heading of Galois theory.

[10]Both Abel (1802–1829) and Galois (1811–1832) died young. Abel died of tuberculosis at the age of 26, while Galois was killed in a duel at the age of 20.

0.5 EXERCISES

Access end-of-section exercises online at **www.webassign.net** 

0.6 Solving Miscellaneous Equations

Equations often arise in calculus that are not polynomial equations of low degree. Many of these complicated-looking equations can be solved easily if you remember the following, which we used in the previous section:

Solving an Equation of the Form $P \cdot Q = 0$

If a product is equal to 0, then at least one of the factors must be 0. That is, if $P \cdot Q = 0$, then either $P = 0$ or $Q = 0$.

Quick Examples

1. $x^5 - 4x^3 = 0$

$x^3(x^2 - 4) = 0$ — Factor the left-hand side.

Either $x^3 = 0$ or $x^2 - 4 = 0$ — Either $P = 0$ or $Q = 0$.

$x = 0, 2$ or -2. — Solve the individual equations.

2. $(x^2 - 1)(x + 2) + (x^2 - 1)(x + 4) = 0$

$(x^2 - 1)[(x + 2) + (x + 4)] = 0$ — Factor the left-hand side.

$(x^2 - 1)(2x + 6) = 0$

Either $x^2 - 1 = 0$ or $2x + 6 = 0$ — Either $P = 0$ or $Q = 0$.

$x = -3, -1,$ or 1. — Solve the individual equations.

EXAMPLE 1 Solving by Factoring

Solve $12x(x^2-4)^5(x^2+2)^6+12x(x^2-4)^6(x^2+2)^5=0$.

Solution

Again, we start by factoring the left-hand side:

$$\begin{aligned}&12x(x^2-4)^5(x^2+2)^6+12x(x^2-4)^6(x^2+2)^5\\&\quad=12x(x^2-4)^5(x^2+2)^5[(x^2+2)+(x^2-4)]\\&\quad=12x(x^2-4)^5(x^2+2)^5(2x^2-2)\\&\quad=24x(x^2-4)^5(x^2+2)^5(x^2-1).\end{aligned}$$

Setting this equal to 0, we get:

$$24x(x^2-4)^5(x^2+2)^5(x^2-1)=0,$$

which means that at least one of the factors of this product must be zero. Now it certainly cannot be the 24, but it could be the x: $x=0$ is one solution. It could also be that

$$(x^2-4)^5=0$$

or

$$x^2-4=0,$$

which has solutions $x=\pm 2$. Could it be that $(x^2+2)^5=0$? If so, then $x^2+2=0$, but this is impossible because $x^2+2\geq 2$, no matter what x is. Finally, it could be that $x^2-1=0$, which has solutions $x=\pm 1$. This gives us five solutions to the original equation:

$$x=-2,-1,0,1,\text{ or }2.$$

EXAMPLE 2 Solving by Factoring

Solve $(x^2-1)(x^2-4)=10$.

Solution Watch out! You may be tempted to say that $x^2-1=10$ or $x^2-4=10$, but this does not follow. If two numbers multiply to give you 10, what must they be? There are lots of possibilities: 2 and 5, 1 and 10, −500,000 and −0.00002 are just a few. The fact that the left-hand side is factored is nearly useless to us if we want to solve this equation. What we will have to do is multiply out, bring the 10 over to the left, and hope that we can factor what we get. Here goes:

$$\begin{aligned}x^4-5x^2+4&=10\\x^4-5x^2-6&=0\\(x^2-6)(x^2+1)&=0\end{aligned}$$

(Here we used a sometimes useful trick that we mentioned in Section 3: We treated x^2 like x and x^4 like x^2, so factoring x^4-5x^2-6 is essentially the same as factoring x^2-5x-6.) *Now* we are allowed to say that one of the factors must be 0: $x^2-6=0$ has solutions $x=\pm\sqrt{6}=\pm 2.449\ldots$ and $x^2+1=0$ has no real solutions. Therefore, we get exactly two solutions, $x=\pm\sqrt{6}=\pm 2.449\ldots$.

To solve equations involving rational expressions, the following rule is very useful.

Solving an Equation of the Form $P/Q = 0$

$$\text{If } \frac{P}{Q} = 0, \text{ then } P = 0.$$

How else could a fraction equal 0? If that is not convincing, multiply both sides by Q (which cannot be 0 if the quotient is defined).

Quick Example

$$\frac{(x+1)(x+2)^2 - (x+1)^2(x+2)}{(x+2)^4} = 0$$

$(x+1)(x+2)^2 - (x+1)^2(x+2) = 0$ — If $\frac{P}{Q} = 0$, then $P = 0$.

$(x+1)(x+2)[(x+2) - (x+1)] = 0$ — Factor.

$(x+1)(x+2)(1) = 0$

Either $x + 1 = 0$ or $x + 2 = 0$,

$x = -1$ or $x = -2$

$x = -1$ — $x = -2$ does not make sense in the original equation: it makes the denominator 0. So it is not a solution and $x = -1$ is the only solution.

EXAMPLE 3 Solving a Rational Equation

Solve $1 - \dfrac{1}{x^2} = 0$.

Solution Write 1 as $\frac{1}{1}$, so that we now have a difference of two rational expressions:

$$\frac{1}{1} - \frac{1}{x^2} = 0.$$

To combine these we can put both over a common denominator of x^2, which gives

$$\frac{x^2 - 1}{x^2} = 0.$$

Now we can set the numerator, $x^2 - 1$, equal to zero. Thus,

$$x^2 - 1 = 0$$

so

$$(x - 1)(x + 1) = 0,$$

giving $x = \pm 1$.

➡ **Before we go on...** This equation could also have been solved by writing

$$1 = \frac{1}{x^2}$$

and then multiplying both sides by x^2. ■

EXAMPLE 4 Another Rational Equation

Solve $\frac{2x-1}{x} + \frac{3}{x-2} = 0.$

Solution We *could* first perform the addition on the left and then set the top equal to 0, but here is another approach. Subtracting the second expression from both sides gives

$$\frac{2x-1}{x} = \frac{-3}{x-2}$$

Cross-multiplying [multiplying both sides by both denominators—that is, by $x(x-2)$] now gives

$$(2x-1)(x-2) = -3x$$

so

$$2x^2 - 5x + 2 = -3x.$$

Adding $3x$ to both sides gives the quadratic equation

$$2x^2 - 2x + 1 = 0.$$

The discriminant is $(-2)^2 - 4 \cdot 2 \cdot 1 = -4 < 0$, so we conclude that there is no real solution.

➡ **Before we go on...** Notice that when we said that $(2x-1)(x-2) = -3x$, we were *not* allowed to conclude that $2x - 1 = -3x$ or $x - 2 = -3x$. ■

EXAMPLE 5 A Rational Equation with Radicals

Solve $\frac{\left(2x\sqrt{x+1} - \frac{x^2}{\sqrt{x+1}}\right)}{x+1} = 0.$

Solution Setting the top equal to 0 gives

$$2x\sqrt{x+1} - \frac{x^2}{\sqrt{x+1}} = 0.$$

This still involves fractions. To get rid of the fractions, we could put everything over a common denominator ($\sqrt{x+1}$) and then set the top equal to 0, or we could multiply the whole equation by that common denominator in the first place to clear fractions. If we do the second, we get

$$2x(x+1) - x^2 = 0$$
$$2x^2 + 2x - x^2 = 0$$
$$x^2 + 2x = 0.$$

Factoring,

$$x(x+2) = 0$$

so either $x = 0$ or $x + 2 = 0$, giving us $x = 0$ or $x = -2$. Again, one of these is not really a solution. The problem is that $x = -2$ cannot be substituted into $\sqrt{x+1}$, because we would then have to take the square root of -1, and we are not allowing ourselves to do that. Therefore, $x = 0$ is the only solution.

0.6 EXERCISES

Access end-of-section exercises online at **www.webassign.net**

0.7 The Coordinate Plane

Q: *Just what is the xy-plane?*

A: The *xy*-plane is an infinite flat surface with two perpendicular lines, usually labeled the ***x*-axis** and ***y*-axis**. These axes are calibrated as shown in Figure 2. (Notice also how the plane is divided into four **quadrants**.)

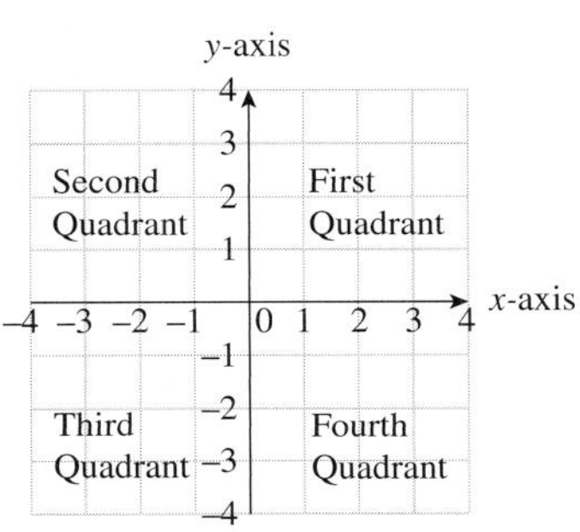

Figure 2

Thus, the *xy*-plane is nothing more than a very large—in fact, infinitely large—flat surface. The purpose of the axes is to allow us to locate specific positions, or **points**, on the plane, with the use of **coordinates**. (If Captain Picard wants to have himself beamed to a specific location, he must supply its coordinates, or he's in trouble.)

Q: *So how do we use coordinates to locate points?*

A: The rule is simple. Each point in the plane has two coordinates, an ***x*-coordinate** and a ***y*-coordinate**. These can be determined in two ways:

1. The *x*-coordinate measures a point's distance to the right or left of the *y*-axis. It is positive if the point is to the right of the axis, negative if it is to the left of the axis, and 0 if it is on the axis. The *y*-coordinate measures a point's distance above or below the *x*-axis. It is positive if the point is above the axis, negative if it is below the axis, and 0 if it is on the axis. Briefly, the *x*-coordinate tells us the *horizontal* position (distance left or right), and the *y*-coordinate tells us the *vertical* position (height).
2. Given a point P, we get its *x*-coordinate by drawing a vertical line from P and seeing where it intersects the *x*-axis. Similarly, we get the *y*-coordinate by extending a horizontal line from P and seeing where it intersects the *y*-axis.

This way of assigning coordinates to points in the plane is often called the system of **Cartesian** coordinates, in honor of the mathematician and philosopher René Descartes (1596–1650), who was the first to use them extensively.

Here are a few examples to help you review coordinates.

EXAMPLE 1 Coordinates of Points

a. Find the coordinates of the indicated points. (See Figure 3. The grid lines are placed at intervals of one unit.)

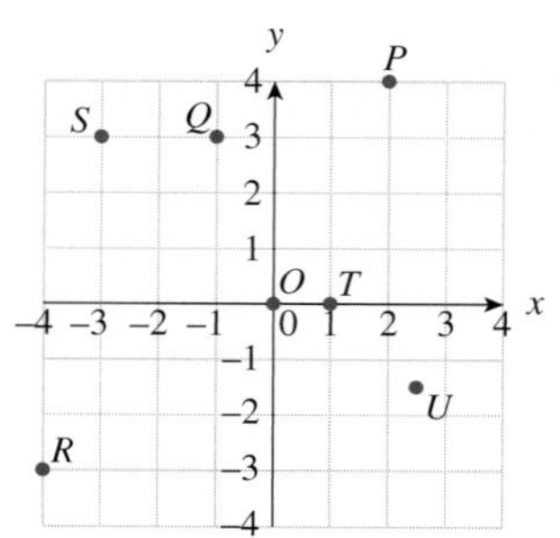

Figure 3

b. Locate the following points in the xy-plane.

$$A(2, 3),\ B(-4, 2),\ C(3, -2.5),\ D(0, -3),\ E(3.5, 0),\ F(-2.5, -1.5)$$

Solution

a. Taking them in alphabetical order, we start with the origin O. This point has height zero and is also zero units to the right of the y-axis, so its coordinates are (0, 0). Turning to P, dropping a vertical line gives $x = 2$ and extending a horizontal line gives $y = 4$. Thus, P has coordinates (2, 4). For practice, determine the coordinates of the remaining points, and check your work against the list that follows:

$$Q(-1, 3),\ R(-4, -3),\ S(-3, 3),\ T(1, 0),\ U(2.5, -1.5)$$

b. In order to locate the given points, we start at the origin (0, 0), and proceed as follows. (See Figure 4.)

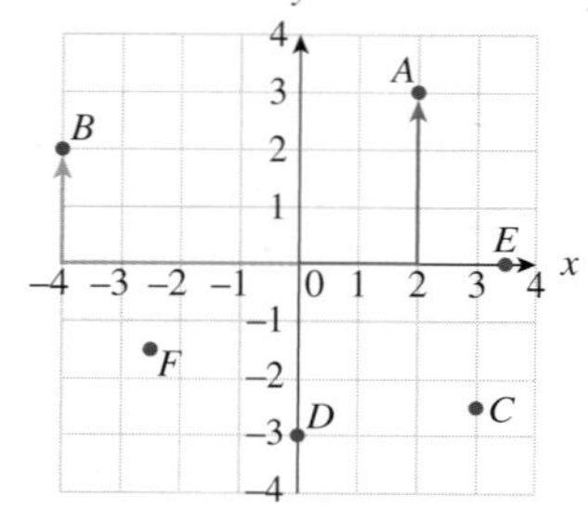

Figure 4

To locate A, we move 2 units to the right and 3 up, as shown.

To locate B, we move -4 units to the right (that is, 4 to the *left*) and 2 up, as shown.

To locate C, we move 3 units right and 2.5 down.

We locate the remaining points in a similar way.

The Graph of an Equation

One of the more surprising developments of mathematics was the realization that equations, which are algebraic objects, can be represented by graphs, which are geometric objects. The kinds of equations that we have in mind are equations in x and y, such as

$$y = 4x - 1, \quad 2x^2 - y = 0, \quad y = 3x^2 + 1, \quad y = \sqrt{x - 1}.$$

The **graph** of an equation in the two variables x and y consists of all points (x, y) in the plane whose coordinates are solutions of the equation.

EXAMPLE 2 Graph of an Equation

Obtain the graph of the equation $y - x^2 = 0$.

Solution We can solve the equation for y to obtain $y = x^2$. Solutions can then be obtained by choosing values for x and then computing y by squaring the value of x, as shown in the following table:

x	−3	−2	−1	0	1	2	3
$y = x^2$	9	4	1	0	1	4	9

Plotting these points (x, y) gives the following picture (left side of Figure 5), suggesting the graph on the right in Figure 5.

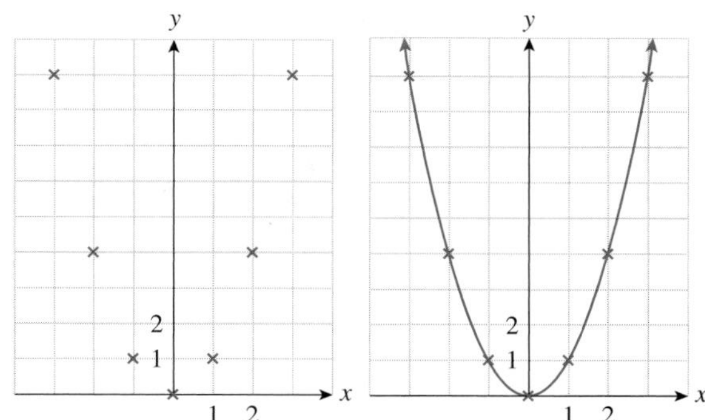

Figure 5

Distance

The distance between two points in the xy-plane can be expressed as a function of their coordinates, as follows:

Distance Formula

The distance between the points $P(x_1, y_1)$ and $Q(x_2, y_2)$ is

$$d = \sqrt{(x_2 - x_1)^2 + (y_2 - y_1)^2} = \sqrt{(\Delta x)^2 + (\Delta y)^2}.$$

Derivation

The distance d is shown in the figure below.

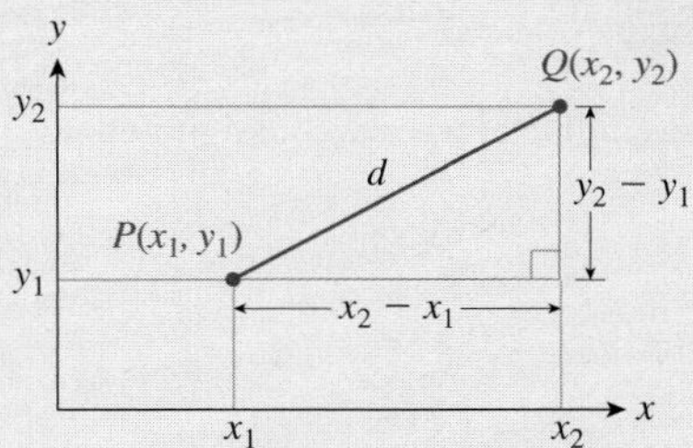

By the Pythagorean theorem applied to the right triangle shown, we get

$$d^2 = (x_2 - x_1)^2 + (y_2 - y_1)^2.$$

Taking square roots (d is a distance, so we take the positive square root), we get the distance formula. Notice that if we switch x_1 with x_2 or y_1 with y_2, we get the same result.

Quick Examples

1. The distance between the points $(3, -2)$ and $(-1, 1)$ is
$$d = \sqrt{(-1-3)^2 + (1+2)^2} = \sqrt{25} = 5.$$

2. The distance from (x, y) to the origin $(0, 0)$ is
$$d = \sqrt{(x-0)^2 + (y-0)^2} = \sqrt{x^2 + y^2}.$$ Distance to the origin

The set of all points (x, y) whose distance from the origin $(0, 0)$ is a fixed quantity r is a circle centered at the origin with radius r. From the second Quick Example, we get the following equation for the circle centered at the origin with radius r:

$$\sqrt{x^2 + y^2} = r. \qquad \text{Distance from the origin} = r.$$

Squaring both sides gives the following equation:

Equation of the Circle of Radius *r* Centered at the Origin

$$x^2 + y^2 = r^2$$

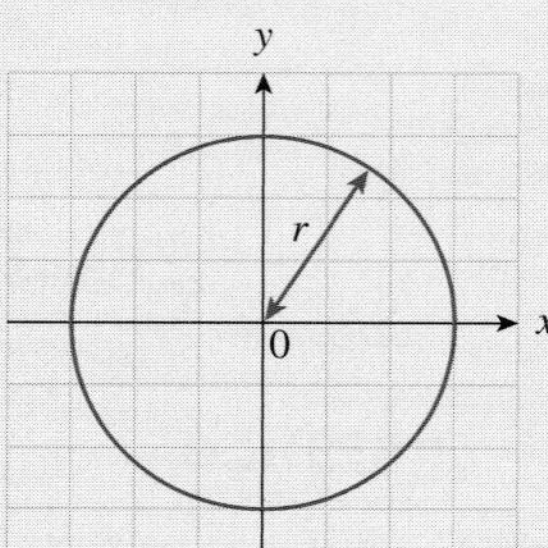

Quick Examples

1. The circle of radius 1 centered at the origin has equation $x^2 + y^2 = 1$.
2. The circle of radius 2 centered at the origin has equation $x^2 + y^2 = 4$.

0.7 EXERCISES

Access end-of-section exercises online at **www.webassign.net** ENHANCED WebAssign

1

Functions and Applications

Website

www.WanerMath.com

At the Website you will find:

- Section-by-section tutorials, including game tutorials with randomized quizzes
- A detailed chapter summary
- A true/false quiz
- Additional review exercises
- Graphers, Excel tutorials, and other resources
- The following extra topic:

  New Functions from Old: Scaled and Shifted Functions

Case Study Modeling Spending on Internet Advertising

You are the new director of *Impact Advertising Inc.'s* Internet division, which has enjoyed a steady 0.25% of the Internet advertising market. You have drawn up an ambitious proposal to expand your division in light of your anticipation that Internet advertising will continue to skyrocket. The VP in charge of Financial Affairs feels that current projections (based on a linear model) do not warrant the level of expansion you propose. **How can you persuade the VP that those projections do not fit the data convincingly?**

Jeff Titcomb/Photographer's Choice / Getty Images

Introduction

To analyze recent trends in spending on Internet advertising and to make reasonable projections, we need a mathematical model of this spending. Where do we start? To apply mathematics to real-world situations like this, we need a good understanding of basic mathematical concepts. Perhaps the most fundamental of these concepts is that of a function: a relationship that shows how one quantity depends on another. Functions may be described numerically and, often, algebraically. They can also be described graphically—a viewpoint that is extremely useful.

The simplest functions—the ones with the simplest formulas and the simplest graphs—are linear functions. Because of their simplicity, they are also among the most useful functions and can often be used to model real-world situations, at least over short periods of time. In discussing linear functions, we will meet the concepts of slope and rate of change, which are the starting point of the mathematics of change.

algebra Review
For this chapter, you should be familiar with real numbers and intervals. To review this material, see **Chapter 0.**

In the last section of this chapter, we discuss *simple linear regression*: construction of linear functions that best fit given collections of data. Regression is used extensively in applied mathematics, statistics, and quantitative methods in business. The inclusion of regression utilities in computer spreadsheets like Excel® makes this powerful mathematical tool readily available for anyone to use.

1.1 Functions from the Numerical, Algebraic, and Graphical Viewpoints

The following table gives the approximate number of Facebook users at various times since its establishment early in 2004.[1]

Year t (Since start of 2004)	0	1	2	3	4	5	6
Facebook Members n (Millions)	0	1	5.5	12	58	150	450

Let's write $n(0)$ for the number of members (in millions) at time $t = 0$, $n(1)$ for the number at time $t = 1$, and so on (we read $n(0)$ as "n of 0"). Thus, $n(0) = 0$, $n(1) = 1$, $n(2) = 5.5$, . . . , $n(6) = 450$. In general, we write $n(t)$ for the number of members (in millions) at time t. We call n a **function** of the variable t, meaning that for each value of t between 0 and 6, n gives us a single corresponding number $n(t)$ (the number of members at that time).

In general, we think of a function as a way of producing new objects from old ones. The functions we deal with in this text produce new numbers from old numbers. The numbers we have in mind are the *real* numbers, including not only positive and negative integers and fractions but also numbers like $\sqrt{2}$ or π. (See Chapter 0 for more on real numbers.) For this reason, the functions we use are called **real-valued functions of a real variable**. For example, the function n takes the year since the start of 2004 as input and returns the number of Facebook members as output (Figure 1).

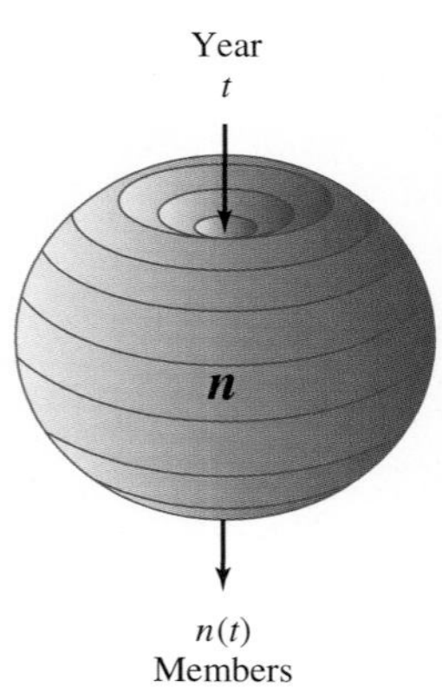

Figure 1

[1]Sources: www.facebook.com, www.insidefacebook.com.

The variable t is called the **independent variable**, while n is called the **dependent variable** as its value depends on t. A function may be specified in several different ways. Here, we have specified the function n **numerically** by giving the values of the function for a number of values of the independent variable, as in the preceding table.

Q: *For which values of t does it make sense to ask for n(t)? In other words, for which years t is the function n defined?*

A: Because $n(t)$ refers to the number of members from the start of 2004 to the start of 2010, $n(t)$ is defined when t is any number between 0 and 6, that is, when $0 \le t \le 6$. Using interval notation (see Chapter 0), we can say that $n(t)$ is defined when t is in the interval [0, 6].

The set of values of the independent variable for which a function is defined is called its **domain** and is a necessary part of the definition of the function. Notice that the preceding table gives the value of $n(t)$ at only some of the infinitely many possible values in the domain [0, 6]. The domain of a function is not always specified explicitly; if no domain is specified for the function f, we take the domain to be the largest set of numbers x for which $f(x)$ makes sense. This "largest possible domain" is sometimes called the **natural domain**.

The previous Facebook data can also be represented on a graph by plotting the given pairs of numbers $(t, n(t))$ in the xy-plane. (See Figure 2. We have connected successive points by line segments.) In general, the **graph** of a function f consists of all points $(x, f(x))$ in the plane with x in the domain of f.

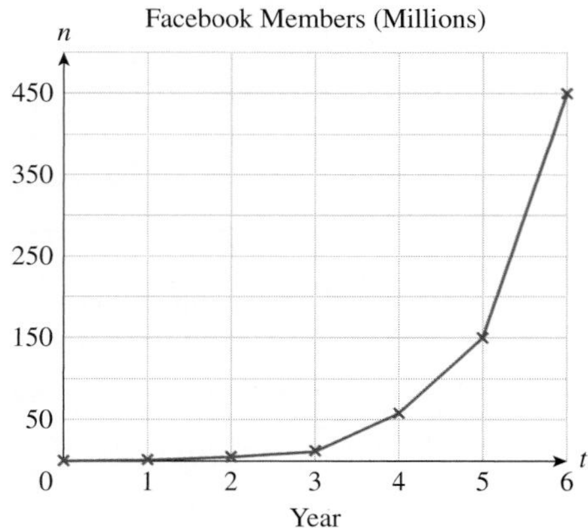

Figure 2

In Figure 2 we specified the function n **graphically** by using a graph to display its values. Suppose now that we had only the graph without the table of data. We could use the graph to find approximate values of n. For instance, to find $n(5)$ from the graph, we do the following:

1. Find the desired value of t at the bottom of the graph ($t = 5$ in this case).
2. Estimate the height (n-coordinate) of the corresponding point on the graph (around 150 in this case).

Thus, $n(5) \approx 150$ million members.*

* In a graphically defined function, we can never know the y-coordinates of points exactly; no matter how accurately a graph is drawn, we can obtain only *approximate* values of the coordinates of points. That is why we have been using the word *estimate* rather than *calculate* and why we say $n(5) \approx 150$ rather than $n(5) = 150$.

In some cases we may be able to use an algebraic formula to calculate the function, and we say that the function is specified **algebraically**. These are not the only ways in which a function can be specified; for instance, it could also be specified **verbally**, as in "Let $n(t)$ be the number of Facebook members, in millions, t years since the start of 2004."* Notice that any function can be represented graphically by plotting the points $(x, f(x))$ for a number of values of x in its domain.

* Specifying a function verbally in this way is useful for understanding what the function is doing, but it gives no numerical information.

Here is a summary of the terms we have just introduced.

Functions

A **real-valued function f of a real-valued variable x** assigns to each real number x in a specified set of numbers, called the **domain** of f, a unique real number $f(x)$, read "f of x." The variable x is called the **independent variable**, and f is called the **dependent variable**. A function is usually specified **numerically** using a table of values, **graphically** using a graph, or **algebraically** using a formula. The **graph of a function** consists of all points $(x, f(x))$ in the plane with x in the domain of f.

Quick Examples

1. **A function specified numerically:** Take $c(t)$ to be the world emission of carbon dioxide in year t since 2000, represented by the following table:[2]

t (Year Since 2000)	$c(t)$ (Billion Metric Tons of CO_2)
0	24
5	28
10	31
15	33
20	36
25	38
30	41

The domain of c is [0, 30], the independent variable is t, the number of years since 2000, and the dependent variable is c, the world production of carbon dioxide in a given year. Some values of c are:

$c(0) = 24$ 24 billion metric tons of CO_2 were produced in 2000.

$c(10) = 31$ 31 billion metric tons of CO_2 were produced in 2010.

$c(30) = 41$ 41 billion metric tons of CO_2 were projected to be produced in 2030.

[2] Figures for 2015 and later are projections. Source: Energy Information Administration (EIA) (www.eia.doe.gov)

Graph of *c*: Plotting the pairs $(t, c(t))$ gives the following graph:

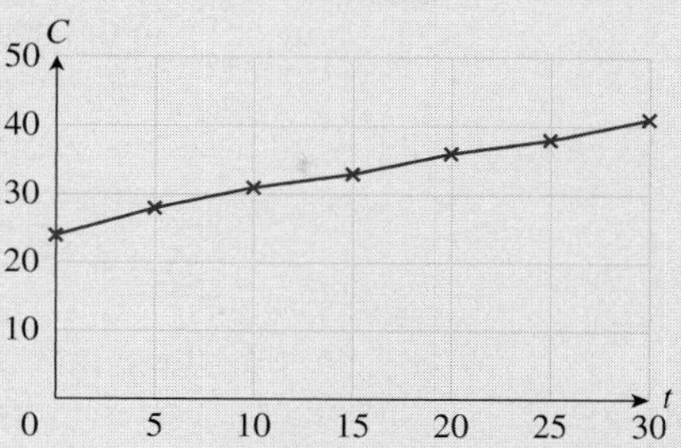

2. **A function specified graphically:** Take $m(t)$ to be the median U.S. home price in thousands of dollars, t years since 2000, as represented by the following graph:[3]

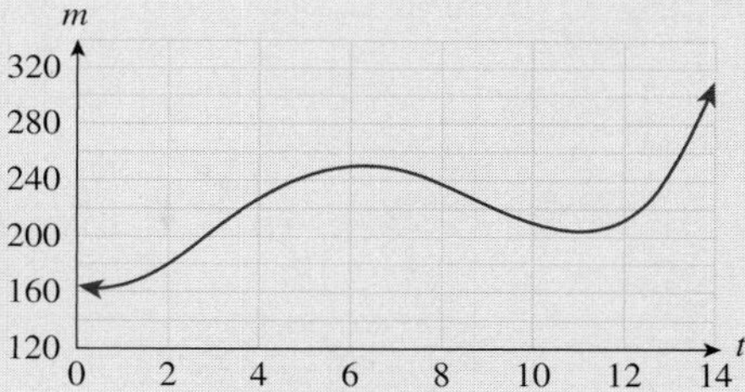

The domain of m is [0, 14], the independent variable is t, the number of years since 2000, and the dependent variable is m, the median U.S. home price in thousands of dollars. Some values of m are:

$m(2) \approx 180$ The median home price in 2002 was about $180,000.

$m(10) \approx 210.$ The median home price in 2010 was about $210,000.

3. **A function specified algebraically:** Let $f(x) = \frac{1}{x}$. The function f is specified algebraically. The independent variable is x and the dependent variable is f. The natural domain of f consists of all real numbers except zero because $f(x)$ makes sense for all values of x other than $x = 0$. Some specific values of f are

$$f(2) = \frac{1}{2} \qquad f(3) = \frac{1}{3} \qquad f(-1) = \frac{1}{-1} = -1$$

$f(0)$ is not defined because 0 is not in the domain of f.

[3]Source for data through end of 2010: www.zillow.com/local-info.

4. The graph of a function: Let $f(x) = x^2$, with domain the set of all real numbers. To draw the graph of f, first choose some convenient values of x in the domain and compute the corresponding y-coordinates $f(x)$:

x	−3	−2	−1	0	1	2	3
$f(x) = x^2$	9	4	1	0	1	4	9

Plotting these points $(x, f(x))$ gives the picture on the left, suggesting the graph on the right.*

* If you plot more points, you will find that they lie on a smooth curve as shown. That is why we did not use line segments to connect the points.

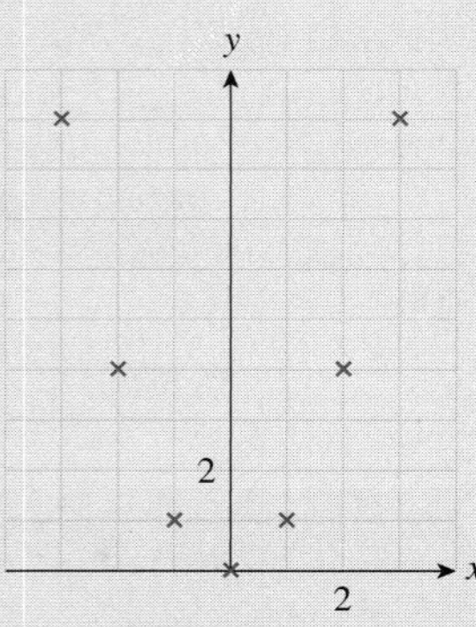

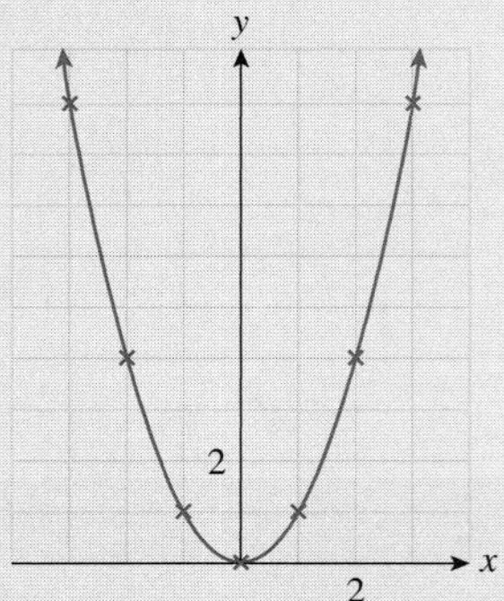

(This particular curve happens to be called a **parabola**, and its lowest point, at the origin, is called its **vertex**.)

EXAMPLE 1 iPod Sales

The total number of iPods sold by Apple up to the end of year x can be approximated by

$$f(x) = 4x^2 + 16x + 2 \text{ million iPods } (0 \le x \le 6),$$

where $x = 0$ represents 2003.[4]

a. What is the domain of f? Compute $f(0)$, $f(2)$, $f(4)$, and $f(6)$. What do these answers tell you about iPod sales? Is $f(-1)$ defined?

b. Compute $f(a)$, $f(-b)$, $f(a+h)$, and $f(a) + h$ assuming that the quantities a, $-b$, and $a + h$ are in the domain of f.

c. Sketch the graph of f. Does the shape of the curve suggest that iPod sales were accelerating or decelerating?

Solution

a. The domain of f is the set of numbers x with $0 \le x \le 6$—that is, the interval $[0, 6]$. If we substitute 0 for x in the formula for $f(x)$, we get

$$f(0) = 4(0)^2 + 16(0) + 2 = 2.$$

By the end of 2003 approximately 2 million iPods had been sold.

[4]Source for data: Apple quarterly earnings reports at www.apple.com/investor/.

Similarly,

$$f(2) = 4(2)^2 + 16(2) + 2 = 50$$ By the end of 2005 approximately 50 million iPods had been sold.

$$f(4) = 4(4)^2 + 16(4) + 2 = 130$$ By the end of 2007 approximately 130 million iPods had been sold.

$$f(6) = 4(6)^2 + 16(6) + 2 = 242.$$ By the end of 2009 approximately 242 million iPods had been sold.

As -1 is not in the domain of f, $f(-1)$ is not defined.

b. To find $f(a)$ we substitute a for x in the formula for $f(x)$ to get

$$f(a) = 4a^2 + 16a + 2.$$ Substitute a for x.

Similarly,

$$f(-b) = 4(-b)^2 + 16(-b) + 2$$ Substitute $-b$ for x.

$$= 4b^2 - 16b + 2$$ $(-b)^2 = b^2$

$$f(a+h) = 4(a+h)^2 + 16(a+h) + 2$$ Substitute $(a+h)$ for x.

$$= 4(a^2 + 2ah + h^2) + 16a + 16h + 2$$ Expand.

$$= 4a^2 + 8ah + 4h^2 + 16a + 16h + 2$$

$$f(a) + h = 4a^2 + 16a + 2 + h.$$ Add h to $f(a)$.

Note how we placed parentheses around the quantities at which we evaluated the function. If we tried to do without any of these parentheses we would likely get an error:

Correct expression: $f(a+h) = 4(a+h)^2 + 16(a+h) + 2.$ ✓

NOT $4a + h^2 + 16a + h + 2x$

Also notice the distinction between $f(a+h)$ and $f(a) + h$: To find $f(a+h)$, we replace x by the quantity $(a+h)$; to find $f(a) + h$ we add h to $f(a)$.

c. To draw the graph of f we plot points of the form $(x, f(x))$ for several values of x in the domain of f. Let us use the values we computed in part (a):

x	0	2	4	6
$f(x) = 4x^2 + 16x + 2$	2	50	130	242

Graphing these points gives the graph shown in Figure 3, suggesting the curve shown on the right.

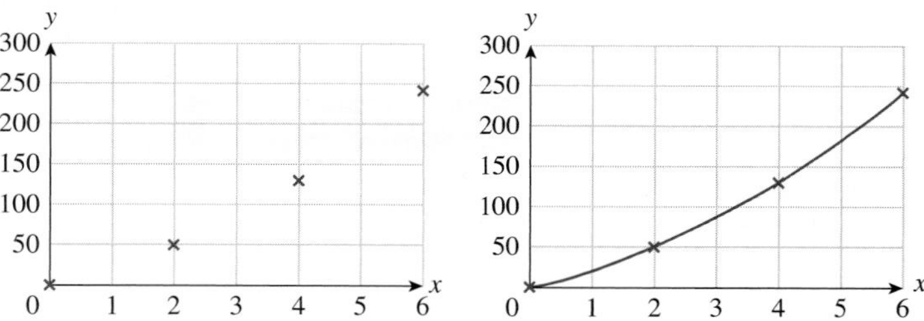

Figure 3

The graph becomes more steep as we move from left to right, suggesting that iPod sales were accelerating.

using Technology

See the Technology Guides at the end of the chapter for detailed instructions on how to obtain the table of values and graph in Example 1 using a TI-83/84 Plus or Excel. Here is an outline:

TI-83/84 Plus
Table of values:
`Y1=4X^2+16X+2`
[2ND] [TABLE].
Graph: [WINDOW];
Xmin = 0, Xmax = 6
[ZOOM] [0].
[More details on page 90.]

Spreadsheet
Table of values: Headings x and $f(x)$ in A1–B1; x-values 0, 2, 4, 6 in A2–A5.
`=4*A2^2+16*A2+2`
in B2; copy down through B5.
Graph: Highlight A1 through B5 and insert a Scatter chart. [More details on page 95.]

Website
www.WanerMath.com
Go to the Function Evaluator and Grapher under Online Utilities, and enter
`4x^2+16x+2`
for y_1. To obtain a table of values, enter the x-values 0, 1, 2, 3 in the Evaluator box, and press "Evaluate" at the top of the box. Graph: Set Xmin = 0, Xmax = 6, and press "Plot Graphs".

➡ **Before we go on...** The following table compares the value of f in Example 1 with the actual sales figures:

x	0	2	4	6
$f(x) = 4x^2 + 16x + 2$	2	50	130	242
Actual iPod Sales (Millions)	2	32	141	240

The actual figures are only stated here for (some) integer values of x; for instance, $x = 4$ gives the total sales up to the end of 2007. But what were, for instance, the sales through June of 2008 ($x = 4.5$)? This is where our formula comes in handy: We can use the formula for f to **interpolate**—that is, to find sales at values of x other than those between values that are stated:

$$f(4.5) = 4(4.5)^2 + 16(4.5) + 2 = 155 \text{ million iPods}.$$

We can also use the formula to **extrapolate**—that is, to predict sales at values of x *outside* the domain—say, for $x = 6.5$ (that is, sales through June 2009):

$$f(6.5) = 4(6.5)^2 + 16(6.5) + 2 = 275 \text{ million iPods}.$$

As a general rule, extrapolation is far less reliable than interpolation: Predicting the future from current data is difficult, especially given the vagaries of the marketplace.

We call the algebraic function f an **algebraic model** of iPod sales because it uses an algebraic formula to model—or mathematically represent (approximately)—the annual sales. The particular kind of algebraic model we used is called a **quadratic model**. (See the end of this section for the names of some commonly used models.) ■

Functions and Equations

Instead of using the usual "function notation" to specify a function, as in, say,

$$f(x) = 4x^2 + 16x + 2, \qquad \text{Function notation}$$

we could have specified it by an equation by replacing $f(x)$ by y:

$$y = 4x^2 + 16x + 2 \qquad \text{Equation notation}$$

(the choice of the letter y is a convention, but any letter will do).

Technically, $y = 4x^2 + 16x + 2$ is an equation and not a function. However, an equation of this type, $y = \textit{Expression in } x$, can be thought of as "specifying y as a function of x." When we specify a function in this way, the variable x is the independent variable and y is the dependent variable.

We could also write the above function as $f = 4x^2 + 16x + 2$, in which case the dependent variable would be f.

Quick Example

If the cost to manufacture x items is given by the "cost function"* C specified by

$$C(x) = 40x + 2{,}000, \qquad \text{Cost function}$$

we could instead write

$$C = 40x + 2{,}000 \qquad \text{Cost equation}$$

and think of C, the cost, as a function of x.

* We will discuss cost functions more fully in the next section.

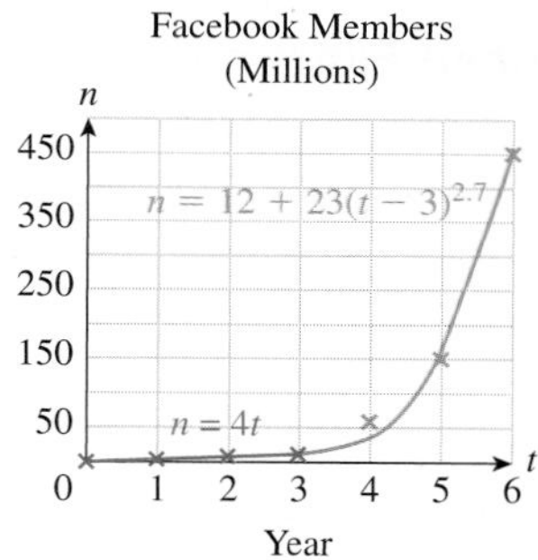

Figure 4

> Function notation and equation notation, sometimes using the same letter for the function name and the dependent variable, are often used interchangeably. It is important to be able to switch back and forth between function notation and equation notation, and we shall do so when it is convenient.

Look again at the graph of the number of Facebook users in Figure 2. From year 0 through year 3 the membership appears to increase more-or-less linearly (that is, the graph is almost a straight line), but then curves upward quite sharply from year 3 to year 6. This behavior can be modeled by using two different functions: one for the interval [0, 3] and another for the interval [3, 6] (see Figure 4).

A function specified by two or more different formulas like this is called a **piecewise-defined function**.

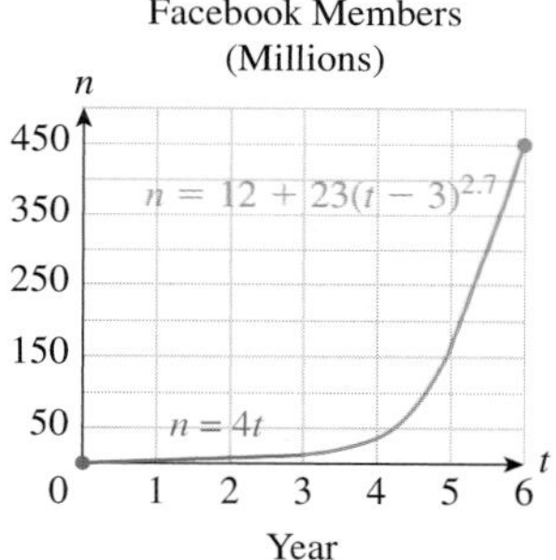

Figure 5

EXAMPLE 2 A Piecewise-Defined Function: Facebook Membership

The number $n(t)$ of Facebook members can be approximated by the following function of time t in years ($t = 0$ represents January 2004):

$$n(t) = \begin{cases} 4t & \text{if } 0 \le t \le 3 \\ 12 + 23(t-3)^{2.7} & \text{if } 3 < t \le 6 \end{cases} \quad \text{million members.}$$

What was the approximate membership of Facebook in January 2005, January 2007, and June 2009? Sketch the graph of n by plotting several points.

Solution We evaluate the given function at the corresponding values of t:

Jan. 2005 ($t = 1$): $n(1) = 4(1) = 4$ — Use the first formula because $0 \le t \le 3$.

Jan. 2007 ($t = 3$): $n(3) = 4(3) = 12$ — Use the first formula because $0 \le t \le 3$.

June 2009 ($t = 5.5$): $n(5.5) = 12 + 23(5.5 - 3)^{2.7} \approx 285.$ — Use the second formula because $3 < t \le 6$.

Thus, the number of Facebook members was approximately 4 million in January 2005, 12 million in January 2007, and 285 million in June 2009.

To sketch the graph of n we use a table of rounded values of $n(t)$ (some of which we have already calculated above), plot the points, and connect them to sketch the graph:

t	0	1	2	3	4	5	6
$n(t)$	0	4	8	12	35	161	459

First Formula (t = 0 to 3) — Second Formula (t = 4 to 6)

The graph (Figure 5) has the following features:

1. The first formula (the line) is used for $0 \le t \le 3$.
2. The second formula (ascending curve) is used for $3 < t \le 6$.
3. The domain is [0, 6], so the graph is cut off at $t = 0$ and $t = 6$.
4. The heavy solid dots at the ends indicate the endpoints of the domain.

using Technology

See the Technology Guides at the end of the chapter for detailed instructions on how to obtain the table of values and graph in Example 2 using a TI-83/84 Plus or Excel. Here is an outline:

TI-83/84 Plus

Table of values:

```
Y1=(X≤3)*(4X)+(X>3)*
(12+23*abs(X-3)^2.7)
```

[2ND] [TABLE].

Graph: [WINDOW]; Xmin = 0, Xmax = 6; [ZOOM] [0].

[More details on page 90.]

Spreadsheet

Table of values: Headings t and $n(t)$ in A1–B1; t-values 0, 1, . . . , 6 in A2–A8.

```
=(A2<=3)*(4*A2)+(A2>3)*
(12+23*abs(A2-3)^2.7)
```

in B2; copy down through B8.

Graph: Highlight A1 through B8 and insert a Scatter chart. [More details on page 96.]

Website
www.WanerMath.com
Go to the Function Evaluator and Grapher under Online Utilities, and enter

```
(x≤3)*(4x)+(x>3)*
  (12+23*abs(x-3)^2.7)
```

for y_1. To obtain a table of values, enter the x-values 0, 1, . . . , 6 in the Evaluator box, and press "Evaluate." at the top of the box. Graph: Set Xmin = 0, Xmax = 6, and press "Plot Graphs."

EXAMPLE 3 More Complicated Piecewise-Defined Functions

Let f be the function specified by

$$f(x) = \begin{cases} -1 & \text{if } -4 \le x < -1 \\ x & \text{if } -1 \le x \le 1 \\ x^2 - 1 & \text{if } 1 < x \le 2 \end{cases}.$$

a. What is the domain of f? Find $f(-2)$, $f(-1)$, $f(0)$, $f(1)$, and $f(2)$.
b. Sketch the graph of f.

Solution

a. The domain of f is $[-4, 2]$, because $f(x)$ is specified only when $-4 \le x \le 2$.

$f(-2) = -1$ We used the first formula because $-4 \le x < -1$.
$f(-1) = -1$ We used the second formula because $-1 \le x \le 1$.
$f(0) = 0$ We used the second formula because $-1 \le x \le 1$.
$f(1) = 1$ We used the second formula because $-1 \le x \le 1$.
$f(2) = 2^2 - 1 = 3$ We used the third formula because $1 < x \le 2$.

b. To sketch the graph by hand, we first sketch the three graphs $y = -1$, $y = x$, and $y = x^2 - 1$, and then use the appropriate portion of each (Figure 6).

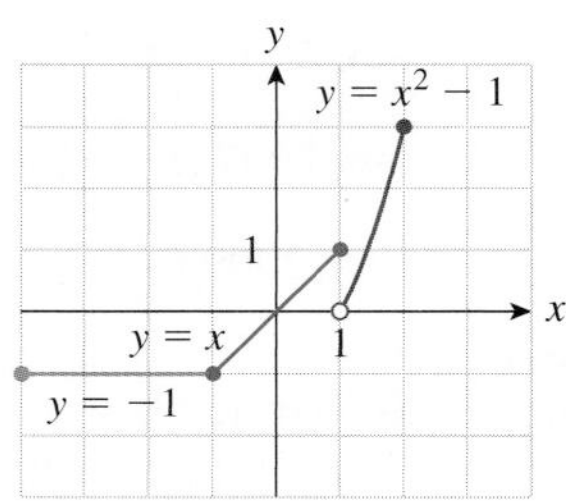

Figure 6

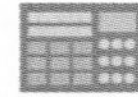
using Technology
For the function in Example 3, use the following technology formula (see the technology discussion for Example 2):

```
(X<-1)*(-1)
 +(-1≤X)*(X≤1)*X
 +(1<X)*(X^2-1)
```

Note that solid dots indicate points on the graph, whereas the open dots indicate points not on the graph. For example, when $x = 1$, the inequalities in the formula tell us that we are to use the middle formula (x) rather than the bottom one ($x^2 - 1$). Thus, $f(1) = 1$, not 0, so we place a solid dot at (1, 1) and an open dot at (1, 0).

Vertical Line Test

Every point in the graph of a function has the form $(x, f(x))$ for some x in the domain of f. Because f assigns a *single* value $f(x)$ to each value of x in the domain, it follows that, in the graph of f, there should be only one y corresponding to any such value of x—namely, $y = f(x)$. In other words, *the graph of a function cannot contain two or more points with the same x-coordinate—that is, two or more points on the same vertical line*. On the other hand, a vertical line at a value of x not in the domain will not contain any points in the graph. This gives us the following rule.

Vertical-Line Test

For a graph to be the graph of a function, every vertical line must intersect the graph in *at most* one point.

Quick Examples

As illustrated below, only graph B passes the vertical line test, so only graph B is the graph of a function.

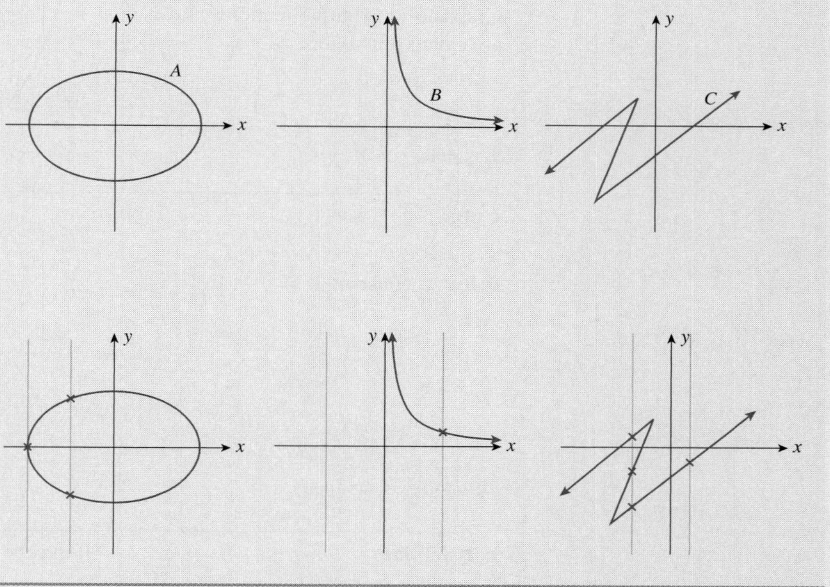

Table 1 lists some common types of functions that are often used to model real world situations.

Table 1 A Compendium of Functions and Their Graphs

Type of Function	*Examples*	
Linear $f(x) = mx + b$ m, b constant Graphs of linear functions are straight lines. The quantity m is the **slope** of the line; the quantity b is the ***y*-intercept** of the line. [See Section 1.3.]	$y = x$ 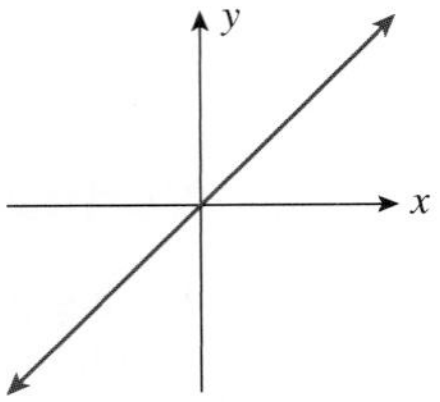	$y = -2x + 2$ 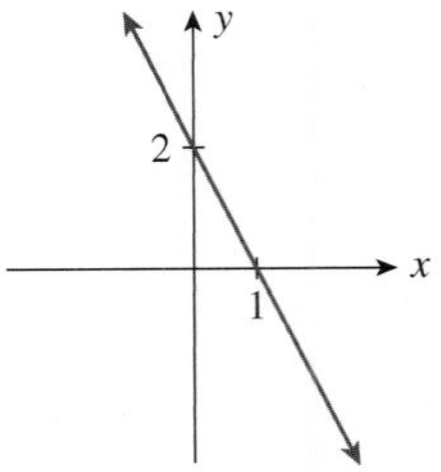
Technology formulas:	`x`	`-2*x+2`
Quadratic $f(x) = ax^2 + bx + c$ a, b, c constant $(a \neq 0)$ Graphs of quadratic functions are called **parabolas**.	$y = x^2$ 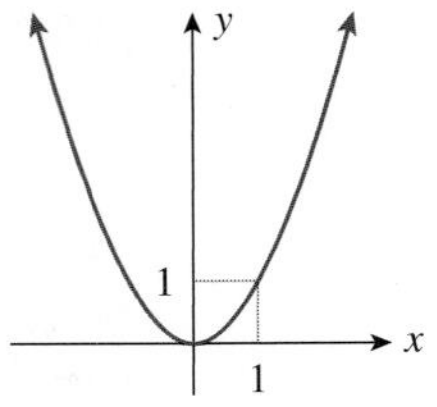	$y = -2x^2 + 2x + 4$ 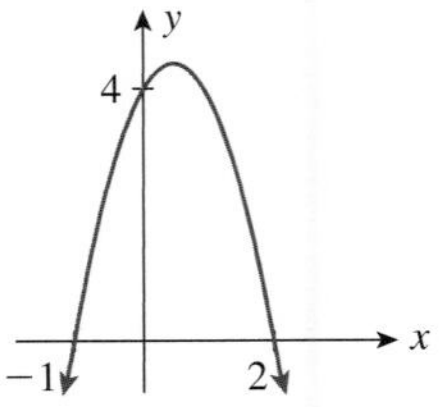
Technology formulas:	`x^2`	`-2*x^2+2*x+4`
Cubic $f(x) = ax^3 + bx^2 + cx + d$ a, b, c, d constant $(a \neq 0)$	$y = x^3$ 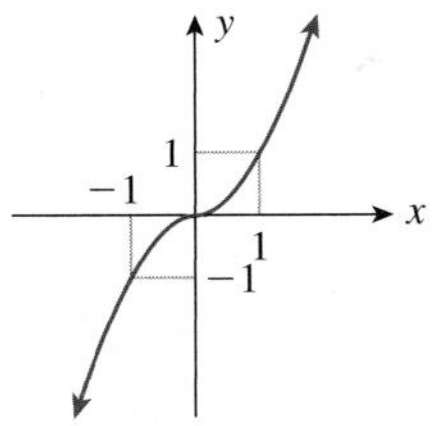	$y = -x^3 + 3x^2 + 1$ 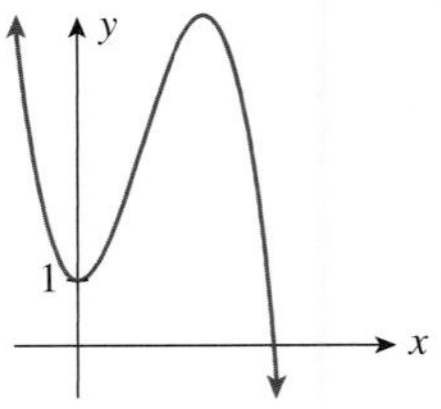
Technology formulas:	`x^3`	`-x^3+3*x^2+1`
Polynomial $f(x) = ax^n + bx^{n-1} + \ldots + rx + s$ $a, b, \ldots, r, s$ constant (includes all of the above functions)	All the above, and $f(x) = x^6 - 2x^5 - 2x^4 + 4x^2$ 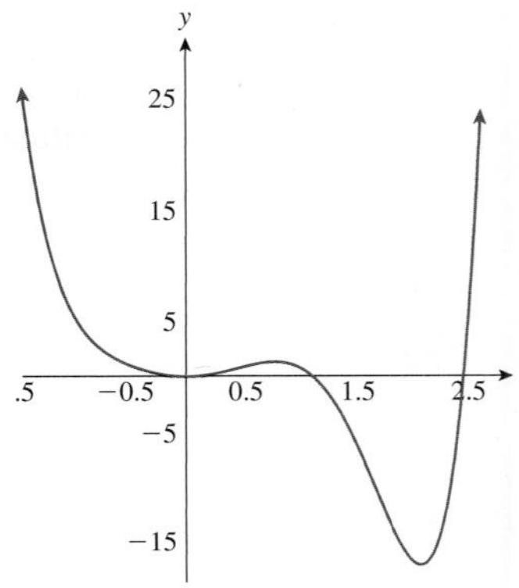	
Technology formula:	`x^6-2x^5-2x^4+4x^2`	

Table 1 (*Continued*)

Type of Function	*Examples*	
Exponential $f(x) = Ab^x$ A, b constant ($b > 0$ and $b \neq 1$) The y-coordinate is multiplied by b every time x increases by 1.	$y = 2^x$ 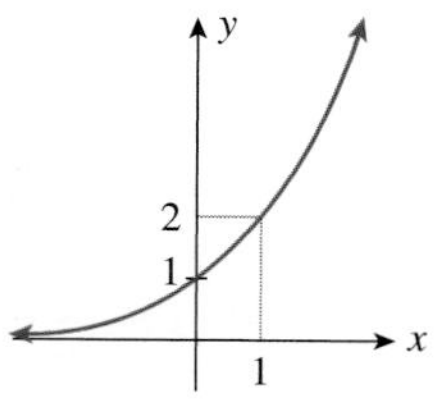 y is doubled every time x increases by 1.	$y = 4(0.5)^x$ 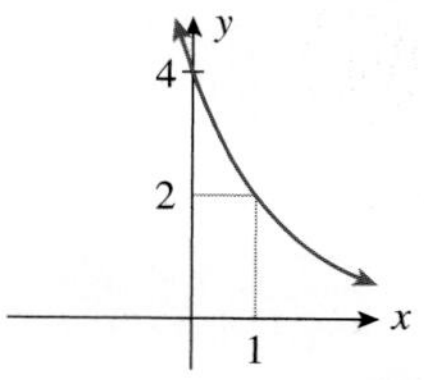 y is halved every time x increases by 1.
Technology formulas:	`2^x`	`4*0.5^x`
Rational $f(x) = \dfrac{P(x)}{Q(x)}$; $P(x)$ and $Q(x)$ polynomials The graph of $y = 1/x$ is a **hyperbola**. The domain excludes zero because $1/0$ is not defined.	$y = \dfrac{1}{x}$ 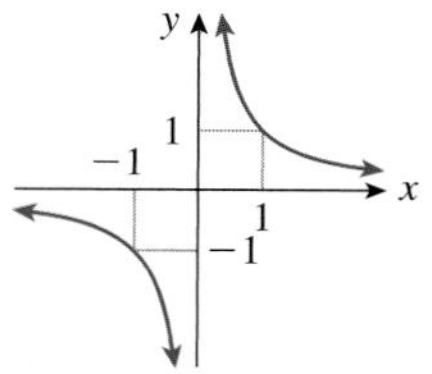	$y = \dfrac{x}{x-1}$ 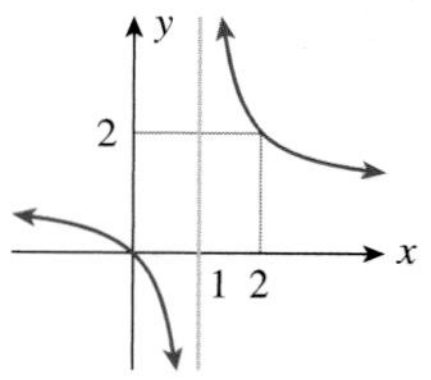
Technology formulas:	`1/x`	`x/(x-1)`
Absolute value For x positive or zero, the graph of $y = \|x\|$ is the same as that of $y = x$. For x negative or zero, it is the same as that of $y = -x$.	$y = \|x\|$ 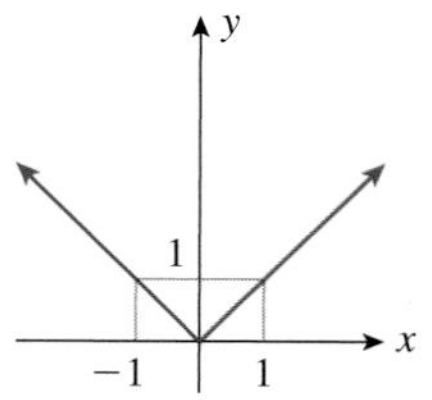	$y = \|2x + 2\|$ 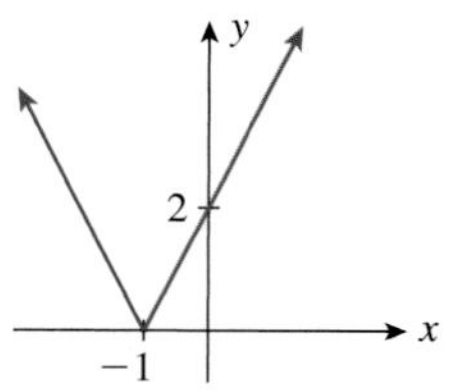
Technology formulas:	`abs(x)`	`abs(2*x+2)`
Square Root The domain of $y = \sqrt{x}$ must be restricted to the nonnegative numbers, because the square root of a negative number is not real. Its graph is the top half of a horizontally oriented parabola.	$y = \sqrt{x}$ 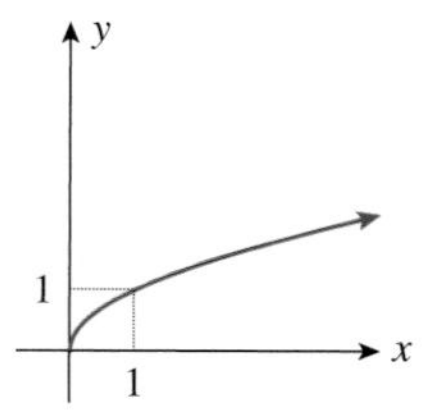	$y = \sqrt{4x - 2}$ 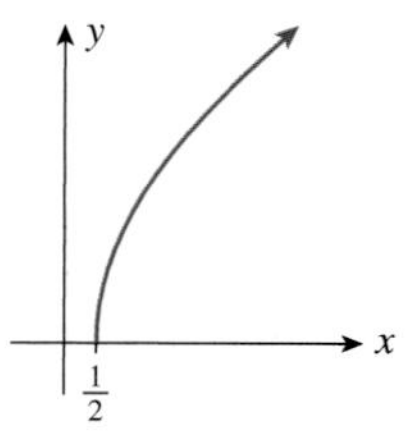
Technology formulas:	`x^0.5` or `√(x)`	`(4*x-2)^0.5` or `√(4*x-2)`

Go to the Website and follow the path

Online Text

→ New Functions from Old: Scaled and Shifted Functions

where you will find complete online interactive text, examples, and exercises on scaling and translating the graph of a function by changing the formula.

Functions and models other than linear ones are called **nonlinear**.

1.1 EXERCISES

Access end-of-section exercises online at **www.webassign.net** 

1.2 Functions and Models

The functions we used in Examples 1 and 2 in Section 1.1 are **mathematical models** of real-life situations, because they model, or represent, situations in mathematical terms.

Mathematical Modeling

To mathematically model a situation means to represent it in mathematical terms. The particular representation used is called a **mathematical model** of the situation. Mathematical models do not always represent a situation perfectly or completely. Some (like Example 1 of Section 1.1) represent a situation only approximately, whereas others represent only some aspects of the situation.

Quick Examples

1. The temperature is now 10°F and increasing by 20° per hour.
Model: $T(t) = 10 + 20t$ (t = time in hours, T = temperature)

2. I invest $1,000 at 5% interest compounded quarterly. Find the value of the investment after t years.
Model: $A(t) = 1{,}000\left(1 + \frac{0.05}{4}\right)^{4t}$ (This is the compound interest formula we will study in Example 6.)

3. I am fencing a rectangular area whose perimeter is 100 ft. Find the area as a function of the width x.
Model: Take y to be the length, so the perimeter is

$$100 = x + y + x + y = 2(x + y).$$

This gives

$$x + y = 50.$$

Thus the length is $y = 50 - x$, and the area is

$$A = xy = x(50 - x).$$

4. You work 8 hours a day Monday through Friday, 5 hours on Saturday, and have Sunday off. Model the number of hours you work as a function of the day of the week n, with $n = 1$ being Sunday.
Model: Take $f(n)$ to be the number of hours you work on the nth day of the week, so

$$f(n) = \begin{cases} 0 & \text{if } n = 1 \\ 8 & \text{if } 2 \le n \le 6 \\ 5 & \text{if } n = 7 \end{cases}.$$

Note that the domain of f is $\{1, 2, 3, 4, 5, 6, 7\}$—a discrete set rather than a continuous interval of the real line.

5. The function

$$f(x) = 4x^2 + 16x + 2 \text{ million iPods sold } (x = \text{years since 2003})$$

in Example 1 of Section 1.1 is a model of iPod sales.

6. The function

$$n(t) = \begin{cases} 4t & \text{if } 0 \le t \le 3 \\ 12 + 23(t-3)^{2.7} & \text{if } 3 < t \le 6 \end{cases} \text{ million members}$$

(t = years since January 2004) in Example 2 of Section 1.1 is a model of Facebook membership.

Types of Models

Quick Examples 1–4 are **analytical models**, obtained by analyzing the situation being modeled, whereas Quick Examples 5 and 6 are **curve-fitting models**, obtained by finding mathematical formulas that approximate observed data. All the models except for Quick Example 4 are **continuous models**, defined by functions whose domains are intervals of the real line, whereas Quick Example 4 is a **discrete model** as its domain is a discrete set, as mentioned above. Discrete models are used extensively in probability and statistics.

Cost, Revenue, and Profit Models

EXAMPLE 1 Modeling Cost: Cost Function

As of August 2010, Yellow Cab Chicago's rates amounted to \$2.05 on entering the cab plus \$1.80 for each mile.[5]

a. Find the cost C of an x-mile trip.

b. Use your answer to calculate the cost of a 40-mile trip.

c. What is the cost of the second mile? What is the cost of the tenth mile?

d. Graph C as a function of x.

Solution

a. We are being asked to find how the cost C depends on the length x of the trip, or to find C as a function of x. Here is the cost in a few cases:

Cost of a 1-mile trip: $C = 1.80(1) + 2.05 = 3.85$ 1 mile at \$1.80 per mile plus \$2.05

Cost of a 2-mile trip: $C = 1.80(2) + 2.05 = 5.65$ 2 miles at \$1.80 per mile plus \$2.05

Cost of a 3-mile trip: $C = 1.80(3) + 2.05 = 7.45$ 3 miles at \$1.80 per mile plus \$2.05

Do you see the pattern? The cost of an x-mile trip is given by the linear function

$$C(x) = 1.80x + 2.05.$$

[5]According to their Web site at www.yellowcabchicago.com.

Notice that the cost function is a sum of two terms: The **variable cost** $1.80x$, which depends on x, and the **fixed cost** 2.05, which is independent of x:

$$\text{Cost} = \text{Variable Cost} + \text{Fixed Cost}.$$

The quantity 1.80 by itself is the incremental cost per mile; you might recognize it as the *slope* of the given linear function. In this context we call 1.80 the **marginal cost**. You might recognize the fixed cost 2.05 as the *C-intercept* of the given linear function.

b. We can use the formula for the cost function to calculate the cost of a 40-mile trip as

$$C(40) = 1.80(40) + 2.05 = \$74.05.$$

c. To calculate the cost of the second mile, we *could* proceed as follows:

Find the cost of a 1-mile trip: $C(1) = 1.80(1) + 2.05 = \$3.85.$

Find the cost of a 2-mile trip: $C(2) = 1.80(2) + 2.05 = \$5.65.$

Therefore, the cost of the second mile is $\$5.65 - \$3.85 = \$1.80.$

But notice that this is just the marginal cost. In fact, the marginal cost is the cost of each additional mile, so we could have done this more simply:

$$\text{Cost of second mile} = \text{Cost of tenth mile} = \text{Marginal cost} = \$1.80.$$

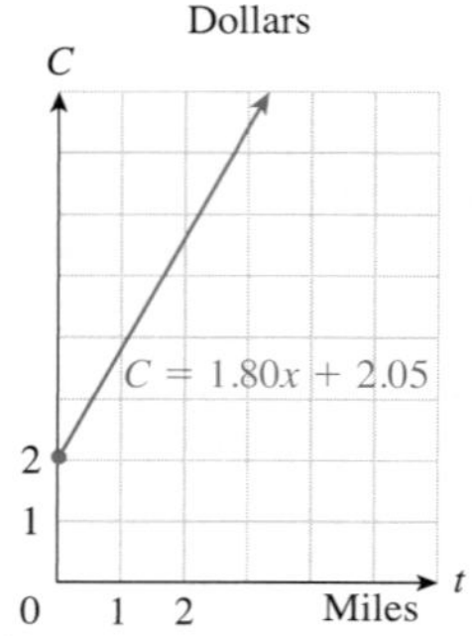

Figure 7

d. Figure 7 shows the graph of the cost function, which we can interpret as a *cost vs. miles* graph. The fixed cost is the starting height on the left, while the marginal cost is the slope of the line: It rises 1.80 units per unit of x. (See Section 1.3 for a discussion of properties of straight lines.)

Before we go on... The cost function in Example 1 is an example of an *analytical model:* We derived the form of the cost function from a knowledge of the cost per mile and the fixed cost.

As we discussed in Section 1.1, we can specify the cost function in Example 1 using equation notation:

$$C = 1.80x + 2.05. \qquad \text{Equation notation}$$

Here, the independent variable is x, and the dependent variable is C. (This is the notation we have used in Figure 7. Remember that we will often switch between function and equation notation when it is convenient to do so.) ■

Here is a summary of some terms we used in Example 1, along with an introduction to some new terms:

Cost, Revenue, and Profit Functions

A **cost function** specifies the cost C as a function of the number of items x. Thus, $C(x)$ is the cost of x items, and has the form

$$\text{Cost} = \text{Variable cost} + \text{Fixed cost}$$

where the variable cost is a function of x and the fixed cost is a constant. A cost function of the form

$$C(x) = mx + b$$

is called a **linear cost function**; the variable cost is mx and the fixed cost is b. The slope m, the **marginal cost**, measures the incremental cost per item.

The **revenue**, or **net sales**, resulting from one or more business transactions is the total income received. If $R(x)$ is the revenue from selling x items at a price of m each, then R is the linear function $R(x) = mx$ and the selling price m can also be called the **marginal revenue**.

The **profit**, or **net income**, on the other hand, is what remains of the revenue when costs are subtracted. If the profit depends linearly on the number of items, the slope m is called the **marginal profit**. Profit, revenue, and cost are related by the following formula.

$$\text{Profit} = \text{Revenue} - \text{Cost}$$
$$P(x) = R(x) - C(x).^{*}$$

* We say that the profit function P is the **difference** between the revenue and cost functions, and express this fact as a formula about functions: $P = R - C$. (We will discuss this further when we talk about the algebra of functions at the end of this section.)

If the profit is negative, say $-\$500$, we refer to a **loss** (of \$500 in this case). To **break even** means to make neither a profit nor a loss. Thus, breakeven occurs when $P = 0$, or

$$R(x) = C(x). \qquad \text{Breakeven}$$

The **break-even point** is the number of items x at which breakeven occurs.

Quick Example

If the daily cost (including operating costs) of manufacturing x T-shirts is $C(x) = 8x + 100$, and the revenue obtained by selling x T-shirts is $R(x) = 10x$, then the daily profit resulting from the manufacture and sale of x T-shirts is

$$P(x) = R(x) - C(x) = 10x - (8x + 100) = 2x - 100.$$

Breakeven occurs when $P(x) = 0$, or $x = 50$.

EXAMPLE 2 Cost, Revenue, and Profit

The annual operating cost of *YSport* Fitness gym is estimated to be

$$C(x) = 100{,}000 + 160x - 0.2x^2 \text{ dollars} \qquad (0 \le x \le 400),$$

where x is the number of members. Annual revenue from membership averages \$800 per member. What is the variable cost? What is the fixed cost? What is the profit function? How many members must *YSport* have to make a profit? What will happen if it has fewer members? If it has more?

Solution The variable cost is the part of the cost function that depends on x:

$$\text{Variable cost} = 160x - 0.2x^2.$$

The fixed cost is the constant term:

$$\text{Fixed cost} = 100{,}000.$$

The annual revenue *YSport* obtains from a single member is \$800. So, if it has x members, it earns an annual revenue of

$$R(x) = 800x.$$

For the profit, we use the formula

$$\begin{aligned} P(x) &= R(x) - C(x) && \text{Formula for profit} \\ &= 800x - (100{,}000 + 160x - 0.2x^2) && \text{Substitute } R(x) \text{ and } C(x). \\ &= -100{,}000 + 640x + 0.2x^2. \end{aligned}$$

To make a profit, *YSport* needs to do better than break even, so let us find the break-even point: the value of x such that $P(x) = 0$. All we have to do is set $P(x) = 0$ and solve for x:

$$-100{,}000 + 640x + 0.2x^2 = 0.$$

Notice that we have a quadratic equation $ax^2 + bx + c = 0$ with $a = 0.2$, $b = 640$, and $c = -100{,}000$. Its solution is given by the quadratic formula:

using Technology

Excel has a feature called "Goal Seek," which can be used to find the point of intersection of the cost and revenue graphs numerically rather than graphically. See the downloadable Excel tutorial for this section at the Website.

$$\begin{aligned} x = \frac{-b \pm \sqrt{b^2 - 4ac}}{2a} &= \frac{-640 \pm \sqrt{640^2 + 4(0.2)(100{,}000)}}{2(0.2)} \\ &\approx \frac{-640 \pm 699.71}{2(0.2)} \\ &\approx 149.3 \text{ or } -3{,}349.3. \end{aligned}$$

We reject the negative solution (as the domain is $[0, 400]$) and conclude that $x \approx 149.3$ members. To make a profit, should *YSport* have 149 members or 150 members? To decide, take a look at Figure 8, which shows two graphs: On the left we see the graph of revenue and cost, and on the right we see the graph of the profit function.

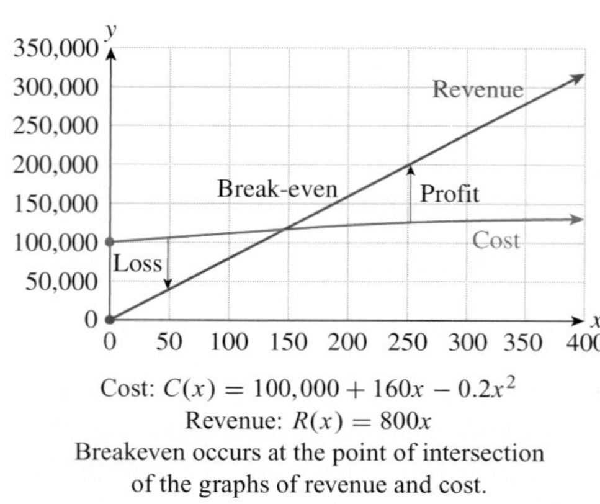

Cost: $C(x) = 100{,}000 + 160x - 0.2x^2$
Revenue: $R(x) = 800x$
Breakeven occurs at the point of intersection of the graphs of revenue and cost.

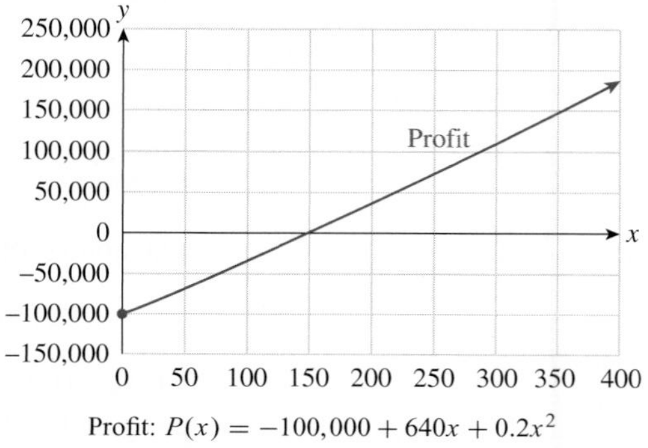

Profit: $P(x) = -100{,}000 + 640x + 0.2x^2$
Breakeven occurs when $P(x) = 0$

Figure 8

For values of x less than the break-even point of 149.3, $P(x)$ is negative, so the company will have a loss. For values of x greater than the break-even point, $P(x)$ is positive, so the company will make a profit. Thus, *YSport Fitness* needs at least 150 members to make a profit. (Note that we rounded 149.3 up to 150 in this case.)

Demand and Supply Models

The demand for a commodity usually goes down as its price goes up. It is traditional to use the letter q for the (quantity of) demand, as measured, for example, in sales. Consider the following example.

EXAMPLE 3 Demand: Private Schools

The demand for private schools in Michigan depends on the tuition cost and can be approximated by

$$q = 77.8p^{-0.11} \text{ thousand students} \qquad (200 \le p \le 2{,}200), \qquad \text{Demand curve}$$

where p is the net tuition cost in dollars.[6]

a. Use technology to plot the demand function.

b. What is the effect on demand if the tuition cost is increased from \$1,000 to \$1,500?

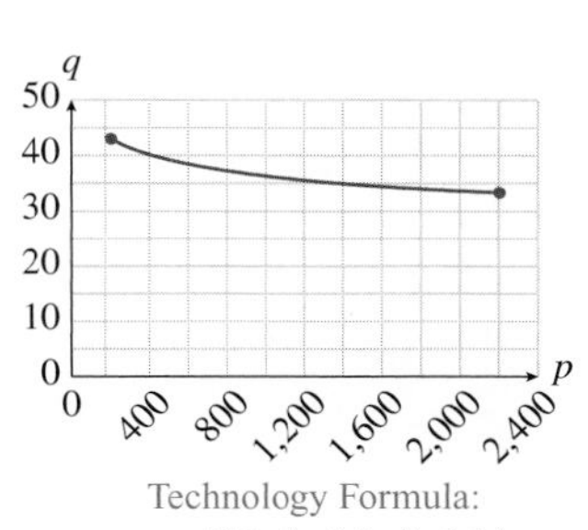

Technology Formula:
y = 77.8x^(-0.11)

Figure 9

Solution

a. The demand function is given by $q(p) = 77.8p^{-0.11}$. Its graph is known as a **demand curve** (Figure 9).

b. The demand at tuition costs of \$1,000 and \$1,500 is

$$q(1{,}000) = 77.8(1{,}000)^{-0.11} \approx 36.4 \text{ thousand students}$$
$$q(1{,}500) = 77.8(1{,}500)^{-0.11} \approx 34.8 \text{ thousand students.}$$

The change in demand is therefore

$$q(1{,}500) - q(1{,}000) \approx 34.8 - 36.4 = -1.6 \text{ thousand students.}$$

We have seen that a demand function gives the number of items consumers are willing to buy at a given price, and a higher price generally results in a lower demand. However, as the price rises, suppliers will be more inclined to produce these items (as opposed to spending their time and money on other products), so supply will generally rise. A **supply function** gives q, the number of items suppliers are willing to make available for sale*, as a function of p, the price per item.

* Although a bit confusing at first, it is traditional to use the same letter q for the quantity of supply and the quantity of demand, particularly when we want to compare them, as in the next example.

Demand, Supply, and Equilibrium Price

A **demand equation** or **demand function** expresses demand q (the number of items demanded) as a function of the unit price p (the price per item). A **supply equation** or **supply function** expresses supply q (the number of items a supplier is willing to make available) as a function of the unit price p (the price per item). It is usually the case that demand decreases and supply increases as the unit price increases.

[6] The tuition cost is net cost: tuition minus tax credit. The model is based on data in "The Universal Tuition Tax Credit: A Proposal to Advance Personal Choice in Education," Patrick L. Anderson, Richard McLellan, J.D., Joseph P. Overton, J.D., Gary Wolfram, Ph.D., Mackinac Center for Public Policy, www.mackinac.org/

Demand and supply are said to be in **equilibrium** when demand equals supply. The corresponding values of p and q are called the **equilibrium price** and **equilibrium demand**. To find the equilibrium price, determine the unit price p where the demand and supply curves cross (sometimes we can determine this value analytically by setting demand equal to supply and solving for p). To find the equilibrium demand, evaluate the demand (or supply) function at the equilibrium price.

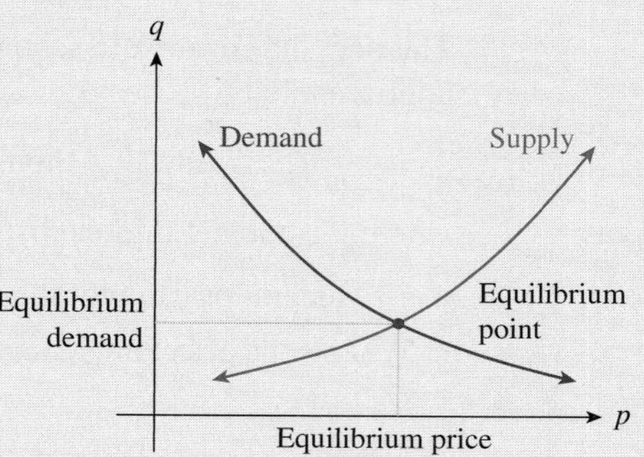

Quick Example

If the demand for your exclusive T-shirts is $q = -20p + 800$ shirts sold per day and the supply is $q = 10p - 100$ shirts per day, then the equilibrium point is obtained when demand = supply:

$$-20p + 800 = 10p - 100$$
$$30p = 900, \text{ giving } p = \$30.$$

The equilibrium price is therefore \$30 and the equilibrium demand is $q = -20(30) + 800 = 200$ shirts per day. What happens at prices other than the equilibrium price is discussed in Example 4.

Note In economics it is customary to plot the independent variable (price) on the vertical axis and the dependent variable (demand or supply) on the horizontal axis, but in this book we follow the usual mathematical convention for all graphs and plot the independent variable on the horizontal axis.

EXAMPLE 4 Demand, Supply, and Equilibrium Price

Continuing with Example 3, suppose that private school institutions are willing to create private schools to accommodate

$$q = 30.4 + 0.006p \text{ thousand students} \qquad (200 \le p \le 2{,}200) \qquad \text{Supply curve}$$

who pay a net tuition of p dollars.

a. Graph the demand curve of Example 3 and the supply curve given here on the same set of axes. Use your graph to estimate, to the nearest \$100, the tuition at which the demand equals the supply. Approximately how many students will be accommodated at that price, known as the **equilibrium price**?

b. What happens if the tuition is higher than the equilibrium price? What happens if it is lower?

c. Estimate the shortage or surplus of openings at private schools if tuition is set at \$1,200.

Solution

a. Figure 10 shows the graphs of demand $q = 77.8p^{-0.11}$ and supply $q = 30.4 + 0.006p$. (See the margin note for a brief description of how to plot them.)

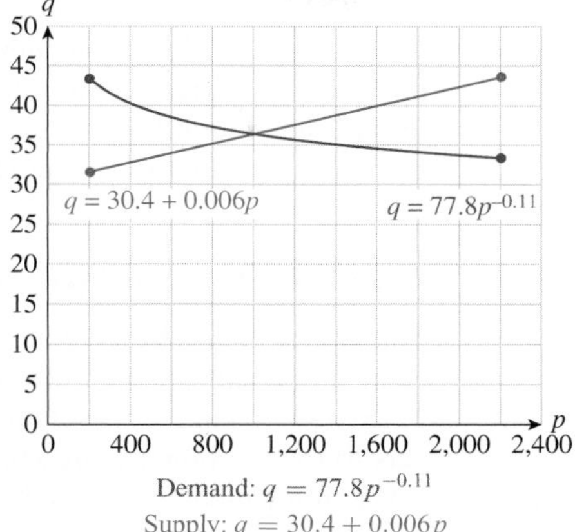

Figure 10

The lines cross close to $p = \$1{,}000$, so we conclude that demand = supply when $p \approx \$1{,}000$ (to the nearest \$100). This is the (approximate) equilibrium tuition price. At that price, we can estimate the demand or supply at around

$$\text{Demand: } q = 77.8(1{,}000)^{-0.11} \approx 36.4$$

$$\text{Supply: } q = 30.4 + 0.006(1{,}000) = 36.4 \qquad \text{Demand = Supply at equilibrium}$$

or 36,400 students.

b. Take a look at Figure 11, which shows what happens if schools charge more or less than the equilibrium price.

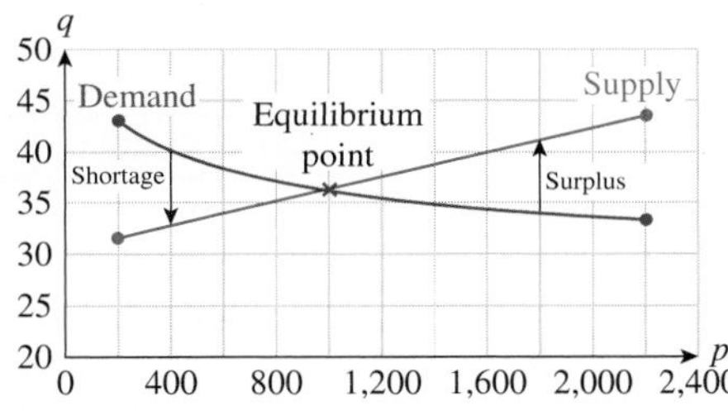

Figure 11

If tuition is, say, \$1,800, then the supply will be larger than demand and there will be a surplus of available openings at private schools. Similarly, if tuition is less—say \$400—then the supply will be less than the demand, and there will be a shortage of available openings.

c. The discussion in part (b) shows that if tuition is set at \$1,200 there will be a surplus of available openings. To estimate that number, we calculate the projected demand and supply when $p = \$1{,}200$:

$$\text{Demand: } q = 77.8(1{,}200)^{-0.11} \approx 35.7 \text{ thousand seats}$$

$$\text{Supply: } q = 30.4 + 0.006(1{,}200) = 37.6 \text{ thousand seats}$$

$$\text{Surplus} = \text{Supply} - \text{Demand} \approx 37.6 - 35.7 = 1.9 \text{ thousand seats.}$$

So, there would be a surplus of around 1,900 available seats.

using Technology

See the Technology Guides at the end of the chapter for detailed instructions on how to obtain the table of values and graph in Example 4 using a TI-83/84 Plus or Excel. Here is an outline:

TI-83/84 Plus

Graphs:
`Y1=77.8*X^(-0.11)`
`Y2=30.4+0.006*X`
[2ND] [TABLE] Graph:
Xmin = 200, Xmax = 2200;
[ZOOM] [0] [More details on page 91.]

Spreadsheet

Headings p, Demand, Supply in A1–C1; p-values 200, 300, ..., 2200 in A2-A22.
`=77.8*A2^(-0.11)` in B2
`=30.4+0.006*A2` in C2
Copy down through C22.
Highlight A1–C22; insert Scatter chart. [More details on page 96.]

 **Website**

www.WanerMath.com
Go to the Function Evaluator and Grapher under Online Utilities, and enter
`77.8*x^(-0.11)` for y_1 and
`30.4+0.006*x` for y_2.
Graph: Set Xmin = 200, Xmax = 2200, and press "Plot Graphs".

Before we go on... We just saw in Example 4 that if tuition is less than the equilibrium price there will be a shortage. If schools were to raise their tuition toward the equilibrium, they would create and fill more openings and increase revenue, because it is the supply equation—and not the demand equation—that determines what one can sell below the equilibrium price. On the other hand, if they were to charge more than the equilibrium price, they will be left with a possibly costly surplus of unused openings (and will want to lower tuition to reduce the surplus). Prices tend to move toward the equilibrium, so supply tends to equal demand. When supply equals demand, we say that the market **clears**. ■

Modeling Change over Time

Things around us change with time. Thus, there are many quantities, such as your income or the temperature in Honolulu, that are natural to think of as functions of time. Example 1 on page 40 (on iPod sales) and Example 2 on page 43 (on Facebook membership) in Section 1.1 are models of change over time. Both of those models are curve-fitting models: We used algebraic functions to approximate observed data.

Note We usually use the independent variable t to denote time (in seconds, hours, days, years, etc.). If a quantity q changes with time, then we can regard q as a function of t. ■

In the next example we are asked to select from among several curve-fitting models for given data.

EXAMPLE 5 Model Selection: Sales

The following table shows annual sales, in billions of dollars, by Nike from 2005 through 2010:[7]

Year	2005	2006	2007	2008	2009	2010
Sales ($ billion)	13.5	15	16.5	18.5	19	19

Take t to be the number of years since 2005, and consider the following four models:

(1) $s(t) = 14 + 1.2t$ Linear model

(2) $s(t) = 13 + 2.2t - 0.2t^2$ Quadratic model

(3) $s(t) = 14(1.07^t)$ Exponential model

(4) $s(t) = \dfrac{19.5}{1 + 0.48(1.8^{-t})}$ Logistic model

a. Which models fit the data significantly better than the rest?

b. Of the models you selected in part (a), which gives the most reasonable prediction for 2013?

[7]Figures are rounded. Source: http://invest.nike.com.

Solution

a. The following table shows the original data together with the values, rounded to the nearest 0.5, for all four models:

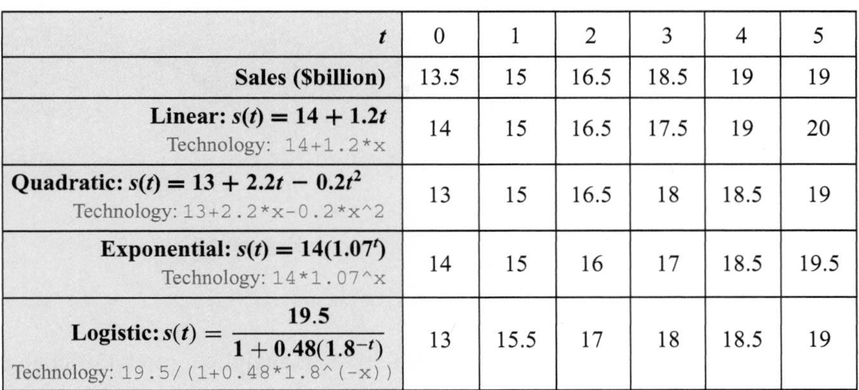

t	0	1	2	3	4	5
Sales ($billion)	13.5	15	16.5	18.5	19	19
Linear: $s(t) = 14 + 1.2t$ Technology: `14+1.2*x`	14	15	16.5	17.5	19	20
Quadratic: $s(t) = 13 + 2.2t - 0.2t^2$ Technology: `13+2.2*x-0.2*x^2`	13	15	16.5	18	18.5	19
Exponential: $s(t) = 14(1.07^t)$ Technology: `14*1.07^x`	14	15	16	17	18.5	19.5
Logistic: $s(t) = \dfrac{19.5}{1 + 0.48(1.8^{-t})}$ Technology: `19.5/(1+0.48*1.8^(-x))`	13	15.5	17	18	18.5	19

Notice that all the models give values that seem reasonably close to the actual sales values. However, the quadratic and logistic curves seem to model their behavior more accurately than the others (see Figure 12).

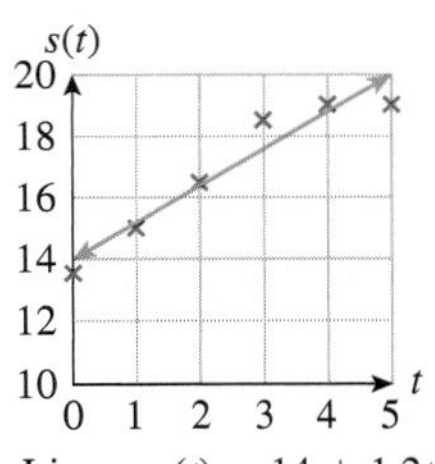

Linear: $s(t) = 14 + 1.2t$

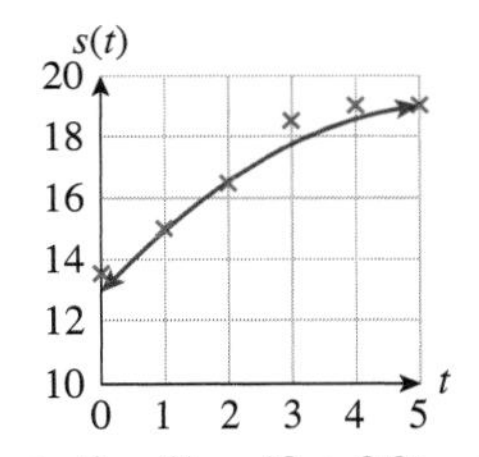

Quadratic: $s(t) = 13 + 2.2t - 0.2t^2$

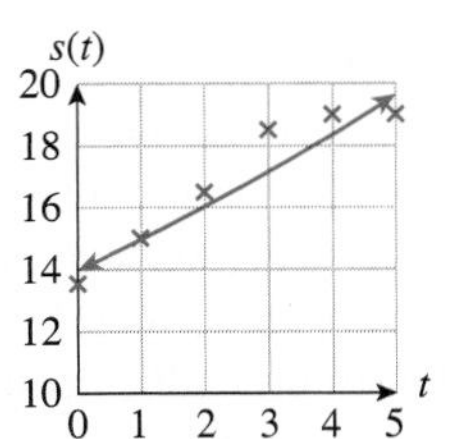

Exponential: $s(t) = 14(1.07^t)$

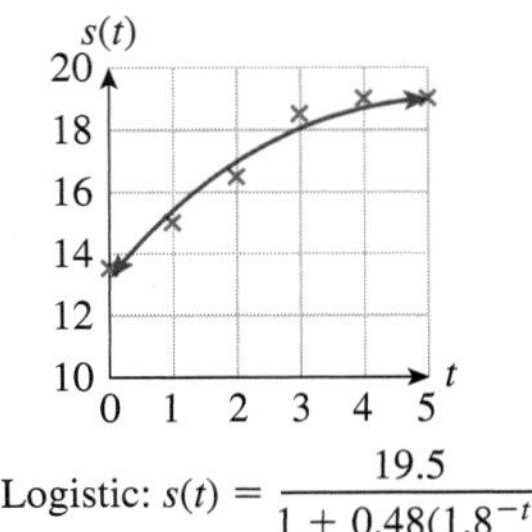

Logistic: $s(t) = \dfrac{19.5}{1 + 0.48(1.8^{-t})}$

Figure 12

We therefore conclude that the quadratic and logistic models fit the data significantly better than the others.

b. Although the quadratic and logistic models both appear to fit the data well, they do not both extrapolate to give reasonable predictions for 2013:

$$\text{Quadratic Model: } s(8) = 13 + 2.2(8) - 0.2(8)^2 = 17.8$$

$$\text{Logistic Model: } s(8) = \frac{19.5}{1 + 0.48(1.8^{-8})} \approx 19.4.$$

using Technology

See the Technology Guides at the end of the chapter for detailed instructions on how to obtain the table and graphs in Example 5 using a TI-83/84 Plus or Excel. Here is an outline:

TI-83/84 Plus
STAT EDIT; enter the values of t in L_1, and $r(t)$ in L_2.
Plotting the points: ZOOM 9
Adding a curve:
Turn on Plot1 in Y= screen. Then (for the second curve)
`Y1=13+2.2*X-0.2*X^2`
Press GRAPH [More details on page 91.]

Spreadsheet
Table of values: Headings t, Sales, $s(t)$ in A1–C1;
t-values in A2-A7; sales values in B2–B7. Formula in C2:
`=13+2.2*A2-0.2*A2^2`
(second model)
Copy down through C7
Graph: Highlight A1–C7; insert Scatter chart. [More details on page 97.]

Website
www.WanerMath.com
In the Function Evaluator and Grapher utility, enter the data and model(s) as shown below. Set xMin = 0, xMax = 5 and press "Plot Graphs".

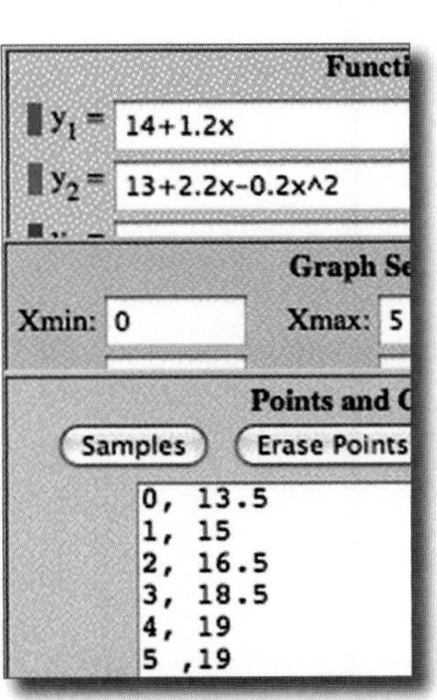

Notice that the quadratic model predicts a significant *decline* in sales whereas the logistic model predicts a more reasonable modest increase. This discrepancy can be seen quite dramatically in Figure 13.

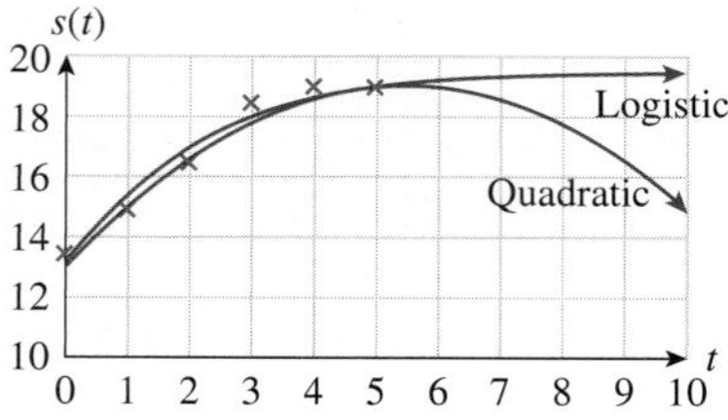

Figure 13

We now derive an analytical model of change over time based on the idea of **compound interest**. Suppose you invest \$500 (the **present value**) in an investment account with an annual yield of 15%, and the interest is reinvested at the end of every year (we say that the interest is **compounded** or **reinvested** once a year). Let t represent the number of years since you made the initial \$500 investment. Each year, the investment is worth 115% (or 1.15 times) of its value the previous year. The **future value** A of your investment changes over time t, so we think of A as a function of t. The following table illustrates how we can calculate the future value for several values of t:

t	0	1	2	3
FutureValue $A(t)$	500	575	661.25	760.44
	A	$500(1.15)$	$500(1.15)^2$	$500(1.15)^3$
		×1.15	×1.15	×1.15

Thus, $A(t) = 500(1.15)^t$. A traditional way to write this formula is

$$A(t) = P(1 + r)^t,$$

where P is the present value ($P = 500$) and r is the annual interest rate ($r = 0.15$).

If, instead of compounding the interest once a year, we compound it every three months (four times a year), we would earn one quarter of the interest ($r/4$ of the current investment) every three months. Because this would happen $4t$ times in t years, the formula for the future value becomes

$$A(t) = P\left(1 + \frac{r}{4}\right)^{4t}.$$

Compound Interest

If an amount (**present value**) P is invested for t years at an annual rate of r, and if the interest is compounded (reinvested) m times per year, then the **future value** A is

$$A(t) = P\left(1 + \frac{r}{m}\right)^{mt}.$$

A special case is **interest compounded once a year**:

$$A(t) = P(1 + r)^t.$$

Quick Example

If \$2,000 is invested for two and a half years in a mutual fund with an annual yield of 12.6% and the earnings are reinvested each month, then $P = 2{,}000$, $r = 0.126$, $m = 12$, and $t = 2.5$, which gives

$$A(2.5) = 2{,}000\left(1 + \frac{0.126}{12}\right)^{12\times 2.5}$$ `2000*(1+0.126/12)^(12*2.5)`

$$= 2{,}000(1.0105)^{30} = \$2{,}736.02.$$

EXAMPLE 6 Compound Interest: Investments

Consider the scenario in the preceding Quick Example: You invest \$2,000 in a mutual fund with an annual yield of 12.6% and the interest is reinvested each month.

a. Find the associated exponential model.

b. T Use a table of values to estimate the year during which the value of your investment reaches \$5,000.

c. Use a graph to confirm your answer in part (b).

Solution

a. Apply the formula

$$A(t) = P\left(1 + \frac{r}{m}\right)^{mt}$$

with $P = 2{,}000$, $r = 0.126$, and $m = 12$. We get

$$A(t) = 2{,}000\left(1 + \frac{0.126}{12}\right)^{12t}$$

$$= 2{,}000(1.0105)^{12t}.$$ `2000*(1+0.126/12)^(12*t)`

This is the exponential model. (What would happen if we left out the last set of parentheses in the technology formula?)

b. We need to find the value of t for which $A(t) = \$5{,}000$, so we need to solve the equation

$$5{,}000 = 2{,}000(1.0105)^{12t}.$$

In Section 2.3 we will learn how to use logarithms to do this algebraically, but we can answer the question now using a graphing calculator, a spreadsheet, or the Function Evaluator and Grapher utility at the Website. Just enter the model and compute the balance at the end of several years. Here are examples of tables obtained using three forms of technology:

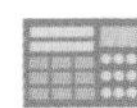 **using Technology**

TI-83/84 Plus

`Y1=2000(1+0.126/12)^(12X)`

2ND TABLE

Spreadsheet

Headings *t* and *A* in A1–B1; *t*-values 0–11 in A2–A13.

`=2000*(1+0.126/12)^(12*A2)`

in B2; copy down through B13.

 **Website**

www.WanerMath.com

Go to the Function Evaluator and Grapher under Online Utilities, and enter

`2000(1+0.126/12)^(12x)`

for y_1. Scroll down to the Evaluator, enter the values 0–11 under *x*-values and press "Evaluate."

X	Y_1
5	3742.9
6	4242.7
7	4809.3
8	5451.5
9	6179.5
10	7004.7
11	7940.1

Y_1=5451.50618802

TI-83/84 Plus

	A	B
1	t	A
2	0	$ 2,000.00
3	1	$ 2,267.07
4	2	$ 2,569.81
5	3	$ 2,912.98
6	4	$ 3,301.97
7	5	$ 3,742.91
8	6	$ 4,242.72
9	7	$ 4,809.29
10	8	$ 5,451.51
11	9	$ 6,179.49

Excel

x-Values	y_1-Values
3	2912.98
4	3301.97
5	3742.91
6	4242.72
7	4809.29
8	5451.51

Website

Because the balance first exceeds \$5,000 at $t = 8$ (the end of year 8), your investment has reached \$5,000 during year 8.

c. Figure 14 shows the graph of $A(t) = 2{,}000(1.0105)^{12t}$ together with the horizontal line $y = 5{,}000$. The graphs cross between $t = 7$ and $t = 8$, confirming that year 8 is the first year during which the value of the investment reaches \$5,000.

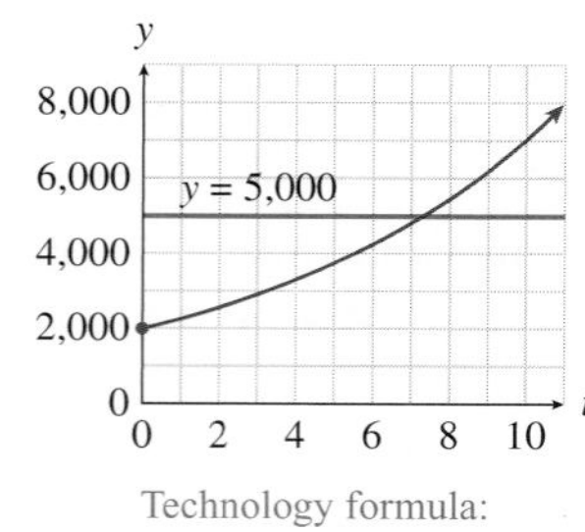

Technology formula:
2000*1.0105^(12*x)

Figure 14

The compound interest examples we saw above are instances of **exponential growth:** a quantity whose magnitude is an increasing exponential function of time. The decay of unstable radioactive isotopes provides instances of **exponential decay:** a quantity whose magnitude is a *decreasing* exponential function of time. For example, carbon 14, an unstable isotope of carbon, decays exponentially to nitrogen. Because carbon 14 decay is extremely slow, it has important applications in the dating of fossils.

EXAMPLE 7 Exponential Decay: Carbon Dating

The amount of carbon 14 remaining in a sample that originally contained A grams is approximately

$$C(t) = A(0.999879)^t,$$

where t is time in years.

a. What percentage of the original amount remains after one year? After two years?

b. Graph the function C for a sample originally containing 50 g of carbon 14, and use your graph to estimate how long, to the nearest 1,000 years, it takes for half the original carbon 14 to decay.

c. A fossilized plant unearthed in an archaeological dig contains 0.50 g of carbon 14 and is known to be 50,000 years old. How much carbon 14 did the plant originally contain?

Solution

Notice that the given model is exponential as it has the form $f(t) = Ab^t$. (See page 47.)

a. At the start of the first year, $t = 0$, so there are

$$C(0) = A(0.999879)^0 = A \text{ grams.}$$

At the end of the first year, $t = 1$, so there are

$$C(1) = A(0.999879)^1 = 0.999879A \text{ grams;}$$

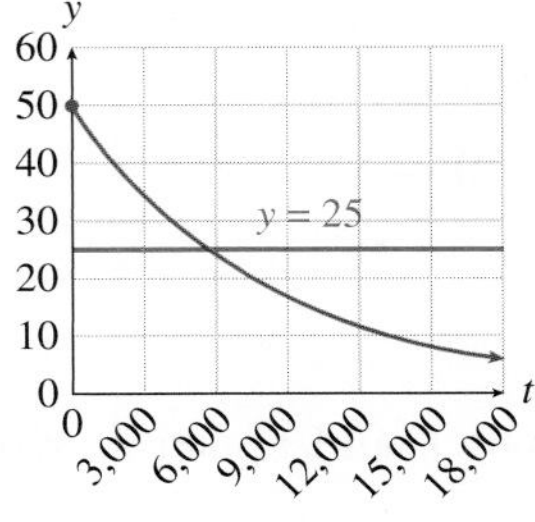

Technology formula:
50*0.999879^x

Figure 15

that is, 99.9879% of the original amount remains. After the second year, the amount remaining is

$$C(2) = A(0.999879)^2 \approx 0.999758A \text{ grams},$$

or about 99.9758% of the original sample.

b. For a sample originally containing 50 g of carbon 14, $A = 50$, so $C(t) = 50(0.999879)^t$. Its graph is shown in Figure 15. We have also plotted the line $y = 25$ on the same graph. The graphs intersect at the point where the original sample has decayed to 25 g: about $t = 6{,}000$ years.

c. We are given the following information: $C = 0.50$, $A =$ the unknown, and $t = 50{,}000$. Substituting gives

$$0.50 = A(0.999879)^{50{,}000}.$$

Solving for A gives

$$A = \frac{0.5}{0.999879^{50{,}000}} \approx 212 \text{ grams}.$$

Thus, the plant originally contained 212 g of carbon 14.

Before we go on...

The formula we used for A in Example 7(c) has the form

$$A(t) = \frac{C}{0.999879^t},$$

which gives the original amount of carbon 14 t years ago in terms of the amount C that is left now. A similar formula can be used in finance to find the present value, given the future value. ■

Algebra of Functions

If you look back at some of the functions considered in this section, you will notice that we frequently constructed them by combining simpler or previously constructed functions. For instance:

Quick Example 3 on page 48:	Area = Width × Length: $A(x) = x(50 - x)$
Example 1:	Cost = Variable Cost + Fixed Cost: $C(x) = 1.80x + 2.05$
Quick Example on page 51:	Profit = Revenue − Cost: $P(x) = 10x - (8x + 100)$.

Let us look a little more deeply at each of the above examples:

Area Example: $A(x) = \text{Width} \times \text{Length} = x(50 - x)$:
Think of the width and length as separate functions of x:

$$\text{Width: } W(x) = x; \qquad \text{Length: } L(x) = 50 - x$$

so that

$$A(x) = W(x)L(x). \qquad \text{Area = Width × Length}$$

We say that the area function A is the **product of the functions** W and L, and we write

$$A = WL. \qquad A \text{ is the product of the functions } W \text{ and } L.$$

To calculate $A(x)$, we multiply $W(x)$ by $L(x)$.

Cost Example: $C(x) =$ Variable Cost + Fixed Cost $= 1.80x + 2.05$:
Think of the variable and fixed costs as separate functions of x:

Variable Cost: $V(x) = 1.80x$;   Fixed Cost: $F(x) = 2.05$*

* F is called a constant function as its value, 2.05, is the same for every value of x.

so that

$$C(x) = V(x) + F(x). \qquad \text{Cost = Variable Cost + Fixed Cost}$$

We say that the cost function C is the **sum of the functions V and F**, and we write

$$C = V + F. \qquad C \text{ is the sum of the functions } V \text{ and } F.$$

To calculate $C(x)$, we add $V(x)$ to $F(x)$.

Profit Example: $P(x) =$ Revenue − Cost $= 10x - (8x + 100)$:
Think of the revenue and cost as separate functions of x:

Revenue: $R(x) = 10x$;   Cost: $C(x) = 8x + 100$

so that

$$P(x) = R(x) - C(x). \qquad \text{Profit = Revenue − Cost}$$

We say that the profit function P is the **difference between the functions R and C**, and we write

$$P = R - C. \qquad P \text{ is the difference of the functions } R \text{ and } C.$$

To calculate $P(x)$, we subtract $C(x)$ from $R(x)$.

Algebra of Functions

If f and g are real-valued functions of the real variable x, then we define their **sum s, difference d, product p,** and **quotient q** as follows:

$s = f + g$ is the function specified by $s(x) = f(x) + g(x)$.
$d = f - g$ is the function specified by $d(x) = f(x) - g(x)$.
$p = fg$ is the function specified by $p(x) = f(x)g(x)$.
$q = \dfrac{f}{g}$ is the function specified by $q(x) = \dfrac{f(x)}{g(x)}$.

Also, if f is as above and c is a constant (real number), then we define the associated **constant multiple m of f** by

$m = cf$ is the function specified by $m(x) = cf(x)$.

Note on Domains

In order for any of the expressions $f(x) + g(x)$, $f(x) - g(x)$, $f(x)g(x)$, or $f(x)/g(x)$ to make sense, x must be simultaneously in the domains of both f and g. Further, for the quotient, the denominator $g(x)$ cannot be zero. Thus, we specify the domains of these functions as follows:

Domain of $f + g$, $f - g$, and fg: All real numbers x simultaneously in the domains of f and g

Domain of f/g: All real numbers x simultaneously in the domains of f and g such that $g(x) \neq 0$

Domain of cf: Same as the domain of f

Quick Examples

1. If $f(x) = x^2 - 1$ and $g(x) = \sqrt{x}$ with domain $[0, +\infty)$, then the sum s of f and g has domain $[0, +\infty)$ and is specified by $s(x) = f(x) + g(x) = x^2 - 1 + \sqrt{x}$.
2. If $f(x) = x^2 - 1$ and $c = 3$, then the associated constant multiple m of f is specified by $m(x) = 3f(x) = 3(x^2 - 1)$.
3. If there are $N = 1{,}000t$ Mars shuttle passengers in year t who pay a total cost of $C = 40{,}000 + 800t$ million dollars, then the cost per passenger is given by the quotient of the two functions,

$$\text{Cost per passenger} = q(t) = \frac{C(t)}{N(t)} = \frac{40{,}000 + 800t}{1{,}000t} \text{ million dollars per passenger.}$$

The largest possible domain of C/N is $(0, +\infty)$, as the quotient is not defined if $t = 0$.

1.2 EXERCISES

Access end-of-section exercises online at **www.webassign.net** ENHANCED WebAssign

1.3 Linear Functions and Models

Linear functions are among the simplest functions and are perhaps the most useful of all mathematical functions.

Linear Function

A **linear function** is one that can be written in the form

$$f(x) = mx + b \qquad \text{Function form}$$

or

$$y = mx + b \qquad \text{Equation form}$$

where m and b are fixed numbers. (The names m and b are traditional.*)

Quick Example

$f(x) = 3x - 1$

$y = 3x - 1$

* Actually, c is sometimes used instead of b. As for m, there has even been some research into the question of its origin, but no one knows exactly why the letter m is used.

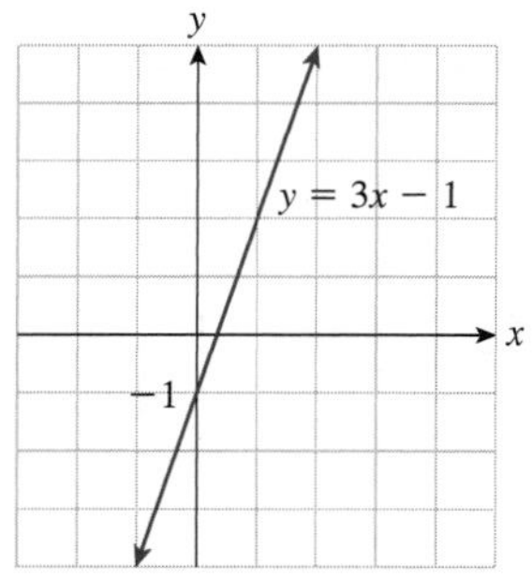

Figure 16

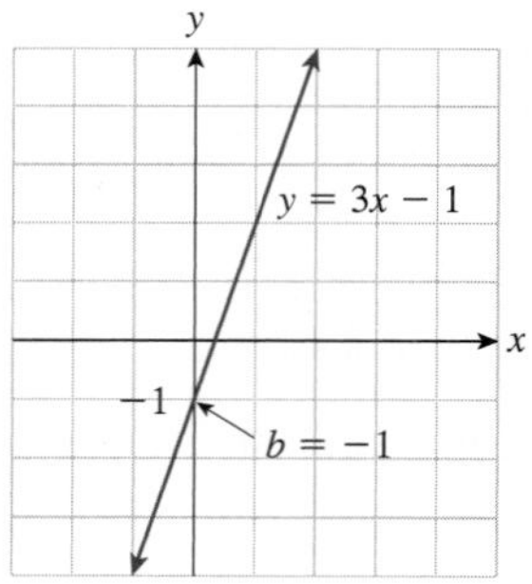

y-intercept $= b = -1$
Graphically, b is the y-intercept of the graph.

Figure 17

Linear Functions from the Numerical and Graphical Point of View

The following table shows values of $y = 3x - 1$ ($m = 3, b = -1$) for some values of x:

x	−4	−3	−2	−1	0	1	2	3	4
y	−13	−10	−7	−4	−1	2	5	8	11

Its graph is shown in Figure 16.

Looking first at the table, notice that setting $x = 0$ gives $y = -1$, the value of b.

Numerically, b is the value of y when x = 0.

On the graph, the corresponding point $(0, -1)$ is the point where the graph crosses the y-axis, and we say that $b = -1$ is the ***y*-intercept** of the graph (Figure 17).

What about m? Looking once again at the table, notice that y increases by $m = 3$ units for every increase of 1 unit in x. This is caused by the term $3x$ in the formula: for every increase of 1 in x we get an increase of $3 \times 1 = 3$ in y.

Numerically, y increases by m units for every 1-unit increase of x.

Likewise, for every increase of 2 in x we get an increase of $3 \times 2 = 6$ in y. In general, if x increases by some amount, y will increase by three times that amount. We write:

$$\text{Change in } y = 3 \times \text{Change in } x.$$

The Change in a Quantity: Delta Notation

If a quantity q changes from q_1 to q_2, the **change in q** is just the difference:

$$\begin{aligned}\text{Change in } q &= \text{Second value} - \text{First value} \\ &= q_2 - q_1.\end{aligned}$$

Mathematicians traditionally use Δ (delta, the Greek equivalent of the Roman letter D) to stand for change, and write the change in q as Δq.

$$\Delta q = \text{Change in } q = q_2 - q_1$$

Quick Examples

1. If x is changed from 1 to 3, we write

$$\Delta x = \text{Second value} - \text{First value} = 3 - 1 = 2.$$

2. Looking at our linear function, we see that, when x changes from 1 to 3, y changes from 2 to 8. So,

$$\Delta y = \text{Second value} - \text{First value} = 8 - 2 = 6.$$

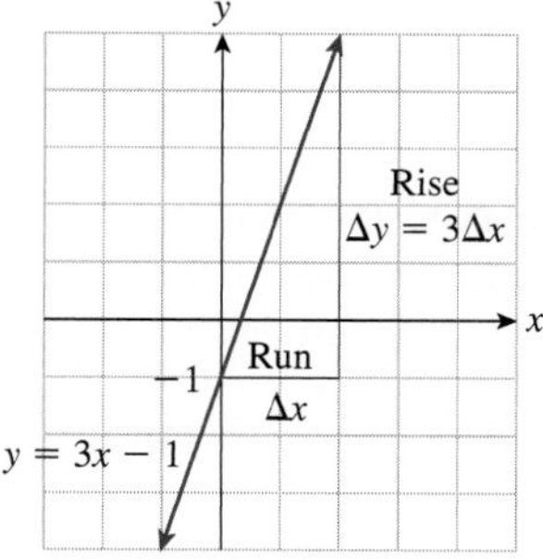

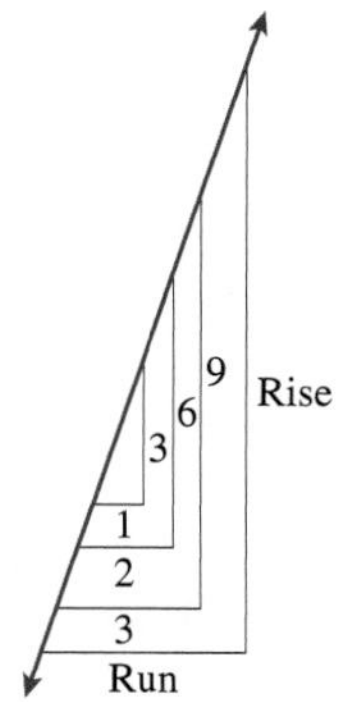

Slope $= m = 3$
Graphically, m is the slope of the graph.

Figure 18

Using delta notation, we can now write, for our linear function,

$$\Delta y = 3\Delta x \qquad \text{Change in } y = 3 \times \text{Change in } x.$$

or

$$\frac{\Delta y}{\Delta x} = 3.$$

Because the value of y increases by exactly 3 units for every increase of 1 unit in x, the graph is a straight line rising by 3 units for every 1 unit we go to the right. We say that we have a **rise** of 3 units for each **run** of 1 unit. Because the value of y changes by $\Delta y = 3\Delta x$ units for every change of Δx units in x, in general we have a rise of $\Delta y = 3\Delta x$ units for each run of Δx units (Figure 18). Thus, we have a rise of 6 for a run of 2, a rise of 9 for a run of 3, and so on. So, $m = 3$ is a measure of the steepness of the line; we call m the **slope of the line**:

$$\text{Slope} = m = \frac{\Delta y}{\Delta x} = \frac{\text{Rise}}{\text{Run}}.$$

In general (replace the number 3 by a general number m), we can say the following.

The Roles of *m* and *b* in the Linear Function *f*(*x*) = *mx* + *b*

Role of *m*

Numerically If $y = mx + b$, then y changes by m units for every 1-unit change in x. A change of Δx units in x results in a change of $\Delta y = m\Delta x$ units in y. Thus,

$$m = \frac{\Delta y}{\Delta x} = \frac{\text{Change in } y}{\text{Change in } x}.$$

Graphically m is the slope of the line $y = mx + b$:

$$m = \frac{\Delta y}{\Delta x} = \frac{\text{Rise}}{\text{Run}} = \text{Slope}.$$

For positive m, the graph rises m units for every 1-unit move to the right, and rises $\Delta y = m\Delta x$ units for every Δx units moved to the right. For negative m, the graph drops $|m|$ units for every 1-unit move to the right, and drops $|m|\Delta x$ units for every Δx units moved to the right.

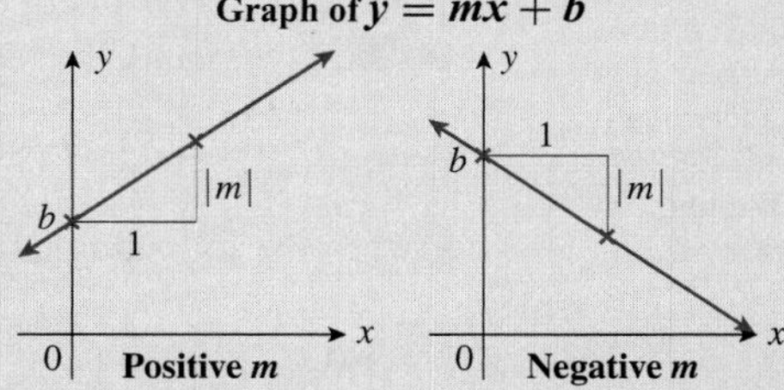

Role of *b*

Numerically When $x = 0$, $y = b$.

Graphically b is the y-intercept of the line $y = mx + b$.

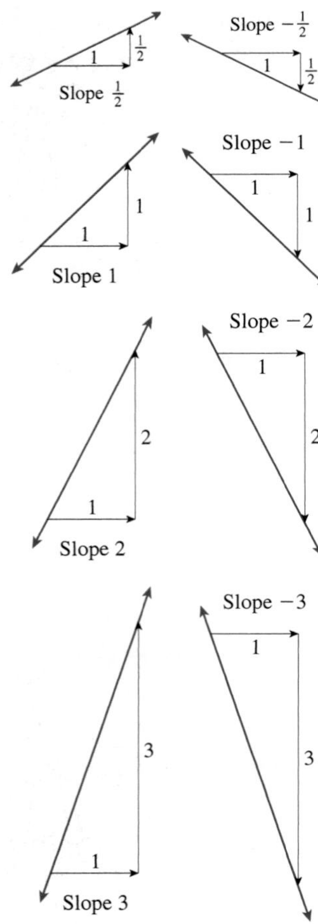

Figure 19

Quick Examples

1. $f(x) = 2x + 1$ has slope $m = 2$ and y-intercept $b = 1$. To sketch the graph, we start at the y-intercept $b = 1$ on the y-axis, and then move 1 unit to the right and up $m = 2$ units to arrive at a second point on the graph. Now connect the two points to obtain the graph on the left.

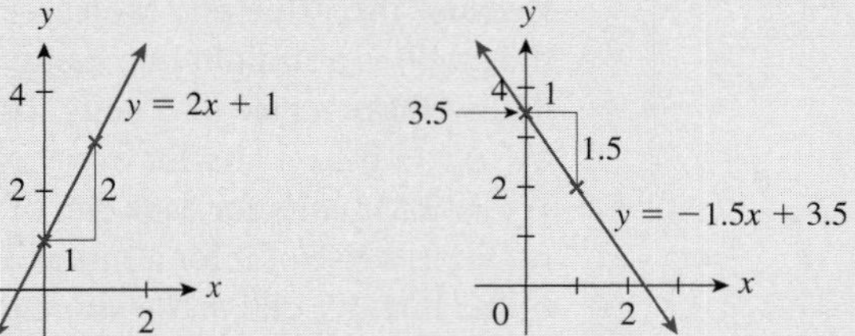

2. The line $y = -1.5x + 3.5$ has slope $m = -1.5$ and y-intercept $b = 3.5$. Because the slope is negative, the graph (above right) goes *down* 1.5 units for every 1 unit it moves to the right.

It helps to be able to picture what different slopes look like, as in Figure 19. Notice that the larger the absolute value of the slope, the steeper is the line.

EXAMPLE 1 Recognizing Linear Data Numerically and Graphically

Which of the following two tables gives the values of a linear function? What is the formula for that function?

x	0	2	4	6	8	10	12
$f(x)$	3	−1	−3	−6	−8	−13	−15

x	0	2	4	6	8	10	12
$g(x)$	3	−1	−5	−9	−13	−17	−21

Solution The function f cannot be linear: If it were, we would have $\Delta f = m\Delta x$ for some fixed number m. However, although the change in x between successive entries in the table is $\Delta x = 2$ each time, the change in f is not the same each time. Thus, the ratio $\Delta f/\Delta x$ is not the same for every successive pair of points.

On the other hand, the ratio $\Delta g/\Delta x$ is the same each time, namely,

$$\frac{\Delta g}{\Delta x} = \frac{-4}{2} = -2$$

as we see in the following table:

x	0	2	4	6	8	10	12
$g(x)$	3	−1	−5	−9	−13	−17	−21

Δx	$2-0=2$	$4-2=2$	$6-4=2$	$8-6=2$	$10-8=2$	$12-10=2$
Δg	$-1-3$ $=-4$	$-5-(-1)$ $=-4$	$-9-(-5)$ $=-4$	$-13-(-9)$ $=-4$	$-17-(-13)$ $=-4$	$-21-(-17)$ $=-4$

using Technology

See the Technology Guides at the end of the chapter for detailed instructions on how to obtain a table with the successive quotients $m = \Delta y/\Delta x$ for the functions f and g in Example 1 using a TI-83/84 Plus or Excel. These tables show at a glance that f is not linear. Here is an outline:

TI-83/84 Plus

STAT EDIT; Enter values of x and $f(x)$ in lists L_1 and L_2.
Highlight the heading L_3 then enter the following formula (including the quotes)

```
"ΔList(L2)/ΔList(L1)"
```

[More details on page 92.]

Spreadsheet
Enter headings x, $f(x)$, Df/Dx in cells A1–C1, and the corresponding values from one of the tables in cells A2–B8. Enter
`=(B3-B2)/(A3-A2)`
in cell C2, and copy down through C8.
[More details on page 98.]

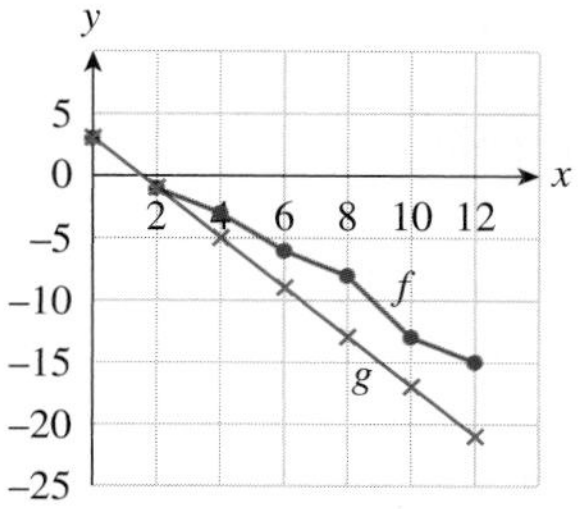

Figure 20

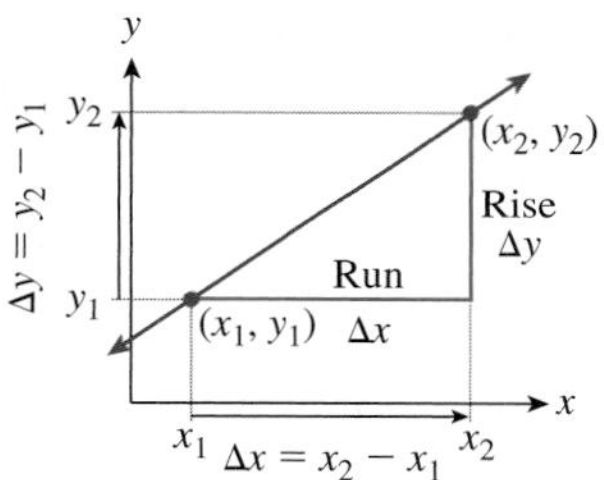

Figure 21

Thus, g is linear with slope $m = -2$. By the table, $g(0) = 3$, hence $b = 3$. Thus,

$$g(x) = -2x + 3. \quad \text{Check that this formula gives the values in the table.}$$

If you graph the points in the tables defining f and g above, it becomes easy to see that g is linear and f is not; the points of g lie on a straight line (with slope -2), whereas the points of f do not lie on a straight line (Figure 20).

Finding a Linear Equation from Data

If we happen to know the slope and y-intercept of a line, writing down its equation is straightforward. For example, if we know that the slope is 3 and the y-intercept is -1, then the equation is $y = 3x - 1$. Sadly, the information we are given is seldom so convenient. For instance, we may know the slope and a point other than the y-intercept, two points on the line, or other information. We therefore need to know how to use the information we are given to obtain the slope and the intercept.

Computing the Slope

We can always determine the slope of a line if we are given two (or more) points on the line, because any two points—say (x_1, y_1) and (x_2, y_2)—determine the line, and hence its slope. To compute the slope when given two points, recall the formula

$$\text{Slope} = m = \frac{\text{Rise}}{\text{Run}} = \frac{\Delta y}{\Delta x}.$$

To find its slope, we need a run Δx and corresponding rise Δy. In Figure 21, we see that we can use $\Delta x = x_2 - x_1$, the change in the x-coordinate from the first point to the second, as our run, and $\Delta y = y_2 - y_1$, the change in the y-coordinate, as our rise. The resulting formula for computing the slope is given in the box.

Computing the Slope of a Line

We can compute the slope m of the line through the points (x_1, y_1) and (x_2, y_2) using

$$m = \frac{\Delta y}{\Delta x} = \frac{y_2 - y_1}{x_2 - x_1}.$$

Quick Examples

1. The slope of the line through $(x_1, y_1) = (1, 3)$ and $(x_2, y_2) = (5, 11)$ is

$$m = \frac{\Delta y}{\Delta x} = \frac{y_2 - y_1}{x_2 - x_1} = \frac{11 - 3}{5 - 1} = \frac{8}{4} = 2.$$

Notice that we can use the points in the reverse order: If we take $(x_1, y_1) = (5, 11)$ and $(x_2, y_2) = (1, 3)$, we obtain the same answer:

$$m = \frac{\Delta y}{\Delta x} = \frac{y_2 - y_1}{x_2 - x_1} = \frac{3 - 11}{1 - 5} = \frac{-8}{-4} = 2.$$

2. The slope of the line through $(x_1, y_1) = (1, 2)$ and $(x_2, y_2) = (2, 1)$ is

$$m = \frac{\Delta y}{\Delta x} = \frac{y_2 - y_1}{x_2 - x_1} = \frac{1 - 2}{2 - 1} = \frac{-1}{1} = -1.$$

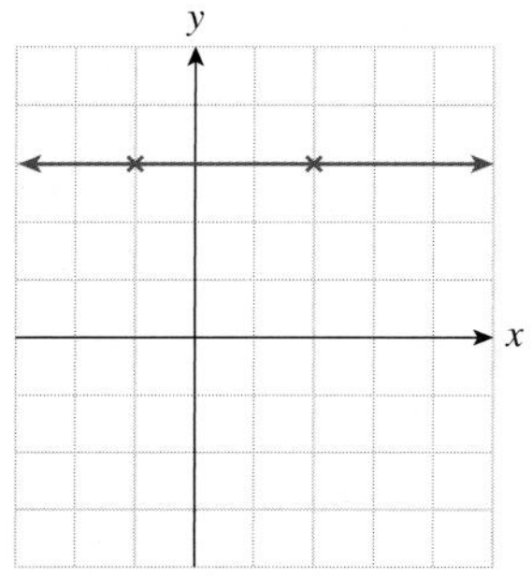

Figure 22

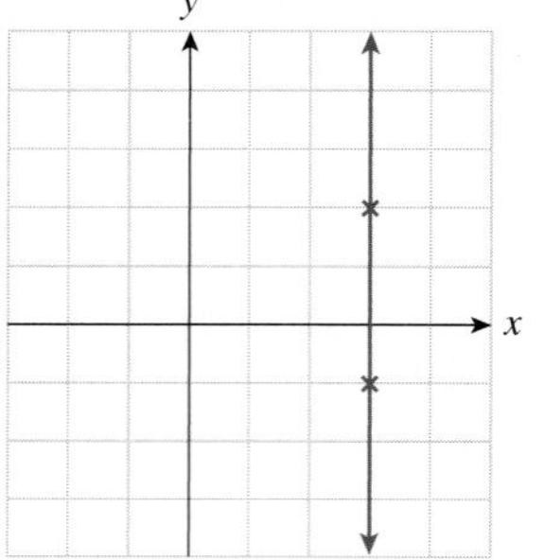

Vertical lines have undefined slope.

Figure 23

3. The slope of the line through (2, 3) and (−1, 3) is

$$m = \frac{\Delta y}{\Delta x} = \frac{y_2 - y_1}{x_2 - x_1} = \frac{3-3}{-1-2} = \frac{0}{-3} = 0.$$

A line of slope 0 has zero rise, so is a *horizontal* line, as shown in Figure 22.

4. The line through (3, 2) and (3, −1) has slope

$$m = \frac{\Delta y}{\Delta x} = \frac{y_2 - y_1}{x_2 - x_1} = \frac{-1-2}{3-3} = \frac{-3}{0},$$

which is undefined. The line passing through these points is *vertical*, as shown in Figure 23.

Computing the *y*-Intercept

Once we know the slope m of a line, and also the coordinates of a point (x_1, y_1), then we can calculate its y-intercept b as follows: The equation of the line must be

$$y = mx + b,$$

where b is as yet unknown. To determine b we use the fact that the line must pass through the point (x_1, y_1), so that (x_1, y_1) satisfies the equation $y = mx + b$. In other words,

$$y_1 = mx_1 + b.$$

Solving for b gives

$$b = y_1 - mx_1.$$

In summary:

Computing the *y*-Intercept of a Line

The y-intercept of the line passing through (x_1, y_1) with slope m is

$$b = y_1 - mx_1.$$

Quick Example

The line through (2, 3) with slope 4 has

$$b = y_1 - mx_1 = 3 - (4)(2) = -5.$$

Its equation is therefore

$$y = mx + b = 4x - 5.$$

EXAMPLE 2 Finding Linear Equations

Find equations for the following straight lines.

a. Through the points (1, 2) and (3, −1)

b. Through (2, −2) and parallel to the line $3x + 4y = 5$

c. Horizontal and through (−9, 5)

d. Vertical and through (−9, 5)

using Technology

See the Technology Guides at the end of the chapter for detailed instructions on how to obtain the slope and intercept in Example 2(a) using a TI-83/84 Plus or a spreadsheet. Here is an outline:

TI-83/84 Plus

STAT EDIT; Enter values of x and y in lists L_1 and L_2.

Slope: Home screen

```
(L2(2)-L2(1))/(L1(2)-L1(1))→M
```

Intercept: Home screen

```
L2(1)-M*L1(1)
```

[More details on page 93.]

Spreadsheet

Enter headings x, y, m, b, in cells A1–D1, and the values (x, y) in cells A2–B3. Enter

```
=(B3-B2)/(A3-A2)
```

in cell C2, and

```
=B2-C2*A2
```

in cell D2.

[More details on page 98.]

Solution

a. To write down the equation of the line, we need the slope m and the y-intercept b.

- ***Slope*** Because we are given two points on the line, we can use the slope formula:

$$m = \frac{y_2 - y_1}{x_2 - x_1} = \frac{-1-2}{3-1} = -\frac{3}{2}.$$

- ***Intercept*** We now have the slope of the line, $m = -3/2$, and also a point—we have two to choose from, so let us choose $(x_1, y_1) = (1, 2)$. We can now use the formula for the y-intercept:

$$b = y_1 - mx_1 = 2 - \left(-\frac{3}{2}\right)(1) = \frac{7}{2}.$$

Thus, the equation of the line is

$$y = -\frac{3}{2}x + \frac{7}{2}. \qquad y = mx + b$$

b. Proceeding as before,

- ***Slope*** We are not given two points on the line, but we are given a parallel line. We use the fact that *parallel lines have the same slope.* (Why?) We can find the slope of $3x + 4y = 5$ by solving for y and then looking at the coefficient of x:

$$y = -\frac{3}{4}x + \frac{5}{4} \qquad \text{To find the slope, solve for } y.$$

so the slope is $-3/4$.

- ***Intercept*** We now have the slope of the line, $m = -3/4$, and also a point $(x_1, y_1) = (2, -2)$. We can now use the formula for the y-intercept:

$$b = y_1 - mx_1 = -2 - \left(-\frac{3}{4}\right)(2) = -\frac{1}{2}.$$

Thus, the equation of the line is

$$y = -\frac{3}{4}x - \frac{1}{2}. \qquad y = mx + b$$

c. We are given a point: $(-9, 5)$. Furthermore, we are told that the line is horizontal, which tells us that the slope is $m = 0$. Therefore, all that remains is the calculation of the y-intercept:

$$b = y_1 - mx_1 = 5 - (0)(-9) = 5$$

so the equation of the line is

$$y = 5. \qquad y = mx + b$$

d. We are given a point: $(-9, 5)$. This time, we are told that the line is vertical, which means that the slope is undefined. Thus, we can't express the equation of the line in the form $y = mx + b$. (This formula makes sense only when the slope m of the line is defined.) What can we do? Well, here are some points on the desired line:

$$(-9, 1), (-9, 2), (-9, 3), \ldots$$

so $x = -9$ and $y = \textit{anything}$. If we simply say that $x = -9$, then these points are all solutions, so the equation is $x = -9$.

Applications: Linear Models

Using linear functions to describe or approximate relationships in the real world is called **linear modeling**.

Recall from Section 1.2 that a **cost function** specifies the cost C as a function of the number of items x.

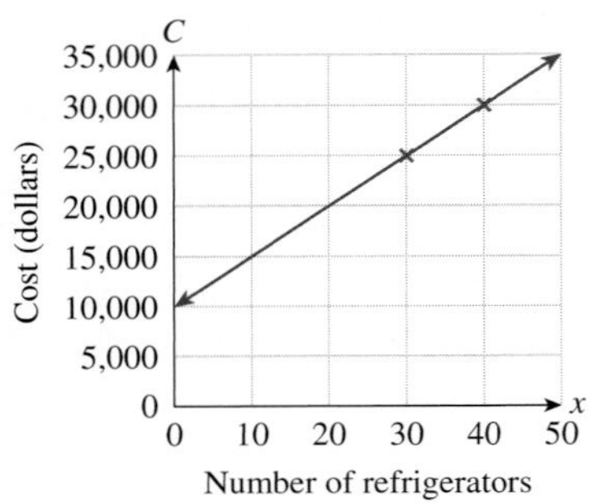

Figure 24

EXAMPLE 3 Linear Cost Function from Data

The manager of the FrozenAir Refrigerator factory notices that on Monday it cost the company a total of \$25,000 to build 30 refrigerators and on Tuesday it cost \$30,000 to build 40 refrigerators. Find a linear cost function based on this information. What is the daily fixed cost, and what is the marginal cost?

Solution We are seeking the cost C as a linear function of x, the number of refrigerators sold:

$$C = mx + b.$$

We are told that $C = 25{,}000$ when $x = 30$, and this amounts to being told that (30, 25,000) is a point on the graph of the cost function. Similarly, (40, 30,000) is another point on the line (Figure 24).

We can use the two points on the line to construct the linear cost equation:

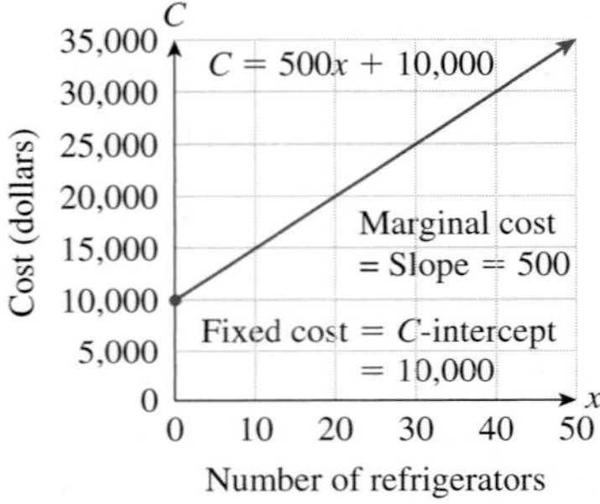

Figure 25

- ***Slope*** $m = \dfrac{C_2 - C_1}{x_2 - x_1} = \dfrac{30{,}000 - 25{,}000}{40 - 30} = 500$ — C plays the role of y.
- ***Intercept*** $b = C_1 - mx_1 = 25{,}000 - (500)(30) = 10{,}000.$ — We used the point $(x_1, C_1) = (30, 25{,}000)$.

The linear cost function is therefore

$$C(x) = 500x + 10{,}000.$$

Because $m = 500$ and $b = 10{,}000$ the factory's fixed cost is \$10,000 each day, and its marginal cost is \$500 per refrigerator. (See page 51 in Section 1.2.) These are illustrated in Figure 25.

➡ **Before we go on...** Recall that, in general, the slope m measures the number of units of change in y per 1-unit change in x, so it is measured in units of y per unit of x:

Units of Slope = Units of y per unit of x.

In Example 3, y is the cost C, measured in dollars, and x is the number of items, measured in refrigerators. Hence,

Units of Slope = Units of y per Unit of x = Dollars per refrigerator.

The y-intercept b, being a value of y, is measured in the same units as y. In Example 3, b is measured in dollars. ■

using Technology

To obtain the cost equation for Example 3 with technology, apply the Technology note for Example 2(a) to the given points (30, 25,000) and (40, 30,000) on the graph of the cost equation.

In Section 1.2 we saw that a **demand function** specifies the demand q as a function of the price p per item.

EXAMPLE 4 Linear Demand Function from Data

You run a small supermarket and must determine how much to charge for Hot'n'Spicy brand baked beans. The following chart shows weekly sales figures (the demand) for Hot'n'Spicy at two different prices.

Price/Can	\$0.50	\$0.75
Demand (cans sold/week)	400	350

a. Model these data with a linear demand function. (See Example 4 in Section 1.2.)

b. How do we interpret the slope and q-intercept of the demand function?

Solution

a. Recall that a demand equation—or demand function—expresses demand q (in this case, the number of cans of beans sold per week) as a function of the unit price p (in this case, price per can). We model the demand using the two points we are given: (0.50, 400) and (0.75, 350).

using Technology

To obtain the demand equation for Example 4 with technology, apply the Technology note for Example 2(a) to the given points (0.50, 400) and (0.75, 350) on the graph of the demand equation.

$$\textbf{\textit{Slope:}} \qquad m = \frac{q_2 - q_1}{p_2 - p_1} = \frac{350 - 400}{0.75 - 0.50} = \frac{-50}{0.25} = -200$$

$$\textbf{\textit{Intercept:}} \ b = q_1 - mp_1 = 400 - (-200)(0.50) = 500$$

So, the demand equation is

$$q = -200p + 500. \qquad q = mp + b$$

b. The key to interpreting the slope in a demand equation is to recall (see the "Before we go on" note at the end of Example 3) that we measure the slope in *units of y per unit of x*. Here, $m = -200$, and the units of m are units of q per unit of p, or the number of cans sold per dollar change in the price. Because m is negative, we see that the number of cans sold decreases as the price increases. We conclude that the weekly sales will drop by 200 cans per $1 increase in the price.

To interpret the q-intercept, recall that it gives the q-coordinate when $p = 0$. Hence it is the number of cans the supermarket can "sell" every week if it were to give them away.*

* Does this seem realistic? Demand is not always unlimited if items were given away. For instance, campus newspapers are sometimes given away, and yet piles of them are often left untaken. Also see the "Before we go on" discussion at the end of this example.

Before we go on...

Q: *Just how reliable is the linear model used in Example 4?*

A: The *actual* demand graph could in principle be obtained by tabulating demand figures for a large number of different prices. If the resulting points were plotted on the *pq* plane, they would probably suggest a curve and not a straight line. However, if you looked at a small enough portion of any curve, you could closely *approximate* it by a straight line. In other words, *over a small range of values of p, a linear model is accurate*. Linear models of real-world situations are generally reliable only for small ranges of the variables. (This point will come up again in some of the exercises.)

■

The next example illustrates modeling change over time t with a linear function of t.

EXAMPLE 5 Modeling Change Over Time: Growth of Sales

The worldwide market for portable navigation devices was expected to grow from 50 million units in 2007 to around 530 million units in 2015.[8]

[8] Sales were expected to grow to more than 500 million in 2015 according to a January 2008 press release by Telematics Research Group. Source: www.telematicsresearch.com.

a. Use this information to model annual worldwide sales of portable navigation devices as a linear function of time t in years since 2007. What is the significance of the slope?

b. Use the model to predict when annual sales of mobile navigation devices will reach 440 million units.

Solution

a. Since we are interested in worldwide sales s of portable navigation devices as a function of time, we take time t to be the independent coordinate (playing the role of x) and the annual sales s, in million of units, to be the dependent coordinate (in the role of y). Notice that 2007 corresponds to $t = 0$ and 2015 corresponds to $t = 8$, so we are given the coordinates of two points on the graph of sales s as a function of time t: (0, 50) and (8, 530). We model the sales using these two points:

$$m = \frac{s_2 - s_1}{t_2 - t_1} = \frac{530 - 50}{8 - 0} = \frac{480}{8} = 60$$

$$b = s_1 - mt_1 = 50 - (60)(0) = 50$$

So, $s = 60t + 50$ million units. $\quad s = mt + b$

The slope m is measured in units of s per unit of t; that is, millions of devices per year, and is thus the *rate of change of annual sales*. To say that $m = 60$ is to say that annual sales are increasing at a rate of 60 million devices per year.

b. Our model of annual sales as a function of time is

$$s = 60t + 50 \text{ million units.}$$

Annual sales of mobile portable devices will reach 440 million when $s = 440$, or

$$440 = 60t + 50$$

Solving for t, $\quad 60t = 440 - 50 = 390$

$$t = \frac{390}{60} = 6.5 \text{ years},$$

which is midway through 2013. Thus annual sales are expected to reach 440 million midway through 2013.

using Technology

To use technology to obtain s as a function of t in Example 5, apply the Technology note for Example 2(a) to the points (0, 50) and (8, 530) on its graph.

EXAMPLE 6 Velocity

You are driving down the Ohio Turnpike, watching the mileage markers to stay awake. Measuring time in hours after you see the 20-mile marker, you see the following markers each half hour:

Time (h)	0	0.5	1	1.5	2
Marker (mi)	20	47	74	101	128

Find your location s as a function of t, the number of hours you have been driving. (The number s is also called your **position** or **displacement**.)

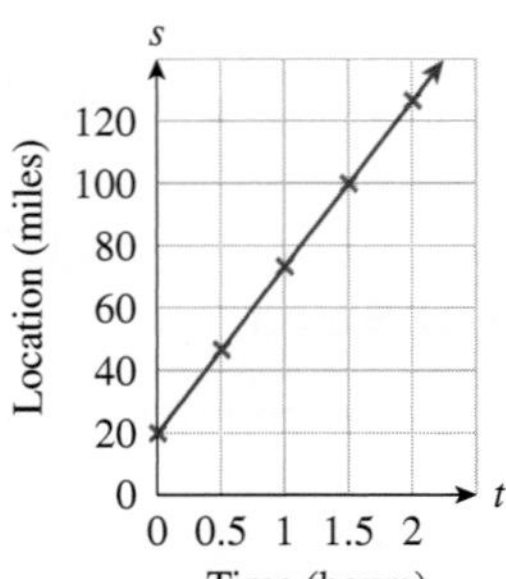

Figure 26

Solution

If we plot the location s versus the time t, the five markers listed give us the graph in Figure 26. These points appear to lie along a straight line. We can verify this by calculating how far you traveled in each half hour. In the first half hour, you traveled $47 - 20 = 27$ miles. In the second half hour you traveled $74 - 47 = 27$ miles also. In fact, you traveled exactly 27 miles each half hour. The points we plotted lie on a straight line that rises 27 units for every 0.5 unit we go to the right, for a slope of $27/0.5 = 54$.

To get the equation of that line, notice that we have the s-intercept, which is the starting marker of 20. Thus, the equation of s as a function of time t is

$$s(t) = 54t + 20. \quad \text{We used } s \text{ in place of } y \text{ and } t \text{ in place of } x.$$

using Technology

To use technology to obtain s as a function of t in Example 6, apply the Technology note for Example 2(a) to the points (0, 20) and (1, 74) on its graph.

Notice the significance of the slope: For every hour you travel, you drive a distance of 54 miles. In other words, you are traveling at a constant velocity of 54 mph. We have uncovered a very important principle:

In the graph of displacement versus time, velocity is given by the slope.

Linear Change over Time

If a quantity q is a linear function of time t,

$$q = mt + b,$$

then the slope m measures the **rate of change** of q, and b is the quantity at time $t = 0$, the **initial quantity**. If q represents the position of a moving object, then the rate of change is also called the **velocity**.

Units of *m* and *b*

The units of measurement of m are units of q per unit of time; for instance, if q is income in dollars and t is time in years, then the rate of change m is measured in dollars per year.

The units of b are units of q; for instance, if q is income in dollars and t is time in years, then b is measured in dollars.

Quick Example

If the accumulated revenue from sales of your video game software is given by $R = 2{,}000t + 500$ dollars, where t is time in years from now, then you have earned \$500 in revenue so far, and the accumulated revenue is increasing at a rate of \$2,000 per year.

Examples 3–6 share the following common theme.

General Linear Models

If $y = mx + b$ is a linear model of changing quantities x and y, then the slope m is the rate at which y is increasing per unit increase in x, and the y-intercept b is the value of y that corresponds to $x = 0$.

Units of *m* and *b*

The slope m is measured in units of y per unit of x, and the intercept b is measured in units of y.

Quick Example

If the number n of spectators at a soccer game is related to the number g of goals your team has scored so far by the equation $n = 20g + 4$, then you can expect 4 spectators if no goals have been scored and 20 additional spectators per additional goal scored.

FAQs

What to Use as *x* and *y*, and How to Interpret a Linear Model

Q: *In a problem where I must find a linear relationship between two quantities, which quantity do I use as x and which do I use as y?*

A: The key is to decide which of the two quantities is the independent variable, and which is the dependent variable. Then use the independent variable as x and the dependent variable as y. In other words, *y depends on x.*

Here are examples of phrases that convey this information, usually of the form *Find y [dependent variable] in terms of x [independent variable]:*

- Find the cost in terms of the number of items. y = Cost, x = # Items
- How does color depend on wavelength? y = Color, x = Wavelength

If no information is conveyed about which variable is intended to be independent, then you can use whichever is convenient.

Q: *How do I interpret a general linear model $y = mx + b$?*

A: The key to interpreting a linear model is to remember the units we use to measure m and b:

The slope m is measured in units of y per unit of x; the intercept b is measured in units of y.

For instance, if $y = 4.3x + 8.1$ and you know that x is measured in feet and y in kilograms, then you can already say, "y is 8.1 kilograms when $x = 0$ feet, and increases at a rate of 4.3 kilograms per foot" without knowing anything more about the situation!

1.3 EXERCISES

Access end-of-section exercises online at **www.webassign.net**

1.4 Linear Regression

We have seen how to find a linear model given two data points: We find the equation of the line that passes through them. However, we often have more than two data points, and they will rarely all lie on a single straight line, but may often come close to doing so. The problem is to find the line coming *closest* to passing through all of the points.

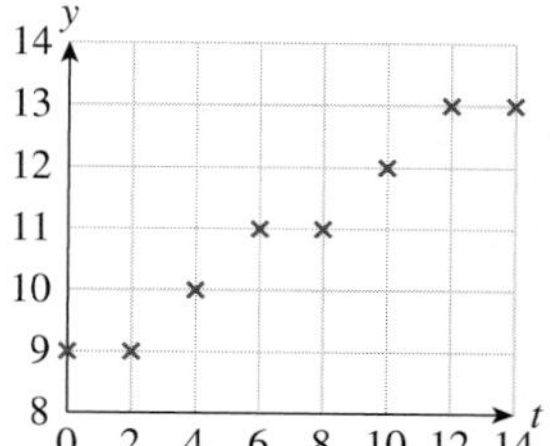
Figure 27(a)

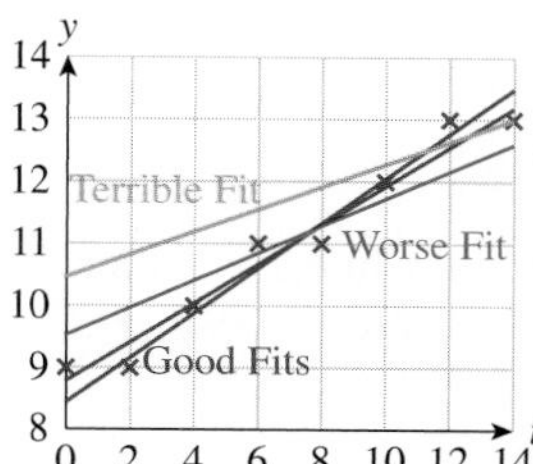

Figure 27(b)

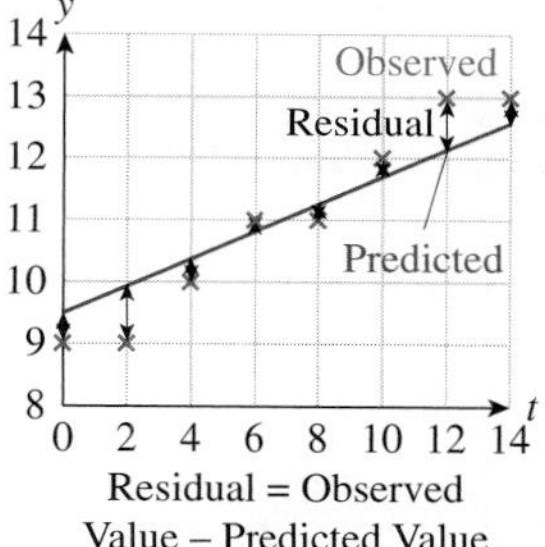

Figure 28

* Why not add the absolute values of the residuals instead? Mathematically, using the squares rather than the absolute values results in a simpler and more elegant solution. Further, using the squares always results in a *single* best-fit line in cases where the *x*-coordinates are all different, whereas this is not the case if we use absolute values.

Suppose, for example, that we are conducting research for a company interested in expanding into Mexico. Of interest to us would be current and projected growth in that country's economy. The following table shows past and projected per capita gross domestic product (GDP)[9] of Mexico for 2000–2014.[10]

Year t ($t = 0$ represents 2000)	0	2	4	6	8	10	12	14
Per Capita GDP y ($1,000)	9	9	10	11	11	12	13	13

A plot of these data suggests a roughly linear growth of the GDP (Figure 27(a)). These points suggest a roughly linear relationship between t and y, although they clearly do not all lie on a single straight line. Figure 27(b) shows the points together with several lines, some fitting better than others. Can we precisely measure which lines fit better than others? For instance, which of the two lines labeled as "good" fits in Figure 27(b) models the data more accurately? We begin by considering, for each value of t, the difference between the actual GDP (the **observed value**) and the GDP predicted by a linear equation (the **predicted value**). The difference between the predicted value and the observed value is called the **residual**.

$$\text{Residual} = \text{Observed Value} - \text{Predicted Value}$$

On the graph, the residuals measure the vertical distances between the (observed) data points and the line (Figure 28) and they tell us how far the linear model is from predicting the actual GDP.

The more accurate our model, the smaller the residuals should be. We can combine all the residuals into a single measure of accuracy by adding their *squares*. (We square the residuals in part to make them all positive.*) The sum of the squares of the residuals is called the **sum-of-squares error, SSE**. Smaller values of SSE indicate more accurate models.

Here are some definitions and formulas for what we have been discussing.

Observed and Predicted Values

Suppose we are given a collection of data points $(x_1, y_1), \ldots, (x_n, y_n)$. The n quantities $y_1, y_2, \ldots, y_n$ are called the **observed *y*-values**. If we model these data with a linear equation

$$\hat{y} = mx + b, \qquad \hat{y} \text{ stands for "estimated } y\text{" or "predicted } y\text{."}$$

then the y-values we get by substituting the given x-values into the equation are called the **predicted *y*-values**:

$$\hat{y}_1 = mx_1 + b \qquad \text{Substitute } x_1 \text{ for } x.$$
$$\hat{y}_2 = mx_2 + b \qquad \text{Substitute } x_2 \text{ for } x.$$
$$\ldots$$
$$\hat{y}_n = mx_n + b. \qquad \text{Substitute } x_n \text{ for } x.$$

[9] The GDP is a measure of the total market value of all goods and services produced within a country.

[10] Data are approximate and/or projected. Sources: CIA World Factbook/www.indexmundi.com, www.economist.com.

Quick Example

Consider the three data points (0, 2), (2, 5), and (3, 6). The observed y-values are $y_1 = 2$, $y_2 = 5$, and $y_3 = 6$. If we model these data with the equation $\hat{y} = x + 2.5$, then the predicted values are:

$$\hat{y}_1 = x_1 + 2.5 = 0 + 2.5 = 2.5$$
$$\hat{y}_2 = x_2 + 2.5 = 2 + 2.5 = 4.5$$
$$\hat{y}_3 = x_3 + 2.5 = 3 + 2.5 = 5.5.$$

Residuals and Sum-of-Squares Error (SSE)

If we model a collection of data $(x_1, y_1), \ldots, (x_n, y_n)$ with a linear equation $\hat{y} = mx + b$, then the **residuals** are the n quantities (Observed Value – Predicted Value):

$$(y_1 - \hat{y}_1), (y_2 - \hat{y}_2), \ldots, (y_n - \hat{y}_n).$$

The **sum-of-squares error (SSE)** is the sum of the squares of the residuals:

$$\text{SSE} = (y_1 - \hat{y}_1)^2 + (y_2 - \hat{y}_2)^2 + \cdots + (y_n - \hat{y}_n)^2.$$

Quick Example

For the data and linear approximation given above, the residuals are:

$$y_1 - \hat{y}_1 = 2 - 2.5 = -0.5$$
$$y_2 - \hat{y}_2 = 5 - 4.5 = 0.5$$
$$y_3 - \hat{y}_3 = 6 - 5.5 = 0.5$$

and so $\text{SSE} = (-0.5)^2 + (0.5)^2 + (0.5)^2 = 0.75$.

using Technology

See the Technology Guides at the end of the chapter for detailed instructions on how to obtain the tables and graphs in Example 1 using a TI-83/84 Plus or a spreadsheet. Here is an outline:

TI-83/84 Plus

STAT EDIT
Values of t in L_1, and y in L_2.
Predicted y: Highlight L_3. Enter `0.5*L1+8`
Squares of residuals: Highlight L_4. Enter `(L2-L3)^2`
SSE: Home screen `sum(L4)`
Graph: `Y1=0.5X+8`
Y = screen: Turn on Plot 1 ZOOM (STAT) [More details on page 93.]

Spreadsheet

Headings t, y, y-hat, Residual^2, m, b, and SSE in A1–F1.
t-values in A2–A9, y-values in B2–B9; 0.25 for m and 9 for b in E2–F2
Predicted y: `=$E$2*A2+$F$2` in C2 and copy down to C9.
Squares of residuals: `=(B2-C2)^2` in D2 and copy down to D9.
SSE: `=SUM(D2:D9)` in G2
Graph: Highlight A1–C9. Insert a Scatter chart.
[More details on page 98.]

EXAMPLE 1 Computing SSE

Using the data above on the GDP in Mexico, compute SSE for the linear models $y = 0.5t + 8$ and $y = 0.25t + 9$. Which model is the better fit?

Solution We begin by creating a table showing the values of t, the observed (given) values of y, and the values predicted by the first model.

Year t	Observed y	Predicted $\hat{y} = 0.5t + 8$
0	9	8
2	9	9
4	10	10
6	11	11
8	11	12
10	12	13
12	13	14
14	13	15

We now add two new columns for the residuals and their squares.

Year t	Observed y	Predicted $\hat{y} = 0.5t + 8$	Residual $y - \hat{y}$	Residual^2 $(y - \hat{y})^2$
0	9	8	$9 - 8 = 1$	$1^2 = 1$
2	9	9	$9 - 9 = 0$	$0^2 = 0$
4	10	10	$10 - 10 = 0$	$0^2 = 0$
6	11	11	$11 - 11 = 0$	$0^2 = 0$
8	11	12	$11 - 12 = -1$	$(-1)^2 = 1$
10	12	13	$12 - 13 = -1$	$(-1)^2 = 1$
12	13	14	$13 - 14 = -1$	$(-1)^2 = 1$
14	13	15	$13 - 15 = -2$	$(-2)^2 = 4$

SSE, the sum of the squares of the residuals, is then the sum of the entries in the last column,

$$\text{SSE} = 8.$$

Repeating the process using the second model, $0.25t + 9$, yields the following table:

Year t	Observed y	Predicted $\hat{y} = 0.25t + 9$	Residual $y - \hat{y}$	Residual^2 $(y - \hat{y})^2$
0	9	9	$9 - 9 = 0$	$0^2 = 0$
2	9	9.5	$9 - 9.5 = -0.5$	$(-0.5)^2 = 0.25$
4	10	10	$10 - 10 = 0$	$0^2 = 0$
6	11	10.5	$11 - 10.5 = 0.5$	$0.5^2 = 0.25$
8	11	11	$11 - 11 = 0$	$0^2 = 0$
10	12	11.5	$12 - 11.5 = 0.5$	$0.5^2 = 0.25$
12	13	12	$13 - 12 = 1$	$1^2 = 1$
14	13	12.5	$13 - 12.5 = 0.5$	$0.5^2 = 0.25$

This time, $\text{SSE} = 2$ and so the second model is a better fit.

Figure 29 shows the data points and the two linear models in question.

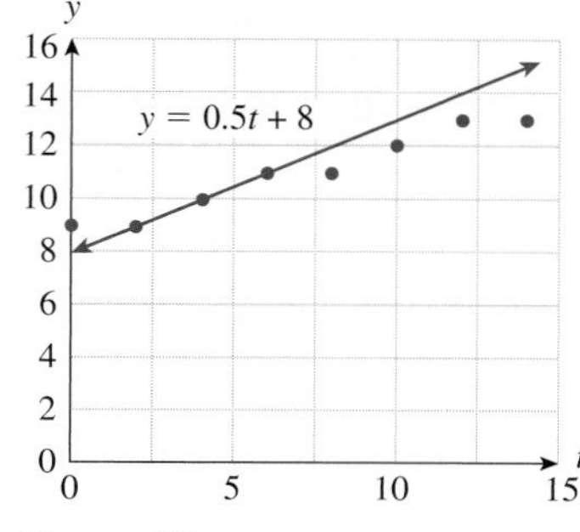

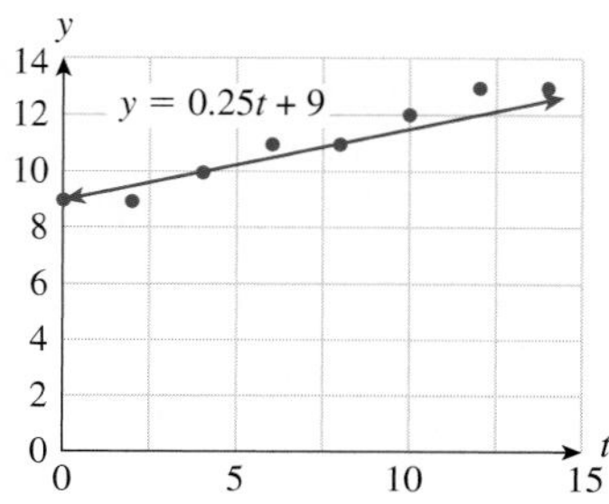

Figure 29

Before we go on...

Q: *It seems clear from the figure that the second model in Example 1 gives a better fit. Why bother to compute SSE to tell me this?*

A: The difference between the two models we chose is so great that it is clear from the graphs which is the better fit. However, if we used a third model with $m = 0.25$ and $b = 9.1$, then its graph would be almost indistinguishable from that of the second, but a slightly better fit as measured by SSE = 1.68.

■

Among all possible lines, there ought to be one with the least possible value of SSE—that is, the greatest possible accuracy as a model. The line (and there is only one such line) that minimizes the sum of the squares of the residuals is called the **regression line**, the **least-squares line**, or the **best-fit line**.

To find the regression line, we need a way to find values of m and b that give the smallest possible value of SSE. As an example, let us take the second linear model in the example above. We said in the "Before we go on" discussion that increasing b from 9 to 9.1 had the desirable effect of decreasing SSE from 2 to 1.68. We could then increase m to 0.26, further reducing SSE to 1.328. Imagine this as a kind of game: Alternately alter the values of m and b by small amounts until SSE is as small as you can make it. This works, but is extremely tedious and time-consuming.

Fortunately, there is an algebraic way to find the regression line. Here is the calculation. To justify it rigorously requires calculus of several variables or linear algebra.

Regression Line

The **regression line (least squares line, best-fit line)** associated with the points $(x_1, y_1), (x_2, y_2), \ldots, (x_n, y_n)$ is the line that gives the minimum SSE. The regression line is

$$y = mx + b,$$

where m and b are computed as follows:

$$m = \frac{n(\sum xy) - (\sum x)(\sum y)}{n(\sum x^2) - (\sum x)^2}$$

$$b = \frac{\sum y - m(\sum x)}{n}$$

n = number of data points.

The quantities m and b are called the **regression coefficients**.

Here, "$\sum$" means "the sum of." Thus, for example,

$$\sum x = \text{Sum of the } x\text{-values} = x_1 + x_2 + \cdots + x_n$$

$$\sum xy = \text{Sum of products} = x_1y_1 + x_2y_2 + \cdots + x_ny_n$$

$$\sum x^2 = \text{Sum of the squares of the } x\text{-values} = x_1^2 + x_2^2 + \cdots + x_n^2.$$

On the other hand,

$$\left(\sum x\right)^2 = \text{Square of } \sum x = \text{Square of the sum of the } x\text{-values.}$$

EXAMPLE 2 Per Capita Gross Domestic Product in Mexico

In Example 1 we considered the following data on the per capita gross domestic product (GDP) of Mexico:

Year x (x = 0 represents 2000)	0	2	4	6	8	10	12	14
Per Capita GDP y (\$1,000)	9	9	10	11	11	12	13	13

Find the best-fit linear model for these data and use the model to predict the per capita GDP in Mexico in 2016.

Solution Let's organize our work in the form of a table, where the original data are entered in the first two columns and the bottom row contains the column sums.

	x	y	xy	x^2
	0	9	0	0
	2	9	18	4
	4	10	40	16
	6	11	66	36
	8	11	88	64
	10	12	120	100
	12	13	156	144
	14	13	182	196
$\sum$ (Sum)	56	88	670	560

Because there are $n = 8$ data points, we get

$$m = \frac{n(\sum xy) - (\sum x)(\sum y)}{n(\sum x^2) - (\sum x)^2} = \frac{8(670) - (56)(88)}{8(560) - (56)^2} \approx 0.321$$

and

$$b = \frac{\sum y - m(\sum x)}{n} \approx \frac{88 - (0.321)(56)}{8} \approx 8.75.$$

So, the regression line is

$$y = 0.321x + 8.75.$$

To predict the per capita GDP in Mexico in 2016 we substitute $x = 16$ and get $y \approx 14$, or \$14,000 per capita.

Figure 30 shows the data points and the regression line (which has SSE ≈ 0.643; a lot lower than in Example 1).

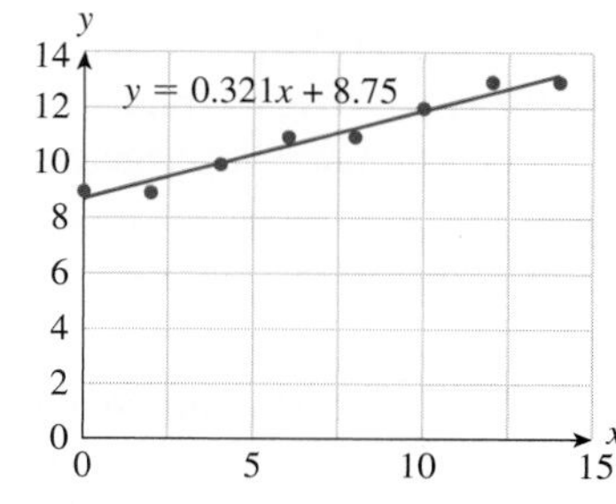

Figure 30

using Technology

See the Technology Guides at the end of the chapter for detailed instructions on how to obtain the regression line and graph in Example 2 using a TI-83/84 Plus or a spreadsheet. Here is an outline:

TI-83/84 Plus
STAT EDIT
Values of x in L_1, and y in L_2.
Regression equation: STAT CALC option #4: `LinReg(ax+b)`
Graph: Y= VARS 5 EQ 1, then ZOOM 9
[More details on page 94.]

Spreadsheet
x-values in A2–A9, y-values in B2–B9
Graph: Highlight A2–B9. Insert a Scatter Chart.
Regression line: Add a linear trendline. [More details on page 99.]

Website
www.WanerMath.com
The following two utilities will calculate and plot regression lines (link to either from Math Tools for Chapter 1):
Simple Regression
Function Evaluator and Grapher

Coefficient of Correlation

If all the data points do not lie on one straight line, we would like to be able to measure how closely they can be approximated by a straight line. Recall that SSE measures the sum of the squares of the deviations from the regression line; therefore it constitutes a measurement of what is called "goodness of fit." (For instance, if SSE $= 0$, then all the points lie on a straight line.) However, SSE depends on the units we use to measure y, and also on the number of data points (the more data points we use, the larger SSE tends to be). Thus, while we can (and do) use SSE to compare the goodness of fit of two lines to the same data, we cannot use it to compare the goodness of fit of one line to one set of data with that of another to a different set of data.

To remove this dependency, statisticians have found a related quantity that can be used to compare the goodness of fit of lines to different sets of data. This quantity, called the **coefficient of correlation** or **correlation coefficient**, and usually denoted r, is between -1 and 1. The closer r is to -1 or 1, the better the fit. For an *exact* fit, we would have $r = -1$ (for a line with negative slope) or $r = 1$ (for a line with positive slope). For a bad fit, we would have r close to 0. Figure 31 shows several collections of data points with least-squares lines and the corresponding values of r.

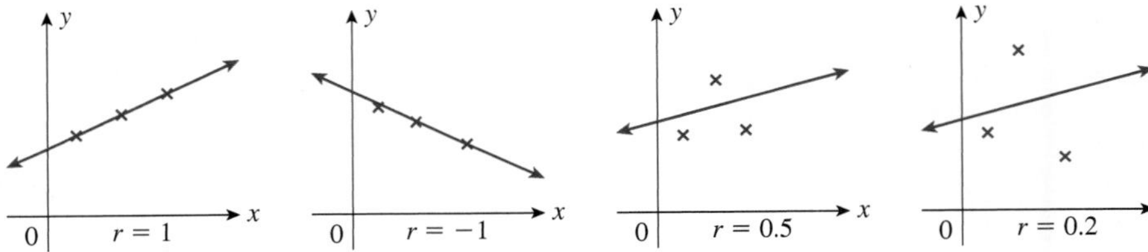

Figure 31

Correlation Coefficient

The coefficient of correlation of the n data points $(x_1, y_1), (x_2, y_2), \ldots, (x_n, y_n)$ is

$$r = \frac{n(\sum xy) - (\sum x)(\sum y)}{\sqrt{n(\sum x^2) - (\sum x)^2} \cdot \sqrt{n(\sum y^2) - (\sum y)^2}}.$$

It measures how closely the data points $(x_1, y_1), (x_2, y_2), \ldots, (x_n, y_n)$ fit the regression line. (The value r^2 is sometimes called the **coefficient of determination**.)

Interpretation

- If r is positive, the regression line has positive slope; if r is negative, the regression line has negative slope.
- If $r = 1$ or -1, then all the data points lie exactly on the regression line; if it is close to ± 1, then all the data points are close to the regression line.
- On the other hand, if r is not close to ± 1, then the data points are not close to the regression line, so the fit is not a good one. As a general rule of thumb, a value of $|r|$ less than around 0.8 indicates a poor fit of the data to the regression line.

using Technology

See the Technology Guides at the end of the chapter for detailed instructions on how to obtain the correlation coefficient in Example 3 using a TI-83/84 Plus or a spreadsheet. Here is an outline:

TI-83/84 Plus

[2ND] [CATALOG] DiagnosticOn

Then [STAT] CALC option #4: LinReg(ax+b) [More details on page 95.]

Spreadsheet

Add a trendline and select the option to "Display R-squared value on chart."

[More details and other alternatives on page 100.]

Website

www.WanerMath.com

The following two utilities will show regression lines and also r^2 (link to either from Math Tools for Chapter 1):

Simple Regression

Function Evaluator and Grapher

EXAMPLE 3 Computing the Coefficient of Correlation

Find the correlation coefficient for the data in Example 2. Is the regression line a good fit?

Solution The formula for r requires $\sum x, \sum x^2, \sum xy, \sum y$, and $\sum y^2$. We have all of these except for $\sum y^2$, which we find in a new column as shown.

	x	y	xy	x^2	y^2
	0	9	0	0	81
	2	9	18	4	81
	4	10	40	16	100
	6	11	66	36	121
	8	11	88	64	121
	10	12	120	100	144
	12	13	156	144	169
	14	13	182	196	169
$\sum$ (Sum)	56	88	670	560	986

Substituting these values into the formula, we get

$$r = \frac{n(\sum xy) - (\sum x)(\sum y)}{\sqrt{n(\sum x^2) - (\sum x)^2} \cdot \sqrt{n(\sum y^2) - (\sum y)^2}}$$

$$= \frac{8(670) - (56)(88)}{\sqrt{8(560) - 56^2} \cdot \sqrt{8(986) - 88^2}}$$

$$\approx 0.982.$$

As r is close to 1, the fit is a fairly good one; that is, the original points lie nearly along a straight line, as can be confirmed from the graph in Example 2.

1.4 EXERCISES

Access end-of-section exercises online at **www.webassign.net**

CHAPTER 1 REVIEW

KEY CONCEPTS

WM Website www.WanerMath.com

Go to the Website at www.WanerMath.com to find a comprehensive and interactive Web-based summary of Chapter 1.

1.1 Functions from the Numerical, Algebraic, and Graphical Viewpoints

Real-valued function f of a real-valued variable x, domain *p. 38*

Independent and dependent variables *p. 38*

Graph of the function f *p. 38*

Numerically specified function *p. 38*

Graphically specified function *p. 38*

Algebraically defined function *p. 38*

Piecewise-defined function *p. 43*

Vertical line test *p. 45*

Common types of algebraic functions and their graphs *p. 46*

1.2 Functions and Models

Mathematical model *p. 48*

Analytical model *p. 49*

Curve-fitting model *p. 49*

Cost, revenue, and profit; marginal cost, revenue, and profit; break-even point *pp. 50–51*

Demand, supply, and equilibrium price *pp. 53–54*

Selecting a model *p. 56*

Compound interest *p. 58*

Exponential growth and decay *p. 60*

Algebra of functions (sum, difference, product, quotient) *p. 62*

1.3 Linear Functions and Models

Linear function $f(x) = mx + b$ *p. 63*

Change in q: $\Delta q = q_2 - q_1$ *p. 64*

Slope of a line:

$$m = \frac{\Delta y}{\Delta x} = \frac{\text{Change in } y}{\text{Change in } x} \quad p.\ 65$$

Interpretations of m *p. 65*

Interpretation of b: y-intercept *p. 65*

Recognizing linear data *p. 66*

Computing the slope of a line *p. 67*

Slopes of horizontal and vertical lines *p. 68*

Computing the y-intercept *p. 68*

Linear modeling *p. 69*

Linear cost *p. 70*

Linear demand *p. 70*

Linear change over time; rate of change; velocity *p. 73*

General linear models *p. 73*

1.4 Linear Regression

Observed and predicted values *p. 75*

Residuals and sum-of-squares error (SSE) *p. 76*

Regression line (least-squares line, best-fit line) *p. 78*

Correlation coefficient; coefficient of determination *p. 80*

REVIEW EXERCISES

In Exercises 1–4, use the graph of the function f to find approximations of the given values.

1.

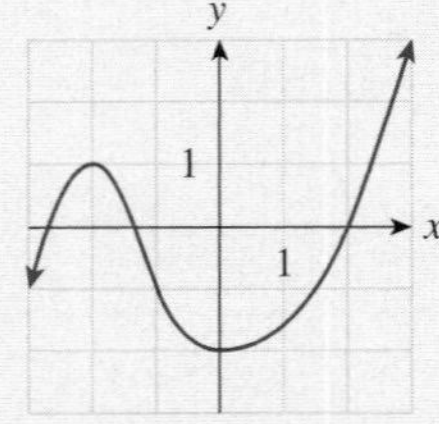

a. $f(-2)$ **b.** $f(0)$
c. $f(2)$ **d.** $f(2) - f(-2)$

2.

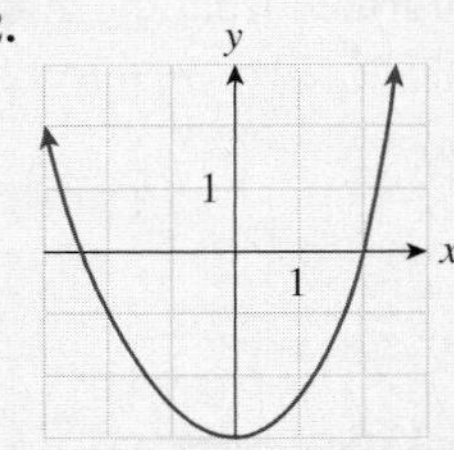

a. $f(-2)$ **b.** $f(0)$
c. $f(2)$ **d.** $f(2) - f(-2)$

3.

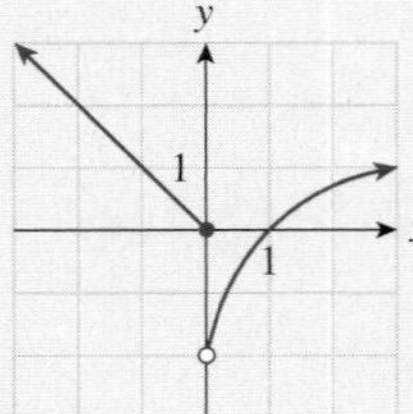

a. $f(-1)$ **b.** $f(0)$
c. $f(1)$ **d.** $f(1) - f(-1)$

4.

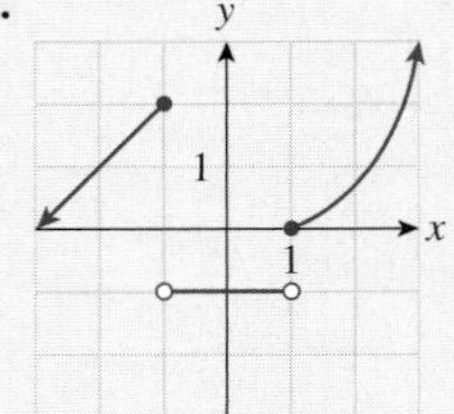

a. $f(-1)$ **b.** $f(0)$
c. $f(1)$ **d.** $f(1) - f(-1)$

In Exercises 5–8, graph the given function or equation.

5. $y = -2x + 5$

6. $2x - 3y = 12$

7. $y = \begin{cases} \frac{1}{2}x & \text{if } -1 \le x \le 1 \\ x - 1 & \text{if } 1 < x \le 3 \end{cases}$

8. $f(x) = 4x - x^2$ with domain $[0, 4]$

In Exercises 9–14, decide whether the specified values come from a linear, quadratic, exponential, or absolute value function.

9.

x	−2	0	1	2	4
$f(x)$	4	2	1	0	2

10.

x	−2	0	1	2	4
$g(x)$	−5	−3	−2	−1	1

11.

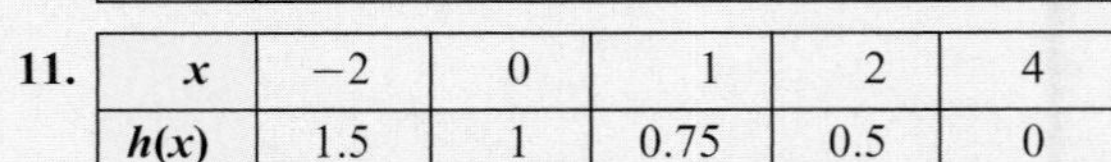

x	−2	0	1	2	4
$h(x)$	1.5	1	0.75	0.5	0

12.

x	−2	0	1	2	4
$k(x)$	0.25	1	2	4	16

13.

x	−2	0	1	2	4
$u(x)$	0	4	3	0	−12

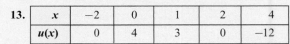

14.

x	-2	0	1	2	4
$w(x)$	32	8	4	2	0.5

In Exercises 15–22, find the equation of the specified line.

15. Through (3, 2) with slope -3

16. Through $(-2, 4)$ with slope -1

17. Through $(1, -3)$ and (5, 2)

18. Through $(-1, 2)$ and (1, 0)

19. Through (1, 2) parallel to $x - 2y = 2$

20. Through $(-3, 1)$ parallel to $-2x - 4y = 5$

21. With slope 4 crossing $2x - 3y = 6$ at its x-intercept

22. With slope 1/2 crossing $3x + y = 6$ at its x-intercept

In Exercises 23 and 24, determine which of the given lines better fits the given points.

23. $(-1, 1), (1, 2), (2, 0)$; $y = -x/2 + 1$ or $y = -x/4 + 1$

24. $(-2, -1), (-1, 1), (0, 1), (1, 2), (2, 4), (3, 3)$; $y = x + 1$ or $y = x/2 + 1$

In Exercises 25 and 26, find the line that best fits the given points and compute the correlation coefficient.

25. $(-1, 1), (1, 2), (2, 0)$

26. $(-2, -1), (-1, 1), (0, 1), (1, 2), (2, 4), (3, 3)$

APPLICATIONS: OHaganBooks.com

27. ***Web Site Traffic*** John Sean O'Hagan is CEO of the online bookstore OHaganBooks.com and notices that, since the establishment of the company Web site six years ago ($t = 0$), the number of visitors to the site has grown quite dramatically, as indicated by the following table:

Year t	0	1	2	3	4	5	6
Web site Traffic $V(t)$ (visits/day)	100	300	1,000	3,300	10,500	33,600	107,400

a. Graph the function V as a function of time t. Which of the following types of function seem to fit the curve best: linear, quadratic, or exponential?

b. Compute the ratios $\frac{V(1)}{V(0)}, \frac{V(2)}{V(1)}, \ldots,$ and $\frac{V(6)}{V(5)}$. What do you notice?

c. Use the result of part (b) to predict Web site traffic next year (to the nearest 100).

28. ***Publishing Costs*** Marjory Maureen Duffin is CEO of publisher Duffin House, a major supplier of paperback titles to OHaganBooks.com. She notices that publishing costs over the past five years have varied considerably as indicated by the following table, which shows the average cost to the company of publishing a paperback novel (t is time in years, and the current year is $t = 5$):

Year t	0	1	2	3	4	5
Cost $C(t)$	\$5.42	\$5.10	\$5.00	\$5.12	\$5.40	\$5.88

a. Graph the function C as a function of time t. Which of the following types of function seem to fit the curve best: linear, quadratic, or exponential?

b. Compute the differences $C(1) - C(0), C(2) - C(1), \ldots,$ and $C(5) - C(4)$, rounded to one decimal place. What do you notice?

c. Use the result of part (b) to predict the cost of producing a paperback novel next year.

29. ***Web Site Stability*** John O'Hagan is considering upgrading the Web server equipment at OHaganBooks.com because of frequent crashes. The tech services manager has been monitoring the frequency of crashes as a function of Web site traffic (measured in thousands of visits per day) and has obtained the following model:

$$c(x) = \begin{cases} 0.03x + 2 & \text{if } 0 \le x \le 50 \\ 0.05x + 1 & \text{if } x > 50 \end{cases}$$

where $c(x)$ is the average number of crashes in a day in which there are x thousand visitors.

a. On average, how many times will the Web site crash on a day when there are 10,000 visits? 50,000 visits? 100,000 visits?

b. What does the coefficient 0.03 tell you about the Web site's stability?

c. Last Friday, the Web site went down 8 times. Estimate the number of visits that day.

30. ***Book Sales*** As OHaganBooks.com has grown in popularity, the sales manager has been monitoring book sales as a function of the Web site traffic (measured in thousands of visits per day) and has obtained the following model:

$$s(x) = \begin{cases} 1.55x & \text{if } 0 \le x \le 100 \\ 1.75x - 20 & \text{if } 100 < x \le 250 \end{cases}$$

where $s(x)$ is the average number of books sold in a day in which there are x thousand visitors.

a. On average, how many books per day does the model predict OHaganBooks.com will sell when it has 60,000 visits in a day? 100,000 visits in a day? 160,000 visits in a day?

b. What does the coefficient 1.75 tell you about book sales?

c. According to the model, approximately how many visitors per day will be needed in order to sell an average of 300 books per day?

31. ***New Users*** The number of registered users at OHaganBooks.com has increased substantially over the past few months. The following table shows the number of new users registering each month for the past six months:

Month t	1	2	3	4	5	6
New Users (thousands)	12.5	37.5	62.5	72.0	74.5	75.0

a. Which of the following models best approximates the data?

(A) $n(t) = \dfrac{300}{4 + 100(5^{-t})}$ **(B)** $n(t) = 13.3t + 8.0$

(C) $n(t) = -2.3t^2 + 30.0t - 3.3$

(D) $n(t) = 7(3^{0.5t})$

b. What do each of the above models predict for the number of new users in the next few months: rising, falling, or leveling off?

32. ***Purchases*** OHaganBooks.com has been promoting a number of books published at Duffin House. The following table shows the number of books purchased each month from Duffin House for the past five months:

Month t	1	2	3	4	5
Purchases (books)	1,330	520	520	1,340	2,980

a. Which of the following models best approximates the data?

(A) $n(t) = \dfrac{3{,}000}{1 + 12(2^{-t})}$ **(B)** $n(t) = \dfrac{2{,}000}{4.2 - 0.7t}$

(C) $n(t) = 300(1.6^t)$

(D) $n(t) = 100(4.1t^2 - 20.4t + 29.5)$

b. What do each of the above models predict for the number of new users in the next few months: rising, falling, leveling off, or something else?

33. ***Internet Advertising*** Several months ago. John O'Hagan investigated the effect on the popularity of OHaganBooks.com of placing banner ads at well-known Internet portals. The following model was obtained from available data:

$v(c) = -0.000005c^2 + 0.085c + 1{,}750$ new visits per day

where c is the monthly expenditure on banner ads.

a. John O'Hagan is considering increasing expenditure on banner ads from the current level of \$5,000 to \$6,000 per month. What will be the resulting effect on Web site popularity?

b. According to the model, would the Web site popularity continue to grow at the same rate if he continued to raise expenditure on advertising \$1,000 each month? Explain.

c. Does this model give a reasonable prediction of traffic at expenditures larger than \$8,500 per month? Why?

34. ***Production Costs*** Over at Duffin House, Marjory Duffin is trying to decide on the size of the print runs for the best-selling new fantasy novel *Larry Plotter and the Simplex Method*. The following model shows a calculation of the total cost to produce a million copies of the novel, based on an analysis of setup and storage costs:

$$c(n) = 0.0008n^2 - 72n + 2{,}000{,}000 \text{ dollars}$$

where n is the print run size (the number of books printed in each run).

a. What would be the effect on cost if the run size was increased from 20,000 to 30,000?

b. Would increasing the run size in further steps of 10,000 result in the same changes in the total cost? Explain.

c. What approximate run size would you recommend that Marjoy Duffin use for a minimum cost?

35. ***Internet Advertising*** When OHaganBooks.com actually went ahead and increased Internet advertising from \$5,000 per month to \$6,000 per month (see Exercise 33) it was noticed that the number of new visits increased from an estimated 2,050 per day to 2,100 per day. Use this information to construct a linear model giving the average number v of new visits per day as a function of the monthly advertising expenditure c.

a. What is the model?

b. Based on the model, how many new visits per day could be anticipated if OHaganBooks.com budgets \$7,000 per month for Internet advertising?

c. The goal is to eventually increase the number of new visits to 2,500 per day. Based on the model, how much should be spent on Internet advertising in order to accomplish this?

36. ***Production Costs*** When Duffin House printed a million copies of *Larry Plotter and the Simplex Method* (see Exercise 34), it used print runs of 20,000, which cost the company \$880,000. For the sequel, *Larry Plotter and the Simplex Method, Phase 2* it used print runs of 40,000 which cost the company \$550,000. Use this information to construct a linear model giving the production cost c as a function of the run size n.

a. What is the model?

b. Based on the model, what would print runs of 25,000 have cost the company?

c. Marjory Duffin has decided to budget \$418,000 for production of the next book in the *Simplex Method* series. Based on the model, how large should the print runs be to accomplish this?

37. ***Recreation*** John O'Hagan has just returned from a sales convention at Puerto Vallarta, Mexico where, in order to win a bet he made with Marjory Duffin (Duffin House was also at the convention), he went bungee jumping at a nearby mountain retreat. The bungee cord he used had the property that a person weighing 70 kg would drop a total distance of 74.5 meters, while a 90 kg person would drop 93.5 meters. Express the distance d a jumper drops as a linear function of the jumper's weight w. John OHagan dropped 90 m. What was his approximate weight?

38. ***Crickets*** The mountain retreat near Puerto Vallarta was so quiet at night that all one could hear was the chirping of the snowy tree crickets. These crickets behave in a rather interesting way: The rate at which they chirp depends linearly on the temperature. Early in the evening, John O'Hagan counted 140 chirps/minute and noticed that the temperature was 80°F. Later in the evening the temperature dropped to 75°F, and the chirping slowed down to 120 chirps/minute. Express the temperature T as a function of the rate of chirping r. The temperature that night dropped to a low of 65°F. At approximately what rate were the crickets chirping at that point?

39. ***Break-Even Analysis*** OHaganBooks.com has recently decided to start selling music albums online through a service it calls *o'Tunes*.[11] Users pay a fee to download an entire music album. Composer royalties and copyright fees cost an average of \$5.50 per album, and the cost of operating and maintaining *o'Tunes* amounts to \$500 per week. The company is currently charging customers \$9.50 per album.

a. What are the associated (weekly) cost, revenue, and profit functions?
b. How many albums must be sold per week in order to make a profit?
c. If the charge is lowered to \$8.00 per album, how many albums must be sold per week in order to make a profit?

40. ***Break-Even Analysis*** OHaganBooks.com also generates revenue through its *o'Books* e-book service. Author royalties and copyright fees cost the company an average of \$4 per novel, and the monthly cost of operating and maintaining the service amounts to \$900 per month. The company is currently charging readers \$5.50 per novel.

a. What are the associated cost, revenue, and profit functions?
b. How many novels must be sold per month in order to break even?
c. If the charge is lowered to \$5.00 per novel, how many books must be sold in order to break even?

41. ***Demand and Profit*** In order to generate a profit from its new *o'Tunes* service, OHaganBooks.com needs to know how the demand for music albums depends on the price it charges. During the first week of the service, it was charging \$7 per album, and sold 500. Raising the price to \$9.50 had the effect of lowering demand to 300 albums per week.

a. Use the given data to construct a linear demand equation.
b. Use the demand equation you constructed in part (a) to estimate the demand if the price was raised to \$12 per album.

[11]The (highly original) name was suggested to John O'Hagan by Marjory Duffin over cocktails one evening.

c. Using the information on cost given in Exercise 39, determine which of the three prices (\$7, \$9.50 and \$12) would result in the largest weekly profit, and the size of that profit.

42. ***Demand and Profit*** In order to generate a profit from its *o'Books* e-book service, OHaganBooks.com needs to know how the demand for novels depends on the price it charges. During the first month of the service, it was charging \$10 per novel, and sold 350. Lowering the price to \$5.50 per novel had the effect of increasing demand to 620 novels per month.

a. Use the given data to construct a linear demand equation.
b. Use the demand equation you constructed in part (a) to estimate the demand if the price was raised to \$15 per novel.
c. Using the information on cost given in Exercise 40, determine which of the three prices (\$5.50, \$10 and \$15) would result in the largest profit, and the size of that profit.

43. ***Demand*** OHaganBooks.com has tried selling music albums on *o'Tunes* at a variety of prices, with the following results:

Price	\$8.00	\$8.50	\$10	\$11.50
Demand (Weekly sales)	440	380	250	180

a. Use the given data to obtain a linear regression model of demand.
b. Use the demand model you constructed in part (a) to estimate the demand if the company charged \$10.50 per album. (Round the answer to the nearest album.)

44. ***Demand*** OHaganBooks.com has tried selling novels through *o'Books* at a variety of prices, with the following results:

Price	\$5.50	\$10	\$11.50	\$12
Demand (Monthly sales)	620	350	350	300

a. Use the given data to obtain a linear regression model of demand.
b. Use the demand model you constructed in part (a) to estimate the demand if the company charged \$8 per novel. (Round the answer to the nearest novel.)

Case Study: Modeling Spending on Internet Advertising

Jeff Titcomb/Photographer's Choice / Getty Images

You are the new director of Impact Advertising Inc.'s Internet division, which has enjoyed a steady 0.25% of the Internet advertising market. You have drawn up an ambitious proposal to expand your division in light of your anticipation that Internet advertising will continue to skyrocket. However, upper management sees things differently and, based on the following email, does not seem likely to approve the budget for your proposal.

TO: JCheddar@impact.com (J. R. Cheddar)
CC: CVODoylePres@impact.com (C. V. O'Doyle, CEO)
FROM: SGLombardoVP@impact.com (S. G. Lombardo, VP Financial Affairs)
SUBJECT: Your Expansion Proposal
DATE: May 30, 2014

Hi John:

Your proposal reflects exactly the kind of ambitious planning and optimism we like to see in our new upper management personnel. Your presentation last week was most impressive, and obviously reflected a great deal of hard work and preparation.

I am in full agreement with you that Internet advertising is on the increase. Indeed, our Market Research department informs me that, based on a regression of the most recently available data, Internet advertising revenue in the United States will continue to grow at a rate of approximately $2.7 billion per year. This translates into approximately $6.75 million in increased revenues per year for Impact, given our 0.25% market share. This rate of expansion is exactly what our planned 2015 budget anticipates. Your proposal, on the other hand, would require a budget of approximately *twice* the 2015 budget allocation, even though your proposal provides no hard evidence to justify this degree of financial backing.

At this stage, therefore, I am sorry to say that I am inclined not to approve the funding for your project, although I would be happy to discuss this further with you. I plan to present my final decision on the 2015 budget at next week's divisional meeting.

Regards, Sylvia

Refusing to admit defeat, you contact the Market Research department and request the details of their projections on Internet advertising. They fax you the following information:[12]

Year	2007	2008	2009	2010	2011	2012	2013	2014
Internet Advertising Revenue ($ Billion)	21.2	23.4	22.7	25.8	28.5	32.6	36	40.5

Regression Model: $y = 2.744x + 19.233$ (x = time in years since 2007)

Correlation Coefficient: $r = 0.970$

Now you see where the VP got that $2.7 billion figure: The slope of the regression equation is close to 2.7, indicating a rate of increase of about $2.7 billion per year. Also, the correlation coefficient is very high—an indication that the linear model fits the data well. In view of this strong evidence, it seems difficult to argue that revenues will increase by significantly more than the projected $2.7 billion per year. To get a

[12]The 2011–2014 figures are projections by eMarketer. Source: www.eMarketer.com.

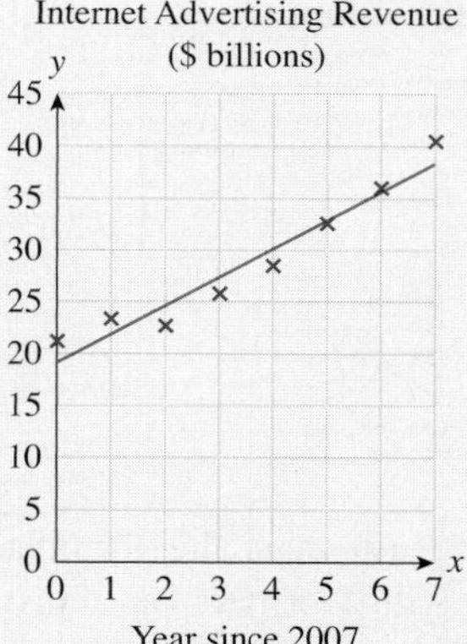

Figure 32

better picture of what's going on, you decide to graph the data together with the regression line in your spreadsheet. What you get is shown in Figure 32. You immediately notice that the data points seem to suggest a curve, and not a straight line. Then again, perhaps the suggestion of a curve is an illusion. Thus there are, you surmise, two possible interpretations of the data:

1. (Your first impression) As a function of time, Internet advertising revenue is nonlinear, and is in fact accelerating (the rate of change is increasing), so a linear model is inappropriate.
2. (Devil's advocate) Internet advertising revenue *is* a linear function of time; the fact that the points do not lie on the regression line is simply a consequence of random factors that do not reflect a long-term trend, such as world events, mergers and acquisitions, short-term fluctuations in economy or the stock market, etc.

You suspect that the VP will probably opt for the second interpretation and discount the graphical evidence of accelerating growth by claiming that it is an illusion: a "statistical fluctuation." That is, of course, a possibility, but you wonder how likely it really is.

For the sake of comparison, you decide to try a regression based on the simplest nonlinear model you can think of—a quadratic function.

$$y = ax^2 + bx + c$$

Your spreadsheet allows you to fit such a function with a click of the mouse. The result is the following.

$$y = 0.3208x^2 + 0.4982x + 21.479 \quad (x = \text{number of years since 2007})$$

$$r = 0.996$$

See Note.*

* Note that this r is *not* the linear correlation coefficient we defined on page 80; what this r measures is how closely the *quadratic* regression model fits the data.

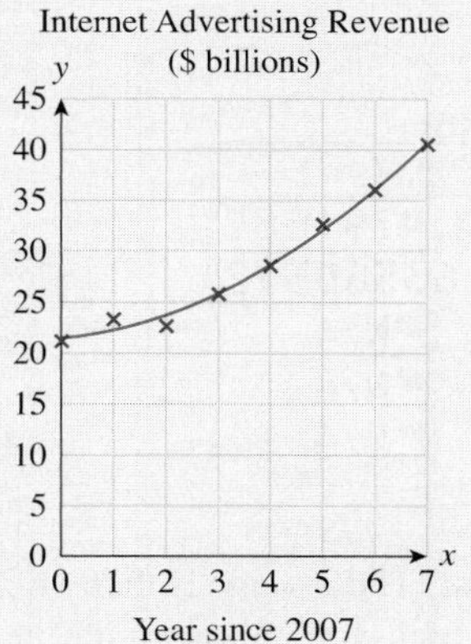

Figure 33

Figure 33 shows the graph of the regression function together with the original data.

Aha! The fit is visually far better, and the correlation coefficient is even higher! Further, the quadratic model predicts 2015 revenue as

$$y = 0.3208(8)^2 + 0.4982(8) + 21.479 \approx \$46.0 \text{ billion},$$

which is \$5.5 billion above the 2014 spending figure in the table above. Given Impact Advertising's 0.25% market share, this translates into an increase in revenues of \$13.75 million, which is about double the estimate predicted by the linear model!

You quickly draft an email to Lombardo, and are about to click "Send" when you decide, as a precaution, to check with a colleague who is knowledgeable in statistics. He tells you to be cautious: The value of r will always tend to increase if you pass from a linear model to a quadratic one because of the increase in "degrees of freedom."† A good way to test whether a quadratic model is more appropriate than a linear one is to compute a statistic called the "p-value" associated with the coefficient of x^2. A low value of p indicates a high degree of confidence that the coefficient of x^2 cannot be zero (see below). Notice that if the coefficient of x^2 *is* zero, then you have a linear model.

† The number of degrees of freedom in a regression model is 1 less than the number of coefficients. For a linear model, it is 1 (there are two coefficients: the slope m and the intercept b), and for a quadratic model it is 2. For a detailed discussion, consult a text on regression analysis.

You can, your colleague explains, obtain the p-value using your spreadsheet as follows (the method we describe here works on all the popular spreadsheets, including *Excel, Google Docs,* and *Open Office Calc*).

First, set up the data in columns, with an extra column for the values of x^2:

◇	A	B	C
1	y	x	x^2
2	21.2	0	0
3	23.4	1	1
4	22.7	2	4
5	25.8	3	9
6	28.5	4	16
7	32.6	5	25
8	36	6	36
9	40.5	7	49

Then, highlight a vacant 5 × 3 block (the block E1:G5 say), type the formula `=LINEST(A2:A9,B2:C9,,TRUE)`, and press Cntl+Shift+Enter (not just Enter!). You will see a table of statistics like the following:

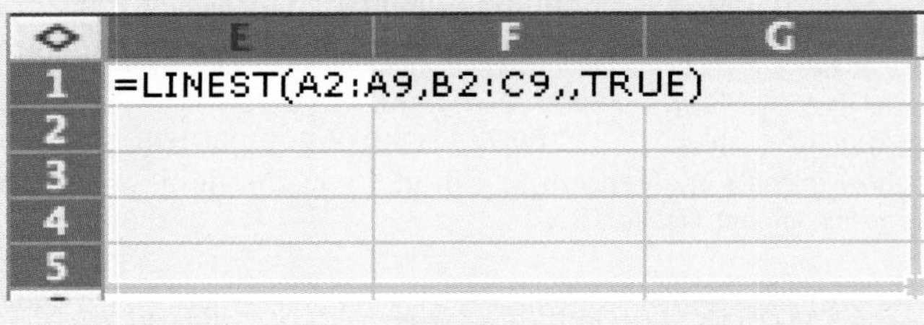

◇	E	F	G
1	=LINEST(A2:A9,B2:C9,,TRUE)		
2			
3			
4			
5			

↓

Cntl+Shift+Enter

◇	E	F	G
1	0.32083333	0.49821429	21.4791667
2	0.05808804	0.42288732	0.63366572
3	0.99157395	0.75290706	#N/A
4	294.198816	5	#N/A
5	333.544405	2.83434524	#N/A

(Notice the coefficients of the quadratic model in the first row.) The p-value is then obtained by the formula `=TDIST(ABS(E1/E2),F4,2)`, which you can compute in any vacant cell. You should get $p \approx 0.00267$.

Q: *What does p actually measure?*

A: *Roughly speaking,* $1 - p \approx 0.997733$ gives the degree of confidence you can have (99.7733%) in asserting that the coefficient of x^2 is not zero. (Technically, p is the probability—allowing for random fluctuation in the data—that, if the coefficient of x^2 were in fact zero, the ratio E1/E2 could be as large as it is.)

In short, you can go ahead and send your email with almost 100% confidence!

EXERCISES

Suppose you are given the following data for the spending on Internet advertising in a hypothetical country in which Impact Advertising also has a 0.25% share of the market.

Year	2010	2011	2012	2013	2014	2015	2016
Spending on Advertising ($ Billion)	0	0.3	1.5	2.6	3.4	4.3	5.0

1. Obtain a linear regression model and the correlation coefficient r. (Take t to be time in years since 2010.) According to the model, at what rate is spending on Internet Advertising increasing in this country? How does this translate to annual revenues for Impact Advertising?
2. Use a spreadsheet or other technology to graph the data together with the best-fit line. Does the graph suggest a quadratic model (parabola)?
3. Test your impression in the preceding exercise by using technology to fit a quadratic function and graphing the resulting curve together with the data. Does the graph suggest that the quadratic model is appropriate?
4. Perform a regression analysis using the quadratic model and find the associated p-value. What does it tell you about the appropriateness of a quadratic model?

TECHNOLOGY GUIDE

TI-83/84 Plus Technology Guide

Section 1.1

Example 1(a) and (c) (page 40) The total number of iPods sold by Apple up to the end of year x can be approximated by $f(x) = 4x^2 + 16x + 2$ million iPods $(0 \le x \le 6)$, where $x = 0$ represents 2003. Compute $f(0)$, $f(2)$, $f(4)$, and $f(6)$, and obtain the graph of f.

Solution with Technology

You can use the Y= screen to enter an algebraically defined function.

1. Enter the function in the Y= screen, as

   `Y1 = 4X^2+16X+2`

   or `Y1 = 4X²+16X+2`

   (See Chapter 0 for a discussion of technology formulas.)

2. To evaluate $f(0)$, for example, enter $Y_1(0)$ in the Home screen to evaluate the function Y_1 at 0. Alternatively, you can use the table feature: After entering the function under Y_1, press [2ND] [TBLSET], and set `Indpnt` to `Ask`. (You do this once and for all; it will permit you to specify values for x in the table screen.) Then, press [2ND] [TABLE], and you will be able to evaluate the function at several values of x. Below (top) is a table showing the values requested:

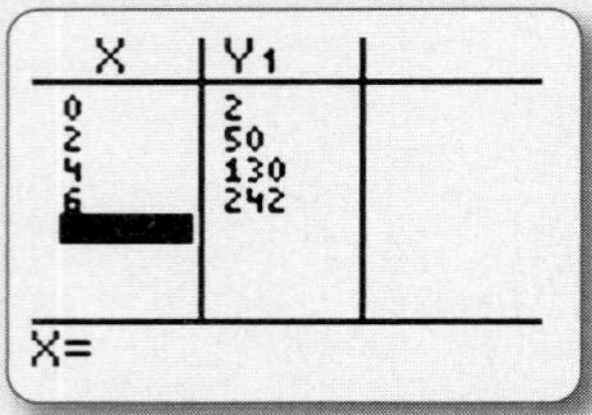

X	Y1
0	2
2	50
4	130
6	242

X=

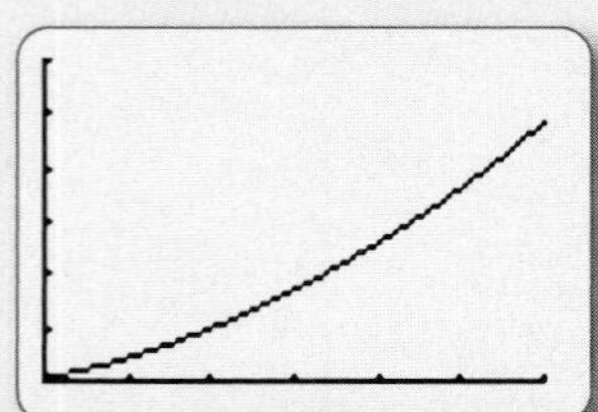

3. To obtain the graph above press [WINDOW], set Xmin = 0, Xmax = 6 (the range of x-values we are interested in), Ymin = 0, Ymax = 300 (we estimated Ymin and Ymax from the corresponding set of y-values in the table) and press [GRAPH] to obtain the curve. Alternatively, you can avoid having to estimate Ymin and Ymax by pressing ZoomFit ([ZOOM] [0]), which automatically sets Ymin and Ymax to the smallest and greatest values of y in the specified range for x.

Example 2 (page 43) The number $n(t)$ of Facebook members can be approximated by the following function of time t in years ($t = 0$ represents January 2004):

$$n(t) = \begin{cases} 4t & \text{if } 0 \le t \le 3 \\ 12 + 23(t-3)^{2.7} & \text{if } 3 < t \le 6 \end{cases}$$

million members.

Obtain a table showing the values $n(t)$ for $t = 0, \ldots, 6$ and also obtain the graph of n.

Solution with Technology

You can enter a piecewise-defined function using the logical inequality operators $<$, $>$, $\le$, and $\ge$, which are found by pressing [2ND] [TEST]:

1. Enter the function n in the Y= screen as:

   `Y1=(X≤3)*(4X)+(X>3)*(12+23*abs(X-3)^2.7)`

   When x is less than or equal to 3, the logical expression `(X≤3)` evaluates to 1 because it is true, and the expression `(X>3)` evaluates to 0 because it is false. The value of the function is therefore given by the expression `(4X)`. When x is greater than 3, the expression `(X≤3)` evaluates to 0 while the expression `(X>3)` evaluates to 1, so the value of the function is given by the expression `(12+23*abs(X-3)^2.7)`. (The reason we use the `abs` in the formula is to prevent an error in evaluating $(x-3)^{2.7}$ when $x < 3$; even though we don't use that formula when $x < 3$, we are in fact evaluating it and multiplying it by zero.)

2. As in Example 1, use the Table feature to compute several values of the function at once by pressing [2ND] [TABLE].

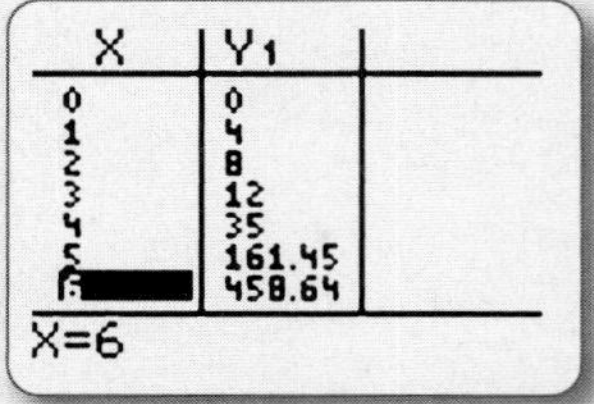

X	Y1
0	0
1	4
2	8
3	12
4	35
5	161.45
6	458.64

X=6

3. To obtain the graph, we proceed as in Example 1: Press [WINDOW], set Xmin = 0, Xmax = 6 (the range of x-values we are interested in), Ymin = 0, Ymax = 500 (see the y-values in the table) and press [GRAPH].

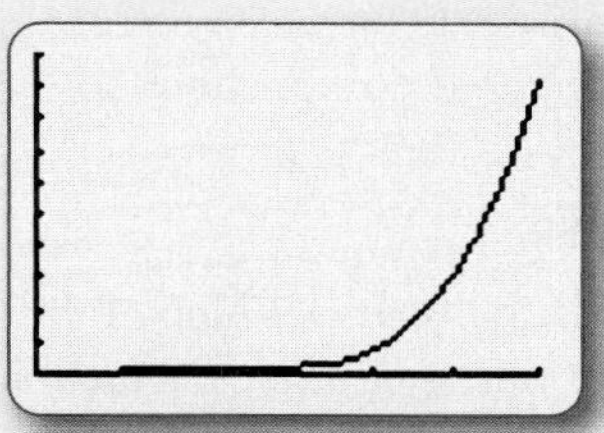

Section 1.2

Example 4(a) (page 54) The demand and supply curves for private schools in Michigan are $q = 77.8p^{-0.11}$ and $q = 30.4 + 0.006p$ thousand students, respectively $(200 \le p \le 2{,}200)$, where p is the net tuition cost in dollars. Graph the demand and supply curves on the same set of axes. Use your graph to estimate, to the nearest \$100, the tuition at which the demand equals the supply (equilibrium price). Approximately how many students will be accommodated at that price?

Solution with Technology

To obtain the graphs of demand and supply:

1. Enter `Y₁=77.8*X^(-0.11)` and `Y₂=30.4+0.006*X` in the "Y=" screen.
2. Press WINDOW, enter Xmin = 200, Xmax = 2200, Ymin = 0, Ymax = 50 and press GRAPH for the graph shown below:

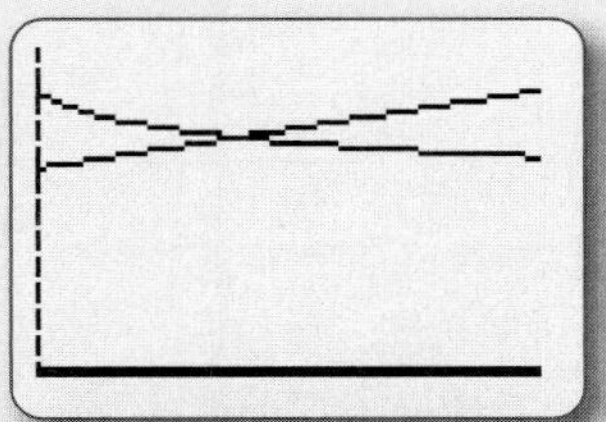

3. To estimate the equilibrium price, press TRACE and use the arrow keys to follow the curve to the approximate point of intersection (around X = 1008) as shown below.

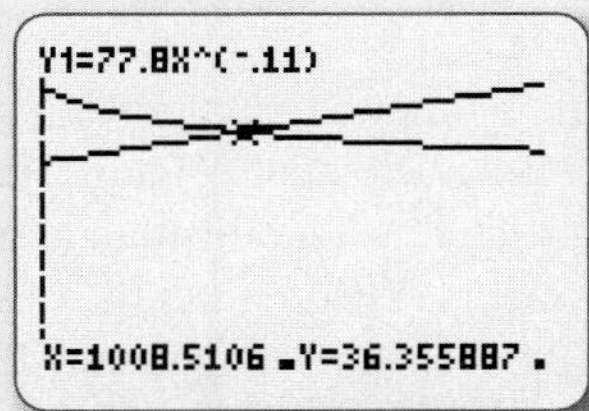

4. For a more accurate estimate, zoom in by pressing ZOOM and selecting Option 1 `ZBox`.
5. Move the curser to a point slightly above and to the left of the intersection, press ENTER, and then move the curser to a point slightly below and to the right and press ENTER again to obtain a box.

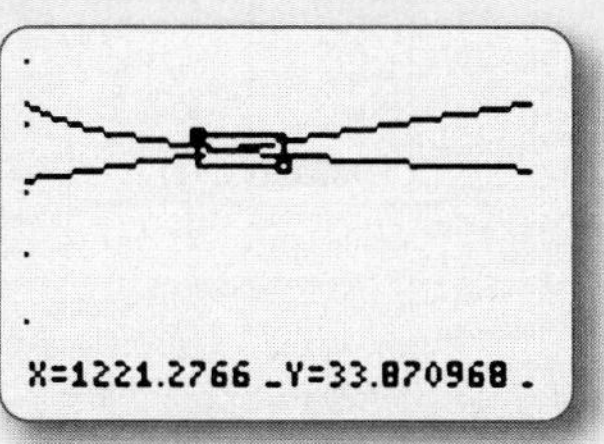

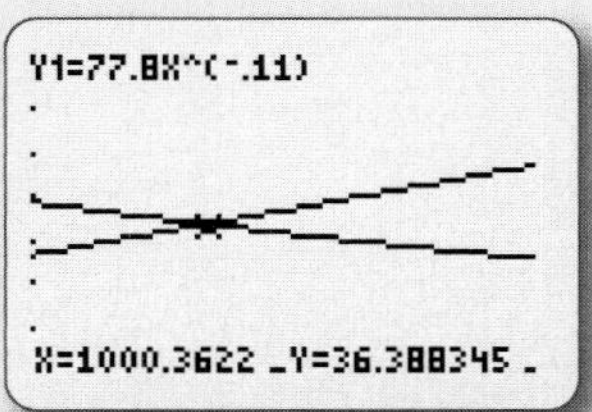

6. Now press ENTER again for a zoomed-in view of the intersection.
7. You can now use TRACE to obtain the intersection coordinates more accurately: X ≈ 1,000, representing a tuition cost of \$1,000. The associated demand is the Y-coordinate: around 36.4 thousand students.

Example 5(a) (page 56) The following table shows annual sales, in billions of dollars, by *Nike* from 2005 through 2010 ($t = 0$ represents 2005):

t	0	1	2	3	4	5
Sales (\$ billion)	13.5	15	16.5	18.5	19	19

Consider the following four models:

(1) $s(t) = 14 + 1.2t$ Linear model

(2) $s(t) = 13 + 2.2t - 0.2t^2$ Quadratic model

(3) $s(t) = 14(1.07^t)$ Exponential model

(4) $s(t) = \dfrac{19.5}{1 + 0.48(1.8^{-t})}$. Logistic model

a. Which models fit the data significantly better than the rest?

b. Of the models you selected in part (a), which gives the most reasonable prediction for 2013?

Solution with Technology

1. First enter the actual revenue data in the stat list editor ([STAT] EDIT) with the values of t in L_1, and the values of $s(t)$ in L_2.

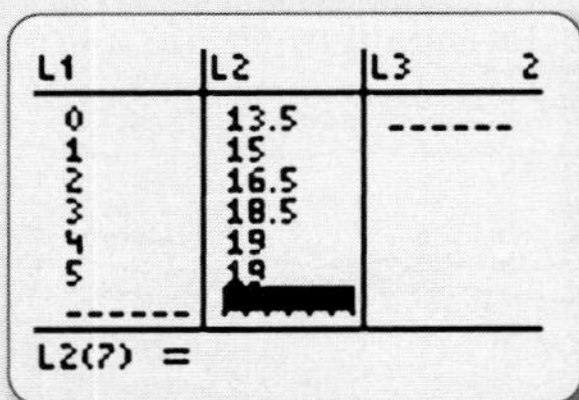

L1
L2
L3
2
0
1
2
3
4
5
13.5
15
16.5
18.5
19
L2(7) =

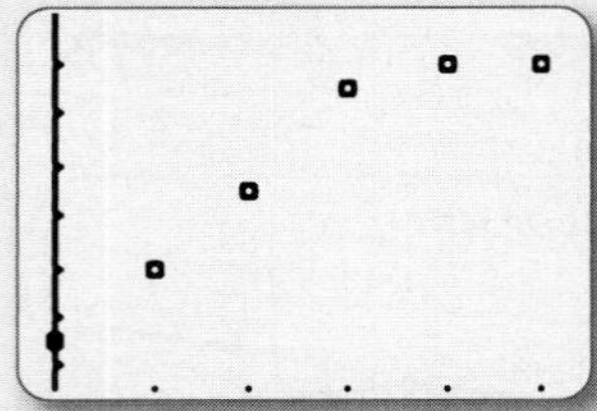

2. Now go to the `Y=` window and turn `Plot1` on by selecting it and pressing [ENTER]. (You can also turn it on in the [2ND] STAT PLOT screen.) Then press ZoomStat ([ZOOM] [9]) to obtain a plot of the points (above).
3. To see any of the four curves plotted along with the points, enter its formula in the Y= screen (for instance, $\texttt{Y}_1\texttt{=13+2.2x-0.2X}^2$ for the second model) and press [GRAPH] (figure on top below).

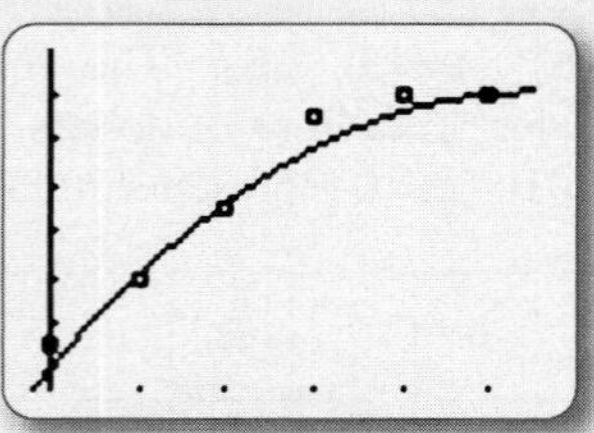

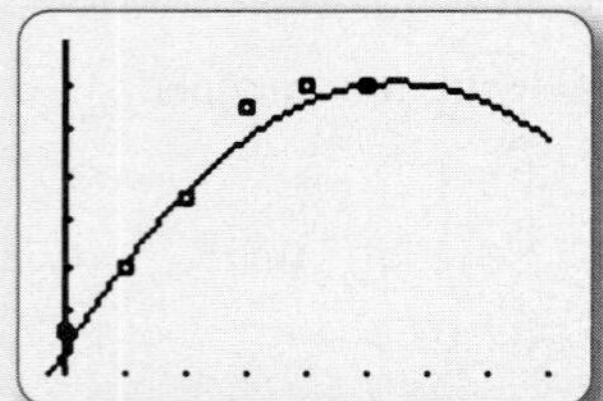

4. To see the extrapolation of the curve to 2013, just change Xmax to 8 (in the [WINDOW] screen) and press [GRAPH] again (lower figure above).
5. Now change Y_1 to see similar graphs for the remaining curves.
6. When you are done, turn `Plot1` off again so that the points you entered do not show up in other graphs.

Section 1.3

Example 1 (page 66) Which of the following two tables gives the values of a linear function? What is the formula for that function?

x	0	2	4	6	8	10	12
$f(x)$	3	−1	−3	−6	−8	−13	−15

x	0	2	4	6	8	10	12
$g(x)$	3	−1	−5	−9	−13	−17	−21

Solution with Technology

We can use the "List" feature in the TI-83/84 Plus to automatically compute the successive quotients $m = \Delta y/\Delta x$ for either f or g as follows:

1. Use the stat list editor ([STAT] EDIT) to enter the values of x and $f(x)$ in the first two columns, called L_1 and L_2, as shown in the screenshot below. (If there is already data in a column you want to use, you can clear it by highlighting the column heading (e.g., L_1) using the arrow key, and pressing [CLEAR] [ENTER].)

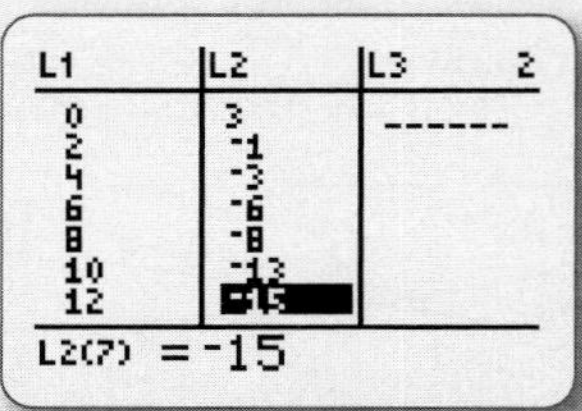

2. Highlight the heading L_3 by using the arrow keys, and enter the following formula (with the quotes, as explained below):

   $\texttt{"}\Delta\texttt{List(L}_2\texttt{)/}\Delta\texttt{List(L}_1\texttt{)"}$ ΔList is found under [2ND] [LIST] OPS. L_1 is [2ND] [1]

   The "ΔList" function computes the differences between successive elements of a list, returning a list with one less element. The formula above then computes the quotients $\Delta y/\Delta x$ in the list L_3 as shown in the following screenshot. As you can see in the third column, $f(x)$ is not linear.

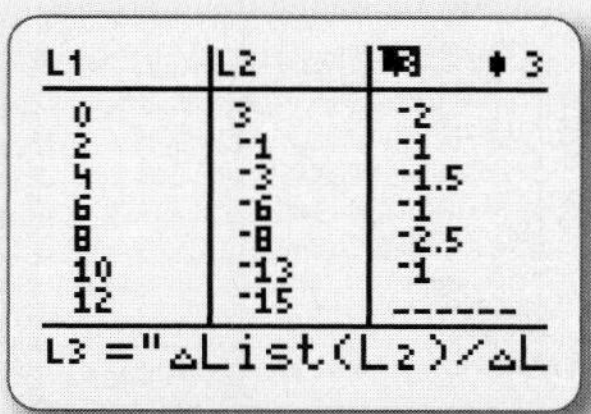

3. To redo the computation for $g(x)$, all you need to do is edit the values of L_2 in the stat list editor. By putting quotes around the formula we used for L_3, we told the calculator to remember the formula, so it automatically recalculates the values.

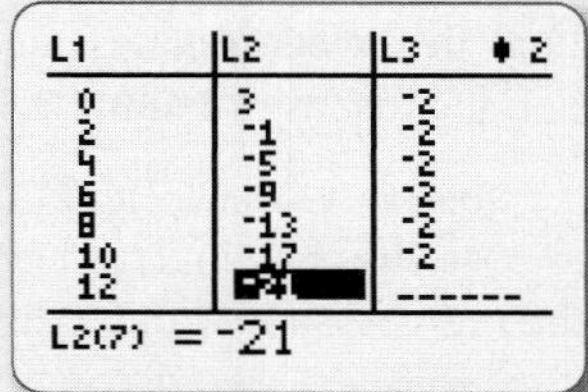

Example 2(a) (page 68) Find the equation of the line through the points (1, 2) and (3, −1).

Solution with Technology

1. Enter the coordinates of the given points in the stat list editor (STAT EDIT) with the values of x in L_1, and the values of y in L_2.

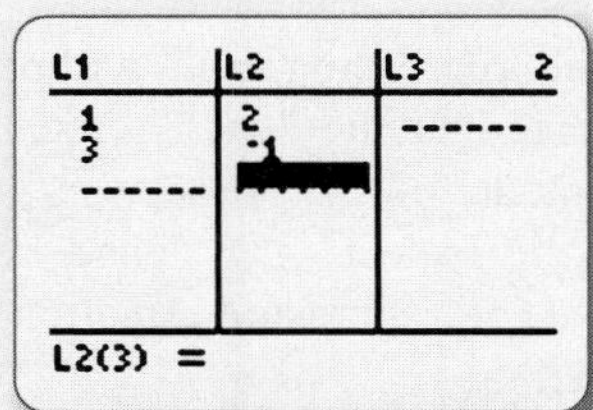

2. To compute the slope, enter the following formula in the Home screen:

   $(L_2(2)-L_2(1))/(L_1(2)-L_1(1))\rightarrow M$

   L_1 and L_2 are under 2ND LIST and the arrow is STO

3. Then, to compute the y-intercept, enter

   $L_2(1)-M*L_1(1)$

Section 1.4

Example 1 (page 76) Using the data on the per capita GDP in Mexico given at the beginning of Section 1.4, compute SSE, the sum-of-squares error, for the linear models $y = 0.5t + 8$ and $y = 0.25t + 9$, and graph the data with the given models.

Solution with Technology

We can use the "List" feature in the TI-83/84 Plus to automate the computation of SSE.

1. Use the stat list editor (STAT EDIT) to enter the given data in the lists L_1 and L_2, as shown in the first screenshot below. (If there is already data in a column you want to use, you can clear it by highlighting the column heading (e.g., L_1) using the arrow key, and pressing CLEAR ENTER.)
2. To compute the predicted values, highlight the heading L_3 using the arrow keys, and enter the following formula for the predicted values (figure on the top below):

   $0.5*L_1+8$ L_1 is 2ND 1

   Pressing ENTER again will fill column 3 with the predicted values (below bottom). Note that only seven of the eight data points can be seen on the screen at one time.

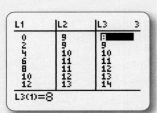

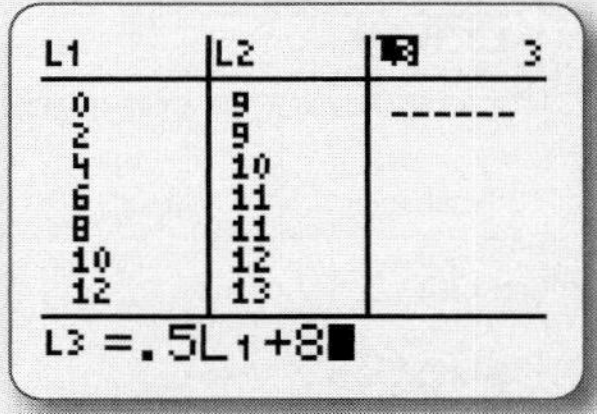

L1 L2 L3 3
0 2 4 6 8 10 12
9 9 10 11 11 12 13
8 9 10 11 12 13 14
L3(1)=8

3. Highlight the heading L_4 and enter the following formula (including the quotes):

   $"(L_2-L_3)\wedge 2"$ Squaring the residuals

TECHNOLOGY GUIDE

4. Pressing ENTER will fill L_4 with the squares of the residuals. (Putting quotes around the formula will allow us to easily check the second model, as we shall see.)

5. To compute SSE, the sum of the entries in L_4, go to the home screen and enter `sum(L4)` (see below; "sum" is under 2ND LIST MATH.)

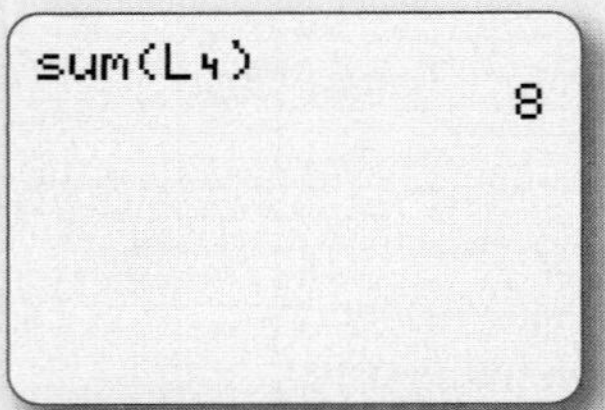

6. To check the second model, go back to the List screen, highlight the heading L_3, enter the formula for the second model, `0.25*L1+9`, and press ENTER. Because we put quotes around the formula for the residuals in L_4, the TI-83/84 Plus will remember the formula and automatically recalculate the values (below top). On the home screen we can again calculate `sum(L4)` to get SSE for the second model (below bottom).

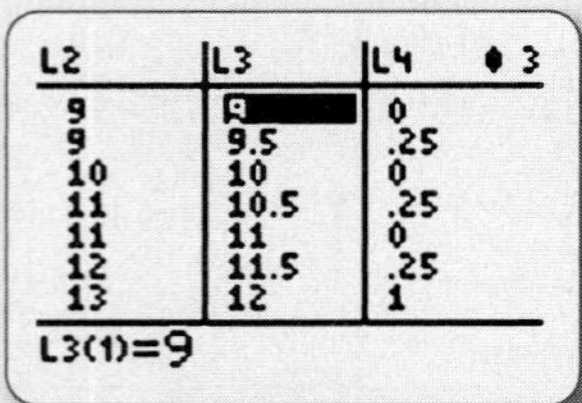

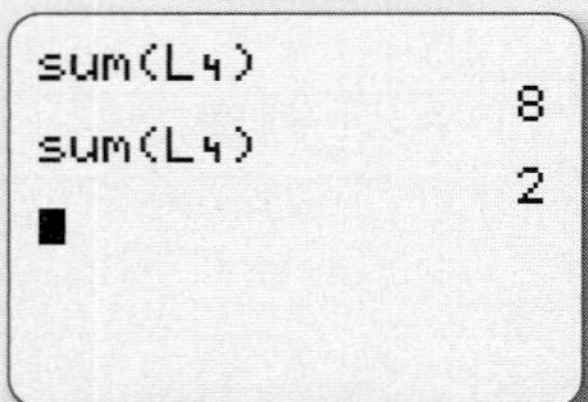

The second model gives a much smaller SSE, so is the better fit.

7. You can also use the TI-83/84 Plus to plot both the original data points and the two lines (see below). Turn `Plot1` on in the STAT PLOT window, obtained by pressing 2ND STAT PLOT. To show the lines, enter them in the "Y=" screen as usual. To obtain a convenient window showing all the points and the lines, press ZOOM and choose 9: `ZoomStat`.

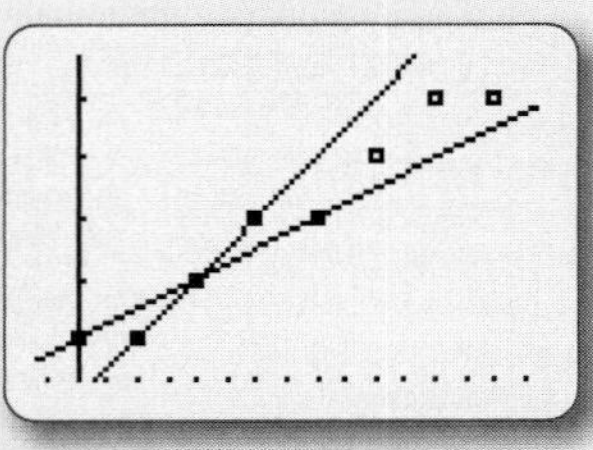

Example 2 (page 79) Use the data on the per capita GDP in Mexico to find the best-fit linear model.

Solution with Technology

1. Enter the data in the TI-83/84 Plus using the List feature, putting the x-coordinates in L_1 and the y-coordinates in L_2, just as in Example 1.

2. Press STAT, select CALC, and choose `#4: LinReg(ax+b)`. Pressing ENTER will cause the equation of the regression line to be displayed in the home screen:

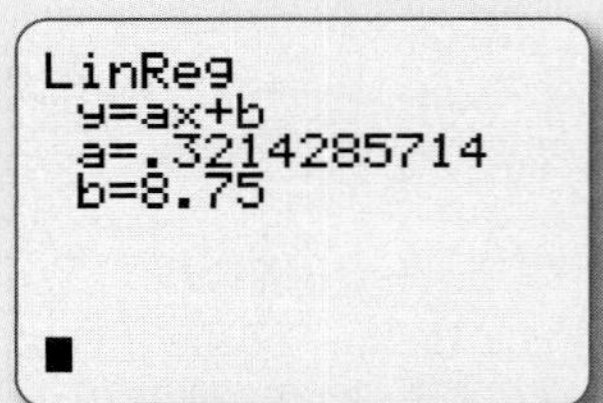

So, the regression line is $y \approx 0.321x + 8.75$.

3. To graph the regression line without having to enter it by hand in the "Y=" screen, press Y=, clear the contents of Y_1, press VARS, choose `#5: Statistics`, select `EQ`, and then choose `#1:RegEQ`. The regression equation will then be entered under Y_1.

4. To simultaneously show the data points, press 2ND STATPLOT and turn `Plot1` on as in Example 1. To obtain a convenient window showing all the points and the line (see below), press ZOOM and choose `#9: ZoomStat`.

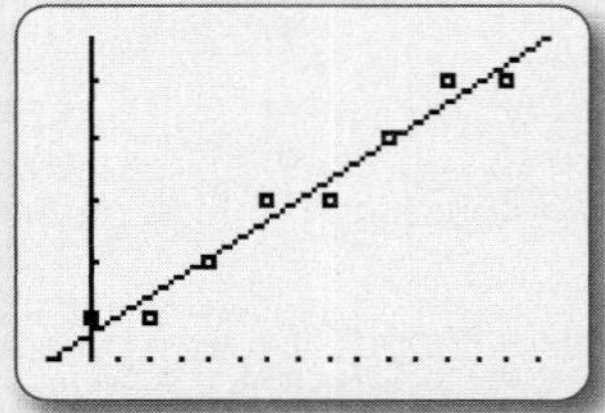

Example 3 (page 81) Find the correlation coefficient for the data in Example 2.

Solution with Technology

To find the correlation coefficient using a TI-83/84 Plus you need to tell the calculator to show you the coefficient at the same time that it shows you the regression line.

1. Press [2ND] [CATALOG] and select `DiagnosticOn` from the list. The command will be pasted to the home screen, and you should then press [ENTER] to execute the command.
2. Once you have done this, the `"LinReg(ax+b)"` command (see the discussion for Example 2) will show you not only a and b, but r and r^2 as well:

SPREADSHEET Technology Guide

Section 1.1

Example 1(a) and (c) (page 40) The total number of iPods sold by Apple up to the end of year x can be approximated by $f(x) = 4x^2 + 16x + 2$ million iPods ($0 \le x \le 6$), where $x = 0$ represents 2003. Compute $f(0)$, $f(2)$, $f(4)$, and $f(6)$, and obtain the graph of f.

Solution with Technology

To create a table of values of f using a spreadsheet:

1. Set up two columns: one for the values of x and one for the values of $f(x)$. Then enter the sequence of values 0, 2, 4, 6 in the x column as shown.

	A	B
1	x	f(x)
2	0	
3	2	
4	4	
5	6	

2. Now we enter a formula for $f(x)$ in cell B2 (below). The technology formula is `4*x^2+16*x+2`. To use this formula in a spreadsheet, we modify it slightly:

`=4*A2^2+16*A2+2` Spreadsheet version of tech formula

Notice that we have preceded the Excel formula by an equals sign (=) and replaced each occurrence of x by the name of the cell holding the value of x (cell A2 in this case).

	A	B	C
1	x	f(x)	
2	0	=4*A2^2+16*A2+2	
3	2		
4	4		
5	6		

Note Instead of typing in the name of the cell "A2" each time, you can simply click on the cell A2, and "A2" will be automatically inserted. ■

3. Now highlight cell B2 and drag the **fill handle** (the little square at the lower right-hand corner of the selection) down until you reach Row 5 as shown below on the top, to obtain the result shown on the bottom.

	A	B	C
1	x	f(x)	
2	0	=4*A2^2+16*A2+2	
3	2		
4	4		
5	6		

	A	B
1	x	f(x)
2	0	2
3	2	50
4	4	130
5	6	242

4. To graph the data, highlight A1 through B5, and insert a "Scatter chart" (the exact method of doing this depends on the specific version of the spreadsheet program). When choosing the style of the chart,

choose a style that shows points connected by lines (if possible) to obtain a graph something like the following:

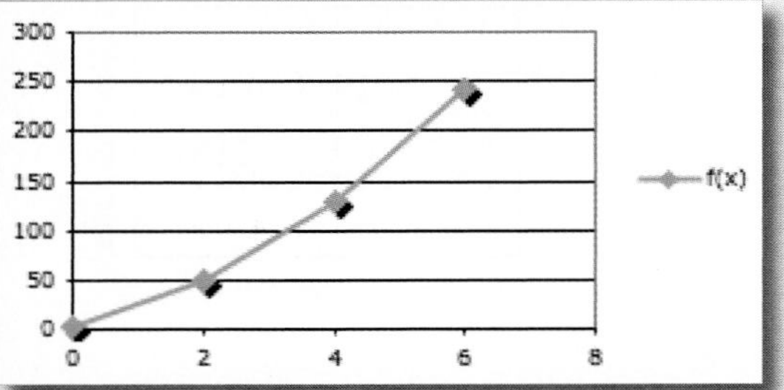

Example 2 (page 43) The number $n(t)$ of Facebook members can be approximated by the following function of time t in years ($t = 0$ represents January 2004):

$$n(t) = \begin{cases} 4t & \text{if } 0 \le t \le 3 \\ 12 + 23(t-3)^{2.7} & \text{if } 3 < t \le 6 \end{cases} \quad \text{million members.}$$

Obtain a table showing the values $n(t)$ for $t = 0, \ldots, 6$ and also obtain the graph of n.

Solution with Technology

We can generate a table of values of $n(t)$ for $t = 0, 1, \ldots, 6$ as follows:

1. Set up two columns; one for the values of t and one for the values of $n(t)$, and enter the values 0, 1, . . . , 6 in the t column as shown in the first screenshot below:
2. We must now enter the formula for n in cell B2. The following formula defines the function n in Excel:

```
=(x<=3)*(4*x)+(x>3)*(12+23*abs
 (x-3)^2.7)
```

When x is less than or equal to 3, the logical expression `(x≤3)` evaluates to 1 because it is true, and the expression `(x>3)` evaluates to 0 because it is false. The value of the function is therefore given by the expression `(4*x)`. When x is greater than 3, the expression `(x≤3)` evaluates to 0 while the expression `(x>3)` evaluates to 1, so the value of the function is given by the expression `(12+23*abs (x-3)^2.7)`. (The reason we use the `abs` in the formula is to prevent an error in evaluating $(x-3)^{2.7}$ when $x < 3$; even though we don't use that formula when $x < 3$, we are in fact evaluating it and multiplying it by zero.) We therefore enter the formula

```
=(A2<=3)*(4*A2)+(A2>3)*(12+23
 *ABS(A2-3)^2.7)
```

in cell B2 and then copy down to cell B8 (below top) to obtain the result shown on the bottom:

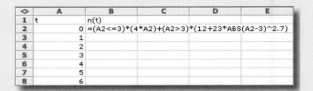

	A	B	C	D	E
1	t	n(t)			
2	0	=(A2<=3)*(4*A2)+(A2>3)*(12+23*ABS(A2-3)^2.7)			
3	1				
4	2				
5	3				
6	4				
7	5				
8	6				

	A	B
1	t	n(t)
2	0	0
3	1	4
4	2	8
5	3	12
6	4	35
7	5	161.454
8	6	458.638

3. To graph the data, highlight A1 through B8, and insert a "Scatter chart" as in Example 1 to obtain the result shown below:

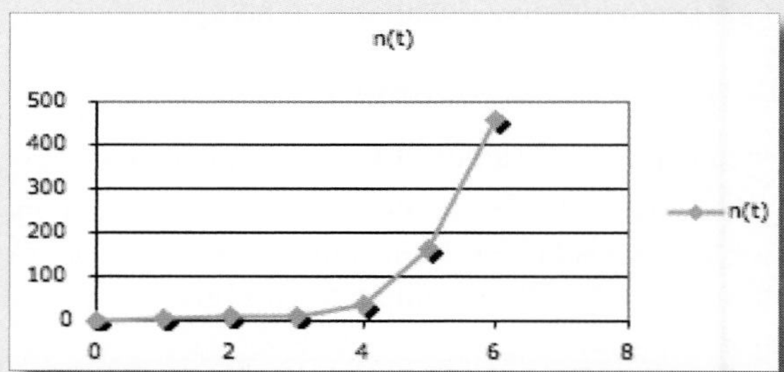

Section 1.2

Example 4(a) (page 54) The demand and supply curves for private schools in Michigan are $q = 77.8p^{-0.11}$ and $q = 30.4 + 0.006p$ thousand students, respectively ($200 \le p \le 2{,}200$), where p is the net tuition cost in dollars. Graph the demand and supply curves on the same set of axes. Use your graph to estimate, to the nearest \$100, the tuition at which the demand equals the supply (equilibrium price). Approximately how many students will be accommodated at that price?

Solution with Technology

To obtain the graphs of demand and supply:

1. Enter the headings p, Demand, and Supply in cells A1–C1 and the p-values 200, 300, . . . , 2,200 in A2–A22.

2. Next, enter the formulas for the demand and supply functions in cells B2 and C2.

   Demand: `=77.8*A2^(-0.11)` in cell B2

   Supply: `=30.4+0.006*A2` in cell C2

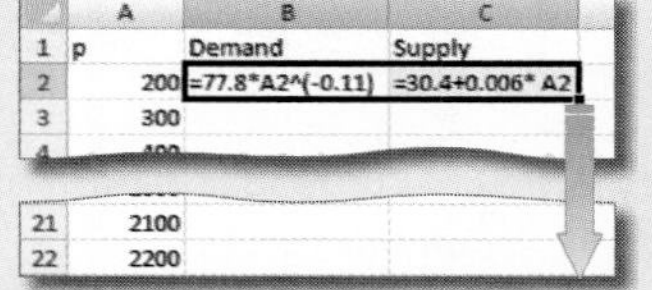

	A	B	C
1	p	Demand	Supply
2	200	=77.8*A2^(-0.11)	=30.4+0.006*A2
3	300		
21	2100		
22	2200		

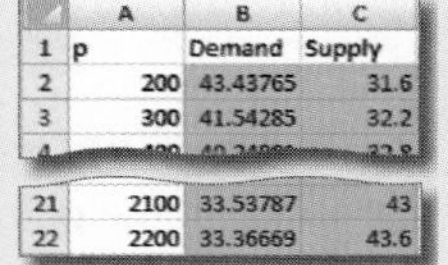

	A	B	C
1	p	Demand	Supply
2	200	43.43765	31.6
3	300	41.54285	32.2
21	2100	33.53787	43
22	2200	33.36669	43.6

3. To graph the data, highlight A1 through C22, and insert a Scatter chart:

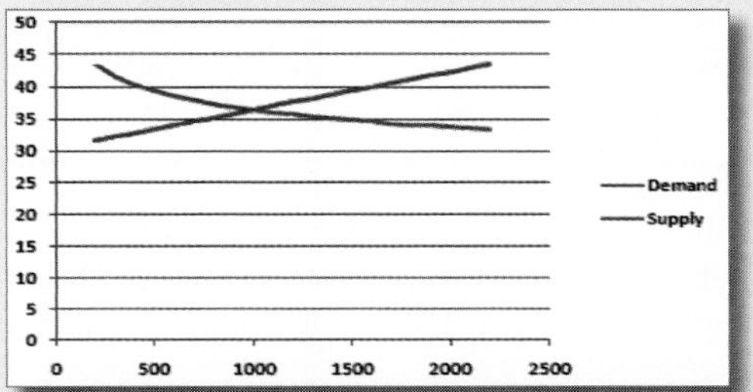

4. If you place the cursor as close as you can get to the intersection point (or just look at the table of values), you will see that the curves cross close to $p = \$1{,}000$ (to the nearest \$100).
5. To more accurately determine where the curves cross, you can narrow down the range of values shown on the x-axis by changing the p-values to 990, 991, . . . , 1010.

Example 5(a) (page 56) The following table shows annual sales, in billions of dollars, by *Nike* from 2005 through 2010 ($t = 0$ represents 2005):

t	0	1	2	3	4	5
Sales ($ billion)	13.5	15	16.5	18.5	19	19

Consider the following four models:

(1) $s(t) = 14t + 1.2t$ Linear model

(2) $s(t) = 13 + 2.2t - 0.2t^2$ Quadratic model

(3) $s(t) = 14(1.07^t)$ Exponential model

(4) $s(t) = \dfrac{19.5}{1 + 0.48(1.8^{-t})}$ Logistic model

a. Which models fit the data significantly better than the rest?

b. Of the models you selected in part (a), which gives the most reasonable prediction for 2013?

Solution with Technology

1. First create a scatter plot of the given data by tabulating the data as shown below, and selecting the Insert tab and choosing a "Scatter" chart:

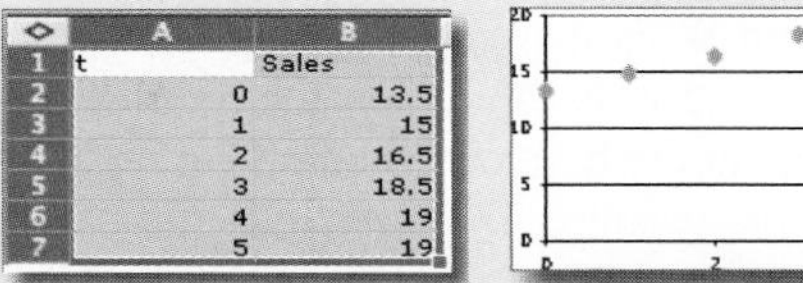

	A	B
1	t	Sales
2	0	13.5
3	1	15
4	2	16.5
5	3	18.5
6	4	19
7	5	19

2. In column C use the formula for the model you are interested in seeing; for example, model (2): `=13+2.2*A2-0.2*A2^2`

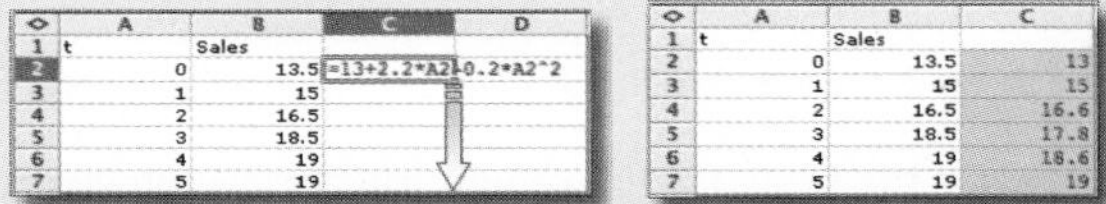

	A	B	C
1	t	Sales	
2	0	13.5	13
3	1	15	15
4	2	16.5	16.6
5	3	18.5	17.8
6	4	19	18.6
7	5	19	19

3. To adjust the graph to include the graph of the model you have added, you need to change the graph data from \$A\$2:\$B\$7 to \$A\$2:\$C\$7 so as to include column C. In Excel you can obtain this by right-clicking on the graph to select "Source Data". In OpenOffice, double-click on the graph and then right-click it to choose "Data Ranges". In Excel, you can also click once on the graph—the effect will be to outline the data you have graphed in columns A and B—and then use the fill handle at the bottom of Column B to extend the selection to Column C as shown:

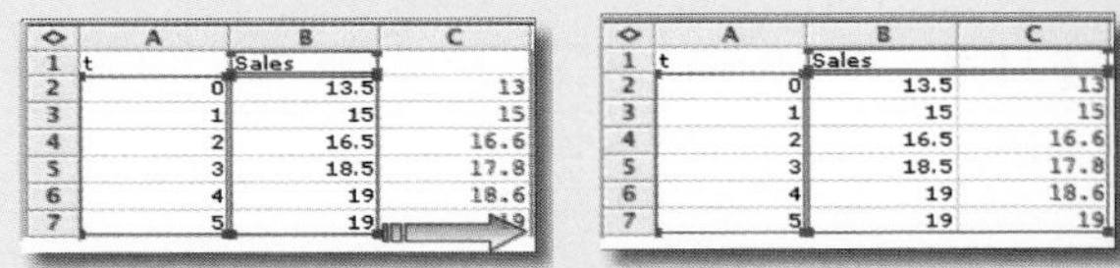

	A	B	C
1	t	Sales	
2	0	13.5	13
3	1	15	15
4	2	16.5	16.6
5	3	18.5	17.8
6	4	19	18.6
7	5	19	19

The graph will now include markers showing the values of both the actual sales and the model you inserted in Column C.

4. Right-click on any of the markers corresponding to column B in the graph (in OpenOffice you would first double-click on the graph), select "Format data series" to add lines connecting the points and remove the markers. The effect will be as shown below, with the model represented by a curve and the actual data points represented by dots:

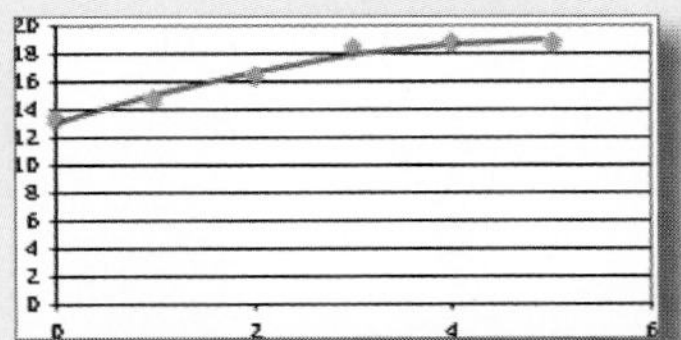

TECHNOLOGY GUIDE

5. To see the extrapolation of the curve to 2013, add the values 6, 7, 8 to Column A. The values of $s(t)$ may automatically be computed in Column C as you type, depending on the spreadsheet. If not, you will need to copy the formula in column C down to C10. (Do not touch Column B, as that contains the observed data up through $t = 5$ only.) Click on the graph, and use the fill handle at the base of Column C to include the new data in the graph:

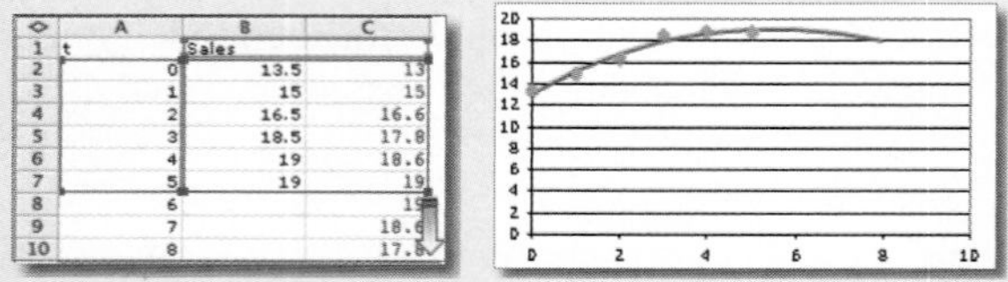

	A	B	C
1	t	Sales	
2	0	13.5	13
3	1	15	15
4	2	16.5	16.6
5	3	18.5	17.8
6	4	19	18.6
7	5	19	19
8	6		19
9	7		18.6
10	8		17.5

6. To see the plots for the remaining curves, change the formula in Column B (and don't forget to copy the new formula down to cell C10 when you do so).

Section 1.3

Example 1 (page 66) Which of the following two tables gives the values of a linear function? What is the formula for that function?

x	0	2	4	6	8	10	12
$f(x)$	3	-1	-3	-6	-8	-13	-15

x	0	2	4	6	8	10	12
$g(x)$	3	-1	-5	-9	-13	-17	-21

Solution with Technology

1. The following worksheet shows how you can compute the successive quotients $m = \Delta y/\Delta x$, and hence check whether a given set of data shows a linear relationship, in which case all the quotients will be the same. (The shading indicates that the formula is to be copied down only as far as cell C7. Why not cell C8?)

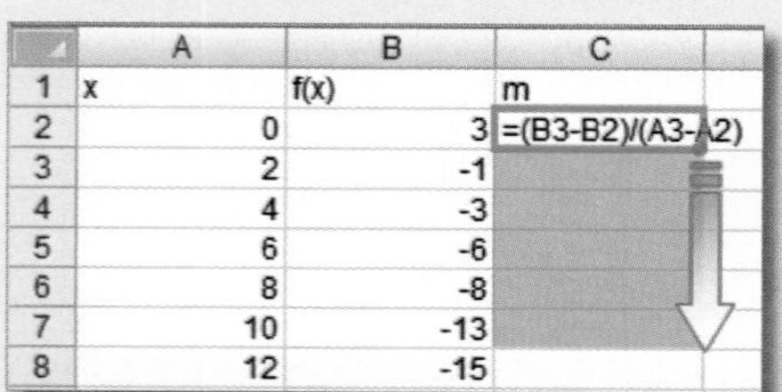

	A	B	C
1	x	f(x)	m
2	0	3	=(B3-B2)/(A3-A2)
3	2	-1	
4	4	-3	
5	6	-6	
6	8	-8	
7	10	-13	
8	12	-15	

2. Here are the results for both f and g.

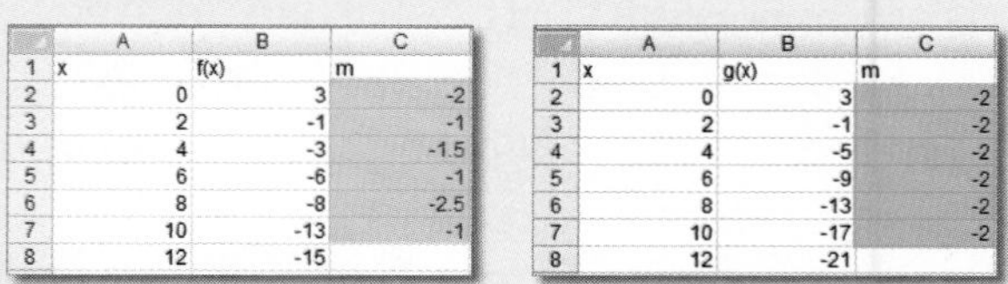

	A	B	C
1	x	f(x)	m
2	0	3	-2
3	2	-1	-1
4	4	-3	-1.5
5	6	-6	-1
6	8	-8	-2.5
7	10	-13	-1
8	12	-15	

	A	B	C
1	x	g(x)	m
2	0	3	-2
3	2	-1	-2
4	4	-5	-2
5	6	-9	-2
6	8	-13	-2
7	10	-17	-2
8	12	-21	

Example 2(a) (page 68) Find the equation of the line through the points (1, 2) and (3, −1).

Solution with Technology

1. Enter the x- and y-coordinates in columns A and B, as shown below on the left.

	A	B
1	x	y
2	1	2
3	3	-1

	A	B	C	D
1	x	y	m	b
2	1	2	=(B3-B2)/(A3-A2)	=B2-C2*A2
3	3	-1	Slope	Intercept

2. Add the headings m and b in C1-D1, and then the formulas for the slope and intercept in C2-D2, as shown above on the right. The result will be as shown below:

	A	B	C	D
1	x	y	m	b
2	1	2	-1.5	3.5
3	3	-1	Slope	Intercept

Section 1.4

Example 1 (page 76) Using the data on the per capita GDP in Mexico given at the beginning of Section 1.4, compute SSE, the sum-of-squares error, for the linear models $y = 0.5t + 8$ and $y = 0.25t + 9$, and graph the data with the given models.

Solution with Technology

1. Begin by setting up your worksheet with the observed data in two columns, t and y, and the predicted data for the first model in the third.

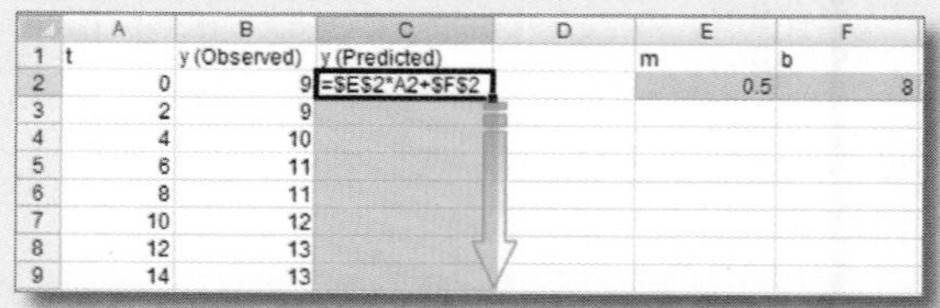

	A	B	C	D	E	F
1	t	y (Observed)	y (Predicted)		m	b
2	0	9	=$E$2*A2+$F$2		0.5	8
3	2	9				
4	4	10				
5	6	11				
6	8	11				
7	10	12				
8	12	13				
9	14	13				

2. Notice that, instead of using the numerical equation for the first model in column C, we used absolute references to the cells containing the slope m and the intercept b. This way, we can switch from one linear model to the next by changing only m and b in cells E2 and F2. (We have deliberately left column D empty in anticipation of the next step.)

3. In column D we compute the squares of the residuals using the Excel formula `=(B2-C2)^2`.

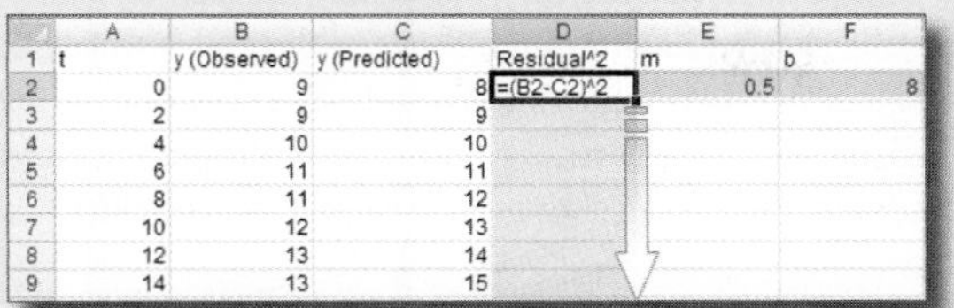

	A	B	C	D	E	F
1	t	y (Observed)	y (Predicted)	Residual^2	m	b
2	0	9	8	=(B2-C2)^2	0.5	8
3	2	9	9			
4	4	10	10			
5	6	11	11			
6	8	11	12			
7	10	12	13			
8	12	13	14			
9	14	13	15			

4. We now compute SSE in cell F4 by summing the entries in column D:

	A	B	C	D	E	F
1	t	y (Observed)	y (Predicted)	Residual^2	m	b
2	0	9	8	1	0.5	8
3	2	9	9	0		
4	4	10	10	0	SSE:	=SUM(D2:D9)
5	6	11	11	0		
6	8	11	12	1		
7	10	12	13	1		
8	12	13	14	1		
9	14	13	15	4		

5. Here is the completed spreadsheet:

	A	B	C	D	E	F
1	t	y (Observed)	y (Predicted)	Residual^2	m	b
2	0	9	8	1	0.5	8
3	2	9	9	0		
4	4	10	10	0	SSE:	8
5	6	11	11	0		
6	8	11	12	1		
7	10	12	13	1		
8	12	13	14	1		
9	14	13	15	4		

6. Changing m to 0.25 and b to 9 gives the sum of squares error for the second model, SSE $= 2$.

	A	B	C	D	E	F
1	t	y (Observed)	y (Predicted)	Residual^2	m	b
2	0	9	9	0	0.25	9
3	2	9	9.5	0.25		
4	4	10	10	0	SSE:	2
5	6	11	10.5	0.25		
6	8	11	11	0		
7	10	12	11.5	0.25		
8	12	13	12	1		
9	14	13	12.5	0.25		

7. To plot both the original data points and each of the two lines, use a scatter plot to graph the data in columns A through C in each of the last two worksheets above.

$y = 0.5t + 8$ $\qquad\qquad$ $y = 0.25t + 9$

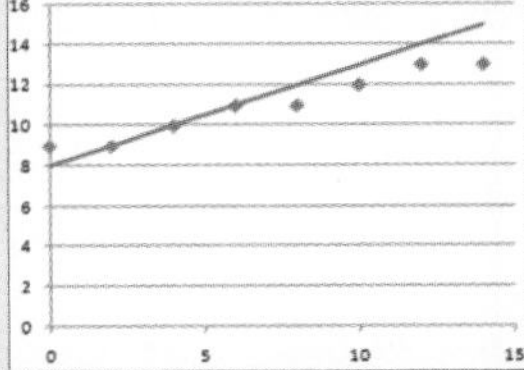

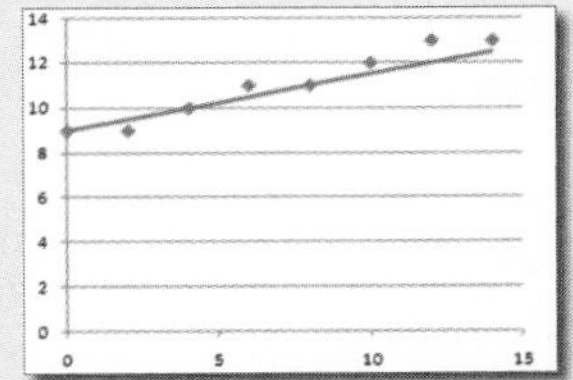

Example 2 (page 79) Use the data on the per capita GDP in Mexico to find the best-fit linear model.

Solution with Technology

Here are two spreadsheet shortcuts for linear regression; one graphical and one based on a spreadsheet formula:

Using a Trendline

1. Start with the original data and insert a scatter plot (below left and right).

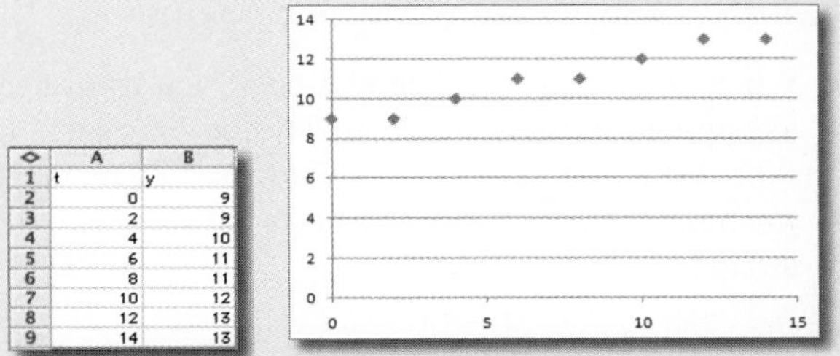

	A	B
1	t	y
2	0	9
3	2	9
4	4	10
5	6	11
6	8	11
7	10	12
8	12	13
9	14	13

2. Insert a "linear trendline", choosing the option to display the equation on the chart. The method for doing so varies from spreadsheet to spreadsheet.[13] In Excel, you can right-click on one of the points in the graph and choose "Add Trendline" (in OpenOffice you would first double-click on the graph). Then, under "Trendline Options", select "Display Equation on chart". The procedure for OpenOffice is almost identical, but you first need to double-click on the graph. The result is shown below.

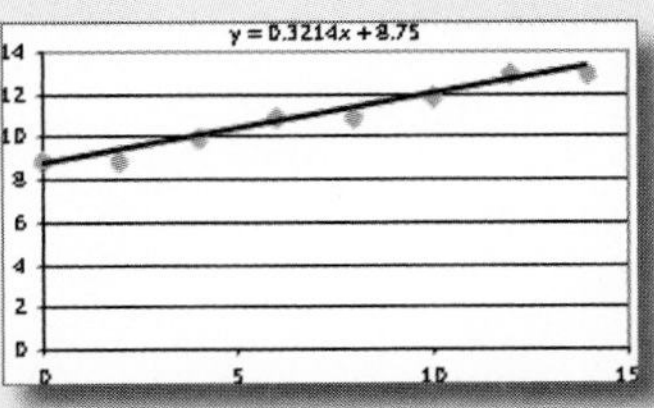

Using a Formula

1. Enter your data as above, and select a block of unused cells two wide and one tall; for example, C2:D2. Then enter the formula

```
=LINEST(B2:B9,A2:A9)
```

[13]At the time of this writing, Google Docs has no trendline feature for its spreadsheet, so you would need to use the formula method.

as shown on the left. Then press Control-Shift-Enter. The result should appear as on the right, with m and b appearing in cells C2 and D2 as shown:

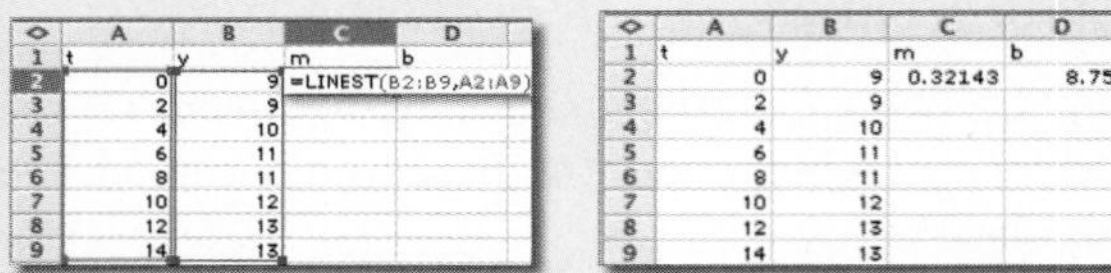

Example 3 (page 81) Find the correlation coefficient for the data in Example 2.

Solution with Technology

1. When you add a trendline to a chart you can select the option "Display R-squared value on chart" to show the value of r^2 on the chart (it is common to examine r^2, which takes on values between 0 and 1, instead of r).

2. Alternatively, the LINEST function we used above in 2 can be used to display quite a few statistics about a best-fit line, including r^2. Instead of selecting a block of cells two wide and one tall as we did in Example 2, we select one two wide and *five* tall. We now enter the requisite LINEST formula with two additional arguments set to "TRUE" as shown, and press Control-Shift-Enter:

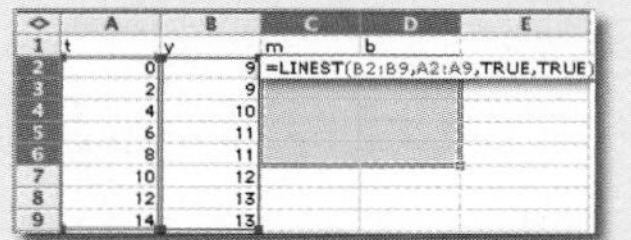

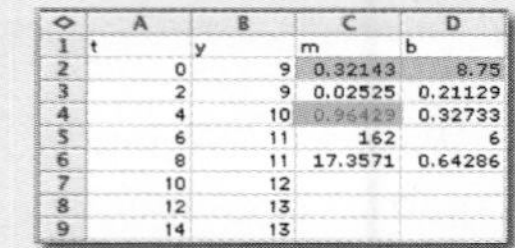

The values of m and b appear in cells C2 and D2 as before, and the value of r^2 in cell C4. (Among the other numbers shown is SSE in cell D6. For the meanings of the remaining numbers shown, do a Web search for "LINEST"; you will see numerous articles, including many that explain all the terms. A good course in statistics wouldn't hurt, either.)

2

Nonlinear Functions and Models

Website
www.WanerMath.com
At the Website you will find:

- Section-by-section tutorials, including game tutorials with randomized quizzes
- A detailed chapter summary
- A true/false quiz
- Additional review exercises
- Graphers, Excel tutorials, and other resources
- The following extra topics:
  - Inverse functions
  - Using and deriving algebraic properties of logarithms

Case Study Checking up on Malthus

In 1798 Thomas R. Malthus (1766–1834) published an influential pamphlet, later expanded into a book, titled *An Essay on the Principle of Population as It Affects the Future Improvement of Society*. One of his main contentions was that population grows geometrically (exponentially), while the supply of resources such as food grows only arithmetically (linearly). Some 200+ years later, you have been asked to check the validity of Malthus's contention. **How do you go about doing so?**

Robert Nickelsberg/Getty Images

Introduction

To see if Malthus was right, we need to see if the data fit the models (linear and exponential) that he suggested or if other models would be better. We saw in Chapter 1 how to fit a linear model. In this chapter we discuss how to construct models that use various *nonlinear* functions.

The nonlinear functions we consider in this chapter are the *quadratic* functions, the simplest nonlinear functions; the *exponential* functions, essential for discussing many kinds of growth and decay, including the growth (and decay) of money in finance and the initial growth of an epidemic; the *logarithmic* functions, needed to fully understand the exponential functions; and the *logistic* functions, used to model growth with an upper limit, such as the spread of an epidemic.

algebra Review
For this chapter, you should be familiar with the algebra reviewed in **Chapter 0, Section 2.**

2.1 Quadratic Functions and Models

In Chapter 1 we studied linear functions. Linear functions are useful, but in real-life applications, they are often accurate for only a limited range of values of the variables. The relationship between two quantities is often best modeled by a curved line rather than a straight line. The simplest function with a graph that is not a straight line is a *quadratic* function.

Quadratic Function

A **quadratic function** of the variable x is a function that can be written in the form

$$f(x) = ax^2 + bx + c \qquad \text{Function form}$$

or

$$y = ax^2 + bx + c \qquad \text{Equation form}$$

where a, b, and c are fixed numbers (with $a \neq 0$).

Quick Examples

1. $f(x) = 3x^2 - 2x + 1$ — $a = 3, b = -2, c = 1$
2. $g(x) = -x^2$ — $a = -1, b = 0, c = 0$
3. $R(p) = -5{,}600p^2 + 14{,}000p$ — $a = -5{,}600, b = 14{,}000, c = 0$

Every quadratic function $f(x) = ax^2 + bx + c$ ($a \neq 0$) has a **parabola** as its graph. Following is a summary of some features of parabolas that we can use to sketch the graph of any quadratic function.*

* We shall not fully justify the formula for the vertex and the axis of symmetry until we have studied some calculus, although it is possible to do so with just algebra.

Features of a Parabola

The graph of $f(x) = ax^2 + bx + c$ $(a \neq 0)$ is a **parabola**. If $a > 0$ the parabola opens upward (concave up) and if $a < 0$ it opens downward (concave down):

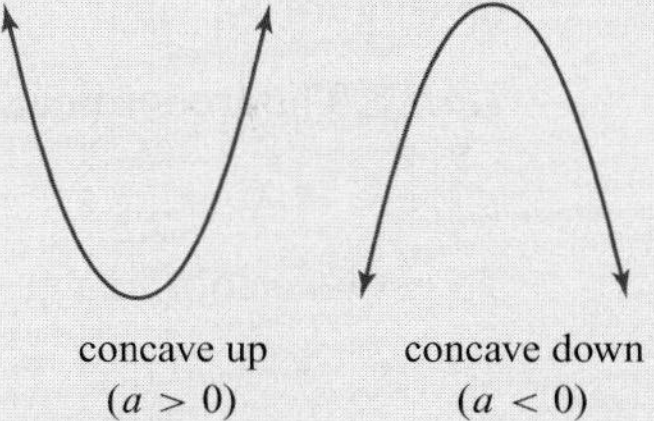

Vertex, Intercepts, and Symmetry

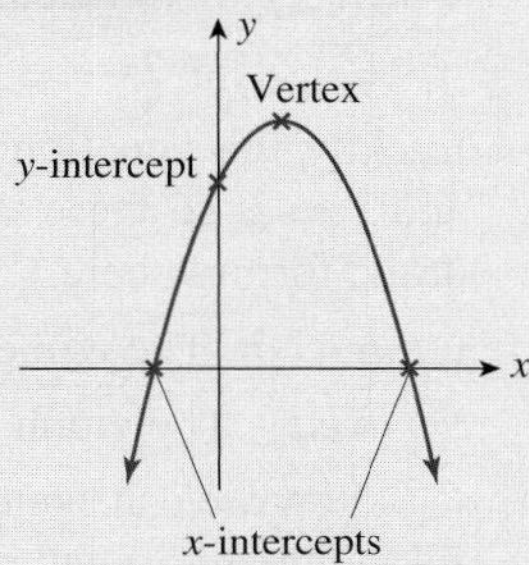

Vertex The vertex is the highest or lowest point of the parabola (see the figure above). Its x-coordinate is $-\dfrac{b}{2a}$. Its y-coordinate is $f\left(-\dfrac{b}{2a}\right)$.

***x*-Intercepts** (if any) These occur when $f(x) = 0$; that is, when

$$ax^2 + bx + c = 0.$$

Solve this equation for x by either factoring or using the quadratic formula. The x-intercepts are

$$x = \frac{-b \pm \sqrt{b^2 - 4ac}}{2a}.$$

If the **discriminant** $b^2 - 4ac$ is positive, there are two x-intercepts. If it is zero, there is a single x-intercept (at the vertex). If it is negative, there are no x-intercepts (so the parabola doesn't touch the x-axis at all).

***y*-Intercept** This occurs when $x = 0$, so

$$y = a(0)^2 + b(0) + c = c.$$

Symmetry The parabola is symmetric with respect to the vertical line through the vertex, which is the line $x = -\dfrac{b}{2a}$.

Note that the x-intercepts can also be written as

$$x = -\frac{b}{2a} \pm \frac{\sqrt{b^2 - 4ac}}{2a},$$

making it clear that they are located symmetrically on either side of the line $x = -b/(2a)$. This partially justifies the claim that the whole parabola is symmetric with respect to this line.

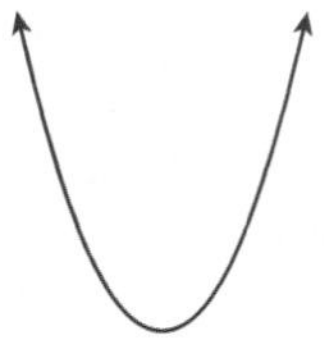

Figure 1

EXAMPLE 1 Sketching the Graph of a Quadratic Function

Sketch the graph of $f(x) = x^2 + 2x - 8$ by hand.

Solution Here, $a = 1$, $b = 2$, and $c = -8$. Because $a > 0$, the parabola is concave up (Figure 1).

Vertex: The x coordinate of the vertex is

$$x = -\frac{b}{2a} = -\frac{2}{2} = -1.$$

To get its y coordinate, we substitute the value of x back into $f(x)$ to get

$$y = f(-1) = (-1)^2 + 2(-1) - 8 = 1 - 2 - 8 = -9.$$

Thus, the coordinates of the vertex are $(-1, -9)$.

x-Intercepts: To calculate the x-intercepts (if any), we solve the equation

$$x^2 + 2x - 8 = 0.$$

Luckily, this equation factors as $(x + 4)(x - 2) = 0$. Thus, the solutions are $x = -4$ and $x = 2$, so these values are the x-intercepts. (We could also have used the quadratic formula here.)

y-Intercept: The y-intercept is given by $c = -8$.

Symmetry: The graph is symmetric around the vertical line $x = -1$.

Now we can sketch the curve as in Figure 2. (As we see in the figure, it is helpful to plot additional points by using the equation $y = x^2 + 2x - 8$, and to use symmetry to obtain others.)

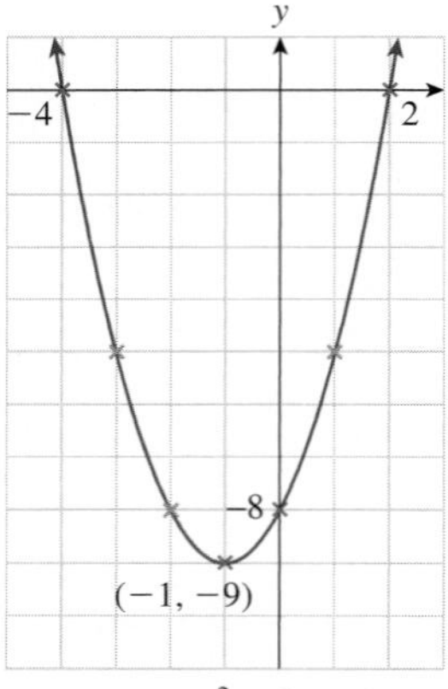

$y = x^2 + 2x - 8$

Figure 2

EXAMPLE 2 One x-Intercept and No x-Intercepts

Sketch the graph of each quadratic function, showing the location of the vertex and intercepts.

a. $f(x) = 4x^2 - 12x + 9$

b. $g(x) = -\frac{1}{2}x^2 + 4x - 12$

Solution

a. We have $a = 4$, $b = -12$, and $c = 9$. Because $a > 0$, this parabola is concave up.

Vertex: $x = -\frac{b}{2a} = \frac{12}{8} = \frac{3}{2}$ x coordinate of vertex

$$y = f\left(\frac{3}{2}\right) = 4\left(\frac{3}{2}\right)^2 - 12\left(\frac{3}{2}\right) + 9 = 0$$ y coordinate of vertex

Thus, the vertex is at the point $(3/2, 0)$.

x-Intercepts:

$$4x^2 - 12x + 9 = 0$$
$$(2x - 3)^2 = 0$$

The only solution is $2x - 3 = 0$, or $x = 3/2$. Note that this coincides with the vertex, which lies on the x-axis.

y-Intercept: $c = 9$

Symmetry: The graph is symmetric around the vertical line $x = 3/2$.

The graph is the narrow parabola shown in Figure 3. (As we remarked in Example 1, plotting additional points and using symmetry helps us obtain an accurate sketch.)

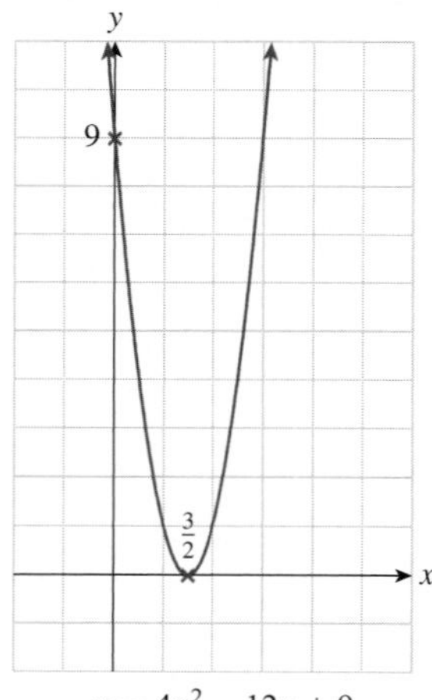

$y = 4x^2 - 12x + 9$

Figure 3

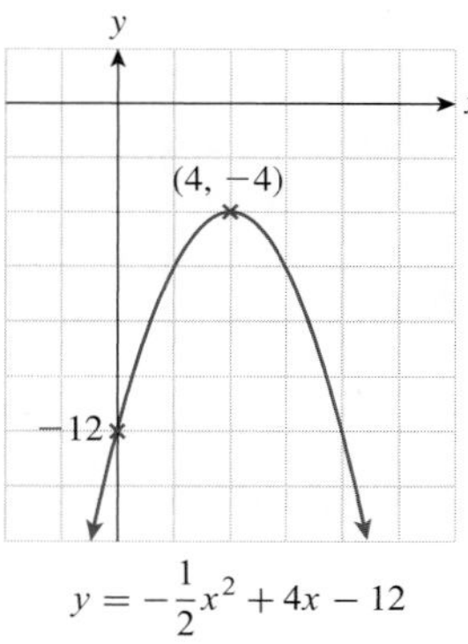

$y = -\frac{1}{2}x^2 + 4x - 12$

Figure 4

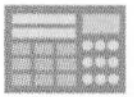 **using Technology**

To automate the computations in Example 2 using a graphing calculator or a spreadsheet, see the Technology Guides at the end of the chapter. Outline:

TI-83/84 Plus

```
Y1=AX^2+BX+C
4→A:-12→B:9→C
WINDOW Xmin=0, Xmax=3
ZOOM 0
```

[More details on page 144.]

Spreadsheet

Enter x values in column A.
Compute the corresponding y values in column B.
Graph the data in columns A and B.
[More details on page 146.]

 **Website**

www.WanerMath.com

In the Function Evaluator and Grapher, enter `4x^2-12x+9` for y_1. For a table of values, enter the various x-values in the Evaluator box, and press "Evaluate".

b. Here, $a = -1/2$, $b = 4$, and $c = -12$. Because $a < 0$, the parabola is concave down. The vertex has x coordinate $-b/(2a) = 4$, with corresponding y coordinate $f(4) = -\frac{1}{2}(4)^2 + 4(4) - 12 = -4$. Thus, the vertex is at $(4, -4)$.

For the x-intercepts, we must solve $-\frac{1}{2}x^2 + 4x - 12 = 0$. If we try to use the quadratic formula, we discover that the discriminant is $b^2 - 4ac = 16 - 24 = -8$. Because the discriminant is negative, there are no solutions of the equation, so there are no x-intercepts.

The y-intercept is given by $c = -12$, and the graph is symmetric around the vertical line $x = 4$.

Because there are no x-intercepts, the graph lies entirely below the x-axis, as shown in Figure 4. (Again, you should plot additional points and use symmetry to ensure that your sketch is accurate.)

APPLICATIONS

Recall that the **revenue** resulting from one or more business transactions is the total payment received. Thus, if q units of some item are sold at p dollars per unit, the revenue resulting from the sale is

$$\text{revenue} = \text{price} \times \text{quantity}$$
$$R = pq.$$

EXAMPLE 3 Demand and Revenue

Alien Publications, Inc. predicts that the demand equation for the sale of its latest illustrated sci-fi novel *Episode 93: Yoda vs. Alien* is

$$q = -2{,}000p + 150{,}000$$

where q is the number of books it can sell each year at a price of $\$p$ per book. What price should Alien Publications, Inc., charge to obtain the maximum annual revenue?

Solution The total revenue depends on the price, as follows:

$$\begin{aligned} R &= pq && \text{Formula for revenue} \\ &= p(-2{,}000p + 150{,}000) && \text{Substitute for } q \text{ from demand equation.} \\ &= -2{,}000p^2 + 150{,}000p. && \text{Simplify.} \end{aligned}$$

We are after the price p that gives the maximum possible revenue. Notice that what we have is a quadratic function of the form $R(p) = ap^2 + bp + c$, where $a = -2{,}000$, $b = 150{,}000$, and $c = 0$. Because a is negative, the graph of the function is a parabola, concave down, so its vertex is its highest point (Figure 5). The p coordinate of the vertex is

$$p = -\frac{b}{2a} = -\frac{150{,}000}{-4{,}000} = 37.5.$$

This value of p gives the highest point on the graph and thus gives the largest value of $R(p)$. We may conclude that Alien Publications, Inc., should charge \$37.50 per book to maximize its annual revenue.

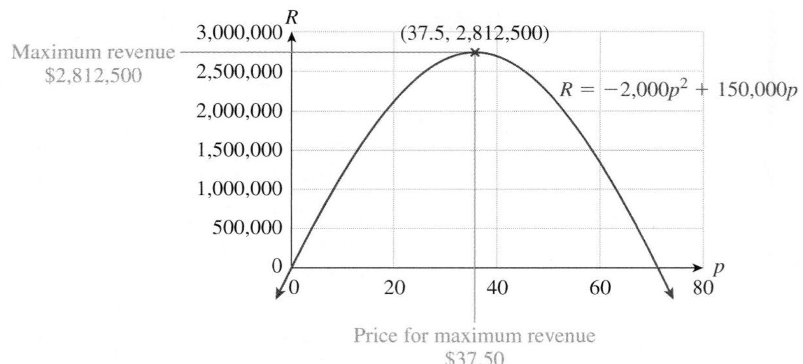

Figure 5

➡ **Before we go on...** You might ask what the maximum annual revenue is for the publisher in Example 3. Because $R(p)$ gives us the revenue at a price of $\$p$, the answer is $R(37.5) = -2{,}000\,(37.5)^2 + 150{,}000(37.5) = 2{,}812{,}500$. In other words, the company will earn total annual revenues from this book amounting to $2,812,500. ■

EXAMPLE 4 Demand, Revenue, and Profit

As the operator of *YSport Fitness* gym, you calculate your demand equation to be

$$q = -0.06p + 84,$$

where q is the number of members in the club and p is the annual membership fee you charge.

a. Your annual operating costs are a fixed cost of $20,000 per year plus a variable cost of $20 per member. Find the annual revenue and profit as functions of the membership price p.

b. At what price should you set the annual membership fee to obtain the maximum revenue? What is the maximum possible revenue?

c. At what price should you set the annual membership fee to obtain the maximum profit? What is the maximum possible profit? What is the corresponding revenue?

Solution

a. The annual revenue is given by

$$\begin{aligned} R &= pq && \text{Formula for revenue} \\ &= p(-0.06p + 84) && \text{Substitute for } q \text{ from demand equation.} \\ &= -0.06p^2 + 84p. && \text{Simplify.} \end{aligned}$$

The annual cost C is given by

$$C = 20{,}000 + 20q. \qquad \text{\$20,000 plus \$20 per member}$$

However, this is a function of q, and not p. To express C as a function of p we substitute for q using the demand equation $q = -0.06p + 84$:

$$\begin{aligned} C &= 20{,}000 + 20(-0.06p + 84) \\ &= 20{,}000 - 1.2p + 1{,}680 \\ &= -1.2p + 21{,}680. \end{aligned}$$

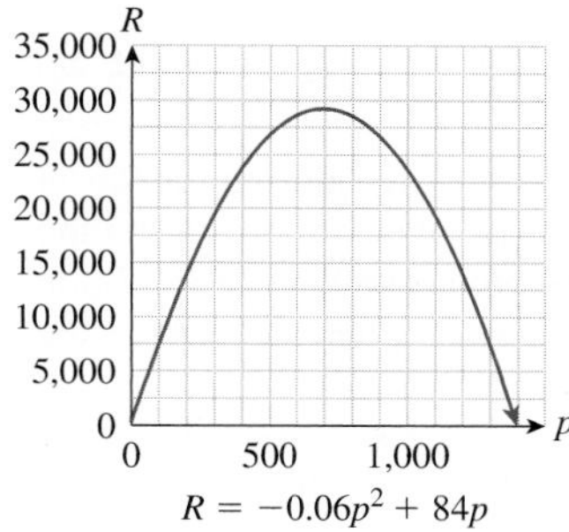

Figure 6

Thus, the profit function is:

$$\begin{aligned} P &= R - C && \text{Formula for profit} \\ &= -0.06p^2 + 84p - (-1.2p + 21{,}680) && \text{Substitute for revenue and cost.} \\ &= -0.06p^2 + 85.2p - 21{,}680. \end{aligned}$$

b. From part (a) the revenue function is given by

$$R = -0.06p^2 + 84p.$$

This is a quadratic function ($a = -0.06, b = 84, c = 0$) whose graph is a concave-down parabola (Figure 6). The maximum revenue corresponds to the highest point of the graph: the vertex, of which the p coordinate is

$$p = -\frac{b}{2a} = -\frac{84}{2(-0.06)} \approx \$700.$$

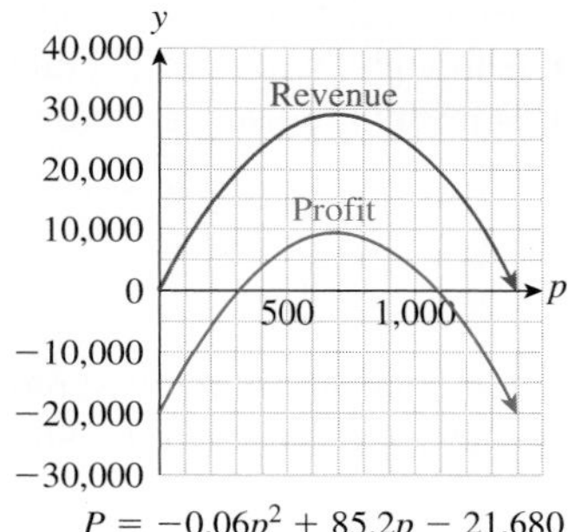

Figure 7

This is the membership fee you should charge for the maximum revenue. The corresponding maximum revenue is given by the y coordinate of the vertex in Figure 6:

$$R(700) = -0.06(700)^2 + 84(700) = \$29{,}400.$$

c. From part (a), the profit function is given by

$$P = -0.06p^2 + 85.2p - 21{,}680.$$

Like the revenue function, the profit function is quadratic ($a = -0.06$, $b = 85.2$, $c = -21{,}680$). Figure 7 shows both the revenue and profit functions. The maximum profit corresponds to the vertex, whose p coordinate is

$$p = -\frac{b}{2a} = -\frac{85.2}{2(-0.06)} \approx \$710.$$

This is the membership fee you should charge for the maximum profit. The corresponding maximum profit is given by the y coordinate of the vertex of the profit curve in Figure 7:

$$P(710) = -0.06(710)^2 + 85.2(710) - 21{,}680 = \$8{,}566.$$

The corresponding revenue is

$$R(710) = -0.06(710)^2 + 84(710) = \$29{,}394,$$

slightly less than the maximum possible revenue of \$29,400.

➡ **Before we go on...** The result of part (c) of Example 4 tells us that the vertex of the profit curve in Figure 7 is slightly to the right of the vertex in the revenue curve. However, the difference is tiny compared to the scale of the graphs, so the graphs appear to be parallel. ■

Q: *Charging \$710 membership brings in less revenue than charging \$700. So why charge \$710?*

A: A membership fee of \$700 does bring in slightly larger revenue than a fee of \$710, but it also brings in a slightly larger membership, which in turn raises the operating expense and has the effect of *lowering* the profit slightly (to \$8,560). In other words, the slightly higher fee, while bringing in less revenue, also lowers the cost, and the net result is a larger profit.

Fitting a Quadratic Function to Data: Quadratic Regression

In Section 1.4 we saw how to fit a regression line to a collection of data points. Here, we see how to use technology to obtain the **quadratic regression curve** associated with a set of points. The quadratic regression curve is the quadratic curve $y = ax^2 + bx + c$ that best fits the data points in the sense that the associated sum-of-squares error (SSE—see Section 1.4) is a minimum. Although there are algebraic methods for obtaining the quadratic regression curve, it is normal to use technology to do this.

EXAMPLE 5 Carbon Dioxide Concentration

The following table shows the annual mean carbon dioxide concentration measured at Mauna Loa Observatory in Hawaii, in parts per million, every 10 years from 1960 through 2010 ($t = 0$ represents 1960).[1]

Year t	0	10	20	30	40	50
PPM CO_2 C	317	326	339	354	369	390

a. Is a linear model appropriate for these data?

b. Find the quadratic model

$$C(t) = at^2 + bt + c$$

that best fits the data.

Solution

a. To see whether a linear model is appropriate, we plot the data points and the regression line using one of the methods of Example 2 in Section 1.4 (Figure 8).

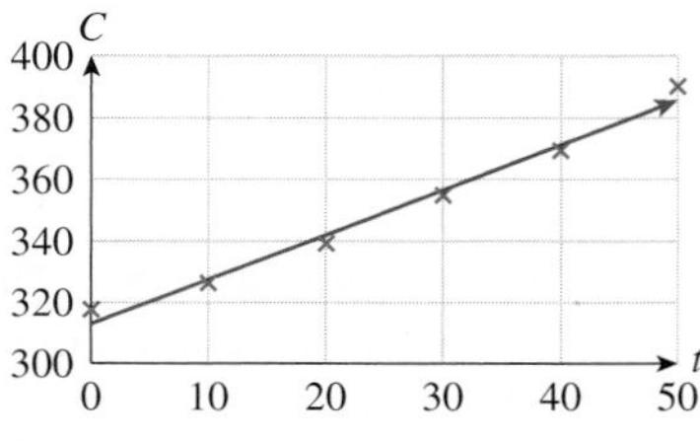

Figure 8

[1]Figures are approximate. Source: U.S. Department of Commerce/National Oceanic and Atmospheric Administration (NOAA) Earth System Research Laboratory, data downloaded from www.esrl.noaa.gov/gmd/ccgg/trends/ on March 13, 2011.

using Technology

For detailed instructions on how to find and graph the regression curve in Example 5 using a graphing calculator or a spreadsheet, see the Technology Guides at the end of the chapter. Outline:

TI-83/84 Plus

[STAT] EDIT values of t in L_1 and values of C in L_2

Regression curve: [STAT] CALC option 5

QuadReg [ENTER]

Graph: [Y=] [VARS] [5] EQ [1], then [ZOOM] [9]

[More details on page 144.]

Spreadsheet

t- and C-values in Columns A and B.

Graph: Highlight data and insert a Scatter chart.

Regression curve: Right-click a datapoint and add polynomial order 2 trendline with option to show equation on chart.

[More details on page 147.]

Website

www.WanerMath.com

In the Simple Regression utility, enter the data in the x- and y-columns and press "y=ax^2+bx+c".

From the graph, we can see that the given data suggest a curve and not a straight line: The observed points are above the regression line at the ends but below in the middle. (We would expect the data points from a linear relation to fall randomly above and below the regression line.)

b. The quadratic model that best fits the data is the quadratic regression model. As with linear regression, there are algebraic formulas to compute a, b, and c, but they are rather involved. However, we exploit the fact that these formulas are built into graphing calculators, spreadsheets, and other technology and obtain the regression curve using technology (see Figure 9):

$$C(t) = 0.012t^2 + 0.85t + 320. \qquad \text{Coefficients rounded to two significant digits}$$

Notice from the graphs that the quadratic regression model appears to give a better fit than the linear regression model. This impression is supported by the values of SSE: For the linear regression model, SSE $\approx$ 58, while for the quadratic regression model, SSE is much smaller, approximately 2.6, indicating a much better fit.

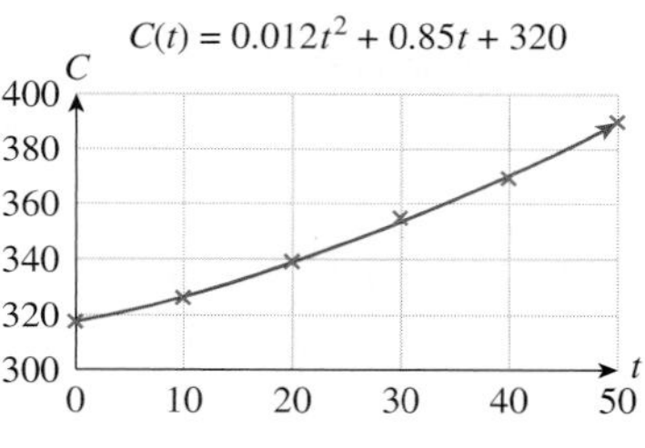

Figure 9

2.1 EXERCISES

Access end-of-section exercises online at **www.webassign.net**

2.2 Exponential Functions and Models

The quadratic functions we discussed in Section 2.1 can be used to model many nonlinear situations. However, exponential functions give better models in some applications, including population growth, radioactive decay, the growth or depreciation of financial investments, and many other phenomena. (We already saw some of these applications in Section 1.2.)

To work effectively with exponential functions, we need to know the laws of exponents. The following list, taken from the algebra review in Chapter 0, gives the laws of exponents we will be using.

The Laws of Exponents

If b and c are positive and x and y are any real numbers, then the following laws hold:

Law	Quick Examples	
1. $b^x b^y = b^{x+y}$	$2^3 2^2 = 2^5 = 32$	$2^{3-x} = 2^3 2^{-x}$
2. $\dfrac{b^x}{b^y} = b^{x-y}$	$\dfrac{4^3}{4^2} = 4^{3-2} = 4^1 = 4$	$3^{x-2} = \dfrac{3^x}{3^2} = \dfrac{3^x}{9}$
3. $\dfrac{1}{b^x} = b^{-x}$	$9^{-0.5} = \dfrac{1}{9^{0.5}} = \dfrac{1}{3}$	$2^{-x} = \dfrac{1}{2^x}$
4. $b^0 = 1$	$(3.3)^0 = 1$	$x^0 = 1$ if $x \neq 0$
5. $(b^x)^y = b^{xy}$	$(2^3)^2 = 2^6 = 64$	$\left(\dfrac{1}{2}\right)^x = (2^{-1})^x = 2^{-x}$
6. $(bc)^x = b^x c^x$	$(4 \cdot 2)^2 = 4^2 2^2 = 64$	$10^x = 5^x 2^x$
7. $\left(\dfrac{b}{c}\right)^x = \dfrac{b^x}{c^x}$	$\left(\dfrac{4}{3}\right)^2 = \dfrac{4^2}{3^2} = \dfrac{16}{9}$	$\left(\dfrac{1}{2}\right)^x = \dfrac{1^x}{2^x} = \dfrac{1}{2^x}$

Here are the functions we will study in this section.

Exponential Function

An **exponential function** has the form

$$f(x) = Ab^x, \qquad \text{Technology: } \texttt{A*b^x}$$

where A and b are constants with $A \neq 0$ and b positive and not equal to 1. We call b the **base** of the exponential function.

Quick Examples

1. $f(x) = 2^x$ — $A = 1, b = 2$; Technology: `2^x`

   $f(1) = 2^1 = 2$ — `2^1`

   $f(-3) = 2^{-3} = \dfrac{1}{8}$ — `2^(-3)`

   $f(0) = 2^0 = 1$ — `2^0`

2. $g(x) = 20(3^x)$ — $A = 20, b = 3$; Technology: `20*3^x`

   $g(2) = 20(3^2) = 20(9) = 180$ — `20*3^2`

   $g(-1) = 20(3^{-1}) = 20\left(\dfrac{1}{3}\right) = 6\dfrac{2}{3}$ — `20*3^(-1)`

3. $h(x) = 2^{-x} = \left(\frac{1}{2}\right)^x$ $A = 1, b = \frac{1}{2}$; Technology: `2^(-x)` or `(1/2)^x`

$h(1) = 2^{-1} = \frac{1}{2}$ `2^(-1)` or `(1/2)^1`

$h(2) = 2^{-2} = \frac{1}{4}$ `2^(-2)` or `(1/2)^2`

4. $k(x) = 3 \cdot 2^{-4x} = 3(2^{-4})^x$ $A = 3, b = 2^{-4}$; Technology: `3*2^(-4*x)`

$k(-2) = 3 \cdot 2^{-4(-2)}$ `3*2^(-4*(-2))`

$= 3 \cdot 2^8 = 3 \cdot 256 = 768$

Exponential Functions from the Numerical and Graphical Points of View

The following table shows values of $f(x) = 3(2^x)$ for some values of x ($A = 3$, $b = 2$):

x	-3	-2	-1	0	1	2	3
$f(x)$	$\frac{3}{8}$	$\frac{3}{4}$	$\frac{3}{2}$	3	6	12	24

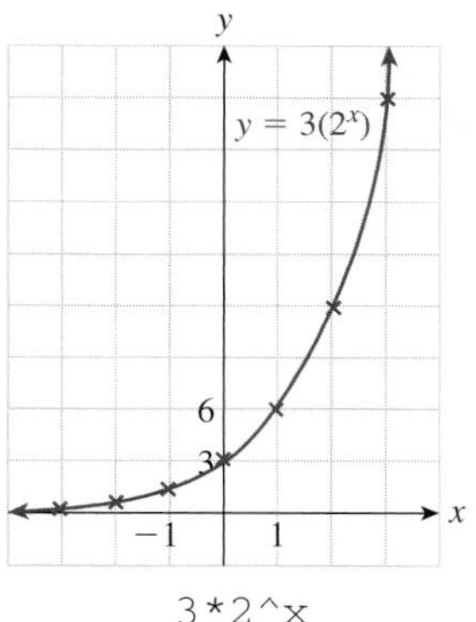

Figure 10

The graph of f is shown in Figure 10.

Notice that the y-intercept is $A = 3$ (obtained by setting $x = 0$). In general:

In the graph of $f(x) = Ab^x$, A is the y-intercept, or the value of y when $x = 0$.

What about b? Notice from the table that the value of y is multiplied by $b = 2$ for every increase of 1 in x. If we decrease x by 1, the y coordinate gets *divided* by $b = 2$.

The value of y is multiplied by b for every one-unit increase of x.

x	-3	-2	-1	0	1	2	3
$f(x)$	$\frac{3}{8}$	$\frac{3}{4}$	$\frac{3}{2}$	3	6	12	24

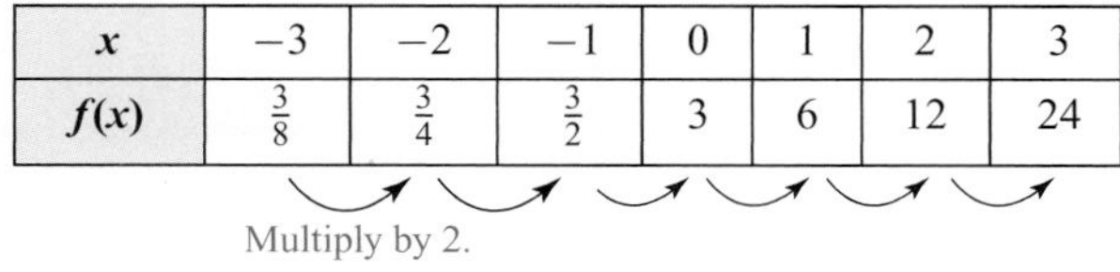

On the graph, if we move one unit to the right from any point on the curve, the y coordinate doubles. Thus, the curve becomes dramatically steeper as the value of x increases. This phenomenon is called **exponential growth**. (See Section 1.2.)

Exponential Function Numerically and Graphically

For the exponential function $f(x) = Ab^x$:

Role of *A*

$f(0) = A$, so A is the y-intercept of the graph of f.

Role of *b*

If x increases by 1, $f(x)$ is multiplied by b.

If x increases by 2, $f(x)$ is multiplied by b^2.

⋮

If x increases by Δx, $f(x)$ is multiplied by $b^{\Delta x}$.

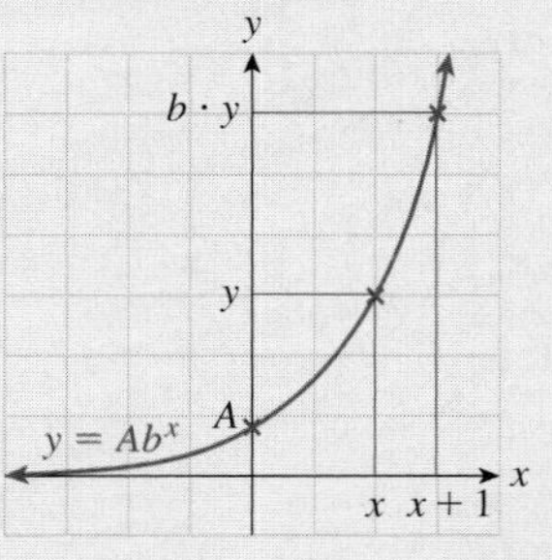

If x increases by 1, y is multiplied by b.

Quick Examples

1. $f_1(x) = 2^x$, $f_2(x) = \left(\frac{1}{2}\right)^x = 2^{-x}$

	A	B	C
1	x	2^x	2^(-x)
2	-3	1/8	8
3	-2	1/4	4
4	-1	1/2	2
5	0	1	1
6	1	2	1/2
7	2	4	1/4
8	3	8	1/8

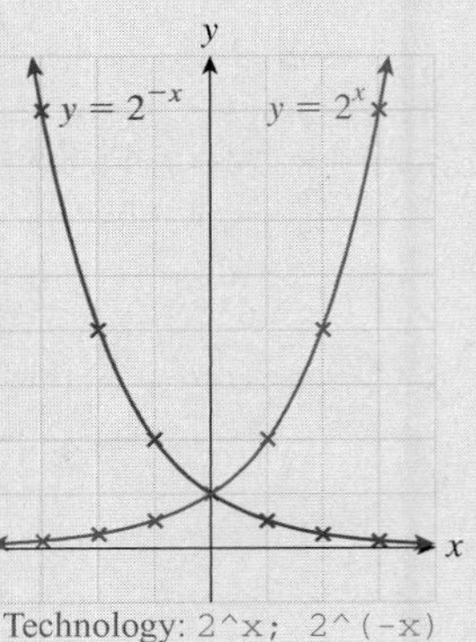

Technology: 2^x; 2^(-x)

When x increases by 1, $f_2(x)$ is multiplied by $\frac{1}{2}$. The function $f_1(x) = 2^x$ illustrates exponential growth, while $f_2(x) = \left(\frac{1}{2}\right)^x$ illustrates the opposite phenomenon: **exponential decay**.

2. $f_1(x) = 2^x$, $f_2(x) = 3^x$, $f_3(x) = 1^x$ (Can you see why f_3 is not an exponential function?)

	A	B	C	D
1	x	2^x	3^x	1^x
2	-3	1/8	1/27	1
3	-2	1/4	1/9	1
4	-1	1/2	1/3	1
5	0	1	1	1
6	1	2	3	1
7	2	4	9	1
8	3	8	27	1

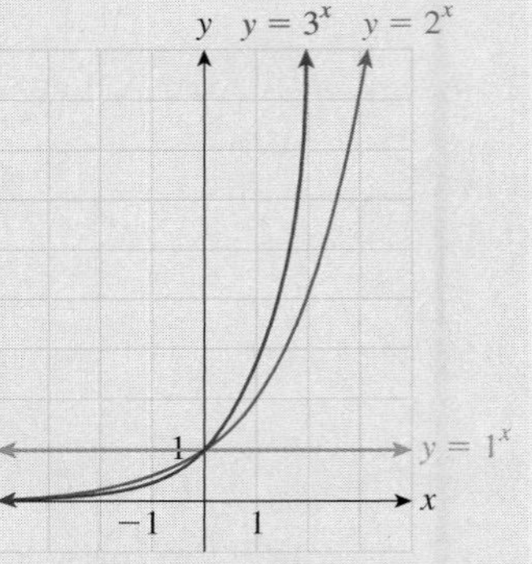

If x increases by 1, 3^x is multiplied by 3. Note also that all the graphs pass through (0, 1). (Why?)

EXAMPLE 1 Recognizing Exponential Data Numerically and Graphically

Some of the values of two functions, f and g, are given in the following table:

x	−2	−1	0	1	2
$f(x)$	−7	−3	1	5	9
$g(x)$	$\frac{2}{9}$	$\frac{2}{3}$	2	6	18

One of these functions is linear, and the other is exponential. Which is which?

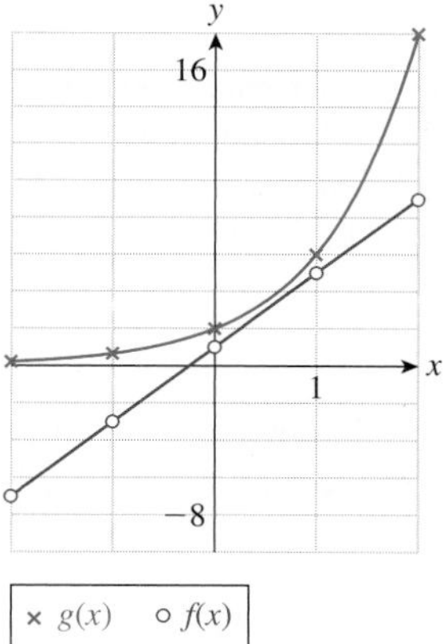

Figure 11

Solution Remember that a linear function increases (or decreases) by the same amount every time x increases by 1. The values of f behave this way: Every time x increases by 1, the value of $f(x)$ increases by 4. Therefore, f is a linear function with a *slope* of 4. Because $f(0) = 1$, we see that

$$f(x) = 4x + 1$$

is a linear formula that fits the data.

On the other hand, every time x increases by 1, the value of $g(x)$ is *multiplied* by 3. Because $g(0) = 2$, we find that

$$g(x) = 2(3^x)$$

is an exponential function fitting the data.

We can visualize the two functions f and g by plotting the data points (Figure 11). The data points for $f(x)$ clearly lie along a straight line, whereas the points for $g(x)$ lie along a curve. The y coordinate of each point for $g(x)$ is 3 times the y coordinate of the preceding point, demonstrating that the curve is an exponential one.

In Section 1.3, we discussed a method for calculating the equation of the line that passes through two given points. In the following example, we show a method for calculating the equation of the exponential curve through two given points.

EXAMPLE 2 Finding the Exponential Curve through Two Points

Find an equation of the exponential curve through (1, 6.3) and (4, 170.1).

Solution We want an equation of the form

$$y = Ab^x \quad (b > 0).$$

Substituting the coordinates of the given points, we get

$$6.3 = Ab^1 \qquad \text{Substitute (1, 6.3).}$$
$$170.1 = Ab^4. \qquad \text{Substitute (4, 170.1).}$$

If we now divide the second equation by the first, we get

$$\frac{170.1}{6.3} = \frac{Ab^4}{Ab} = b^3$$
$$b^3 = 27$$
$$b = 27^{1/3} \qquad \text{Take reciprocal power of both sides.}$$
$$b = 3.$$

Now that we have b, we can substitute its value into the first equation to obtain

$$6.3 = 3A \qquad \text{Substitute } b = 3 \text{ into the equation } 6.3 = Ab^1.$$
$$A = \frac{6.3}{3} = 2.1.$$

We have both constants, $A = 2.1$ and $b = 3$, so the model is

$$y = 2.1(3^x).$$

Example 6 will show how to use technology to fit an exponential function to two or more data points.

APPLICATIONS

Recall some terminology we mentioned earlier: A quantity y experiences **exponential growth** if $y = Ab^t$ with $b > 1$. (Here we use t for the independent variable, thinking of time.) It experiences **exponential decay** if $y = Ab^t$ with $0 < b < 1$. We already saw several examples of exponential growth and decay in Section 1.2.

EXAMPLE 3 Exponential Growth and Decay

a. Compound Interest (See Section 1.2 Example 6.) If \$2,000 is invested in a mutual fund with an annual yield of 12.6% and the earnings are reinvested each month, then the future value after t years is

$$A(t) = P\left(1 + \frac{r}{m}\right)^{mt} = 2{,}000\left(1 + \frac{0.126}{12}\right)^{12t} = 2{,}000(1.0105)^{12t},$$

which can be written as $2{,}000(1.0105^{12})^t$, so $A = 2{,}000$ and $b = 1.0105^{12}$. This is an example of exponential growth, because $b > 1$.

b. Carbon Decay (See Section 1.2 Example 7.) The amount of carbon 14 remaining in a sample that originally contained A grams is approximately

$$C(t) = A(0.999879)^t.$$

This is an instance of exponential decay, because $b < 1$.

➡ **Before we go on...** Refer again to part (a). In Example 6(b) of Section 1.2 we showed how to use technology to answer questions such as the following: "When, to the nearest year, will the value of your investment reach \$5,000?" ■

The next example shows an application to public health.

EXAMPLE 4 Exponential Growth: Epidemics

In the early stages of the AIDS epidemic during the 1980s, the number of cases in the United States was increasing by about 50% every 6 months. By the start of 1983, there were approximately 1,600 AIDS cases in the United States.[2]

a. Assuming an exponential growth model, find a function that predicts the number of people infected t years after the start of 1983.

b. Use the model to estimate the number of people infected by October 1, 1986, and also by the end of that year.

Solution

a. One way of finding the desired exponential function is to reason as follows: At time $t = 0$ (January 1, 1983), the number of people infected was 1,600, so $A = 1{,}600$. Every 6 months, the number of cases increased to 150% of the number 6 months earlier—that is, to 1.50 times that number. Each year, it therefore increased to $(1.50)^2 = 2.25$ times the number one year earlier. Hence, after t years, we need to multiply the original 1,600 by 2.25^t, so the model is

$$y = 1{,}600(2.25^t) \text{ cases.}$$

[2]Data based on regression of the 1982–1986 figures. Source for data: Centers for Disease Control and Prevention. HIV/AIDS Surveillance Report, 2000;12 (No. 2).

Alternatively, if we wish to use the method of Example 2, we need two data points. We are given one point: (0, 1,600). Because y increased by 50% every 6 months, 6 months later it reached $1{,}600 + 800 = 2{,}400$ ($t = 0.5$). This information gives a second point: (0.5, 2,400). We can now apply the method in Example 2 to find the model above.

b. October 1, 1986, corresponds to $t = 3.75$ (because October 1 is 9 months, or $9/12 = 0.75$ of a year after January 1). Substituting this value of t in the model gives

$$y = 1{,}600(2.25^{3.75}) \approx 33{,}481 \text{ cases}$$ `1600*2.25^3.75`

By the end of 1986, the model predicts that

$$y = 1{,}600(2.25^4) = 41{,}006 \text{ cases.}$$

(The actual number of cases was around 41,700.)

➡ **Before we go on...** Increasing the number of cases by 50% every 6 months couldn't continue for very long and this is borne out by observations. If increasing by 50% every 6 months did continue, then by January 2003 ($t = 20$), the number of infected people would have been

$$1{,}600(2.25^{20}) \approx 17{,}700{,}000{,}000$$

a number that is more than 50 times the size of the U.S. population! Thus, although the exponential model is fairly reliable in the early stages of the epidemic, it is unreliable for predicting long-term trends. ■

Epidemiologists use more sophisticated models to measure the spread of epidemics, and these models predict a leveling-off phenomenon as the number of cases becomes a significant part of the total population. We discuss such a model, the **logistic function**, in Section 2.4.

The Number e and More Applications

In nature we find examples of growth that occurs *continuously*, as though "interest" is being added more often than every second or fraction of a second. To model this, we need to see what happens to the compound interest formula of Section 1.2 as we let m (the number of times interest is added per year) become extremely large. Something very interesting does happen: We end up with a more compact and elegant formula than we began with. To see why, let's look at a very simple situation.

Suppose we invest $1 in the bank for 1 year at 100% interest, compounded m times per year. If $m = 1$, then 100% interest is added every year, and so our money doubles at the end of the year. In general, the accumulated capital at the end of the year is

$$A = 1\left(1 + \frac{1}{m}\right)^m = \left(1 + \frac{1}{m}\right)^m.$$ `(1+1/m)^m`

	A	B
1	m	(1+1/m)^m
2	1	2
3	10	2.59374246
4	100	2.704813829
5	1000	2.716923932
6	10000	2.718145927
7	100000	2.718268237
8	1000000	2.718280469
9	10000000	2.718281694
10	100000000	2.718281786
11	1000000000	2.718282031

Now, we are interested in what A becomes for large values of m. On the left is a spreadsheet showing the quantity $\left(1 + \frac{1}{m}\right)^m$ for larger and larger values of m.

Something interesting *does* seem to be happening! The numbers appear to be getting closer and closer to a specific value. In mathematical terminology, we say that the numbers **converge** to a fixed number, 2.71828 . . . , called the **limiting value*** of the quantities $\left(1 + \frac{1}{m}\right)^m$. This number, called e, is one of the most important in mathematics. The number e is irrational, just as the more familiar number π is, so we cannot write down its exact numerical value. To 20 decimal places,

$$e = 2.71828182845904523536. . . .$$

* See Chapter 3 for more on limits.

We now say that, if \$1 is invested for 1 year at 100% interest **compounded continuously**, the accumulated money at the end of that year will amount to $\$e = \$2.72$ (to the nearest cent). But what about the following more general question?

Q: *What about a more general scenario: If we invest an amount \$P for t years at an interest rate of r, compounded continuously, what will be the accumulated amount A at the end of that period?*

A: In the special case above (P, t, and r all equal 1), we took the compound interest formula and let m get larger and larger. We do the same more generally, after a little preliminary work with the algebra of exponentials.

$$A = P\left(1 + \frac{r}{m}\right)^{mt}$$

$$= P\left(1 + \frac{1}{(m/r)}\right)^{mt} \qquad \text{Substituting } \frac{r}{m} = \frac{1}{(m/r)}$$

$$= P\left(1 + \frac{1}{(m/r)}\right)^{(m/r)rt} \qquad \text{Substituting } m = \left(\frac{m}{r}\right)r$$

$$= P\left[\left(1 + \frac{1}{(m/r)}\right)^{(m/r)}\right]^{rt} \qquad \text{Using the rule } a^{bc} = (a^b)^c$$

For continuous compounding of interest, we let m, and hence m/r, get very large. This affects only the term in brackets, which converges to e, and we get the formula

$$A = Pe^{rt}.$$

Q: *How do I obtain powers of e or e itself on a TI-83/84 Plus or in a spreadsheet?*

A: On the TI-83/84 Plus, enter e^x as `e^(x)`, where `e^(` can be obtained by pressing [2ND] [LN]. To obtain the number e on the TI-83/84 Plus, enter `e^(1)`. Spreadsheets have a built-in function called `EXP`; `EXP(x)` gives the value of e^x. To obtain the number e in a spreadsheet, enter `=EXP(1)`.

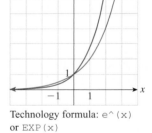

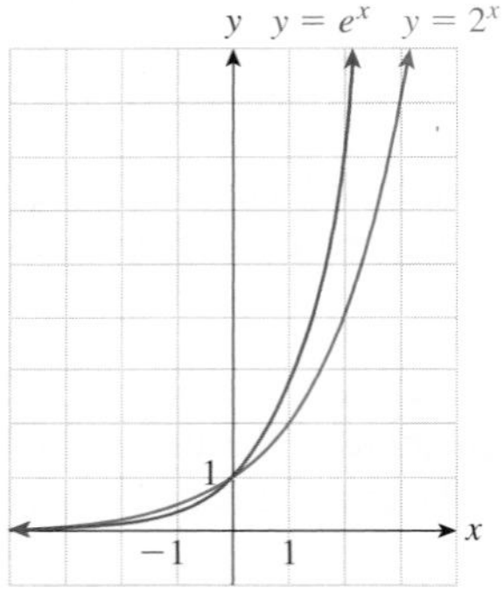

Technology formula: `e^(x)` or `EXP(x)`

Figure 12

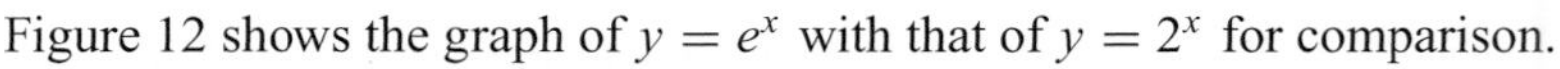

Figure 12 shows the graph of $y = e^x$ with that of $y = 2^x$ for comparison.

The Number e and Continuous Compounding

The number e is the limiting value of the quantities $\left(1 + \frac{1}{m}\right)^m$ as m gets larger and larger, and has the value 2.71828182845904523536 . . .

If \$$P$ is invested at an annual interest rate r compounded continuously, the accumulated amount after t years is

$$A(t) = Pe^{rt}.$$ `P*e^(r*t)` or `P*EXP(r*t)`

Quick Examples

1. If \$100 is invested in an account that bears 15% interest compounded continuously, at the end of 10 years the investment will be worth

$$A(10) = 100e^{(0.15)(10)} = \$448.17.$$ `100*e^(0.15*10)` or `100*EXP(0.15*10)`

2. If \$1 is invested in an account that bears 100% interest compounded continuously, at the end of x years the investment will be worth

$$A(x) = e^x \text{ dollars.}$$

EXAMPLE 5 Continuous Compounding

a. You invest \$10,000 at *Fastrack Savings & Loan,* which pays 6% compounded continuously. Express the balance in your account as a function of the number of years t and calculate the amount of money you will have after 5 years.

b. Your friend has just invested \$20,000 in *Constant Growth Funds,* whose stocks are continuously *declining* at a rate of 6% per year. How much will her investment be worth in 5 years?

c. T During which year will the value of your investment first exceed that of your friend?

Solution

a. We use the continuous growth formula with $P = 10{,}000$, $r = 0.06$, and t variable, getting

$$A(t) = Pe^{rt} = 10{,}000e^{0.06t}.$$

In 5 years,

$$A(5) = 10{,}000e^{0.06(5)}$$
$$= 10{,}000e^{0.3}$$
$$\approx \$13{,}498.59.$$

b. Because the investment is depreciating, we use a negative value for r and take $P = 20{,}000$, $r = -0.06$, and $t = 5$, getting

$$A(t) = Pe^{rt} = 20{,}000e^{-0.06t}$$
$$A(5) = 20{,}000e^{-0.06(5)}$$
$$= 20{,}000e^{-0.3}$$
$$\approx \$14{,}816.36.$$

c. We can answer the question now using a graphing calculator, a spreadsheet, or the Function Evaluator and Grapher tool at the Website. Just enter the exponential models of parts (a) and (b) and create tables to compute the values at the end of several years:

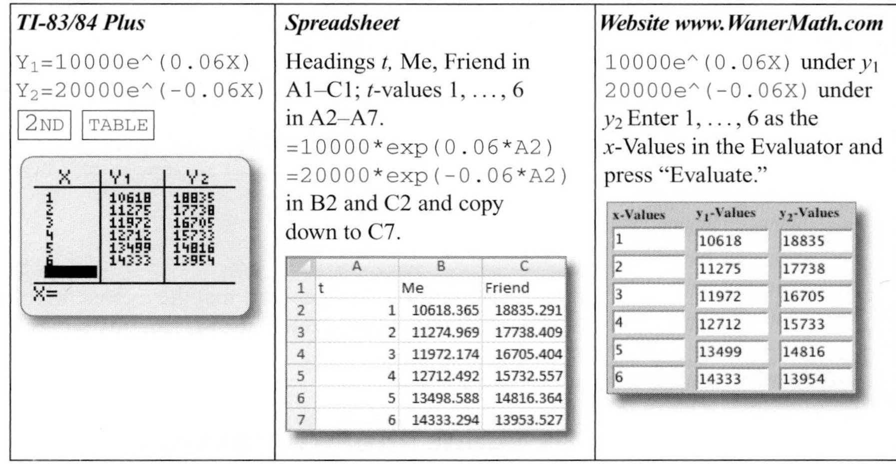

TI-83/84 Plus

`Y1=10000e^(0.06X)`
`Y2=20000e^(-0.06X)`
2ND TABLE

X	Y1	Y2
1	10618	18835
2	11275	17738
3	11972	16705
4	12712	15733
5	13499	14816
6	14333	13954

X=

Spreadsheet

Headings *t,* Me, Friend in A1–C1; *t*-values 1, . . . , 6 in A2–A7.
`=10000*exp(0.06*A2)`
`=20000*exp(-0.06*A2)`
in B2 and C2 and copy down to C7.

	A	B	C
1	t	Me	Friend
2	1	10618.365	18835.291
3	2	11274.969	17738.409
4	3	11972.174	16705.404
5	4	12712.492	15732.557
6	5	13498.588	14816.364
7	6	14333.294	13953.527

Website www.WanerMath.com

`10000e^(0.06X)` under y_1
`20000e^(-0.06X)` under y_2 Enter 1, . . . , 6 as the x-Values in the Evaluator and press "Evaluate."

x-Values	y_1-Values	y_2-Values
1	10618	18835
2	11275	17738
3	11972	16705
4	12712	15733
5	13499	14816
6	14333	13954

From the table, we see that the value of your investment overtakes that of your friend after $t = 5$ (the end of year 5) and before $t = 6$ (the end of year 6). Thus your investment first exceeds that of your friend sometime during year 6.

➡ Before we go on...

Q: *How does continuous compounding compare with monthly compounding?*

A: To repeat the calculation in part (a) of Example 5 using monthly compounding instead of continuous compounding, we use the compound interest formula with $P = 10{,}000$, $r = 0.06$, $m = 12$, and $t = 5$ and find

$$A(5) = 10{,}000(1 + 0.06/12)^{60} \approx \$13{,}488.50.$$

Thus, continuous compounding earns you approximately \$10 more than monthly compounding on a 5-year, \$10,000 investment. This is little to get excited about.

■

If we write the continuous compounding formula $A(t) = Pe^{rt}$ as $A(t) = P(e^r)^t$, we see that $A(t)$ is an exponential function of t, where the base is $b = e^r$, so we have really not introduced a new kind of function. In fact, exponential functions are often written in this way:

Exponential Functions: Alternative Form

We can write any exponential function in the following alternative form:

$$f(x) = Ae^{rx}$$

where A and r are constants. If r is positive, f models exponential growth; if r is negative, f models exponential decay.

Quick Examples

1. $f(x) = 100e^{0.15x}$ — Exponential growth $A = 100, r = 0.15$
2. $f(t) = Ae^{-0.000\,121\,01t}$ — Exponential decay of carbon 14; $r = -0.000\,121\,01$
3. $f(t) = 100e^{0.15t} = 100\left(e^{0.15}\right)^t = 100(1.1618)^t$ — Converting Ae^{rt} to the form Ab^t

We will see in Chapter 4 that the exponential function with base e exhibits some interesting properties when we measure its rate of change, and this is the real mathematical importance of e.

Exponential Regression

Starting with a set of data that suggests an exponential curve, we can use technology to compute the exponential regression curve in much the same way as we did for the quadratic regression curve in Example 5 of Section 2.1.

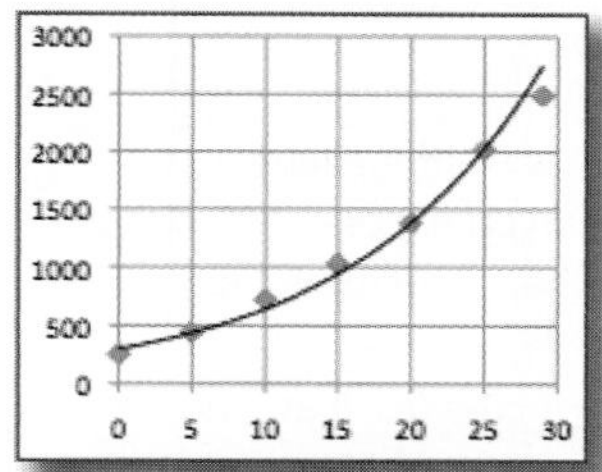

Figure 13

using Technology

See the Technology Guides at the end of the chapter for detailed instructions on how to obtain the regression curve and graph for Example 6 using a graphing calculator or spreadsheet. Outline:

TI-83/84 Plus
STAT EDIT values of t in L_1 and values of C in L_2
Regression curve: STAT CALC option #0 ExpReg ENTER
Graph: Y= VARS 5 EQ 1, then ZOOM 9
[More details on page 145.]

Spreadsheet
t- and C-values in Columns A and B, graph these data. Regression curve: Add exponential Trendline with option to show equation.
[More details on page 148.]

Website
www.WanerMath.com
In the Simple Regression utility, enter the data in the x and y columns and press "y=a(b^x)".

EXAMPLE 6 Exponential Regression: Health Expenditures

The following table shows annual expenditure on health in the U.S. from 1980 through 2009 ($t = 0$ represents 1980).[3]

Year t	0	5	10	15	20	25	29
Expenditure ($ billion)	256	444	724	1,030	1,380	2,020	2,490

a. Find the exponential regression model

$$C(t) = Ab^t$$

for the annual expenditure.

b. Use the regression model to estimate the expenditure in 2002 ($t = 22$; the actual expenditure was approximately $1,640 billion).

Solution

a. We use technology to obtain the exponential regression curve (See Figure 13):

$$C(t) \approx 296(1.08)^t. \qquad \text{Coefficients rounded}$$

b. Using the model $C(t) \approx 296(1.08)^t$ we find that

$$C(22) \approx 296(1.08)^{22} \approx \$1{,}609 \text{ billion},$$

which is close to the actual number of about $1,640 billion.

Before we go on... We said in the preceding section that the regression curve gives the smallest value of the sum-of-squares error, SSE (the sum of the squares of the residuals). However, exponential regression as computed via technology generally minimizes the sum of the squares of the residuals of the *logarithms* (logarithms are discussed in the next section). Using logarithms allows one easily to convert an exponential function into a linear one and then use linear regression formulas. However, in Section 2.4, we will discuss a way of using Excel's Solver to minimize SSE directly, which allows us to find the best-fit exponential curve directly without the need for devices to simplify the mathematics. If we do this, we obtain a very different equation:

$$C(t) \approx 353(1.07)^t.$$

If you plot this function, you will notice that it seems to fit the data more closely than the regression curve. ■

FAQs

When to Use an Exponential Model for Data Points, and when to Use e in Your Model

Q: *Given a set of data points that appear to be curving upward, how can I tell whether to use a quadratic model or an exponential model?*

[3] Data are rounded. Source: U.S. Department of Health & Human Services/Centers for Medicare & Medicaid Services, National Health Expenditure Data, downloaded April 2011 from www.cms.gov.

A: Here are some things to look for:

- Do the data values appear to double at regular intervals? (For example, do the values approximately double every 5 units?) If so, then an exponential model is appropriate. If it takes longer and longer to double, then a quadratic model may be more appropriate.
- Do the values first decrease to a low point and then increase? If so, then a quadratic model is more appropriate.

It is also helpful to use technology to graph both the regression quadratic and exponential curves and to visually inspect the graphs to determine which gives the closest fit to the data.

Q: *We have two ways of writing exponential functions: $f(x) = Ab^x$ and $f(x) = Ae^{rx}$. How do we know which one to use?*

A: The two forms are equivalent, and it is always possible to convert from one form to the other.* So, use whichever form seems to be convenient for a particular situation. For instance, $f(t) = A(3^t)$ conveniently models exponential growth that is tripling every unit of time, whereas $f(t) = Ae^{0.06t}$ conveniently models an investment with continuous compounding at 6%.

* Quick Example 3 on page 118 shows how to convert Ae^{rx} to Ab^x. Conversion from Ab^x to Ae^{rx} involves logarithms: $r = \ln b$.

2.2 EXERCISES

Access end-of-section exercises online at **www.webassign.net** ENHANCED WebAssign

2.3 Logarithmic Functions and Models

Logarithms were invented by John Napier (1550–1617) in the late sixteenth century as a means of aiding calculation. His invention made possible the prodigious hand calculations of astronomer Johannes Kepler (1571–1630), who was the first to describe accurately the orbits and the motions of the planets. Today, computers and calculators have done away with that use of logarithms, but many other uses remain. In particular, the logarithm is used to model real-world phenomena in numerous fields, including physics, finance, and economics.

From the equation

$$2^3 = 8$$

we can see that the power to which we need to raise 2 in order to get 8 is 3. We abbreviate the phrase "the power to which we need to raise 2 in order to get 8" as "$\log_2 8$." Thus, another way of writing the equation $2^3 = 8$ is

$$\log_2 8 = 3. \quad \text{The power to which we need to raise 2 in order to get 8 is 3.}$$

This is read "the base 2 logarithm of 8 is 3" or "the log, base 2, of 8 is 3."

Here is the general definition.

Base *b* Logarithm

The **base *b* logarithm of *x***, $\log_b x$, is the power to which we need to raise b in order to get x. Symbolically,

$$\log_b x = y \quad \text{means} \quad b^y = x.$$

Logarithmic form — *Exponential form*

Quick Examples

1. The following table lists some exponential equations and their equivalent logarithmic forms:

Exponential Form	$10^3 = 1000$	$4^2 = 16$	$3^3 = 27$	$5^1 = 5$	$7^0 = 1$	$4^{-2} = \frac{1}{16}$	$25^{1/2} = 5$
Logarithmic Form	$\log_{10} 1000 = 3$	$\log_4 16 = 2$	$\log_3 27 = 3$	$\log_5 5 = 1$	$\log_7 1 = 0$	$\log_4 \frac{1}{16} = -2$	$\log_{25} 5 = \frac{1}{2}$

2. $\log_3 9$ = the power to which we need to raise 3 in order to get 9. Because $3^2 = 9$, this power is 2, so $\log_3 9 = 2$.
3. $\log_{10} 10{,}000$ = the power to which we need to raise 10 in order to get 10,000. Because $10^4 = 10{,}000$, this power is 4, so $\log_{10} 10{,}000 = 4$.
4. $\log_3 \frac{1}{27}$ is the power to which we need to raise 3 in order to get $\frac{1}{27}$. Because $3^{-3} = \frac{1}{27}$ this power is -3, so $\log_3 \frac{1}{27} = -3$.
5. $\log_b 1 = 0$ for every positive number b other than 1 because $b^0 = 1$.

Note The number $\log_b x$ is defined only if b and x are both positive and $b \neq 1$. Thus, it is impossible to compute, say, $\log_3(-9)$ (because there is no power of 3 that equals -9), or $\log_1(2)$ (because there is no power of 1 that equals 2). ■

Logarithms with base 10 and base e are frequently used, so they have special names and notations.

Common Logarithm, Natural Logarithm

The following are standard abbreviations.

		TI-83/84 Plus & Spreadsheet Formula
Base 10: $\log_{10} x = \log x$	*Common Logarithm*	`log(x)`
Base *e*: $\log_e x = \ln x$	*Natural Logarithm*	`ln(x)`

Quick Examples

Logarithmic Form	**Exponential Form**
1. $\log 10{,}000 = 4$	$10^4 = 10{,}000$
2. $\log 10 = 1$	$10^1 = 10$
3. $\log \frac{1}{10{,}000} = -4$	$10^{-4} = \frac{1}{10{,}000}$
4. $\ln e = 1$	$e^1 = e$
5. $\ln 1 = 0$	$e^0 = 1$
6. $\ln 2 = 0.69314718\ldots$	$e^{0.69314718\ldots} = 2$

Some technologies (such as calculators) do not permit direct calculation of logarithms other than common and natural logarithms. To compute logarithms with other bases with these technologies, we can use the following formula:

Change-of-Base Formula

$$\log_b a = \frac{\log a}{\log b} = \frac{\ln a}{\ln b}$$ Change-of-base formula*

Quick Examples

1. $\log_{11} 9 = \frac{\log 9}{\log 11} \approx 0.91631$ `log(9)/log(11)`
2. $\log_{11} 9 = \frac{\ln 9}{\ln 11} \approx 0.91631$ `ln(9)/ln(11)`
3. $\log_{3.2}\left(\frac{1.42}{3.4}\right) \approx -0.75065$ `log(1.42/3.4)/log(3.2)`

Using Technology to Compute Logarithms

To compute $\log_b x$ using technology, use the following formulas:

TI-83/84 Plus	`log(x)/log(b)`	Example: $\log_2(16)$ is `log(16)/log(2)`
Spreadsheet:	`=LOG(x,b)`	Example: $\log_2(16)$ is `=LOG(16,2)`

* Here is a quick explanation of why this formula works: To calculate $\log_b a$, we ask, "to what power must we raise b to get a?" To check the formula, we try using $\log a/\log b$ as the exponent.

$$b^{\frac{\log a}{\log b}} = (10^{\log b})^{\frac{\log a}{\log b}}$$ (because $b = 10^{\log b}$)
$$= 10^{\log a} = a$$

so this exponent works!

One important use of logarithms is to solve equations in which the unknown is in the exponent.

EXAMPLE 1 Solving Equations with Unknowns in the Exponent

Solve the following equations.

a. $5^{-x} = 125$ **b.** $3^{2x-1} = 6$ **c.** $100(1.005)^{3x} = 200$

Solution

a. Write the given equation $5^{-x} = 125$ in logarithmic form:

$$-x = \log_5 125$$

This gives $x = -\log_5 125 = -3$.

b. In logarithmic form, $3^{2x-1} = 6$ becomes

$$2x - 1 = \log_3 6$$
$$2x = 1 + \log_3 6$$

giving $$x = \frac{1 + \log_3 6}{2} \approx \frac{1 + 1.6309}{2} \approx 1.3155.$$

c. We cannot write the given equation, $100(1.005)^{3x} = 200$, directly in logarithmic form. We must first divide both sides by 100:

$$1.005^{3x} = \frac{200}{100} = 2$$
$$3x = \log_{1.005} 2$$
$$x = \frac{\log_{1.005} 2}{3} \approx \frac{138.9757}{3} \approx 46.3252.$$

Now that we know what logarithms are, we can talk about functions based on logarithms:

Logarithmic Function

A **logarithmic function** has the form

$$f(x) = \log_b x + C \qquad (b \text{ and } C \text{ are constants with } b > 0, b \neq 1)$$

or, alternatively,

$$f(x) = A \ln x + C. \qquad (A, C \text{ constants with } A \neq 0)$$

Quick Examples

1. $f(x) = \log x$
2. $g(x) = \ln x - 5$
3. $h(x) = \log_2 x + 1$
4. $k(x) = 3.2 \ln x + 7.2$

Q: *What is the difference between the two forms of the logarithmic function?*

A: None, really—they're equivalent: We can start with an equation in the first form and use the change-of-base formula to rewrite it:

$$\begin{aligned} f(x) &= \log_b x + C \\ &= \frac{\ln x}{\ln b} + C \qquad \text{Change-of-base formula} \\ &= \left(\frac{1}{\ln b}\right)\ln x + C. \end{aligned}$$

Our function now has the form $f(x) = A \ln x + C$, where $A = 1/\ln b$. We can go the other way as well, to rewrite $A \ln x + C$ in the form $\log_b x + C$.

EXAMPLE 2 Graphs of Logarithmic Functions

a. Sketch the graph of $f(x) = \log_2 x$ by hand.

b. Use technology to compare the graph in part (a) with the graphs of $\log_b x$ for $b = 1/4$, $1/2$, and 4.

Solution

a. To sketch the graph of $f(x) = \log_2 x$ by hand, we begin with a table of values. Because $\log_2 x$ is not defined when $x = 0$, we choose several values of x close to zero and also some larger values, all chosen so that their logarithms are easy to compute:

x	$\frac{1}{8}$	$\frac{1}{4}$	$\frac{1}{2}$	1	2	4	8
$f(x) = \log_2 x$	−3	−2	−1	0	1	2	3

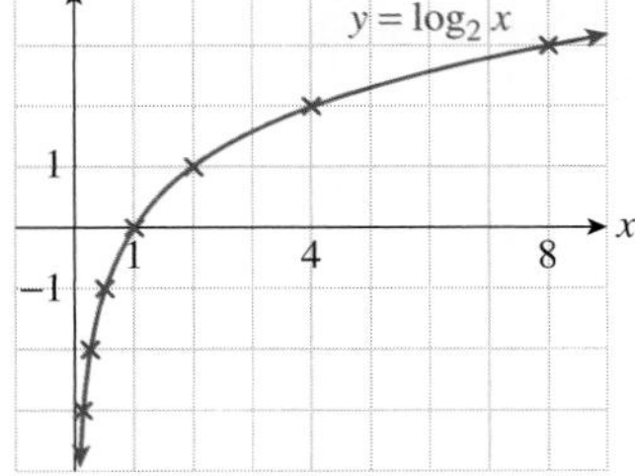

Figure 14

Graphing these points and joining them by a smooth curve gives us Figure 14.

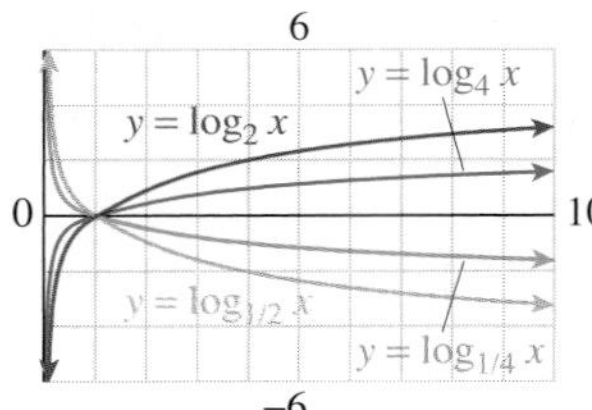

Figure 15

b. We enter the logarithmic functions in graphing utilities as follows (note the use of the change-of-base formula in the TI-83/84 Plus version):

TI-83/84 Plus	**Spreadsheet**
`Y1=log(X)/log(0.25)`	`=LOG(x,0.25)`
`Y2=log(X)/log(0.5)`	`=LOG(x,0.5)`
`Y3=log(X)/log(2)`	`=LOG(x,2)`
`Y4=log(X)/log(4)`	`=LOG(x,4)`

Figure 15 shows the resulting graphs.

➡ **Before we go on...** Notice that the graphs of the logarithmic functions in Example 2 all pass through the point (1, 0). (Why?) Notice further that the graphs of the logarithmic functions with bases less than 1 are upside-down versions of the others. Finally, how are these graphs related to the graphs of exponential functions? ■

Below are some important algebraic properties of logarithms we shall use throughout the rest of this section.

Website
www.WanerMath.com
Follow the path
Chapter 2
→ Using and Deriving Algebraic Properties of Logarithms
to find a list of logarithmic identities and a discussion on where they come from.
Follow the path
Chapter 2
→ Inverse Functions
for a general discussion of inverse functions, including further discussion of the relationship between logarithmic and exponential functions.

Logarithm Identities

The following identities hold for all positive bases $a \neq 1$ and $b \neq 1$, all positive numbers x and y, and every real number r. These identities follow from the laws of exponents.

Identity	**Quick Examples**
1. $\log_b(xy) = \log_b x + \log_b y$	$\log_2 16 = \log_2 8 + \log_2 2$
2. $\log_b\left(\frac{x}{y}\right) = \log_b x - \log_b y$	$\log_2\left(\frac{5}{3}\right) = \log_2 5 - \log_2 3$
3. $\log_b(x^r) = r \log_b x$	$\log_2(6^5) = 5 \log_2 6$
4. $\log_b b = 1$; $\log_b 1 = 0$	$\log_2 2 = 1$; $\ln e = 1$; $\log_{11} 1 = 0$
5. $\log_b\left(\frac{1}{x}\right) = -\log_b x$	$\log_2\left(\frac{1}{3}\right) = -\log_2 3$
6. $\log_b x = \frac{\log_a x}{\log_a b}$	$\log_2 5 = \frac{\log_{10} 5}{\log_{10} 2} = \frac{\log 5}{\log 2}$

Relationship with Exponential Functions
The following two identities demonstrate that the operations of taking the base b logarithm and raising b to a power are *inverse* to each other.

Identity	**Quick Examples**
1. $\log_b(b^x) = x$ *In words:* The power to which you raise b in order to get b^x is x. (!)	$\log_2(2^7) = 7$
2. $b^{\log_b x} = x$ *In words:* Raising b to the power to which it must be raised to get x yields x. (!)	$5^{\log_5 8} = 8$

APPLICATIONS

EXAMPLE 3 Investments: How Long?

Global bonds sold by Mexico are yielding an average of 2.51% per year.[4] At that interest rate, how long will it take a \$1,000 investment to be worth \$1,200 if the interest is compounded monthly?

Solution Substituting $A = 1{,}200$, $P = 1{,}000$, $r = 0.0251$, and $m = 12$ in the compound interest equation gives

$$\begin{aligned} A(t) &= P\left(1 + \frac{r}{m}\right)^{mt} \\ 1{,}200 &= 1{,}000\left(1 + \frac{0.0251}{12}\right)^{12t} \\ &\approx 1{,}000(1.002092)^{12t} \end{aligned}$$

and we must solve for t. We first divide both sides by 1,000, getting an equation in exponential form:

$$1.2 = 1.002092^{12t}.$$

In logarithmic form, this becomes

$$12t = \log_{1.002092}(1.2).$$

We can now solve for t:

$$t = \frac{\log_{1.002092}(1.2)}{12} \approx 7.3 \text{ years.}$$ `log(1.2)/(log(1.002092)*12)`

Thus, it will take approximately 7.3 years for a \$1,000 investment to be worth \$1,200.

➡ **Before we go on...** We can use the logarithm identities to solve the equation

$$1.2 = 1.002092^{12t}$$

that arose in Example 3 (and also more general equations with unknowns in the exponent) by taking the natural logarithm of both sides:

$$\begin{aligned} \ln 1.2 &= \ln(1.002092^{12t}) \\ &= 12t \ln 1.002092. \end{aligned}$$ By Identity 3

We can now solve this for t to get

$$t = \frac{\ln 1.2}{12 \ln 1.002092},$$

which, by the change-of-base formula, is equivalent to the answer we got in Example 3. ■

EXAMPLE 4 Half-Life

a. The weight of carbon 14 that remains in a sample that originally contained A grams is given by

$$C(t) = A(0.999879)^t$$

where t is time in years. Find the **half-life**, the time it takes half of the carbon 14 in a sample to decay.

[4] In 2011 (Bonds maturing 03/03/2015). Source: www.bloomberg.com.

b. Repeat part (a) using the following alternative form of the exponential model in part (a):

$$C(t) = Ae^{-0.000\,121\,01t}$$ See Quick Examples, page 118.

c. Another radioactive material has a half-life of 7,000 years. Find an exponential decay model in the form

$$R(t) = Ae^{-kt}$$

for the amount of undecayed material remaining. (The constant k is called the **decay constant**.)

d. How long will it take for 99.95% of the substance in a sample of the material in part (c) to decay?

Solution

a. We want to find the value of t for which $C(t) =$ the weight of undecayed carbon 14 left $=$ half the original weight $= 0.5A$. Substituting, we get

$$0.5A = A(0.999879)^t.$$

Dividing both sides by A gives

$$0.5 = 0.999879^t \qquad \text{Exponential form}$$
$$t = \log_{0.999879} 0.5 \approx 5{,}728 \text{ years.} \qquad \text{Logarithmic form}$$

b. This is similar to part (a): We want to solve the equation

$$0.5A = Ae^{-0.000\,121\,01t}$$

for t. Dividing both sides by A gives

$$0.5 = e^{-0.000\,121\,01t}.$$

Taking the natural logarithm of both sides gives

$$\ln(0.5) = \ln(e^{-0.000\,121\,01t}) = -0.000\,121\,01t \qquad \text{Identity 3: } \ln(e^a) = a\ln e = a$$
$$t = \frac{\ln(0.5)}{-0.000\,121\,01} \approx 5{,}728 \text{ years,}$$

as we obtained in part (a).

c. This time we are given the half-life, which we can use to find the exponential model $R(t) = Ae^{-kt}$. At time $t = 0$, the amount of radioactive material is

$$R(0) = Ae^0 = A.$$

Because half of the sample decays in 7,000 years, this sample will decay to $0.5A$ grams in 7,000 years ($t = 7{,}000$). Substituting this information gives

$$0.5A = Ae^{-k(7{,}000)}.$$

Canceling A and taking natural logarithms (again using Identity 3) gives

$$\ln(0.5) = -7{,}000k$$

so the decay constant k is

$$k = -\frac{\ln(0.5)}{7{,}000} \approx 0.000\,099\,021.$$

Therefore, the model is

$$R(t) = Ae^{-0.000\,099\,021t}.$$

d. If 99.95% of the substance in a sample has decayed, then the amount of undecayed material left is 0.05% of the original amount, or $0.0005A$. We have

$$\begin{aligned} 0.0005A &= Ae^{-0.000\,099\,021t} \\ 0.0005 &= e^{-0.000\,099\,021t} \\ \ln(0.0005) &= -0.000\,099\,021t \\ t &= \frac{\ln(0.0005)}{-0.000\,099\,021} \approx 76{,}760 \text{ years.} \end{aligned}$$

Before we go on...

Q: *In parts (a) and (b) of Example 4 we were given two different forms of the model for carbon 14 decay. How do we convert an exponential function in one form to the other?*

A: We have already seen (See Quick Example 3 on page 118) how to convert from the form $f(t) = Ae^{rt}$ in part (b) to the form $f(t) = Ab^t$ in part (a). To go the other way, start with the model in part (a), and equate it to the desired form:

$$C(t) = A(0.999\,879)^t = Ae^{rt}.$$

To solve for r, cancel the As and take the natural logarithm of both sides:

$$t\ln(0.999\,879) = rt\ln e = rt$$

so $r = \ln(0.999\,879) \approx -0.000\,121\,007,$

giving

$$C(t) = Ae^{-0.000\,121\,01t}$$

as in part (b).

■

We can use the work we did in parts (b) and (c) of the above example to obtain a formula for the decay constant in an exponential decay model for any radioactive substance when we know its half-life. Write the half-life as t_h. Then the calculation in part (b) gives

$$k = -\frac{\ln(0.5)}{t_h} = \frac{\ln 2}{t_h}. \qquad -\ln(0.5) = -\ln\left(\frac{1}{2}\right) = \ln 2$$

Multiplying both sides by t_h gives us the relationship $t_h k = \ln 2$.

Exponential Decay Model and Half-Life

An **exponential decay function** has the form

$$Q(t) = Q_0 e^{-kt}. \qquad Q_0, k \text{ both positive}$$

Q_0 represents the value of Q at time $t = 0$, and k is the **decay constant.** The decay constant k and half-life t_h for Q are related by

$$t_h k = \ln 2.$$

Quick Examples

1. $Q(t) = Q_0e^{-0.000\,121\,01t}$ is the decay function for carbon 14 (see Example 4b).
2. If $t_h = 10$ years, then $10k = \ln 2$, so $k = \frac{\ln 2}{10} \approx 0.06931$ and the decay model is
$$Q(t) = Q_0e^{-0.06931t}.$$
3. If $k = 0.0123$, then $t_h(0.0123) = \ln 2$, so the half-life is
$$t_h = \frac{\ln 2}{0.0123} \approx 56.35 \text{ years.}$$

We can repeat the analysis above for exponential growth models:

Exponential Growth Model and Doubling Time

An **exponential growth function** has the form

$$Q(t) = Q_0e^{kt}. \qquad Q_0,\ k \text{ both positive}$$

Q_0 represents the value of Q at time $t = 0$, and k is the **growth constant**. The growth constant k and doubling time t_d for Q are related by

$$t_dk = \ln 2.$$

Quick Examples

1. $P(t) = 1{,}000e^{0.05t}$ — \$1,000 invested at 5% annually with interest compounded continuously
2. If $t_d = 10$ years, then $10k = \ln 2$, so $k = \frac{\ln 2}{10} \approx 0.06931$ and the growth model is
$$Q(t) = Q_0e^{0.06931t}.$$
3. If $k = 0.0123$, then $t_d(0.0123) = \ln 2$, so the doubling time is
$$t_d = \frac{\ln 2}{0.0123} \approx 56.35 \text{ years.}$$

using Technology

To obtain the regression curve and graph for Example 5 using a graphing calculator or a spreadsheet, see the Technology Guides at the end of the chapter. Outline:

TI-83/84 Plus
[STAT] EDIT values of t in L_1 and values of S in L_2
Regression curve: [STAT] CALC option #9 LnReg [ENTER]
Graph: [Y=] [VARS] [5] EQ [1], then [ZOOM] [9]
[More details on page 145.]

Spreadsheet
t- and S-values in Columns A and B
Graph: Highlight data and insert a Scatter chart.
Regression curve: Right-click a datapoint and add logarithmic Trendline with option to show equation on chart.
[More details on page 148.]

Website
www.WanerMath.com
In the Function Evaluator and Grapher, enter the data as shown, press "Examples" until the logarithmic model `$1*ln(x)+$2` shows in the first box, and press "Fit Curve".

Logarithmic Regression

If we start with a set of data that suggests a logarithmic curve we can, by repeating the methods from previous sections, use technology to find the logarithmic regression curve $y = \log_b x + C$ approximating the data.

EXAMPLE 5 Research & Development

The following table shows the total spent on research and development by universities and colleges in the U.S., in billions of dollars, for the period 1998–2008 (t is the number of years since 1990).[5]

[5]2008 data is preliminary. Source: National Science Foundation, Division of Science Resources Statistics. 2010. *National Patterns of R&D Resources: 2008 Data Update.* NSF 10-314. Arlington, VA. www.nsf.gov/statistics/nsf10314/.

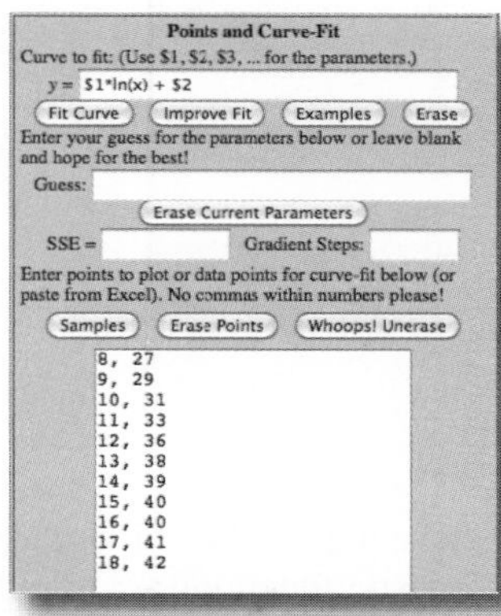

Year t	8	9	10	11	12	13	14	15	16	17	18
Spending ($ billions)	27	29	31	33	36	38	39	40	40	41	42

Find the best-fit logarithmic model of the form

$$S(t) = A \ln t + C$$

and use the model to project total spending on research by universities and colleges in 2012, assuming the trend continues.

Solution We use technology to get the following regression model:

$$S(t) = 19.3 \ln t - 12.8.$$

Coefficients rounded

Because 2012 is represented by $t = 22$, we have

$$S(22) = 19.3 \ln(22) - 12.8 \approx 47.$$

Why did we round the result to two significant digits?

So, research and development spending by universities and colleges is projected to be around $47 billion in 2012.

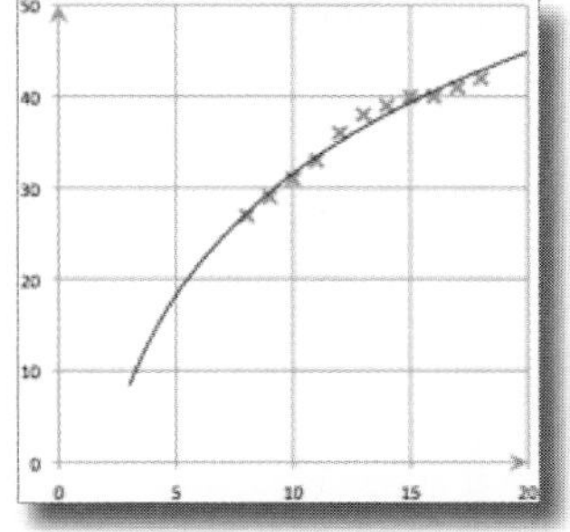

Figure 16

Before we go on... The model in Example 5 seems to give reasonable estimates when we extrapolate forward, but extrapolating backward is quite another matter: The logarithm curve drops sharply to the left of the given range and becomes negative for small values of t (Figure 16). ■

2.3 EXERCISES

Access end-of-section exercises online at **www.webassign.net**

ENHANCED WebAssign

2.4 Logistic Functions and Models

Figure 17 shows wired broadband penetration in the United States as a function of time t in years ($t = 0$ represents 2000).[6]

The left-hand part of the curve in Figure 17, from $t = 2$ to, say, $t = 6$, looks roughly like exponential growth: P behaves (roughly) like an exponential function, with the y-coordinates growing by a factor of around 1.5 per year. Then, as the market starts to become saturated, the growth of P slows and its value approaches a "saturation" point that appears to be around 30%. **Logistic** functions have just this

[6]Broadband penetration is the number of broadband installations divided by the total population. Source for data: Organisation for Economic Co-operation and Development (OECD) Directorate for Science, Technology, and Industry, table of Historical Penetration Rates, June 2010, downloaded April 2011 from www.oecd.org/sti/ict/broadband.

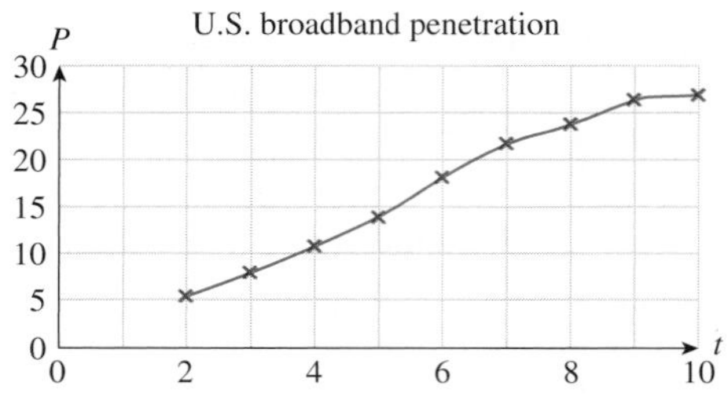

Figure 17

kind of behavior, growing exponentially at first and then leveling off. In addition to modeling the demand for a new technology or product, logistic functions are often used in epidemic and population modeling. In an epidemic, the number of infected people often grows exponentially at first and then slows when a significant proportion of the entire susceptible population is infected and the epidemic has "run its course." Similarly, populations may grow exponentially at first and then slow as they approach the capacity of the available resources.

Logistic Function

A **logistic function** has the form

$$f(x) = \frac{N}{1 + Ab^{-x}}$$

for nonzero constants N, A, and b (A and b positive and $b \neq 1$).

Quick Example

$N = 6$, $A = 2$, $b = 1.1$ gives $f(x) = \dfrac{6}{1 + 2(1.1^{-x})}$ `6/(1+2*1.1^-x)`

$f(0) = \dfrac{6}{1+2} = 2$ The y-intercept is $N/(1 + A)$.

$f(1{,}000) = \dfrac{6}{1 + 2(1.1^{-1{,}000})} \approx \dfrac{6}{1+0} = 6 = N$ When x is large, $f(x) \approx N$.

Graph of a Logistic Function

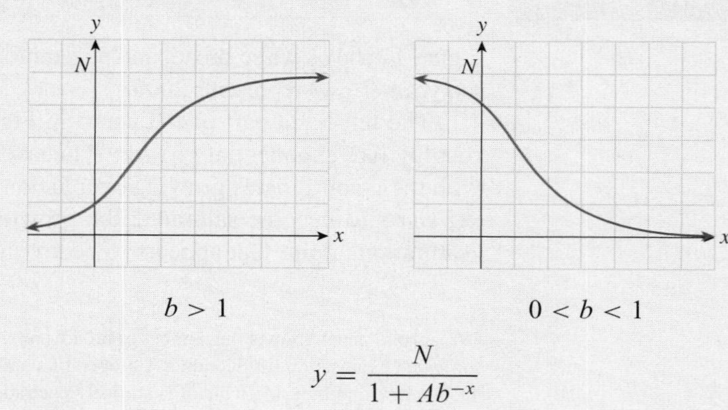

$$y = \frac{N}{1 + Ab^{-x}}$$

Properties of the Logistic Curve $y = \dfrac{N}{1 + Ab^{-x}}$

- The graph is an S-shaped curve sandwiched between the horizontal lines $y = 0$ and $y = N$. N is called the **limiting value** of the logistic curve.
- If $b > 1$ the graph rises; if $b < 1$, the graph falls.
- The y-intercept is $\dfrac{N}{1 + A}$.
- The curve is steepest when $t = \dfrac{\ln A}{\ln b}$. We will see why in Chapter 5.

Note If we write b^{-x} as e^{-kx} (so that $k = \ln b$), we get the following alternative form of the logistic function:

$$f(x) = \frac{N}{1 + Ae^{-kx}}.$$

■

Q: *How does the constant b affect the graph?*

A: To understand the role of b, we first rewrite the logistic function by multiplying top and bottom by b^x:

$$\begin{aligned} f(x) &= \frac{N}{1 + Ab^{-x}} \\ &= \frac{Nb^x}{(1 + Ab^{-x})b^x} \\ &= \frac{Nb^x}{b^x + A} \qquad \text{Because } b^{-x}b^x = 1 \end{aligned}$$

For values of x close to 0, the quantity b^x is close to 1, so the denominator is approximately $1 + A$, giving

$$f(x) \approx \frac{Nb^x}{1 + A} = \left(\frac{N}{1 + A}\right) b^x.$$

In other words, $f(x)$ is approximately exponential with base b for values of x close to 0. Put another way, if x represents time, then initially the logistic function behaves like an exponential function.

To summarize:

Logistic Function for Small *x* and the Role of *b*

For small values of x, we have

$$\frac{N}{1 + Ab^{-x}} \approx \left(\frac{N}{1 + A}\right) b^x.$$

Thus, for small x, the logistic function grows approximately exponentially with base b.

Quick Example

Let

$$f(x) = \frac{50}{1 + 24(3^{-x})}. \qquad N = 50, A = 24, b = 3$$

Then

$$f(x) \approx \left(\frac{50}{1 + 24}\right)(3^x) = 2(3^x)$$

for small values of x. The following figure compares their graphs:

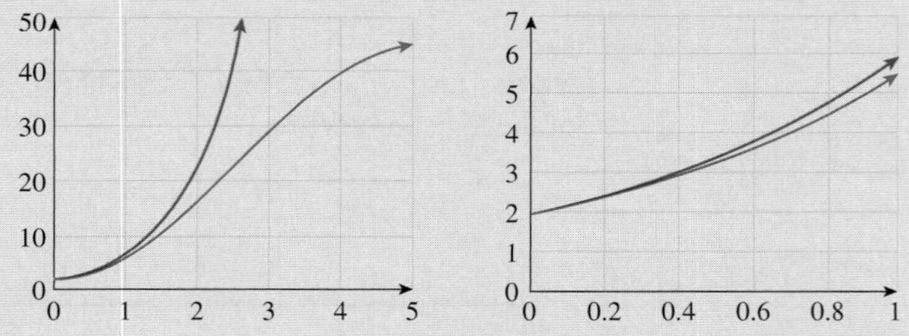

The upper curve is the exponential curve.

Modeling with the Logistic Function

EXAMPLE 1 Epidemics

A flu epidemic is spreading through the U.S. population. An estimated 150 million people are susceptible to this particular strain, and it is predicted that all susceptible people will eventually become infected. There are 10,000 people already infected, and the number is doubling every 2 weeks. Use a logistic function to model the number of people infected. Hence predict when, to the nearest week, 1 million people will be infected.

Solution Let t be time in weeks, and let $P(t)$ be the total number of people infected at time t. We want to express P as a logistic function of t, so that

$$P(t) = \frac{N}{1 + Ab^{-t}}.$$

We are told that, in the long run, 150 million people will be infected, so that

$$N = 150{,}000{,}000. \qquad \text{Limiting value of } P$$

At the current time ($t = 0$), 10,000 people are infected, so

$$10{,}000 = \frac{N}{1 + A} = \frac{150{,}000{,}000}{1 + A}. \qquad \text{Value of } P \text{ when } t = 0$$

Solving for A gives

$$10{,}000(1 + A) = 150{,}000{,}000$$
$$1 + A = 15{,}000$$
$$A = 14{,}999.$$

What about b? At the beginning of the epidemic (t near 0), P is growing approximately exponentially, doubling every 2 weeks. Using the technique of Section 2.2, we find that the exponential curve passing through the points (0, 10,000) and (2, 20,000) is

$$y = 10{,}000(\sqrt{2})^t$$

giving us $b = \sqrt{2}$. Now that we have the constants N, A, and b, we can write down the logistic model:

$$P(t) = \frac{150{,}000{,}000}{1 + 14{,}999(\sqrt{2})^{-t}}.$$

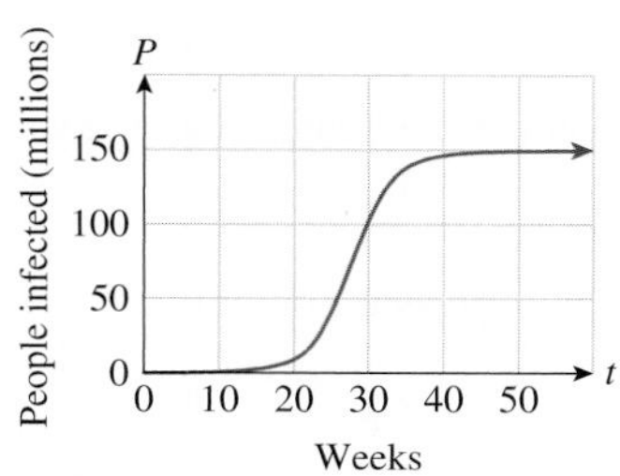

Figure 18

The graph of this function is shown in Figure 18.

Now we tackle the question of prediction: When will 1 million people be infected? In other words: When is $P(t) = 1{,}000{,}000$?

$$\begin{aligned} 1{,}000{,}000 &= \frac{150{,}000{,}000}{1 + 14{,}999(\sqrt{2})^{-t}} \\ 1{,}000{,}000[1 + 14{,}999(\sqrt{2})^{-t}] &= 150{,}000{,}000 \\ 1 + 14{,}999(\sqrt{2})^{-t} &= 150 \\ 14{,}999(\sqrt{2})^{-t} &= 149 \\ (\sqrt{2})^{-t} &= \frac{149}{14{,}999} \\ -t &= \log_{\sqrt{2}}\left(\frac{149}{14{,}999}\right) \approx -13.31 \quad \text{Logarithmic form} \end{aligned}$$

Thus, 1 million people will be infected by about the thirteenth week.

Before we go on... We said earlier that the logistic curve is steepest when $t = \dfrac{\ln A}{\ln b}$. In Example 1, this occurs when $t = \dfrac{\ln 14{,}999}{\ln\sqrt{2}} \approx 28$ weeks into the epidemic. At that time, the number of cases is growing most rapidly (look at the apparent slope of the graph at the corresponding point). ■

Logistic Regression

Let's go back to the data on broadband penetration in the United States with which we began this section and try to determine the long-term percentage of broadband penetration. In order to be able to make predictions such as this, we require a model for the data, so we will need to do some form of regression.

EXAMPLE 2 Broadband Penetration

Here are the data graphed in Figure 17:

Year (t)	2	3	4	5	6	7	8	9	10
Penetration (%) (P)	5.5	7.9	10.9	14.2	18.2	21.9	23.9	26.5	27.1

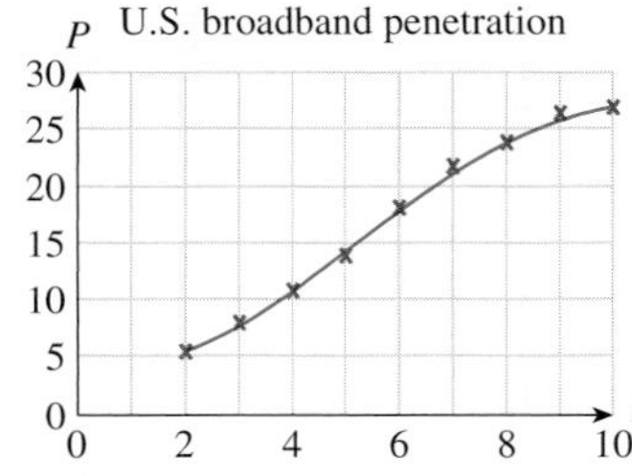

Figure 19

using Technology

See the Technology Guide at the end of the chapter for detailed instructions on how to obtain the regression curve and graph for Example 2 using a graphing calculator or spreadsheet. Outline:

TI-83/84 Plus
[STAT] EDIT values of t in L_1 and values of P in L_2
Regression curve: [STAT] CALC option #B Logistic [ENTER]
Graph: [Y=] [VARS] [5] EQ [1], then [ZOOM] [9] [More details on page 146.]

Spreadsheet
Use the Solver Add-in to obtain the best-fit logistic curve. [More details on page 148.]

Website
www.WanerMath.com
In the Function Evaluator and Grapher, enter the data as shown, press "Examples" until the logistic model `$1/(1+$2*$3^(-x))` shows in the first box, and press "Fit Curve".

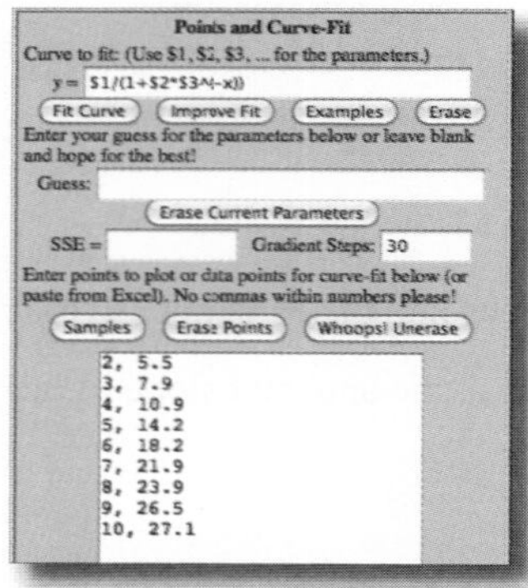

Find a logistic regression curve of the form

$$P(t) = \frac{N}{1 + Ab^{-t}}.$$

In the long term, what percentage of broadband penetration in the United States does the model predict?

Solution We can use technology to obtain the following regression model:

$$P(t) \approx \frac{29.842}{1 + 12.502(1.642)^{-t}}.$$ Coefficients rounded to 3 decimal places

Its graph and the original data are shown in Figure 19. Because $N = 29.842$, this model predicts that, in the long term, the percentage of broadband penetration in the United States will be 29.842%, or about 30%.

➡ **Before we go on...** Logistic regression programs generally estimate all three constants N, A, and b for a model $y = \dfrac{N}{1 + Ab^{-x}}$. However, there are times, as in Example 1, when we already know the limiting value N and require estimates of only A and b. In such cases, we can use technology like Excel Solver to find A and b for the best-fit curve with N fixed. Alternatively, we can use exponential regression to compute estimates of A and b as follows: First rewrite the logistic equation as

$$\frac{N}{y} = 1 + Ab^{-x},$$

so that

$$\frac{N}{y} - 1 = Ab^{-x} = A(b^{-1})^x.$$

This equation gives $N/y - 1$ as an exponential function of x. Thus, if we do exponential regression using the data points $(x, N/y - 1)$, we can obtain estimates for A and b^{-1} (and hence b). This is done in Exercises 35 and 36.

It is important to note that the resulting curve is not the best-fit curve (in the sense of minimizing SSE; see the "Before we go on" discussion on page 119 after Example 6 in Section 2.2) and will be thus be different from that obtained using the method in Example 2. ■

2.4 EXERCISES

Access end-of-section exercises online at **www.webassign.net** ENHANCED WebAssign

CHAPTER 2 REVIEW

KEY CONCEPTS

WW Website www.WanerMath.com

Go to the Website at www.WanerMath.com to find a comprehensive and interactive Web-based summary of Chapter 2.

2.1 Quadratic Functions and Models

A **quadratic function** has the form $f(x) = ax^2 + bx + c$. *p. 102*

The graph of $f(x) = ax^2 + bx + c$ $(a \neq 0)$ is a **parabola.** *p. 103*

The x-coordinate of the **vertex** is $-\frac{b}{2a}$. The y-coordinate is $f\left(-\frac{b}{2a}\right)$. *p. 103*

***x*-intercepts** (if any) occur at

$$x = \frac{-b \pm \sqrt{b^2 - 4ac}}{2a}. \quad p.\ 103$$

The ***y*-intercept** occurs at $y = c$. *p. 103*

The parabola is **symmetric** with respect to the vertical line through the vertex. *p. 103*

Sketching the graph of a quadratic function *p. 104*

Application to maximizing revenue *p. 106*

Application to maximizing profit *p. 106*

Finding the quadratic regression curve *p. 108*

2.2 Exponential Functions and Models

An **exponential function** has the form $f(x) = Ab^x$. *p. 110*

Roles of the constants A and b in an exponential function $f(x) = Ab^x$ *p. 111*

Recognizing exponential data *p. 112*

Finding the exponential curve through two points *p. 113*

Application to compound interest *p. 114*

Application to exponential decay (carbon dating) *p. 114*

Application to exponential growth (epidemics) *p. 114*

The number e and continuous compounding *p. 116*

Alternative form of an exponential function: $f(x) = Ae^{rx}$ *p. 118*

Finding the exponential regression curve *p. 118*

2.3 Logarithmic Functions and Models

The **base *b* logarithm of *x***: $y = \log_b x$ means $b^y = x$ *p. 121*

Common logarithm, $\log x = \log_{10} x$, and **natural logarithm**, $\ln x = \log_e x$ *p. 121*

Change-of-base formula *p. 122*

Solving equations with unknowns in the exponent *p. 122*

A **logarithmic function** has the form $f(x) = \log_b x + C$ or $f(x) = A \ln x + C$. *p. 123*

Graphs of logarithmic functions *p. 123*

Logarithm identities *p. 124*

Application to investments (How long?) *p. 125*

Application to half-life *p. 125*

Exponential decay models and half-life *p. 127*

Exponential growth models and doubling time *p. 128*

Finding the logarithmic regression curve *p. 128*

2.4 Logistic Functions and Models

A **logistic function** has the form

$$f(x) = \frac{N}{1 + Ab^{-x}}. \quad p.\ 130$$

Properties of the logistic curve, point where curve is steepest *p. 131*

Logistic function for small x, the role of b *p. 131*

Application to epidemics *p. 132*

Finding the logistic regression curve *p. 133*

REVIEW EXERCISES

Sketch the graph of the quadratic functions in Exercises 1 and 2, indicating the coordinates of the vertex, the y-intercept, and the x-intercepts (if any).

1. $f(x) = x^2 + 2x - 3$

2. $f(x) = -x^2 - x - 1$

In Exercises 3 and 4, the values of two functions, f and g, are given in a table. One, both, or neither of them may be exponential. Decide which, if any, are exponential, and give the exponential models for those that are.

3.

x	−2	−1	0	1	2
$f(x)$	20	10	5	2.5	1.25
$g(x)$	8	4	2	1	0

4.

x	−2	−1	0	1	2
$f(x)$	8	6	4	2	1
$g(x)$	$\frac{3}{4}$	$\frac{3}{2}$	3	6	12

In Exercises 5 and 6, graph the given pairs of functions on the same set of axes with $-3 \leq x \leq 3$.

5. $f(x) = \frac{1}{2}(3^x)$; $g(x) = \frac{1}{2}(3^{-x})$

6. $f(x) = 2(4^x)$; $g(x) = 2(4^{-x})$

T *On the same set of axes, use technology to graph the pairs of functions in Exercises 7 and 8 for the given range of x. Identify which graph corresponds to which function.*

7. $f(x) = e^x$; $g(x) = e^{0.8x}$; $-3 \leq x \leq 3$

8. $f(x) = 2(1.01)^x$; $g(x) = 2(0.99)^x$; $-100 \leq x \leq 100$

In Exercises 9–14, compute the indicated quantity.

9. A $3,000 investment earns 3% interest, compounded monthly. Find its value after 5 years.

10. A $10,000 investment earns 2.5% interest, compounded quarterly. Find its value after 10 years.

11. An investment earns 3% interest, compounded monthly, and is worth $5,000 after 10 years. Find its initial value.

12. An investment earns 2.5% interest, compounded quarterly, and is worth \$10,000 after 10 years. Find its initial value.

13. A \$3,000 investment earns 3% interest, compounded continuously. Find its value after 5 years.

14. A \$10,000 investment earns 2.5% interest, compounded continuously. Find its value after 10 years.

In Exercises 15–18, find a formula of the form $f(x) = Ab^x$ using the given information.

15. $f(0) = 4.5$; the value of f triples for every half-unit increase in x.

16. $f(0) = 5$; the value of f decreases by 75% for every 1-unit increase in x.

17. $f(1) = 2$, $f(3) = 18$.

18. $f(1) = 10$, $f(3) = 5$.

In Exercises 19–22, use logarithms to solve the given equation for x.

19. $3^{-2x} = 4$

20. $2^{2x^2-1} = 2$

21. $300(10^{3x}) = 315$

22. $P(1+i)^{mx} = A$

On the same set of axes, graph the pairs of functions in Exercises 23 and 24.

23. $f(x) = \log_3 x$; $g(x) = \log_{(1/3)} x$

24. $f(x) = \log x$; $g(x) = \log_{(1/10)} x$

In Exercises 25–28, use the given information to find an exponential model of the form $Q = Q_0e^{-kt}$ or $Q = Q_0e^{kt}$, as appropriate. Round all coefficients to 3 significant digits when rounding is necessary.

25. Q is the amount of radioactive substance with a half-life of 100 years in a sample originally containing 5 g (t is time in years).

26. Q is the number of cats on an island whose cat population was originally 10,000 but is being cut in half every 5 years (t is time in years).

27. Q is the diameter (in cm) of a circular patch of mold on your roommate's damp towel you have been monitoring with morbid fascination. You measured the patch at 2.5 cm across 4 days ago, and have observed that it is doubling in diameter every 2 days (t is time in days).

28. Q is the population of cats on another island whose cat population was originally 10,000 but is doubling every 15 months (t is time in months).

In Exercises 29–32, find the time required, to the nearest 0.1 year, for the investment to reach the desired goal.

29. \$2,000 invested at 4%, compounded monthly; goal: \$3,000

30. \$2,000 invested at 6.75%, compounded daily; goal: \$3,000

31. \$2,000 invested at 3.75%, compounded continuously; goal: \$3,000

32. \$1,000 invested at 100%, compounded quarterly; goal: \$1,200

In Exercises 33–36, find equations for the logistic functions of x with the stated properties.

33. Through (0, 100), initially increasing by 50% per unit of x, and limiting value 900.

34. Initially exponential of the form $y = 5(1.1)^x$ with limiting value 25.

35. Passing through (0, 5) and decreasing from a limiting value of 20 to 0 at a rate of 20% per unit of x when x is near 0.

36. Initially exponential of the form $y = 2(0.8)^x$ with a value close to 10 when $x = -60$.

APPLICATIONS: OHaganBooks.com

37. ***Web Site Traffic*** The daily traffic ("hits per day") at OHaganBooks.com apparently depends on the monthly expenditure on Internet advertising. The following model is based on information collected over the past few months:

$$h = -0.000005c^2 + 0.085c + 1{,}750.$$

Here, h is the average number of hits per day at OHaganBooks.com, and c is the monthly advertising expenditure.

a. According to the model, what monthly advertising expenditure will result in the largest volume of traffic at OHaganBooks.com? What is that volume?

b. In addition to predicting a maximum volume of traffic, the model predicts that the traffic will eventually drop to zero if the advertising expenditure is increased too far. What expenditure (to the nearest dollar) results in no Web site traffic?

c. What feature of the formula for this quadratic model indicates that it will predict an eventual decline in traffic as advertising expenditure increases?

38. ***Broadband Access*** Pablo Pelogrande, a new summer intern at OHaganBooks.com in 2013, was asked by John O'Hagan to research the extent of broadband access in the United States. Pelogrande found some very old data online on broadband access from the start of 2001 to the end of 2003 and used it to construct the following quadratic model of the growth rate of broadband access:

$n(t) = 2t^2 - 6t + 12$ million new American adults with broadband per year

(t is time in years; $t = 0$ represents the start 2000).[7]

a. What is the appropriate domain of n?

b. According to the model, when was the growth rate at a minimum?

c. Does the model predict a zero growth rate at any particular time? If so, when?

d. What feature of the formula for this quadratic model indicates that the growth rate eventually increases?

e. Does the fact that $n(t)$ decreases for $t \leq 1.5$ suggest that the number of broadband users actually declined before June 2001? Explain.

[7]Based on data for 2001–2003. Source for data: Pew Internet and American Life Project data memos dated May 18, 2003 and April 19, 2004, downloaded from www.pewinternet.org.

f. Pelogande extrapolated the model in order to estimate the growth rate at the beginning of 2013 and 2014. What did he find? Comment on the answer.

39. *Revenue and Profit* Some time ago, a consultant formulated the following linear model of demand for online novels:

$$q = -60p + 950$$

where q is the monthly demand for OHaganBooks.com's online novels at a price of p dollars per novel.

a. Use this model to express the monthly revenue as a function of the unit price p, and hence determine the price you should charge for a maximum monthly revenue.

b. Author royalties and copyright fees cost the company an average of $4 per novel, and the monthly cost of operating and maintaining the online publishing service amounts to $900 per month. Express the monthly profit P as a function of the unit price p, and hence determine the unit price you should charge for a maximum monthly profit. What is the resulting profit (or loss)?

40. *Revenue and Profit* Billy-Sean O'Hagan is John O'Hagan's son and a freshman in college. He notices that the demand for the college newspaper was 2,000 copies each week when the paper was given away free of charge, but dropped to 1,000 each week when the college started charging 10¢/copy.

a. Write down the associated linear demand function.

b. Use your demand function to express the revenue as a function of the unit price p, and hence determine the price the college should charge for a maximum revenue. At that price, what is the revenue from sales of one edition of the newspaper?

c. It costs the college 4¢ to produce each copy of the paper, plus an additional fixed cost of $200. Express the profit P as a function of the unit price p, and hence determine the unit price the college should charge for a maximum monthly profit (or minimum loss). What is the resulting profit (or loss)?

41. *Lobsters* Marjory Duffin, CEO of Duffin House, is particularly fond of having steamed lobster at working lunches with executives from OHaganBooks.com and is therefore alarmed by the fact that the yearly lobster harvest from New York's Long Island Sound has been decreasing dramatically since 1997. Indeed, the size of the annual harvest can be approximated by

$$n(t) = 9.1(0.81^t) \text{ million pounds}$$

where t is time in years since 1997.[8]

a. The model tells us that the harvest was ____ million pounds in 1997 and decreasing by ___% each year.

b. What does the model predict for the 2013 harvest?

42. *Stock Prices* In the period immediately following its initial public offering (IPO), OHaganBooks.com's stock was doubling in value every 3 hours. If you bought $10,000 worth of the stock when it was first offered, how much was your stock worth after 8 hours?

[8]Authors' regression model. Source for data: Long Island Sound Study, data downloaded May 2011 from longislandsoundstudy.net/2010/07/lobster-landings/.

43. *Lobsters* (See Exercise 41.) Marjory Duffin has just left John O'Hagan, CEO of OHaganBooks.com, a frantic phone message to the effect that this year's lobster harvest from New York's Long Island Sound is predicted to dip below 200,000 pounds, making that planned lobster working lunch more urgent than ever. What year is it?

44. *Stock Prices* We saw in Exercise 42 that OHaganBooks.com's stock was doubling in value every 3 hours, following its IPO. If you bought $10,000 worth of the stock when it was first offered, how long from the initial offering did it take your investment to reach $50,000?

45. *Lobsters* We saw in Exercise 41 that the Long Island Sound lobster harvest was given by $n(t) = 9.1(0.81^t)$ million pounds t years after 1997. However, in 2010, thanks to the efforts of Duffin House, Inc. it turned around and started increasing by 24% each year.[9] What, to the nearest 10,000 pounds, was the actual size of the harvest in 2013?

46. *Stock Prices* We saw in Exercise 42 that OHaganBooks.com's stock was doubling in value every 3 hours, following its IPO. After 10 hours of trading, the stock turns around and starts losing one third of its value every 4 hours. How long (from the initial offering) will it be before your stock is once again worth $10,000?

47. T ***Lobsters*** The following chart shows some of the data that went into the model in Exercise 41:

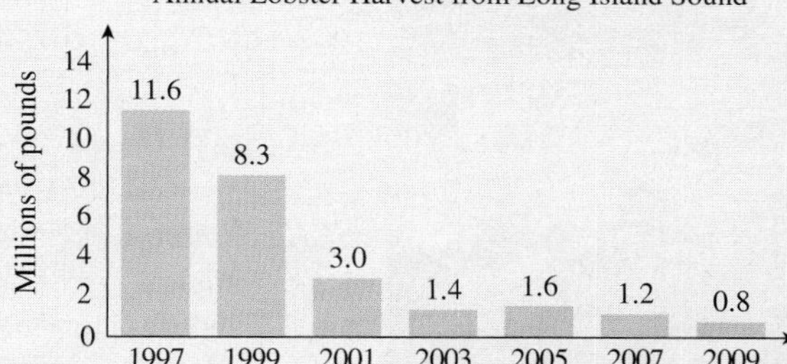

Use these data to obtain an exponential regression curve of the form $n(t) = Ab^t$, with $t = 0$ corresponding to 1997 and coefficients rounded to 2 significant digits.

48. T ***Stock Prices*** The actual stock price of OHaganBooks.com in the hours following its IPO is shown in the following chart:

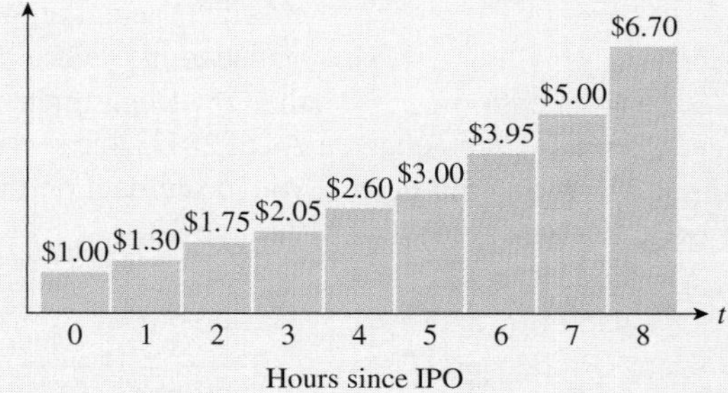

[9]This claim, like Duffin House, is fiction.

Use the data to obtain an exponential regression curve of the form $P(t) = Ab^t$, with $t = 0$ the time in hours since the IPO and coefficients rounded to 3 significant digits. At the end of which hour will the stock price first be above \$10?

49. ***Hardware Life*** *(Based on a question from the GRE economics exam)* To estimate the rate at which new computer hard drives will have to be retired, OHaganBooks.com uses the "survivor curve":

$$L_x = L_0 e^{-x/t}$$

where

L_x = number of surviving hard drives at age x
L_0 = number of hard drives initially
t = average life in years.

All of the following are implied by the curve *except:*

(A) Some of the equipment is retired during the first year of service.
(B) Some equipment survives three average lives.
(C) More than half the equipment survives the average life.
(D) Increasing the average life of equipment by using more durable materials would increase the number surviving at every age.
(E) The number of survivors never reaches zero.

50. ***Sales*** OHaganBooks.com modeled its weekly sales over a period of time with the function

$$s(t) = 6{,}050 + \frac{4{,}470}{1 + 14(1.73^{-t})}$$

as shown in the following graph (t is measured in weeks):

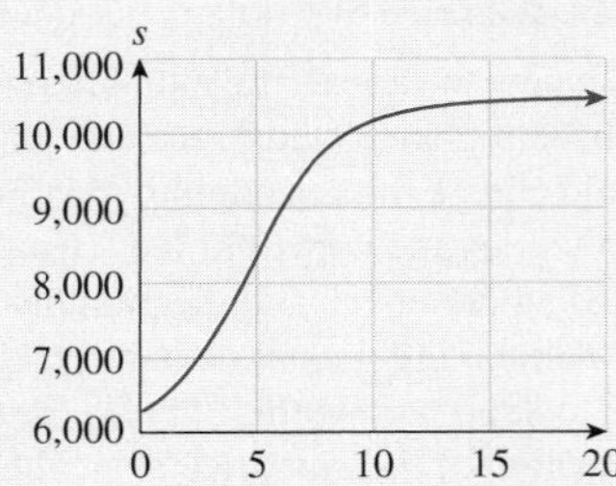

a. As time goes on, it appears that weekly sales are leveling off. At what value are they leveling off?
b. When did weekly sales rise above 10,000?
c. When, to the nearest week, were sales rising most rapidly?

Case Study

Checking up on Malthus

Robert Nickelsberg/Getty Images

In 1798 Thomas R. Malthus (1766–1834) published an influential pamphlet, later expanded into a book, titled *An Essay on the Principle of Population As It Affects the Future Improvement of Society.* One of his main contentions was that population grows geometrically (exponentially) while the supply of resources such as food grows only arithmetically (linearly). This led him to the pessimistic conclusion that population would always reach the limits of subsistence and precipitate famine, war, and ill health, unless population could be checked by other means. He advocated "moral restraint," which includes the pattern of late marriage common in Western Europe at the time and now common in most developed countries and which leads to a lower reproduction rate.

Two hundred years later, you have been asked to check the validity of Malthus's contention. That population grows geometrically, at least over short periods of time, is commonly assumed. That resources grow linearly is more questionable. You decide to check the actual production of a common crop, wheat, in the United States. Agricultural statistics like these are available from the U.S. government on the Internet, through the U.S. Department of Agriculture's National Agricultural Statistics Service (NASS). As of 2011, this service was available at www.nass.usda.gov. Looking through this site, you locate data on the annual production of all wheat in the United States from 1900 through 2010.

Website
www.WanerMath.com
To download an Excel sheet with the data used in the case study, go to Everything for Calculus, scroll down to the case study for Chapter 2, and click on "Wheat Production Data (Excel)".

Year	1900	1901	. . .	2009	2010
Thousands of Bushels	599,315	762,546	. . .	2,218,061	2,208,391

Graphing these data (using Excel, for example), you obtain the graph in Figure 20.

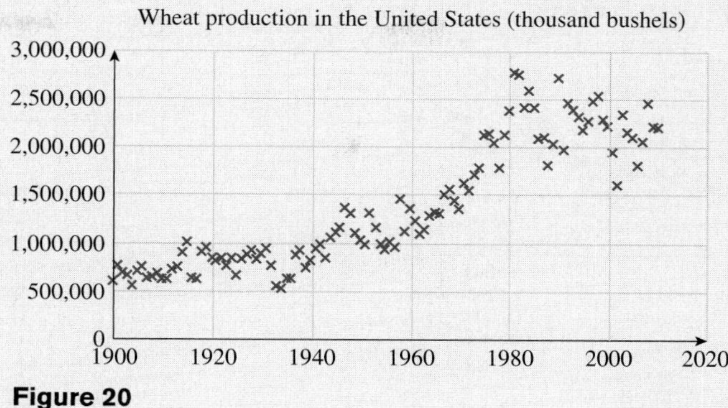

Figure 20

This does not look very linear, particularly in the last half of the twentieth century, but you continue checking the mathematics. Using Excel's built-in linear regression capabilities, you find that the line that best fits these data, shown in Figure 21, has $r^2 = 0.8039$. (Recall the discussion of the correlation coefficient r in Section 1.4. A similar statistic is available for other types of regression as well.)

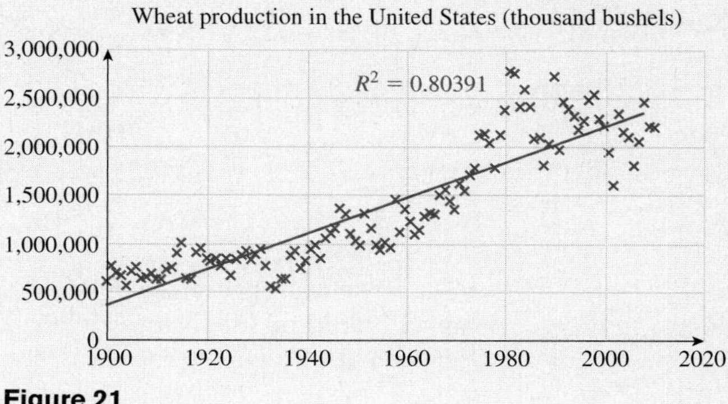

Figure 21

* Recall that the residuals are defined as $y_{\text{Observed}} - y_{\text{Predicted}}$ (see Section 1.4) and are the vertical distances between the observed data points and the regression line.

Although that is a fairly high correlation, you notice that the residuals* are not distributed randomly: The actual wheat production starts out higher than the line, dips below the line from about 1920 to about 1970, then rises above the line, and finally appears to dip below the line around 2002. This behavior seems to suggest a logistic curve or perhaps a cubic curve. On the other hand, it is also possible that the apparent dip at the end of the data is not statistically significant—it could be nothing more than a transitory fluctuation in the wheat production industry—so perhaps we should also consider models that do not bend downward, like exponential and quadratic models.

Following is a comparison of the four proposed models (with coefficients rounded to 3 significant digits). For the independent variable, we used t = time in years since 1900. SSE is the sum-of-squares error.

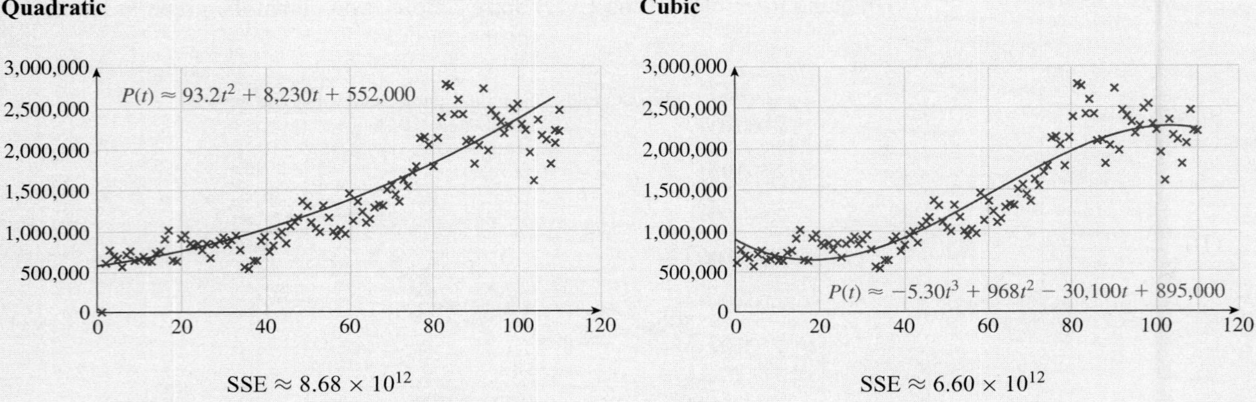

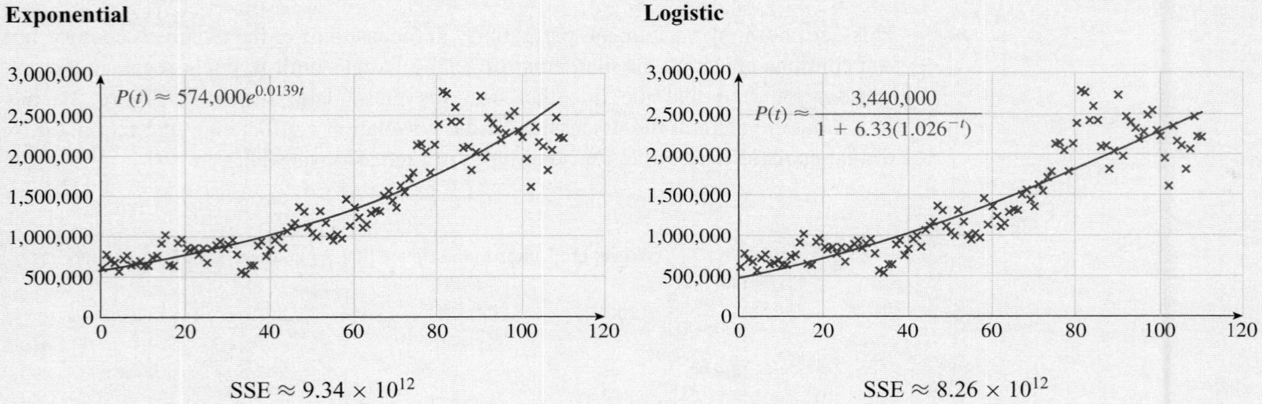

The model that appears to best fit the data seems to be the cubic model; both visually and by virtue of SSE. Notice also that the cubic model predicts a *decrease* in the production of wheat in the near term (see Figure 22).

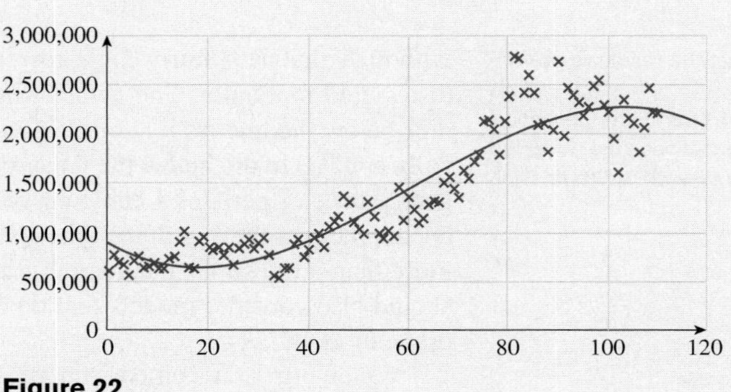

Figure 22

So you prepare a report that documents your findings and concludes that things are even worse than Malthus predicted, at least as far as wheat production in the United States is concerned: The supply is deceasing while the population is still increasing more-or-less exponentially. (See Exercise 75 in Section 2.2.)

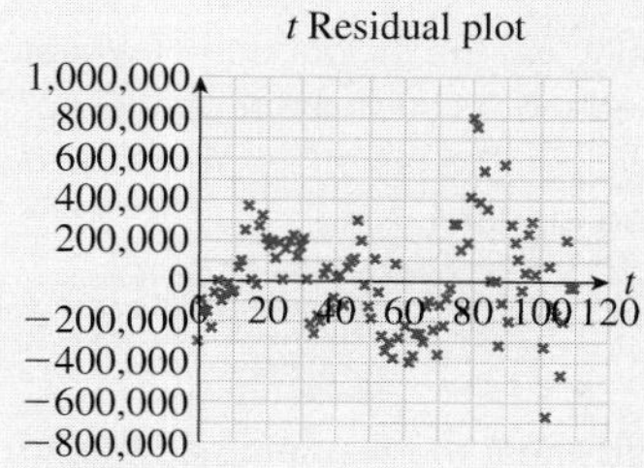

Figure 23

You are about to hit "Send," which will dispatch copies of your report to a significant number of people on whom the success of your career depends, when you notice something strange about the pattern of data in Figure 22: The observed data points appear to hug the regression curve quite closely for small values of t, but appear to become more and more scattered as the value of t increases. In the language of residuals, the residuals are small for small values of t but then tend to get larger with increasing t. Figure 23 shows a plot of the residuals that shows this trend even more clearly.

This reminds you vaguely of something that came up in your college business statistics course, so you consult the textbook from that class that (fortunately) you still own and discover that a pattern of residuals with increasing magnitude suggests that, instead of modeling y versus t directly, you instead model $\ln y$ versus t. (The residuals for large values of t will then be scaled down by the logarithm.)

Figure 24 shows the resulting plot together with the regression line (what we call the "linear transformed model").

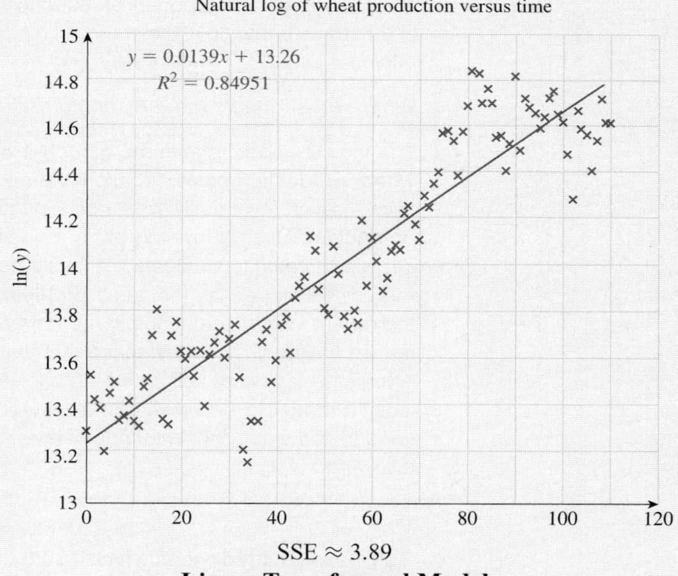

Figure 24

Notice that this time, the regression patterns no longer suggest an obvious curve. Further, they no longer appear to grow with increasing t. Although SSE is dramatically lower than the values for the earlier models, the contrast is a false one; the units of y are now different, and comparing SSE with that of the earlier models is like comparing apples and oranges. While SSE depends on the units of measurement used, the coefficient of determination r^2 discussed in Section 1.4 is independent of the units used. A similar statistic is available for other types of regression as well, as well as something called "adjusted r^2."

The value of r^2 for the transformed model is approximately 0.850, while r^2 for the cubic model* is about 0.861, which is fairly close.

* The "adjusted r^2" from statistics that corrects for model size.

Q: *If the cubic model and the linear transformed model have similar values of r^2, how do I decide which is more appropriate?*

A: The cubic model, if extrapolated, predicts unrealistically that the production of wheat will plunge in the near future, but the linear transformed model sees the recent drop-off as just one of several market fluctuations that show up in the residuals. You should therefore favor the more reasonable linear transformed model.

Q: *The linear transformed model gives us ln y versus t. What does it say about y versus t?*

A: Accurately write down the equation of the transformed linear model, being careful to replace y by $\ln y$:

$$\ln y = 0.0139t + 13.26.$$

Rewriting this in exponential form gives

$$\begin{aligned} y &= e^{0.0139t+13.26} \\ &= e^{13.26}e^{0.0139t} \\ &\approx 574{,}000e^{0.0139t}, \quad \text{Coefficients rounded to 3 digits} \end{aligned}$$

which is exactly the exponential model we found earlier! (In fact, using the natural logarithm transformation is the standard method of computing the regression exponential curve.)

Q: *What of the logistic model; should that not be the most realistic?*

A: The logistic model seems as though it *ought* to be the most appropriate, because wheat production cannot reasonably be expected to continue increasing exponentially forever; eventually resource limitations must lead to a leveling off of wheat production. Such a leveling off, if it occurred before the population started to level off, would seem to vindicate Malthus's pessimistic predictions. However, the logistic regression model has the same problem as the cubic model: It is trying to interpret the recent large fluctuations in the data as evidence of leveling off, but we really do not yet have significant evidence that that is occurring. Wheat production—even if it is logistic—appears still in the early (exponential) stage of growth. In general, for a logistic model to be reliable in its prediction of the leveling-off value N, we would need to see significant evidence of leveling off in the data. (See, however, Exercise 2 following.)

You now conclude that wheat production for the past 100 years is better described as increasing exponentially than linearly, contradicting Malthus, and moreover that it shows no sign of leveling off as yet.

EXERCISES

1. Use the wheat production data starting at 1950 to construct the exponential regression model in two ways: directly, and using a linear transformed model as above. (Round coefficients to 3 digits.) Compare the growth constant k of your model with that of the exponential model based on the data from 1900 on. How would you interpret the difference?
2. Compute the least-squares logistic model for the data in the preceding exercise. (Round coefficients to 3 significant digits.) At what level does it predict that wheat production will level off? (Note on using Excel Solver for logistic regression: Before running Solver, press Options in the Solver window and turn "Automatic Scaling" on. This adjusts the algorithm for the fact that the constants A, N, and b have vastly different orders of magnitude.) Give

two graphs: one showing the data with the exponential regression model, and the other showing the data with the logistic regression model. Which model gives a better fit visually? Justify your observation by computing SSE directly for both models. Comment on your answer in terms of Malthus's assertions.

3. Find the production figures for another common crop grown in the United States. Compare the linear, quadratic, cubic, exponential, and logistic models. What can you conclude?
4. Below are the census figures for the population of the U.S. (in thousands) from 1820 to 2010.[10] Compare the linear, quadratic, and exponential models. What can you conclude?

Year	1820	1830	1840	1850	1860	1870	1880	1890	1900	1910
Population (1000s)	9,638	12,861	17,063	23,192	31,443	38,558	50,189	62,980	76,212	92,228
Year	1920	1930	1940	1950	1960	1970	1980	1990	2000	2010
Population (1000s)	106,022	123,203	132,165	151,326	179,323	203,302	226,542	248,710	281,422	308,746

[10]Source: Bureau of the Census, U.S. Department of Commerce.

TECHNOLOGY GUIDE

TI-83/84 Plus Technology Guide

Section 2.1

Example 2 (page 104) Sketch the graph of each quadratic function, showing the location of the vertex and intercepts.

a. $f(x) = 4x^2 - 12x + 9$ **b.** $g(x) = -\frac{1}{2}x^2 + 4x - 12$

Solution with Technology

We will do part (a).

1. Start by storing the coefficients a, b, c using

   `4→A:-12→B:9→C`

   STO> gives the arrow ALPHA . gives the colon

2. Save your quadratic as Y_1 using the Y = screen:

   `Y1=AX^2+BX+C`

3. To obtain the x-coordinate of the vertex, enter its formula as shown below on the left.

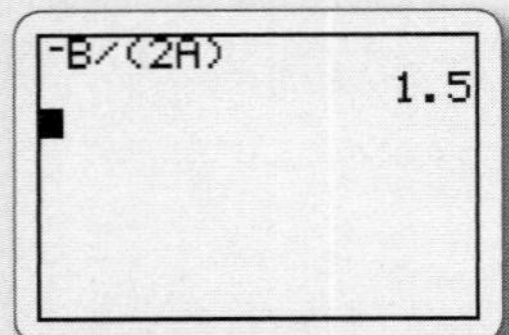

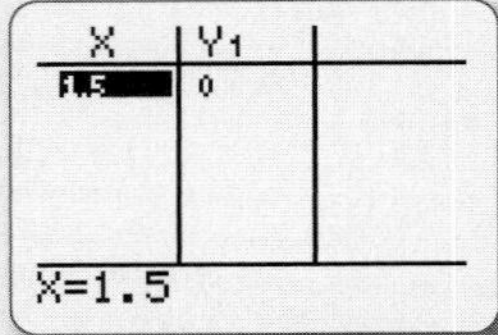

4. The y-coordinate of the vertex can be obtained from the Table screen by entering $x = 1.5$ as shown above on the right. (If you can't enter values of x, press 2ND TBLSET, and set `Indpnt` to Ask.) From the table, we see that the vertex is at the point (1.5, 0).
5. To obtain the x-intercepts, enter the quadratic formula on the home screen as shown:

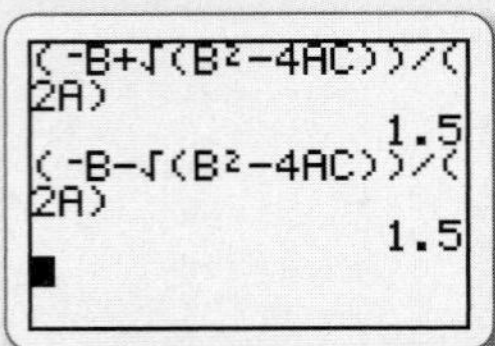

   Because both intercepts agree, we conclude that the graph intersects the x-axis on a single point (at the vertex).

6. To graph the function, we need to select good values for Xmin and Xmax. In general, we would like our graph to show the vertex as well as all the intercepts. To see the vertex, make sure that its x coordinate (1.5) is between Xmin and Xmax. To see the x-intercepts, make sure that they are also between Xmin and Xmax. To see the y-intercept, make sure that $x = 0$ is between Xmin and Xmax. Thus, to see everything, choose Xmin and Xmax so that the interval [xMin, xMax] contains the x coordinate of the vertex, the x-intercepts, and 0. For this example, we can choose an interval like [−1, 3].
7. Once xMin and xMax are chosen, you can obtain convenient values of yMin and yMax by pressing ZOOM and selecting the option `ZoomFit`. (Make sure that your quadratic equation is entered in the `Y=` screen before doing this!)

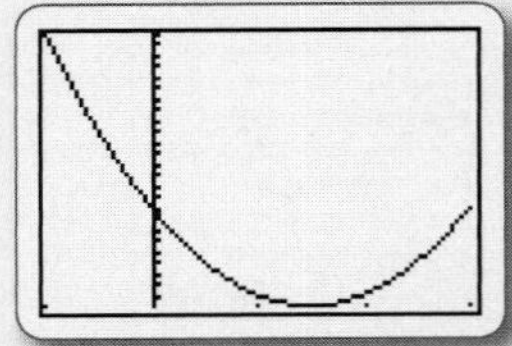

Example 5 (page 108) The following table shows the annual mean carbon dioxide concentration measured at Mauna Loa Observatory in Hawaii, in parts per million, every 10 years from 1960 through 2010 ($t = 0$ represents 1960).

Year t	0	10	20	30	40	50
PPM CO_2 C	317	326	339	354	369	390

Find the quadratic model

$$C(t) = at^2 + bt + c$$

that best fits the data.

Solution with Technology

1. Using STAT EDIT enter the data with the x-coordinates (values of t) in L_1 and the y-coordinates (values of C) in L_2, just as in Section 1.4:

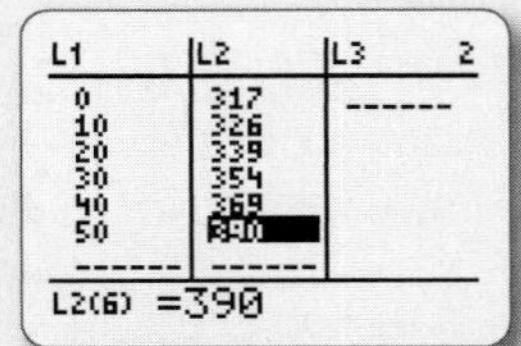

2. Press STAT, select CALC, and choose option #5 QuadReg. Pressing ENTER gives the quadratic regression curve in the home screen:

$$y \approx 0.01214x^2 + 0.8471x + 316.9.$$ Coefficients rounded to 4 decimal places

3. Now go to the Y= window and turn Plot1 on by selecting it and pressing ENTER. (You can also turn it on in the 2ND STAT PLOT screen.)
4. Next, enter the regression equation in the Y= screen by pressing Y=, clearing out whatever function is there, and pressing VARS 5 and selecting EQ option #1: RegEq as shown below left.

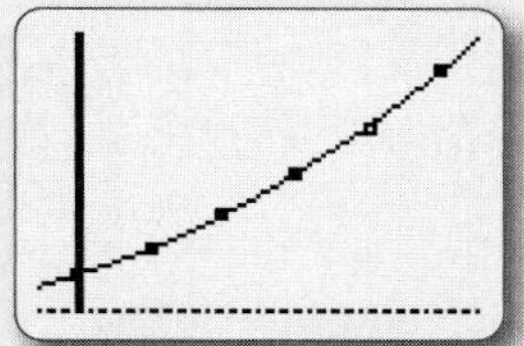

5. To obtain a convenient window showing all the points and the lines, press ZOOM and choose option #9: ZoomStat as shown above on the right.

Note When you are done viewing the graph, it is a good idea to turn Plot1 off again to avoid errors in graphing or data points showing up in your other graphs. ■

Section 2.2

Example 6(a) (page 119) The following table shows annual expenditure on health in the U.S. from 1980 through 2009 ($t = 0$ represents 1980).

Year t	0	5	10	15	20	25	29
Expenditure ($ billion)	256	444	724	1,030	1,380	2,020	2,490

Find the exponential regression model

$$C(t) = Ab^t$$

for the annual expenditure.

Solution with Technology

This is very similar to Example 5 in Section 2.1 (see the Technology Guide for Section 2.1):

1. Use STAT EDIT to enter the table of values.

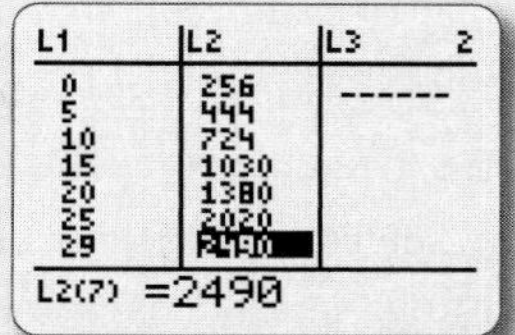

2. Press STAT, select CALC, and choose option #0 ExpReg. Pressing ENTER gives the exponential regression curve in the home screen:

$$C(t) \approx 296.25(1.0798)^t.$$ Coefficients rounded

3. To graph the points and regression line in the same window, turn Plot1 on (see the Technology Guide for Example 5 in Section 2.1) and enter the regression equation in the Y= screen by pressing Y=, clearing out whatever function is there, and pressing VARS 5 and selecting EQ option #1: RegEq. Then press ZOOM and choose option #9: ZoomStat to see the graph.

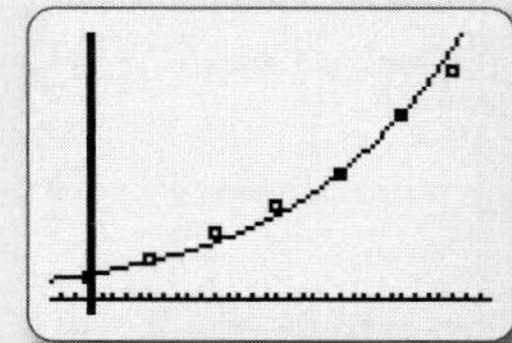

Note When you are done viewing the graph, it is a good idea to turn Plot1 off again to avoid errors in graphing or data points showing up in your other graphs. ■

Section 2.3

Example 5 (page 128) The following table shows the total spent on research and development by universities and colleges in the U.S., in billions of dollars, for the period 1998–2008 (t is the year since 1990).

Year t	8	9	10	11	12	13
Spending ($ billions)	27	29	31	33	36	38
Year t	14	15	16	17	18	
Spending ($ billions)	39	40	40	41	42	

Find the best-fit logarithmic model of the form

$$S(t) = A \ln t + C$$

and use the model to project total spending on research by universities and colleges in 2012, assuming the trend continues.

TECHNOLOGY GUIDE

Solution with Technology

This is very similar to Example 5 in Section 2.1 (see the Technology Guide for Section 2.1):

1. Use STAT EDIT to enter the table of values.
2. Press STAT, select CALC, and choose option #9 LnReg. Pressing ENTER gives the logarithmic regression curve in the home screen:

$$S(t) \approx 19.25 \ln t - 12.78. \qquad \text{Coefficients rounded}$$

3. To graph the points and regression line in the same window, turn Plot1 on (see the Technology Guide for Example 5 in Section 2.1) and enter the regression equation in the Y= screen by pressing Y=, clearing out whatever function is there, and pressing VARS 5 and selecting EQ option #1: RegEq. Then press ZOOM and choose option #9: ZoomStat to see the graph.

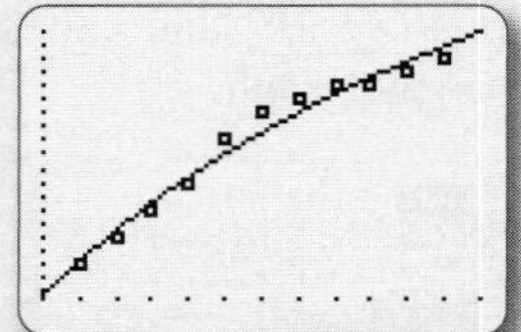

Section 2.4

Example 2 (page 133) The following table shows wired broadband penetration in the United States as a function of time t in years ($t = 0$ represents 2000).

Year (t)	2	3	4	5	6	7	8	9	10
Penetration (%) (P)	5.5	7.9	10.9	14.2	18.2	21.9	23.9	26.5	27.1

Find a logistic regression curve of the form

$$P(t) = \frac{N}{1 + Ab^{-t}}.$$

Solution with Technology

This is very similar to Example 5 in Section 2.1 (see the Technology Guide for Section 2.1):

1. Use STAT EDIT to enter the table of values.
2. Press STAT, select CALC, and choose option #B Logistic. Pressing ENTER gives the logistic regression curve in the home screen:

$$P(t) \approx \frac{29.842}{1 + 12.502e^{-0.49592t}}. \qquad \text{Coefficients rounded}$$

This is not exactly the form we are seeking, but we can convert it to that form by writing

$$e^{-0.49592t} = (e^{0.49592})^{-t} \approx 1.642^{-t}$$

so

$$P(t) \approx \frac{29.842}{1 + 12.502(1.642)^{-t}}.$$

3. To graph the points and regression line in the same window, turn Plot1 on (see the Technology Guide for Example 5 in Section 2.1) and enter the regression equation in the Y= screen by pressing Y=, clearing out whatever function is there, and pressing VARS 5 and selecting EQ option #1: RegEq. Then press ZOOM and choose option #9: ZoomStat to see the graph.

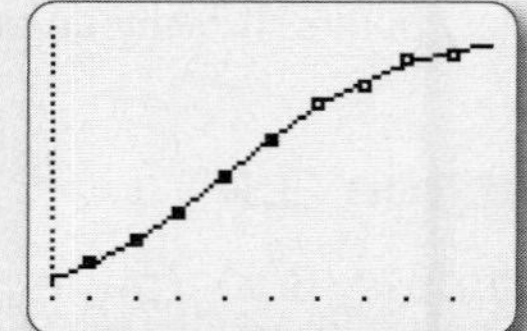

SPREADSHEET Technology Guide

Section 2.1

Example 2 (page 104) Sketch the graph of each quadratic function, showing the location of the vertex and intercepts.

a. $f(x) = 4x^2 - 12x + 9$

b. $g(x) = -\frac{1}{2}x^2 + 4x - 12$

Solution with Technology

We can set up a worksheet so that all we have to enter are the coefficients a, b, and c, and a range of x-values for the graph. Here is a possible layout that will plot 101 points using the coefficients for part (a).

1. First, we compute the x coordinates:

	A	B	C	D	E
1	x	y	a	4	
2	=$D$4		b	-12	
3	=A2+$D$6		c	9	
4			Xmin	-10	
5			Xmax	10	
6			Delta X	=(D5-D4)/100	
101					
102					

2. To add the y coordinates, we use the technology formula

```
a*x^2+b*x+c
```

replacing a, b, and c with (absolute) references to the cells containing their values.

	A	B	C	D
1	x	y	a	4
2	-10	=$D$1*A2^2+$D$2*A2+$D$3	b	-12
3	-9.8		c	9
4	-9.6		Xmin	-10
5	-9.4		Xmax	10
6	-9.2		Delta X	0.2
101				
102	10			

3. Graphing the data in columns A and B gives the graph shown here:

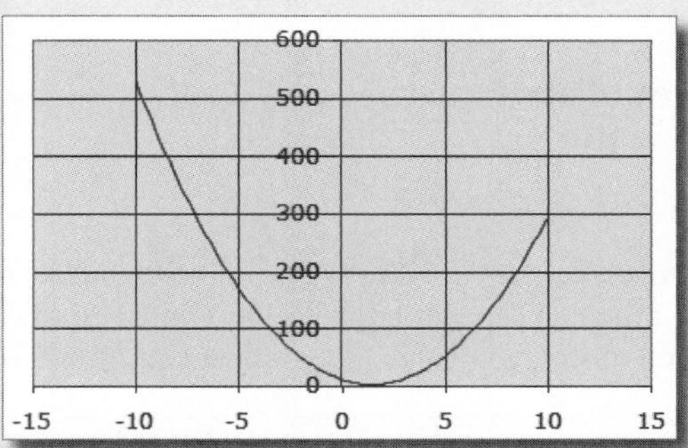

$$y = 4x^2 - 12x + 9$$

4. We can go further and compute the exact coordinates of the vertex and intercepts:

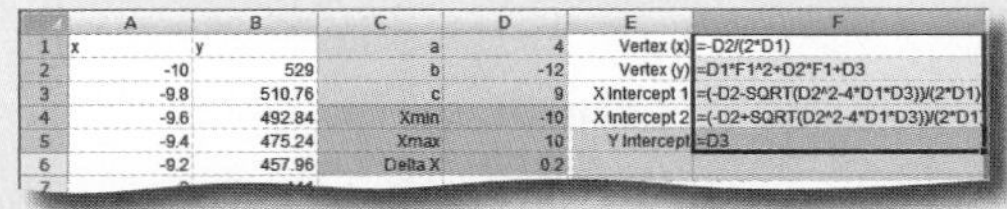

	A	B	C	D	E	F
1	x	y	a	4	Vertex (x)	=-D2/(2*D1)
2	-10	529	b	-12	Vertex (y)	=D1*F1^2+D2*F1+D3
3	-9.8	510.76	c	9	X Intercept 1	=(-D2-SQRT(D2^2-4*D1*D3))/(2*D1)
4	-9.6	492.84	Xmin	-10	X Intercept 2	=(-D2+SQRT(D2^2-4*D1*D3))/(2*D1
5	-9.4	475.24	Xmax	10	Y Intercept	=D3
6	-9.2	457.96	Delta X	0.2		

The completed sheet should look like this:

	A	B	C	D	E	F
1	x	y	a	4	Vertex (x)	1.5
2	-10	529	b	-12	Vertex (y)	0
3	-9.8	510.76	c	9	X Intercept 1	1.5
4	-9.6	492.84	Xmin	-10	X Intercept 2	1.5
5	-9.4	475.24	Xmax	10	Y Intercept	9
6	-9.2	457.96	Delta X	0.2		

We can now save this sheet as a template to handle all quadratic functions. For instance, to do part (b), we just change the values of a, b, and c in column D to $a = -1/2$, $b = 4$, and $c = -12$.

Example 5 (page 108) The following table shows the annual mean carbon dioxide concentration measured at Mauna Loa Observatory in Hawaii, in parts per million, every 10 years from 1960 through 2010 ($t = 0$ represents 1960).

Year t	0	10	20	30	40	50
PPM CO_2 C	317	326	339	354	369	390

Find the quadratic model

$$C(t) = at^2 + bt + c$$

that best fits the data.

Solution with Technology

As in Section 1.4, Example 2, we start with a scatter plot of the original data, and add a trendline:

1. Start with the original data and a "Scatter plot" (see Section 1.2 Example 5).

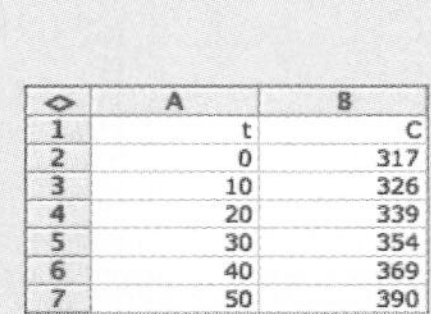

	A	B
1	t	C
2	0	317
3	10	326
4	20	339
5	30	354
6	40	369
7	50	390

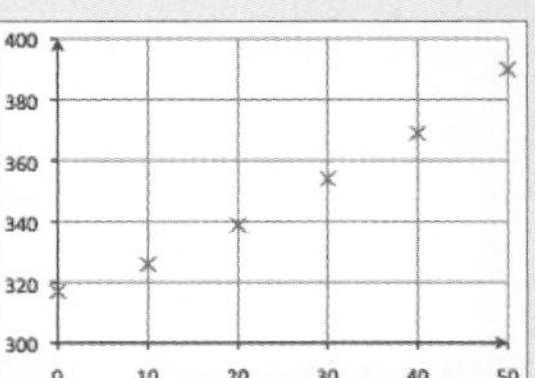

2. Add a quadratic trend line. (As of the time of this writing, among the common spreadsheets only Excel has the ability to add a polynomial trendline.) Right-click on any data point in the chart and select "Add Trendline," then select a "Polynomial" type of order 2 and check the option to "Display Equation on chart."

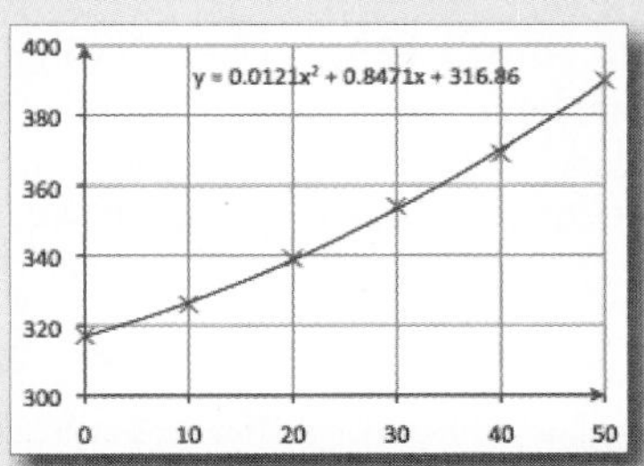

Section 2.2

Example 6(a) (page 119) The following table shows annual expenditure on health in the U.S. from 1980 through 2009 ($t = 0$ represents 1980).

Year t	0	5	10	15	20	25	29
Expenditure ($ billion)	256	444	724	1,030	1,380	2,020	2,490

Find the exponential regression model

$$C(t) = Ab^t$$

for the annual expenditure.

Solution with Technology

This is very similar to Example 5 in Section 2.1 (see the Technology Guide for Section 2.1):

1. Start with a "Scatter plot" of the observed data.
2. Add an exponential trendline:[11] The details vary from spreadsheet to spreadsheet—in OpenOffice, first double-click on the graph. Right-click on any data point in the chart and select "Add Trendline," then select an "Exponential" type and check the option to "Display Equation on chart."

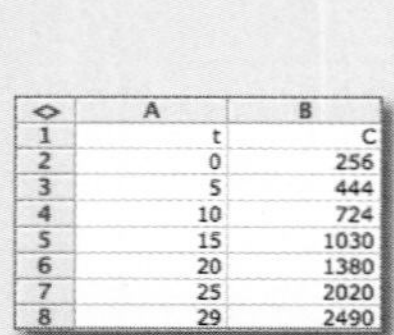

	A	B
1	t	C
2	0	256
3	5	444
4	10	724
5	15	1030
6	20	1380
7	25	2020
8	29	2490

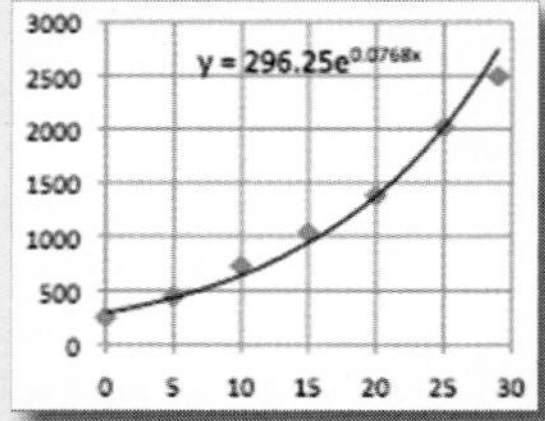

Notice that the regression curve is given in the form Ae^{kt} rather than Ab^t. To transform it, write

$$296.25e^{0.0768t} = 296.25(e^{0.0768})^t$$
$$\approx 296.25(1.0798)^t. \qquad e^{0.0768} \approx 1.0798$$

Section 2.3

Example 5 (page 128) The following table shows the total spent on research and development by universities and colleges in the U.S., in billions of dollars, for the period 1998–2008 (t is the year since 1990).

Year t	8	9	10	11	12	13
Spending ($ billions)	27	29	31	33	36	38
Year t	14	15	16	17	18	
Spending ($ billions)	39	40	40	41	42	

Find the best-fit logarithmic model of the form

$$S(t) = A \ln t + C$$

and use the model to project total spending on research by universities and colleges in 2012, assuming the trend continues.

Solution with Technology

This is very similar to Example 5 in Section 2.1 (see the Technology Guide for Section 2.1): We start, as usual, with a "Scatter plot" of the observed data and add a Logarithmic trendline. Here is the result:

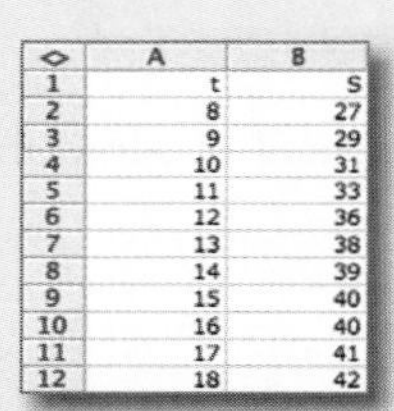

	A	B
1	t	S
2	8	27
3	9	29
4	10	31
5	11	33
6	12	36
7	13	38
8	14	39
9	15	40
10	16	40
11	17	41
12	18	42

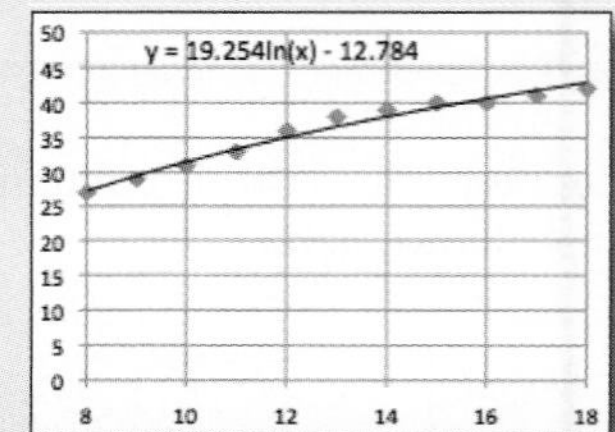

Section 2.4

Example 2 (page 133) The following table shows wired broadband penetration in the United States as a function of time t in years ($t = 0$ represents 2000).

Year (t)	2	3	4	5	6	7	8	9	10
Penetration (%) (P)	5.5	7.9	10.9	14.2	18.2	21.9	23.9	26.5	27.1

Find a logistic regression curve of the form

$$P(t) = \frac{N}{1 + Ab^{-t}}.$$

Solution with Technology

At the time of this writing, available spreadsheets did not have a built-in logistic regression calculation, so we use an alternative method that works for any type of regression curve. The Solver included with Windows versions of Excel and some Mac versions can find logistic regression curves while the Solver included with some other spreadsheets is not yet capable of this, so the instructions here are specific to Excel.

[11] At the time of this writing, Google Docs has no trendline feature for its spreadsheet.

1. First use rough estimates for N, A, and b, and compute the sum-of-squares error (SSE; see Section 1.4) directly:

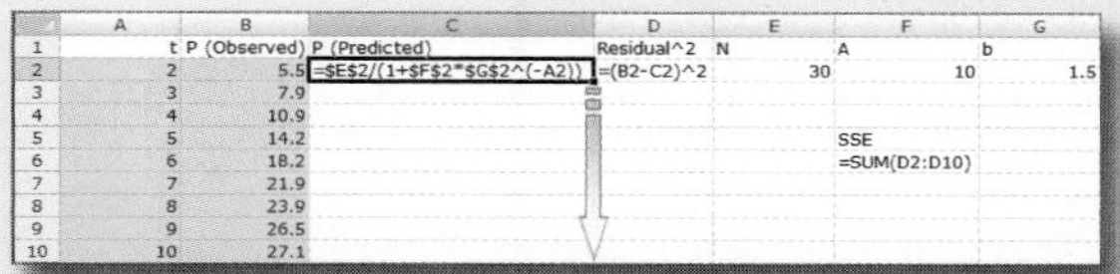

	A	B	C	D	E	F	G
1	t	P (Observed)	P (Predicted)	Residual^2	N	A	b
2	2	5.5	=$E$2/(1+$F$2*$G$2^(-A2))	=(B2-C2)^2	30	10	1.5
3	3	7.9					
4	4	10.9					
5	5	14.2				SSE	
6	6	18.2				=SUM(D2:D10)	
7	7	21.9					
8	8	23.9					
9	9	26.5					
10	10	27.1					

Cells E2:G2 contain our initial rough estimates of N, A, and b. For N, we used 30 (notice that the y-coordinates do appear to level off around 30). For A, we used the fact that the y-intercept is $N/(1+A)$ and the y-intercept appears to be approximately 3. In other words,

$$3 = \frac{30}{1+A}.$$

Because a very rough estimate is all we are after, using $A = 10$ will do just fine. For b, we chose 1.5 as the values of P appear to be increasing by around 50% per year initially (again, this is rough).

2. Cell C2 contains the formula for $P(t)$, and the square of the resulting residual is computed in D2.

3. Cell F6 will contain SSE. The completed spreadsheet should look like this:

	A	B	C	D	E	F	G
1	t	P (Observed)	P (Predicted)	Residual^2	N	A	b
2	2	5.5	5.510204082	0.000104123	30	10	1.5
3	3	7.9	7.570093458	0.108838326			
4	4	10.9	10.08298755	0.66750934			
5	5	14.2	12.94849023	1.566276702		SSE	
6	6	18.2	15.97516435	4.949893654		31.13446345	
7	7	21.9	18.92414191	8.855731375			
8	8	23.9	21.57987063	5.383000302			
9	9	26.5	23.80720074	7.251167845			
10	10	27.1	25.56639582	2.35194178			

The best-fit curve will result from values of N, A, and b that give a minimum value for SSE. We shall use Excel's "Solver," found in the "Analysis" group on the "Data" tab. (If "Solver" does not appear in the Analysis group, you will have to install the Solver Add-in using the Excel Options dialogue.) The Solver dialogue box with the necessary fields completed to solve the problem looks like this:

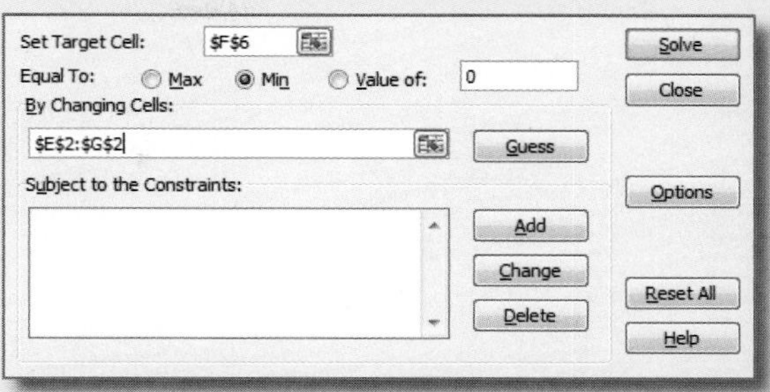

- The Target Cell refers to the cell that contains SSE.
- "Min" is selected because we are minimizing SSE.
- "Changing Cells" are obtained by selecting the cells that contain the current values of N, A, and b.

4. When you have filled in the values for the three items above, press "Solve" and tell Solver to Keep Solver Solution when done. You will find $N \approx 29.842$, $A \approx 12.501$, and $b \approx 1.642$ so that

$$P(t) \approx \frac{29.842}{1 + 12.501(1.642)^{-t}}.$$

If you use a scatter plot to graph the data in columns A, B and C, you will obtain the following graph:

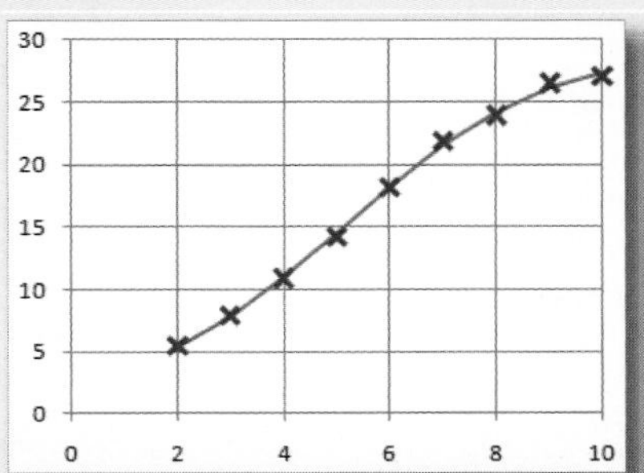

3

Introduction to the Derivative

Website
www.WanerMath.com
At the Website you will find:

- Section-by-section tutorials, including game tutorials with randomized quizzes
- A detailed chapter summary
- A true/false quiz
- Additional review exercises
- Graphers, Excel tutorials, and other resources
- The following extra topics:
  - Sketching the graph of the derivative
  - Continuity and differentiability

Case Study Reducing Sulfur Emissions

The Environmental Protection Agency (EPA) wants to formulate a policy that will encourage utilities to reduce sulfur emissions. Its goal is to reduce annual emissions of sulfur dioxide by a total of 10 million tons from the current level of 25 million tons by imposing a fixed charge for every ton of sulfur released into the environment per year. The EPA has some data showing the marginal cost to utilities of reducing sulfur emissions. As a consultant to the EPA, you must determine the amount to be charged per ton of sulfur emissions in light of these data.

Norbert Schaefer/CORBIS

Introduction

In the world around us, everything is changing. The mathematics of change is largely about the rate of change: how fast and in which direction the change is occurring. Is the Dow Jones average going up, and if so, how fast? If I raise my prices, how many customers will I lose? If I launch this missile, how fast will it be traveling after two seconds, how high will it go, and where will it come down?

We have already discussed the concept of rate of change for linear functions (straight lines), where the slope measures the rate of change. But this works only because a straight line maintains a constant rate of change along its whole length. Other functions rise faster here than there—or rise in one place and fall in another—so that the rate of change varies along the graph. The first achievement of calculus is to provide a systematic and straightforward way of calculating (hence the name) these rates of change. To describe a changing world, we need a language of change, and that is what calculus is.

The history of calculus is an interesting story of personalities, intellectual movements, and controversy. Credit for its invention is given to two mathematicians: Isaac Newton (1642–1727) and Gottfried Leibniz (1646–1716). Newton, an English mathematician and scientist, developed calculus first, probably in the 1660s. We say "probably" because, for various reasons, he did not publish his ideas until much later. This allowed Leibniz, a German mathematician and philosopher, to publish his own version of calculus first, in 1684. Fifteen years later, stirred up by nationalist fervor in England and on the continent, controversy erupted over who should get the credit for the invention of calculus. The debate got so heated that the Royal Society (of which Newton and Leibniz were both members) set up a commission to investigate the question. The commission decided in favor of Newton, who happened to be president of the society at the time. The consensus today is that both mathematicians deserve credit because they came to the same conclusions working independently. This is not really surprising: Both built on well-known work of other people, and it was almost inevitable that someone would put it all together at about that time.

algebra Review
For this chapter, you should be familiar with the algebra reviewed in **Chapter 0, Section 2.**

3.1 Limits: Numerical and Graphical Viewpoints

Rates of change are calculated by derivatives, but an important part of the definition of the derivative is something called a **limit**. Arguably, much of mathematics since the eighteenth century has revolved around understanding, refining, and exploiting the idea of the limit. The basic idea is easy, but getting the technicalities right is not.

Evaluating Limits Numerically

Start with a very simple example: Look at the function $f(x) = 2 + x$ and ask: What happens to $f(x)$ as x approaches 3? The following table shows the value of $f(x)$ for values of x close to and on either side of 3:

x approaching 3 from the left → ← x approaching 3 from the right

x	2.9	2.99	2.999	2.9999	3	3.0001	3.001	3.01	3.1
$f(x) = 2 + x$	4.9	4.99	4.999	4.9999		5.0001	5.001	5.01	5.1

We have left the entry under 3 blank to emphasize that when calculating the limit of $f(x)$ as x *approaches* 3, we are not interested in its value when x *equals* 3.

Notice from the table that the closer x gets to 3 from either side, the closer $f(x)$ gets to 5. We write this as

$$\lim_{x\to 3} f(x) = 5. \qquad \text{The limit of } f(x), \text{ as } x \text{ approaches 3, equals 5.}$$

Q: *Why all the fuss? Can't we simply substitute $x = 3$ and avoid having to use a table?*

A: This happens to work for *some* functions, but not for *all* functions. The following example illustrates this point.

EXAMPLE 1 Estimating a Limit Numerically

Use a table to estimate the following limits:

a. $\displaystyle\lim_{x\to 2} \frac{x^3-8}{x-2}$ **b.** $\displaystyle\lim_{x\to 0} \frac{e^{2x}-1}{x}$

Solution

a. We cannot simply substitute $x = 2$, because the function $f(x) = \dfrac{x^3-8}{x-2}$ is not defined at $x = 2$. (Why?)* Instead, we use a table of values as we did above, with x approaching 2 from both sides.

* However, if you factor $x^3 - 8$, you will find that $f(x)$ can be simplified to a function that *is* defined at $x = 2$. This point will be discussed (and this example redone) in Section 3.3. The function in part (b) cannot be simplified by factoring.

x approaching 2 from the left → ← x approaching 2 from the right

x	1.9	1.99	1.999	1.9999	2	2.0001	2.001	2.01	2.1
$f(x) = \dfrac{x^3-8}{x-2}$	11.41	11.9401	11.9940	11.9994		12.0006	12.0060	12.0601	12.61

We notice that as x approaches 2 from either side, $f(x)$ appears to be approaching 12. This suggests that the limit is 12, and we write

$$\lim_{x\to 2} \frac{x^3-8}{x-2} = 12.$$

b. The function $g(x) = \dfrac{e^{2x}-1}{x}$ is not defined at $x = 0$ (nor can it even be simplified to one which *is* defined at $x = 0$). In the following table, we allow x to approach 0 from both sides:

x approaching 0 from the left → ← x approaching 0 from the right

x	−0.1	−0.01	−0.001	−0.0001	0	0.0001	0.001	0.01	0.1
$g(x) = \dfrac{e^{2x}-1}{x}$	1.8127	1.9801	1.9980	1.9998		2.0002	2.0020	2.0201	2.2140

The table suggests that $\displaystyle\lim_{x\to 0} \frac{e^{2x}-1}{x} = 2.$

using Technology

To automate the computations in Example 1 using a graphing calculator or a spreadsheet, see the Technology Guides at the end of the chapter. Outline for part (a):

TI-83/84 Plus

Home screen: `Y1=(X^3-8)/(X-2)`

[2ND] [TBLSET] `Indpnt` set to `Ask`

[2ND] [TABLE] Enter some values of x from the example: 1.9, 1.99, 1.999 . . .

[More details on page 213.]

Spreadsheet
Headings x, $f(x)$ in A1–B1 and again in C1–D1.
In A2–A5 enter 1.9, 1.99, 1.999, 1.9999.
In C1–C5 enter 2.1, 2.01, 2.001, 2.0001. Enter
`=(A2^3-8)/(A2-2)`
in B2 and copy down to B5. Copy and paste the same formula in D2–D5.
[More details on page 215.]

Website
www.WanerMath.com
In the Function Evaluator and Grapher, enter `(x^3-8)/(x-2)` for y_1. For a table of values, enter the various x-values in the Evaluator box, and press "Evaluate".

Before we go on... Although the table *suggests* that the limit in Example 1 part (b) is 2, it by no means establishes that fact conclusively. It is *conceivable* (though not in fact the case here) that putting $x = 0.000000087$ could result in, say, $g(x) = 426$. Using a table can only *suggest* a value for the limit. In the next two sections we shall discuss algebraic techniques to allow us to actually *calculate* limits. ■

Before we continue, let us make a more formal definition.

Definition of a Limit

If $f(x)$ approaches the number L as x approaches (but is not equal to) a from both sides, then we say that $f(x)$ **approaches L as $x \to a$** ("x approaches a") or that the **limit** of $f(x)$ as $x \to a$ is L. More precisely, *we can make $f(x)$ be as close to L as we like by choosing any x sufficiently close to (but not equal to) a on either side.* We write

$$\lim_{x \to a} f(x) = L$$

or

$$f(x) \to L \text{ as } x \to a.$$

If $f(x)$ *fails* to approach *a single fixed number* as x approaches a from both sides, then we say that $f(x)$ **has no limit** as $x \to a$, or

$$\lim_{x \to a} f(x) \textbf{ does not exist.}$$

Quick Examples

1. $\lim_{x \to 3}(2 + x) = 5$ — See discussion before Example 1.
2. $\lim_{x \to -2}(3x) = -6$ — As x approaches -2, $3x$ approaches -6.
3. $\lim_{x \to 0}(x^2 - 2x + 1)$ exists. — In fact, the limit is 1.
4. $\lim_{x \to 5} \frac{1}{x} = \frac{1}{5}$ — As x approaches 5, $\frac{1}{x}$ approaches $\frac{1}{5}$.
5. $\lim_{x \to 2} \frac{x^3 - 8}{x - 2} = 12$ — See Example 1. (We cannot just put $x = 2$ here.)

(For examples where the limit does not exist, see Example 2.)

Notes

1. It is important that $f(x)$ approach the same number as x approaches a from *both sides*. For instance, if $f(x)$ approaches 5 for $x = 1.9, 1.99, 1.999, \ldots$, but approaches 4 for $x = 2.1, 2.01, 2.001, \ldots$, then the limit as $x \to 2$ does not exist. (See Example 2 for such a situation.)
2. It may happen that $f(x)$ does not approach any fixed number at all as $x \to a$ from either side. In this case, we also say that the limit does not exist.
3. If a happens to be an endpoint of the domain of f, then the function is only defined on one side of a, and so the limit as $x \to a$ does not exist. For example, the natural domain of $f(x) = \sqrt{x}$ is $[0, +\infty)$, so the limit of $f(x)$ as $x \to 0$ does not exist. The appropriate kind of limit to consider in such a case is a **one-sided limit** (see Example 2). ■

The next example gives instances in which a stated limit does not exist.

EXAMPLE 2 Limits Do Not Always Exist

Do the following limits exist?

a. $\lim_{x \to 0} \frac{1}{x^2}$ **b.** $\lim_{x \to 0} \frac{|x|}{x}$ **c.** $\lim_{x \to 2} \frac{1}{x-2}$ **d.** $\lim_{x \to 1} \sqrt{x-1}$

Solution

a. Here is a table of values for $f(x) = \frac{1}{x^2}$, with x approaching 0 from both sides.

x approaching 0 from the left → ← *x* approaching 0 from the right

x	−0.1	−0.01	−0.001	−0.0001	0	0.0001	0.001	0.01	0.1
$f(x) = \frac{1}{x^2}$	100	10,000	1,000,000	100,000,000		100,000,000	1,000,000	10,000	100

The table shows that as x gets closer to zero on either side, $f(x)$ gets larger and larger **without bound**—that is, if you name any number, no matter how large, $f(x)$ will be even larger than that if x is sufficiently close to 0. Because $f(x)$ is not approaching any real number, we conclude that $\lim_{x \to 0} \frac{1}{x^2}$ does not exist. Because $f(x)$ is becoming arbitrarily large, we also say that $\lim_{x \to 0} \frac{1}{x^2}$ **diverges to** $+\infty$, or just

$$\lim_{x \to 0} \frac{1}{x^2} = +\infty.$$

Note This is not meant to imply that the limit exists; the symbol $+\infty$ does not represent any real number. We write $\lim_{x \to a} f(x) = +\infty$ to indicate two things: (1) The limit does not exist and (2) the function gets large without bound as x approaches a. ■

b. Here is a table of values for $f(x) = \frac{|x|}{x}$, with x approaching 0 from both sides.

x approaching 0 from the left → ← *x* approaching 0 from the right

x	−0.1	−0.01	−0.001	−0.0001	0	0.0001	0.001	0.01	0.1
$f(x) = \frac{\lvert x \rvert}{x}$	−1	−1	−1	−1		1	1	1	1

The table shows that $f(x)$ does not approach the same limit as x approaches 0 from both sides. There appear to be two *different* limits: the limit as we approach 0 from the left and the limit as we approach from the right. We write

$$\lim_{x \to 0^-} f(x) = -1$$

read as "the limit as x approaches 0 from the left (or from below) is −1" and

$$\lim_{x \to 0^+} f(x) = 1$$

read as "the limit as x approaches 0 from the right (or from above) is 1." These are called the **one-sided limits** of $f(x)$. In order for f to have a **two-sided limit**, the two one-sided limits must be equal. Because they are not, we conclude that $\lim_{x \to 0} f(x)$ does not exist.

c. Near $x = 2$, we have the following table of values for $f(x) = \dfrac{1}{x-2}$:

	x approaching 2 from the left →					← x approaching 2 from the right			
x	1.9	1.99	1.999	1.9999	2	2.0001	2.001	2.01	2.1
$f(x) = \dfrac{1}{x-2}$	−10	−100	−1,000	−10,000		10,000	1,000	100	10

Because $f(x)$ is approaching no (single) real number as $x \to 2$, we see that $\lim_{x\to 2} \dfrac{1}{x-2}$ does not exist. Notice also that $\dfrac{1}{x-2}$ diverges to $+\infty$ as $x \to 2$ from the positive side (right half of the table) and to $-\infty$ as $x \to 2$ from the left (left half of the table). In other words,

$$\lim_{x\to 2^-} \frac{1}{x-2} = -\infty$$

$$\lim_{x\to 2^+} \frac{1}{x-2} = +\infty$$

$$\lim_{x\to 2} \frac{1}{x-2} \text{ does not exist.}$$

d. The natural domain of $f(x) = \sqrt{x-1}$ is $[1, +\infty)$, as $f(x)$ is defined only when $x \geq 1$. Thus we cannot evaluate $f(x)$ if x is to the left of 1. This means that

$$\lim_{x\to 1^-} \sqrt{x-1} \text{ does not exist,}$$

and our table looks like this:

		← x approaching 1 from the right				
x	1	1.00001	1.0001	1.001	1.01	1.1
$f(x) = \sqrt{x-1}$		0.0032	0.0100	0.0316	0.1000	0.3162

The values suggest that

$$\lim_{x\to 1^+} \sqrt{x-1} = 0.$$

In fact, we can obtain this limit by substituting $x = 1$ in the formula for $f(x)$ (see the comments after the example).

Q: *In Example 2(d) (and also in some of the Quick Examples before that) we could find a limit of an algebraically specified function by simply substituting the value of x in the formula for f(x). Does this always work?*

A: Short answer: Yes, when it makes sense. If the function is specified by a *single* algebraic formula and if $x = a$ is in the domain of *f*, then the limit can be obtained by substituting. We will say more about this when we discuss the algebraic approach to limits in Section 3.3. Remember, however, that, by definition, the limit of a function as $x \to a$ has nothing to do with its value at $x = a$, but rather its values for *x close to a*.

Q: *If f(x) is undefined when x = a, then the limit does not exist, right?*

A: Wrong. If $f(a)$ is not defined, then the limit may or may not exist. Example 1 shows instances where the limit *does* exist, and Example 2 shows instances where it does not. Again, the limit of a function as $x \to a$ has nothing to do with its value at $x = a$, but rather its values for x *close to a*.

In another useful kind of limit, we let x approach either $+\infty$ or $-\infty$, by which we mean that we let x get arbitrarily large or let x become an arbitrarily large negative number. The next example illustrates this.

EXAMPLE 3 Limits at Infinity

Use a table to estimate: **a.** $\lim_{x \to +\infty} \dfrac{2x^2 - 4x}{x^2 - 1}$ and **b.** $\lim_{x \to -\infty} \dfrac{2x^2 - 4x}{x^2 - 1}$.

Solution

a. By saying that x is "approaching $+\infty$," we mean that x is getting larger and larger without bound, so we make the following table:

x approaching $+\infty$ →

x	10	100	1,000	10,000	100,000
$f(x) = \dfrac{2x^2 - 4x}{x^2 - 1}$	1.6162	1.9602	1.9960	1.9996	2.0000

(Note that we are only approaching $+\infty$ from the left because we can hardly approach it from the right!) What seems to be happening is that $f(x)$ is approaching 2. Thus we write

$$\lim_{x \to +\infty} f(x) = 2.$$

b. Here, x is approaching $-\infty$, so we make a similar table, this time with x assuming negative values of greater and greater magnitude (read this table from right to left):

← x approaching $-\infty$

x	−100,000	−10,000	−1,000	−100	−10
$f(x) = \dfrac{2x^2 - 4x}{x^2 - 1}$	2.0000	2.0004	2.0040	2.0402	2.4242

Once again, $f(x)$ is approaching 2. Thus, $\lim_{x \to -\infty} f(x) = 2$.

Estimating Limits Graphically

We can often estimate a limit from a graph, as the next example shows.

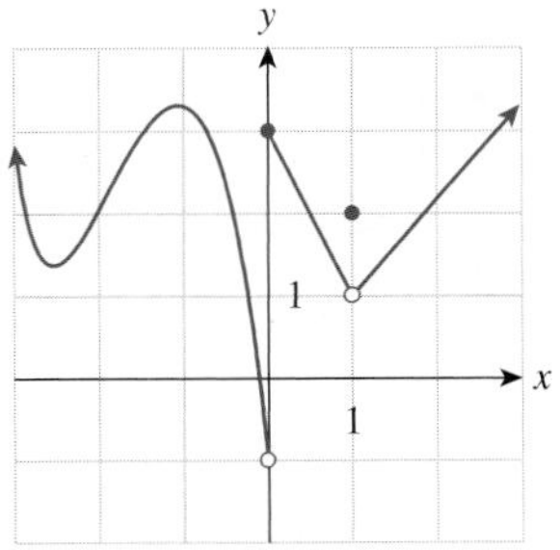

EXAMPLE 4 Estimating Limits Graphically

The graph of a function f is shown in Figure 1. (Recall that the solid dots indicate points on the graph, and the hollow dots indicate points not on the graph.) From the graph, analyze the following limits.

Figure 1

a. $\lim_{x \to -2} f(x)$ **b.** $\lim_{x \to 0} f(x)$ **c.** $\lim_{x \to 1} f(x)$ **d.** $\lim_{x \to +\infty} f(x)$

Solution Since we are given only a graph of f, we must analyze these limits graphically.

a. Imagine that Figure 1 was drawn on a graphing calculator equipped with a trace feature that allows us to move a cursor along the graph and see the coordinates as we go. To simulate this, place a pencil point on the graph to the left of $x = -2$, and move it along the curve so that the x-coordinate approaches -2. (See Figure 2.) We evaluate the limit numerically by noting the behavior of the y-coordinates.*

However, we can see directly from the graph that the y-coordinate approaches 2. Similarly, if we place our pencil point to the right of $x = -2$ and move it to the left, the y coordinate will approach 2 from that side as well (Figure 3). Therefore, as x approaches -2 from either side, $f(x)$ approaches 2, so

$$\lim_{x \to -2} f(x) = 2.$$

* For a visual animation of this process, look at the online tutorial for this section at the Website.

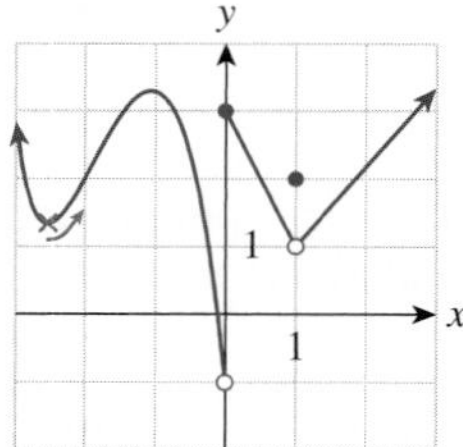

Figure 2

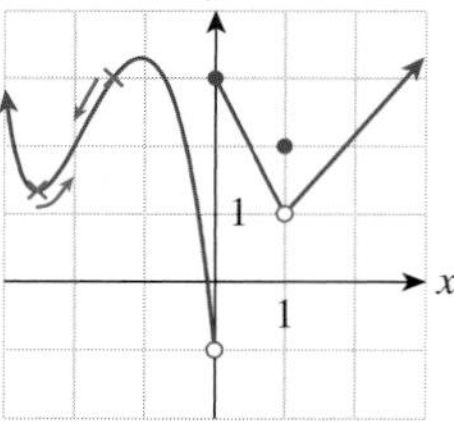

Figure 3

b. This time we move our pencil point toward $x = 0$. Referring to Figure 4, if we start from the left of $x = 0$ and approach 0 (by moving right), the y-coordinate approaches -1. However, if we start from the right of $x = 0$ and approach 0 (by moving left), the y-coordinate approaches 3. Thus (see Example 2),

$$\lim_{x \to 0^-} f(x) = -1$$

and

$$\lim_{x \to 0^+} f(x) = 3.$$

Because these limits are not equal, we conclude that

$$\lim_{x \to 0} f(x) \text{ does not exist.}$$

In this case there is a "break" in the graph at $x = 0$, and we say that the function is **discontinuous** at $x = 0$ (see Section 3.2).

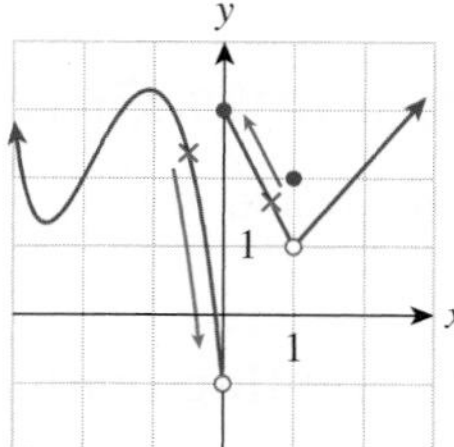

Figure 4

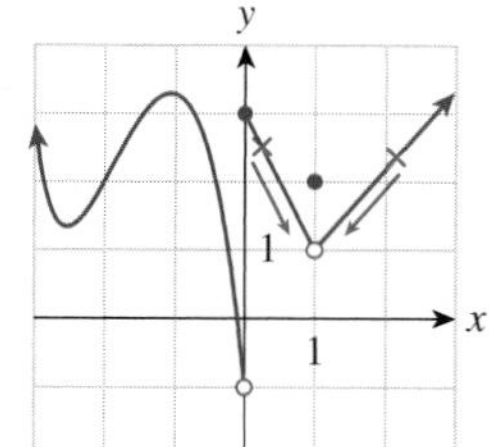

Figure 5

c. Once more we think about a pencil point moving along the graph with the x-coordinate this time approaching $x = 1$ from the left and from the right (Figure 5).

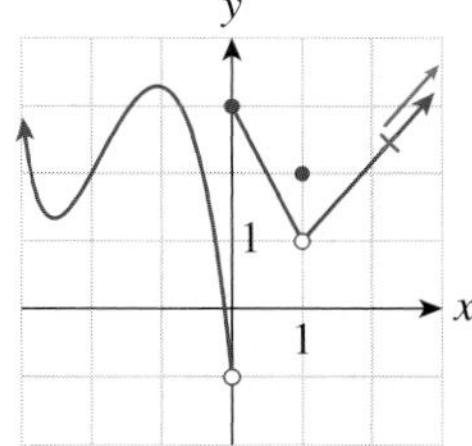

Figure 6

As the x-coordinate of the point approaches 1 from either side, the y-coordinate approaches 1 also. Therefore,

$$\lim_{x \to 1} f(x) = 1.$$

d. For this limit, x is supposed to approach infinity. We think about a pencil point moving along the graph further and further to the right as shown in Figure 6.

As the x-coordinate gets larger, the y-coordinate also gets larger and larger without bound. Thus, $f(x)$ diverges to $+\infty$:

$$\lim_{x \to +\infty} f(x) = +\infty.$$

Similarly,

$$\lim_{x \to -\infty} f(x) = +\infty.$$

Before we go on... In Example 4(c) $\lim_{x \to 1} f(x) = 1$ but $f(1) = 2$ (why?). Thus, $\lim_{x \to 1} f(x) \neq f(1)$. In other words, the limit of $f(x)$ as x *approaches* 1 is not the same as the value of f *at* $x = 1$. Always keep in mind that when we evaluate a limit as $x \to a$, *we do not care about the value of the function at* $x = a$. We only care about the value of $f(x)$ as x *approaches* a. In other words, $f(a)$ may or may not equal $\lim_{x \to a} f(x)$. ■

Here is a summary of the graphical method we used in Example 4, together with some additional information:

Evaluating Limits Graphically

To decide whether $\lim_{x \to a} f(x)$ exists and to find its value if it does:

1. Draw the graph of $f(x)$ by hand or with graphing technology.
2. Position your pencil point (or the Trace cursor) on a point of the graph to the right of $x = a$.
3. Move the point *along the graph* toward $x = a$ from the right and read the y-coordinate as you go. The value the y-coordinate approaches (if any) is the limit $\lim_{x \to a^+} f(x)$.
4. Repeat Steps 2 and 3, this time starting from a point on the graph to the left of $x = a$, and approaching $x = a$ along the graph from the left. The value the y-coordinate approaches (if any) is $\lim_{x \to a^-} f(x)$.
5. If the left and right limits both exist and have the same value L, then $\lim_{x \to a} f(x) = L$. Otherwise, the limit does not exist. The value $f(a)$ has no relevance whatsoever.
6. To evaluate $\lim_{x \to +\infty} f(x)$, move the pencil point toward the far right of the graph and estimate the value the y-coordinate approaches (if any). For $\lim_{x \to -\infty} f(x)$, move the pencil point toward the far left.
7. If $x = a$ happens to be an endpoint of the domain of f, then only a one-sided limit is possible at $x = a$. For instance, if the domain is $(-\infty, 4]$, then $\lim_{x \to 4^-} f(x)$ may exist, but neither $\lim_{x \to 4^+} f(x)$ nor $\lim_{x \to 4} f(x)$ exists.

In the next example we use both the numerical and graphical approaches.

EXAMPLE 5 Infinite Limit

Does $\lim_{x \to 0^+} \frac{1}{x}$ exist?

Solution

Numerical Method Because we are asked for only the right-hand limit, we need only list values of x approaching 0 from the right.

← x approaching 0 from the right

x	0	0.0001	0.001	0.01	0.1
$f(x) = \frac{1}{x}$		10,000	1,000	100	10

What seems to be happening as x approaches 0 from the right is that $f(x)$ is increasing without bound, as in Example 4(d). That is, if you name any number, no matter how large, $f(x)$ will be even larger than that if x is sufficiently close to zero. Thus, the limit diverges to $+\infty$, so

$$\lim_{x \to 0^+} \frac{1}{x} = +\infty$$

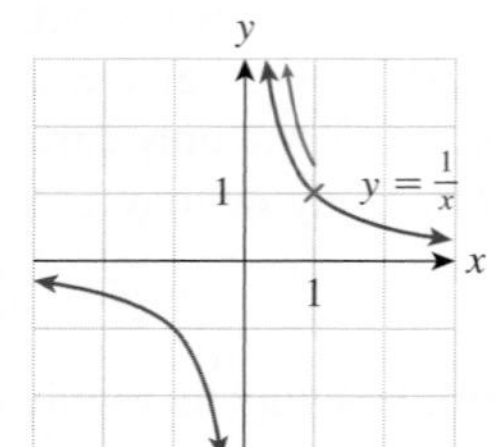

Figure 7

Graphical Method Recall that the graph of $f(x) = \frac{1}{x}$ is the standard hyperbola shown in Figure 7. The figure also shows the pencil point moving so that its x-coordinate approaches 0 from the right. Because the point moves along the graph, it is forced to go higher and higher. In other words, its y-coordinate becomes larger and larger, approaching $+\infty$. Thus, we conclude that

$$\lim_{x \to 0^+} \frac{1}{x} = +\infty.$$

➡ **Before we go on...** In Example 5 you should check that

$$\lim_{x \to 0^-} \frac{1}{x} = -\infty. \qquad \frac{1}{x} \text{ diverges to } -\infty \text{ as } x \to 0^-.$$

Also, check that

$$\lim_{x \to +\infty} \frac{1}{x} = \lim_{x \to -\infty} \frac{1}{x} = 0. \blacksquare$$

APPLICATION

EXAMPLE 6 Broadband Penetration

Wired broadband penetration in the United States can be modeled by

$$P(t) = \frac{29.842}{1 + 12.502(1.642)^{-t}} \quad (t \geq 0),$$

where t is time in years since 2000.[1]

[1] See Example 2 in Section 2.4. Broadband penetration is the number of broadband installations divided by the total population. Source for data: Organization for Economic Cooperation and Development (OECD) Directorate for Science, Technology, and Industry, table of Historical Penetration Rates, June 2010, downloaded April 2011 from www.oecd.org/sti/ict/broadband.

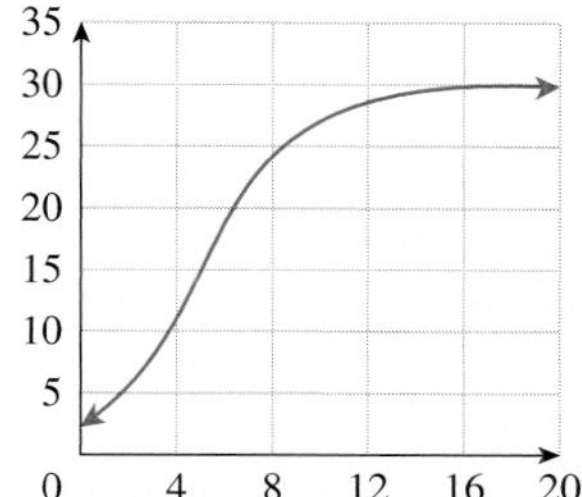

Figure 8

a. Estimate $\lim_{t\to+\infty} P(t)$ and interpret the answer.

b. Estimate $\lim_{t\to 0^+} P(t)$ and interpret the answer.

Solution

a. Figure 8 shows a plot of $P(t)$ for $0 \le t \le 20$. Using either the numerical or the graphical approach, we find

$$\lim_{t\to+\infty} P(t) = \lim_{t\to+\infty} \frac{29.842}{1 + 12.502(1.642)^{-t}} \approx 30.$$

(The actual limit is 29.842. Why?) Thus, in the long term (as t gets larger and larger), broadband penetration in the United States is expected to approach 30%; that is, the number of installations is expected to approach 30% of the total population.

b. The limit here is

$$\lim_{t\to 0^+} P(t) = \lim_{t\to 0^+} \frac{29.842}{1 + 12.502(1.642)^{-t}} \approx 2.21.$$

(Notice that in this case, we can simply put $t = 0$ to evaluate this limit.) Thus, the closer t gets to 0 (representing 2000) from the right, the closer $P(t)$ gets to 2.21%, meaning that, in 2000, broadband penetration was about 2.2% of the population.

FAQs

Determining when a Limit Does or Does Not Exist

Q: *If I substitute $x = a$ in the formula for a function and find that the function is defined there, it means that $\lim_{x\to a} f(x)$ exists and equals $f(a)$, right?*

A: Correct, provided the function is specified by *a single algebraic formula* and is not, say, piecewise-defined. We shall say more about this in the next two sections.

Q: *If I substitute $x = a$ in the formula for a function and find that the function is* not *defined there, it means that $\lim_{x\to a} f(x)$ does not exist, right?*

A: Wrong. The limit may still exist, as in Example 1, or may not exist, as in Example 2. In general, whether or not $\lim_{x\to a} f(x)$ exists has nothing to do with $f(a)$, but rather the values of f when x is *very close to, but not equal to a.*

Q: *Is there a quick and easy way of telling from a graph whether $\lim_{x\to a} f(x)$ exists?*

A: Yes. If you cover up the portion of the graph corresponding to $x = a$ and it appears as though the visible part of the graph could be made into a continuous curve by filling in a suitable point at $x = a$, then the limit exists. (The "suitable point" need not be $(a, f(a))$.) Otherwise, it does not. Try this method with the curves in Example 4.

3.1 EXERCISES

Access end-of-section exercises online at **www.webassign.net**

3.2 Limits and Continuity

In the last section we saw examples of graphs that had various kinds of "breaks" or "jumps." For instance, in Example 4 we looked at the graph in Figure 9. This graph appears to have breaks, or **discontinuities**, at $x = 0$ and at $x = 1$. At $x = 0$ we saw that $\lim_{x\to 0} f(x)$ does not exist because the left- and right-hand limits are not the same. Thus, the discontinuity at $x = 0$ seems to be due to the fact that the limit does not exist there. On the other hand, at $x = 1$, $\lim_{x\to 1} f(x)$ *does* exist (it is equal to 1), but is not equal to $f(1) = 2$.

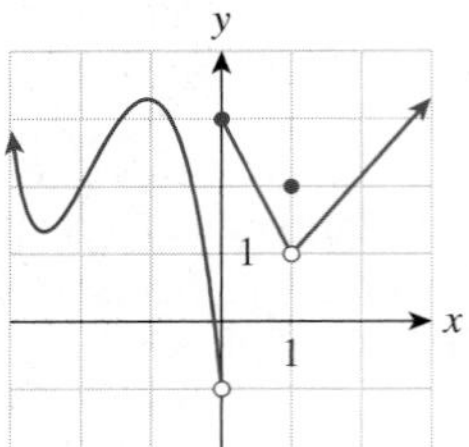

Figure 9

Thus, we have identified two kinds of discontinuity:

1. Points where the limit of the function does not exist.

   $x = 0$ in Figure 9 because $\lim_{x\to 0} f(x)$ does not exist.

2. Points where the limit exists but does not equal the value of the function.

   $x = 1$ in Figure 9 because $\lim_{x\to 1} f(x) = 1 \neq f(1)$

On the other hand, there is no discontinuity at, say, $x = -2$, where we find that $\lim_{x\to -2} f(x)$ exists and equals 2 and $f(-2)$ is also equal to 2. In other words,

$$\lim_{x\to -2} f(x) = 2 = f(-2).$$

The point $x = -2$ is an example of a point where f is **continuous**. (Notice that you can draw the portion of the graph near $x = -2$ without lifting your pencil from the paper.) Similarly, f is continuous at *every* point other than $x = 0$ and $x = 1$. Here is the mathematical definition.

Continuous Function

Let f be a function and let a be a number in the domain of f. Then f is **continuous at a** if

a. $\lim_{x\to a} f(x)$ exists, and

b. $\lim_{x\to a} f(x) = f(a)$.

If the number a is an endpoint of the domain, we will understand that the limit is the left or right limit, as appropriate.

The function f is said to be **continuous on its domain** if it is continuous at each point in its domain.

If f is not continuous at a particular a in its domain, we say that f is **discontinuous** at a or that f has a **discontinuity** at a. Thus, a discontinuity can occur at $x = a$ if either

a. $\lim_{x\to a} f(x)$ does not exist, or

b. $\lim_{x\to a} f(x)$ exists but is not equal to $f(a)$.

Quick Examples

1. The function shown in Figure 9 is continuous at $x = -1$ and $x = 2$. It is discontinuous at $x = 0$ and $x = 1$, and so is not continuous on its domain.
2. The function $f(x) = x^2$ is continuous on its domain. (Think of its graph, which contains no breaks.)
3. The function shown in Figure 10 is continuous on its domain. In particular, it is continuous at the left endpoint $x = -1$ of its domain, because $\lim_{x\to -1^+} f(x) = 1 = f(-1)$ (recall that we use only a one-sided limit at an endpoint of the domain).

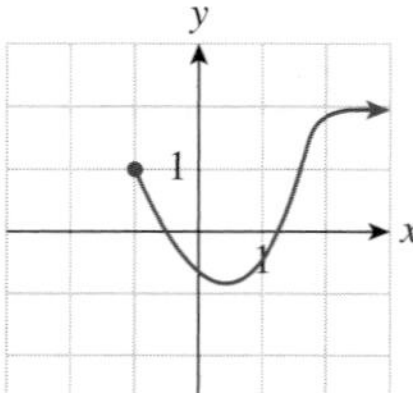

Figure 10

4. The function f whose graph is shown on the left in the following figure is continuous on its domain. (Although the graph breaks at $x = 2$, that is not a point of its domain.) The function g whose graph is shown on the right is not continuous on its domain because it has a discontinuity at $x = 2$. (Here, $x = 2$ is a point of the domain of g.)

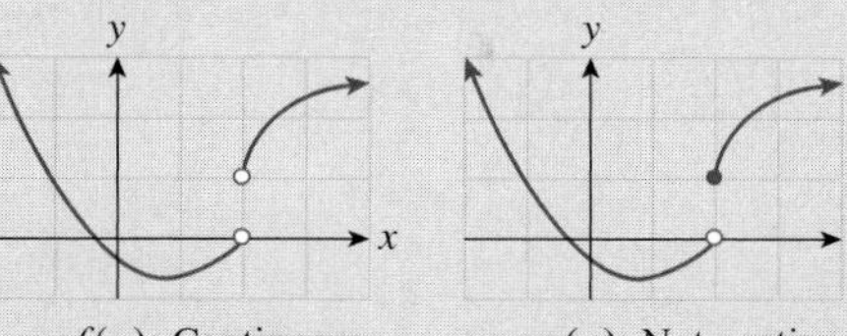

$y = f(x)$: Continuous on its domain

$y = g(x)$: Not continuous on its domain

Note Continuity and discontinuity of a function are defined only for points in a function's domain; a function cannot be continuous at a point not in its domain, and it cannot be discontinuous there either. So, if a is not in the domain of f—that is, if $f(a)$ is not defined—then it is meaningless to talk about whether f is continuous or discontinuous at a. ■

EXAMPLE 1 Continuous and Discontinuous Functions

Which of the following functions are continuous on their domains?

a. $h(x) = \begin{cases} x + 3 & \text{if } x \le 1 \\ 5 - x & \text{if } x > 1 \end{cases}$

b. $k(x) = \begin{cases} x + 3 & \text{if } x \le 1 \\ 1 - x & \text{if } x > 1 \end{cases}$

c. $f(x) = \dfrac{1}{x}$

d. $g(x) = \begin{cases} \dfrac{1}{x} & \text{if } x \ne 0 \\ 0 & \text{if } x = 0 \end{cases}$

Solution

a and **b.** The graphs of h and k are shown in Figure 11.

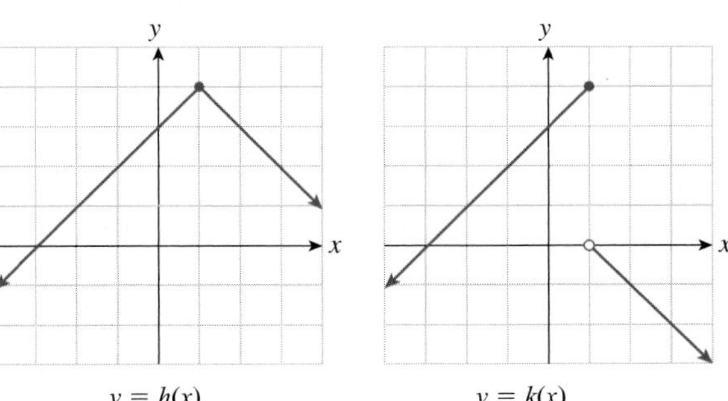

$y = h(x)$ $y = k(x)$

Figure 11

Even though the graph of h is made up of two different line segments, it is continuous at every point of its domain, including $x = 1$ because

$$\lim_{x\to 1} h(x) = 4 = h(1).$$

On the other hand, $x = 1$ is also in the domain of k, but $\lim_{x\to 1} k(x)$ does not exist. Thus, k is discontinuous at $x = 1$ and thus not continuous on its domain.

using Technology

We can use technology to draw (approximate) graphs of the functions in Example 1(a), (b), and (c). Here are the technology formulas that will work for the TI-83/84 Plus, spreadsheets, and Website function evaluator and grapher. (In the TI-83/84 Plus, replace <= by ≤. In spreadsheets, replace x by a cell reference and insert an equals sign in front of the formula.)

a. `(x+3)*(x<=1)`
`+(5-x)*(x>1)`

b. `(x+3)*(x<=1)`
`+(1-x)*(x>1)`

c. `(1/x)`

Observe in each case how technology handles the breaks in the curves.

c and **d.** The graphs of f and g are shown in Figure 12.

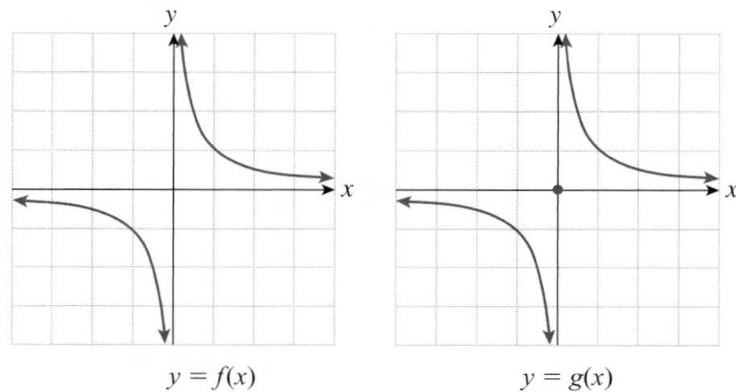

Figure 12

The domain of f consists of all real numbers except 0 and f is continuous at all such numbers. (Notice that 0 is not in the domain of f, so the question of continuity at 0 does not arise.) Thus, f is continuous on its domain.

The function g, on the other hand, has its domain expanded to include 0, so we now need to check whether g is continuous at 0. From the graph, it is easy to see that g is discontinuous there because $\lim_{x\to 0} g(x)$ does not exist. Thus, g is not continuous on its domain because it is discontinuous at 0.

➡ Before we go on...

Q: *Wait a minute! How can a function like $f(x) = 1/x$ be continuous when its graph has a break in it?*

A: We are not claiming that f is continuous *at every real number.* What we are saying is that f is continuous *on its domain;* the break in the graph occurs at a point not in the domain of f. In other words, f is continuous on the set of all nonzero real numbers; it is not continuous on the set of *all* real numbers because it is not even defined on that set.

■

EXAMPLE 2 Continuous Except at a Point

In each case, say what, if any, value of $f(a)$ would make f continuous at a.

a. $f(x) = \dfrac{x^3 - 8}{x - 2}$; $a = 2$ **b.** $f(x) = \dfrac{e^{2x} - 1}{x}$; $a = 0$ **c.** $f(x) = \dfrac{|x|}{x}$; $a = 0$

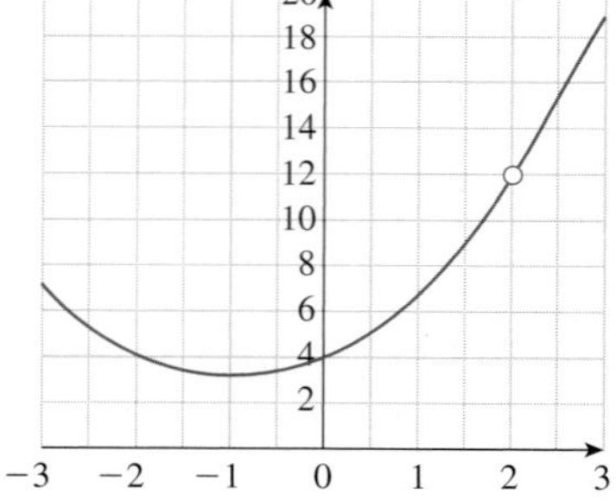

Figure 13

Solution

a. In Figure 13 we see the graph of $f(x) = \dfrac{x^3 - 8}{x - 2}$. The point corresponding to $x = 2$ is missing because f is not (yet) defined there. (Your graphing utility will probably miss this subtlety and render a continuous curve. See the technology note in the margin.) To turn f into a function that is continuous at $x = 2$, we need to "fill in the gap" so as to obtain a continuous curve. Since the graph suggests that the missing point is (2, 12), let us define $f(2) = 12$.

Does f now become continuous if we take $f(2) = 12$? From the graph, or Example 1(a) of Section 3.1,

$$\lim_{x\to 2} f(x) = \lim_{x\to 2} \frac{x^3 - 8}{x - 2} = 12,$$

which is now equal to $f(2)$. Thus, $\lim_{x\to 2} f(x) = f(2)$, showing that f is now continuous at $x = 2$.

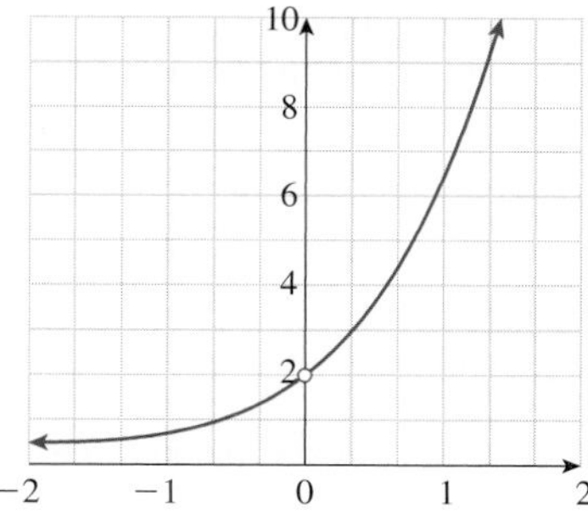

Figure 14

using Technology

It is instructive to see how technology handles the functions in Example 2. Here are the technology formulas that will work for the TI-83/84 Plus, spreadsheets, and Website function evaluator and grapher. (In spreadsheets, replace `x` by a cell reference and insert an equal sign in front of the formula.)

a. `(x^3-8)/(x-2)`

b. `(e^(2x)-1)/x`

Spreadsheet:
`=(exp(2*A2)-1)/A2`

c. `abs(x)/x`

In each case, compare the graph rendered by technology with the corresponding figure in Example 2.

b. In Example 1(b) of the preceding section, we saw that

$$\lim_{x\to 0} f(x) = \lim_{x\to 0} \frac{e^{2x}-1}{x} = 2,$$

and so, as in part (a), we must define $f(0) = 2$. This is confirmed by the graph, shown in Figure 14.

c. We considered the function $f(x) = |x|/x$ in Example 2 in Section 3.1. Its graph is shown in Figure 15.

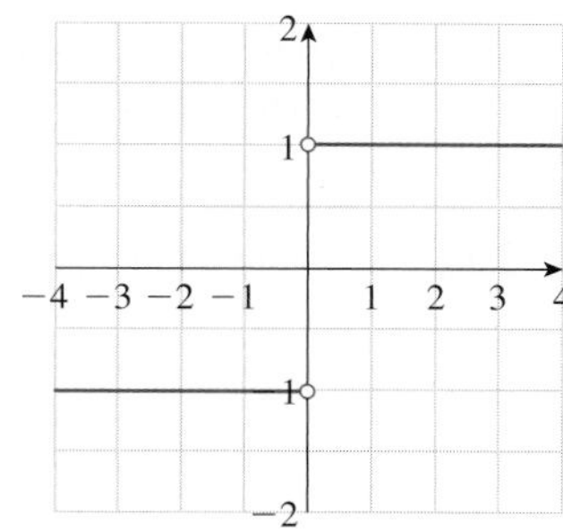

Figure 15

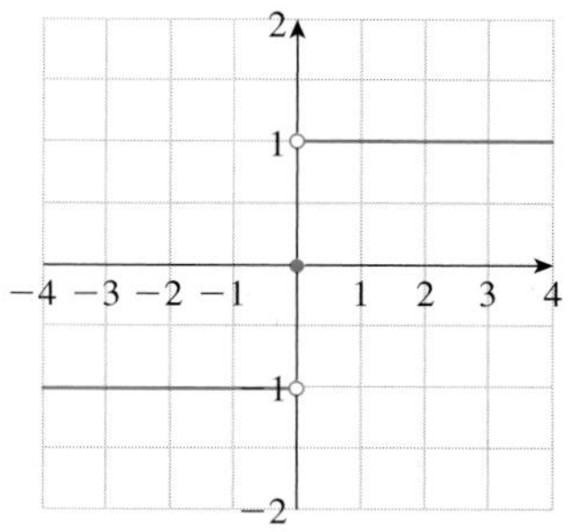

Figure 16

Now we encounter a problem: No matter how we try to fill in the gap at $x = 0$, the result will be a discontinuous function. For example, setting $f(0) = 0$ will result in the discontinuous function shown in Figure 16. We conclude that it is impossible to assign any value to $f(0)$ to turn f into a function that is continuous at $x = 0$.

We can also see this result algebraically: In Example 2 of Section 3.1, we saw that $\lim_{x\to 0} \frac{|x|}{x}$ does not exist. Thus, the resulting function will fail to be continuous at 0, no matter how we define $f(0)$.

A function not defined at an isolated point is said to have a **singularity** at that point. The function in part (a) of Example 2 has a singularity at $x = 2$, and the functions in parts (b) and (c) have singularities at $x = 0$. The functions in parts (a) and (b) have **removable singularities** because we can make these functions continuous at $x = a$ by properly defining $f(a)$. The function in part (c) has an **essential singularity** because we cannot make f continuous at $x = a$ just by defining $f(a)$ properly.

3.2 EXERCISES

Access end-of-section exercises online at **www.webassign.net** ENHANCED WebAssign

3.3 Limits and Continuity: Algebraic Viewpoint

Although numerical and graphical estimation of limits is effective, the estimates these methods yield may not be perfectly accurate. The algebraic method, when it can be used, will always yield an exact answer. Moreover, algebraic analysis of a function often enables us to take a function apart and see "what makes it tick."

Let's start with the function $f(x) = 2 + x$ and ask: What happens to $f(x)$ as x approaches 3? To answer this algebraically, notice that as x gets closer and closer to 3, the quantity $2 + x$ must get closer and closer to $2 + 3 = 5$. Hence,

$$\lim_{x \to 3} f(x) = \lim_{x \to 3}(2 + x) = 2 + 3 = 5$$

Q: *Is that all there is to the algebraic method? Just substitute $x = a$?*

A: Under certain circumstances: Notice that by substituting $x = 3$ we *evaluated the function at $x = 3$*. In other words, we relied on the fact that

$$\lim_{x \to 3} f(x) = f(3).$$

In Section 3.2 we said that a function satisfying this equation is *continuous* at $x = 3$.

Thus,

If we know that the function f is continuous at a point a, we can compute $\lim_{x \to a} f(x)$ by simply substituting $x = a$ into $f(x)$.

To use this fact, we need to know how to recognize continuous functions when we see them. Geometrically, they are easy to spot: A function is continuous at $x = a$ if its graph has no break at $x = a$. Algebraically, a large class of functions are known to be continuous on their domains—those, roughly speaking, that are *specified by a single formula*.

We can be more precise: A **closed-form function** is any function that can be obtained by combining constants, powers of x, exponential functions, radicals, logarithms, absolute values, trigonometric functions (and some other functions we do not encounter in this text) into a *single* mathematical formula by means of the usual arithmetic operations and composition of functions. (They can be as complicated as we like.)

Closed-Form Functions

A function is **written in closed form** if it is specified by combining constants, powers of x, exponential functions, radicals, logarithms, absolute values, trigonometric functions (and some other functions we do not encounter in this text) into a *single* mathematical formula by means of the usual arithmetic operations and composition of functions. A **closed-form function** is any function that can be written in closed form.

Quick Examples

1. $3x^2 - |x| + 1$, $\dfrac{\sqrt{x^2 - 1}}{6x - 1}$, $e^{-\frac{4x^2-1}{x}}$, and $\sqrt{\log_3(x^2 - 1)}$ are written in closed form, so they are all closed-form functions.

2. $f(x) = \begin{cases} -1 & \text{if } x \le -1 \\ x^2 + x & \text{if } -1 < x \le 1 \\ 2 - x & \text{if } 1 < x \le 2 \end{cases}$ is not written in closed-form because $f(x)$ is not expressed by a *single* mathematical formula.*

* It is possible to rewrite some piecewise-defined functions in closed form (using a single formula), but not this particular function, so $f(x)$ is not a closed-form function.

What is so special about closed-form functions is the following theorem.

Theorem 3.1 Continuity of Closed-Form Functions

Every closed-form function is continuous on its domain. Thus, if f is a closed-form function and $f(a)$ is defined, then $\lim_{x\to a} f(x)$ exists, and equals $f(a)$. (When a is an endpoint of the domain of f, we understand the limit to be the appropriate one-sided limit.)

Quick Example

$f(x) = 1/x$ is a closed-form function, and its natural domain consists of all real numbers except 0. Thus, f is continuous at every nonzero real number. That is,

$$\lim_{x\to a} \frac{1}{x} = \frac{1}{a}$$

provided $a \neq 0$.

Mathematics majors spend a great deal of time studying the proof of this theorem. We ask you to accept it without proof.

EXAMPLE 1 Limit of a Closed-Form Function

Evaluate the following limits algebraically:

a. $\lim_{x\to 1} \frac{x^3 - 8}{x - 2}$ **b.** $\lim_{x\to 2} \frac{x^3 - 8}{x - 2}$.

Solution

a. First, notice that $(x^3 - 8)/(x - 2)$ is a closed-form function because it is specified by a single algebraic formula. Also, $x = 1$ is in the domain of this function. Therefore the theorem applies, and

$$\lim_{x\to 1} \frac{x^3 - 8}{x - 2} = \frac{1^3 - 8}{1 - 2} = 7.$$

b. Although $(x^3 - 8)/(x - 2)$ is a closed-form function, $x = 2$ is not in its domain. Thus, the theorem does not apply and we cannot obtain the limit by substitution. However—and this is the key to finding such limits—*some preliminary algebraic simplification will allow us to obtain a closed-form function with $x = 2$ in its domain*. To do this, notice first that the numerator can be factored as

$$x^3 - 8 = (x - 2)(x^2 + 2x + 4).$$

Thus,

$$\frac{x^3 - 8}{x - 2} = \frac{(x - 2)(x^2 + 2x + 4)}{x - 2} = x^2 + 2x + 4.$$

Once we have canceled the offending $(x - 2)$ in the denominator, we are left with a closed-form function *with 2 in its domain*. Thus,

$$\lim_{x\to 2} \frac{x^3 - 8}{x - 2} = \lim_{x\to 2}(x^2 + 2x + 4)$$

$$= 2^2 + 2(2) + 4 = 12. \quad \text{Substitute } x = 2.$$

This confirms the answer we found numerically in Example 1 in Section 3.1.

➡ **Before we go on...** Notice that in Example 1(b) before simplification, the substitution $x = 2$ yields

$$\frac{x^3 - 8}{x - 2} = \frac{8 - 8}{2 - 2} = \frac{0}{0}.$$

Worse than the fact that 0/0 is undefined, it also conveys absolutely no information as to what the limit might be. (The limit turned out to be 12!) We therefore call the expression 0/0 an **indeterminate form**. Once simplified, the function became $x^2 + 2x + 4$, which, upon the substitution $x = 2$, yielded 12—no longer an indeterminate form. In general, we have the following rule of thumb:

If the substitution $x = a$ yields the indeterminate form 0/0, try simplifying by the method in Example 1.

We will say more about indeterminate forms in Example 2. ■

Q: *There is something suspicious about Example 1(b). If 2 was not in the domain before simplifying but was in the domain after simplifying, we must have changed the function, right?*

A: Correct. In fact, when we said that

$$\frac{x^3 - 8}{x - 2} = x^2 + 2x + 4$$

Domain excludes 2 Domain includes 2

we were lying a little bit. What we really meant is that these two expressions are equal *where both are defined*. The functions $(x^3 - 8)/(x - 2)$ and $x^2 + 2x + 4$ are different functions. The difference is that $x = 2$ is not in the domain of $(x^3 - 8)/(x - 2)$ and is in the domain of $x^2 + 2x + 4$. Since $\lim_{x\to 2} f(x)$ explicitly *ignores* any value that f may have at 2, this does not affect the limit. From the point of view of the limit at 2, these functions *are* equal. In general we have the following rule.

Functions with Equal Limits

If $f(x) = g(x)$ for all x except possibly $x = a$, then

$$\lim_{x\to a} f(x) = \lim_{x\to a} g(x).$$

Quick Example

$$\frac{x^2 - 1}{x - 1} = x + 1 \text{ for all } x \text{ except } x = 1.$$

Write $\frac{x^2 - 1}{x - 1}$ as $\frac{(x + 1)(x - 1)}{x - 1}$ and cancel the $(x - 1)$.

Therefore,

$$\lim_{x\to 1} \frac{x^2 - 1}{x - 1} = \lim_{x\to 1}(x + 1) = 1 + 1 = 2.$$

Q: *How do we find $\lim_{x\to a} f(x)$ when $x = a$ is not in the domain of the function f and we cannot simplify the given function to make a a point of the domain?*

A: In such a case, it might be necessary to analyze the function by some other method, such as numerically or graphically. However, if we do not obtain the indeterminate form 0/0 upon substitution, we can often say what the limit is, as the following example shows.

EXAMPLE 2 Limit of a Closed-Form Function at a Point Not in Its Domain: The Determinate Form *k*/0

Evaluate the following limits, if they exist:

a. $\lim_{x\to 1^+} \dfrac{x^2 - 4x + 1}{x - 1}$ **b.** $\lim_{x\to 1} \dfrac{x^2 - 4x + 1}{x - 1}$ **c.** $\lim_{x\to 1} \dfrac{x^2 - 4x + 1}{x^2 - 2x + 1}$

Solution

a. Although the function $f(x) = \dfrac{x^2 - 4x + 1}{x - 1}$ is a closed-form function, $x = 1$ is not in its domain. Notice that substituting $x = 1$ gives

$$\frac{x^2 - 4x + 1}{x - 1} = \frac{1^2 - 4 + 1}{1 - 1} = \frac{-2}{0}$$

The **determinate** form $\frac{k}{0}$

which, although not defined, conveys important information to us: As x gets closer and closer to 1, the numerator approaches -2 and the denominator gets closer and closer to 0. Now, if we divide a number close to -2 by a number close to zero, we get a number of large absolute value; for instance,

$$\frac{-2.1}{0.0001} = -21{,}000 \quad \text{and} \quad \frac{-2.1}{-0.0001} = 21{,}000$$

$$\frac{-2.01}{0.00001} = -201{,}000 \quad \text{and} \quad \frac{-2.01}{-0.00001} = 201{,}000.$$

(Compare Example 5 in Section 3.1.) In our limit for part (a), x is approaching 1 from the right, so the denominator $x - 1$ is positive (as x is to the right of 1). Thus we have the scenario illustrated previously on the left, and we can conclude that

$$\lim_{x\to 1^+} \frac{x^2 - 4x + 1}{x - 1} = -\infty.$$

Think of this as $\frac{-2}{0^+} = -\infty$.

b. This time, x could be approaching 1 from either side. We already have, from part (a)

$$\lim_{x\to 1^+} \frac{x^2 - 4x + 1}{x - 1} = -\infty.$$

The same reasoning we used in part (a) gives

$$\lim_{x\to 1^-} \frac{x^2 - 4x + 1}{x - 1} = +\infty$$

Think of this as $\frac{-2}{0^-} = +\infty$.

because now the denominator is negative and still approaching zero while the numerator still approaches -2 and therefore is also negative. (See the numerical calculations above on the right.) Because the left and right limits do not agree, we conclude that

$$\lim_{x\to 1} \frac{x^2 - 4x + 1}{x - 1} \text{ does not exist.}$$

c. First notice that the denominator factors:

$$\lim_{x\to 1}\frac{x^2-4x+1}{x^2-2x+1}=\lim_{x\to 1}\frac{x^2-4x+1}{(x-1)^2}.$$

As x approaches 1, the numerator approaches -2 as before, and the denominator approaches 0. However, this time, the denominator $(x-1)^2$, being a square, is ≥ 0, regardless of from which side x is approaching 1. Thus, the entire function is negative as x approaches 1, and

$$\lim_{x\to 1}\frac{x^2-4x+1}{(x-1)^2}=-\infty. \qquad \frac{-2}{0^+}=-\infty$$

➡ **Before we go on...** In general, the determinate forms $\frac{k}{0^+}$ and $\frac{k}{0^-}$ will always yield $\pm\infty$, with the sign depending on the sign of the overall expression as $x \to a$. (When we write the form $\frac{k}{0}$ we always mean $k \neq 0$.) This and other determinate forms are discussed further after Example 4.

Figure 17 shows the graphs of $\frac{x^2-4x+1}{x-1}$ and $\frac{x^2-4x+1}{(x-1)^2}$ from Example 2. You should check that results we obtained above agree with a geometric analysis of these graphs near $x = 1$.

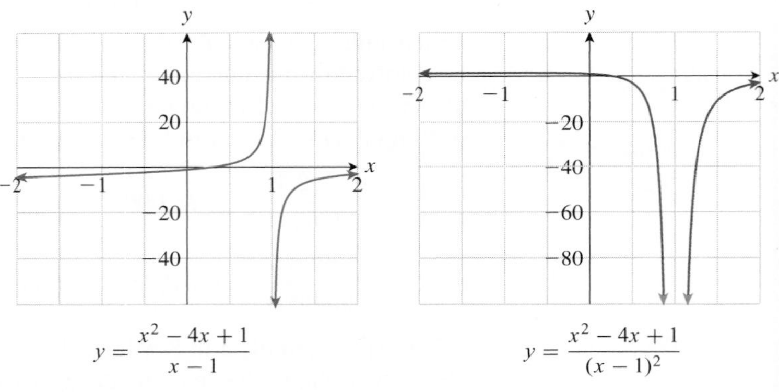

Figure 17

■

We can also use algebraic techniques to analyze functions that are not given in closed form.

EXAMPLE 3 Functions Not Written in Closed Form

For which values of x are the following piecewise defined functions continuous?

a. $f(x)=\begin{cases} x^2+2 & \text{if } x<1 \\ 2x-1 & \text{if } x\geq 1 \end{cases}$ **b.** $g(x)=\begin{cases} x^2-x+1 & \text{if } x\leq 0 \\ 1-x & \text{if } 0<x\leq 1 \\ x-3 & \text{if } x>1 \end{cases}$

Solution

a. The function $f(x)$ is given in closed form over the intervals $(-\infty, 1)$ and $[1, +\infty)$. At $x = 1$, $f(x)$ suddenly switches from one closed-form formula to another, so

$x = 1$ is the only place where there is a potential problem with continuity. To investigate the continuity of $f(x)$ at $x = 1$, let's calculate the limit there:

$$\begin{aligned} \lim_{x \to 1^-} f(x) &= \lim_{x \to 1^-} (x^2 + 2) && f(x) = x^2 + 2 \text{ for } x < 1. \\ &= (1)^2 + 2 = 3 && x^2 + 2 \text{ is closed-form.} \\ \lim_{x \to 1^+} f(x) &= \lim_{x \to 1^+} (2x - 1) && f(x) = 2x - 1 \text{ for } x > 1. \\ &= 2(1) - 1 = 1. && 2x - 1 \text{ is closed-form.} \end{aligned}$$

Because the left and right limits are different, $\lim_{x \to 1} f(x)$ does not exist, and so $f(x)$ is discontinuous at $x = 1$.

b. The only potential points of discontinuity for $g(x)$ occur at $x = 0$ and $x = 1$:

$$\lim_{x \to 0^-} g(x) = \lim_{x \to 0^-} (x^2 - x + 1) = 1$$

$$\lim_{x \to 0^+} g(x) = \lim_{x \to 0^+} (1 - x) = 1.$$

Thus, $\lim_{x \to 0} g(x) = 1$. Further, $g(0) = 0^2 - 0 + 1 = 1$ from the formula, and so

$$\lim_{x \to 0} g(x) = g(0),$$

which shows that $g(x)$ is continuous at $x = 0$. At $x = 1$ we have

$$\lim_{x \to 1^-} g(x) = \lim_{x \to 1^-} (1 - x) = 0$$

$$\lim_{x \to 1^+} g(x) = \lim_{x \to 1^+} (x - 3) = -2$$

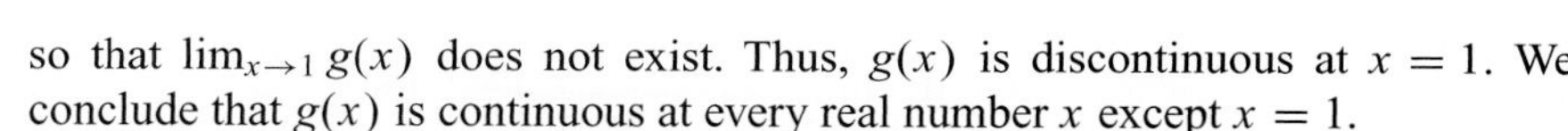

so that $\lim_{x \to 1} g(x)$ does not exist. Thus, $g(x)$ is discontinuous at $x = 1$. We conclude that $g(x)$ is continuous at every real number x except $x = 1$.

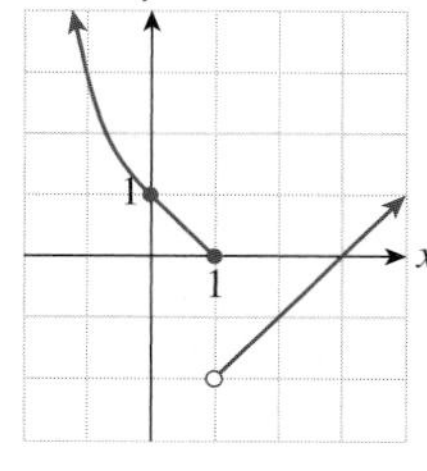

$y = g(x)$

Figure 18

➡ **Before we go on...** Figure 18 shows the graph of g from Example 3(b). Notice how the discontinuity at $x = 1$ shows up as a break in the graph, whereas at $x = 0$ the two pieces "fit together" at the point (0, 1). ■

Limits at Infinity

Let's look once again at some limits similar to those in Examples 3 and 6 in Section 3.1.

EXAMPLE 4 Limits at Infinity

Compute the following limits, if they exist:

a. $\displaystyle\lim_{x \to +\infty} \frac{2x^2 - 4x}{x^2 - 1}$ **b.** $\displaystyle\lim_{x \to -\infty} \frac{2x^2 - 4x}{x^2 - 1}$

c. $\displaystyle\lim_{x \to +\infty} \frac{-x^3 - 4x}{2x^2 - 1}$ **d.** $\displaystyle\lim_{x \to +\infty} \frac{2x^2 - 4x}{5x^3 - 3x + 5}$

e. $\displaystyle\lim_{t \to +\infty} (e^{0.1t} - 20)$ **f.** $\displaystyle\lim_{t \to +\infty} \frac{80}{1 + 2.2(3.68)^{-t}}$

Solution **a** and **b.** While calculating the values for the tables used in Example 3 in Section 3.1, you might have noticed that the highest power of x in both the numerator and denominator dominated the calculations. For instance, when $x = 100{,}000$, the term $2x^2$ in the numerator has the value of 20,000,000,000, whereas the term $4x$ has the comparatively insignificant value of 400,000. Similarly, the term x^2 in the denominator overwhelms the term -1. In other words, for large values of x (or negative values with large magnitude),

$$\frac{2x^2 - 4x}{x^2 - 1} \approx \frac{2x^2}{x^2} = 2.$$ Use only the highest powers top and bottom.

Therefore,

$$\lim_{x\to\pm\infty} \frac{2x^2 - 4x}{x^2 - 1} = \lim_{x\to\pm\infty} \frac{2x^2}{x^2} = \lim_{x\to\pm\infty} 2 = 2.$$

The procedure of using only the highest powers of x to compute the limit is stated formally and justified after this example.

c. Applying the previous technique of looking only at highest powers gives

$$\lim_{x\to+\infty} \frac{-x^3 - 4x}{2x^2 - 1} = \lim_{x\to+\infty} \frac{-x^3}{2x^2}$$ Use only the highest powers top and bottom.

$$= \lim_{x\to+\infty} \frac{-x}{2}.$$ Simplify.

As x gets large, $-x/2$ gets large in magnitude but negative, so the limit is

$$\lim_{x\to+\infty} \frac{-x}{2} = -\infty.$$ $\frac{-\infty}{2} = -\infty$ (See below.)

d. $$\lim_{x\to+\infty} \frac{2x^2 - 4x}{5x^3 - 3x + 5} = \lim_{x\to+\infty} \frac{2x^2}{5x^3}$$ Use only the highest powers top and bottom.

$$= \lim_{x\to+\infty} \frac{2}{5x}.$$

As x gets large, $2/(5x)$ gets close to zero, so the limit is

$$\lim_{x\to+\infty} \frac{2}{5x} = 0.$$ $\frac{2}{\infty} = 0$ (See below.)

e. Here we do not have a ratio of polynomials. However, we know that, as t becomes large and positive, so does $e^{0.1t}$, and hence also $e^{0.1t} - 20$. Thus,

$$\lim_{t\to+\infty} (e^{0.1t} - 20) = +\infty$$ $e^{+\infty} = +\infty$ (See below.)

f. As $t \to +\infty$, the term $(3.68)^{-t} = \frac{1}{3.68^t}$ in the denominator, being 1 divided by a very large number, approaches zero. Hence the denominator $1 + 2.2(3.68)^{-t}$ approaches $1 + 2.2(0) = 1$ as $t \to +\infty$. Thus,

$$\lim_{t\to+\infty} \frac{80}{1 + 2.2(3.68)^{-t}} = \frac{80}{1 + 2.2(0)} = 80$$ $(3.68)^{-\infty} = 0$ (See below.)

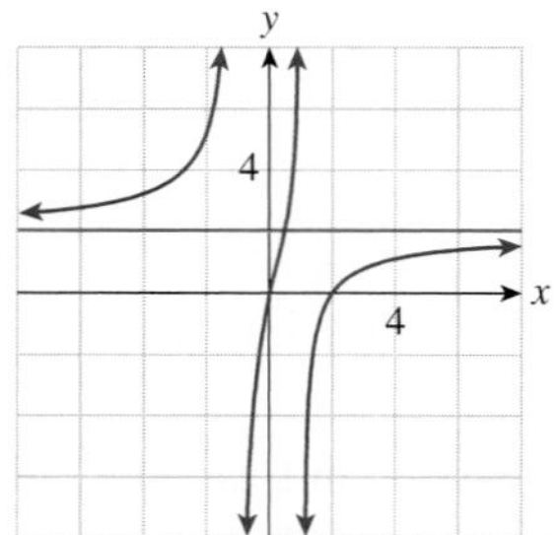

Figure 19

➡ **Before we go on...** Let's now look at the graph of the function $\frac{2x^2 - 4x}{x^2 - 1}$ in parts (a) and (b) of Example 4. We say that the graph of f has a **horizontal asymptote** at $y = 2$ because of the limits we have just calculated. This means that the graph approaches the horizontal line $y = 2$ far to the right or left (in this case, to both the right and left). Figure 19 shows the graph of f together with the line $y = 2$.

The graph reveals some additional interesting information: as $x \to 1^+$, $f(x) \to -\infty$, and as $x \to 1^-$, $f(x) \to +\infty$. Thus,

$$\lim_{x\to 1} f(x) \text{ does not exist.}$$

See if you can determine what happens as $x \to -1$.

If you graph the functions in parts (d) and (f) of Example 4, you will again see a horizontal asymptote. Do the limits in parts (c) and (e) show horizontal asymptotes? ■

It is worthwhile looking again at what we did in each of the limits in Example 4:

a and **b.** We saw that $\dfrac{2x^2-4x}{x^2-1} \approx \dfrac{2x^2}{x^2}$, and then we canceled the x^2. Notice that, before we cancel, letting x approach $\pm\infty$ in the numerator and denominator yields the ratio $\dfrac{\infty}{\infty}$, which, like $\dfrac{0}{0}$, is another *indeterminate form*, and indicates to us that further work is needed—in this case cancellation—before we can write down the limit.

c. We obtained $\dfrac{-x^3-4x}{2x^2-1} \approx \dfrac{-x^3}{2x^2}$, which results in another indeterminate form, $\dfrac{-\infty}{\infty}$, as $x \to +\infty$. Cancellation of the x^2 gave us $\dfrac{-x}{2}$, resulting in the *determinate* form $\dfrac{-\infty}{2} = -\infty$ (a very large number divided by 2 is again a very large number).

d. Here, $\dfrac{2x^2-4x}{5x^3-3x+5} \approx \dfrac{2x^2}{5x^3} = \dfrac{2}{5x}$, and the cancellation step turns the indeterminate form $\dfrac{\infty}{\infty}$ into the determinate form $\dfrac{2}{\infty} = 0$ (dividing 2 by a very large number yields a very small number).

e. We reasoned that e raised to a large positive number is large and positive. Putting $t = +\infty$ gives us the determinate form $e^{+\infty} = +\infty$.

f. Here we reasoned that 3.68 raised to a large *negative* number is close to zero. Putting $t = -\infty$ gives us the determinate form $3.68^{-\infty} = 1/3.68^{+\infty} = 1/\infty = 0$ (see (d)).

In parts (a)–(d) of Example 4, $f(x)$ was a **rational function**: a quotient of polynomial functions. We calculated the limit of $f(x)$ at $\pm\infty$ by ignoring all powers of x in both the numerator and denominator except for the largest. Following is a theorem that justifies this procedure.

Theorem 3.2 Evaluating the Limit of a Rational Function at $\pm\infty$

If $f(x)$ has the form

$$f(x) = \frac{c_n x^n + c_{n-1}x^{n-1} + \cdots + c_1 x + c_0}{d_m x^m + d_{m-1}x^{m-1} + \cdots + d_1 x + d_0}$$

with the c_i and d_i constants ($c_n \neq 0$ and $d_m \neq 0$), then we can calculate the limit of $f(x)$ as $x \to \pm\infty$ by ignoring all powers of x except the highest in both the numerator and denominator. Thus,

$$\lim_{x\to\pm\infty} f(x) = \lim_{x\to\pm\infty} \frac{c_n x^n}{d_m x^m}.$$

Quick Examples

(See Example 4.)

1. $\lim_{x\to+\infty} \frac{2x^2-4x}{x^2-1} = \lim_{x\to+\infty} \frac{2x^2}{x^2} = \lim_{x\to+\infty} 2 = 2$

2. $\lim_{x\to+\infty} \frac{-x^3-4x}{2x^2-1} = \lim_{x\to+\infty} \frac{-x^3}{2x^2} = \lim_{x\to+\infty} \frac{-x}{2} = -\infty$

3. $\lim_{x\to+\infty} \frac{2x^2-4x}{5x^3-3x+5} = \lim_{x\to+\infty} \frac{2x^2}{5x^3} = \lim_{x\to+\infty} \frac{2}{5x} = 0$

Proof Our function $f(x)$ is a polynomial of degree n divided by a polynomial of degree m. If n happens to be larger than m, then dividing the top and bottom by the largest power x^n of x gives

$$\begin{aligned} f(x) &= \frac{c_nx^n + c_{n-1}x^{n-1} + \cdots + c_1x + c_0}{d_mx^m + d_{m-1}x^{m-1} + \cdots + d_1x + d_0} \\ &= \frac{c_nx^n/x^n + c_{n-1}x^{n-1}/x^n + \cdots + c_1x/x^n + c_0/x^n}{d_mx^m/x^n + d_{m-1}x^{m-1}/x^n + \cdots + d_1x/x^n + d_0/x^n}. \end{aligned}$$

Canceling powers of x in each term and remembering that $n > m$ leaves us with

$$f(x) = \frac{c_n + c_{n-1}/x + \cdots + c_1/x^{n-1} + c_0/x^n}{d_m/x^{n-m} + d_{m-1}/x^{n-m+1} + \cdots + d_1/x^{n-1} + d_0/x^n}.$$

As $x \to \pm\infty$, all the terms shown in red approach 0, so we can ignore them in taking the limit. (The first term in the denominator happens to approach 0 as well, but we retain it for convenience.) Thus,

$$\lim_{x\to\pm\infty} f(x) = \lim_{x\to\pm\infty} \frac{c_n}{d_m/x^{n-m}} = \lim_{x\to\pm\infty} \frac{c_nx^n}{d_mx^m},$$

as required. The cases when n is smaller than m and $m = n$ are proved similarly by dividing top and bottom by the largest power of x in each case.

Some Determinate and Indeterminate Forms

The following table brings these ideas together with our observations in Example 2.

Some Determinate and Indeterminate Forms

$\frac{0}{0}$ and $\pm\frac{\infty}{\infty}$ are **indeterminate**; evaluating limits in which these arise requires simplification or further analysis.†

The following are **determinate** forms for any nonzero number k:

$\frac{k}{0^\pm} = \pm\infty$	$\frac{k}{\text{Small}} = \text{Big*}$ (See Example 2.)
$k(\pm\infty) = \pm\infty$	$k \times \text{Big} = \text{Big*}$
$k \pm \infty = \pm\infty$	$k \pm \text{Big} = \pm\text{Big}$

† Some other indeterminate forms are: $\pm\infty \cdot 0$, $\infty - \infty$, and 1^∞. (These are not discussed in this text, but see the Communication and Reasoning exercises for this section.)

$$\pm\frac{\infty}{k} = \pm\infty \qquad \frac{\text{Big}}{k} = \text{Big}^*$$

$$\pm\frac{k}{\infty} = 0 \qquad \frac{k}{\text{Big}} = \text{Small}$$

and, if $k > 1$, then

$$k^{+\infty} = +\infty \qquad k^{\text{Big positive}} = \text{Big}$$

$$k^{-\infty} = 0. \qquad k^{\text{Big negative}} = \text{Small}$$

*The sign gets switched in these forms if k is negative.

Quick Examples

1. $\lim_{x\to 0} \frac{60}{2x^2} = +\infty$ $\qquad \frac{k}{0^+} = +\infty$
2. $\lim_{x\to -1^-} \frac{2x-6}{x+1} = +\infty$ $\qquad \frac{-8}{0^-} = +\infty$
3. $\lim_{x\to -\infty} 3x - 5 = -\infty$ $\qquad 3(-\infty) - 5 = -\infty - 5 = -\infty$
4. $\lim_{x\to +\infty} \frac{2x}{60} = +\infty$ $\qquad \frac{2(\infty)}{60} = \infty$
5. $\lim_{x\to -\infty} \frac{60}{2x} = 0$ $\qquad \frac{60}{2(-\infty)} = 0$
6. $\lim_{x\to +\infty} \frac{60x}{2x} = 30$ $\qquad \frac{\infty}{\infty}$ is indeterminate but we can cancel.
7. $\lim_{x\to -\infty} \frac{60}{e^x - 1} = \frac{60}{0-1} = -60$ $\qquad e^{-\infty} = 0$

FAQs

Strategy for Evaluating Limits Algebraically

Q: *Is there a systematic way to evaluate a limit $\lim_{x\to a} f(x)$ algebraically?*

A: The following approach is often successful:

Case 1: *a* Is a Finite Number (Not $\pm\infty$)

1. Decide whether f is a closed-form function. If it is not, then find the left and right limits at the values of x where the function changes from one formula to another.
2. If f *is* a closed-form function, try substituting $x = a$ in the formula for $f(x)$. Then one of the following three things may happen:

   $f(a)$ is defined. Then $\lim_{x\to a} f(x) = f(a)$.

   $f(a)$ is not defined and has the indeterminate form 0/0. Try to simplify the expression for f to cancel one of the terms that gives 0.

   $f(a)$ is not defined and has one of the determinate forms listed above in the above table. Use the table to determine the limit as in the Quick Examples.

Case 2: $a = \pm\infty$

Remember that we can use the determinate forms $k^{+\infty} = \infty$ and $k^{-\infty} = 0$ if $k > 1$. Further, if the given function is a polynomial or ratio of polynomials, use the technique of Example 4: Focus only on the highest powers of x and then simplify to obtain either a number L, in which case the limit exists and equals L, or one of the determinate forms $\pm\infty/k = \pm\infty$ or $\pm k/\infty = 0$.

There is another technique for evaluating certain difficult limits, called *l'Hospital's rule,* but this uses derivatives, so we'll have to wait to discuss it until Section 4.1.

3.3 EXERCISES

Access end-of-section exercises online at **www.webassign.net**

3.4 Average Rate of Change

Calculus is the mathematics of change, inspired largely by observation of continuously changing quantities around us in the real world. As an example, the Consumer Price Index (CPI) C increased from 211 points in January 2009 to 220 points in January 2011.[2] As we saw in Chapter 1, the **change** in this index can be measured as the difference:

$$\Delta C = \text{Second value} - \text{First value} = 220 - 211 = 9 \text{ points.}$$

(The fact that the CPI increased is reflected in the positive sign of the change.) The kind of question we will concentrate on is *how fast* the CPI was changing. Because C increased by 9 points in 2 years, we say it averaged a $9/2 = 4.5$ point rise each year. (It actually rose 5 points the first year and 4 the second, giving an average rise of 4.5 points each year.)

Alternatively, we might want to measure this rate in points per month rather than points per year. Because C increased by 9 points in 24 months, it increased at an average rate of $9/24 = 0.375$ points per month.

In both cases, we obtained the average rate of change by dividing the change by the corresponding length of time:

$$\text{Average rate of change} = \frac{\text{Change in } C}{\text{Change in time}} = \frac{9}{2} = 4.50 \text{ points per year}$$

$$\text{Average rate of change} = \frac{\text{Change in } C}{\text{Change in time}} = \frac{9}{24} = 0.375 \text{ points per month.}$$

[2]Figures are approximate. Source: Bureau of Labor Statistics www.bls.gov.

Average Rate of Change of a Function Numerically and Graphically

EXAMPLE 1 Standard and Poor's 500

The following table lists the approximate value of Standard and Poor's 500 stock market index (S&P) during the period 2005–2011[3] ($t = 5$ represents 2005):

t (year)	5	6	7	8	9	10	11
$S(t)$ (points)	1,200	1,300	1,400	1,400	900	1,150	1,300

a. What was the average rate of change in the S&P over the 2-year period 2005–2007 (the period $5 \le t \le 7$ or $[5, 7]$ in interval notation); over the 4-year period 2005–2009 (the period $5 \le t \le 9$ or $[5, 9]$); and over the period $[6, 11]$?

b. Graph the values shown in the table. How are the rates of change reflected in the graph?

Solution

a. During the 2-year period [5, 7], the S&P changed as follows:

Start of the period ($t = 5$):	$S(5) = 1{,}200$
End of the period ($t = 7$):	$S(7) = 1{,}400$
Change during the period [5, 7]:	$S(7) - S(5) = 200$

Thus, the S&P increased by 200 points in 2 years, giving an average rate of change of $200/2 = 100$ points per year. We can write the calculation this way:

$$\begin{aligned}\text{Average rate of change of } S &= \frac{\text{Change in } S}{\text{Change in } t}\\ &= \frac{\Delta S}{\Delta t}\\ &= \frac{S(7) - S(5)}{7 - 5}\\ &= \frac{1{,}400 - 1{,}200}{7 - 5} = \frac{200}{2} = 100 \text{ points per year.}\end{aligned}$$

Interpreting the result: During the period [5, 7] (that is, 2005–2007), the S&P increased at an average rate of 100 points per year.

Similarly, the average rate of change during the period [5, 9] was

$$\begin{aligned}\text{Average rate of change of } S &= \frac{\Delta S}{\Delta t} = \frac{S(9) - S(5)}{9 - 5} = \frac{900 - 1{,}200}{9 - 5}\\ &= \frac{-300}{4} = -75 \text{ points per year.}\end{aligned}$$

Interpreting the result: During the period [5, 9] (that is, 2005–2009), the S&P *decreased* at an average rate of 75 points per year.

[3]The values are approximate values at the start of the given year. Source: http://finance.google.com.

Finally, during the period [6, 11], the average change was

$$\text{Average rate of change of } S = \frac{\Delta S}{\Delta t} = \frac{S(11) - S(6)}{11 - 6} = \frac{1{,}300 - 1{,}300}{11 - 6}$$

$$= \frac{0}{5} = 0 \text{ points per year.}$$

Interpreting the result: During the period [6, 11] the average rate of change of the S&P was zero points per year (even though its value did fluctuate during that period).

b. In Chapter 1, we saw that the rate of change of a quantity that changes linearly with time is measured by the slope of its graph. However, the S&P index does not change linearly with time. Figure 20 shows the data plotted two different ways: (a) as a bar chart and (b) as a piecewise linear graph. Bar charts are more commonly used in the media, but Figure 20(b) on the right illustrates the changing index more clearly.

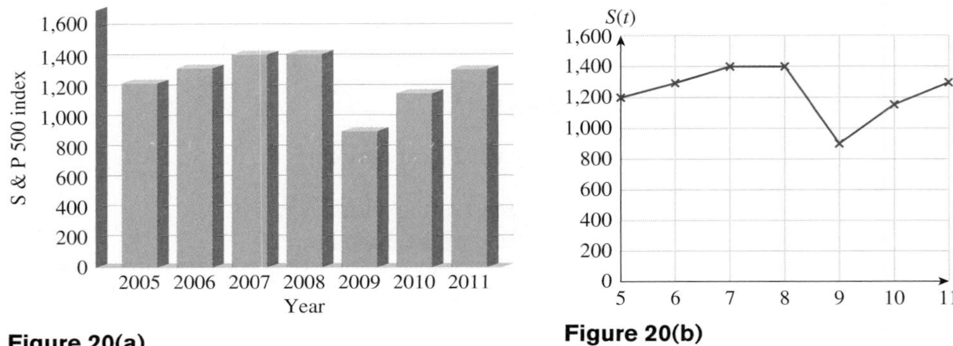

Figure 20(a) **Figure 20(b)**

We saw in part (a) that the average rate of change of S over the interval [5, 9] is the ratio

$$\text{Average rate of change of } S = \frac{\Delta S}{\Delta t} = \frac{S(9) - S(5)}{9 - 5} = -75 \text{ points per year.}$$

Notice that this rate of change is also the slope of the line through P and Q shown in Figure 21, and we can estimate this slope directly from the graph as shown.

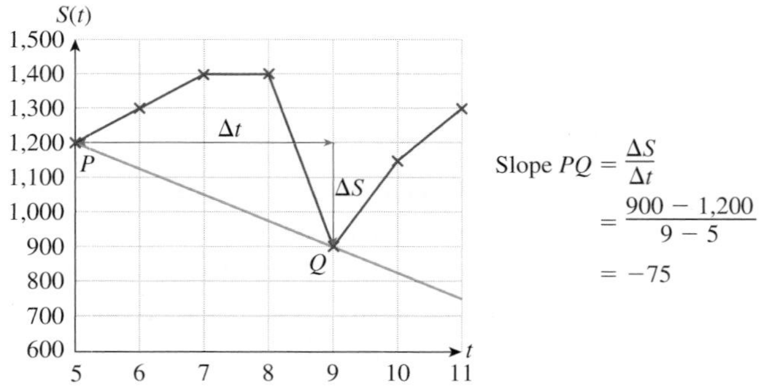

Figure 21

Average Rate of Change as Slope: The average rate of change of the S&P over the interval [5, 9] is the slope of the line passing through the points on the graph where $t = 5$ and $t = 9$.

Similarly, the average rates of change of the S&P over the intervals [5, 7] and [6, 11] are the slopes of the lines through pairs of corresponding points.

Here is the formal definition of the average rate of change of a function over an interval.

Change and Average Rate of Change of f over $[a, b]$: Difference Quotient

The **change** in $f(x)$ over the interval $[a, b]$ is

$$\begin{aligned}\text{Change in } f &= \Delta f \\ &= \text{Second value} - \text{First value} \\ &= f(b) - f(a).\end{aligned}$$

The **average rate of change** of $f(x)$ over the interval $[a, b]$ is

$$\begin{aligned}\text{Average rate of change of } f &= \frac{\text{Change in } f}{\text{Change in } x} \\ &= \frac{\Delta f}{\Delta x} = \frac{f(b) - f(a)}{b - a} \\ &= \text{Slope of line through points } P \text{ and } Q \\ &\quad \text{(see figure).}\end{aligned}$$

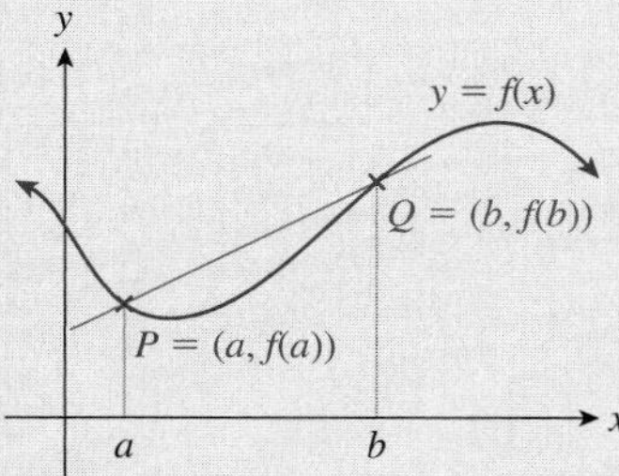

Average rate of change = Slope of PQ

We also call this average rate of change the **difference quotient** of f over the interval $[a, b]$. (It is the *quotient* of the *differences* $f(b) - f(a)$ and $b - a$.) A line through two points of a graph like P and Q is called a **secant line** of the graph.

Units

The units of the change Δf in f are the units of $f(x)$.
The units of the average rate of change of f are units of $f(x)$ per unit of x.*

* The average rate of change is a slope, and so it is measured in the same units as the slope: units of y (or $f(x)$) per unit of x.

Quick Example

If $f(3) = -1$ billion dollars, $f(5) = 0.5$ billion dollars, and x is measured in years, then the change and average rate of change of f over the interval $[3, 5]$ are given by

$$\text{Change in } f = f(5) - f(3) = 0.5 - (-1) = 1.5 \text{ billion dollars}$$

$$\begin{aligned}\text{Average rate of change} &= \frac{f(5) - f(3)}{5 - 3} = \frac{0.5 - (-1)}{2} \\ &= 0.75 \text{ billion dollars/year.}\end{aligned}$$

Alternative Formula: Average Rate of Change of *f* over $[a, a + h]$

(Replace b in the formula for the average rate of change by $a + h$.) The average rate of change of f over the interval $[a, a + h]$ is

$$\text{Average rate of change of } f = \frac{f(a + h) - f(a)}{h}. \qquad \text{Replace } b \text{ by } a + h.$$

In Example 1 we saw that the average rate of change of a quantity can be estimated directly from a graph. Here is an example that further illustrates the graphical approach.

EXAMPLE 2 Freon 22 Production

Figure 22 shows the number of tons $f(t)$ of ozone-layer-damaging Freon 22 (chlorodifluoromethane) produced annually in developing countries for the period 2000–2010 (t is time in years, and $t = 0$ represents 2000).[4]

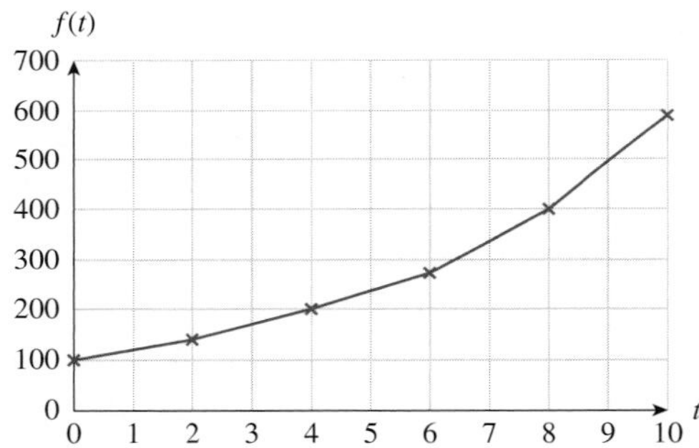

Figure 22

a. Use the graph to estimate the average rate of change of $f(t)$ with respect to t over the interval [4, 8] and interpret the result.

b. Over which 2-year period(s) was the average rate of change of Freon 22 production the greatest?

c. Multiple choice: For the period of time under consideration, Freon 22 production was

(A) increasing at a faster and faster rate.
(B) increasing at a slower and slower rate.
(C) decreasing at a faster and faster rate.
(D) decreasing at a slower and slower rate.

Solution

a. The average rate of change of f over the interval [4, 8] is given by the slope of the line through the points P and Q shown in Figure 23.

---

[4]Figures from 2007 on were projected. Source: *New York Times*, February 23, 2007, p. C1.

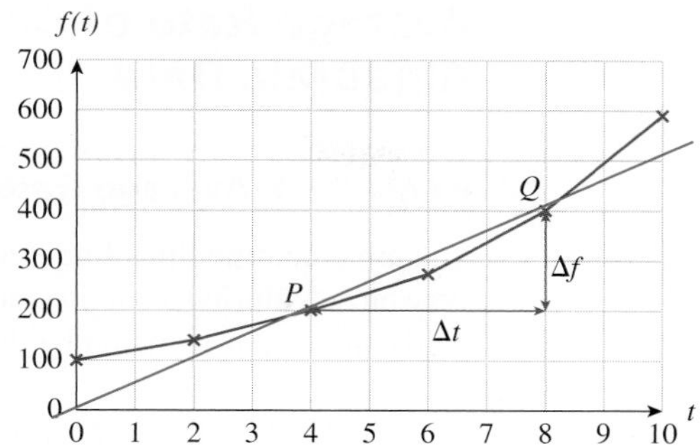

Figure 23

From the figure,

$$\text{Average rate of change of } f = \frac{\Delta f}{\Delta t} = \text{slope } PQ \approx \frac{400 - 200}{8 - 4} = \frac{200}{4} = 50.$$

Thus, the rate of change of f over the interval [4, 8] is approximately 50.

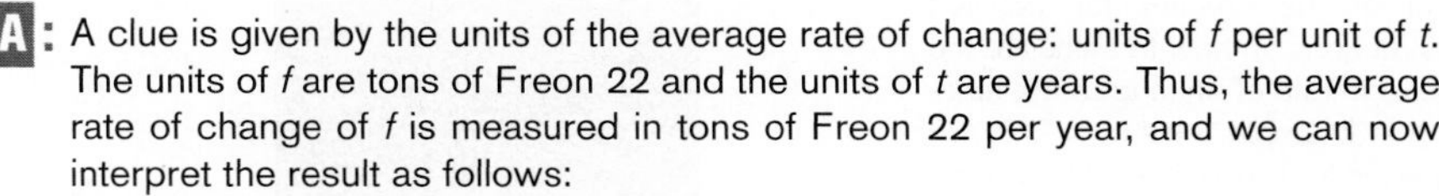

Q: *How do we interpret the result?*

A: A clue is given by the units of the average rate of change: units of f per unit of t. The units of f are tons of Freon 22 and the units of t are years. Thus, the average rate of change of f is measured in tons of Freon 22 per year, and we can now interpret the result as follows:

Interpreting the average rate of change: Annual production of Freon 22 was increasing at an average rate of 50 tons of Freon 22 per year from 2004 to 2008.

b. The rates of change of Freon 22 production over successive 2-year periods are given by the slopes of the individual line segments that make up the graph in Figure 22. Thus, the greatest average rate of change in a single 2-year period corresponds to the segment(s) with the largest slope. If you look at the figure, you will notice that the segment corresponding to [8, 10] is the steepest. Thus, the average rate of change of Freon 22 production was largest over the 2-year period from 2008 to 2010.

c. Looking again at the figure, notice that the graph rises as we go from left to right; that is, the value of the function (Freon 22 production) is increasing with increasing t. At the same time, the fact that the curve bends up (is concave up) with increasing t tells us that the successive slopes get steeper, and so the average rates of change increase as well (Choice (A)).

➡ **Before we go on...** Notice in Example 2 that we do not get exact answers from a graph; the best we can do is *estimate* the rates of change: Was the exact answer to part (a) closer to 49 or 51? Two people can reasonably disagree about results read from a graph, and you should bear this in mind when you check the answers to the exercises. ■

Perhaps the most sophisticated way to compute the average rate of change of a quantity is through the use of a mathematical formula or model for the quantity in question.

Average Rate of Change of a Function Using Algebraic Data

EXAMPLE 3 Average Rate of Change from a Formula

You are a commodities trader and you monitor the price of gold on the spot market very closely during an active morning. Suppose you find that the price of an ounce of gold can be approximated by the function

$$G(t) = 5t^2 - 85t + 1{,}762 \qquad (7.5 \le t \le 10.5),$$

where t is time in hours. (See Figure 24. $t = 8$ represents 8:00 am.)

Looking at the graph on the right, we can see that the price of gold was falling at the beginning of the time period, but by $t = 8.5$ the fall had slowed to a stop, whereupon the market turned around and the price began to rise more and more rapidly toward the end of the period. What was the average rate of change of the price of gold over the $1\frac{1}{2}$-hour period starting at 8:00 am (the interval [8, 9.5] on the t-axis)?

using Technology

See the Technology Guides at the end of the chapter for detailed instructions on how to calculate the average rate of change of the function in Example 3 using a TI-83/84 Plus or a spreadsheet. Here is an outline:

TI-83/84 Plus

`Y1=5X^2-85X+1762`

Home screen: `(Y1(9.5)-Y1(8))/(9.5-8)`

[More details on page 213.]

Spreadsheet

Headings t, $G(t)$, Rate of Change in A1–C1

t-values 8, 9.5 in A2–A3

`=5*A2^2-85*A2+1762`

in B2, copied down to B3

`=(B3-B2)/(A3-A2)` in C2.

[More details on page 215.]

Source: www.kitco.com

$G(t)$ graph: vertical axis 1,395 to 1,425; horizontal axis t from 7.5 to 10.5

$G(t) = 5t^2 - 85t + 1{,}762$

Figure 24

Solution We have

$$\text{Average rate of change of } G \text{ over } [8, 9.5] = \frac{\Delta G}{\Delta t} = \frac{G(9.5) - G(8)}{9.5 - 8}.$$

From the formula for $G(t)$, we find

$$G(9.5) = 5(9.5)^2 - 85(9.5) + 1{,}762 = 1{,}405.75$$
$$G(8) = 5(8)^2 - 85(8) + 1{,}762 = 1{,}402.$$

Thus, the average rate of change of G is given by

$$\frac{G(9.5) - G(8)}{9.5 - 8} = \frac{1{,}405.75 - 1{,}402}{1.5} = \frac{3.75}{1.5} = \$2.50 \text{ per hour.}$$

In other words, the price of gold increased at an average rate of \$2.50 per hour over the $1\frac{1}{2}$-hour period.

EXAMPLE 4 T Rates of Change over Shorter Intervals

Continuing with Example 3, use technology to compute the average rate of change of

$$G(t) = 5t^2 - 85t + 1{,}762 \qquad (7.5 \le t \le 10.5)$$

over the intervals $[8, 8 + h]$, where $h = 1, 0.1, 0.01, 0.001$, and 0.0001. What do the answers tell you about the price of gold?

Solution

We use the "alternative" formula

$$\text{Average rate of change of } G \text{ over } [a, a+h] = \frac{G(a+h) - G(a)}{h}$$

so

$$\text{Average rate of change of } G \text{ over } [8, 8+h] = \frac{G(8+h) - G(8)}{h}.$$

Let us calculate this average rate of change for some of the values of h listed:

using Technology

Example 4 is the kind of example where the use of technology can make a huge difference. See the Technology Guides at the end of the chapter to find out how to do the above computations almost effortlessly using a TI-83/84 Plus or a spreadsheet. Here is an outline:

TI-83/84 Plus

`Y1=5X^2-85X+1762`

Home screen:

```
(Y1(8+1)-Y1(8))/1
(Y1(8+0.1)-Y1(8))/0.1
(Y1(8+0.01)-Y1(8))/0.01
```

etc.

[More details on page 213.]

Spreadsheet

Headings *a*, *h*, *t*, *G*(*t*), Rate of Change in A1–E1

`8` in A2, `1` in B2,

`=A2` in C2, `=A2+B2` in C3

`=5*C2^2-85*C2+1762` in D2; copy down to D3

`=(D3-D2)/(C3-C2)` in E2

[More details on page 215.]

$h = 1$: $G(8+h) = G(8+1) = G(9) = 5(9)^2 - 85(9) + 1{,}762 = 1{,}402$

$$G(8) = 5(8)^2 - 85(8) + 1{,}762 = 1{,}402$$

$$\text{Average rate of change of } G = \frac{G(9) - G(8)}{1} = \frac{1{,}402 - 1{,}402}{1} = 0$$

$h = 0.1$: $G(8+h) = G(8+0.1) = G(8.1) = 5(8.1)^2 - 85(8.1) + 1{,}762$

$$= 1{,}401.55$$

$$G(8) = 5(8)^2 - 85(8) + 1{,}762 = 1{,}402$$

$$\text{Average rate of change of } G = \frac{G(8.1) - G(8)}{0.1} = \frac{1{,}401.55 - 1{,}402}{0.1} = \frac{-0.45}{0.1}$$

$$= -4.5$$

$h = 0.01$: $G(8+h) = G(8+0.01) = G(8.01) = 5(8.01)^2 - 85(8.01) + 1{,}762$

$$= 1{,}401.9505$$

$$G(8) = 5(8)^2 - 85(8) + 1{,}762 = 1{,}402$$

$$\text{Average rate of change of } G = \frac{G(8.01) - G(8)}{0.01} = \frac{1{,}401.9505 - 1{,}402}{0.01} = \frac{-0.0495}{0.01}$$

$$= -4.95$$

Continuing in this way, we get the values in the following table:

h	1	0.1	0.01	0.001	0.0001
Ave. Rate of Change $\dfrac{G(8+h) - G(8)}{h}$	0	−4.5	−4.95	−4.995	−4.9995

Each value is an average rate of change of G. For example, the value corresponding to $h = 0.01$ is -4.95, which tells us:

> *Over the interval* [8, 8.01] *the price of gold was decreasing at an average rate of* \$4.95 *per hour.*

In other words, during the first one hundredth of an hour (or 36 seconds) starting at $t = $ 8:00 am, the price of gold was decreasing at an average rate of \$4.95 per hour. Put another way, in those 36 seconds, the price of gold decreased at a rate that, if continued, would have produced a decrease of \$4.95 in the price of gold during the next hour. We will return to this example at the beginning of Section 3.5.

FAQs

Recognizing When and How to Compute the Average Rate of Change and How to Interpret the Answer

Q: *How do I know, by looking at the wording of a problem, that it is asking for an average rate of change?*

A: If a problem does not ask for an average rate of change directly, it might do so indirectly, as in "On average, how fast is quantity q increasing?"

Q: *If I know that a problem calls for computing an average rate of change, how should I compute it? By hand or using technology?*

A: All the computations can be done by hand, but when hand calculations are not called for, using technology might save time.

Q: *Lots of problems ask us to "interpret" the answer. How do I do that for questions involving average rates of change?*

A: The *units* of the average rate of change are often the key to interpreting the results:

The units of the average rate of change of $f(x)$ are units of $f(x)$ per unit of x.

Thus, for instance, if $f(x)$ is the cost, in dollars, of a trip of x miles in length, and the average rate of change of f is calculated to be 10, then the units of the average rate of change are dollars per mile, and so we can interpret the answer by saying that the cost of a trip rises an average of \$10 for each additional mile.

EXERCISES

Access end-of-section exercises online at **www.webassign.net**

3.5 Derivatives: Numerical and Graphical Viewpoints

In Example 4 of Section 3.4, we looked at the average rate of change of the function $G(t) = 5t^2 - 85t + 1{,}762$ approximating the price of gold on the spot market over smaller and smaller intervals of time. We obtained the following table showing the average rates of change of G over the intervals $[8, 8 + h]$ for successively smaller values of h:

h getting smaller; interval $[8, 8 + h]$ getting smaller →

h	1	0.1	0.01	0.001	0.0001
Ave. Rate of Change over $[8, 8 + h]$	0	−4.5	−4.95	−4.995	−4.9995

Rate of change approaching −\$5 per hour →

The average rates of change of the price of gold over smaller and smaller periods of time, starting at the instant $t = 8$ (8:00 am), appear to be getting closer and closer to −\$5 per hour. As we look at these shrinking periods of time, we are getting closer to looking at what happens at the *instant* $t = 8$. So it seems reasonable to say that the average rates of change are approaching the **instantaneous rate of change** at $t = 8$, which the table suggests is −\$5 per hour. This is how fast the price of gold was changing *exactly* at 8:00 am.

At $t = 8$, *the instantaneous rate of change of* $G(t)$ *is* -5.

We express this fact mathematically by writing $G'(8) = -5$ (which we read as "G prime of 8 equals −5"). Thus

$G'(8) = -5$ *means that, at* $t = 8$, *the instantaneous rate of change of* $G(t)$ *is* -5.

The process of letting h get smaller and smaller is called taking the **limit** as h approaches 0 (as you recognize if you've done the sections on limits). As in the sections on limits, we write $h \to 0$ as shorthand for "h approaches 0." Thus, taking the limit of the average rates of change as $h \to 0$ gives us the instantaneous rate of change.

Q: *All these intervals $[8, 8 + h]$ are intervals to the right of 8. What about small intervals to the left of 8, such as [7.9, 8]?*

A: We can compute the average rate of change of our function for such intervals by choosing h to be negative ($h = -0.1, -0.01$, etc.) and using the same difference quotient formula we used for positive h:

$$\text{Average rate of change of } G \text{ over } [8 + h, 8] = \frac{G(8) - G(8 + h)}{8 - (8 + h)}$$

Here are the results we get using negative h:

h getting closer to 0; interval $[8 + h, 8]$ getting smaller →

h	−1	−0.1	−0.01	−0.001	−0.0001
Ave. Rate of Change over $[8 + h, 8]$	−10	−5.5	−5.05	−5.005	−5.0005

Rate of change approaching −\$5 per hour →

Notice that the average rates of change are again getting closer and closer to −5 as h approaches 0, suggesting once again that the instantaneous rate of change is −\$5 per hour.

Instantaneous Rate of Change of f(x) at x = a: Derivative

The **instantaneous rate of change** of $f(x)$ at $x = a$ is defined as

$$f'(a) = \lim_{h \to 0} \frac{f(a+h) - f(a)}{h}.$$

f prime of a equals the limit, as h approaches 0, of the ratio $\frac{f(a+h) - f(a)}{h}$.

The quantity $f'(a)$ is also called the **derivative of f(x) at x = a.** Finding the derivative of f is called **differentiating *f***.

Units

The units of $f'(a)$ are the same as the units of the average rate of change: units of f per unit of x.

Quick Examples

1. If $f(x) = 5x^2 - 85x + 1{,}762$, then the two tables on the previous page suggest that

$$f'(8) = \lim_{h \to 0} \frac{f(8+h) - f(8)}{h} = -5.$$

2. If $f(t)$ is the number of insects in your dorm room at time t hours, and you know that $f(3) = 5$ and $f'(3) = 8$, this means that, at time $t = 3$ hours, there are 5 insects in your room, and this number is growing at an instantaneous rate of 8 insects per hour.

IMPORTANT NOTES

1. Sections 3.1–3.3 discuss limits in some detail. If you have not (yet) covered those sections, you can trust to your intuition.
2. The formula for the derivative tells us that the instantaneous rate of change is the limit of the average rates of change $[f(a+h) - f(a)]/h$ over smaller and smaller intervals. Thus, the value of $f'(a)$ can be approximated by computing the average rate of change for smaller and smaller values of h, both positive and negative.
3. If a happens to be an endpoint of the domain of f, then $f'(a)$ does not exist, as then $[f(a+h) - f(a)]/h$ only has a one-sided limit as $h \to 0$, and so

$$\lim_{h \to 0} \frac{f(a+h) - f(a)}{h} \text{ does not exist.*}$$

4. In this section we will only *approximate* derivatives. In Section 3.6 we will begin to see how we find the *exact* values of derivatives.
5. $f'(a)$ is a number we can calculate, or at least approximate, for various values of a, as we have done in the earlier example. Since $f'(a)$ depends on the value of a, we can think of f' as *a function of a*. (We return to this idea at the end of this section.) An old name for f' is "the function *derived from f*," which has been shortened to the *derivative* of *f*.
6. It is because f' is a function that we sometimes refer to $f'(a)$ as "the derivative of *f evaluated at a*," or the "derivative of $f(x)$ evaluated at $x = a$."

* One could define the derivative at an endpoint by instead using the associated one-sided limit; for instance, if a is a left endpoint of the domain of f, then one could define

$$f'(a) = \lim_{h \to 0^+} \frac{f(a+h) - f(a)}{h}$$

as we did in previous editions of this book. However, in this edition we have decided to follow the usual convention and say that the derivative at an endpoint does not exist.

It may happen that the average rates of change $[f(a+h) - f(a)]/h$ do not approach any fixed number at all as h approaches zero, or that they approach one number on the intervals using positive h, and another on those using negative h. If this happens, $\lim_{h\to 0}[f(a+h) - f(a)]/h$ does not exist, and we say that f is **not differentiable** at $x = a$, or **$f'(a)$ does not exist**. When the limit *does* exist, we say that f is **differentiable** at the point $x = a$, or **$f'(a)$ exists**. It is comforting to know that all polynomials and exponential functions are differentiable at *every* point. On the other hand, certain functions are not differentiable. Examples are $f(x) = |x|$ and $f(x) = x^{1/3}$, neither of which is differentiable at $x = 0$. (See Section 4.1.)

EXAMPLE 1 Instantaneous Rate of Change: Numerically and Graphically

The air temperature one spring morning, t hours after 7:00 am, was given by the function $f(t) = 50 + 0.1t^4$ degrees Fahrenheit $(0 \le t \le 4)$.

a. How fast was the temperature rising at 9:00 am?

b. How is the instantaneous rate of change of temperature at 9:00 am reflected in the graph of temperature vs. time?

Solution

a. We are being asked to find the instantaneous rate of change of the temperature at $t = 2$, so we need to find $f'(2)$. To do this we examine the average rates of change

$$\frac{f(2+h) - f(2)}{h} \qquad \text{Average rate of change = difference quotient}$$

for values of h approaching 0. Calculating the average rate of change over $[2, 2+h]$ for $h = 1$, 0.1, 0.01, 0.001, and 0.0001 we get the following values (rounded to four decimal places):*

* We can quickly compute these values using technology as in Example 4 in Section 3.4. (See the Technology Guides at the end of the chapter.)

h	1	0.1	0.01	0.001	0.0001
Average Rate of Change Over $[2, 2+h]$	6.5	3.4481	3.2241	3.2024	3.2002

Here are the values we get using negative values of h:

h	−1	−0.1	−0.01	−0.001	−0.0001
Average Rate of Change Over $[2+h, 2]$	1.5	2.9679	3.1761	3.1976	3.1998

The average rates of change are clearly approaching the number 3.2, so we can say that $f'(2) = 3.2$. Thus, at 9:00 in the morning, the temperature was rising at the rate of 3.2 degrees per hour.

b. We saw in Section 3.4 that the average rate of change of f over an interval is the slope of the secant line through the corresponding points on the graph of f. Figure 25 illustrates this for the intervals $[2, 2+h]$ with $h = 1$, 0.5, and 0.1.

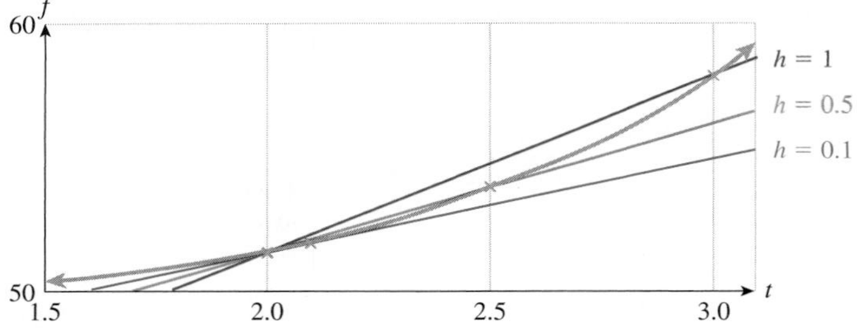

Figure 25

All three secant lines pass though the point $(2, f(2)) = (2, 51.6)$ on the graph of f. Each of them passes through a second point on the curve (the second point is different for each secant line) and this second point gets closer and closer to (2, 51.6) as h gets closer to 0. What seems to be happening is that the secant lines are getting closer and closer to a line that just touches the curve at (2, 51.6): the **tangent line** at (2, 51.6), shown in Figure 26.

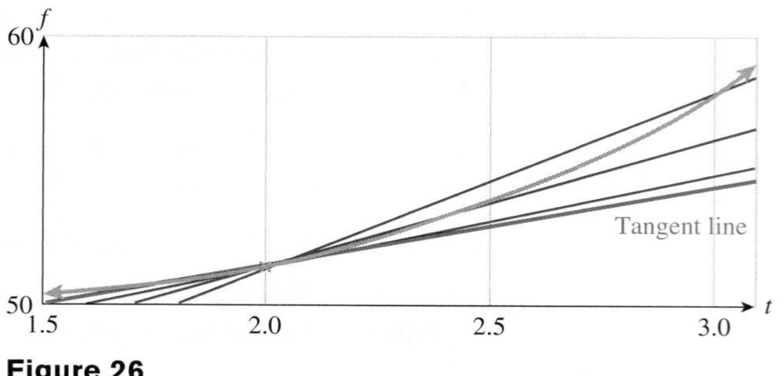

Figure 26

Q: *What is the slope of this tangent line?*

A: Because the slopes of the secant lines are getting closer and closer to 3.2, and because the secant lines are approaching the tangent line, the tangent line must have slope 3.2. In other words,

At the point on the graph where $x = 2$, the slope of the tangent line is $f'(2)$.

Q: *What is the difference between $f(2)$ and $f'(2)$?*

A: An important question. Briefly, $f(2)$ is the *value of* f when $t = 2$, while $f'(2)$ is the *rate at which f is changing* when $t = 2$. Here,

$$f(2) = 50 + 0.1(2)^4 = 51.6 \text{ degrees.}$$

Thus, at 9:00 am ($t = 2$), the temperature was 51.6 degrees. On the other hand,

$$f'(2) = 3.2 \text{ degrees per hour.}$$ Units of slope are units of f per unit of t.

This means that, at 9:00 am ($t = 2$), the temperature was increasing at a rate of 3.2 degrees per hour.

Because we have been talking about tangent lines, we should say more about what they *are*. A tangent line to a *circle* is a line that touches the circle in just one point. A tangent line gives the circle "a glancing blow," as shown in Figure 27.

For a smooth curve other than a circle, a tangent line may touch the curve at more than one point, or pass through it (Figure 28).

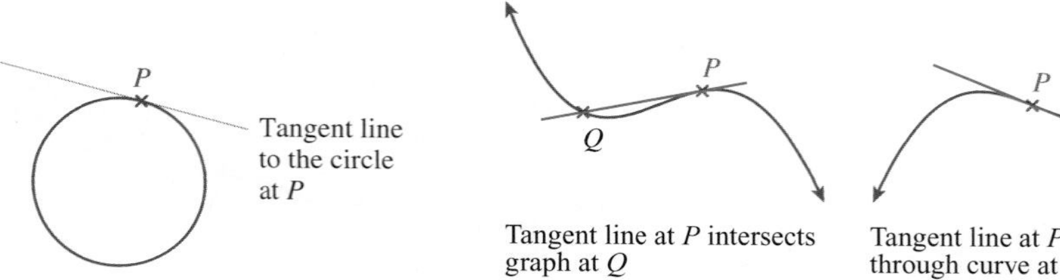

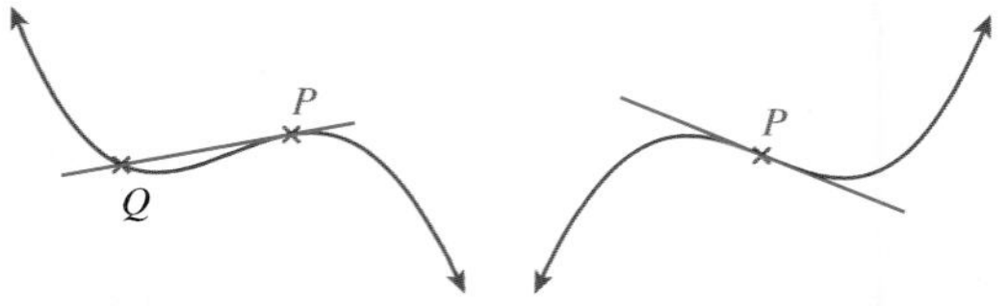

Tangent line at P intersects graph at Q

Tangent line at P passes through curve at P

Figure 27

Figure 28

However, all tangent lines have the following interesting property in common: If we focus on a small portion of the curve very close to the point P—in other words, if we "zoom in" to the graph near the point P—the curve will appear almost straight, and almost indistinguishable from the tangent line (Figure 29).

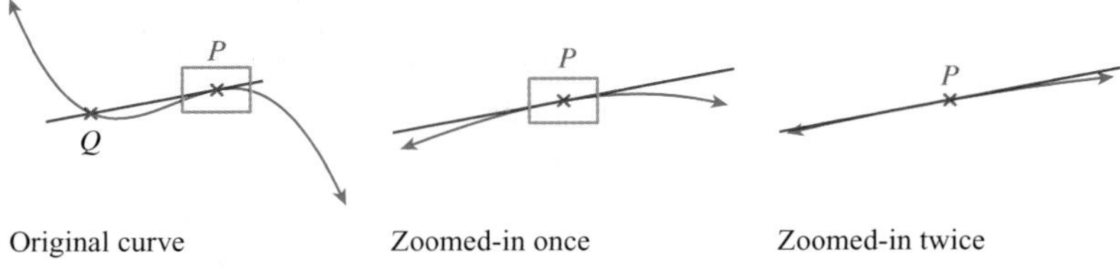

Figure 29

You can check this property by zooming in on the curve shown in Figures 25 and 26 in the previous example near the point where $x = 2$.

Secant and Tangent Lines

The *slope of the secant line* through the points on the graph of f where $x = a$ and $x = a + h$ is given by the average rate of change, or difference quotient,

$$m_{\text{sec}} = \text{slope of secant} = \text{average rate of change} = \frac{f(a+h) - f(a)}{h}.$$

The *slope of the tangent line* through the point on the graph of f where $x = a$ is given by the instantaneous rate of change, or derivative

$$m_{\text{tan}} = \text{slope of tangent} = \text{instantaneous rate of change} = \text{derivative}$$
$$= f'(a) = \lim_{h \to 0} \frac{f(a+h) - f(a)}{h},$$

assuming the limit exists.

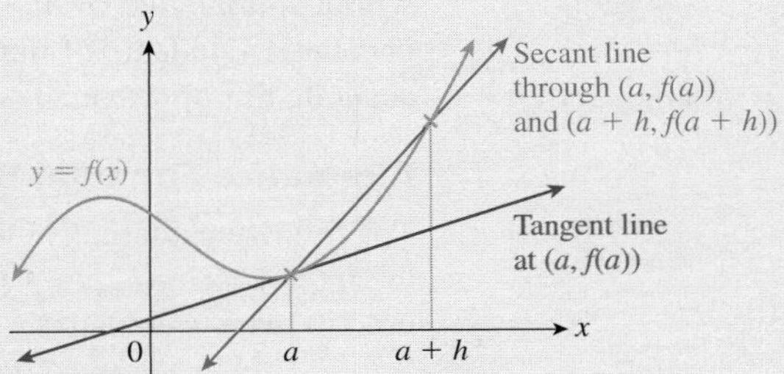

Quick Example

In the following graph, the tangent line at the point where $x = 2$ has slope 3. Therefore, the derivative at $x = 2$ is 3. That is, $f'(2) = 3$.

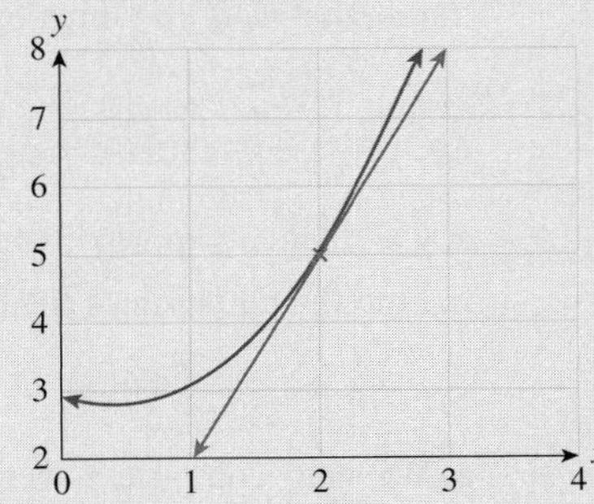

Note It might happen that the tangent line is vertical at some point or does not exist at all. These are the cases in which f is not differentiable at the given point. (See Section 3.6.) ■

We can now give a more precise definition of what we mean by the tangent line to a point P on the graph of f at a given point: The **tangent line** to the graph of f at the point $P(a, f(a))$ is the straight line passing through P with slope $f'(a)$.

Quick Approximation of the Derivative

Q: *Do we always need to make tables of difference quotients as above in order to calculate an approximate value for the derivative? That seems like a large amount of work just to get an approximation.*

A: We can usually *approximate* the value of the derivative by using a single, small value of h. In the example above, the value $h = 0.0001$ would have given a pretty good approximation. The problems with using a fixed value of h are that (1) we do not get an *exact* answer, only an *approximation* of the derivative, and (2) how good an approximation it is depends on the function we're differentiating.* However, with most of the functions we'll be considering, setting $h = 0.0001$ does give us a good approximation.

* In fact, no matter how small the value we decide to use for h, it is possible to craft a function f for which the difference quotient at a is not even close to $f'(a)$.

Calculating a Quick Approximation of the Derivative

We can calculate an approximate value of $f'(a)$ by using the formula

$$f'(a) \approx \frac{f(a+h) - f(a)}{h} \qquad \text{Rate of change over } [a,\ a+h]$$

with a small value of h. The value $h = 0.0001$ works for most examples we encounter (students of numerical methods study the question of exactly how accurate this approximation is).

Alternative Formula: The Balanced Difference Quotient

The following alternative formula, which measures the rate of change of f over the interval $[a - h, a + h]$, often gives a more accurate result, and is the one used in many calculators:

$$f'(a) \approx \frac{f(a+h) - f(a-h)}{2h}. \qquad \text{Rate of change over } [a-h,\ a+h]$$

Note For the quick approximations to be valid, the function f must be differentiable; that is, $f'(a)$ must exist. ■

EXAMPLE 2 Quick Approximation of the Derivative

a. Calculate an approximate value of $f'(1.5)$ if $f(x) = x^2 - 4x$.

b. Find the equation of the tangent line at the point on the graph where $x = 1.5$.

Solution

a. We shall compute both the ordinary difference quotient and the balanced difference quotient.

Ordinary Difference Quotient: Using $h = 0.0001$, the ordinary difference quotient is:

$$f'(1.5) \approx \frac{f(1.5 + 0.0001) - f(1.5)}{0.0001} \qquad \text{Ordinary difference quotient}$$

$$= \frac{f(1.5001) - f(1.5)}{0.0001}$$

$$= \frac{(1.5001^2 - 4 \times 1.5001) - (1.5^2 - 4 \times 1.5)}{0.0001} = -0.9999.$$

This answer is accurate to 0.0001; in fact, $f'(1.5) = -1$.

Graphically, we can picture this approximation as follows: Zoom in on the curve using the window $1.5 \le x \le 1.5001$ and measure the slope of the secant line joining both ends of the curve segment. Figure 30 shows close-up views of the curve and tangent line near the point P in which we are interested, the third view being the zoomed-in view used for this approximation.

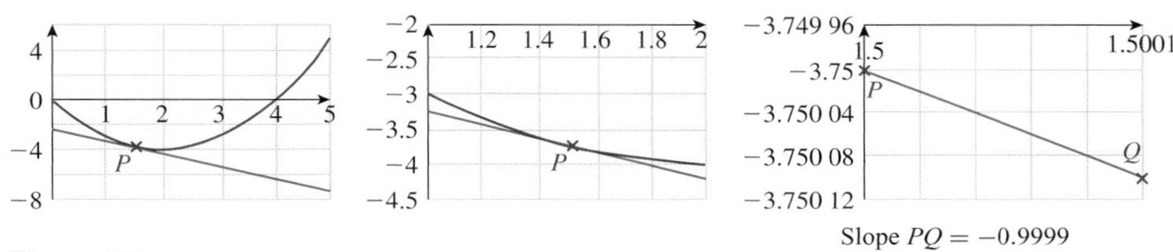

Figure 30

* The balanced difference quotient always gives the exact derivative for a quadratic function.

using Technology

See the Technology Guides at the end of the chapter to find out how to calculate the quick approximations to the derivative in Example 2 using a TI-83/84 Plus or a spreadsheet. Here is an outline:

TI-83/84 Plus

`Y1=X^2-4*X`

Home screen:

`(Y1(1.5001)-Y1(1.5))/0.0001`

`(Y1(1.5001)-Y1(1.4999))/0.0002`

[More details on page 214.]

Spreadsheet

Headings a, h, x, $f(x)$, Diff Quotient, Balanced Diff Quotient in A1–F1

`1.5` in A2, `0.0001` in B2, `=A2-B2` in C2, `=A2` in C3, `=A2+B2` in C4

`=C2^2-4*C2` in D2; copy down to D4

`=(D3-D2)/(C3-C2)` in E2

`=(D4-D2)/(C4-C2)` in E3

[More details on page 216.]

Notice that in the third window the tangent line and curve are indistinguishable. Also, the point P in which we are interested is on the left edge of the window.

Balanced Difference Quotient: For the balanced difference quotient, we get

$$f'(1.5) \approx \frac{f(1.5 + 0.0001) - f(1.5 - 0.0001)}{2(0.0001)} \qquad \text{Balanced difference quotient}$$

$$= \frac{f(1.5001) - f(1.4999)}{0.0002}$$

$$= \frac{(1.5001^2 - 4 \times 1.5001) - (1.4999^2 - 4 \times 1.4999)}{0.0002} = -1.$$

This balanced difference quotient gives the exact answer in this case!* Graphically, it is as though we have zoomed in using a window that puts the point P in the *center* of the screen (Figure 31) rather than at the left edge.

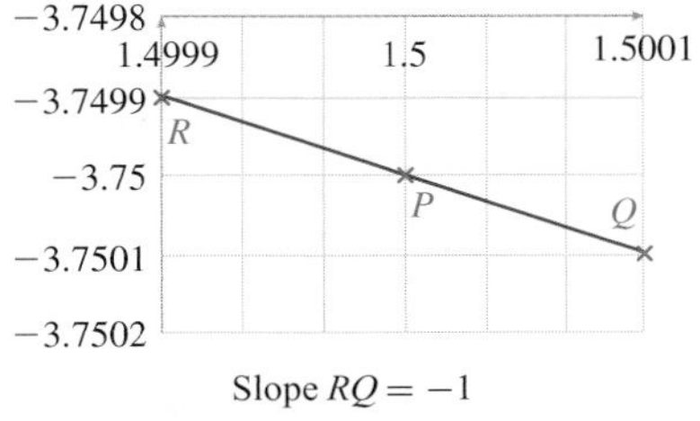

Figure 31

b. We find the equation of the tangent line from a point on the line and its slope, as we did in Chapter 1:

- ***Point*** $(1.5, f(1.5)) = (1.5, -3.75)$.
- ***Slope*** $m = f'(1.5) = -1$. Slope of the tangent line = derivative.

The equation is

$$y = mx + b,$$

where $m = -1$ and $b = y_1 - mx_1 = -3.75 - (-1)(1.5) = -2.25$. Thus, the equation of the tangent line is

$$y = -x - 2.25.$$

Q: *Why can't we simply put $h = 0.000\,000\,000\,000\,000\,000\,01$ for an incredibly accurate approximation to the instantaneous rate of change and be done with it?*

A: This approach would certainly work if you were patient enough to do the (thankless) calculation by hand! However, doing it with the help of technology—even an ordinary calculator—will cause problems: The issue is that calculators and spreadsheets represent numbers with a maximum number of significant digits (15 in the case of Excel). As the value of h gets smaller, the value of $f(a+h)$ gets closer and closer to the value of $f(a)$. For example, if $f(x) = 50 + 0.1x^4$, Excel might compute

$$\begin{aligned} &f(2 + 0.000\,000\,000\,000\,1) - f(2) \\ &\quad = 51.600\,000\,000\,000\,3 - 51.6 \qquad \text{Rounded to 15 digits} \\ &\quad = 0.000\,000\,000\,000\,3 \end{aligned}$$

and the corresponding difference quotient would be 3, not 3.2 as it should be. If h gets even smaller, Excel will not be able to distinguish between $f(a+h)$ and $f(a)$ at all, in which case it will compute 0 for the rate of change. This loss in accuracy when subtracting two very close numbers is called **subtractive error**.

Thus, there is a trade-off in lowering the value of h: Smaller values of h yield *mathematically* more accurate approximations of the derivative, but if h gets too small, subtractive error becomes a problem and decreases the accuracy of computations that use technology.

Leibniz *d* Notation

We introduced the notation $f'(x)$ for the derivative of f at x, but there is another interesting notation. We have written the average rate of change as

$$\text{Average rate of change} = \frac{\Delta f}{\Delta x}. \qquad \frac{\text{Change in } f}{\text{Change in } x}$$

As we use smaller and smaller values for Δx, we approach the instantaneous rate of change, or derivative, for which we also have the notation df/dx, due to Leibniz:

$$\text{Instantaneous rate of change} = \lim_{\Delta x \to 0} \frac{\Delta f}{\Delta x} = \frac{df}{dx}.$$

That is, df/dx is just another notation for $f'(x)$. Do not think of df/dx as an actual quotient of two numbers: Remember that we only use an actual quotient $\Delta f/\Delta x$ to *approximate* the value of df/dx.

In Example 3, we apply the quick approximation method of estimating the derivative.

EXAMPLE 3 Velocity

My friend Eric, an enthusiastic baseball player, claims he can "probably" throw a ball upward at a speed of 100 feet per second (ft/s).* Our physicist friends tell us that its height s (in feet) t seconds later would be $s = 100t - 16t^2$. Find its average velocity over the interval [2, 3] and its instantaneous velocity exactly 2 seconds after Eric throws it.

* Eric's claim is difficult to believe; 100 ft/s corresponds to around 68 mph, and professional pitchers can throw *forward* at about 100 mph.

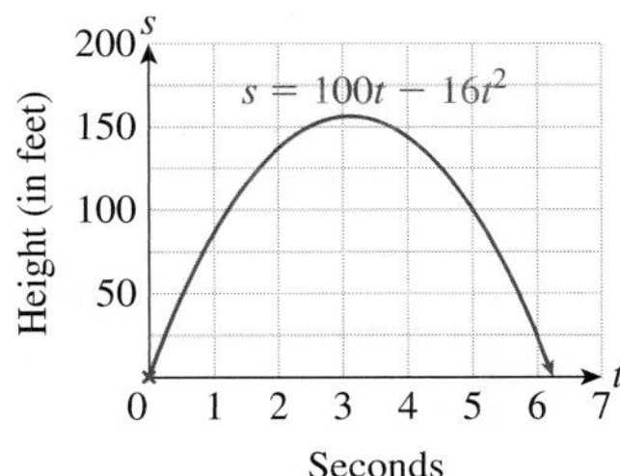

Figure 32

Solution The graph of the ball's height as a function of time is shown in Figure 32. Asking for the velocity is really asking for the rate of change of height with respect to time. (Why?) Consider average velocity first. To compute the **average velocity** of the ball from time 2 to time 3, we first compute the change in height:

$$\Delta s = s(3) - s(2) = 156 - 136 = 20 \text{ ft}.$$

Since it rises 20 feet in $\Delta t = 1$ second, we use the defining formula *speed* = *distance/time* to get the average velocity:

$$\text{Average velocity} = \frac{\Delta s}{\Delta t} = \frac{20}{1} = 20 \text{ ft/sec}.$$

from time $t = 2$ to $t = 3$. This is just the difference quotient, so

The average velocity is the average rate of change of height.

To get the **instantaneous velocity** at $t = 2$, we find the instantaneous rate of change of height. In other words, we need to calculate the derivative ds/dt at $t = 2$. Using the balanced quick approximation described earlier, we get

$$\begin{aligned}\frac{ds}{dt} &\approx \frac{s(2 + 0.0001) - s(2 - 0.0001)}{2(0.0001)} \\ &= \frac{s(2.0001) - s(1.9999)}{0.0002} \\ &= \frac{100(2.0001) - 16(2.0001)^2 - (100(1.9999) - 16(1.9999)^2)}{0.0002} \\ &= 36 \text{ ft/sec}.\end{aligned}$$

In fact, this happens to be the exact answer; the instantaneous velocity at $t = 2$ is exactly 36 ft/sec. (Try an even smaller value of h to persuade yourself.)

➡ **Before we go on...** If we repeat the calculation in Example 3 at time $t = 5$, we get

$$\frac{ds}{dt} = -60 \text{ ft/sec}.$$

The negative sign tells us that the ball is *falling* at a rate of 60 feet per second at time $t = 5$. (How does the fact that it is falling at $t = 5$ show up on the graph?) ■

The preceding example gives another interpretation of the derivative.

Average and Instantaneous Velocity

For an object moving in a straight line with position $s(t)$ at time t, the **average velocity** from time t to time $t+h$ is the average rate of change of position with respect to time:

$$v_{ave} = \frac{s(t+h)-s(t)}{h} = \frac{\Delta s}{\Delta t}.$$

Average velocity = Average rate of change of position

The **instantaneous velocity** at time t is

$$v = \lim_{h\to 0} \frac{s(t+h)-s(t)}{h} = \frac{ds}{dt}.$$

Instantaneous velocity = Instantaneous rate of change of position

In other words, *instantaneous velocity is the derivative of position with respect to time.*

Here is one last comment on Leibniz notation. In Example 3, we could have written the velocity either as s' or as ds/dt, as we chose to do. To write the answer to the question, that the velocity at $t = 2$ sec was 36 ft/sec, we can write either

$$s'(2) = 36$$

or

$$\left.\frac{ds}{dt}\right|_{t=2} = 36.$$

The notation "$|_{t=2}$" is read "evaluated at $t = 2$." Similarly, if $y = f(x)$, we can write the instantaneous rate of change of f at $x = 5$ in either functional notation as

$$f'(5)$$

The derivative of f, evaluated at $x = 5$

or in Leibniz notation as

$$\left.\frac{dy}{dx}\right|_{x=5}.$$

The derivative of y, evaluated at $x = 5$

The latter notation is obviously more cumbersome than the functional notation $f'(5)$, but the notation dy/dx has compensating advantages. You should practice using both notations.

The Derivative Function

The derivative $f'(x)$ is a number we can calculate, or at least approximate, for various values of x. Because $f'(x)$ depends on the value of x, we may think of f' as a function of x. This function is the **derivative function**.

Derivative Function

If f is a function, its **derivative function** f' is the function whose value $f'(x)$ is the derivative of f at x. Its domain is the set of all x at which f is differentiable. Equivalently, f' associates to each x the slope of the tangent to the graph of the function f at x, or the rate of change of f at x. The formula for the derivative function is

$$f'(x) = \lim_{h\to 0} \frac{f(x+h)-f(x)}{h}.$$

Derivative function

Quick Examples

1. Let $f(x) = 3x - 1$. The graph of f is a straight line that has slope 3 everywhere. In other words, $f'(x) = 3$ for every choice of x; that is, f' is a constant function.

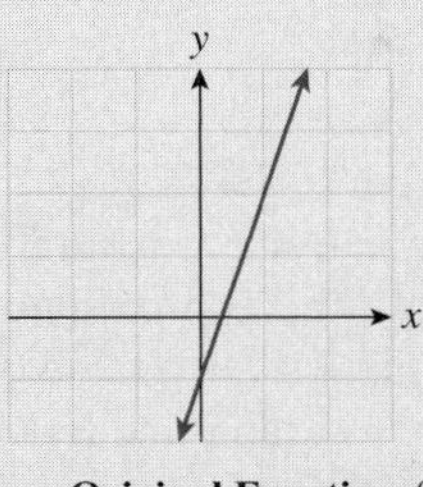

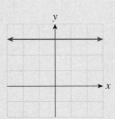

Original Function f
$f(x) = 3x - 1$

Derivative Function f'
$f'(x) = 3$

2. Given the graph of a function f, we can get a rough sketch of the graph of f' by estimating the slope of the tangent to the graph of f at several points, as illustrated below.*

* This method is discussed in detail on the Website at Online Text → Sketching the Graph of the Derivative.

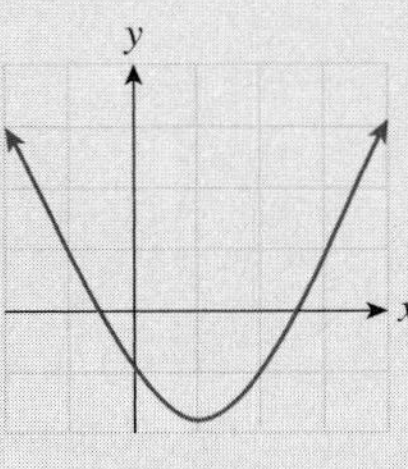

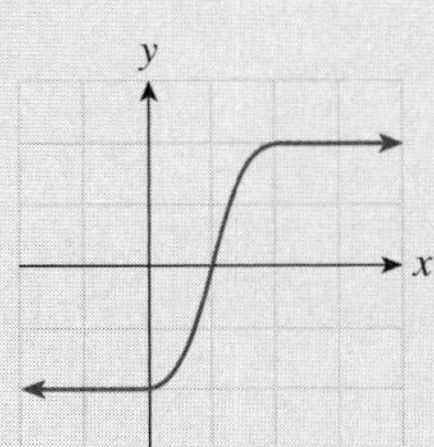

Original Function f
$y = f(x)$

Derivative Function f'
$y = f'(x)$

For x between -2 and 0, the graph of f is linear with slope -2. As x increases from 0 to 2, the slope increases from -2 to 2. For x larger than 2, the graph of f is linear with slope 2. (Notice that, when $x = 1$, the graph of f has a horizontal tangent, so $f'(1) = 0$.)

3. Look again at the graph on the left in Quick Example 2. When $x < 1$ the derivative $f'(x)$ is negative, so the graph has negative slope and f is **decreasing**; its values are going down as x increases. When $x > 1$ the derivative $f'(x)$ is positive, so the graph has positive slope and f is **increasing**; its values are going up as x increases.

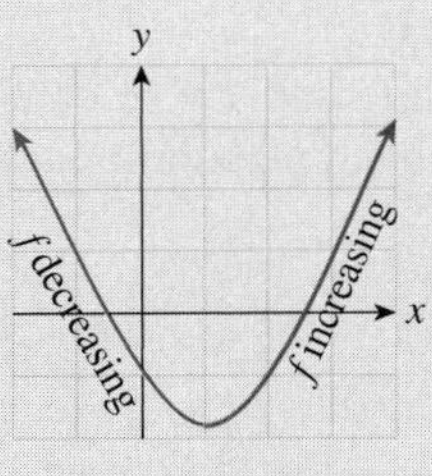

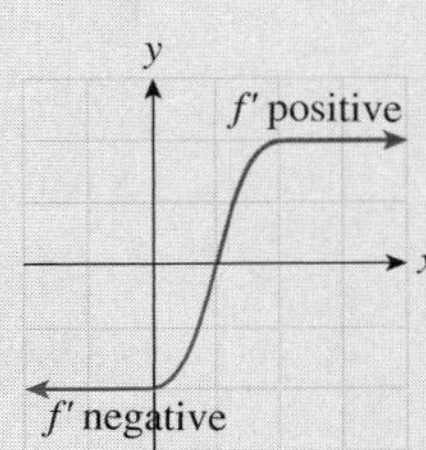

f decreasing for $x < 1$
f increasing for $x > 1$

f' negative for $x < 1$
f' positive for $x > 1$

The following example shows how we can use technology to graph the (approximate) derivative of a function, where it exists.

EXAMPLE 4 Graphing the Derivative with Technology

Use technology to graph the derivative of $f(x) = -2x^2 + 6x + 5$ for values of x starting at -5.

Solution The TI-83/84 Plus has a built-in function that approximates the derivative, and we can use it to graph the derivative of a function. In a spreadsheet, we need to create the approximation using one of the quick approximation formulas and we can then graph a table of its values. See the technology note in the margin to find out how to graph the derivative (Figure 33) using the Website graphing utility, the TI-83/84 Plus, and a spreadsheet.

using Technology

See the Technology Guides at the end of the chapter to find out how to obtain a table of values of and graph the derivative in Example 4 using a TI-83/84 Plus or a spreadsheet. Here is an outline:

TI-83/84 Plus

`Y1=-2X^2+6X+5`
`Y2=nDeriv(Y1,X,X)`
[More details on page 214.]

Spreadsheet

Value of h in E2
Values of x from A2 down increasing by h
`-2*A2^2+6*A2+5` from B2 down
`=(B3-B2)/$E$2` from C2 down
Insert scatter chart using columns A and C [More details on page 216.]

Website

www.WanerMath.com

Web grapher:

Online Utilities→ Function Evaluator and Grapher

Enter `deriv(-2*x^2+6*x+5)` for y_1. Alternatively, enter `-2*x^2+6*x+5` for y_1 and `deriv(y1)` for y_2.

Excel grapher:

Student Home→ Online Utilities→ Excel First and Second Derivative Graphing Utility

Function: `-2*x^2+6*x+5`

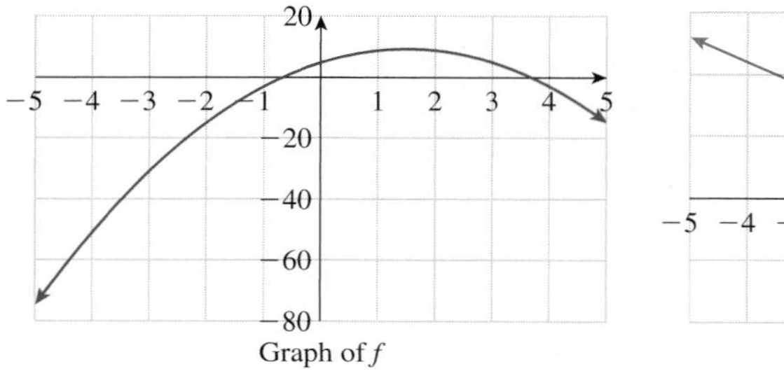

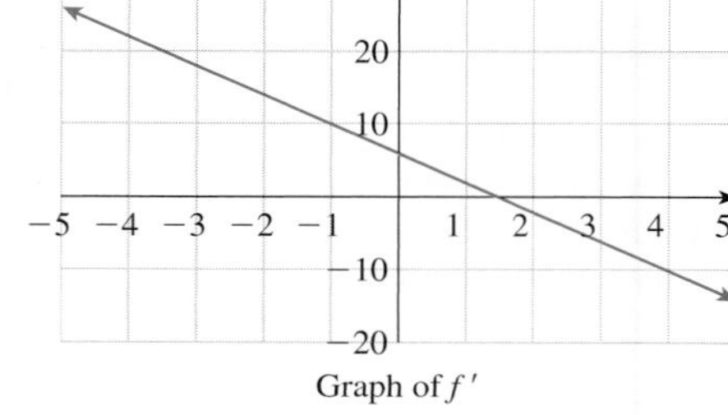

Figure 33

We said that f' records the slope of (the tangent line to) the function f at each point. Notice that the graph of f' confirms that the slope of the graph of f is decreasing as x increases from -5 to 5. Note also that the graph of f reaches a high point at $x = 1.5$ (the vertex of the parabola). At that point, the slope of the tangent is zero; that is, $f'(1.5) = 0$, as we see in the graph of f'.

EXAMPLE 5 An Application: Broadband Penetration

Wired broadband penetration in the United States can be modeled by

$$P(t) = \frac{29.842}{1 + 12.502(1.642)^{-t}} \qquad (0 \le t \le 12),$$

where t is time in years since 2000.[5] Graph both P and its derivative, and determine when broadband penetration was growing most rapidly.

Solution Using one of the methods in Example 4, we obtain the graphs shown in Figure 34.

[5] Broadband penetration is the number of broadband installations divided by the total population. Source for data: Organisation for Economic Co-operation and Development (OECD) Directorate for Science, Technology, and Industry, table of Historical Penetration Rates, June 2010, downloaded April 2011 from www.oecd.org/sti/ict/broadband.

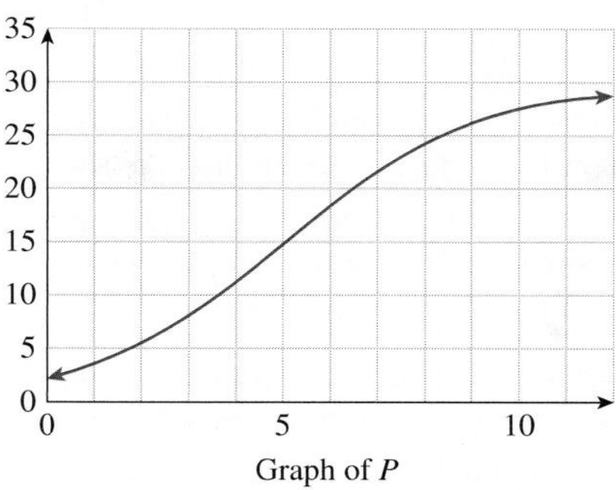

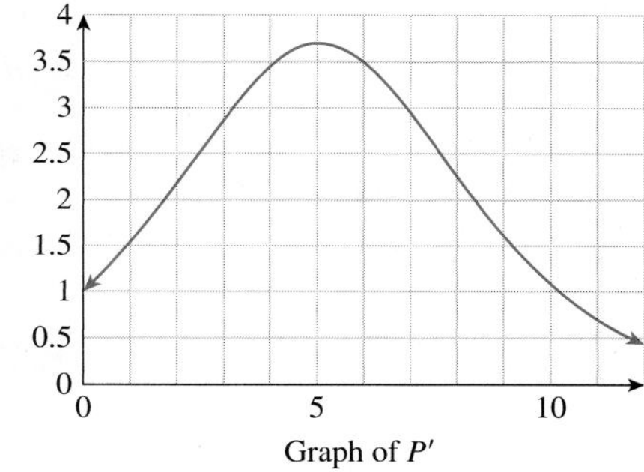

Figure 34

From the graph on the right, we see that P' reaches a peak somewhere near $t = 5$ (the beginning of 2005). Recalling that P' measures the *slope* of the graph of P, we can conclude that the graph of P is steepest near $t = 5$, indicating that, according to the model, broadband penetration was growing most rapidly at the start of 2005. Notice that this is not so easy to see directly on the graph of P.

To determine the point of maximum growth more accurately, we can zoom in on the graph of P' using the range $4.5 \le t \le 5.5$ (Figure 35).

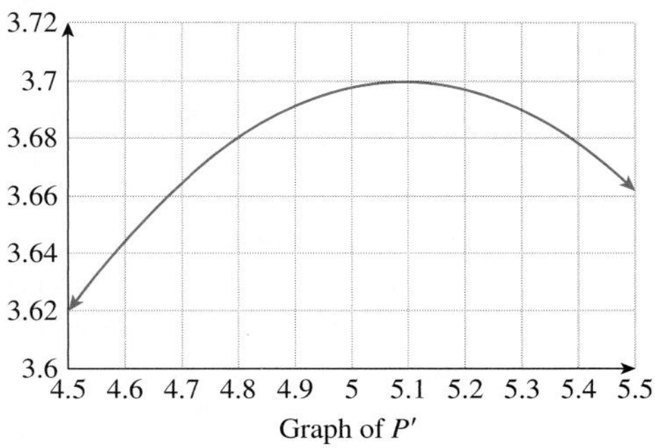

Figure 35

We can now see that P' reaches its highest point around $t = 5.1$, so we conclude that broadband penetration was growing most rapidly in early 2005.

➡ **Before we go on...** Besides helping us to determine the point of maximum growth, the graph of P' in Example 5 gives us a great deal of additional information. As just one example, in Figure 35 we can see that the maximum value of P' is about 3.7, indicating that broadband penetration grew at a fastest rate of about 3.7 percentage points per year. ■

Website
www.WanerMath.com
At the Website you can find the following optional interactive online sections:
- **Continuity and Differentiability**
- **Sketching the Graph of the Derivative**

You can find these sections by following
Everything for Calculus → Chapter 3 (Online Sections)

FAQs

Recognizing When and How to Compute the Instantaneous Rate of Change

Q: *How do I know, by looking at the wording of a problem, that it is asking for an instantaneous rate of change?*

A: If a problem does not ask for an instantaneous rate of change directly, it might do so indirectly, as in "How fast is quantity q increasing?" or "Find the rate of increase of q."

Q: *If I know that a problem calls for estimating an instantaneous rate of change, how should I estimate it: with a table showing smaller and smaller values of h, or by using a quick approximation?*

A: For most practical purposes, a quick approximation is accurate enough. Use a table showing smaller and smaller values of *h* when you would like to check the accuracy.

Q: *Which should I use in computing a quick approximation: the balanced difference quotient or the ordinary difference quotient?*

A: In general, the balanced difference quotient gives a more accurate answer.

3.5 EXERCISES

Access end-of-section exercises online at **www.webassign.net** 

3.6 Derivatives: Algebraic Viewpoint

In Section 3.5 we saw how to estimate the derivative of a function using numerical and graphical approaches. In this section we use an algebraic approach that will give us the *exact value* of the derivative, rather than just an approximation, when the function is specified algebraically.

This algebraic approach is quite straightforward: Instead of subtracting numbers to estimate the average rate of change over smaller and smaller intervals, we subtract algebraic expressions. Our starting point is the definition of the derivative in terms of the difference quotient:

$$f'(a) = \lim_{h\to 0} \frac{f(a+h) - f(a)}{h}.$$

EXAMPLE 1 Calculating the Derivative at a Point Algebraically

Let $f(x) = x^2$. Use the definition of the derivative to compute $f'(3)$ algebraically.

Solution Substituting $a = 3$ into the definition of the derivative, we get:

$$f'(3) = \lim_{h\to 0} \frac{f(3+h) - f(3)}{h}$$ Formula for the derivative

$$= \lim_{h\to 0} \frac{\overbrace{(3+h)^2}^{f(3+h)} - \overbrace{3^2}^{f(3)}}{h}$$ Substitute for $f(3)$ and $f(3+h)$.

$$= \lim_{h\to 0} \frac{(9+6h+h^2) - 9}{h}$$ Expand $(3+h)^2$.

$$= \lim_{h\to 0} \frac{6h+h^2}{h}$$ Cancel the 9.

$$= \lim_{h\to 0} \frac{h(6+h)}{h}$$ Factor out h.

$$= \lim_{h\to 0} (6+h).$$ Cancel the h.

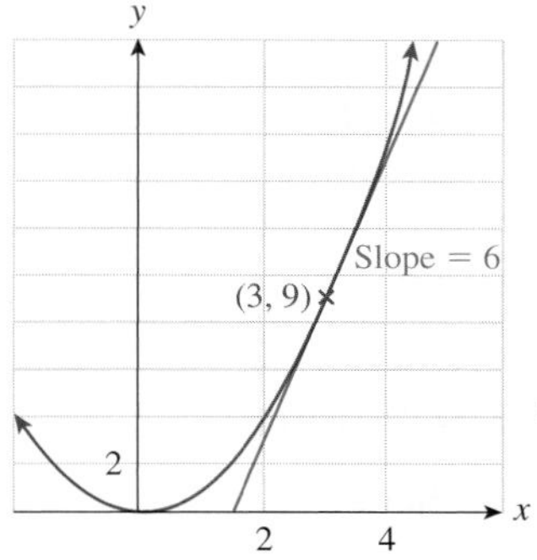

Figure 36

Now we let h approach 0. As h gets closer and closer to 0, the sum $6 + h$ clearly gets closer and closer to $6 + 0 = 6$. Thus,

$$f'(3) = \lim_{h \to 0} (6 + h) = 6. \qquad \text{As } h \to 0, (6 + h) \to 6$$

(Calculations of limits like this are discussed and justified more fully in Sections 3.2 and 3.3.)

Before we go on... We did the following calculation in Example 1: If $f(x) = x^2$, then $f'(3) = 6$. In other words, the tangent to the graph of $y = x^2$ at the point (3, 9) has slope 6 (Figure 36). ■

There is nothing very special about $a = 3$ in Example 1. Let's try to compute $f'(x)$ for general x.

EXAMPLE 2 Calculating the Derivative Function Algebraically

Let $f(x) = x^2$.

a. Use the definition of the derivative to compute $f'(x)$ algebraically.
b. Use the answer to evaluate $f'(3)$.

Solution

a. Once again, our starting point is the definition of the derivative in terms of the difference quotient:

$$f'(x) = \lim_{h \to 0} \frac{f(x+h) - f(x)}{h} \qquad \text{Formula for the derivative}$$

$$= \lim_{h \to 0} \frac{\overbrace{(x+h)^2}^{f(x+h)} - \overbrace{x^2}^{f(x)}}{h} \qquad \text{Substitute for } f(x) \text{ and } f(x+h).$$

$$= \lim_{h \to 0} \frac{(x^2 + 2xh + h^2) - x^2}{h} \qquad \text{Expand } (x+h)^2.$$

$$= \lim_{h \to 0} \frac{2xh + h^2}{h} \qquad \text{Cancel the } x^2.$$

$$= \lim_{h \to 0} \frac{h(2x + h)}{h} \qquad \text{Factor out } h.$$

$$= \lim_{h \to 0} (2x + h). \qquad \text{Cancel the } h.$$

Now we let h approach 0. As h gets closer and closer to 0, the sum $2x + h$ clearly gets closer and closer to $2x + 0 = 2x$. Thus,

$$f'(x) = \lim_{h \to 0} (2x + h) = 2x.$$

This is the derivative function.

b. Now that we have a *formula* for the derivative of f, we can obtain $f'(a)$ for any value of a we choose by simply evaluating f' there. For instance,

$$f'(3) = 2(3) = 6$$

as we saw in Example 1.

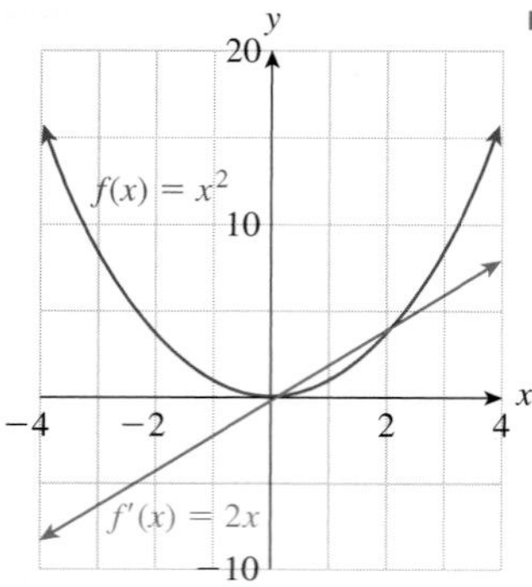

Figure 37

➡ **Before we go on...** The graphs of $f(x) = x^2$ and $f'(x) = 2x$ from Example 2 are familiar. Their graphs are shown in Figure 37.

When $x < 0$, the parabola slopes downward, which is reflected in the fact that the derivative $2x$ is negative there. When $x > 0$, the parabola slopes upward, which is reflected in the fact that the derivative is positive there. The parabola has a horizontal tangent line at $x = 0$, reflected in the fact that $2x = 0$ there. ■

EXAMPLE 3 More Computations of Derivative Functions

Compute the derivative $f'(x)$ for each of the following functions:

a. $f(x) = x^3$ **b.** $f(x) = 2x^2 - x$ **c.** $f(x) = \dfrac{1}{x}$

Solution

a. $f'(x) = \lim_{h\to 0} \dfrac{f(x+h) - f(x)}{h}$ Derivative formula

$= \lim_{h\to 0} \dfrac{\overbrace{(x+h)^3}^{f(x+h)} - \overbrace{x^3}^{f(x)}}{h}$ Substitute for $f(x)$ and $f(x+h)$.

$= \lim_{h\to 0} \dfrac{(x^3 + 3x^2h + 3xh^2 + h^3) - x^3}{h}$ Expand $(x+h)^3$.

$= \lim_{h\to 0} \dfrac{3x^2h + 3xh^2 + h^3}{h}$ Cancel the x^3.

$= \lim_{h\to 0} \dfrac{h(3x^2 + 3xh + h^2)}{h}$ Factor out h.

$= \lim_{h\to 0} (3x^2 + 3xh + h^2)$ Cancel the h.

$= 3x^2.$ Let h approach 0.

b. $f'(x) = \lim_{h\to 0} \dfrac{f(x+h) - f(x)}{h}$ Derivative formula

$= \lim_{h\to 0} \dfrac{(\overbrace{2(x+h)^2 - (x+h)}^{f(x+h)}) - \overbrace{(2x^2 - x)}^{f(x)}}{h}$ Substitute for $f(x)$ and $f(x+h)$.

$= \lim_{h\to 0} \dfrac{(2x^2 + 4xh + 2h^2 - x - h) - (2x^2 - x)}{h}$ Expand.

$= \lim_{h\to 0} \dfrac{4xh + 2h^2 - h}{h}$ Cancel the $2x^2$ and x.

$= \lim_{h\to 0} \dfrac{h(4x + 2h - 1)}{h}$ Factor out h.

$= \lim_{h\to 0} (4x + 2h - 1)$ Cancel the h.

$= 4x - 1.$ Let h approach 0.

c. $$f'(x) = \lim_{h\to 0} \frac{f(x+h) - f(x)}{h}$$ Derivative formula

$$= \lim_{h\to 0} \frac{\left[\overbrace{\frac{1}{x+h}}^{f(x+h)} - \overbrace{\frac{1}{x}}^{f(x)}\right]}{h}$$ Substitute for $f(x)$ and $f(x+h)$.

$$= \lim_{h\to 0} \frac{\left[\frac{x-(x+h)}{(x+h)x}\right]}{h}$$ Subtrac… …ctions.

$$= \lim_{h\to 0} \frac{1}{h}\left[\frac{x-(x+h)}{(x+h)x}\right]$$ Dividing by h = … by $1/h$. …ng

$$= \lim_{h\to 0} \left[\frac{-h}{h(x+h)x}\right]$$ Simplify.

$$= \lim_{h\to 0} \left[\frac{-1}{(x+h)x}\right]$$ Cancel the h.

$$= \frac{-1}{x^2}$$ Let h approach 0.

In Example 4, we redo Example 3 of Section 3.5, this time getting an exact, rather than approximate, answer.

EXAMPLE 4 Velocity

My friend Eric, an enthusiastic baseball player, claims he can "probably" throw a ball upward at a speed of 100 feet per second (ft/sec). Our physicist friends tell us that its height s (in feet) t seconds later would be $s(t) = 100t - 16t^2$. Find the ball's instantaneous velocity function and its velocity exactly 2 seconds after Eric throws it.

Solution The instantaneous velocity function is the derivative ds/dt, which we calculate as follows:

$$\frac{ds}{dt} = \lim_{h\to 0} \frac{s(t+h) - s(t)}{h}.$$

Let us compute $s(t+h)$ and $s(t)$ separately:

$$\begin{aligned} s(t) &= 100t - 16t^2 \\ s(t+h) &= 100(t+h) - 16(t+h)^2 \\ &= 100t + 100h - 16(t^2 + 2th + h^2) \\ &= 100t + 100h - 16t^2 - 32th - 16h^2. \end{aligned}$$

Therefore,

$$\begin{aligned} \frac{ds}{dt} &= \lim_{h\to 0} \frac{s(t+h) - s(t)}{h} \\ &= \lim_{h\to 0} \frac{100t + 100h - 16t^2 - 32th - 16h^2 - (100t - 16t^2)}{h} \\ &= \lim_{h\to 0} \frac{100h - 32th - 16h^2}{h} = \lim_{h\to 0} \frac{h(100 - 32t - 16h)}{h} \\ &= \lim_{h\to 0} (100 - 32t - 16h) = 100 - 32t \text{ ft/sec}. \end{aligned}$$

Thus, the velocity exactly 2 seconds after Eric throws it is

$$\left.\frac{ds}{dt}\right|_{t=2} = 100 - 32(2) = 36 \text{ ft/sec}.$$

This verifies the accuracy of the approximation we made in Section 3.5.

➡ **Before we go on...** From the derivative function in Example 4, we can now describe the behavior of the velocity of the ball: Immediately on release ($t = 0$) the ball is traveling at 100 feet per second upward. The ball then slows down; precisely, it loses 32 feet per second of speed every second. When, exactly, does the velocity become zero and what happens after that? ■

Q: *Do we always have to calculate the limit of the difference quotient to find a formula for the derivative function?*

A: As it turns out, no. In Section 4.1 we will start to look at shortcuts for finding derivatives that allow us to bypass the definition of the derivative in many cases.

A Function Not Differentiable at a Point

Recall from Section 3.5 that a function is **differentiable** at a point a if $f'(a)$ exists; that is, if the difference quotient $[f(a+h) - f(a)]/h$ approaches a fixed value as h approaches 0. In Section 3.5, we mentioned that the function $f(x) = |x|$ is not differentiable at $x = 0$. In Example 5, we find out why.

EXAMPLE 5 A Function Not Differentiable at 0

Numerically, graphically, and algebraically investigate the differentiability of the function $f(x) = |x|$ at the points **(a)** $x = 1$ and **(b)** $x = 0$.

Solution

a. We compute

$$f'(1) = \lim_{h\to 0} \frac{f(1+h) - f(1)}{h} = \lim_{h\to 0} \frac{|1+h| - 1}{h}.$$

Numerically, we can make tables of the values of the average rate of change $(|1+h| - 1)/h$ for h positive or negative and approaching 0:

h	1	0.1	0.01	0.001	0.0001
Average Rate of Change Over [1, 1 + h]	1	1	1	1	1

h	−1	−0.1	−0.01	−0.001	−0.0001
Average Rate of Change Over [1 + h, 1]	1	1	1	1	1

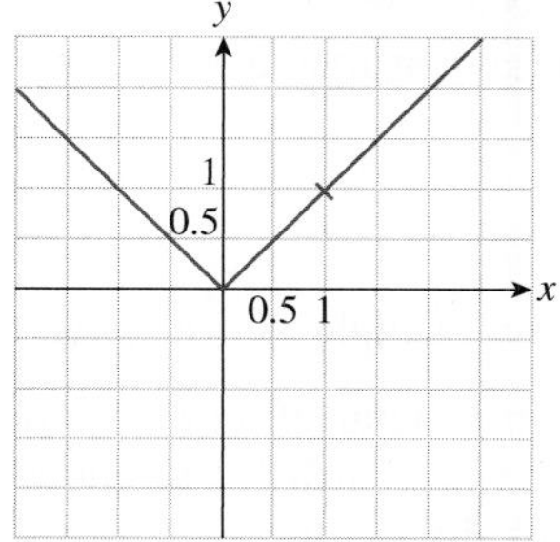

Figure 38

From these tables it appears that $f'(1)$ is equal to 1. We can verify that algebraically: For h that is sufficiently small, $1 + h$ is positive (even if h is negative) and so

$$
\begin{aligned}
f'(1) &= \lim_{h \to 0} \frac{1 + h - 1}{h} \\
&= \lim_{h \to 0} \frac{h}{h} && \text{Cancel the 1s.} \\
&= \lim_{h \to 0} 1 && \text{Cancel the } h. \\
&= 1.
\end{aligned}
$$

Graphically, we are seeing the fact that the tangent line at the point (1, 1) has slope 1 because the graph is a straight line with slope 1 near that point (Figure 38).

b.
$$
\begin{aligned}
f'(0) &= \lim_{h \to 0} \frac{f(0 + h) - f(0)}{h} \\
&= \lim_{h \to 0} \frac{|0 + h| - 0}{h} \\
&= \lim_{h \to 0} \frac{|h|}{h}
\end{aligned}
$$

If we make tables of values in this case we get the following:

h	1	0.1	0.01	0.001	0.0001
Average Rate of Change over [0, 0 + *h*]	1	1	1	1	1

h	−1	−0.1	−0.01	−0.001	−0.0001
Average Rate of Change over [0 + *h*, 0]	−1	−1	−1	−1	−1

For the limit and hence the derivative $f'(0)$ to exist, the average rates of change should approach the same number for both positive and negative h. Because they do not, f is not differentiable at $x = 0$. We can verify this conclusion algebraically: If h is positive, then $|h| = h$, and so the ratio $|h|/h$ is 1, regardless of how small h is. Thus, according to the values of the difference quotients with $h > 0$, the limit should be 1. On the other hand if h is negative, then $|h| = -h$ (positive) and so $|h|/h = -1$, meaning that the limit should be -1. Because the limit cannot be both -1 and 1 (it must be a single number for the derivative to exist), we conclude that $f'(0)$ does not exist.

To see what is happening graphically, take a look at Figure 39, which shows zoomed-in views of the graph of f near $x = 0$.

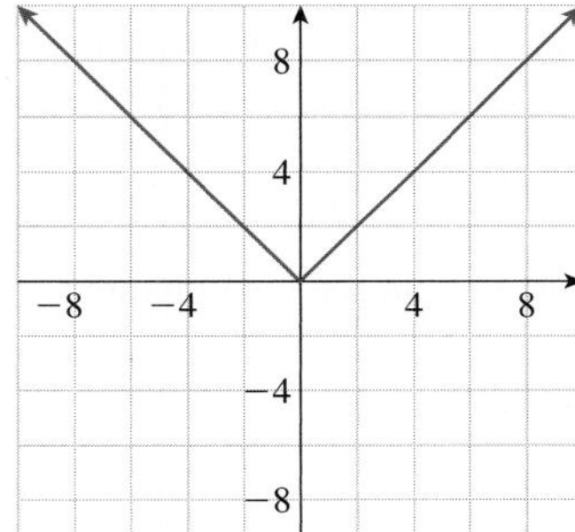

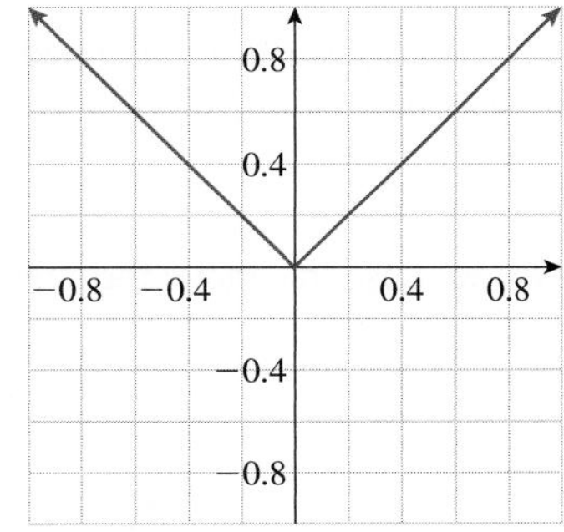

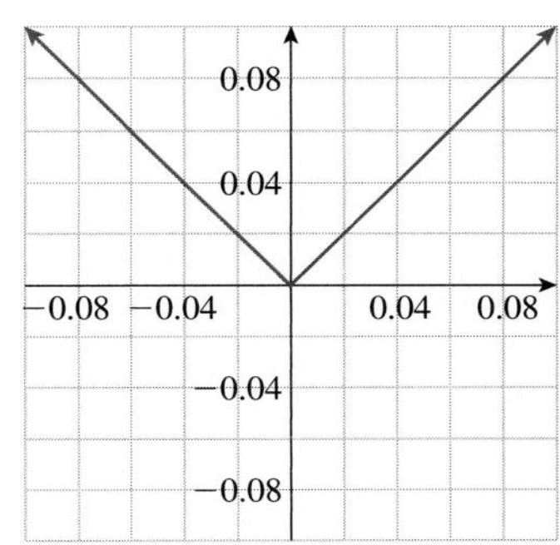

Figure 39

No matter what scale we use to view the graph, it has a sharp corner at $x = 0$ and hence has no tangent line there. Since there is no tangent line at $x = 0$, the function is not differentiable there.

➡ **Before we go on...** Notice that $|x| = \begin{cases} -x & \text{if } x < 0 \\ x & \text{if } x \geq 0 \end{cases}$ is an example of a piecewise-linear function whose graph comes to a point at $x = 0$. In general, if $f(x)$ is any piecewise linear function whose graph comes to a point at $x = a$, it will be nondifferentiable at $x = a$ for the same reason that $|x|$ fails to be differentiable at $x = 0$.

If we repeat the computation in Example 5(a) using any nonzero value for a in place of 1, we see that f is differentiable there as well. If a is positive, we find that $f'(a) = 1$ and, if a is negative, $f'(a) = -1$. In other words, the derivative function is

$$f'(x) = \begin{cases} -1 & \text{if } x < 0 \\ 1 & \text{if } x > 0 \end{cases}.$$

Immediately to the left of $x = 0$, we see that $f'(x) = -1$, immediately to the right, $f'(x) = 1$; and when $x = 0$, $f'(x)$ is not defined. ■

Q: *So does that mean there is no single formula for the derivative of $|x|$?*

A: Actually, there is a convenient formula. Consider the ratio $\frac{|x|}{x}$. If x is positive, then $|x| = x$, so $\frac{|x|}{x} = \frac{x}{x} = 1$. On the other hand, if x is negative, then $|x| = -x$, so $\frac{|x|}{x} = \frac{-x}{x} = -1$. In other words,

$$\frac{|x|}{x} = \begin{cases} -1 & \text{if } x < 0 \\ 1 & \text{if } x > 0 \end{cases},$$

which is exactly the formula we obtained for $f'(x)$. We have therefore obtained a convenient closed-form formula for the derivative of |x|!

Derivative of |x|

If $f(x) = |x|$, then $f'(x) = \frac{|x|}{x}$.

Note that $|x|/x$ is not defined if $x = 0$, reflecting the fact that $f'(x)$ does not exist when $x = 0$.

We will use the above formula extensively in the next chapter.

FAQ

Computing Derivatives Algebraically

Q: *The algebraic computation of $f'(x)$ seems to require a number of steps. How do I remember what to do, and when?*

A: If you examine the computations in the examples above, you will find the following pattern:

1. Write out the formula for $f'(x)$, as the limit of the difference quotient, then substitute $f(x+h)$ and $f(x)$.
2. Expand and simplify the *numerator* of the expression, but not the denominator.
3. After simplifying the numerator, factor out an h to cancel with the h in the denominator. If h does not factor out of the numerator, you might have made an error. (A frequent error is a wrong sign.)
4. After canceling the h, you should be able to see what the limit is by letting $h \to 0$.

3.6 EXERCISES

Access end-of-section exercises online at **www.webassign.net**

CHAPTER 3 REVIEW

KEY CONCEPTS

 Website www.WanerMath.com

Go to the Website at www.WanerMath.com to find a comprehensive and interactive Web-based summary of Chapter 3.

3.1 Limits: Numerical and Graphical Viewpoints

$\lim_{x\to a} f(x) = L$ means that $f(x)$ approaches L as x approaches a *p. 154*

What it means for a limit to exist *p. 155*

Limits at infinity *p. 157*

Estimating limits graphically *p. 157*

Interpreting limits in real-world situations *p. 160*

3.2 Limits and Continuity

f is continuous at a if $\lim_{x\to a} f(x)$ exists and $\lim_{x\to a} f(x) = f(a)$ *p. 162*

Discontinuous, continuous on domain *p. 162*

Determining whether a given function is continuous *p. 163*

3.3 Limits and Continuity: Algebraic Viewpoint

Closed-form function *p. 166*

Limits of closed form functions *p. 167*

Simplifying to obtain limits *p. 167*

The indeterminate form 0/0 *p. 168*

The determinate form $k/0$ *p. 169*

Limits of piecewise-defined functions *p. 170*

Limits at infinity *p. 171*

Determinate and indeterminate forms *p. 174–175*

3.4 Average Rate of Change

Average rate of change of $f(x)$ over $[a, b]$: $\dfrac{\Delta f}{\Delta x} = \dfrac{f(b) - f(a)}{b - a}$ *p. 179*

Average rate of change as slope of the secant line *p. 179*

Computing the average rate of change from a graph *p. 180*

Computing the average rate of change from a formula *p. 182*

Computing the average rate of change over short intervals $[a, a + h]$ *p. 183*

3.5 Derivatives: Numerical and Graphical Viewpoints

Instantaneous rate of change of $f(x)$ (derivative of f at a);

$f'(a) = \lim_{h\to 0} \dfrac{f(a + h) - f(a)}{h}$ *p. 186*

The derivative as slope of the tangent line *p. 189*

Quick approximation of the derivative *p. 190*

$\dfrac{d}{dx}$ Notation *p. 192*

The derivative as velocity *p. 193*

Average and instantaneous velocity *p. 194*

The derivative function *p. 194*

Graphing the derivative function with technology *p. 196*

3.6 Derivatives: Algebraic Viewpoint

Derivative at the point $x = a$:

$f'(a) = \lim_{h\to 0} \dfrac{f(a + h) - f(a)}{h}$ *p. 198*

Derivative function:

$f'(x) = \lim_{h\to 0} \dfrac{f(x + h) - f(x)}{h}$ *p. 199*

Examples of the computation of $f'(x)$ *p. 200*

$f(x) = |x|$ is not differentiable at $x = 0$ *p. 202*

REVIEW EXERCISES

T indicates exercises that must be solved using technology

Numerically *estimate whether the limits in Exercises 1–4 exist. If a limit does exist, give its approximate value.*

1. $\lim_{x\to 3} \dfrac{x^2 - x - 6}{x - 3}$ **2.** $\lim_{x\to 3} \dfrac{x^2 - 2x - 6}{x - 3}$

3. $\lim_{x\to -1} \dfrac{|x + 1|}{x^2 - x - 2}$ **4.** $\lim_{x\to -1} \dfrac{|x + 1|}{x^2 + x - 2}$

In Exercises 5 and 6, the graph of a function f is shown. Graphically determine whether the given limits exist. If a limit does exist, give its approximate value.

5.

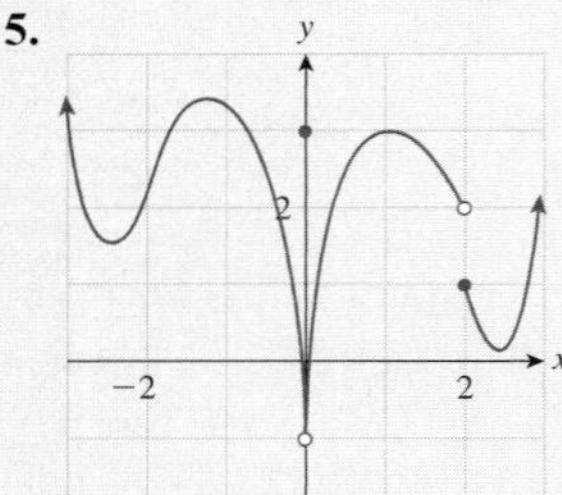

a. $\lim_{x\to 0} f(x)$ **b.** $\lim_{x\to 1} f(x)$ **c.** $\lim_{x\to 2} f(x)$

6.

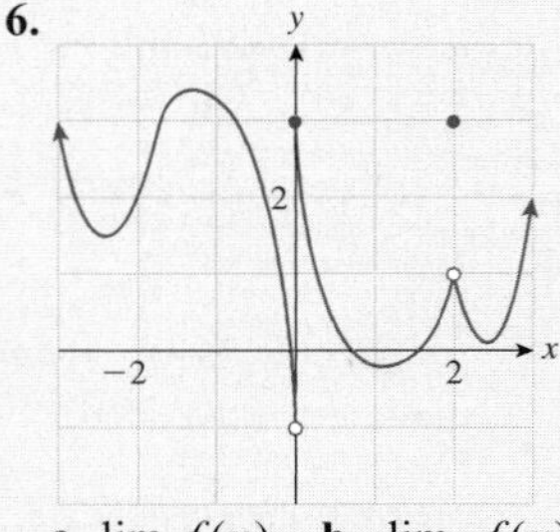

a. $\lim_{x\to 0} f(x)$ **b.** $\lim_{x\to -2} f(x)$ **c.** $\lim_{x\to 2} f(x)$

Calculate the limits in Exercises 7–30 algebraically. If a limit does not exist, say why.

7. $\lim_{x\to -2} \dfrac{x^2}{x - 3}$ **8.** $\lim_{x\to 3} \dfrac{x^2 - 9}{2x - 6}$

9. $\lim_{x\to -2} \dfrac{x^2 - 4}{x^3 + 2x^2}$ **10.** $\lim_{x\to -1} \dfrac{x^2 - 9}{2x - 6}$

11. $\lim_{x\to 0} \dfrac{x}{2x^2 - x}$ **12.** $\lim_{x\to 1} \dfrac{x^2 - 9}{x - 1}$

13. $\lim_{x\to -1} \dfrac{x^2 + 3x}{x^2 - x - 2}$ **14.** $\lim_{x\to -1^+} \dfrac{x^2 + 1}{x^2 + 3x + 2}$

15. $\lim_{x\to 8} \dfrac{x^2 - 6x - 16}{x^2 - 9x + 8}$ **16.** $\lim_{x\to 4} \dfrac{x^2 + 3x}{x^2 - 8x + 16}$

17. $\lim_{x\to 4} \dfrac{x^2 + 8}{x^2 - 2x - 8}$ **18.** $\lim_{x\to 6} \dfrac{x^2 - 5x - 6}{x^2 - 36}$

19. $\lim_{x\to 1/2} \dfrac{x^2 + 8}{4x^2 - 4x + 1}$ **20.** $\lim_{x\to 1/2} \dfrac{x^2 + 3x}{2x^2 + 3x - 1}$

21. $\lim_{x\to +\infty} \dfrac{10x^2 + 300x + 1}{5x^3 + 2}$ **22.** $\lim_{x\to +\infty} \dfrac{2x^4 + 20x^3}{1{,}000x^6 + 6}$

23. $\lim_{x\to -\infty} \dfrac{x^2 - x - 6}{x - 3}$ **24.** $\lim_{x\to +\infty} \dfrac{x^2 - x - 6}{4x^2 - 3}$

25. $\lim\limits_{t\to+\infty} \dfrac{-5}{5+5.3(3^{2t})}$

26. $\lim\limits_{t\to+\infty} \left(3+\dfrac{2}{e^{4t}}\right)$

27. $\lim\limits_{x\to+\infty} \dfrac{2}{5+4e^{-3x}}$

28. $\lim\limits_{x\to+\infty} (4e^{3x}+12)$

29. $\lim\limits_{t\to+\infty} \dfrac{1+2^{-3t}}{1+5.3e^{-t}}$

30. $\lim\limits_{x\to-\infty} \dfrac{8+0.5^x}{2-3^{2x}}$

In Exercises 31–34, find the average rate of change of the given function over the interval $[a, a+h]$ for $h = 1, 0.01,$ and 0.001. (Round answers to four decimal places.) Then estimate the slope of the tangent line to the graph of the function at a.

31. $f(x) = \frac{1}{x+1};\ a = 0$

32. $f(x) = x^x;\ a = 2$

33. $f(x) = e^{2x};\ a = 0$

34. $f(x) = \ln(2x);\ a = 1$

*In Exercises 35–38, you are given the graph of a function with four points marked. Determine at which (if any) of these points the derivative of the function is: **(a)** -1, **(b)** 0, **(c)** 1, and **(d)** 2.*

35.

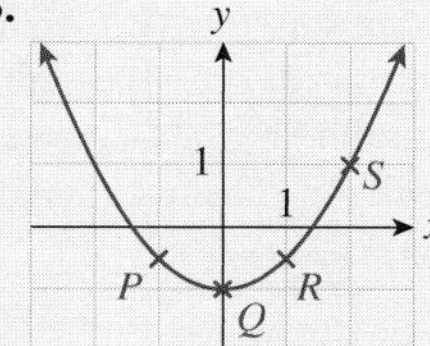

36.

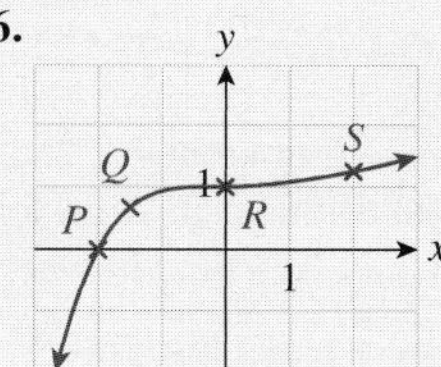

37.

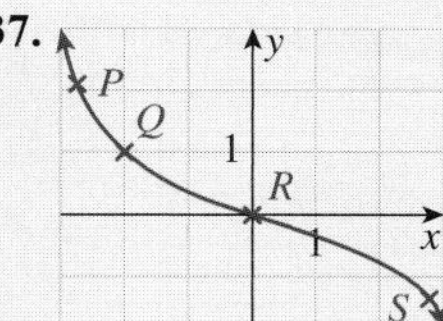

38.

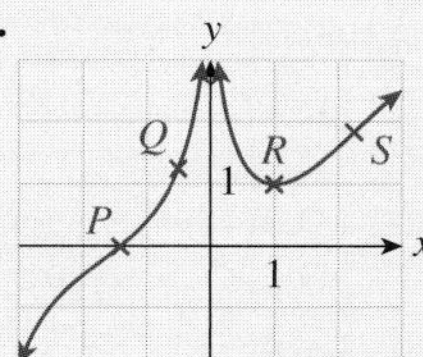

39. Let f have the graph shown.

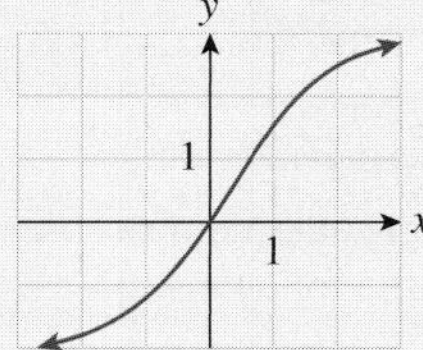

Select the correct answer.

a. The average rate of change of f over the interval [0, 2] is
(A) greater than $f'(0)$. **(B)** less than $f'(0)$.
(C) approximately equal to $f'(0)$.

b. The average rate of change of f over the interval $[-1, 1]$ is
(A) greater than $f'(0)$. **(B)** less than $f'(0)$.
(C) approximately equal to $f'(0)$.

c. Over the interval [0, 2], the instantaneous rate of change of f is
(A) increasing. **(B)** decreasing.
(C) neither increasing nor decreasing.

d. Over the interval $[-2, 2]$, the instantaneous rate of change of f is
(A) increasing, then decreasing.
(B) decreasing, then increasing.
(C) approximately constant.

e. When $x = 2$, $f(x)$ is
(A) approximately 1 and increasing at a rate of about 2.5 units per unit of x.
(B) approximately 1.2 and increasing at a rate of about 1 unit per unit of x.
(C) approximately 2.5 and increasing at a rate of about 0.5 units per unit of x.
(D) approximately 2.5 and increasing at a rate of about 2.5 units per unit of x.

40. Let f have the graph shown.

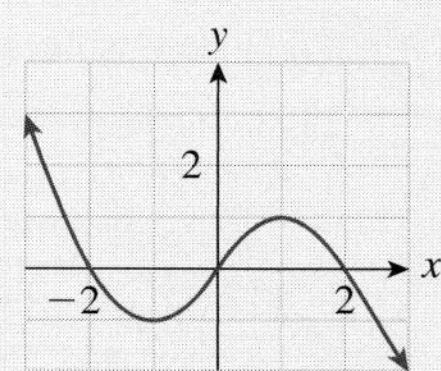

Select the correct answer.

a. The average rate of change of f over the interval [0, 1] is
(A) greater than $f'(0)$. **(B)** less than $f'(0)$.
(C) approximately equal to $f'(0)$.

b. The average rate of change of f over the interval [0, 2] is
(A) greater than $f'(1)$. **(B)** less than $f'(1)$.
(C) approximately equal to $f'(1)$.

c. Over the interval $[-2, 0]$, the instantaneous rate of change of f is
(A) increasing. **(B)** decreasing.
(C) neither increasing nor decreasing.

d. Over the interval $[-2, 2]$, the instantaneous rate of change of f is
(A) increasing, then decreasing.
(B) decreasing, then increasing.
(C) approximately constant.

e. When $x = 0$, $f(x)$ is
(A) approximately 0 and increasing at a rate of about 1.5 units per unit of x.
(B) approximately 0 and decreasing at a rate of about 1.5 units per unit of x.
(C) approximately 1.5 and neither increasing nor decreasing.
(D) approximately 0 and neither increasing nor decreasing.

In Exercises 41–44, use the definition of the derivative to calculate the derivative of each of the given functions algebraically.

41. $f(x) = x^2 + x$

42. $f(x) = 3x^2 - x + 1$

43. $f(x) = 1 - \dfrac{2}{x}$

44. $f(x) = \dfrac{1}{x} + 1$

T *In Exercises 45–48, use technology to graph the derivative of the given function. In each case, choose a range of x-values and y-values that shows the interesting features of the graph.*

45. $f(x) = 10x^5 + \dfrac{1}{2}x^4 - x + 2$

46. $f(x) = \dfrac{10}{x^5} + \dfrac{1}{2x^4} - \dfrac{1}{x} + 2$

47. $f(x) = 3x^3 + 3\sqrt[3]{x}$

48. $f(x) = \dfrac{2}{x^{2.1}} - \dfrac{x^{0.1}}{2}$

APPLICATIONS: OHaganBooks.com

49. ***Stock Investments*** OHaganBooks.com CEO John O'Hagan has terrible luck with stocks. The following graph shows the value of Fly-By-Night Airlines stock that he bought acting on a "hot tip" from Marjory Duffin (CEO of Duffin House publishers and a close business associate):

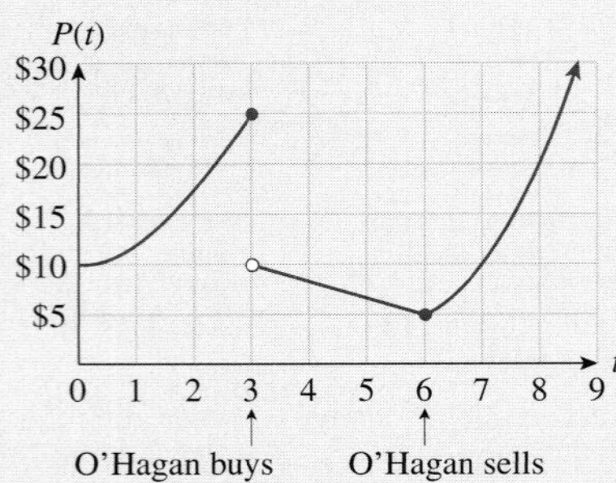

a. Compute $P(3)$, $\lim_{t\to 3^-} P(t)$ and $\lim_{t\to 3^+} P(t)$. Does $\lim_{t\to 3} P(t)$ exist? Interpret your answers in terms of Fly-By-Night stock.

b. Is P continuous at $t = 6$? Is P differentiable at $t = 6$? Interpret your answers in terms of Fly-By-Night stock.

50. ***Stock Investments*** John O'Hagan's golf partner Juan Robles seems to have had better luck with his investment in Gapple Gomputer Inc. stocks as shown in the following graph:

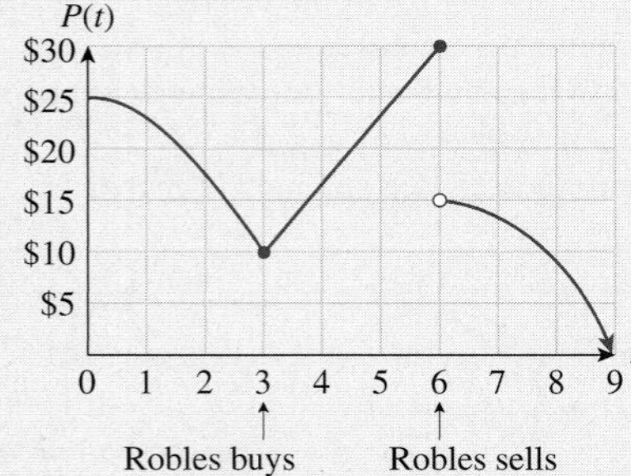

a. Compute $P(6)$, $\lim_{t\to 6^-} P(t)$ and $\lim_{t\to 6^+} P(t)$. Does $\lim_{t\to 6} P(t)$ exist? Interpret your answers in terms of Gapple stock.

b. Is P continuous at $t = 3$? Is P differentiable at $t = 3$? Interpret your answers in terms of Gapple stock.

51. ***Real Estate*** Marjory Duffin has persuaded John O'Hagan to consider investing a portion of OHaganBooks.com profits in real estate, now that the real estate market seems to have bottomed out. A real estate broker friend of hers emailed her the following (somewhat optimistic) graph from brokersadvocacy.com:[6]

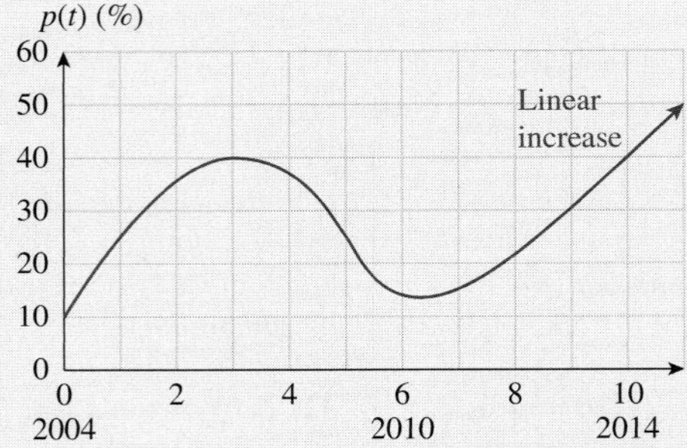

Here, $p(t)$ is the home price percentage over the 2003 level.

a. Assuming the trend shown in the graph were to continue indefinitely, estimate $\lim_{t\to 3} p(t)$ and $\lim_{t\to +\infty} p(t)$ and interpret the results.

b. Estimate $\lim_{t\to +\infty} p'(t)$ and interpret the result.

52. ***Advertising Costs*** OHaganBooks.com has (on further advice from Marjory Duffin) mounted an aggressive online marketing strategy. The following graph shows the weekly cost of this campaign for the six-week period since the start of July (t is time in weeks):

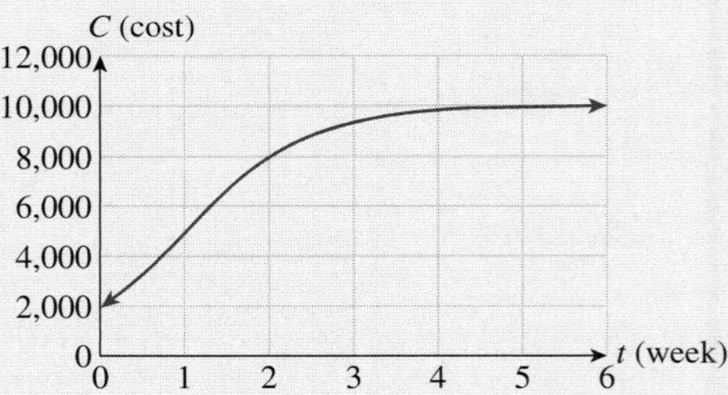

a. Assuming the trend shown in the graph were to continue indefinitely, estimate $\lim_{t\to 2} C(t)$ and $\lim_{t\to +\infty} C(t)$ and interpret the results.

b. Estimate $\lim_{t\to +\infty} C'(t)$ and interpret the result.

53. ***Sales*** Since the start of July, OHaganBooks.com has seen its weekly sales increase, as shown in the following table:

Week	1	2	3	4	5	6
Sales (books)	6,500	7,000	7,200	7,800	8,500	9,000

a. What was the average rate of increase of weekly sales over this entire period?

b. During which 1-week interval(s) did the rate of increase of sales exceed the average rate?

c. During which 2-week interval(s) did the weekly sales rise at the highest average rate, and what was that average rate?

[6] Authors' note: As of August 2012, brokersadvocacy.com is unregistered.

54. ***Rising Sea Level*** Marjory Duffin recently purchased a beachfront condominium in New York and is now in a panic, having just seen some disturbing figures about rising sea levels (sea levels as measured in New York relative to the 1900 level).[7]

Year since 1900	0	25	50	75	100	125
Sea Level (mm)	0	60	140	240	310	390

a. What was the average rate of increase of the sea level over this entire period?
b. During which 25-year interval(s) did the rate of increase of the sea level exceed the average rate?
c. Marjory Duffin's condominium is about 2 meters above sea level. Using the average rate of change from part (a), estimate how long she has before the sea rises to her condominium.

55. ***Real Estate*** The following graph (see Exercise 51) shows the home price index chart emailed to Marjory Duffin by a real estate broker:

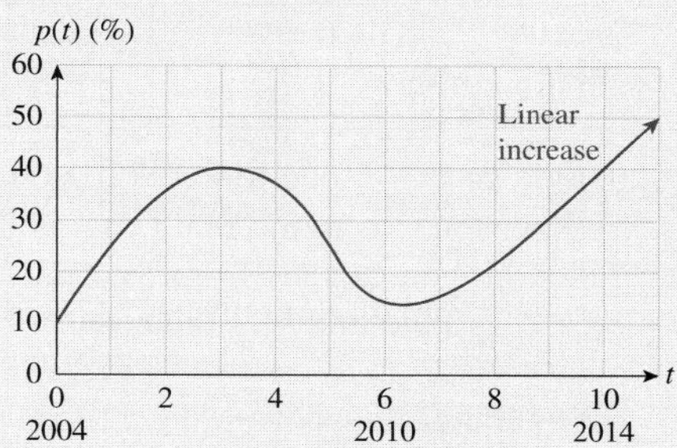

Use the graph to answer the following questions:
a. What was the average rate of change of the index over the 10-year period beginning 2004?
b. What was the average rate of change of the index over the period [3, 10]?
c. Which of the following is correct? Over the period [4, 6],
(A) The rate of change of the index increased.
(B) The rate of change of the index increased and then decreased.
(C) The rate of change of the index decreased.
(D) The rate of change of the index decreased and then increased.

56. ***Advertising Costs*** The following graph (see Exercise 52) shows the weekly cost of OHaganBooks.com's online ad campaign for the 6-week period since the start of July (t is time in weeks).

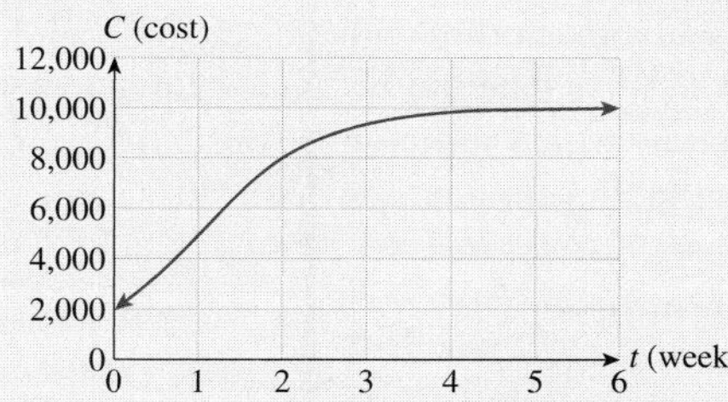

Use the graph to answer the following questions:
a. What was the average rate of change of cost over the entire six-week period?
b. What was the average rate of change of cost over the period [2, 6]?
c. Which of the following is correct? Over the period [2, 6],
(A) The rate of change of cost increased and the cost increased.
(B) The rate of change of cost decreased and the cost increased.
(C) The rate of change of cost increased and the cost decreased.
(D) The rate of change of cost decreased and the cost decreased.

57. ***Sales*** OHaganBooks.com fits the curve

$$w(t) = 36t^2 + 250t + 6{,}240 \quad (0 \le t \le 6)$$

to its weekly sales figures from Exercise 53, as shown in the following graph:

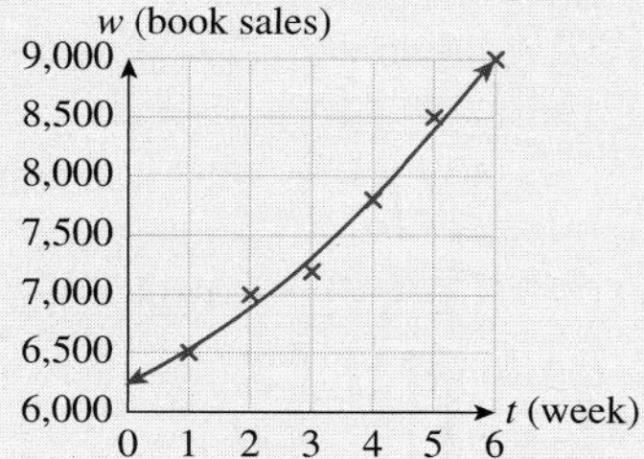

a. Compute the derivative function $w'(t)$.
b. According to the model, what was the rate of increase of sales at the beginning of the second week ($t = 1$)?
c. If we extrapolate the model, what would be the rate of increase of weekly sales at the beginning of the 8th week ($t = 7$)?

58. ***Sea Levels*** Marjory Duffin fit the curve

$$s(t) = 0.002t^2 + 3t - 6.4 \quad (0 \le t \le 125)$$

to her sea level figures from Exercise 54, as shown in the following graph:

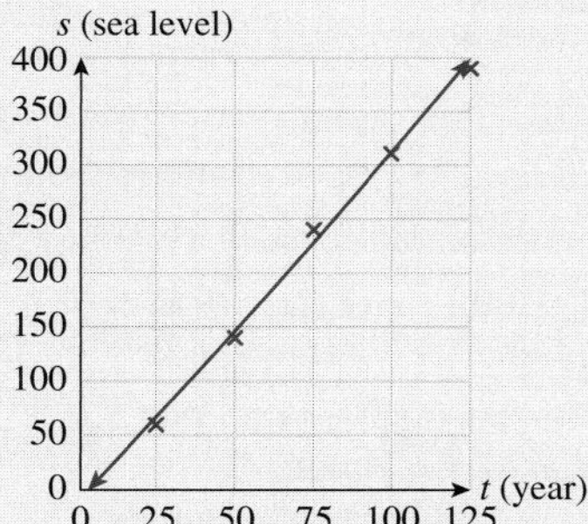

a. Compute the derivative function $s'(t)$.
b. According to the model, what was the rate of increase of the sea level in 2000 ($t = 100$)?
c. If we extrapolate the model, what would be the rate of increase of the sea level in 2100 ($t = 200$)?

[7]The 2025 level is a projection. Source: New England Integrated Science & Assessment (www.neisa.unh.edu/Climate/index.html).

Case Study

Reducing Sulfur Emissions

Norbert Schaefer/CORBIS

The Environmental Protection Agency (EPA) wishes to formulate a policy that will encourage utilities to reduce sulfur emissions. Its goal is to reduce annual emissions of sulfur dioxide by a total of 10 million tons from the current level of 25 million tons by imposing a fixed charge for every ton of sulfur released into the environment per year. As a consultant to the EPA, you must determine the amount to be charged per ton of sulfur emissions.

You would like first to know the cost to the utility industry of reducing sulfur emissions. In other words, you would like to have a cost function of the form

$$C(q) = \text{Cost of removing } q \text{ tons of sulfur dioxide.}$$

Unfortunately, you do not have such a function handy. You do, however, have the following data, which show the *marginal* cost (that is, the *rate of change* of cost) to the utility industry of reducing sulfur emissions at several levels of reduction.[8]

Reduction (tons) q	8,000,000	10,000,000	12,000,000
Marginal Cost (\$ per ton) $C'(q)$	270	360	779

C

Cost of reducing sulfur emissions by q tons

0 5 10 15 q

Million tons of sulfur reduction

Figure 40

The table tells you that $C'(8{,}000{,}000) = \$270$ per ton, $C'(10{,}000{,}000) = \$360$ per ton, and $C'(12{,}000{,}000) = \$779$ per ton. Recalling that $C'(q)$ is the slope of the tangent to the graph of the cost function, you can see from the table that this slope is positive and increasing as q increases, so the graph of the cost function has the general shape shown in Figure 40.

Notice that the slope (additional cost) is increasing as you move to the right, so the utility industry has no cost incentive to reduce emissions further, as it costs the industry significantly more per ton for each additional ton of sulfur it removes. What you would like—if the goal of reducing total emissions by 10 million tons is to be reached—is that, somehow, the imposition of a fixed charge for every ton of sulfur dioxide released will *alter* the form of the cost curve so that it has the general shape shown in Figure 41. In this ideal curve, the cost D to utilities is lowest at a reduction level of 10 million tons, so if the utilities act to minimize cost, they can be expected to reduce emissions by 10 million tons, which is exactly the EPA goal! From the graph, you can see that the tangent line to the curve at the point where $q = 10$ million tons is horizontal, and thus has zero slope: $D'(10{,}000{,}000) = \$0$ per ton. Further, the slope $D'(q)$ is negative for values of q to the left of 10 million tons and positive for values to the right.

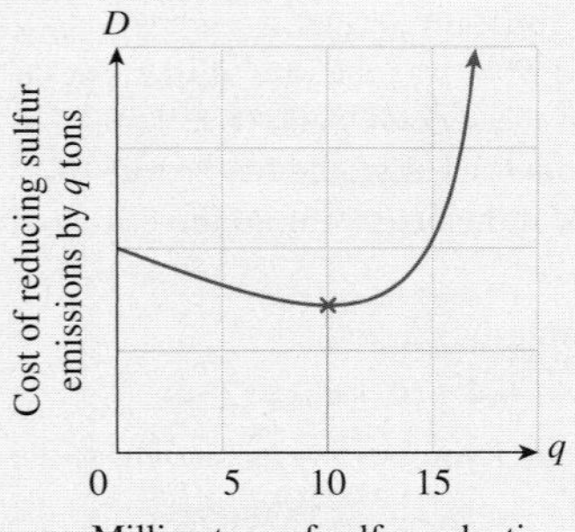

Figure 41

So, how much should the EPA charge per ton of sulfur released into the environment? Suppose the EPA charges $\$k$ per ton, so that

$$\text{Emission charge to utilities} = k \times \text{Sulfur emissions.}$$

It is your job to calculate k. Because you are working with q as the independent variable, you decide that it would be best to formulate the emission charge as a function of q. However, q represents the amount by which sulfur emissions have been reduced from the original 25 million tons; that is, the amount by which sulfur emissions are *lower than* the original 25 million tons:

$$q = 25{,}000{,}000 - \text{Sulfur emissions.}$$

[8] These figures were produced in a computerized study of reducing sulfur emissions from the 1980 level by the given amounts. Source: Congress of the United States, Congressional Budget Office, *Curbing Acid Rain: Cost, Budget and Coal Market Effects* (Washington, DC: Government Printing Office, 1986): xx, xxii, 23, 80.

Thus, the total annual emission charge to the utilities is

$$k \times \text{Sulfur emissions} = k(25{,}000{,}000 - q) = 25{,}000{,}000k - kq.$$

This results in a total cost to the utilities of

$$\begin{aligned}\text{Total cost} &= \text{Cost of reducing emissions} + \text{Emission charge}\\ D(q) &= C(q) + 25{,}000{,}000k - kq.\end{aligned}$$

You now recall from calculus that the derivative of a sum of two functions is the sum of their derivatives (you will see why in Section 4.1*), so the derivative of D is given by

* This statement makes intuitive sense: For instance, if C is changing at a rate of 3 units per second and D is changing at a rate of 2 units per second, then their sum is changing at a rate of $3 + 2 = 5$ units per second.

$$D'(q) = \text{Derivative of } C + \text{Derivative of } (25{,}000{,}000k - kq).$$

The function $y = 25{,}000{,}000k - kq$ is a linear function of q with slope $-k$ and intercept $25{,}000{,}000k$. Thus its derivative is just its slope: $-k$. Therefore:

$$D'(q) = C'(q) - k.$$

Remember that you want

$$D'(10{,}000{,}000) = 0.$$

Thus,

$$C'(10{,}000{,}000) - k = 0.$$

Referring to the table, you see that

$$360 - k = 0$$

or

$$k = \$360 \text{ per ton}.$$

In other words, all you need to do is set the emission charge at $k = \$360$ per ton of sulfur emitted. Further, to ensure that the resulting curve will have the general shape shown in Figure 40, you would like to have $D'(q)$ negative for $q < 10{,}000{,}000$ and positive for $q > 10{,}000{,}000$. To check this, write

$$\begin{aligned}D'(q) &= C'(q) - k\\ &= C'(q) - 360\end{aligned}$$

and refer to the table to obtain

$$D'(8{,}000{,}000) = 270 - 360 = -90 < 0 \checkmark$$

and

$$D'(12{,}000{,}000) = 779 - 360 = 419 > 0 \checkmark$$

Thus, based on the given data, the resulting curve will have the shape you require. You therefore inform the EPA that an annual emissions charge of \$360 per ton of sulfur released into the environment will create the desired incentive: to reduce sulfur emissions by 10 million tons per year.

One week later, you are informed that this charge would be unrealistic because the utilities cannot possibly afford such a cost. You are asked whether there is an alternative plan that accomplishes the 10-million-ton reduction goal and yet is cheaper to the utilities by \$5 billion per year. You then look at your expression for the emission charge

$$25{,}000{,}000k - kq$$

and notice that, if you decrease this amount by \$5 billion, the derivative will not change at all because it will still have the same slope (only the intercept is affected). Thus, you propose the following revised formula for the emission charge:

$$\begin{aligned} &25{,}000{,}000k - kq - 5{,}000{,}000{,}000 \\ &= 25{,}000{,}000(360) - 360q - 5{,}000{,}000{,}000 \\ &= 4{,}000{,}000{,}000 - 360q. \end{aligned}$$

At the expected reduction level of 10 million tons, the total amount paid by the utilities will then be

$$4{,}000{,}000{,}000 - 360(10{,}000{,}000) = \$400{,}000{,}000.$$

Thus, your revised proposal is the following: Impose an annual emissions charge of \$360 per ton of sulfur released into the environment and hand back \$5 billion in the form of subsidies. The effect of this policy will be to cause the utilities industry to reduce sulfur emissions by 10 million tons per year and will result in \$400 million in annual revenues to the government.

Notice that this policy also provides an incentive for the utilities to search for cheaper ways to reduce emissions. For instance, if they lowered costs to the point where they could achieve a reduction level of 12 million tons, they would have a total emission charge of

$$4{,}000{,}000{,}000 - 360(12{,}000{,}000) = -\$320{,}000{,}000.$$

The fact that this is negative means that the government would be paying the utilities industry \$320 million more in annual subsidies than the industry is paying in per ton emission charges.

EXERCISES

1. Excluding subsidies, what should the annual emission charge be if the goal is to reduce sulfur emissions by 8 million tons?

2. Excluding subsidies, what should the annual emission charge be if the goal is to reduce sulfur emissions by 12 million tons?

3. What is the *marginal emission charge* (derivative of emission charge) in your revised proposal (as stated before the exercise set)? What is the relationship between the marginal cost of reducing sulfur emissions before emissions charges are implemented and the marginal emission charge, at the optimal reduction under your revised proposal?

4. We said that the revised policy provided an incentive for utilities to find cheaper ways to reduce emissions. How would $C(q)$ have to change to make 12 million tons the optimum reduction?

5. What change in $C(q)$ would make 8 million tons the optimum reduction?

6. If the scenario in Exercise 5 took place, what would the EPA have to do in order to make 10 million tons the optimal reduction once again?

7. Due to intense lobbying by the utility industry, you are asked to revise the proposed policy so that the utility industry will pay no charge if sulfur emissions are reduced by the desired 10 million tons. How can you accomplish this?

8. Suppose that instead of imposing a fixed charge per ton of emission, you decide to use a sliding scale, so that the total charge to the industry for annual emissions of x tons will be $\$kx^2$ for some k. What must k be to again make 10 million tons the optimum reduction? HINT [The derivative of kx^2 is $2kx$.]

TI-83/84 Plus Technology Guide

Section 3.1

Example 1 (page 153) Use a table to estimate the following limits.

a. $\lim_{x\to 2} \dfrac{x^3-8}{x-2}$ **b.** $\lim_{x\to 0} \dfrac{e^{2x}-1}{x}$

Solution with Technology

On the TI-83/84 Plus, use the table feature to automate these computations as follows:

1. Define `Y1=(X^3-8)/(X-2)` for part (a) or `Y1=(e^(2X)-1)/X` for part (b).
2. Press [2ND] [TABLE] to list its values for the given values of x. (If the calculator does not allow you to enter values of x, press [2ND] [TBLSET] and set `Indpnt` to `Ask`).

   Here is the table showing some of the values for part (a):

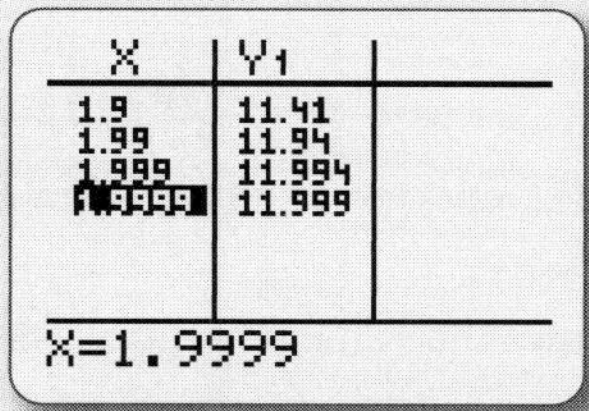

X	Y1
1.9	11.41
1.99	11.94
1.999	11.994
1.9999	11.999

X=1.9999

3. For part (b) use `Y1=(e^(2X)-1)/X` and use values of x approaching 0 from either side.

Section 3.4

Example 3 (page 182) The price of an ounce of gold can be approximated by the function

$$G(t) = 5t^2 - 85t + 1{,}762 \quad (7.5 \le t \le 10.5)$$

where t is time in hours. ($t = 8$ represents 8:00 am.) What was the average rate of change of the price of gold over the $1\frac{1}{2}$-hour period starting at 8:00 am (the interval [8, 9.5] on the t-axis)?

Solution with Technology

On the TI-83/84 Plus:

1. Enter the function G as `Y1` (using X for t):

   `Y1=5X^2-85X+1762`

2. Now find the average rate of change over [8, 9.5] by evaluating the following on the home screen:

   `(Y1(9.5)-Y1(8))/(9.5-8)`

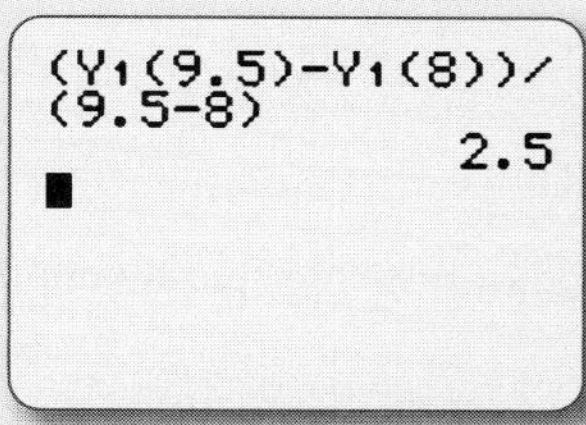

As shown on the screen, the average rate is of change is 2.5.

Example 4 (page 183) Continuing with Example 3, use technology to compute the average rate of change of

$$G(t) = 5t^2 - 85t + 1{,}762 \qquad (7.5 \le t \le 10.5)$$

over the intervals $[8, 8+h]$, where $h = 1$, 0.1, 0.01, 0.001, and 0.0001.

Solution with Technology

1. As in Example 4, enter the function G as `Y1` (using X for t):

   `Y1=5X^2-85X+1762`

2. Now find the average rate of change for $h = 1$ by evaluating, on the home screen,

   `(Y1(8+1)-Y1(8))/1,`

   which gives 0.

3. To evaluate for $h = 0.1$, recall the expression using [2ND] [ENTER] and then change the 1, both places it occurs, to 0.1, getting

   `(Y1(8+0.1)-Y1(8))/0.1,`

   which gives -4.95.

4. Continuing, we can evaluate the average rate of change for all the desired values of h:

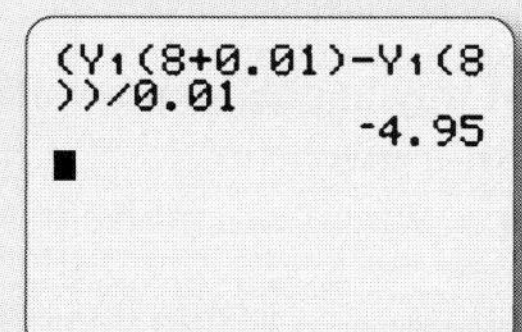

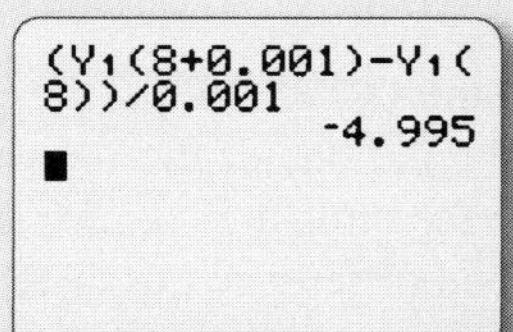

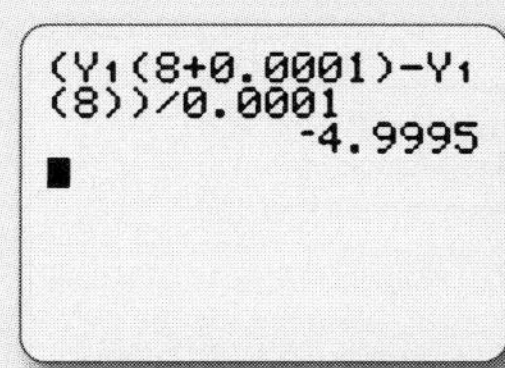

Section 3.5

Example 2 (page 190) Calculate an approximate value of $f'(1.5)$ if $f(x) = x^2 - 4x$, and then find the equation of the tangent line at the point on the graph where $x = 1.5$.

Solution with Technology

1. In the TI-83/84 Plus, enter the function f as `Y1`

```
Y1=X^2-4*X
```

2. Go to the home screen to compute the approximations:

```
(Y1(1.5001)-Y1(1.5))/0.0001
```
Usual difference quotient

```
(Y1(1.5001)-Y1(1.4999))/0.0002
```
Balanced difference quotient

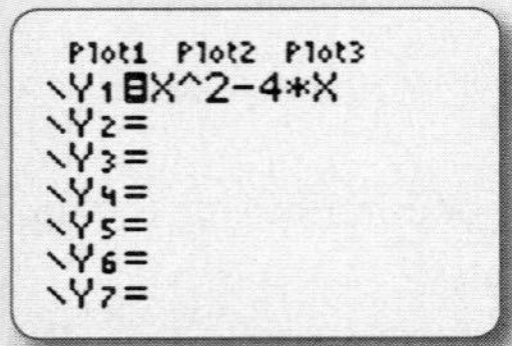

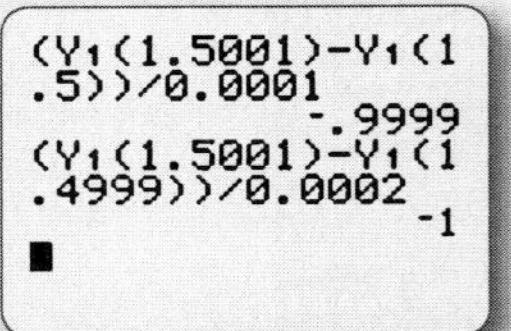

From the display on the right, we find that the difference quotient quick approximation is -0.9999 and the balanced difference quotient quick approximation is -1, which is in fact the exact value of $f'(1.5)$. See the discussion in the text for the calculation of the equation of the tangent line.

Example 4 (page 196) Use technology to graph the derivative of $f(x) = -2x^2 + 6x + 5$ for values of x in starting at -5.

Solution with Technology

On the TI-83/84 Plus, the easiest way to obtain quick approximations of the derivative of a given function is to use the built-in `nDeriv` function, which calculates balanced difference quotients.

1. On the `Y=` screen, first enter the function:

```
Y1=-2X^2+6X+5
```

2. Then set

```
Y2=nDeriv(Y1,X,X)
```
For `nDeriv` press [MATH] [8]

which is the TI-83/84 Plus's approximation of $f'(x)$ (see figure on the left below). Alternatively, we can enter the balanced difference quotient directly:

```
Y2=(Y1(X+0.001)-Y1(X-0.001))/0.002
```

(The TI-83/84 Plus uses $h = 0.001$ by default in the balanced difference quotient when calculating `nDeriv`, but this can be changed by giving a value of h as a fourth argument, like `nDeriv(Y1,X,X,0.0001)`.) To see a table of approximate values of the derivative, we press [2ND] [TABLE] and choose a collection of values for x (shown on the right below):

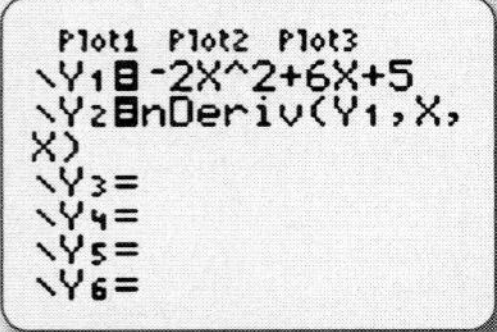

X	Y1	Y2
-3	-31	18
-2	-15	14
-1	-3	10
0	5	6
1	9	2
2	9	-2
3	5	-6

X=3

Here, `Y1` shows the value of f and `Y2` shows the values of f'.

To graph the function or its derivative, we can graph `Y1` or `Y2` in a window showing the given domain $[-5, 5]$:

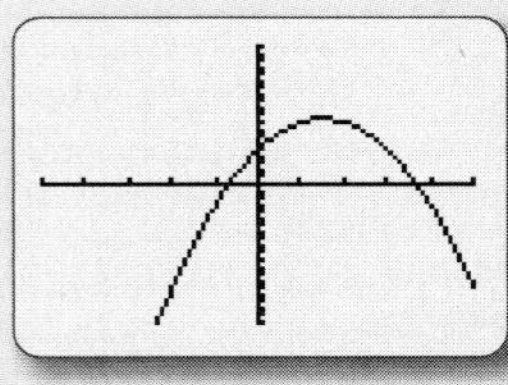

Graph of f

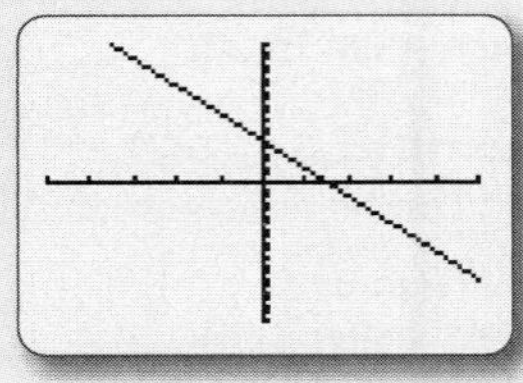

Graph of f'

SPREADSHEET Technology Guide

Section 3.1

Example 1 (page 153) Use a table to estimate the following limits.

a. $\lim\limits_{x \to 2} \dfrac{x^3 - 8}{x - 2}$ **b.** $\lim\limits_{x \to 0} \dfrac{e^{2x} - 1}{x}$

Solution with Technology

1. Set up your spreadsheet to duplicate the table in part (a) as follows:

	A	B	C	D
1	x	f(x)	x	f(x)
2	1.9	=(A2^3-8)/(A2-2)	2.1	
3	1.99		2.01	
4	1.999		2.001	
5	1.9999		2.0001	

↓

	A	B	C	D
1	x	f(x)	x	f(x)
2	1.9	11.41	2.1	12.61
3	1.99	11.9401	2.01	12.0601
4	1.999	11.994001	2.001	12.006001
5	1.9999	11.99940001	2.0001	12.00060001

(The formula in cell B2 is copied to columns B and D as indicated by the shading.) The values of $f(x)$ will be calculated in columns B and D.

2. For part (b), use the formula `=(EXP(2*A2)-1)/A2` in cell B2 and, in columns A and C, use values of x approaching 0 from either side.

Section 3.4

Example 3 (page 182) The price of an ounce of gold can be approximated by the function

$$G(t) = 5t^2 + 85t + 1{,}762 \quad (7.5 \le t \le 10.5)$$

where t is time in hours. ($t = 8$ represents 8:00 am.) What was the average rate of change of the price of gold over the $1\frac{1}{2}$-hour period starting at 8:00 am (the interval [8, 9.5] on the t-axis)?

Solution with Technology

To use a spreadsheet to compute the average rate of change of G:

1. Start with two columns, one for values of t and one for values of $G(t)$, which you enter using the formula for G:

```
=5*A2^2-85*A2+1762
```

	A	B
1	t	G(t)
2	8	=5*A2^2-85*A2+1762
3	9.5	

2. Next, calculate the average rate of change as shown here:

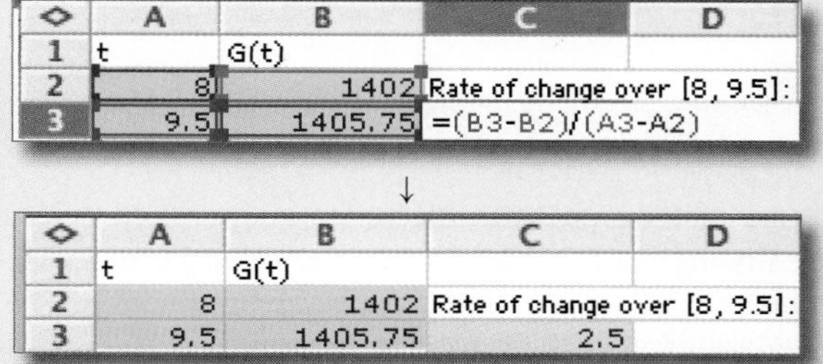

	A	B	C	D
1	t	G(t)		
2	8	1402	Rate of change over [8, 9.5]:	
3	9.5	1405.75	=(B3-B2)/(A3-A2)	

↓

	A	B	C	D
1	t	G(t)		
2	8	1402	Rate of change over [8, 9.5]:	
3	9.5	1405.75	2.5	

In Example 4, we describe another, more versatile Excel template for computing rates of change.

Example 4 (page 183) Continuing with Example 3, use technology to compute the average rate of change of

$$G(t) = 5t^2 + 85t + 1{,}762 \qquad (7.5 \le t \le 10.5)$$

over the intervals $[8, 8 + h]$, where $h = 1$, 0.1, 0.01, 0.001, and 0.0001.

Solution with Technology

The template we can use to compute the rates of change is an extension of what we used in Example 3:

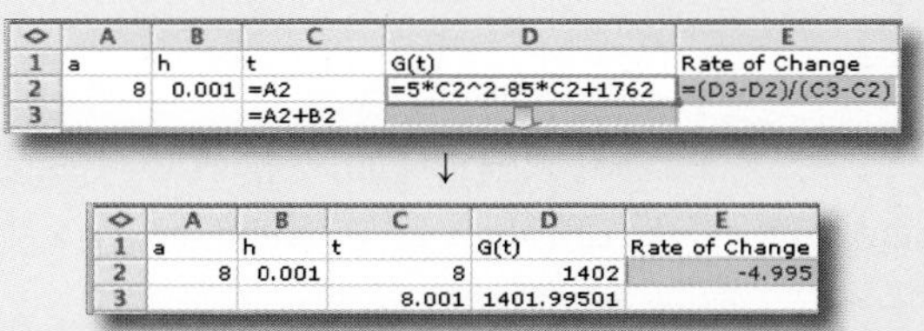

	A	B	C	D	E
1	a	h	t	G(t)	Rate of Change
2	8	0.001	=A2	=5*C2^2-85*C2+1762	=(D3-D2)/(C3-C2)
3			=A2+B2		

↓

	A	B	C	D	E
1	a	h	t	G(t)	Rate of Change
2	8	0.001	8	1402	-4.995
3			8.001	1401.99501	

1. Column C contains the values $t = a$ and $t = a + h$ we are using for the independent variable.
2. The formula in cell E2 is the average-rate-of-change formula $\Delta G/\Delta t$. Entering the different values $h = 1$, 0.1, 0.01, 0.001, and 0.0001 in cell B2 gives the results shown in Example 4.

TECHNOLOGY GUIDE

Section 3.5

Example 2 (page 190) Calculate an approximate value of $f'(1.5)$ if $f(x) = x^2 - 4x$, and then find the equation of the tangent line at the point on the graph where $x = 1.5$.

Solution with Technology

You can compute both the difference quotient and the balanced difference quotient approximations in a spreadsheet using the following extension of the worksheet in Example 4 in Section 3.4:

	A	B	C	D	E	F
1	a	h	x	f(x)	Diff Quotients	Balanced Diff Quotient
2	1.5	0.0001	=A2-B2	=C2^2-4*C2	=(D3-D2)/(C3-C2)	=(D4-D2)/(C4-C2)
3			=A2			
4			=A2+B2			

Notice that we get two difference quotients in column E. The first uses $h = -0.0001$ while the second uses $h = 0.0001$ and is the one we use for our quick approximation. The balanced quotient is their average (column F). The results are as follows:

	A	B	C	D	E	F
1	a	h	x	f(x)	Diff Quotients	Balanced Diff Quotient
2	1.5	0.0001	1.4999	-3.7499	-1.0001	-1
3			1.5	-3.75	-0.9999	
4			1.5001	-3.7501		

From the results shown above, we find that the difference quotient quick approximation is -0.9999 and that the balanced difference quotient quick approximation is -1, which is in fact the exact value of $f'(1.5)$. See the discussion in the text for the calculation of the equation of the tangent line.

Example 4 (page 196) Use technology to graph the derivative of $f(x) = -2x^2 + 6x + 5$ for values of x starting at -5.

Solution with Technology

1. Start with a table of values for the function f:

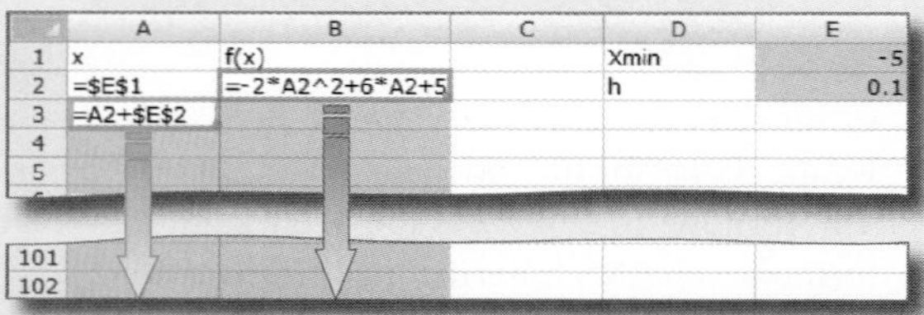

	A	B	C	D	E
1	x	f(x)		Xmin	-5
2	=$E$1	=-2*A2^2+6*A2+5		h	0.1
3	=A2+$E$2				
4					
5					
101					
102					

2. Next, compute approximate derivatives in Column C:

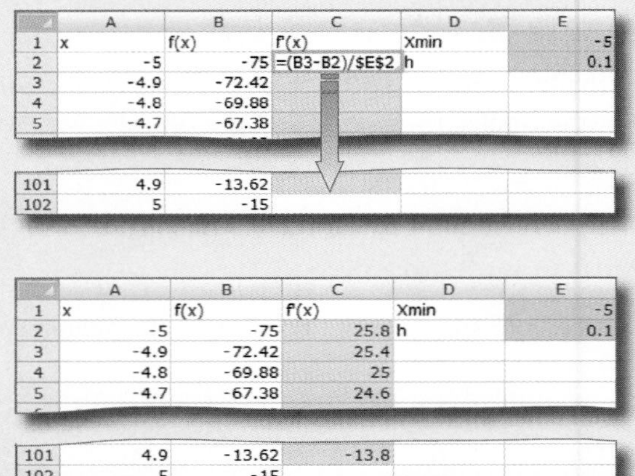

	A	B	C	D	E
1	x	f(x)	f'(x)	Xmin	-5
2	-5	-75	=(B3-B2)/$E$2	h	0.1
3	-4.9	-72.42			
4	-4.8	-69.88			
5	-4.7	-67.38			
101	4.9	-13.62			
102	5	-15			

	A	B	C	D	E
1	x	f(x)	f'(x)	Xmin	-5
2	-5	-75	25.8	h	0.1
3	-4.9	-72.42	25.4		
4	-4.8	-69.88	25		
5	-4.7	-67.38	24.6		
101	4.9	-13.62	-13.8		
102	5	-15			

You cannot paste the difference quotient formula into cell C102. (Why?) Notice that this worksheet uses the ordinary difference quotients, $[f(x+h) - f(x)]/h$. If you prefer, you can use balanced difference quotients $[f(x+h) - f(x-h)]/(2h)$, in which case cells C2 and C102 would both have to be left blank.

We now graph the function and the derivative on different graphs as follows:

1. First, graph the function f in the usual way, using Columns A and B.
2. Make a copy of this graph and click on it once. Columns A and B should be outlined, indicating that these are the columns used in the graph.
3. By dragging from the center of the bottom edge of the box, move the Column B box over to Column C as shown:

	A	B	C
96	4.4	-7.32	-11.8
97	4.5	-8.5	-12.2
98	4.6	-9.72	-12.6
99	4.7	-10.98	-13
100	4.8	-12.28	-13.4
101	4.9	-13.62	-13.8
102	5	-15	

↓

	A	B	C
96	4.4	-7.32	-11.8
97	4.5	-8.5	-12.2
98	4.6	-9.72	-12.6
99	4.7	-10.98	-13
100	4.8	-12.28	-13.4
101	4.9	-13.62	-13.8
102	5	-15	

The graph will then show the derivative (Columns A and C):

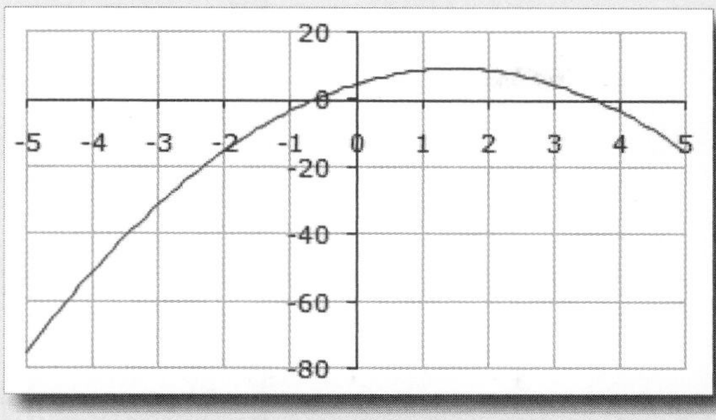

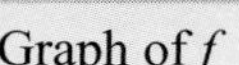
Graph of f

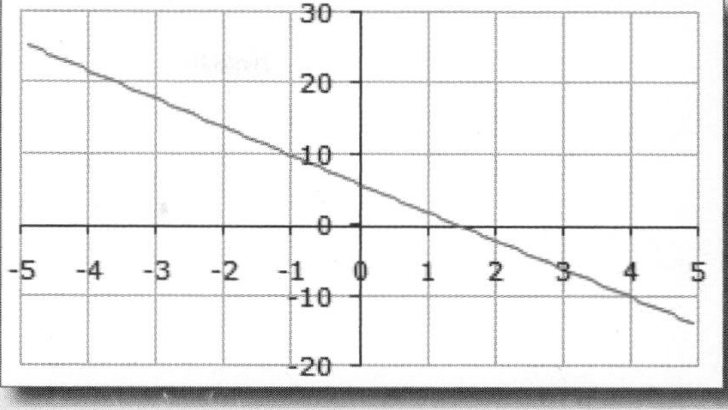

Graph of f'

4

Techniques of Differentiation with Applications

Website

www.WanerMath.com

At the Website you will find:

- Section-by-section tutorials, including game tutorials with randomized quizzes
- A detailed chapter summary
- A true/false quiz
- Additional review exercises
- Graphers, Excel tutorials, and other resources
- The following extra topic:

    Linear Approximation and Error Estimation

Case Study Projecting Market Growth

You are on the board of directors at Fullcourt Academic Press. The sales director of the high school division has just burst into your office with a proposal for an expansion strategy based on the assumption that the number of graduates from private high schools in the U.S. will grow at a rate of at least 4,000 per year through the year 2015. Because the figures actually appear to be leveling off, you are suspicious about this estimate. You would like to devise a model that predicts this trend before tomorrow's scheduled board meeting. **How do you go about doing this?**

Yuri Arcurs/Shutterstock

Introduction

In Chapter 3 we studied the concept of the derivative of a function, and we saw some of the applications for which derivatives are useful. However, computing the derivative of a function algebraically seemed to be a time-consuming process, forcing us to restrict attention to fairly simply functions.

In this chapter we develop shortcut techniques that will allow us to write down the derivative of a function directly without having to calculate any limit. These techniques will also enable us to differentiate any closed-form function—that is, any function, no matter how complicated, that can be specified by a formula involving powers, radicals, absolute values, exponents, and logarithms. (In Chapter 9 we will discuss how to add trigonometric functions to this list.) We also show how to find the derivatives of functions that are only specified *implicitly*—that is, functions for which we are not given an explicit formula for y in terms of x but only an equation relating x and y.

algebra Review

For this chapter, you should be familiar with the algebra reviewed in **Chapter 0, Sections 3 and 4.**

4.1 Derivatives of Powers, Sums, and Constant Multiples

Up to this point we have approximated derivatives using difference quotients, and we have done exact calculations using the definition of the derivative as the limit of a difference quotient. In general, we would prefer to have an exact calculation, and it is also very useful to have a formula for the derivative function when we can find one. However, the calculation of a derivative as a limit is often tedious, so it would be nice to have a quicker method. We discuss the first of the shortcut rules in this section. By the end of this chapter, we will be able to find fairly quickly the derivative of almost any function we can write.

Shortcut Formula: The Power Rule

If you look at Examples 2 and 3 in Section 3.6, you may notice a pattern:

$$f(x) = x^2 \quad \Rightarrow \quad f'(x) = 2x$$
$$f(x) = x^3 \quad \Rightarrow \quad f'(x) = 3x^2.$$

This pattern generalizes to any power of x:

Theorem 4.1 The Power Rule

If n is any constant and $f(x) = x^n$, then

$$f'(x) = nx^{n-1}.$$

Quick Examples

1. If $f(x) = x^2$, then $f'(x) = 2x^1 = 2x$.
2. If $f(x) = x^3$, then $f'(x) = 3x^2$.
3. If $f(x) = x$, rewrite as $f(x) = x^1$, so $f'(x) = 1x^0 = 1$.
4. If $f(x) = 1$, rewrite as $f(x) = x^0$, so $f'(x) = 0x^{-1} = 0$.

Website
www.WanerMath.com
At the Website you can find a proof of the power rule by following:
Everything for Calculus
→ Chapter 4
→ Proof of the Power Rule

The proof of the power rule involves first studying the case when n is a positive integer, and then studying the cases of other types of exponents (negative integer, rational number, irrational number). You can find a proof at the Website.

EXAMPLE 1 Using the Power Rule for Negative and Fractional Exponents

Calculate the derivatives of the following:

a. $f(x) = \frac{1}{x}$ **b.** $f(x) = \frac{1}{x^2}$ **c.** $f(x) = \sqrt{x}$

Solution

a. Rewrite* as $f(x) = x^{-1}$. Then $f'(x) = (-1)x^{-2} = -\frac{1}{x^2}$.

b. Rewrite as $f(x) = x^{-2}$. Then $f'(x) = (-2)x^{-3} = -\frac{2}{x^3}$.

c. Rewrite as $f(x) = x^{0.5}$. Then $f'(x) = 0.5x^{-0.5} = \frac{0.5}{x^{0.5}}$. Alternatively, rewrite $f(x)$ as $x^{1/2}$, so that $f'(x) = \frac{1}{2}x^{-1/2} = \frac{1}{2x^{1/2}} = \frac{1}{2\sqrt{x}}$.

* See Section 0.2 in the Precalculus Review to brush up on negative and fractional exponents. Pay particular attention to rational, radical, and exponent forms.

By rewriting the given functions in Example 1 before taking derivatives, we converted them from **rational** or **radical form** (as in, say, $\frac{1}{x^2}$ and $\sqrt{x}$) to **exponent form** (as in x^{-2} and $x^{0.5}$; see the Precalculus Review, Section 0.2) to enable us to use the power rule. (See the Caution below.)

Caution

We cannot apply the power rule to terms in the denominators or under square roots. For example:

1. The derivative of $\frac{1}{x^2}$ is **NOT** $\frac{1}{2x}$; it is $-\frac{2}{x^3}$. See Example 1(b).
2. The derivative of $\sqrt{x^3}$ is **NOT** $\sqrt{3x^2}$; it is $1.5x^{0.5}$. Rewrite $\sqrt{x^3}$ as $x^{3/2}$ or $x^{1.5}$ and apply the power rule.

Some of the derivatives in Example 1 are very useful to remember, so we summarize them in Table 1. We suggest that you add to this table as you learn more derivatives. It is *extremely* helpful to remember the derivatives of common functions such as $1/x$ and $\sqrt{x}$, even though they can be obtained by using the power rule as in the above example.

Table 1 Table of Derivative Formulas

$f(x)$	$f'(x)$
1	0
x	1
x^2	$2x$
x^3	$3x^2$
x^n	nx^{n-1}
$\frac{1}{x}$	$-\frac{1}{x^2}$
$\frac{1}{x^2}$	$-\frac{2}{x^3}$
$\sqrt{x}$	$\frac{1}{2\sqrt{x}}$

Another Notation: Differential Notation

Here is a useful notation based on the "d-notation" we discussed in Section 3.5. **Differential notation** is based on an abbreviation for the phrase "the derivative with respect to x." For example, we learned that if $f(x) = x^3$, then $f'(x) = 3x^2$. When we say "$f'(x) = 3x^2$," we mean the following:

The derivative of x^3 with respect to x equals $3x^2$.

You may wonder why we sneaked in the words "with respect to x." All this means is that the variable of the function is x, and not any other variable.* Because we use the phrase "the derivative with respect to x" often, we use the following abbreviation.

* This may seem odd in the case of $f(x) = x^3$ because there are no other variables to worry about. But in expressions like st^3 that involve variables other than x, it is necessary to specify just what the variable of the function is. This is the same reason that we write "$f(x) = x^3$" rather than just "$f = x^3$."

Differential Notation; Differentiation

$\dfrac{d}{dx}$ means "the derivative with respect to x."

Thus, $\dfrac{d}{dx}[f(x)]$ is the same thing as $f'(x)$, the derivative of $f(x)$ with respect to x. If y is a function of x, then the derivative of y with respect to x is

$$\frac{d}{dx}(y) \qquad \text{or, more compactly,} \qquad \frac{dy}{dx}.$$

To **differentiate** a function $f(x)$ with respect to x means to take its derivative with respect to x.

Quick Examples

In Words	Formula
1. The derivative with respect to x of x^3 is $3x^2$.	$\dfrac{d}{dx}(x^3) = 3x^2$
2. The derivative with respect to t of $\dfrac{1}{t}$ is $-\dfrac{1}{t^2}$.	$\dfrac{d}{dt}\left(\dfrac{1}{t}\right) = -\dfrac{1}{t^2}$
3. If $y = x^4$, then $\dfrac{dy}{dx} = 4x^3$.	
4. If $u = \dfrac{1}{t^2}$, then $\dfrac{du}{dt} = -\dfrac{2}{t^3}$.	

Notes

1. $\dfrac{dy}{dx}$ is Leibniz's notation for the derivative we discussed in Section 3.5. (See the discussion before Example 3 there.)

2. Leibniz notation illustrates units nicely: Units of $\dfrac{dy}{dx}$ are units of y per unit of x.

3. We can (and often do!) use different kind of brackets or parentheses in Liebniz notation; for instance, $\dfrac{d}{dx}[x^3]$, $\dfrac{d}{dx}(x^3)$, and $\dfrac{d}{dx}\{x^3\}$ all mean the same thing (and equal $3x^2$). ■

The Rules for Sums and Constant Multiples

We can now find the derivatives of more complicated functions, such as polynomials, using the following rules. If f and g are functions and if c is a constant, we saw in Section 1.2 how to obtain the **sum**, $\boldsymbol{f + g}$, **difference**, $\boldsymbol{f - g}$, and **constant multiple**, $\boldsymbol{cf}$.

Theorem 4.2 Derivatives of Sums, Differences, and Constant Multiples

If f and g are any two differentiable functions, and if c is any constant, then the sum, $f+g$, the difference, $f-g$, and the constant multiple, cf, are differentiable, and

$$[f \pm g]'(x) = f'(x) \pm g'(x) \qquad \text{Sum Rule}$$

$$[cf]'(x) = cf'(x). \qquad \text{Constant Multiple Rule}$$

In Words:

- The derivative of a sum is the sum of the derivatives, and the derivative of a difference is the difference of the derivatives.
- The derivative of c times a function is c times the derivative of the function.

Differential Notation:

$$\frac{d}{dx}[f(x) \pm g(x)] = \frac{d}{dx}f(x) \pm \frac{d}{dx}g(x)$$

$$\frac{d}{dx}[cf(x)] = c\frac{d}{dx}f(x)$$

Quick Examples

1. $\frac{d}{dx}(x^2 - x^4) = \frac{d}{dx}(x^2) - \frac{d}{dx}(x^4) = 2x - 4x^3$

2. $\frac{d}{dx}(7x^3) = 7\frac{d}{dx}(x^3) = 7(3x^2) = 21x^2$

   In other words, we multiply the coefficient (7) by the exponent (3), and then decrease the exponent by 1.

3. $\frac{d}{dx}(12x) = 12\frac{d}{dx}(x) = 12(1) = 12$

   In other words, the derivative of a constant times x is that constant.

4. $\frac{d}{dx}(-x^{0.5}) = \frac{d}{dx}[(-1)x^{0.5}] = (-1)\frac{d}{dx}(x^{0.5}) = (-1)(0.5)x^{-0.5}$
   $= -0.5x^{-0.5}$

5. $\frac{d}{dx}(12) = \frac{d}{dx}[12(1)] = 12\frac{d}{dx}(1) = 12(0) = 0.$

   In other words, the derivative of a constant is zero.

6. If my company earns twice as much (annual) revenue as yours and the derivative of your revenue function is the curve on the left, then the derivative of my revenue function is the curve on the right.

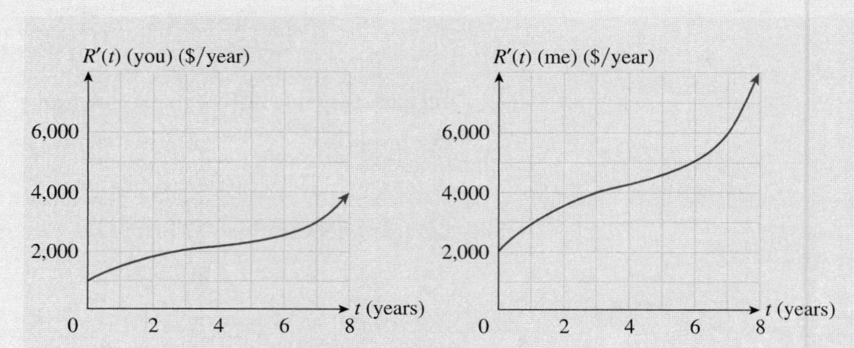

7. Suppose that a company's revenue R and cost C are changing with time. Then so is the profit, $P(t) = R(t) - C(t)$, and the rate of change of the profit is

$$P'(t) = R'(t) - C'(t).$$

In words: *The derivative of the profit is the derivative of revenue minus the derivative of cost.*

Proof of the Sum Rule

By the definition of the derivative of a function,

$$\begin{aligned}
\frac{d}{dx}[f(x) + g(x)] &= \lim_{h\to 0} \frac{[f(x+h) + g(x+h)] - [f(x) + g(x)]}{h} \\
&= \lim_{h\to 0} \frac{[f(x+h) - f(x)] + [g(x+h) - g(x)]}{h} \\
&= \lim_{h\to 0} \left[\frac{f(x+h) - f(x)}{h} + \frac{g(x+h) - g(x)}{h} \right] \\
&= \lim_{h\to 0} \frac{f(x+h) - f(x)}{h} + \lim_{h\to 0} \frac{g(x+h) - g(x)}{h} \\
&= \frac{d}{dx}[f(x)] + \frac{d}{dx}[g(x)].
\end{aligned}$$

The next-to-last step uses a property of limits: The limit of a sum is the sum of the limits. Think about why this should be true. The last step uses the definition of the derivative again (and the fact that the functions are differentiable).

The proof of the rule for constant multiples is similar.

EXAMPLE 2 Combining the Sum and Constant Multiple Rules, and Dealing with x in the Denominator

Find the derivatives of the following:

a. $f(x) = 3x^2 + 2x - 4$

b. $f(x) = \dfrac{2x}{3} - \dfrac{6}{x} + \dfrac{2}{3x^{0.2}} - \dfrac{x^4}{2}$

c. $f(x) = \dfrac{|x|}{4} + \dfrac{1}{2\sqrt{x}}$

Solution

a. $\frac{d}{dx}(3x^2 + 2x - 4) = \frac{d}{dx}(3x^2) + \frac{d}{dx}(2x - 4)$ Rule for sums

$= \frac{d}{dx}(3x^2) + \frac{d}{dx}(2x) - \frac{d}{dx}(4)$ Rule for differences

$= 3(2x) + 2(1) - 0$ See Quick Example 2.

$= 6x + 2$

b. Notice that f has x and powers of x in the denominator. We deal with these terms the same way we did in Example 1, by rewriting them in exponent form (that is, in the form constant × power of x; see Section 0.2 in the Precalculus Review):

$f(x) = \frac{2x}{3} - \frac{6}{x} + \frac{2}{3x^{0.2}} - \frac{x^4}{2}$ Rational form

$= \frac{2}{3}x - 6x^{-1} + \frac{2}{3}x^{-0.2} - \frac{1}{2}x^4.$ Exponent form

We are now ready to take the derivative:

$f'(x) = \frac{2}{3}(1) - 6(-1)x^{-2} + \frac{2}{3}(-0.2)x^{-1.2} - \frac{1}{2}(4x^3)$

$= \frac{2}{3} + 6x^{-2} - \frac{0.4}{3}x^{-1.2} - 2x^3$ Exponent form

$= \frac{2}{3} + \frac{6}{x^2} - \frac{0.4}{3x^{1.2}} - 2x^3.$ Rational form

c. Rewrite $f(x)$ using exponent form as follows:

$f(x) = \frac{|x|}{4} + \frac{1}{2\sqrt{x}}$ Rational form

$= \frac{1}{4}|x| + \frac{1}{2}x^{-1/2}.$ Exponent form

Now recall from the end of Section 3.6 that the derivative of $|x|$ is $\frac{|x|}{x}$. Thus,

$f'(x) = \frac{1}{4}\frac{|x|}{x} + \frac{1}{2}\left(\frac{-1}{2}x^{-3/2}\right)$

$= \frac{|x|}{4x} - \frac{1}{4}x^{-3/2}$ Simplify

$= \frac{|x|}{4x} - \frac{1}{4x^{3/2}}.$ Rational form

Notice that in Example 2(a) we had three terms in the expression for $f(x)$, not just two. By applying the rule for sums and differences twice, we saw that the derivative of a sum or difference of three terms is the sum or difference of the derivatives of the terms. (One of those terms had zero derivative, so the final answer had only two terms.) In fact, the derivative of a sum or difference of any number of terms is the sum or difference of the derivatives of the terms. Put another way, to take the derivative of a sum or difference of any number of terms, we take derivatives term by term.

Note Nothing forces us to use only x as the independent variable when taking derivatives (although it is traditional to give x preference). For instance, part (a) in Example 2 can be rewritten as

$$\frac{d}{dt}(3t^2 + 2t - 4) = 6t + 2 \qquad \frac{d}{dt} \text{ means "derivative with respect to } t\text{."}$$

or

$$\frac{d}{du}(3u^2 + 2u - 4) = 6u + 2. \qquad \frac{d}{du} \text{ means "derivative with respect to } u\text{."}$$ ■

In the previous examples, we saw instances of the following important facts. (Think about these graphically to see why they must be true.)

The Derivative of a Constant Times *x* and the Derivative of a Constant

If c is any constant, then:

Rule	Quick Examples	
$\frac{d}{dx}(cx) = c$	$\frac{d}{dx}(6x) = 6$	$\frac{d}{dx}(-x) = -1$
$\frac{d}{dx}(c) = 0$	$\frac{d}{dx}(5) = 0$	$\frac{d}{dx}(\pi) = 0$

In Section 3.5 we pointed out that the derivative of a function cannot exist at an endpoint of its domain, as the defining limit does not exist there (see the "Important Notes" after the definition of instantaneous rate of change). Thus, for instance, $f(x) = \sqrt{x}$ and $g(x) = x^{1/4}$ are not differentiable at the endpoint $x = 0$ of their domains. In Example 5 of Section 3.6 we saw that $h(x) = |x|$ also fails to be differentiable at $x = 0$, even though $x = 0$ is not an endpoint of its domain (the domain of h is the set of all real numbers). In the next example we see how to spot the non-differentiability at a point of this and other functions simply by looking at the formulas for their derivatives.

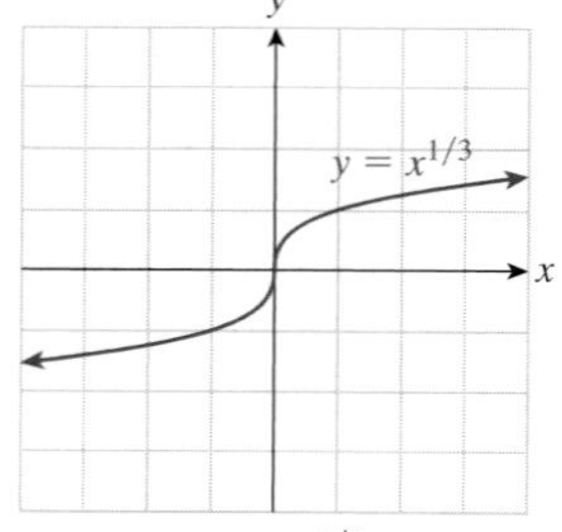

$f(x) = x^{1/3}$

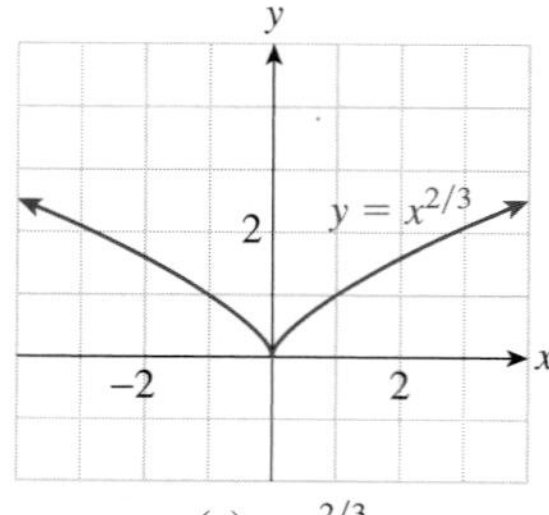

$g(x) = x^{2/3}$

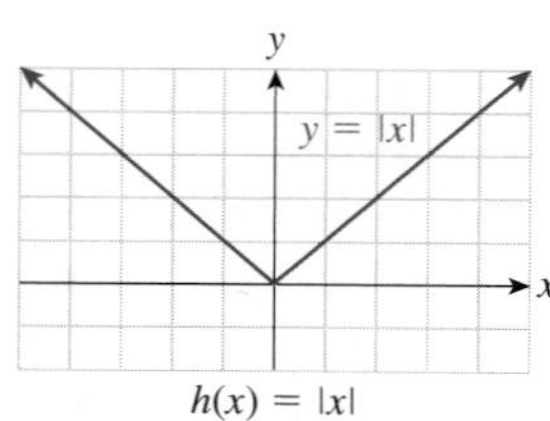

$h(x) = |x|$

Figure 1

EXAMPLE 3 Functions Not Differentiable at a Point

Find the natural domains of the derivatives of $f(x) = x^{1/3}$, $g(x) = x^{2/3}$, and $h(x) = |x|$.

Solution Let's look at the derivatives of the three functions given:

$$f(x) = x^{1/3}, \text{ so } \quad f'(x) = \frac{1}{3}x^{-2/3} = \frac{1}{3x^{2/3}}.$$

$$g(x) = x^{2/3}, \text{ so } \quad g'(x) = \frac{2}{3}x^{-1/3} = \frac{2}{3x^{1/3}}.$$

$$h(x) = |x|, \text{ so } h'(x) = \frac{|x|}{x}.$$

The derivatives of all three functions are defined only for nonzero values of x, and their natural domains consist of all real numbers except 0. Thus, the derivatives f', g', and h' do not exist at $x = 0$. In other words, these functions are not differentiable at $x = 0$. If we look at Figure 1 we notice why these functions fail to be differentiable at $x = 0$: The graph of f has a vertical tangent line at 0. Because a vertical line

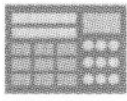

using Technology

If you try to graph the function $f(x) = x^{2/3}$ using the format

```
X^(2/3)
```

you may get only the right-hand portion of the graph of g in Figure 1 because graphing utilities are (often) not programmed to raise negative numbers to fractional exponents. (However, many will handle `X^(1/3)` correctly, as a special case they recognize.) To avoid this difficulty, you can take advantage of the identity

$$x^{2/3} = (x^2)^{1/3}$$

so that it is always a nonnegative number that is being raised to a fractional exponent. Thus, use the format

```
(X^2)^(1/3)
```

to obtain both portions of the graph.

has undefined slope, the derivative is undefined at that point. The graphs of g and h come to a sharp point at 0, where it is not meaningful to speak about the slope of the tangent line; therefore, the derivatives of g and h are not defined there. (In the case of g, where the sharp point is called a *cusp,* a vertical tangent line would seem appropriate, but as in the case of f, its slope is undefined.)

You can also detect this nondifferentiability by computing some difference quotients numerically, as we did for h in Section 3.6.

APPLICATION

EXAMPLE 4 Gold Price

You are a commodities trader and you monitor the price of gold on the spot market very closely during an active morning. Suppose you find that the price of an ounce of gold can be approximated by the function

$$G(t) = 5t^2 - 85t + 1{,}762 \qquad (7.5 \le t \le 10.5),$$

where t is time in hours. (See Figure 2. $t = 8$ represents 8:00 am.)

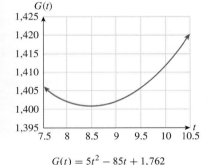

Source: www.kitco.com

$G(t) = 5t^2 - 85t + 1{,}762$

Figure 2

a. According to the model, how fast was the price of gold changing at 8:00 am?

b. According to the model, the price of gold

(A) increased at a faster and faster rate
(B) increased at a slower and slower rate
(C) decreased at a faster and faster rate
(D) decreased at a slower and slower rate

between 7:30 and 8:30 am.

Solution

a. Differentiating the given function with respect to t gives

$$G'(t) = 10t - 85.$$

Because 8:00 am corresponds to $t = 8$, we obtain

$$G'(8) = 10(8) - 85 = -5.$$

The units of the derivative are dollars per hour, so we conclude that, at 8:00 am, the price of gold was dropping at a rate of \$5 per hour.

b. From the graph, we can see that, between 7:30 and 8:30 am (the interval [7.5, 8.5]), the price of gold was decreasing. Also from the graph, we see that the slope of the tangent becomes less and less negative as t increases, so the price of gold is decreasing at a slower and slower rate (choice (D)).

We can also see this algebraically from the derivative, $G'(t) = 10t - 85$: For values of t less than 8.5, $G'(t)$ is negative; that is, the rate of change of G is negative, so the price of gold is decreasing. Further, as t increases, $G'(t)$ becomes less and less negative, so the price of gold is decreasing at a slower and slower rate, confirming that choice (D) is the correct one.

An Application to Limits: L'Hospital's Rule

The limits that caused us some trouble in Sections 3.1–3.3 are those of the form $\lim_{x\to a} f(x)$ in which substituting $x = a$ gave us an indeterminate form, such as

$$\lim_{x\to 2} \frac{x^3 - 8}{x - 2} \qquad \text{Substituting } x = 2 \text{ yields } \tfrac{0}{0}.$$

$$\lim_{x\to +\infty} \frac{2x - 4}{x - 1}. \qquad \text{Substituting } x = +\infty \text{ yields } \tfrac{\infty}{\infty}.$$

L'Hospital's rule* gives us an alternate way of computing limits such as these without the need to do any preliminary simplification. It also allows us to compute some limits for which algebraic simplification does not work.

* Guillaume François Antoine, Marquis de l'Hospital (1661–1704) wrote the first textbook on calculus, *Analyse des infiniment petits pour l'intelligence des lignes courbes,* in 1692. The rule now known as l'Hospital's rule appeared first in this book.

Theorem 4.3 L'Hospital's Rule

If f and g are two differentiable functions such that substituting $x = a$ in the expression $\dfrac{f(x)}{g(x)}$ gives the indeterminate form $\dfrac{0}{0}$ or $\dfrac{\infty}{\infty}$, then

$$\lim_{x\to a} \frac{f(x)}{g(x)} = \lim_{x\to a} \frac{f'(x)}{g'(x)}.$$

That is, we can replace $f(x)$ and $g(x)$ with their *derivatives* and try again to take the limit.

Quick Examples

1. Substituting $x = 2$ in $\dfrac{x^3 - 8}{x - 2}$ yields $\dfrac{0}{0}$. Therefore, l'Hospital's rule applies and

$$\lim_{x\to 2} \frac{x^3 - 8}{x - 2} = \lim_{x\to 2} \frac{3x^2}{1} = \frac{3(2)^2}{1} = 12.$$

2. Substituting $x = +\infty$ in $\dfrac{2x - 4}{x - 1}$ yields $\dfrac{\infty}{\infty}$. Therefore, l'Hospital's rule applies and

$$\lim_{x\to +\infty} \frac{2x - 4}{x - 1} = \lim_{x\to +\infty} \frac{2}{1} = 2.$$

The proof of l'Hospital's rule is beyond the scope of this text.†

† A proof of l'Hospital's rule can be found in most advanced calculus textbooks.

EXAMPLE 5 Applying L'Hospital's Rule

Check whether l'Hospital's rule applies to each of the following limits. If it does, use it to evaluate the limit. Otherwise, use some other method to evaluate the limit.

a. $\lim_{x\to 1} \dfrac{x^2-2x+1}{4x^3-3x^2-6x+5}$ **b.** $\lim_{x\to +\infty} \dfrac{2x^2-4x}{5x^3-3x+5}$

c. $\lim_{x\to 1} \dfrac{x-1}{x^3-3x^2+3x-1}$ **d.** $\lim_{x\to 1} \dfrac{x}{x^3-3x^2+3x-1}$

Solution

a. Setting $x = 1$ yields

$$\frac{1-2+1}{4-3-6+5} = \frac{0}{0}.$$

Therefore, l'Hospital's rule applies and

$$\lim_{x\to 1} \frac{x^2-2x+1}{4x^3-3x^2-6x+5} = \lim_{x\to 1} \frac{2x-2}{12x^2-6x-6}.$$

We are left with a closed-form function. However, we cannot substitute $x = 1$ to find the limit because the function $(2x-2)/(12x^2-6x-6)$ is still not defined at $x = 1$. In fact, if we set $x = 1$, we again get $0/0$. Thus, l'Hospital's rule applies again, and

$$\lim_{x\to 1} \frac{2x-2}{12x^2-6x-6} = \lim_{x\to 1} \frac{2}{24x-6}.$$

Once again we have a closed-form function, but this time it is defined when $x = 1$, giving

$$\frac{2}{24-6} = \frac{1}{9}.$$

Thus,

$$\lim_{x\to 1} \frac{x^2-2x+1}{4x^3-3x^2-6x+5} = \frac{1}{9}.$$

b. Setting $x = +\infty$ yields $\dfrac{\infty}{\infty}$, so

$$\lim_{x\to +\infty} \frac{2x^2-4x}{5x^3-3x+5} = \lim_{x\to +\infty} \frac{4x-4}{15x^2-3}.$$

Setting $x = +\infty$ again yields $\dfrac{\infty}{\infty}$, so we can apply the rule again to obtain

$$\lim_{x\to +\infty} \frac{4x-4}{15x^2-3} = \lim_{x\to +\infty} \frac{4}{30x}.$$

Note that we cannot apply l'Hospital's rule a third time because setting $x = +\infty$ yields the *determinate* form $4/\infty = 0$ (see the discussion at the end of Section 3.3). Thus, the limit is 0.

c. Setting $x = 1$ yields $0/0$ so, by l'Hospital's rule,

$$\lim_{x\to 1}\frac{x-1}{x^3-3x^2+3x-1} = \lim_{x\to 1}\frac{1}{3x^2-6x+3}.$$

We are left with a closed-form function that is still not defined at $x = 1$. Further, l'Hospital's rule no longer applies because putting $x = 1$ yields the determinate form $1/0$. To investigate this limit, we refer to the discussion at the end of Section 3.3 and find

$$\lim_{x\to 1}\frac{1}{3x^2-6x+3} = \lim_{x\to 1}\frac{1}{3(x-1)^2} = +\infty. \qquad \frac{1}{0^+} = +\infty$$

d. Setting $x = 1$ in the expression yields the determinate form $1/0$, so l'Hospital's rule does not apply here. Using the methods of Section 3.3 again, we find that the limit does not exist.

FAQs

Using the Rules and Recognizing when a Function Is Not Differentiable

Q: I would *like* to say that the derivative of $5x^2 - 8x + 4$ is just $10x - 8$ without having to go through all that stuff about derivatives of sums and constant multiples. Can I simply forget about all the rules and write down the answer?

A: We developed the rules for sums and constant multiples precisely for that reason: so that we could simply write down a derivative without having to think about it too hard. So, you are perfectly justified in simply writing down the derivative without going through the rules, but bear in mind that what you are really doing is applying the power rule, the rule for sums, and the rule for multiples over and over.

Q: Is there a way of telling from its formula whether a function f is not differentiable at a point?

A: Here are some indicators to look for in the formula for f:

- The absolute value of some expression; f may not be differentiable at points where that expression is zero.

  Example: $f(x) = 3x^2 - |x - 4|$ is not differentiable at $x = 4$.

- A fractional power smaller than 1 of some expression; f may not be differentiable at points where that expression is zero.

  Example: $f(x) = (x^2 - 16)^{2/3}$ is not differentiable at $x = \pm 4$.

4.1 EXERCISES

Access end-of-section exercises online at **www.webassign.net** ENHANCED WebAssign

4.2 A First Application: Marginal Analysis

In Chapter 1, we considered linear *cost functions* of the form $C(x) = mx + b$, where C is the total cost, x is the number of items, and m and b are constants. The slope m is the *marginal cost*. It measures the *cost of one more item*. Notice that the derivative of $C(x) = mx + b$ is $C'(x) = m$. In other words, for a linear cost function, *the marginal cost is the derivative of the cost function.*

In general, we make the following definition.

Marginal Cost

Recall from Section 1.2 that a **cost function** C specifies the total cost as a function of the number of items x, so that $C(x)$ is the total cost of x items. The **marginal cost function** is the derivative C' of the cost function C. Thus, $C'(x)$ measures the rate of change of cost with respect to x.

Units

The units of marginal cost are units of cost (dollars, say) per item.

Interpretation

*We interpret $C'(x)$ as the approximate cost of one more item.**

* See Example 1.

Quick Example

If $C(x) = 400x + 1,000$ dollars, then the marginal cost function is $C'(x) = \$400$ per item (a constant).

EXAMPLE 1 Marginal Cost

Suppose that the cost in dollars to manufacture portable music players is given by

$$C(x) = 150,000 + 20x - 0.0001x^2$$

where x is the number of music players manufactured.† Find the marginal cost function C' and use it to estimate the cost of manufacturing the 50,001st music player.

† The term $0.0001x^2$ may reflect a cost saving for high levels of production, such as a bulk discount in the cost of electronic components.

Solution Since

$$C(x) = 150,000 + 20x - 0.0001x^2$$

the marginal cost function is

$$C'(x) = 20 - 0.0002x.$$

The units of $C'(x)$ are units of C (dollars) per unit of x (music players). Thus, $C'(x)$ is measured in dollars per music player.

The cost of the 50,001st music player is the amount by which the total cost would rise if we increased production from 50,000 music players to 50,001. Thus, we need to know the rate at which the total cost rises as we increase production. This rate of change is measured by the derivative, or marginal cost, which we just computed. At $x = 50{,}000$, we get

$$C'(50{,}000) = 20 - 0.0002(50{,}000) = \$10 \text{ per music player.}$$

In other words, we estimate that the 50,001st music player will cost approximately \$10.

Before we go on... In Example 1, the marginal cost is really only an *approximation* to the cost of the 50,001st music player:

$$\begin{aligned} C'(50{,}000) &\approx \frac{C(50{,}001) - C(50{,}000)}{1} && \text{Set } h = 1 \text{ in the definition of the derivative.} \\ &= C(50{,}001) - C(50{,}000) \\ &= \text{cost of the 50,001st music player} \end{aligned}$$

The exact cost of the 50,001st music player is

$$\begin{aligned} C(50{,}001) - C(50{,}000) &= [150{,}000 + 20(50{,}001) - 0.0001(50{,}001)^2] \\ &\quad - [150{,}000 + 20(50{,}000) - 0.0001(50{,}000)^2] \\ &= \$9.9999 \end{aligned}$$

So, the marginal cost is a good approximation to the actual cost.

Graphically, we are using the tangent line to approximate the cost function near a production level of 50,000. Figure 3 shows the graph of the cost function together with the tangent line at $x = 50{,}000$. Notice that the tangent line is essentially indistinguishable from the graph of the function for some distance on either side of 50,000.

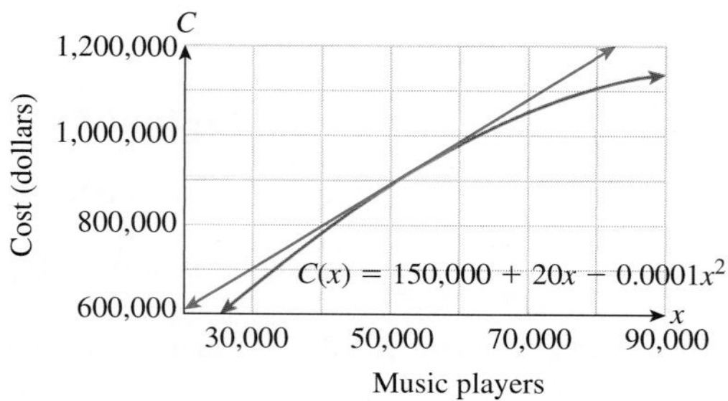

Figure 3

Notes

1. In general, the difference quotient $[C(x+h) - C(x)]/h$ gives the **average cost per item** to produce h more items at a current production level of x items. (Why?)
2. Notice that $C'(x)$ is much easier to calculate than $[C(x+h) - C(x)]/h$. (Try it.) ■

We can extend the idea of marginal cost to include other functions we discussed in Section 1.2, like revenue and profit:

Marginal Revenue and Profit

Recall that a **revenue** or **profit function** specifies the total revenue R or profit P as a function of the number of items x. The derivatives, R' and P', of these functions are called the **marginal revenue** and **marginal profit** functions. They measure the rate of change of revenue and profit with respect to the number of items.

Units

The units of marginal revenue and profit are the same as those of marginal cost: dollars (or euros, pesos, etc.) per item.

Interpretation

We interpret $R'(x)$ and $P'(x)$ as the approximate revenue and profit from the sale of one more item.

EXAMPLE 2 Marginal Revenue and Profit

You operate an *iPod* refurbishing service (a typical refurbished iPod might have a custom color case with blinking lights and a personalized logo). The cost to refurbish x iPods in a month is calculated to be

$$C(x) = 0.25x^2 + 40x + 1{,}000 \text{ dollars.}$$

You charge customers \$80 per iPod for the work.

a. Calculate the marginal revenue and profit functions. Interpret the results.

b. Compute the revenue and profit, and also the marginal revenue and profit, if you have refurbished 20 units this month. Interpret the results.

c. For which value of x is the marginal profit zero? Interpret your answer.

Solution

a. We first calculate the revenue and profit functions:

$$\begin{aligned} R(x) &= 80x && \text{Revenue = Price × Quantity} \\ P(x) &= R(x) - C(x) && \text{Profit = Revenue − Cost} \\ &= 80x - (0.25x^2 + 40x + 1{,}000) \\ P(x) &= -0.25x^2 + 40x - 1{,}000. \end{aligned}$$

The marginal revenue and profit functions are then the derivatives:

$$\begin{aligned} \text{Marginal revenue} &= R'(x) = 80 \\ \text{Marginal profit} &= P'(x) = -0.5x + 40. \end{aligned}$$

Interpretation: $R'(x)$ gives the approximate revenue from the refurbishing of one more item, and $P'(x)$ gives the approximate profit from the refurbishing of one more item. Thus, if x iPods have been refurbished in a month, you will earn a revenue of \$80 and make a profit of approximately $\$(-0.5x + 40)$ if you refurbish one more that month.

Notice that the marginal revenue is a constant, so you earn the same revenue (\$80) for each iPod you refurbish. However, the marginal profit, $\$(-0.5x + 40)$, decreases as x increases, so your additional profit is about 50¢ less for each additional iPod you refurbish.

b. From part (a), the revenue, profit, marginal revenue, and marginal profit functions are

$$R(x) = 80x$$
$$P(x) = -0.25x^2 + 40x - 1{,}000$$
$$R'(x) = 80$$
$$P'(x) = -0.5x + 40$$

Because you have refurbished $x = 20$ iPods this month, $x = 20$, so

$$R(20) = 80(20) = \$1{,}600 \quad \text{Total revenue from 20 iPods}$$
$$P(20) = -0.25(20)^2 + 40(20) - 1{,}000 = -\$300 \quad \text{Total profit from 20 iPods}$$
$$R'(20) = \$80 \text{ per unit} \quad \text{Approximate revenue from the 21st iPod}$$
$$P'(20) = -0.5(20) + 40 = \$30 \text{ per unit} \quad \text{Approximate profit from the 21st iPod}$$

Interpretation: If you refurbish 20 iPods in a month, you will earn a total revenue of \$160 and a profit of −\$300 (indicating a loss of \$300). Refurbishing one more iPod that month will earn you an additional revenue of \$80 and an additional profit of about \$30.

c. The marginal profit is zero when $P'(x) = 0$:

$$-0.5x + 40 = 0$$
$$x = \frac{40}{0.5} = 80 \text{ iPods}$$

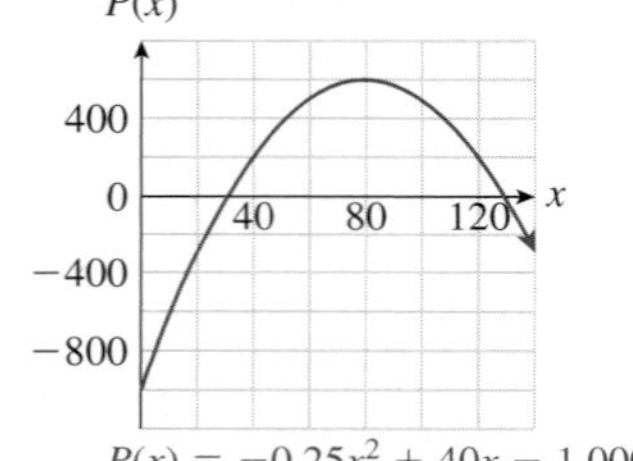

$P(x) = -0.25x^2 + 40x - 1{,}000$

Figure 4

Thus, if you refurbish 80 iPods in a month, refurbishing one more will get you (approximately) zero additional profit. To understand this further, let us take a look at the graph of the profit function, shown in Figure 4. Notice that the graph is a parabola (the profit function is quadratic) with vertex at the point $x = 80$, where $P'(x) = 0$, so the profit is a maximum at this value of x.

➡ **Before we go on...** In general, setting $P'(x) = 0$ and solving for x will always give the exact values of x for which the profit peaks as in Figure 4, assuming there is such a value. We recommend that you graph the profit function to check whether the profit is indeed a maximum at such a point. ■

EXAMPLE 3 Marginal Product

A consultant determines that *Precision Manufacturers'* annual profit (in dollars) is given by

$$P(n) = -200{,}000 + 400{,}000n - 4{,}600n^2 - 10n^3 \qquad (10 \le n \le 50),$$

where n is the number of assembly-line workers it employs.

a. Compute $P'(n)$. $P'(n)$ is called the **marginal product** at the employment level of n assembly-line workers. What are its units?

b. Calculate $P(20)$ and $P'(20)$, and interpret the results.

c. Precision Manufacturers currently employs 20 assembly-line workers and is considering laying off some of them. What advice would you give the company's management?

Solution

a. Taking the derivative gives

$$P'(n) = 400{,}000 - 9{,}200n - 30n^2.$$

The units of $P'(n)$ are profit (in dollars) per worker.

b. Substituting into the formula for $P(n)$, we get

$$P(20) = -200{,}000 + 400{,}000(20) - 4{,}600(20)^2 - 10(20)^3 = \$5{,}880{,}000.$$

Thus, Precision Manufacturers will make an annual profit of $5,880,000 if it employs 20 assembly-line workers. On the other hand,

$$P'(20) = 400{,}000 - 9{,}200(20) - 30(20)^2 = \$204{,}000/\text{worker}.$$

Thus, at an employment level of 20 assembly-line workers, annual profit is increasing at a rate of $204,000 per additional worker. In other words, if the company were to employ one more assembly-line worker, its annual profit would increase by approximately $204,000.

c. Because the marginal product is positive, profits will increase if the company increases the number of workers and will decrease if it decreases the number of workers, so your advice would be to hire additional assembly-line workers. Downsizing the assembly-line workforce would reduce its annual profits.

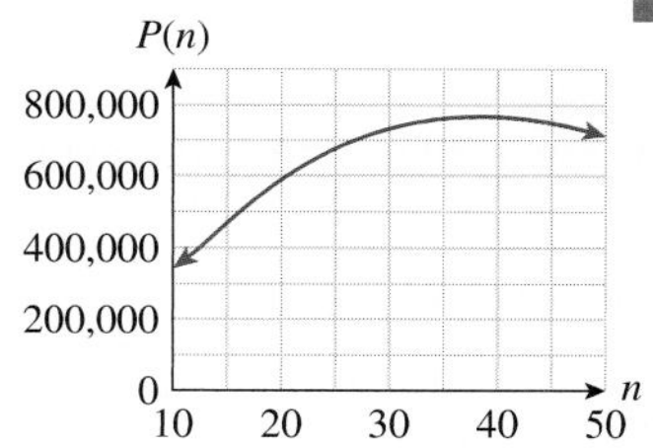

Figure 5

Before we go on... In Example 3, it would be interesting for Precision Manufacturers to ascertain how many additional assembly-line workers it should hire to obtain the *maximum* annual profit. Taking our cue from Example 2, we suspect that such a value of n would correspond to a point where $P'(n) = 0$. Figure 5 shows the graph of P, and on it we see that the highest point of the graph is indeed a point where the tangent line is horizontal; that is, $P'(n) = 0$, and occurs somewhere between $n = 35$ and 40. To compute this value of n more accurately, set $P'(n) = 0$ and solve for n:

$$P'(n) = 400{,}000 - 9{,}200n - 30n^2 = 0 \quad \text{or} \quad 40{,}000 - 920n - 3n^2 = 0.$$

We can now obtain n using the quadratic formula:

$$n = \frac{-b \pm \sqrt{b^2 - 4ac}}{2a} = \frac{920 \pm \sqrt{920^2 - 4(-3)(40{,}000)}}{2(-3)}$$

$$= \frac{920 \pm \sqrt{1{,}326{,}400}}{-6} \approx -345.3 \text{ or } 38.6.$$

The only meaningful solution is the positive one, $n \approx 38.6$ workers, and we conclude that the company should employ between 38 and 39 assembly-line workers for a maximum profit. To see which gives the larger profit, 38 or 39, we check:

$$P(38) = \$7{,}808{,}880$$

while

$$P(39) = \$7{,}810{,}210.$$

This tells us that the company should employ 39 assembly-line workers for a maximum profit. Thus, instead of laying off any of its 20 assembly-line workers, the company should hire 19 additional assembly-line workers for a total of 39. ■

Average Cost

EXAMPLE 4 Average Cost

Suppose the cost in dollars to manufacture portable music players is given by

$$C(x) = 150{,}000 + 20x - 0.0001x^2$$

where x is the number of music players manufactured. (This is the cost equation we saw in Example 1.)

a. Find the average cost per music player if 50,000 music players are manufactured.

b. Find a formula for the average cost per music player if x music players are manufactured. This function of x is called the **average cost function, $\bar{C}(x)$**.

Solution

a. The total cost of manufacturing 50,000 music players is given by

$$C(50{,}000) = 150{,}000 + 20(50{,}000) - 0.0001(50{,}000)^2$$
$$= \$900{,}000.$$

Because 50,000 music players cost a total of \$900,000 to manufacture, the average cost of manufacturing one music player is this total cost divided by 50,000:

$$\bar{C}(50{,}000) = \frac{900{,}000}{50{,}000} = \$18.00 \text{ per music player.}$$

Thus, if 50,000 music players are manufactured, each music player costs the manufacturer an average of \$18.00 to manufacture.

b. If we replace 50,000 by x, we get the general formula for the average cost of manufacturing x music players:

$$\bar{C}(x) = \frac{C(x)}{x}$$
$$= \frac{1}{x}(150{,}000 + 20x - 0.0001x^2)$$
$$= \frac{150{,}000}{x} + 20 - 0.0001x. \quad \text{Average cost function}$$

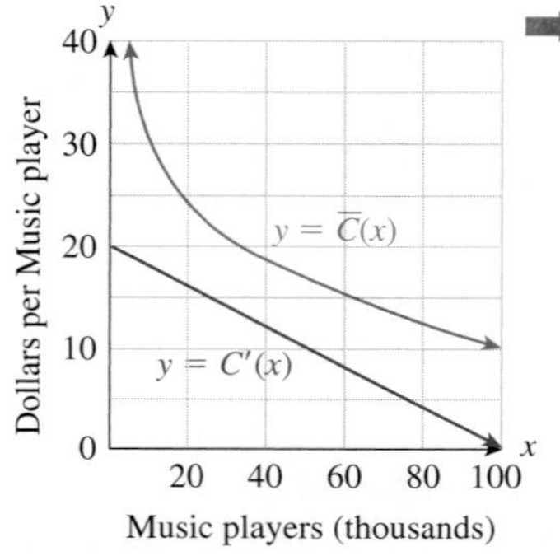

Figure 6

Before we go on... Average cost and marginal cost convey different but related information. The average cost $\bar{C}(50{,}000) = \$18$ that we calculated in Example 4 is the cost per item of manufacturing the first 50,000 music players, whereas the marginal cost $C'(50{,}000) = \$10$ that we calculated in Example 1 gives the (approximate) cost of manufacturing the *next* music player. Thus, according to our calculations, the first 50,000 music players cost an average of \$18 to manufacture, but it costs only about \$10 to manufacture the next one. Note that the marginal cost at a production level of 50,000 music players is lower than the average cost. This means that the average cost to manufacture CDs is going down with increasing volume. (Think about why.)

Figure 6 shows the graphs of average and marginal cost. Notice how the decreasing marginal cost seems to pull the average cost down with it. ■

To summarize:

Average Cost

Given a cost function C, the **average cost** of the first x items is given by

$$\bar{C}(x) = \frac{C(x)}{x}.$$

The average cost is distinct from the **marginal cost** $C'(x)$, which tells us the approximate cost of the *next* item.

Quick Example

For the cost function $C(x) = 20x + 100$ dollars

Marginal Cost $= C'(x) =$ \$20 per additional item.

Average Cost $= \bar{C}(x) = \dfrac{C(x)}{x} = \dfrac{20x + 100}{x} = \$(20 + 100/x)$ per item.

4.2 EXERCISES

Access end-of-section exercises online at **www.webassign.net** ENHANCED WebAssign

4.3 The Product and Quotient Rules

We know how to find the derivatives of functions that are sums of powers, such as polynomials. In general, if a function is a sum or difference of functions whose derivatives we know, then we know how to find its derivative. But what about *products and quotients* of functions whose derivatives we know? For instance, how do we calculate the derivative of something like $x^2/(x + 1)$? The derivative of $x^2/(x + 1)$ is not, as one might suspect, $2x/1 = 2x$. That calculation is based on an assumption that the derivative of a quotient is the quotient of the derivatives. But it is easy to see that this assumption is false: For instance, the derivative of $1/x$ is not $0/1 = 0$, but $-1/x^2$. Similarly, the derivative of a product is not the product of the derivatives: For instance, the derivative of $x = 1 \cdot x$ is not $0 \cdot 1 = 0$, but 1.

To identify the correct method of computing the derivatives of products and quotients, let's look at a simple example. We know that the daily revenue resulting from the sale of q items per day at a price of p dollars per item is given by the product, $R = pq$ dollars. Suppose you are currently selling wall posters on campus. At this time your daily sales are 50 posters, and sales are increasing at a rate of 4 per day. Furthermore, you are currently charging \$10 per poster, and you are also raising the price at a rate of \$2 per day. Let's use this information to estimate how fast your daily revenue is increasing. In other words, let us estimate the rate of change, dR/dt, of the revenue R.

There are two contributions to the rate of change of daily revenue: the increase in daily sales and the increase in the unit price. We have

$$\frac{dR}{dt} \text{ due to increasing price: } \quad \$2 \text{ per day} \times 50 \text{ posters} = \$100 \text{ per day}$$

$$\frac{dR}{dt} \text{ due to increasing sales: } \quad \$10 \text{ per poster} \times 4 \text{ posters per day} = \$40 \text{ per day.}$$

Thus, we estimate the daily revenue to be increasing at a rate of \$100 + \$40 = \$140 per day. Let us translate what we have said into symbols:

$$\frac{dR}{dt} \text{ due to increasing price: } \quad \frac{dp}{dt} \times q$$

$$\frac{dR}{dt} \text{ due to increasing sales: } \quad p \times \frac{dq}{dt}.$$

Thus, the rate of change of revenue is given by

$$\frac{dR}{dt} = \frac{dp}{dt}q + p\frac{dq}{dt}.$$

Because $R = pq$, we have discovered the following rule for differentiating a product:

$$\frac{d}{dt}(pq) = \frac{dp}{dt}q + p\frac{dq}{dt}.$$

The derivative of a product is the derivative of the first times the second, plus the first times the derivative of the second.

This rule and a similar rule for differentiating quotients are given next, and also a discussion of how these results are proved rigorously.

Product Rule

If f and g are differentiable functions of x, then so is their product fg, and

$$\frac{d}{dx}[f(x)g(x)] = f'(x)g(x) + f(x)g'(x).$$

Product Rule in Words

The derivative of a product is the derivative of the first times the second, plus the first times the derivative of the second.

Quick Example

Let $f(x) = x^2$ and $g(x) = 3x - 1$. Because f and g are both differentiable functions of x, so is their product fg, and its derivative is

$$\frac{d}{dx}[x^2(3x-1)] = \underset{\uparrow\text{ Derivative of first}}{2x} \cdot \underset{\uparrow\text{ Second}}{(3x-1)} + \underset{\uparrow\text{ First}}{x^2} \cdot \underset{\uparrow\text{ Derivative of second}}{(3)}.$$

Quotient Rule

If f and g are differentiable functions of x, then so is their quotient f/g, and

$$\frac{d}{dx}\left(\frac{f(x)}{g(x)}\right) = \frac{f'(x)g(x) - f(x)g'(x)}{[g(x)]^2}.$$

provided $g(x) \neq 0$.*

Quotient Rule in Words

The derivative of a quotient is the derivative of the top times the bottom, minus the top times the derivative of the bottom, all over the bottom squared.

* If $g(x)$ is zero, then the quotient $f(x)/g(x)$ is not defined in the first place.

Quick Example

Let $f(x) = x^3$ and $g(x) = x^2 - 1$. Because f and g are both differentiable functions of x, so is their quotient f/g, and its derivative is

Derivative of top ↓ Bottom ↓ Top ↓ Derivative of bottom ↓

$$\frac{d}{dx}\left(\frac{x^3}{x^2-1}\right) = \frac{3x^2(x^2-1) - x^3 \cdot 2x}{(x^2-1)^2},$$

↑ Bottom squared

provided $x \neq 1$ or -1.

Notes

1. Don't try to remember the rules by the symbols we have used, but remember them in words. (The slogans are easy to remember, even if the terms are not precise.)
2. One more time: *The derivative of a product is* NOT *the product of the derivatives, and the derivative of a quotient is* NOT *the quotient of the derivatives.* To find the derivative of a product, you must use the product rule, and to find the derivative of a quotient, you must use the quotient rule.* ■

* Leibniz made this mistake at first, too, so you would be in good company if you forgot to use the product or quotient rule.

Q: *Wait a minute! The expression $2x^3$ is a product, and we already know that its derivative is $6x^2$. Where did we use the product rule?*

A: To differentiate functions such as $2x^3$, we have used the rule from Section 4.1:

The derivative of c times a function is c times the derivative of the function.

However, the product rule gives us the same result:

Derivative of first ↓ Second ↓ First ↓ Derivative of second ↓

$$\frac{d}{dx}(2x^3) = (0)(x^3) + (2)(3x^2) = 6x^2 \qquad \text{Product rule}$$

$$\frac{d}{dx}(2x^3) = (2)(3x^2) = 6x^2 \qquad \text{Derivative of a constant times a function}$$

We do not recommend that you use the product rule to differentiate functions such as $2x^3$; continue to use the simpler rule when one of the factors is a constant.

Derivation of the Product Rule

Before we look at more examples of using the product and quotient rules, let's see why the product rule is true. To calculate the derivative of the product $f(x)g(x)$ of two differentiable functions, we go back to the definition of the derivative:

$$\frac{d}{dx}[f(x)g(x)] = \lim_{h\to 0} \frac{f(x+h)g(x+h) - f(x)g(x)}{h}.$$

We now rewrite this expression so that we can evaluate the limit: Notice that the numerator reflects a simultaneous change in f [from $f(x)$ to $f(x+h)$] and g [from

$g(x)$ to $g(x+h)$]. To separate the two effects, we add and subtract a quantity in the numerator that reflects a change in only one of the functions:

$$\frac{d}{dx}[f(x)g(x)] = \lim_{h\to 0}\frac{f(x+h)g(x+h) - f(x)g(x)}{h}$$

$$= \lim_{h\to 0}\frac{f(x+h)g(x+h) - f(x)g(x+h) + f(x)g(x+h) - f(x)g(x)}{h}$$ We subtracted and added the quantity* $f(x)g(x+h)$.

$$= \lim_{h\to 0}\frac{[f(x+h) - f(x)]\,g(x+h) + f(x)[g(x+h) - g(x)]}{h}$$ Common factors

$$= \lim_{h\to 0}\left(\frac{f(x+h)-f(x)}{h}\right)g(x+h) + \lim_{h\to 0} f(x)\left(\frac{g(x+h)-g(x)}{h}\right)$$ Limit of sum

$$= \lim_{h\to 0}\left(\frac{f(x+h)-f(x)}{h}\right)\lim_{h\to 0} g(x+h) + \lim_{h\to 0} f(x)\lim_{h\to 0}\left(\frac{g(x+h)-g(x)}{h}\right)$$ Limit of product

Now we already know the following four limits:

$$\lim_{h\to 0}\frac{f(x+h)-f(x)}{h} = f'(x)$$ Definition of derivative of f; f is differentiable.

$$\lim_{h\to 0}\frac{g(x+h)-g(x)}{h} = g'(x)$$ Definition of derivative of g; g is differentiable.

$$\lim_{h\to 0} g(x+h) = g(x)$$ If g is differentiable, it must be continuous.†

$$\lim_{h\to 0} f(x) = f(x)$$ Limit of a constant

Putting these limits into the one we're calculating, we get

$$\frac{d}{dx}[f(x)g(x)] = f'(x)g(x) + f(x)g'(x)$$

which is the product rule.

* Adding an appropriate form of zero is an age-old mathematical ploy.

† For a proof of the fact that, if g is differentiable, it must be continuous, go to the Website and follow the path

Everything for Calculus
→ Chapter 4
→ Continuity and Differentiability

EXAMPLE 1 Using the Product Rule

Compute the following derivatives.

a. $\frac{d}{dx}[(x^{3.2}+1)(1-x)]$ Simplify the answer.

b. $\frac{d}{dx}[(x+1)(x^2+1)(x^3+1)]$ Do not expand the answer.

c. $\frac{d}{dx}\left[\frac{x|x|}{2}\right]$

 **Website**
www.WanerMath.com
The quotient rule can be proved in a very similar way. Go to the Website and follow the path

Everything for Calculus
→ Chapter 4
→ Proof of Quotient Rule

Solution

a. We can do the calculation in two ways.

Using the Product Rule:

$$\frac{d}{dx}[(x^{3.2}+1)(1-x)] = \underbrace{(3.2x^{2.2})}_{\text{Derivative of first}}\underbrace{(1-x)}_{\text{Second}} + \underbrace{(x^{3.2}+1)}_{\text{First}}\underbrace{(-1)}_{\text{Derivative of second}}$$

$$= 3.2x^{2.2} - 3.2x^{3.2} - x^{3.2} - 1$$ Expand the answer.

$$= -4.2x^{3.2} + 3.2x^{2.2} - 1$$

Not Using the Product Rule: First, expand the given expression.

$$(x^{3.2}+1)(1-x) = -x^{4.2}+x^{3.2}-x+1$$

Thus,

$$\begin{aligned}\frac{d}{dx}[(x^{3.2}+1)(1-x)] &= \frac{d}{dx}(-x^{4.2}+x^{3.2}-x+1)\\ &= -4.2x^{3.2}+3.2x^{2.2}-1\end{aligned}$$

In this example the product rule saves us little or no work, but in later sections we shall see examples that can be done in no other way. Learn how to use the product rule now!

b. Here we have a product of *three* functions, not just two. We can find the derivative by using the product rule twice:

$$\begin{aligned}&\frac{d}{dx}[(x+1)(x^2+1)(x^3+1)]\\ &= \frac{d}{dx}(x+1)\cdot[(x^2+1)(x^3+1)] + (x+1)\cdot\frac{d}{dx}[(x^2+1)(x^3+1)]\\ &= (1)(x^2+1)(x^3+1)+(x+1)[(2x)(x^3+1)+(x^2+1)(3x^2)]\\ &= (1)(x^2+1)(x^3+1)+(x+1)(2x)(x^3+1)+(x+1)(x^2+1)(3x^2)\end{aligned}$$

We can see here a more general product rule:

$$(fgh)' = f'gh + fg'h + fgh'$$

Notice that every factor has a chance to contribute to the rate of change of the product. There are similar formulas for products of four or more functions.

c. First write $\frac{x|x|}{2}$ as $\frac{1}{2}x|x|$.

$$\frac{d}{dx}\left[\frac{1}{2}x|x|\right] = \frac{1}{2}\frac{d}{dx}[x|x|] \qquad \text{Constant multiple rule}$$

$$= \frac{1}{2}\left((1)\cdot|x| + x\cdot\frac{|x|}{x}\right) \qquad \text{Recall that } \frac{d}{dx}|x| = \frac{|x|}{x}.$$

$$= \frac{1}{2}(|x|+|x|) \qquad \text{Cancel the } x.$$

$$= \frac{1}{2}(2|x|) = |x| \qquad \text{See the note.*}$$

* Notice that we have found a function whose derivative is $|x|$; namely $x|x|/2$. Notice also that the derivation we gave assumes that $x \neq 0$ because we divided by x in the third step. However, one can verify, using the definition of the derivative as a limit, that $x|x|/2$ is differentiable at $x = 0$ as well, and that its derivative at $x = 0$ is 0, implying that the formula $\frac{d}{dx}(x|x|/2) = |x|$ is valid for all values of x, including 0.

EXAMPLE 2 Using the Quotient Rule

Compute the derivatives **a.** $\frac{d}{dx}\left[\frac{1-3.2x^{-0.1}}{x+1}\right]$ **b.** $\frac{d}{dx}\left[\frac{(x+1)(x+2)}{x-1}\right]$

Solution

Derivative of top ↓ Bottom ↓ Top ↓ Derivative of bottom ↓

a. $$\frac{d}{dx}\left[\frac{1-3.2x^{-0.1}}{x+1}\right] = \frac{(0.32x^{-1.1})(x+1)-(1-3.2x^{-0.1})(1)}{(x+1)^2}$$

↑ Bottom squared

$$= \frac{0.32x^{-0.1} + 0.32x^{-1.1} - 1 + 3.2x^{-0.1}}{(x+1)^2} \quad \text{Expand the numerator.}$$

$$= \frac{3.52x^{-0.1} + 0.32x^{-1.1} - 1}{(x+1)^2}$$

b. Here we have both a product and a quotient. Which rule do we use, the product or the quotient rule? Here is a way to decide. Think about how we would calculate, step by step, the value of $(x+1)(x+2)/(x-1)$ for a specific value of x—say $x = 11$. Here is how we would probably do it:

1. Calculate $(x+1)(x+2) = (11+1)(11+2) = 156$.
2. Calculate $x - 1 = 11 - 1 = 10$.
3. Divide 156 by 10 to get 15.6.

Now ask: *What was the last operation we performed?* The last operation we performed was division, so we can regard the whole expression as a *quotient*—that is, as $(x+1)(x+2)$ *divided by* $(x-1)$. Therefore, we should use the quotient rule.

The first thing the quotient rule tells us to do is to take the derivative of the numerator. Now, the numerator is a product, so we must use the product rule to take its derivative. Here is the calculation:

$$\frac{d}{dx}\left[\frac{(x+1)(x+2)}{x-1}\right] = \frac{\overbrace{[(1)(x+2) + (x+1)(1)]}^{\text{Derivative of top}}\overbrace{(x-1)}^{\text{Bottom}} - \overbrace{[(x+1)(x+2)]}^{\text{Top}}\overbrace{(1)}^{\text{Derivative of bottom}}}{\underbrace{(x-1)^2}_{\text{Bottom squared}}}$$

$$= \frac{(2x+3)(x-1) - (x+1)(x+2)}{(x-1)^2}$$

$$= \frac{x^2 - 2x - 5}{(x-1)^2}$$

What is important is to determine the *order of operations* and, in particular, to determine the last operation to be performed. Pretending to do an actual calculation reminds us of the order of operations; we call this technique the **calculation thought experiment**.

➡ **Before we go on...** We used the quotient rule in Example 2 because the function was a quotient; we used the product rule to calculate the derivative of the numerator because the numerator was a product. Get used to this: Differentiation rules usually must be used in combination.

Here is another way we could have done this problem: Our calculation thought experiment could have taken the following form.

1. Calculate $(x+1)/(x-1) = (11+1)/(11-1) = 1.2$.
2. Calculate $x + 2 = 11 + 2 = 13$.
3. Multiply 1.2 by 13 to get 15.6.

We would have then regarded the expression as a *product*—the product of the factors $(x+1)/(x-1)$ and $(x+2)$—and used the product rule instead. We can't escape the quotient rule, however: We need to use it to take the derivative of the first factor, $(x+1)/(x-1)$. Try this approach for practice and check that you get the same answer. ■

Calculation Thought Experiment

The **calculation thought experiment** is a technique to determine whether to treat an algebraic expression as a product, quotient, sum, or difference. Given an expression, consider the steps you would use in computing its value. If the last operation is multiplication, treat the expression as a product; if the last operation is division, treat the expression as a quotient; and so on.

Quick Examples

1. $(3x^2 - 4)(2x + 1)$ can be computed by first calculating the expressions in parentheses and then multiplying. Because the last step is multiplication, we can treat the expression as a product.
2. $\dfrac{2x - 1}{x}$ can be computed by first calculating the numerator and denominator and then dividing one by the other. Because the last step is division, we can treat the expression as a quotient.
3. $x^2 + (4x - 1)(x + 2)$ can be computed by first calculating x^2, then calculating the product $(4x - 1)(x + 2)$, and finally adding the two answers. Thus, we can treat the expression as a sum.
4. $(3x^2 - 1)^5$ can be computed by first calculating the expression in parentheses and then raising the answer to the fifth power. Thus, we can treat the expression as a power. (We shall see how to differentiate powers of expressions in Section 4.4.)
5. The expression $(x + 1)(x + 2)/(x - 1)$ can be treated as either a quotient or a product: We can write it as a quotient: $\dfrac{(x + 1)(x + 2)}{x - 1}$ or as a product: $(x + 1)\left(\dfrac{x + 2}{x - 1}\right)$. (See Example 2(b).)

EXAMPLE 3 Using the Calculation Thought Experiment

Find $\dfrac{d}{dx}\left[6x^2 + 5\left(\dfrac{x}{x - 1}\right)\right]$.

Solution The calculation thought experiment tells us that the expression we are asked to differentiate can be treated as a *sum*. Because the derivative of a sum is the sum of the derivatives, we get

$$\frac{d}{dx}\left[6x^2 + 5\left(\frac{x}{x - 1}\right)\right] = \frac{d}{dx}(6x^2) + \frac{d}{dx}\left[5\left(\frac{x}{x - 1}\right)\right].$$

In other words, we must take the derivatives of $6x^2$ and $5\left(\dfrac{x}{x - 1}\right)$ separately and then add the answers. The derivative of $6x^2$ is $12x$. There are two ways of taking the derivative of $5\left(\dfrac{x}{x - 1}\right)$: We could either first multiply the expression $\left(\dfrac{x}{x - 1}\right)$ by 5 to get $\left(\dfrac{5x}{x - 1}\right)$ and then take its derivative using the quotient rule, or we could pull the 5 out, as we do next.

$$\frac{d}{dx}\left[6x^2+5\left(\frac{x}{x-1}\right)\right]=\frac{d}{dx}(6x^2)+\frac{d}{dx}\left[5\left(\frac{x}{x-1}\right)\right]\qquad\text{Derivative of sum}$$

$$=12x+5\frac{d}{dx}\left(\frac{x}{x-1}\right)\qquad\text{Constant}\times\text{Function}$$

$$=12x+5\left(\frac{(1)(x-1)-(x)(1)}{(x-1)^2}\right)\qquad\text{Quotient rule}$$

$$=12x+5\left(\frac{-1}{(x-1)^2}\right)$$

$$=12x-\frac{5}{(x-1)^2}$$

APPLICATIONS

In the next example, we return to a scenario similar to the one discussed at the start of this section.

EXAMPLE 4 Applying the Product and Quotient Rules: Revenue and Average Cost

Sales of your newly launched miniature wall posters for college dorms, *iMiniPosters,* are really taking off. (Those old-fashioned large wall posters no longer fit in today's "downsized" college dorm rooms.) Monthly sales to students at the start of this year were 1,500 iMiniPosters, and since that time, sales have been increasing by 300 posters each month, even though the price you charge has also been going up.

a. The price you charge for iMiniPosters is given by

$$p(t)=10+0.05t^2 \text{ dollars per poster,}$$

where t is time in months since the start of January of this year. Find a formula for the monthly revenue, and then compute its rate of change at the beginning of March.

b. The number of students who purchase iMiniPosters in a month is given by

$$n(t)=800+0.2t,$$

where t is as in part (a). Find a formula for the average number of posters each student buys, and hence estimate the rate at which this number was growing at the beginning of March.

Solution

a. To compute monthly revenue as a function of time t, we use

$$R(t)=p(t)q(t).\qquad\text{Revenue = Price}\times\text{Quantity}$$

We already have a formula for $p(t)$. The function $q(t)$ measures sales, which were 1,500 posters/month at time $t=0$, and were rising by 300 per month:

$$q(t)=1{,}500+300t.$$

Therefore, the formula for revenue is

$$R(t)=p(t)q(t)$$

$$R(t)=(10+0.05t^2)(1{,}500+300t).$$

Rather than expand this expression, we shall leave it as a product so that we can use the product rule in computing its rate of change:

$$\begin{aligned} R'(t) &= p'(t)q(t) + p(t)q'(t) \\ &= [0.10t][1{,}500 + 300t] + [10 + 0.05t^2][300]. \end{aligned}$$

Because the beginning of March corresponds to $t = 2$, we have

$$\begin{aligned} R'(2) &= [0.10(2)][1{,}500 + 300(2)] + [10 + 0.05(2)^2][300] \\ &= (0.2)(2{,}100) + (10.2)(300) = \$3{,}480 \text{ per month.} \end{aligned}$$

Therefore, your monthly revenue was increasing at a rate of \$3,480 per month at the beginning of March.

b. The average number of posters sold to each student is

$$k(t) = \frac{\text{Number of posters}}{\text{Number of students}} = \frac{q(t)}{n(t)} = \frac{1{,}500 + 300t}{800 + 0.2t}.$$

The rate of change of $k(t)$ is computed with the quotient rule:

$$\begin{aligned} k'(t) &= \frac{q'(t)n(t) - q(t)n'(t)}{n(t)^2} \\ &= \frac{(300)(800 + 0.2t) - (1{,}500 + 300t)(0.2)}{(800 + 0.2t)^2} \end{aligned}$$

so that

$$\begin{aligned} k'(2) &= \frac{(300)[800 + 0.2(2)] - [1{,}500 + 300(2)](0.2)}{[800 + 0.2(2)]^2} \\ &= \frac{(300)(800.4) - (2{,}100)(0.2)}{800.4^2} \approx 0.37 \text{ posters/student per month.} \end{aligned}$$

Therefore, the average number of posters sold to each student was increasing at a rate of about 0.37 posters/student per month.

4.3 EXERCISES

Access end-of-section exercises online at **www.webassign.net**

4.4 The Chain Rule

We can now find the derivatives of expressions involving powers of x combined using addition, subtraction, multiplication, and division, but we still cannot take the derivative of an expression like $(3x + 1)^{0.5}$. For this we need one more rule. The function $h(x) = (3x + 1)^{0.5}$ is not a sum, difference, product, or quotient. To find out what it is, we can use the calculation thought experiment and think about the last operation we would perform in calculating $h(x)$.

1. Calculate $3x + 1$.

2. Take the 0.5 power (square root) of the answer.

The last operation is "take the 0.5 power." We do not yet have a rule for finding the derivative of the 0.5 power of a quantity other than x.

There is a way to build $h(x) = (3x+1)^{0.5}$ out of two simpler functions: $u(x) = 3x + 1$ (the function that corresponds to the first step in the calculation above) and $f(x) = x^{0.5}$ (the function that corresponds to the second step):

$$\begin{aligned} h(x) &= (3x+1)^{0.5} \\ &= [u(x)]^{0.5} && u(x) = 3x+1 \\ &= f(u(x)). && f(x) = x^{0.5} \end{aligned}$$

We say that h is the **composite** of f and u. We read $f(u(x))$ as "f of u of x."

To compute $h(1)$, say, we first compute $3 \cdot 1 + 1 = 4$ and then take the square root of 4, giving $h(1) = 2$. To compute $f(u(1))$ we follow exactly the same steps: First compute $u(1) = 4$ and then $f(u(1)) = f(4) = 2$. We always compute $f(u(x))$ from the inside out: Given x, first compute $u(x)$ and then $f(u(x))$.

Now, f and u are functions *whose derivatives we know.* The *chain rule* allows us to use our knowledge of the derivatives of f and u to find the derivative of $f(u(x))$. For the purposes of stating the rule, let us avoid some of the nested parentheses by abbreviating $u(x)$ as u. Thus, we write $f(u)$ instead of $f(u(x))$ and remember that u is a function of x.

Chain Rule

If f is a differentiable function of u and u is a differentiable function of x, then the composite $f(u)$ is a differentiable function of x, and

$$\frac{d}{dx}[f(u)] = f'(u)\frac{du}{dx}. \qquad \text{Chain rule}$$

In words *The derivative of f(quantity) is the derivative of f, evaluated at that quantity, times the derivative of the quantity.*

Quick Examples

In the Quick Examples that follow, u, "the quantity," is some (unspecified) differentiable function of x.

1. Take $f(u) = u^2$. Then

$$\frac{d}{dx}(u^2) = 2u\frac{du}{dx}. \qquad \text{Because } f'(u) = 2u$$

The derivative of a quantity squared is two times the quantity, times the derivative of the quantity.

2. Take $f(u) = u^{0.5}$. Then

$$\frac{d}{dx}(u^{0.5}) = 0.5u^{-0.5}\frac{du}{dx}. \qquad \text{Because } f'(u) = 0.5u^{-0.5}$$

The derivative of a quantity raised to the 0.5 is 0.5 times the quantity raised to the −0.5, times the derivative of the quantity.

As the quick examples illustrate, for every power of a function u whose derivative we know, we now get a "generalized" differentiation rule. The following table gives more examples.

Original Rule	*Generalized Rule*	*In Words*
$\frac{d}{dx}(x^2) = 2x$	$\frac{d}{dx}(u^2) = 2u\frac{du}{dx}$	*The derivative of a quantity squared is twice the quantity, times the derivative of the quantity.*
$\frac{d}{dx}(x^3) = 3x^2$	$\frac{d}{dx}(u^3) = 3u^2\frac{du}{dx}$	*The derivative of a quantity cubed is 3 times the quantity squared, times the derivative of the quantity.*
$\frac{d}{dx}\left(\frac{1}{x}\right) = -\frac{1}{x^2}$	$\frac{d}{dx}\left(\frac{1}{u}\right) = -\frac{1}{u^2}\frac{du}{dx}$	*The derivative of 1 over a quantity is negative 1 over the quantity squared, times the derivative of the quantity.*
Power Rule	***Generalized Power Rule***	***In Words***
$\frac{d}{dx}(x^n) = nx^{n-1}$	$\frac{d}{dx}(u^n) = nu^{n-1}\frac{du}{dx}$	*The derivative of a quantity raised to the n is n times the quantity raised to the n − 1, times the derivative of the quantity.*
$\frac{d}{dx}\lvert x\rvert = \frac{\lvert x\rvert}{x}$	$\frac{d}{dx}\lvert u\rvert = \frac{\lvert u\rvert}{u}\frac{du}{dx}$	*The derivative of the absolute value of a quantity is the absolute value of the quantity divided by the quantity, times the derivative of the quantity.*

To motivate the chain rule, let us see why it is true in the special case when $f(u) = u^3$, where the chain rule tells us that

$$\frac{d}{dx}(u^3) = 3u^2\frac{du}{dx}. \qquad \text{Generalized power rule with } n = 3$$

But we could have done this using the product rule instead:

$$\frac{d}{dx}(u^3) = \frac{d}{dx}(u \cdot u \cdot u) = \frac{du}{dx}u \cdot u + u\frac{du}{dx}u + u \cdot u\frac{du}{dx} = 3u^2\frac{du}{dx},$$

which gives us the same result. A similar argument works for $f(u) = u^n$, where $n = 2, 3, 4, \ldots$. We can then use the quotient rule and the chain rule for positive powers to verify the generalized power rule for *negative* powers as well. For the case of a general differentiable function *f*, the proof of the chain rule is beyond the scope of this book, but you can find one on the Website by following the path

Website → Everything for Calculus → Chapter 4 → Proof of Chain Rule.

EXAMPLE 1 Using the Chain Rule

Compute the following derivatives.

a. $\frac{d}{dx}[(2x^2 + x)^3]$ **b.** $\frac{d}{dx}[(x^3 + x)^{100}]$ **c.** $\frac{d}{dx}\sqrt{3x + 1}$ **d.** $\frac{d}{dx}\lvert 4x^2 - x\rvert$

Solution

a. Using the calculation thought experiment, we see that the last operation we would perform in calculating $(2x^2 + x)^3$ is that of *cubing*. Thus we think of $(2x^2 + x)^3$ as *a quantity cubed*. There are two similar methods we can use to calculate its derivative.

Method 1: Using the formula We think of $(2x^2+x)^3$ as u^3, where $u=2x^2+x$. By the formula,

$$\frac{d}{dx}(u^3) = 3u^2\frac{du}{dx}. \qquad \text{Generalized power rule}$$

Now substitute for u:

$$\frac{d}{dx}[(2x^2+x)^3] = 3(2x^2+x)^2\frac{d}{dx}(2x^2+x)$$
$$= 3(2x^2+x)^2(4x+1)$$

Method 2: Using the verbal form If we prefer to use the verbal form, we get:

The derivative of $(2x^2+x)$ *cubed is three times* $(2x^2+x)$ *squared, times the derivative of* $(2x^2+x)$.

In symbols,

$$\frac{d}{dx}[(2x^2+x)^3] = 3(2x^2+x)^2(4x+1),$$

as we obtained above.

b. First, the calculation thought experiment: If we were computing $(x^3+x)^{100}$, the last operation we would perform is *raising a quantity to the power* 100. Thus we are dealing with *a quantity raised to the power* 100, and so we must again use the generalized power rule. According to the verbal form of the generalized power rule, the derivative of a quantity raised to the power 100 is 100 times that quantity to the power 99, times the derivative of that quantity. In symbols,

$$\frac{d}{dx}[(x^3+x)^{100}] = 100(x^3+x)^{99}(3x^2+1).$$

c. We first rewrite the expression $\sqrt{3x+1}$ as $(3x+1)^{0.5}$ and then use the generalized power rule as in parts (a) and (b):

The derivative of a quantity raised to the 0.5 *is* 0.5 *times the quantity raised to the* −0.5, *times the derivative of the quantity.*

Thus,

$$\frac{d}{dx}[(3x+1)^{0.5}] = 0.5(3x+1)^{-0.5}\cdot 3 = 1.5(3x+1)^{-0.5}.$$

d. The calculation thought experiment tells us that $|4x^2-x|$ is the absolute value of a quantity, so we use the generalized rule for absolute values (above):

$$\frac{d}{dx}|u| = \frac{|u|}{u}\frac{du}{dx}, \text{ or, in words,}$$

The derivative of the absolute value of a quantity is the absolute value of the quantity divided by the quantity, times the derivative of the quantity.

Thus,

$$\frac{d}{dx}|4x^2-x| = \frac{|4x^2-x|}{4x^2-x}\cdot(8x-1). \qquad \frac{d}{dx}|u| = \frac{|u|}{u}\frac{du}{dx}$$

➡ **Before we go on...** The following are examples of common errors in solving Example 1(b):

$$\text{“}\frac{d}{dx}[(x^3+x)^{100}] = 100(3x^2+1)^{99}\text{”} \qquad ✗ \ \textit{WRONG!}$$

$$\text{“}\frac{d}{dx}[(x^3+x)^{100}] = 100(x^3+x)^{99}.\text{”} \qquad ✗ \ \textit{WRONG!}$$

Remember that the generalized power rule says that the derivative of a quantity to the power 100 is 100 times *that same quantity* raised to the power 99, *times the derivative of that quantity.* ■

Q: *It seems that there are now two formulas for the derivative of an nth power:*

1. $\frac{d}{dx}[x^n] = nx^{n-1}$
2. $\frac{d}{dx}[u^n] = nu^{n-1}\frac{du}{dx}$.

Which one do I use?

A: Formula 1 is actually a special case of Formula 2: Formula 1 is the original power rule, which applies only to a power of x. For instance, it applies to x^{10}, but it does not apply to $(2x+1)^{10}$ because the quantity that is being raised to a power is not x. Formula 2 applies to a power of any *function of* x, such as $(2x+1)^{10}$. It can even be used in place of the original power rule. For example, if we take $u = x$ in Formula 2, we obtain

$$\frac{d}{dx}[x^n] = nx^{n-1}\frac{dx}{dx}$$

$$= nx^{n-1}. \quad \text{The derivative of } x \text{ with respect to } x \text{ is } 1.$$

Thus, the generalized power rule really *is* a generalization of the original power rule, as its name suggests.

EXAMPLE 2 More Examples Using the Chain Rule

Find: **a.** $\frac{d}{dx}[(2x^5+x^2-20)^{-2/3}]$ **b.** $\frac{d}{dx}\left[\frac{1}{\sqrt{x+2}}\right]$ **c.** $\frac{d}{dx}\left[\frac{1}{x^2+x}\right]$

Solution Each of the given functions is, or can be rewritten as, a power of a function whose derivative we know. Thus, we can use the method of Example 1.

a. $\frac{d}{dx}[(2x^5+x^2-20)^{-2/3}] = -\frac{2}{3}(2x^5+x^2-20)^{-5/3}(10x^4+2x)$

b. $\frac{d}{dx}\left[\frac{1}{\sqrt{x+2}}\right] = \frac{d}{dx}(x+2)^{-1/2} = -\frac{1}{2}(x+2)^{-3/2}\cdot 1 = -\frac{1}{2(x+2)^{3/2}}$

c. $\frac{d}{dx}\left[\frac{1}{x^2+x}\right] = \frac{d}{dx}(x^2+x)^{-1} = -(x^2+x)^{-2}(2x+1) = -\frac{2x+1}{(x^2+x)^2}$

➡ **Before we go on...** In Example 2(c), we could have used the quotient rule instead of the generalized power rule. We can think of the quantity $1/(x^2+x)$ in two different ways using the calculation thought experiment:

1. As 1 divided by something—in other words, as a quotient

2. As something raised to the -1 power

Of course, we get the same derivative using either approach. ■

We now look at some more complicated examples.

EXAMPLE 3 Harder Examples Using the Chain Rule

Find $\frac{dy}{dx}$ in each case. **a.** $y = [(x+1)^{-2.5} + 3x]^{-3}$ **b.** $y = (x+10)^3\sqrt{1-x^2}$

Solution

a. The calculation thought experiment tells us that the last operation we would perform in calculating y is raising the quantity $[(x+1)^{-2.5} + 3x]$ to the power -3. Thus, we use the generalized power rule.

$$\frac{dy}{dx} = -3[(x+1)^{-2.5} + 3x]^{-4}\frac{d}{dx}[(x+1)^{-2.5} + 3x]$$

We are not yet done; we must still find the derivative of $(x+1)^{-2.5} + 3x$. Finding the derivative of a complicated function in several steps helps to keep the problem manageable. Continuing, we have

$$\frac{dy}{dx} = -3[(x+1)^{-2.5} + 3x]^{-4}\frac{d}{dx}[(x+1)^{-2.5} + 3x]$$

$$= -3[(x+1)^{-2.5} + 3x]^{-4}\left(\frac{d}{dx}[(x+1)^{-2.5}] + \frac{d}{dx}(3x)\right).$$ Derivative of a sum

Now we have two derivatives left to calculate. The second of these we know to be 3, and the first is the derivative of a quantity raised to the -2.5 power. Thus

$$\frac{dy}{dx} = -3[(x+1)^{-2.5} + 3x]^{-4}[-2.5(x+1)^{-3.5}\cdot 1 + 3].$$

b. The expression $(x+10)^3\sqrt{1-x^2}$ is a product, so we use the product rule:

$$\frac{d}{dx}[(x+10)^3\sqrt{1-x^2}] = \left(\frac{d}{dx}[(x+10)^3]\right)\sqrt{1-x^2} + (x+10)^3\left(\frac{d}{dx}\sqrt{1-x^2}\right)$$

$$= 3(x+10)^2\sqrt{1-x^2} + (x+10)^3\frac{1}{2\sqrt{1-x^2}}(-2x)$$

$$= 3(x+10)^2\sqrt{1-x^2} - \frac{x(x+10)^3}{\sqrt{1-x^2}}.$$

APPLICATIONS

The next example is a new way of looking at Example 3 from Section 4.2.

EXAMPLE 4 Marginal Product

A consultant determines that *Precision Manufacturers'* annual profit (in dollars) is given by

$$P = -200{,}000 + 4{,}000q - 0.46q^2 - 0.00001q^3,$$

where q is the number of surgical lasers it sells each year. The consultant also informs Precision that the number of surgical lasers it can manufacture each year depends on the number n of assembly-line workers it employs according to the equation

$$q = 100n.$$ Each worker contributes 100 lasers per year.

Use the chain rule to find the marginal product $\frac{dP}{dn}$.

Solution We could calculate the marginal product by substituting the expression for q in the expression for P to obtain P as a function of n (as given in Example 3 from Section 4.2) and then finding dP/dn. Alternatively—and this will simplify the calculation—we can use the chain rule. To see how the chain rule applies, notice that P is a function of q, where q in turn is given as a function of n. By the chain rule,

$$\frac{dP}{dn} = P'(q)\frac{dq}{dn} \qquad \text{Chain rule}$$

$$= \frac{dP}{dq}\frac{dq}{dn}. \qquad \text{Notice how the "quantities" } dq \text{ appear to cancel.}$$

Now we compute

$$\frac{dP}{dq} = 4{,}000 - 0.92q - 0.00003q^2$$

and $$\frac{dq}{dn} = 100.$$

Substituting into the equation for $\frac{dP}{dn}$ gives

$$\frac{dP}{dn} = (4{,}000 - 0.92q - 0.00003q^2)(100)$$

$$= 400{,}000 - 92q - 0.003q^2.$$

Notice that the answer has q as a variable. We can express dP/dn as a function of n by substituting $100n$ for q:

$$\frac{dP}{dn} = 400{,}000 - 92(100n) - 0.003(100n)^2$$

$$= 400{,}000 - 9{,}200n - 30n^2.$$

The equation

$$\frac{dP}{dn} = \frac{dP}{dq}\frac{dq}{dn}$$

in the example above is an appealing way of writing the chain rule because it suggests that the "quantities" dq cancel. In general, we can write the chain rule as follows.

Chain Rule in Differential Notation

If y is a differentiable function of u, and u is a differentiable function of x, then

$$\frac{dy}{dx} = \frac{dy}{du}\frac{du}{dx}. \qquad \text{The terms } du \text{ cancel.}$$

Notice how the units of measurement also cancel:

$$\frac{\text{Units of } y}{\text{Units of } x} = \frac{\text{Units of } y \ \cancel{\text{Units of } u}}{\cancel{\text{Units of } u} \ \text{Units of } x}.$$

Quick Examples

1. If $y = u^3$, where $u = 4x + 1$, then

$$\frac{dy}{dx} = \frac{dy}{du}\frac{du}{dx} = 3u^2 \cdot 4 = 12u^2 = 12(4x + 1)^2.$$

2. If $q = 43p^2$, where p (and hence q also) is a differentiable function of t, then

$$\frac{dq}{dt} = \frac{dq}{dp}\frac{dp}{dt}$$

$$= 86p\frac{dp}{dt}.$$ p is not specified, so we leave dp/dt as is.

EXAMPLE 5 Marginal Revenue

Suppose a company's weekly revenue R is given as a function of the unit price p, and p in turn is given as a function of weekly sales q (by means of a demand equation). If

$$\left.\frac{dR}{dp}\right|_{q=1,000} = \$40 \text{ per } \$1 \text{ increase in price}$$

and

$$\left.\frac{dp}{dq}\right|_{q=1,000} = -\$20 \text{ per additional item sold per week}$$

find the marginal revenue when sales are 1,000 items per week.

Solution The marginal revenue is $\frac{dR}{dq}$. By the chain rule, we have

$$\frac{dR}{dq} = \frac{dR}{dp}\frac{dp}{dq}.$$ Units: Revenue per item
= Revenue per \$1 price increase × price increase per additional item.

Because we are interested in the marginal revenue at a demand level of 1,000 items per week, we have

$$\left.\frac{dR}{dq}\right|_{q=1,000} = (40)(-20) = -\$800 \text{ per additional item sold.}$$

Thus, if the price is lowered to increase the demand from 1,000 to 1,001 items per week, the weekly revenue will drop by approximately \$800.

Look again at the way the terms "du" appeared to cancel in the differential formula $\frac{dy}{dx} = \frac{dy}{du}\frac{du}{dx}$. In fact, the chain rule tells us more:

* The notion of "thinking of x as a function of y" will be made more precise in Section 4.6.

Manipulating Derivatives in Differential Notation

1. Suppose y is a function of x. Then, thinking of x as a function of y (as, for instance, when we can solve for x)* we have

$$\frac{dx}{dy} = \frac{1}{\left(\frac{dy}{dx}\right)}, \text{ provided } \frac{dy}{dx} \neq 0.$$ Notice again how $\frac{dy}{dx}$ behaves like a fraction.

Quick Example

In the demand equation $q = -0.2p - 8$, we have $\frac{dq}{dp} = -0.2$. Therefore,

$$\frac{dp}{dq} = \frac{1}{\left(\frac{dq}{dp}\right)} = \frac{1}{-0.2} = -5.$$

2. Suppose x and y are functions of t. Then, thinking of y as a function of x (as, for instance, when we can solve for t as a function of x, and hence obtain y as a function of x), we have

$$\frac{dy}{dx} = \frac{dy/dt}{dx/dt}.$$ The terms dt appear to cancel.

Quick Example

If $x = 3 - 0.2t$ and $y = 6 + 6t$, then

$$\frac{dy}{dx} = \frac{dy/dt}{dx/dt} = \frac{6}{-0.2} = -30.$$

To see why the above formulas work, notice that the second formula,

$$\frac{dy}{dx} = \frac{\left(\frac{dy}{dt}\right)}{\left(\frac{dx}{dt}\right)}$$

can be written as

$$\frac{dy}{dx}\frac{dx}{dt} = \frac{dy}{dt},$$ Multiply both sides by $\frac{dx}{dt}$.

which is just the differential form of the chain rule. For the first formula, use the second formula with y playing the role of t:

$$\frac{dy}{dx} = \frac{dy/dy}{dx/dy}$$

$$= \frac{1}{dx/dy}.$$ $\frac{dy}{dy} = \frac{d}{dy}[y] = 1$

FAQs

Using the Chain Rule

Q: *How do I decide whether or not to use the chain rule when taking a derivative?*

A: Use the calculation thought experiment (Section 4.3): Given an expression, consider the steps you would use in computing its value.

- If the last step is *raising a quantity to a power,* as in $\left(\frac{x^2-1}{x+4}\right)^4$, then the first step to use is the chain rule (in the form of the generalized power rule):

$$\frac{d}{dx}\left(\frac{x^2-1}{x+4}\right)^4 = 4\left(\frac{x^2-1}{x+4}\right)^3 \frac{d}{dx}\left(\frac{x^2-1}{x+4}\right).$$

Then use the appropriate rules to finish the computation. You may need to again use the calculation thought experiment to decide on the next step (here the quotient rule):

$$= 4\left(\frac{x^2-1}{x+4}\right)^3 \frac{(2x)(x+4)-(x^2-1)(1)}{(x+4)^2}.$$

- If the last step is *division,* as in $\frac{(x^2-1)}{(3x+4)^4}$, then the first step to use is the quotient rule:

$$\frac{d}{dx}\frac{(x^2-1)}{(3x+4)^4} = \frac{(2x)(3x+4)^4-(x^2-1)\frac{d}{dx}(3x+4)^4}{(3x+4)^8}.$$

Then use the appropriate rules to finish the computation (here the chain rule):

$$= \frac{(2x)(3x+4)^4-(x^2-1)[4(3x+4)^3(3)]}{(3x+4)^8}.$$

- If the last step is *multiplication, addition, subtraction, or multiplication by a constant,* then the first rule to use is the product rule, or the rule for sums, differences, or constant multiples as appropriate.

Q: *Every time I compute a derivative, I leave something out. How do I make sure I am really done when taking the derivative of a complicated-looking expression?*

A: Until you are an expert at taking derivatives, the key is to use one rule at a time and write out each step, rather than trying to compute the derivative in a single step. To illustrate this, try computing the derivative of $(x+10)^3\sqrt{1-x^2}$ in Example 3(b) in two ways: First try to compute it in a single step, and then compute it by writing out each step as shown in the example. How do your results compare? For more practice, try Exercises 87 and 88 associated with this section.

4.4 EXERCISES

Access end-of-section exercises online at **www.webassign.net** ENHANCED WebAssign

4.5 Derivatives of Logarithmic and Exponential Functions

At this point, we know how to take the derivative of any algebraic expression in x (involving powers, radicals, and so on). We now turn to the derivatives of logarithmic and exponential functions.

Derivative of the Natural Logarithm

$$\frac{d}{dx}[\ln x] = \frac{1}{x}$$ Recall that $\ln x = \log_e x$.

Quick Examples

1. $\frac{d}{dx}[3 \ln x] = 3 \cdot \frac{1}{x} = \frac{3}{x}$ Derivative of a constant times a function
2. $\frac{d}{dx}[x \ln x] = 1 \cdot \ln x + x \cdot \frac{1}{x}$ Product rule, because $x \ln x$ is a product
   $= \ln x + 1.$

The above simple formula works only for the natural logarithm (the logarithm with base e). For logarithms with bases other than e, we have the following:

Derivative of the Logarithm with Base *b*

$$\frac{d}{dx}[\log_b x] = \frac{1}{x \ln b}$$ Notice that, if $b = e$, we get the same formula as previously.

Quick Examples

1. $\frac{d}{dx}[\log_3 x] = \frac{1}{x \ln 3} \approx \frac{1}{1.0986x}$
2. $\frac{d}{dx}[\log_2(x^4)] = \frac{d}{dx}(4 \log_2 x)$ We used the logarithm identity $\log_b(x^r) = r \log_b x$.
   $= 4 \cdot \frac{1}{x \ln 2} \approx \frac{4}{0.6931x}$

Derivation of the formulas $\frac{d}{dx}[\ln x] = \frac{1}{x}$ and $\frac{d}{dx}[\log_b x] = \frac{1}{x \ln b}$

To compute $\frac{d}{dx}[\ln x]$, we need to use the definition of the derivative. We also use properties of the logarithm to help evaluate the limit.

$$\frac{d}{dx}[\ln x] = \lim_{h\to 0} \frac{\ln(x+h) - \ln x}{h} \quad \text{Definition of the derivative}$$

$$= \lim_{h\to 0} \frac{1}{h}\, [\ln(x+h) - \ln x] \quad \text{Algebra}$$

$$= \lim_{h\to 0} \frac{1}{h} \ln\left(\frac{x+h}{x}\right) \quad \text{Properties of the logarithm}$$

$$= \lim_{h\to 0} \frac{1}{h} \ln\left(1 + \frac{h}{x}\right) \quad \text{Algebra}$$

$$= \lim_{h\to 0} \ln\left(1 + \frac{h}{x}\right)^{1/h} \quad \text{Properties of the logarithm}$$

which we rewrite as

$$\lim_{h\to 0} \ln\left[\left(1 + \frac{1}{(x/h)}\right)^{x/h}\right]^{1/x}.$$

As $h \to 0^+$, the quantity x/h gets large and positive, and so the quantity in brackets approaches e (see the definition of e in Section 2.2), which leaves us with

$$\ln[e]^{1/x} = \frac{1}{x} \ln e = \frac{1}{x}$$

which is the derivative we are after.* What about the limit as $h \to 0^-$? We will glide over that case and leave it for the interested reader to pursue.†

The rule for the derivative of $\log_b x$ follows from the fact that $\log_b x = \ln x / \ln b$.

If we were to take the derivative of the natural logarithm of a *quantity* (a function of x), rather than just x, we would need to use the chain rule:

* We actually used the fact that the logarithm function is continuous when we took the limit.

† Here is an outline of the argument for negative h. Because x must be positive for $\ln x$ to be defined, we find that $x/h \to -\infty$ as $h \to 0^-$, and so we must consider the quantity $(1 + 1/m)^m$ for large *negative* m. It turns out the limit is still e (check it numerically!) and so the computation above still works.

Derivatives of Logarithms of Functions

Original Rule	*Generalized Rule*	*In Words*
$\frac{d}{dx}[\ln x] = \frac{1}{x}$	$\frac{d}{dx}[\ln u] = \frac{1}{u}\frac{du}{dx}$	*The derivative of the natural logarithm of a quantity is 1 over that quantity, times the derivative of that quantity.*
$\frac{d}{dx}[\log_b x] = \frac{1}{x \ln b}$	$\frac{d}{dx}[\log_b u] = \frac{1}{u \ln b}\frac{du}{dx}$	*The derivative of the log to base b of a quantity is 1 over the product of* ln *b and that quantity, times the derivative of that quantity.*

Quick Examples

1. $\frac{d}{dx}\ln[x^2+1] = \frac{1}{x^2+1}\frac{d}{dx}(x^2+1)$ $\quad u = x^2+1$ (See the margin note.§)

$$= \frac{1}{x^2+1}(2x) = \frac{2x}{x^2+1}$$

2. $\frac{d}{dx}\log_2[x^3+x] = \frac{1}{(x^3+x)\ln 2}\frac{d}{dx}(x^3+x)$ $\quad u = x^3+x$

$$= \frac{1}{(x^3+x)\ln 2}(3x^2+1) = \frac{3x^2+1}{(x^3+x)\ln 2}$$

§ If we were to evaluate $\ln(x^2+1)$, the last operation we would perform would be to take the natural logarithm of a quantity. Thus, the calculation thought experiment tells us that we are dealing with ln *of a quantity*, and so we need the generalized logarithm rule as stated above.

EXAMPLE 1 Derivative of Logarithmic Function

Compute the following derivatives:

a. $\frac{d}{dx}[\ln\sqrt{x+1}]$ **b.** $\frac{d}{dx}[\ln[(1+x)(2-x)]]$ **c.** $\frac{d}{dx}[\ln|x|]$

Solution

a. The calculation thought experiment tells us that we have the natural logarithm of a quantity, so

$$\frac{d}{dx}[\ln\sqrt{x+1}] = \frac{1}{\sqrt{x+1}}\frac{d}{dx}\sqrt{x+1} \qquad \frac{d}{dx}\ln u = \frac{1}{u}\frac{du}{dx}$$

$$= \frac{1}{\sqrt{x+1}} \cdot \frac{1}{2\sqrt{x+1}} \qquad \frac{d}{dx}\sqrt{u} = \frac{1}{2\sqrt{u}}\frac{du}{dx}$$

$$= \frac{1}{2(x+1)}.$$

Q: *What happened to the square root?*

A: As with many problems involving logarithms, we could have done this one differently and much more easily if we had simplified the expression $\ln\sqrt{x+1}$ using the properties of logarithms *before* differentiating. Doing this, we get the following:

Part (a) redone by simplifying first:

$$\ln\sqrt{x+1} = \ln(x+1)^{1/2} = \frac{1}{2}\ln(x+1). \qquad \text{Simplify the logarithm first.}$$

Thus,

$$\frac{d}{dx}[\ln\sqrt{x+1}] = \frac{d}{dx}\left[\frac{1}{2}\ln(x+1)\right]$$

$$= \frac{1}{2}\left[\frac{1}{x+1}\right] \cdot 1 = \frac{1}{2(x+1)}.$$

A *lot* easier!

b. This time, we simplify the expression $\ln[(1+x)(2-x)]$ before taking the derivative:

$$\ln[(1+x)(2-x)] = \ln(1+x) + \ln(2-x). \qquad \text{Simplify the logarithm first.}$$

Thus,

$$\frac{d}{dx}[\ln[(1+x)(2-x)]] = \frac{d}{dx}[\ln[(1+x)]] + \frac{d}{dx}[\ln[(2-x)]]$$

$$= \frac{1}{1+x} - \frac{1}{2-x}. \qquad \frac{d}{dx}\ln u = \frac{1}{u}\frac{du}{dx}$$

For practice, try doing this calculation without simplifying first. What other differentiation rule do you need to use?

c. Before we start, we note that $\ln x$ is defined only for positive values of x, so its domain is the set of positive real numbers. The domain of $\ln |x|$, on the other hand, is the set of *all* nonzero real numbers. For example, $\ln|-2| = \ln 2 \approx 0.6931$. For this reason, $\ln |x|$ often turns out to be more useful than the ordinary logarithm function.

$$\frac{d}{dx}[\ln|x|] = \frac{1}{|x|}\frac{d}{dx}|x| \qquad \frac{d}{dx}\ln u = \frac{1}{u}\frac{du}{dx}$$

$$= \frac{1}{|x|}\frac{|x|}{x} \qquad \text{Recall that } \frac{d}{dx}|x| = \frac{|x|}{x}.$$

$$= \frac{1}{x}$$

Before we go on... Figure 7(a) shows the graphs of $y = \ln|x|$ and $y = 1/x$. Figure 7(b) shows the graphs of $y = \ln|x|$ and $y = 1/|x|$. You should be able to see from these graphs why the derivative of $\ln|x|$ is $1/x$ and not $1/|x|$.

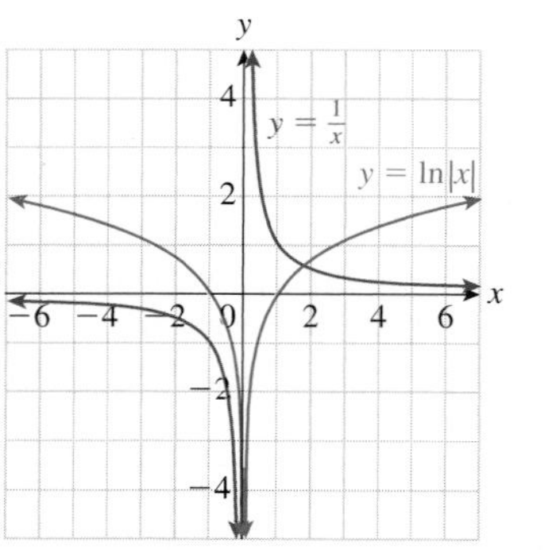

Figure 7(a)

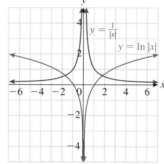

Figure 7(b)

This last example, in conjunction with the chain rule, gives us the following formulas.

Derivative of Logarithms of Absolute Values

Original Rule	*Generalized Rule*	*In Words*
$\frac{d}{dx}[\ln\|x\|] = \frac{1}{x}$	$\frac{d}{dx}[\ln\|u\|] = \frac{1}{u}\frac{du}{dx}$	*The derivative of the natural logarithm of the absolute value of a quantity is 1 over that quantity, times the derivative of that quantity.*
$\frac{d}{dx}[\log_b\|x\|] = \frac{1}{x \ln b}$	$\frac{d}{dx}[\log_b\|u\|] = \frac{1}{u \ln b}\frac{du}{dx}$	*The derivative of the log to base b of the absolute value of a quantity is 1 over the product of* ln *b and that quantity, times the derivative of that quantity.*

Note: Compare the above formulas with those on page 256. They tell us that we can simply ignore the absolute values in $\ln|u|$ or $\log_b|u|$ when taking the derivative.

Quick Examples

1. $$\frac{d}{dx}[\ln|x^2-1|] = \frac{1}{x^2-1}\frac{d}{dx}(x^2-1) \qquad u = x^2 - 1$$
$$= \frac{1}{x^2-1}(2x) = \frac{2x}{x^2-1}$$

2. $$\frac{d}{dx}[\log_2|x^3+x|] = \frac{1}{(x^3+x)\ln 2}\frac{d}{dx}(x^3+x) \qquad u = x^3 + x$$
$$= \frac{1}{(x^3+x)\ln 2}(3x^2+1) = \frac{3x^2+1}{(x^3+x)\ln 2}$$

We now turn to the derivatives of *exponential* functions—that is, functions of the form $f(x) = b^x$. We begin by showing how *not* to differentiate them.

Caution The derivative of b^x is *not* xb^{x-1}. The power rule applies only to *constant* exponents. In this case the exponent is decidedly *not* constant, and so the power rule does not apply.

The following shows the correct way of differentiating b^x, beginning with a special case.

Derivative of e^x

$$\frac{d}{dx}[e^x] = e^x$$

Quick Examples

1. $$\frac{d}{dx}[3e^x] = 3\frac{d}{dx}[e^x] = 3e^x \qquad \text{Constant multiple rule}$$

2. $$\frac{d}{dx}\left[\frac{e^x}{x}\right] = \frac{e^x x - e^x(1)}{x^2} \qquad \text{Quotient rule}$$
$$= \frac{e^x(x-1)}{x^2}$$

* There is another—very simple—function that is its own derivative. What is it?

Thus, e^x has the amazing property that its derivative is itself!* For bases other than e, we have the following generalization:

Derivative of b^x

If b is any positive number, then

$$\frac{d}{dx}[b^x] = b^x \ln b.$$

Note that if $b = e$, we obtain the previous formula.

Quick Example

$$\frac{d}{dx}[3^x] = 3^x \ln 3$$

Derivation of the Formula $\frac{d}{dx}[e^x] = e^x$

† This shortcut is an example of a technique called *logarithmic differentiation*, which is occasionally useful. We will see it again in the next section.

To find the derivative of e^x, we use a shortcut.† Write $g(x) = e^x$. Then

$$\ln g(x) = x.$$

Take the derivative of both sides of this equation to get

$$\frac{g'(x)}{g(x)} = 1$$

or

$$g'(x) = g(x) = e^x.$$

In other words, the exponential function with base e is its own derivative. The rule for exponential functions with other bases follows from the equality $b^x = e^{x \ln b}$ (why?) and the chain rule. (Try it.)

If we were to take the derivative of e raised to a *quantity*, not just x, we would need to use the chain rule, as follows.

Derivatives of Exponentials of Functions

Original Rule	*Generalized Rule*	*In Words*
$\frac{d}{dx}[e^x] = e^x$	$\frac{d}{dx}[e^u] = e^u \frac{du}{dx}$	*The derivative of e raised to a quantity is e raised to that quantity, times the derivative of that quantity.*
$\frac{d}{dx}[b^x] = b^x \ln b$	$\frac{d}{dx}[b^u] = b^u \ln b \frac{du}{dx}$	*The derivative of b raised to a quantity is b raised to that quantity, times* ln *b, times the derivative of that quantity.*

* The calculation thought experiment tells us that we have e raised to a quantity.

Quick Examples

1. $\frac{d}{dx}\left[e^{x^2+1}\right] = e^{x^2+1}\frac{d}{dx}[x^2+1]$ $\quad u = x^2 + 1$ (See margin note.*)

   $= e^{x^2+1}(2x) = 2x\, e^{x^2+1}$

2. $\frac{d}{dx}[2^{3x}] = 2^{3x}\ln 2\frac{d}{dx}[3x]$ $\quad u = 3x$

   $= 2^{3x}(\ln 2)(3) = (3\ln 2)2^{3x}$

3. $\frac{d}{dt}[30e^{1.02t}] = 30e^{1.02t}(1.02) = 30.6e^{1.02t}$ $\quad u = 1.02t$

4. If \$1,000 is invested in an account earning 5% per year compounded continuously, then the rate of change of the account balance after t years is

$$\frac{d}{dt}[1{,}000e^{0.05t}] = 1{,}000(0.05)e^{0.05t} = 50e^{0.05t} \text{ dollars/year.}$$

APPLICATIONS

EXAMPLE 2 Epidemics

In the early stages of the AIDS epidemic during the 1980s, the number of cases in the United States was increasing by about 50% every 6 months. By the start of 1983, there were approximately 1,600 AIDS cases in the United States.[1] Had this trend continued, how many new cases per year would have been occurring by the start of 1993?

Solution To find the answer, we must first model this exponential growth using the methods of Chapter 2. Referring to Example 4 in Section 2.2, we find that t years after the start of 1983 the number of cases is

$$A = 1{,}600(2.25^t).$$

We are asking for the number of new cases each year. In other words, we want the rate of change, dA/dt:

$$\frac{dA}{dt} = 1{,}600(2.25)^t \ln 2.25 \text{ cases per year.}$$

At the start of 1993, $t = 10$, so the number of new cases per year is

$$\left.\frac{dA}{dt}\right|_{t=10} = 1{,}600(2.25)^{10}\ln 2.25 \approx 4{,}300{,}000 \text{ cases per year.}$$

➡ **Before we go on...** In Example 2, the figure for the number of new cases per year is so large because we assumed that exponential growth—the 50% increase every 6 months—would continue. A more realistic model for the spread of a disease is the logistic model. (See Section 2.4, as well as the next example.) ■

[1] Data based on regression of 1982–1986 figures. Source for data: Centers for Disease Control and Prevention. HIV/AIDS Surveillance Report, 2000;12 (No. 2).

EXAMPLE 3 Sales Growth

The sales of the *Cyberpunk II* video game can be modeled by the logistic curve

$$q(t) = \frac{10,000}{1 + 0.5e^{-0.4t}}$$

where $q(t)$ is the total number of units sold t months after its introduction. How fast is the game selling 2 years after its introduction?

Solution We are asked for $q'(24)$. We can find the derivative of $q(t)$ using the quotient rule, or we can first write

$$q(t) = 10,000(1 + 0.5e^{-0.4t})^{-1}$$

and then use the generalized power rule:

$$\begin{aligned} q'(t) &= -10,000(1 + 0.5e^{-0.4t})^{-2}(0.5e^{-0.4t})(-0.4) \\ &= \frac{2,000e^{-0.4t}}{(1 + 0.5e^{-0.4t})^2}. \end{aligned}$$

Thus,

$$q'(24) = \frac{2,000e^{-0.4(24)}}{(1 + 0.5e^{-0.4(24)})^2} \approx 0.135 \text{ units per month.}$$

So, after 2 years, sales are quite slow.

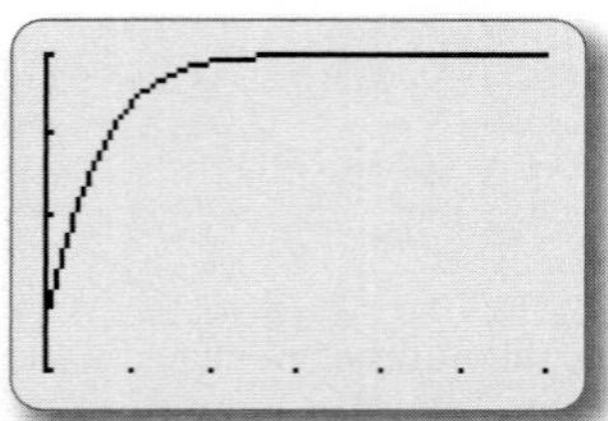

Figure 8

Before we go on... We can check the answer in Example 3 graphically. If we plot the total sales curve for $0 \le t \le 30$ and $6,000 \le q \le 10,000$, on a TI-83/84 Plus, for example, we get the graph shown in Figure 8. Notice that total sales level off at about 10,000 units.* We computed $q'(24)$, which is the slope of the curve at the point with t-coordinate 24. If we zoom in to the portion of the curve near $t = 24$, we obtain the graph shown in Figure 9, with $23 \le t \le 25$ and $9,999 \le q \le 10,000$. The curve is almost linear in this range. If we use the two endpoints of this segment of the curve, (23, 9,999.4948) and (25, 9,999.7730), we can approximate the derivative as

$$\frac{9,999.7730 - 9,999.4948}{25 - 23} = 0.1391$$

which is accurate to two decimal places. ■

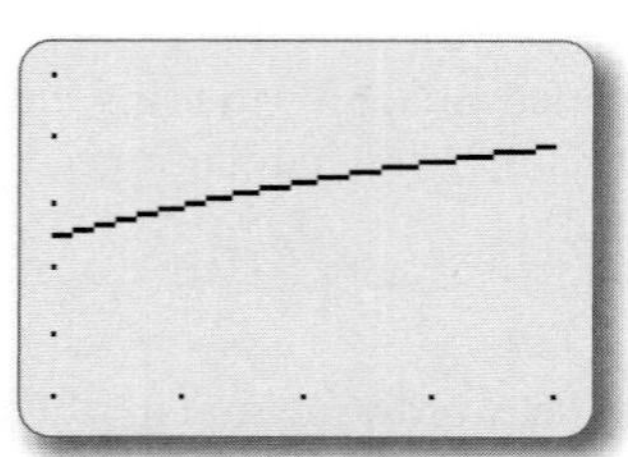

Figure 9

* We can also say this using limits: $\lim_{t \to +\infty} q(t) = 10,000$.

4.5 EXERCISES

Access end-of-section exercises online at **www.webassign.net**

4.6 Implicit Differentiation

Consider the equation $y^5 + y + x = 0$, whose graph is shown in Figure 10.

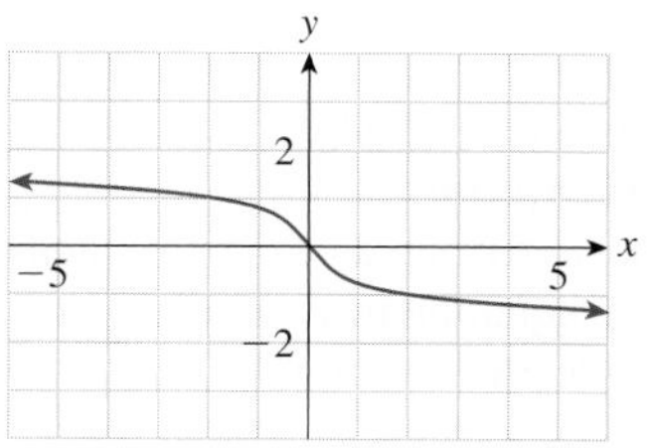

How did we obtain this graph? We did not solve for y as a function of x; that is impossible. In fact, we solved for x in terms of y to find points to plot. Nonetheless, the graph in Figure 10 is the graph of a function because it passes the vertical line test: Every vertical line crosses the graph no more than once, so for each value of x there is no more than one corresponding value of y. Because we cannot solve for y explicitly in terms of x, we say that the equation $y^5 + y + x = 0$ determines y as an **implicit function** of x.

Now, suppose we want to find the slope of the tangent line to this curve at, say, the point $(2, -1)$ (which, you should check, is a point on the curve). In the following example we find, surprisingly, that it is possible to obtain a formula for dy/dx without having to first solve the equation for y.

Figure 10

EXAMPLE 1 Implicit Differentiation

Find $\dfrac{dy}{dx}$, given that $y^5 + y + x = 0$.

Solution We use the chain rule and a little cleverness. Think of y as a function of x and take the derivative with respect to x of both sides of the equation:

$$y^5 + y + x = 0 \qquad \text{Original equation}$$

$$\frac{d}{dx}[y^5 + y + x] = \frac{d}{dx}[0] \qquad \text{Derivative with respect to } x \text{ of both sides}$$

$$\frac{d}{dx}[y^5] + \frac{d}{dx}[y] + \frac{d}{dx}[x] = 0. \qquad \text{Derivative rules}$$

Now we must be careful. The derivative *with respect to x* of y^5 is *not* $5y^4$. Rather, because y is a function of x, we must use the chain rule, which tells us that

$$\frac{d}{dx}[y^5] = 5y^4\frac{dy}{dx}.$$

Thus, we get

$$5y^4\frac{dy}{dx} + \frac{dy}{dx} + 1 = 0.$$

We want to find dy/dx, so we *solve for it*:

$$(5y^4 + 1)\frac{dy}{dx} = -1 \qquad \text{Isolate } dy/dx \text{ on one side.}$$

$$\frac{dy}{dx} = -\frac{1}{5y^4 + 1}. \qquad \text{Divide both sides by } 5y^4 + 1.$$

Before we go on... Note that we should not expect to obtain dy/dx as an explicit function of x if y was not an explicit function of x to begin with. For example, the formula we found for dy/dx in Example 1 is not a function of x because there is a y in

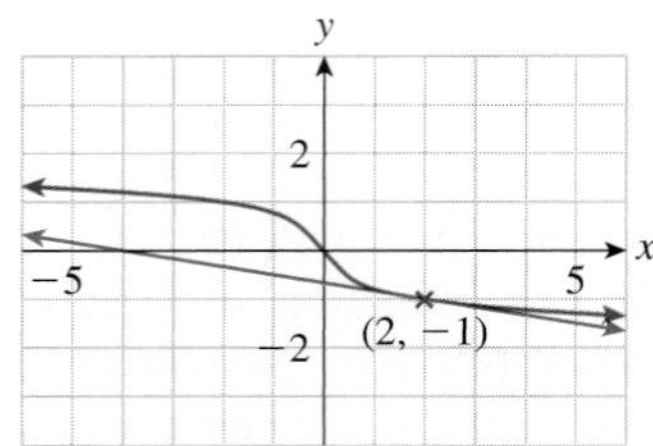

Figure 11

it. However, the result is still useful because we can evaluate the derivative at any point on the graph. For instance, at the point $(2, -1)$ on the graph, we get

$$\frac{dy}{dx} = -\frac{1}{5y^4 + 1} = -\frac{1}{5(-1)^4 + 1} = -\frac{1}{6}.$$

Thus, the slope of the tangent line to the curve $y^5 + y + x = 0$ at the point $(2, -1)$ is $-1/6$. Figure 11 shows the graph and this tangent line. ■

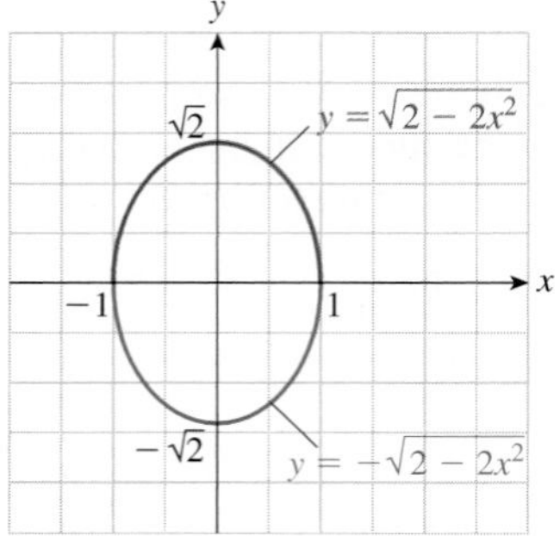

Figure 12

This procedure we just used—differentiating an equation to find dy/dx without first solving the equation for y—is called **implicit differentiation**.

In Example 1 we were given an equation in x and y that determined y as an (implicit) function of x, even though we could not solve for y. But an equation in x and y need not always determine y as a function of x. Consider, for example, the equation

$$2x^2 + y^2 = 2.$$

Solving for y yields $y = \pm\sqrt{2 - 2x^2}$. The $\pm$ sign reminds us that for some values of x there are two corresponding values for y. We can graph this equation by superimposing the graphs of

$$y = \sqrt{2 - 2x^2} \quad \text{and} \quad y = -\sqrt{2 - 2x^2}.$$

The graph, an *ellipse*, is shown in Figure 12.

The graph of $y = \sqrt{2 - 2x^2}$ constitutes the top half of the ellipse, and the graph of $y = -\sqrt{2 - 2x^2}$ constitutes the bottom half.

EXAMPLE 2 Slope of Tangent Line

Refer to Figure 12. Find the slope of the tangent line to the ellipse $2x^2 + y^2 = 2$ at the point $(1/\sqrt{2}, 1)$.

Solution Because $(1/\sqrt{2}, 1)$ is on the top half of the ellipse in Figure 12, we *could* differentiate the function $y = \sqrt{2 - 2x^2}$ to obtain the result, but it is actually easier to apply implicit differentiation to the original equation.

$$2x^2 + y^2 = 2 \qquad \text{Original equation}$$

$$\frac{d}{dx}[2x^2 + y^2] = \frac{d}{dx}[2] \qquad \text{Derivative with respect to } x \text{ of both sides}$$

$$4x + 2y\frac{dy}{dx} = 0$$

$$2y\frac{dy}{dx} = -4x \qquad \text{Solve for } dy/dx.$$

$$\frac{dy}{dx} = -\frac{4x}{2y} = -\frac{2x}{y}$$

To find the slope at $(1/\sqrt{2}, 1)$ we now substitute for x and y:

$$\left.\frac{dy}{dx}\right|_{(1/\sqrt{2}, 1)} = -\frac{2/\sqrt{2}}{1} = -\sqrt{2}.$$

Thus, the slope of the tangent to the ellipse at the point $(1/\sqrt{2}, 1)$ is $-\sqrt{2} \approx -1.414$.

EXAMPLE 3 Tangent Line for an Implicit Function

Find the equation of the tangent line to the curve $\ln y = xy$ at the point where $y = 1$.

Solution First, we use implicit differentiation to find dy/dx:

$$\frac{d}{dx}[\ln y] = \frac{d}{dx}[xy] \qquad \text{Take } d/dx \text{ of both sides.}$$

$$\frac{1}{y}\frac{dy}{dx} = (1)y + x\frac{dy}{dx}. \qquad \text{Chain rule on left, product rule on right}$$

To solve for dy/dx, we bring all the terms containing dy/dx to the left-hand side and all terms not containing it to the right-hand side:

$$\frac{1}{y}\frac{dy}{dx} - x\frac{dy}{dx} = y \qquad \text{Bring the terms with } dy/dx \text{ to the left.}$$

$$\frac{dy}{dx}\left(\frac{1}{y} - x\right) = y \qquad \text{Factor out } dy/dx.$$

$$\frac{dy}{dx}\left(\frac{1 - xy}{y}\right) = y$$

$$\frac{dy}{dx} = y\left(\frac{y}{1 - xy}\right) = \frac{y^2}{1 - xy}. \qquad \text{Solve for } dy/dx.$$

The derivative gives the slope of the tangent line, so we want to evaluate the derivative at the point where $y = 1$. However, the formula for dy/dx requires values for both x and y. We get the value of x by substituting $y = 1$ in the original equation:

$$\ln y = xy$$

$$\ln 1 = x \cdot 1$$

But $\ln 1 = 0$, and so $x = 0$ for this point. Thus,

$$\left.\frac{dy}{dx}\right|_{(0,1)} = \frac{1^2}{1 - (0)(1)} = 1.$$

Therefore, the tangent line is the line through $(x, y) = (0, 1)$ with slope 1, which is

$$y = x + 1.$$

➡ **Before we go on...** Example 3 presents an instance of an implicit function in which it is simply not possible to solve for y. Try it. ■

Sometimes, it is easiest to differentiate a complicated function of x by first taking the logarithm and then using implicit differentiation—a technique called **logarithmic differentiation**.

EXAMPLE 4 Logarithmic Differentiation

Find $\frac{d}{dx}\left[\frac{(x+1)^{10}(x^2+1)^{11}}{(x^3+1)^{12}}\right]$ without using the product or quotient rules.

Solution Write

$$y = \frac{(x+1)^{10}(x^2+1)^{11}}{(x^3+1)^{12}}$$

and then take the natural logarithm of both sides:

$$\ln y = \ln\left[\frac{(x+1)^{10}(x^2+1)^{11}}{(x^3+1)^{12}}\right].$$

We can use properties of the logarithm to simplify the right-hand side:

$$\begin{aligned}\ln y &= \ln(x+1)^{10} + \ln(x^2+1)^{11} - \ln(x^3+1)^{12} \\ &= 10\ln(x+1) + 11\ln(x^2+1) - 12\ln(x^3+1).\end{aligned}$$

Now we can find $\dfrac{dy}{dx}$ using implicit differentiation:

$$\frac{1}{y}\frac{dy}{dx} = \frac{10}{x+1} + \frac{22x}{x^2+1} - \frac{36x^2}{x^3+1} \qquad \text{Take } d/dx \text{ of both sides.}$$

$$\frac{dy}{dx} = y\left(\frac{10}{x+1} + \frac{22x}{x^2+1} - \frac{36x^2}{x^3+1}\right) \qquad \text{Solve for } dy/dx.$$

$$= \frac{(x+1)^{10}(x^2+1)^{11}}{(x^3+1)^{12}}\left(\frac{10}{x+1} + \frac{22x}{x^2+1} - \frac{36x^2}{x^3+1}\right). \qquad \text{Substitute for } y.$$

➡ **Before we go on...** Redo Example 4 using the product and quotient rules (and the chain rule) instead of logarithmic differentiation and compare the answers. Compare also the amount of work involved in both methods. ■

APPLICATION

Productivity usually depends on both labor and capital. Suppose, for example, you are managing a surfboard manufacturing company. You can measure its productivity by counting the number of surfboards the company makes each year. As a measure of labor, you can use the number of employees, and as a measure of capital you can use its operating budget. The so-called *Cobb-Douglas* model uses a function of the form:

$$P = Kx^a y^{1-a}, \qquad \text{Cobb-Douglas model for productivity}$$

where P stands for the number of surfboards made each year, x is the number of employees, and y is the operating budget. The numbers K and a are constants that depend on the particular situation studied, with a between 0 and 1.

EXAMPLE 5 Cobb-Douglas Production Function

The surfboard company you own has the Cobb-Douglas production function

$$P = x^{0.3}y^{0.7}$$

where P is the number of surfboards it produces per year, x is the number of employees, and y is the daily operating budget (in dollars). Assume that the production level P is constant.

a. Find $\dfrac{dy}{dx}$.

b. Evaluate this derivative at $x = 30$ and $y = 10{,}000$, and interpret the answer.

Solution

a. We are given the equation $P = x^{0.3}y^{0.7}$, in which P is constant. We find $\dfrac{dy}{dx}$ by implicit differentiation.

$$0 = \frac{d}{dx}[x^{0.3}y^{0.7}] \qquad d/dx \text{ of both sides}$$

$$0 = 0.3x^{-0.7}y^{0.7} + x^{0.3}(0.7)y^{-0.3}\frac{dy}{dx} \qquad \text{Product and chain rules}$$

$$-0.7x^{0.3}y^{-0.3}\frac{dy}{dx} = 0.3x^{-0.7}y^{0.7} \qquad \text{Bring term with } dy/dx \text{ to left.}$$

$$\frac{dy}{dx} = -\frac{0.3x^{-0.7}y^{0.7}}{0.7x^{0.3}y^{-0.3}} \qquad \text{Solve for } dy/dx.$$

$$= -\frac{3y}{7x}. \qquad \text{Simplify.}$$

b. Evaluating this derivative at $x = 30$ and $y = 10{,}000$ gives

$$\left.\frac{dy}{dx}\right|_{x=30,\ y=10{,}000} = -\frac{3(10{,}000)}{7(30)} \approx -143.$$

To interpret this result, first look at the units of the derivative: We recall that the units of dy/dx are units of y per unit of x. Because y is the daily budget, its units are dollars; because x is the number of employees, its units are employees. Thus,

$$\left.\frac{dy}{dx}\right|_{x=30,\ y=10{,}000} \approx -\$143 \text{ per employee.}$$

Next, recall that dy/dx measures the rate of change of y as x changes. Because the answer is negative, the daily budget to maintain production at the fixed level is decreasing by approximately \$143 per additional employee at an employment level of 30 employees and a daily operating budget of \$10,000. In other words, increasing the workforce by one worker will result in a savings of approximately \$143 per day. Roughly speaking, *a new employee is worth \$143 per day* at the current levels of employment and production.

4.6 EXERCISES

ENHANCED WebAssign

Access end-of-section exercises online at **www.webassign.net**

CHAPTER 4 REVIEW

KEY CONCEPTS

WW Website www.WanerMath.com

Go to the Website at www.WanerMath.com to find a comprehensive and interactive Web-based summary of Chapter 4.

4.1 Derivatives of Powers, Sums, and Constant Multiples

Power Rule: If n is any constant and $f(x) = x^n$, then $f'(x) = nx^{n-1}$. *p. 220*

Using the power rule for negative and fractional exponents *p. 221*

Sums, differences, and constant multiples *p. 223*

Combining the rules *p. 224*

$\frac{d}{dx}(cx) = c, \quad \frac{d}{dx}(c) = 0$ *p. 226*

$f(x) = x^{1/3}$, $g(x) = x^{2/3}$, and $h(x) = |x|$ are not differentiable at $x = 0$. *p. 226*

L'Hospital's rule *p. 228*

4.2 A First Application: Marginal Analysis

Marginal cost function $C'(x)$ *p. 231*

Marginal revenue and profit functions $R'(x)$ and $P'(x)$ *p. 233*

What it means when the marginal profit is zero *p. 234*

Marginal product *p. 234*

Average cost of the first x items:

$\bar{C}(x) = \frac{C(x)}{x}$ *p. 236*

4.3 The Product and Quotient Rules

Product rule: $\frac{d}{dx}[f(x)g(x)] = f'(x)g(x) + f(x)g'(x)$ *p. 238*

Quotient rule: $\frac{d}{dx}\left[\frac{f(x)}{g(x)}\right] = \frac{f'(x)g(x) - f(x)g'(x)}{[g(x)]^2}$ *p. 238*

Using the product rule *p. 240*

Using the quotient rule *p. 241*

Calculation thought experiment *p. 243*

Application to revenue and average cost *p. 244*

4.4 The Chain Rule

Chain rule: $\frac{d}{dx}[f(u)] = f'(u)\frac{du}{dx}$ *p. 246*

Generalized power rule:

$\frac{d}{dx}[u^n] = nu^{n-1}\frac{du}{dx}$ *p. 247*

Using the chain rule *p. 247*

Application to marginal product *p. 250*

Chain rule in differential notation:

$\frac{dy}{dx} = \frac{dy}{du}\frac{du}{dx}$ *p. 251*

Manipulating derivatives in differential notation *p. 253*

4.5 Derivatives of Logarithmic and Exponential Functions

Derivative of the natural logarithm:

$\frac{d}{dx}[\ln x] = \frac{1}{x}$ *p. 255*

Derivative of logarithm with base b:

$\frac{d}{dx}[\log_b x] = \frac{1}{x \ln b}$ *p. 255*

Derivatives of logarithms of functions:

$\frac{d}{dx}[\ln u] = \frac{1}{u}\frac{du}{dx}$

$\frac{d}{dx}[\log_b u] = \frac{1}{u \ln b}\frac{du}{dx}$ *p. 256*

Derivatives of logarithms of absolute values:

$\frac{d}{dx}[\ln|x|] = \frac{1}{x} \quad \frac{d}{dx}[\ln|u|] = \frac{1}{u}\frac{du}{dx}$

$\frac{d}{dx}[\log_b|x|] = \frac{1}{x \ln b}$

$\frac{d}{dx}[\log_b|u|] = \frac{1}{u \ln b}\frac{du}{dx}$ *p. 258*

Derivative of e^x: $\frac{d}{dx}[e^x] = e^x$ *p. 259*

Derivative of b^x: $\frac{d}{dx}[b^x] = b^x \ln b$ *p. 260*

Derivatives of exponential functions *p. 260*

Application to epidemics *p. 261*

Application to sales growth (logistic function) *p. 262*

4.6 Implicit Differentiation

Implicit function of x *p. 263*

Implicit differentiation *p. 263*

Using implicit differentiation *p. 264*

Finding a tangent line *p. 264*

Logarithmic differentiation *p. 265*

REVIEW EXERCISES

In Exercises 1–26, find the derivative of the given function.

1. $f(x) = 10x^5 + \frac{1}{2}x^4 - x + 2$

2. $f(x) = \frac{10}{x^5} + \frac{1}{2x^4} - \frac{1}{x} + 2$

3. $f(x) = 3x^3 + 3\sqrt[3]{x}$

4. $f(x) = \frac{2}{x^{2.1}} - \frac{x^{0.1}}{2}$

5. $f(x) = x + \frac{1}{x^2}$

6. $f(x) = 2x - \frac{1}{x}$

7. $f(x) = \frac{4}{3x} - \frac{2}{x^{0.1}} + \frac{x^{1.1}}{3.2} - 4$

8. $f(x) = \frac{4}{x} + \frac{x}{4} - |x|$

9. $f(x) = e^x(x^2 - 1)$

10. $f(x) = \frac{x^2 + 1}{x^2 - 1}$

11. $f(x) = \frac{|x| + 1}{3x^2 + 1}$

12. $f(x) = (|x| + x)(2 - 3x^2)$

13. $f(x) = (4x - 1)^{-1}$

14. $f(x) = (x + 7)^{-2}$

15. $f(x) = (x^2 - 1)^{10}$

16. $f(x) = \frac{1}{(x^2 - 1)^{10}}$

17. $f(x) = [2 + (x + 1)^{-0.1}]^{4.3}$

18. $f(x) = [(x + 1)^{0.1} - 4x]^{-5.1}$

19. $f(x) = e^x(x^2 + 1)^{10}$

20. $f(x) = \left[\dfrac{x-1}{3x+1}\right]^3$ **21.** $f(x) = \dfrac{3^x}{x-1}$

22. $f(x) = 4^{-x}(x+1)$ **23.** $f(x) = e^{x^2-1}$

24. $f(x) = (x^2+1)e^{x^2-1}$ **25.** $f(x) = \ln(x^2-1)$

26. $f(x) = \dfrac{\ln(x^2-1)}{x^2-1}$

In Exercises 27–34, find all values of x (if any) where the tangent line to the graph of the given equation is horizontal.

27. $y = -3x^2 + 7x - 1$ **28.** $y = 5x^2 - 2x + 1$

29. $y = \dfrac{x}{2} + \dfrac{2}{x}$ **30.** $y = \dfrac{x^2}{2} - \dfrac{8}{x^2}$

31. $y = x - e^{2x-1}$ **32.** $y = e^{x^2}$

33. $y = \dfrac{x}{x+1}$ **34.** $y = \sqrt{x}(x-1)$

In Exercises 35–40, find dy/dx for the given equation.

35. $x^2 - y^2 = x$ **36.** $2xy + y^2 = y$

37. $e^{xy} + xy = 2$ **38.** $\ln\left(\dfrac{y}{x}\right) = y$

39. $y = \dfrac{(2x-1)^4(3x+4)}{(x+1)(3x-1)^3}$ **40.** $y = x^{x-1}3^x$

In Exercises 41–46, find the equation of the tangent line to the graph of the given equation at the specified point.

41. $y = (x^2 - 3x)^{-2}$; $x = 1$ **42.** $y = (2x^2 - 3)^{-3}$; $x = -1$

43. $y = x^2 e^{-x}$; $x = -1$ **44.** $y = \dfrac{x}{1+e^x}$; $x = 0$

45. $xy - y^2 = x^2 - 3$; $(-1, 1)$

46. $\ln(xy) + y^2 = 1$; $(-1, -1)$

APPLICATIONS: OHaganBooks.com

47. ***Sales*** OHaganBooks.com fits the cubic curve

$$w(t) = -3.7t^3 + 74.6t^2 + 135.5t + 6{,}300$$

to its weekly sales figures (see Chapter 3 Review Exercise 57; t is time in weeks), as shown in the following graph:

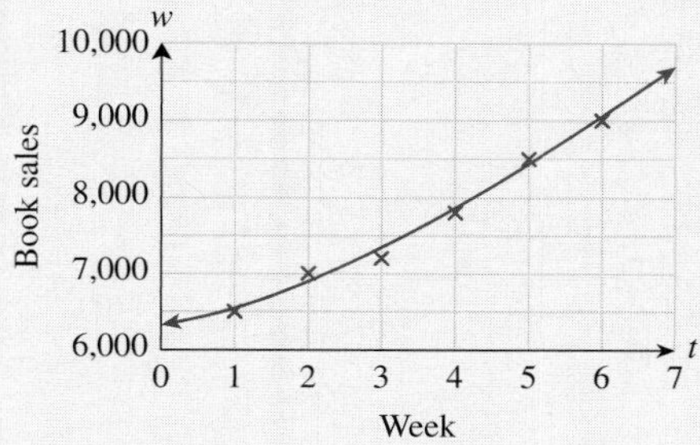

a. According to the cubic model, what was the rate of increase of sales at the beginning of the second week ($t = 1$)? (Round your answer to the nearest unit.)

b. If we extrapolate the model, what would be the rate of increase of weekly sales at the beginning of the eighth week ($t = 7$)?

c. Graph the function w for $0 \le t \le 20$. Would it be realistic to use the function to predict sales through week 20? Why?

d. By examining the graph, say why the choice of a quadratic model would result in radically different long-term predictions of sales.

48. ***Rising Sea Level*** Marjory Duffin is still toying with various models to fit to the New York sea level figures she had seen after purchasing a beachfront condominium in New York (see Chapter 3 Review Exercise 58). Following is a cubic curve she obtained using regression:

$$L(t) = -0.0001t^3 + 0.02t^2 + 2.2t \text{ mm}$$

(t is time in years since 1900). The curve and data are shown in the following graph:

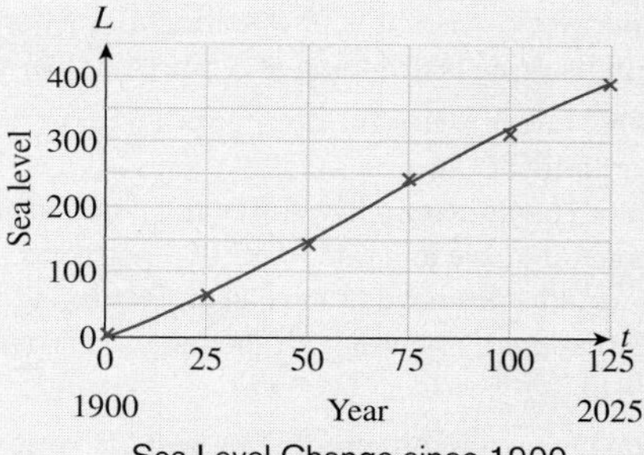

Sea Level Change since 1900

a. According to the cubic model, what was the rate at which the sea level was rising in 2000 ($t = 100$)? (Round your answer to two significant digits.)

b. If we extrapolate the model, what would be the rate at which the sea level is rising in 2025 ($t = 125$)?

c. Graph the function L for $0 \le t \le 200$. Why is it not realistic to use the function to predict the sea level through 2100?

d. James Stewart, a summer intern at Duffin House Publishers, differs. As he puts it, "The cubic curve came from doing regression on the actual data, and thus reflects the actual trend of the data. We can't argue against reality!" Comment on this assertion.

49. ***Cost*** As OHaganBooks.com's sales increase, so do its costs. If we take into account volume discounts from suppliers and shippers, the weekly cost of selling x books is

$$C(x) = -0.00002x^2 + 3.2x + 5{,}400 \text{ dollars.}$$

a. What is the marginal cost at a sales level of 8,000 books per week?

b. What is the average cost per book at a sales level of 8,000 books per week?

c. What is the marginal average cost ($d\bar{C}/dx$) at a sales level of 8,000 books per week?

d. Interpret the results of parts (a)–(c).

50. *Cost* OHaganBooks.com has been experiencing a run of bad luck with its summer college intern program in association with PCU (Party Central University), begun as a result of a suggestion by Marjory Duffin over dinner one evening. The frequent errors in filling orders, charges from movie download sites and dating sites, and beverages spilled on computer equipment have resulted in an estimated weekly cost to the company of

$$C(x) = 25x^2 - 5.2x + 4{,}000 \text{ dollars,}$$

where x is the number of college interns employed.

a. What is the marginal cost at a level of 10 interns?
b. What is the average cost per intern at a level of 10 interns?
c. What is the marginal average cost at a level of 10 interns?
d. Interpret the results of parts (a)–(c).

51. *Revenue* At the moment, OHaganBooks.com is selling 1,000 books per week and its sales are rising at a rate of 200 books per week. Also, it is now selling all its books for \$20 each, but its price is dropping at a rate of \$1 per week.

a. At what rate is OHaganBooks.com's weekly revenue rising or falling?
b. John O'Hagan would like to see the company's weekly revenue increase at a rate of \$5,000 per week. At what rate would sales have to have been increasing to accomplish that goal, assuming all the other information is as given above?

52. *Revenue* Due to ongoing problems with its large college intern program in association with PCU (see Exercise 50), OHaganBooks.com has arranged to transfer its interns to its competitor JungleBooks.com (whose headquarters happens to be across the road) for a small fee. At the moment, it is transferring 5 students per week, and this number is rising at a rate of 4 students per week. Also, it is now charging JungleBooks \$400 per intern, but this amount is decreasing at a rate of \$20 per week.

a. At what rate is OHaganBooks.com's weekly revenue from this transaction rising or falling?
b. Flush with success of the transfer program, John O'Hagan would like to see the company's resulting revenue increase at a rate of \$3,900 per week. At what rate would the transfer of interns have to increase to accomplish that goal, assuming all the other information is as given above?

53. *Percentage Rate of Change of Revenue* The percentage rate of change of a quantity Q is Q'/Q. Why is the percentage rate of change of revenue always equal to the sum of the percentage rates of change of unit price and weekly sales?

54. *P/E Ratios* At the beginning of last week, OHaganBooks .com stock was selling for \$100 per share, rising at a rate of \$50 per year. Its earnings amounted to \$1 per share, rising at a rate of \$0.10 per year. At what rate was its price-to-earnings (P/E) ratio, the ratio of its stock price to its earnings per share, rising or falling?

55. *P/E Ratios* Refer to Exercise 54. Jay Campbell, who recently invested in OHaganBooks.com stock, would have liked to see the P/E ratio increase at a rate of 100 points per year. How fast would the stock have to have been rising, assuming all the other information is as given in Exercise 54?

56. *Percentage Rate of Change of P/E Ratios* Refer to Exercise 54. The percentage rate of change of a quantity Q is Q'/Q. Why is the percentage rate of change of P/E always equal to the percentage rate of change of unit price minus the percentage rate of change of earnings?

57. *Sales* OHaganBooks.com decided that the cubic curve in Exercise 47 was not suitable for extrapolation, so instead it tried

$$s(t) = 6{,}000 + \frac{4{,}500}{1 + e^{-0.55(t-4.8)}}$$

as shown in the following graph:

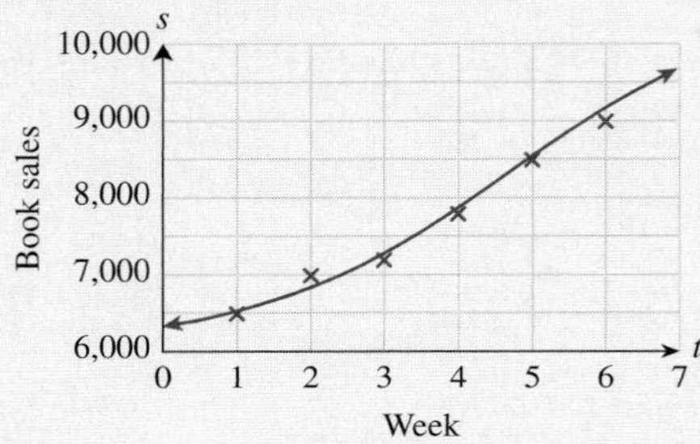

a. Compute $s'(t)$ and use the answer to estimate the rate of increase of weekly sales at the beginning of the seventh week ($t = 6$). (Round your answer to the nearest unit.)
b. Compute $\lim_{t\to+\infty} s'(t)$ and interpret the answer.

58. *Rising Sea Level* Upon some reflection, Marjory Duffin decided that the curve in Exercise 48 was not suitable for extrapolation, so instead she tried

$$L(t) = \frac{418}{1 + 17.2e^{-0.041t}} \qquad (0 \le t \le 125)$$

(t is time in years since 1900) as shown in the following graph:

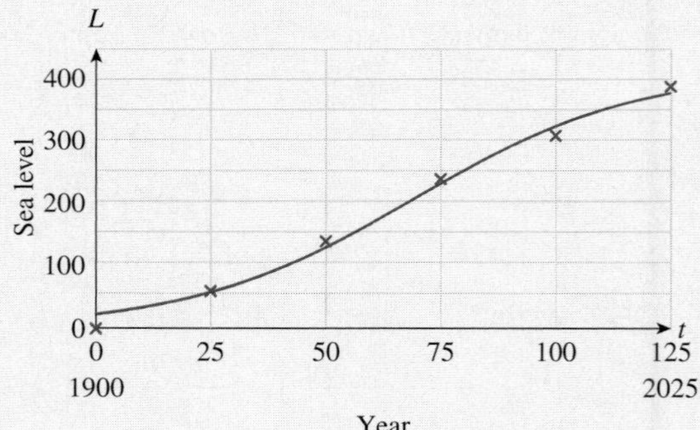

a. Compute $L'(t)$ and use the answer to estimate the rate at which the sea level was rising in 2000 ($t = 100$). (Round your answer to two decimal places.)
b. Compute $\lim_{t\to+\infty} L'(t)$ and interpret the answer.

59. ***Web Site Activity*** The number of "hits" on OHaganBooks .com's Web site was 1,000 per day at the beginning of the year, and was growing at a rate of 5% per week. If this growth rate continued for the whole year (52 weeks), find the rate of increase (in hits per day per week) at the end of the year.

60. ***Web Site Activity*** The number of "hits" on ShadyDownload .net during the summer intern program at OHaganBooks .com was 100 per day at the beginning of the intern program, and was growing at a rate of 15% per day. If this growth rate continued for the duration of the whole summer intern program (85 days), find the rate of increase (in hits per day per day) at the end of the program.

61. ***Demand and Revenue*** The price p that OHaganBooks.com charges for its latest leather-bound gift edition of *The Complete Larry Potter* is related to the demand q in weekly sales by the equation

$$250pq + q^2 = 13{,}500{,}000.$$

Suppose the price is set at \$50, which would make the demand 1,000 copies per week.

a. Using implicit differentiation, compute the rate of change of demand with respect to price, and interpret the result. (Round the answer to two decimal places.)

b. Use the result of part (a) to compute the rate of change of revenue with respect to price. Should the price be raised or lowered to increase revenue?

62. ***Demand and Revenue*** The price p that OHaganBooks.com charges for its latest leather-bound gift edition of *Lord of the Fields* is related to the demand q in weekly sales by the equation

$$100pq + q^2 = 5{,}000{,}000.$$

Suppose the price is set at \$40, which would make the demand 1,000 copies per week.

a. Using implicit differentiation, compute the rate of change of demand with respect to price, and interpret the result. (Round the answer to two decimal places.)

b. Use the result of part (a) to compute the rate of change of revenue with respect to price. Should the price be raised or lowered to increase revenue?

Case Study: Projecting Market Growth

Yuri Arcurs/Shutterstock

You are on the board of directors at *Fullcourt Academic Press,* a major textbook supplier to private schools, and various expansion strategies will be discussed at tomorrow's board meeting. TJM, the sales director of the high school division, has just burst into your office with his last-minute proposal based on data showing the number of private high school graduates in the U.S. each year over the past 18 years:[2]

Year	1995	1996	1997	1998	1999	2000	2001	2002	2003
Graduates (thousands)	245	254	265	273	279	279	285	296	301
Year	2004	2005	2006	2007	2008	2009	2010	2011	2012
Graduates (thousands)	307	307	307	314	314	315	315	316	316

He is asserts that, despite the unspectacular numbers in the past few years, the long-term trend appears to support a basic premise of his proposal for an expansion strategy: that the number of high school seniors in private schools in the U.S. will be growing at a rate of about 4,000 per year through 2015. He points out that the rate of increase predicted by the regression line is approximately 4,080 students per year, supporting his premise.

In order to decide whether to support TJM's proposal at tomorrow's board meeting, you would like to first determine whether the linear regression prediction of around 4,000 students per year is reasonable, especially in view of the more recent figures. You open your spreadsheet and graph the data with the regression line (Figure 13). The data suggest that the number of graduates began to "level off" (in the language of calculus, the *derivative appears to be decreasing*) toward the end of the

Thousands of graduates

$N(t)$: 200, 220, 240, 260, 280, 300, 320, 340

t: 0 2 4 6 8 10 12 14 16 18 (1995 – 2013)

t = Time in years since 1995

Regression Line: $y = 4.08t + 259$

Figure 13

[2]Data through 2011 are National Center for Educational Statistics actual and projected data as of April 2010. Source: National Center for Educational Statistics http://nces.ed.gov/.

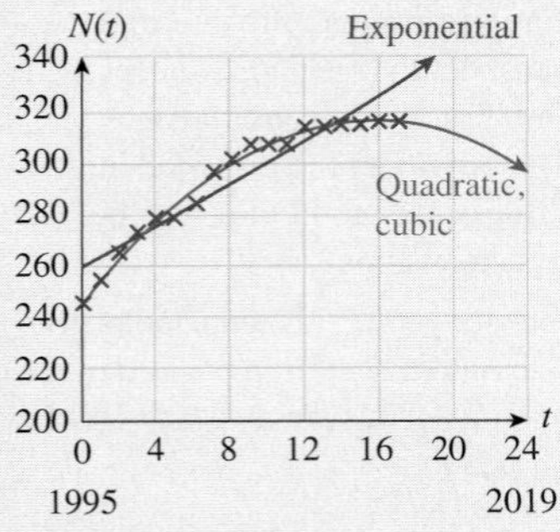

Figure 14

using Technology

You can obtain the best-fit logistic curve using technology.

TI-83/84 Plus

STAT EDIT

Enter 0–17 in L_1 and the values of $N(t)$ in L_2.

STAT CALC and select Option #B `Logistic`

To plot the regression curve with the data: Graph: Y= VARS 5 EQ 1, then ZOOM 9

[See the TI-83/84 Plus Technology Guide for Section 1.4 for more details on plotting the data and curve.]

Website

www.WanerMath.com

In the Function Ealuator and Grapher, enter the data as shown, press "Examples" until the logistic model `$1/(1+$2*$3^(-x))` shows in the first box, and press "Fit Curve". (You might need to press "Fit Curve" several times before it manages to find the curve. It also helps to read the instructions (link near the top of the Web page under the heading).)

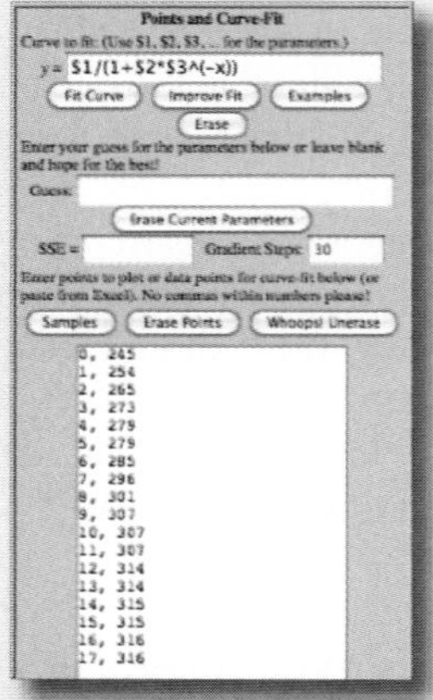

period. Moreover, you recall reading somewhere that the numbers of students in the lower grades have also begun to level off, so it is safe to predict that the slowing of growth in the senior class will continue over the next few years, contrary to what TJM has claimed. In order to make a meaningful prediction, you would really need some precise data about numbers in the lower grades, but the meeting is tomorrow and you would like a quick and easy way of extrapolating the data by "extending the curve to the right."

It would certainly be helpful if you had a mathematical model of the data in Figure 13 that you could use to project the current trend. But what kind of model should you use? The linear model is no good because it does not show any change in the derivative (the derivative of a linear function is constant). In addition, best-fit polynomial and exponential functions do not accurately reflect the leveling off, as you realize after trying to fit a few of them (Figure 14).

You then recall that a logistic curve can model the leveling-off property you desire, and so you try a model of the form

$$N(t) = \frac{M}{1 + Ab^{-t}}.$$

Figure 15 shows the best-fit logistic curve, which has a sum-of-squares error (SSE) of around 109. (See Section 1.4 or any of the regression examples in Chapter 2 for a discussion of SSE.)

$$N(t) = \frac{323.9}{1 + 0.3234(1.186)^{-t}} \qquad \text{SSE} \approx 109$$

Its graph shows the leveling off and also gives more reasonable long-term predictions.

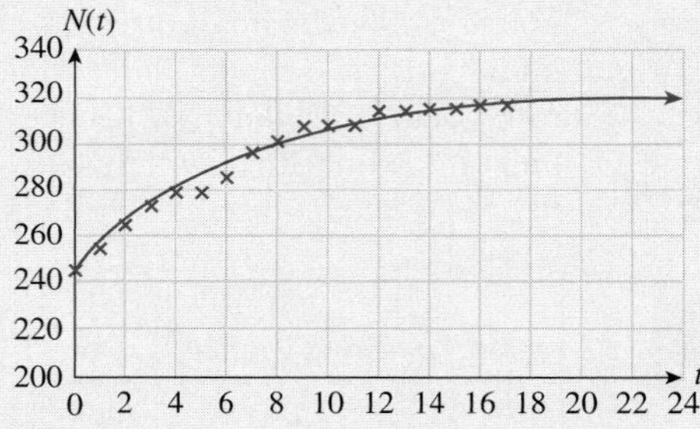

Figure 15

The rate of increase of high school students—pertinent to TJM's report—is given by the derivative, $N'(t)$:

$$N(t) = \frac{M}{1 + Ab^{-t}}$$

$$N'(t) = -\frac{M}{(1 + Ab^{-t})^2}\frac{d}{dt}[1 + Ab^{-t}]$$

$$= \frac{MAb^{-t}\ln b}{(1 + Ab^{-t})^2}.$$

The rate of increase in the number of high school students in 2015 ($t = 20$) is given by

$$N'(20) = \frac{(323.9)(0.3234)(1.186)^{-20}\ln 1.186}{(1 + 0.3234(1.186)^{-20})^2}$$

$$\approx 0.577 \text{ thousand students per year,}$$

or about 580 students per year—far less than the optimistic estimate of 4,000 in the proposal! Thus, TJM's prediction is suspect and further research will have to be done before the board can even consider the proposal.

To reassure yourself, you decide to look for another kind of S-shaped model as a backup. After flipping through a calculus book, you stumble across a function that is slightly more general than the one you have:

$$N(t) = \frac{M}{1 + Ab^{-t}} + C. \qquad \text{Shifted Logistic Curve*}$$

* To find a detailed discussion of scaled and shifted functions, visit the Website and follow Chapter 1 → New Functions from Old: Scaled and Shifted Functions.

The added term C has the effect of shifting the graph up C units. Turning once again to your calculus book (see the discussion of logistic regression in Section 2.4), you see that a best-fit curve is one that minimizes the sum-of-squares error, and you find the best-fit curve by again using the utility on the Website (this time, with the model `$1/(1+$2*$3^(-x))+$4` and initial guess 323.9, 0.3234, 1.186, 0; that is, keeping the current values and setting $c = 0$). You obtain the model

$$N(t) = \frac{135.5}{1 + 1.192(1.268)^{-t}} + 184.6. \qquad \text{SSE} \approx 100$$

The value of SSE has decreased only slightly, and, as seen in Figure 16, the shifted logistic curve seems almost identical to the unshifted curve, but does seem to level off slightly faster (compare the portions of the two curves on the extreme right).

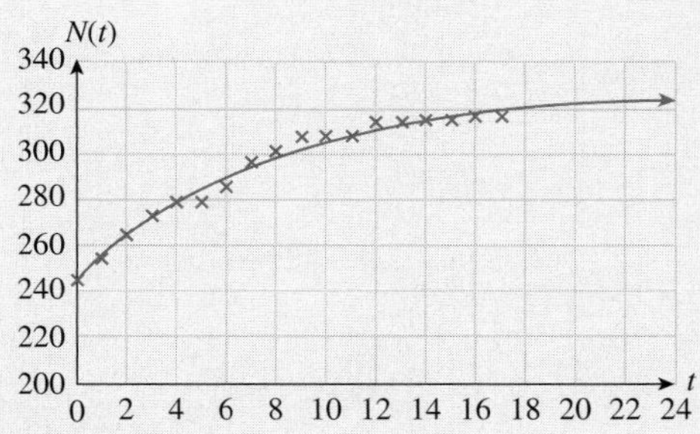

Figure 16

You decide to use the shifted model to obtain another estimate of the projected rate of change in 2015. As the two models differ by a constant, their derivatives are given by the same formula, so you compute

$$N'(20) = \frac{(135.5)(1.192)(1.268)^{-20}\ln 1.268}{(1 + 1.192(1.268)^{-20})^2}$$

$$\approx 0.325 \text{ thousand students per year,}$$

or about 325 students per year, even less than the prediction of the logistic model.

Q: *Why do the two models give very different predictions of the rate of change in 2015?*

A: The long-term prediction in any logistic model is highly sensitive to small changes in the data and/or the model. This is one reason why using regression curve-fitting models to make long-term projections can be a risky undertaking.

Q: *Then what is the point of using any model to project in the first place?*

A: Projections are always tricky as we cannot foresee the future. But a *good* model is not merely one that seems to fit the data well, but rather a model whose structure is based on the situation being modeled. For instance, a *good* model of student graduation rates should take into account such factors as the birth rate, current school populations at all levels, and the relative popularity of private schools as opposed to public schools. It is by using models of this kind that the National Center for Educational Statistics is able to make the projections shown in the data above.

EXERCISES

1. In 1994 there were 246,000 private high school graduates. What do the two logistic models (unshifted and shifted) "predict" for 1994? (Round answer to the nearest 1,000.) Which gives the better prediction?
2. What is the long-term prediction of each of the two models? (Round answer to the nearest 1,000.)
3. Find $\lim_{t \to +\infty} N'(t)$ for both models, and interpret the results.
4. T You receive a last-minute memo from TJM to the effect that, sorry, the 2011 and 2012 figures are not accurate. Use technology to re-estimate M, A, b, and C for the shifted logistic model in the absence of this data and obtain new estimates for the 2011 and 2012 data. What does the new model predict the rate of change in the number of high school seniors will be in 2015?
5. T ***Another Model*** Using the original data, find the best-fit shifted logistic curve of the form

$$N(t) = c + b\frac{a(t-m)}{1+a|t-m|}. \qquad (a, b, c, m \text{ constant})$$

Its graph is shown below:

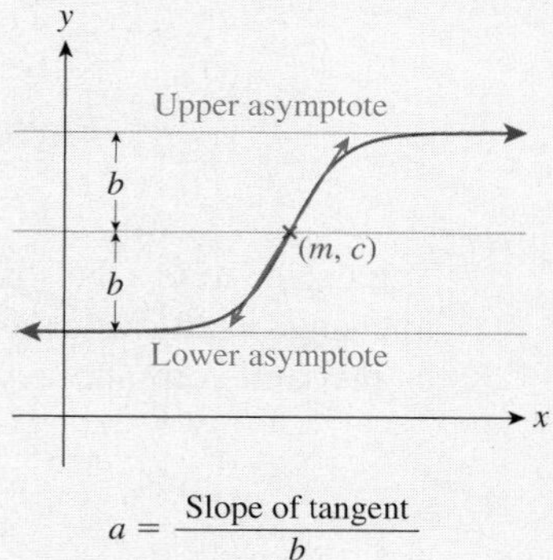

$$a = \frac{\text{Slope of tangent}}{b}$$

(Use the model `$1+$2*$3*(x-$4)/(1+$3*abs(x-$4))` and start with the following values: $a = 0.05$, $b = 160$, $c = 250$, $m = 5$; that is, input `250, 160, 0.05, 5` in the "Guess" field.) Graph the data together with the model. What is SSE? Is the model as accurate a fit as the model used in the text? What does this model predict will be the growth rate of the number of high school graduates in 2015? Comment on the answer. (Round the coefficients in the model and all answers to four decimal places.)

6. T ***Demand for Freon*** The demand for chlorofluorocarbon-12 (CFC-12)—the ozone-depleting refrigerant commonly known as Freon 12 (the name given to it by Du Pont)—has

been declining significantly in response to regulation and concern about the ozone layer. The chart below shows the projected demand for CFC-12 for the period 1994–2005.[3]

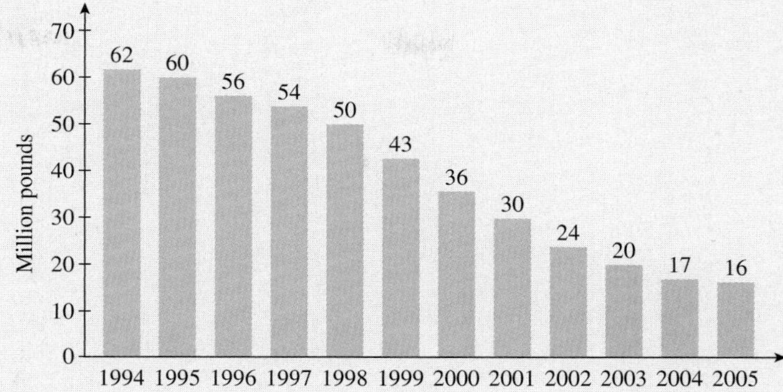

a. Use technology to obtain the best-fit equation of the form

$$N(t) = c + b\frac{a(t - m)}{1 + a|t - m|}, \qquad (a, b, c, m \text{ constant})$$

where t is the number of years since 1990. Use your function to estimate the total demand for CFC-12 from the start of the year 2000 to the start of 2010. [Start with the following values: $a = 1$, $b = -25$, $c = 35$, and $m = 10$, and round your answers to four decimal places.]

b. According to your model, how fast is the demand for Freon declining in 2000?

[3]Source: The Automobile Consulting Group (*New York Times,* December 26, 1993, p. F23). The exact figures were not given, and the chart is a reasonable facsimile of the chart that appeared in *New York Times.*

5

Further Applications of the Derivative

Website

www.WanerMath.com

At the Website you will find:

- Section-by-section tutorials, including game tutorials with randomized quizzes
- A detailed chapter summary
- A true/false quiz
- Additional review exercises
- Graphers, Excel tutorials, and other resources
- The following extra topic:

    Linear Approximation and Error Estimation

Case Study Production Lot Size Management

Your publishing company is planning the production of its latest best seller, which it predicts will sell 100,000 copies each month over the coming year. The book will be printed in several batches of the same number, evenly spaced throughout the year. Each print run has a setup cost of \$5,000, a single book costs \$1 to produce, and monthly storage costs for books awaiting shipment average 1¢ per book. **To meet the anticipated demand at minimum total cost to your company, how many printing runs should you plan?**

SERDAR/Alamy

Introduction

In this chapter we begin to see the power of calculus as an optimization tool. In Chapter 2 we saw how to price an item in order to get the largest revenue when the demand function is linear. Using calculus, we can handle nonlinear functions, which are much more general. In Section 5.1 we show how calculus can be used to solve the problem of finding the values of a variable that lead to a maximum or minimum value of a given function. In Section 5.2 we show how this helps us in various real-world applications.

Another theme in this chapter is that calculus can help us to draw and understand the graph of a function. By the time you have completed the material in Section 5.1, you will be able to locate and sketch some of the important features of a graph, such as where it rises and where it falls. In Section 5.3 we look at the *second derivative,* the derivative of the derivative function, and what it tells us about how the graph *curves*. We also see how the second derivative is used to model the notion of *acceleration*. In Section 5.4 we put a number of ideas together that help to explain what you see in a graph (drawn, for example, using graphing technology) and to locate its most important points.

We also include sections on related rates and elasticity of demand. The first of these (Section 5.5) examines further the concept of the derivative as a rate of change. The second (Section 5.6) returns to the problem of optimizing revenue based on the demand equation, looking at it in a new way that leads to an important idea in economics—elasticity.

algebra Review
For this chapter, you should be familiar with the algebra reviewed in **Chapter 0, sections 5 and 6.**

5.1 Maxima and Minima

Figure 1 shows the graph of a function f whose domain is the closed interval $[a, b]$. A mathematician sees lots of interesting things going on here. There are hills and valleys, and even a small chasm (called a *cusp*) near the center. For many purposes, the important features of this curve are the highs and lows. Suppose, for example, you know that the price of the stock of a certain company will follow this graph during the course of a week. Although you would certainly make a handsome profit if you bought at time a and sold at time b, your best strategy would be to follow the old adage to "buy low and sell high," buying at all the lows and selling at all the highs.

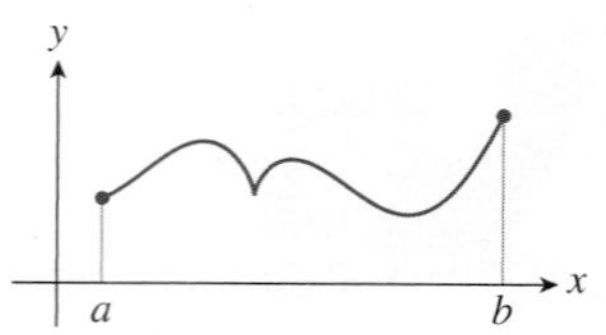

Figure 1

Figure 2 shows the graph once again with the highs and lows marked. Mathematicians have names for these points: the highs (at the x-values p, r, and b) are referred to as **relative maxima**, and the lows (at the x-values a, q, and s) are referred to as **relative minima**. Collectively, these highs and lows are referred to as **relative extrema**. (A point of language: The singular forms of the plurals *minima, maxima,* and *extrema* are *minimum, maximum,* and *extremum.*)

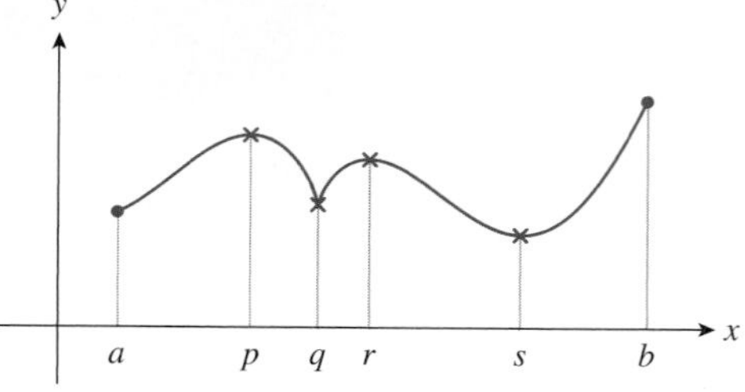

Figure 2

Why do we refer to these points as relative extrema? Take a look at the point corresponding to $x = r$. It is the highest point of the graph *compared to other points nearby*. If you were an extremely nearsighted mountaineer standing at the point

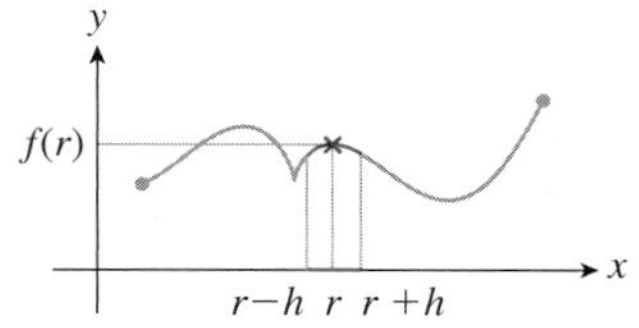

Figure 3

where $x = r$, you would *think* that you were at the highest point of the graph, not being able to see the distant peaks at $x = p$ and $x = b$.

Let's translate into mathematical terms. We are talking about the heights of various points on the curve. The height of the curve at $x = r$ is $f(r)$, so we are saying that $f(r)$ is greater than or equal to $f(x)$ for every x near r. In other words, *$f(r)$ is the greatest value that $f(x)$ has for all choices of x between $r - h$ and $r + h$* for some (possibly small) h. (See Figure 3.)

We can phrase the formal definition as follows.

Relative Extrema

f has a **relative maximum** at $x = r$ if there is some interval $(r - h, r + h)$ (even a very small one) for which $f(r) \geq f(x)$ for all x in $(r - h, r + h)$ for which $f(x)$ is defined.

f has a **relative minimum** at $x = r$ if there is some interval $(r - h, r + h)$ (even a very small one) for which $f(r) \leq f(x)$ for all x in $(r - h, r + h)$ for which $f(x)$ is defined.

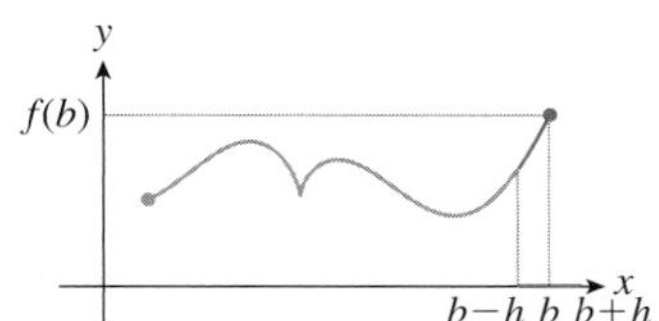

Figure 4

Quick Examples

In Figure 2, f has the following relative extrema:

1. Relative maxima at p and r.
2. A relative maximum at b. (See Figure 4.) Note that $f(x)$ is not defined for $x > b$. However, $f(b) \geq f(x)$ for every x in the interval $(b - h, b + h)$ *for which $f(x)$ is defined*—that is, for every x in $(b - h, b]$.*
3. Relative minima at a, q, and s.

* Our definition of relative extremum allows f to have a relative extremum at an endpoint of its domain; the definitions used in some books do not. In view of examples like the stock market investing strategy mentioned above, we find it more useful to allow endpoints as relative extrema.

Looking carefully at Figure 2, we can see that the lowest point on the whole graph is where $x = s$ and the highest point is where $x = b$. This means that $f(s)$ is the least value of f on the whole domain of f (the interval $[a, b]$) and $f(b)$ is the greatest value. We call these the *absolute* minimum and maximum.

Absolute Extrema

f has an **absolute maximum** at $x = r$ if $f(r) \geq f(x)$ for every x in the domain of f.

f has an **absolute minimum** at $x = r$ if $f(r) \leq f(x)$ for every x in the domain of f.

Quick Examples

1. In Figure 2, f has an absolute maximum at b and an absolute minimum at s.
2. If $f(x) = x^2$, then $f(x) \geq f(0)$ for every real number x. Therefore, $f(x) = x^2$ has an absolute minimum at $x = 0$. (See Figure 5.)
3. Generalizing (2), every quadratic function $f(x) = ax^2 + bx + c$ has an absolute extremum at its vertex $x = -b/(2a)$; an absolute minimum if $a > 0$, and an absolute maximum if $a < 0$.

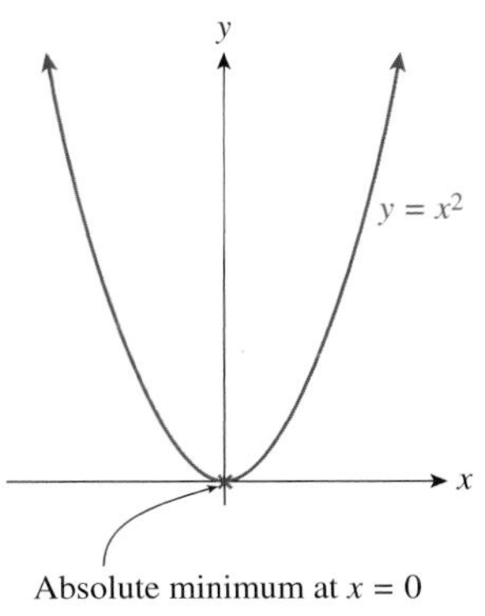

Figure 5

Note If f has an absolute extremum at $x = r$, then it automatically satisfies the requirement for a *relative* extremum there as well; take $h = 1$ (or any other value) in the definition of relative extremum. Thus, absolute extrema are special types of relative extrema. ■

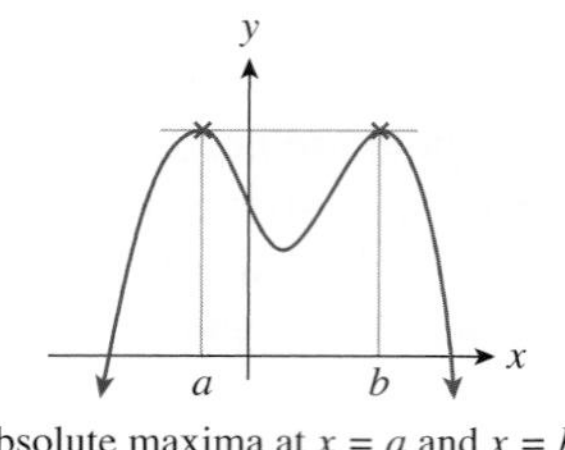

Absolute maxima at $x = a$ and $x = b$

Figure 6

Some graphs have no absolute extrema at all (think of the graph of $y = x$), while others might have an absolute minimum but no absolute maximum (like $y = x^2$), or vice versa. When f does have an absolute maximum, there is only one absolute maximum *value* of f, but this value may occur at different values of x, and similarly for absolute minima. (See Figure 6.)

Q: *At how many different values of x can f take on its absolute maximum value?*

A: An extreme case is that of a constant function; because we use $\geq$ in the definition of absolute maximum, a constant function has an absolute maximum (and minimum) at every point in its domain.

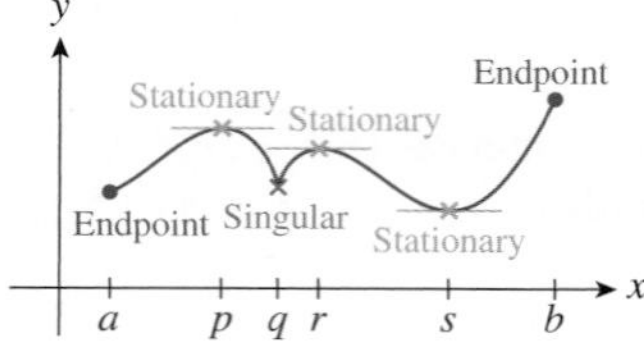

Figure 7

Now, how do we go about locating extrema? In many cases we can get a good idea by using graphing technology to zoom in on a maximum or minimum and approximate its coordinates. However, calculus gives us a way to find the exact locations of the extrema and at the same time to understand why the graph of a function behaves the way it does. In fact, it is often best to combine the powers of graphing technology with those of calculus, as we shall see.

In Figure 7 we see the graph from Figure 1 once more, but we have labeled each extreme point as one of three types. Notice that two extrema occur at endpoints and the others at **interior points;** that is, points other than endpoints. Let us look first at the extrema occurring at interior points: At the points labeled "Stationary," the tangent lines to the graph are horizontal, and so have slope 0, so f' (which gives the slope) is 0. Any time $f'(x) = 0$, we say that f has a **stationary point** at x because the rate of change of f is zero there. We call an extremum that occurs at a stationary point a **stationary extremum.** In general, to find the exact location of each stationary point, we need to solve the equation $f'(x) = 0$. Note that stationary points are always interior points, as f' is not defined at endpoints of the domain.

There is a relative minimum in Figure 7 at $x = q$, but there is no horizontal tangent there. In fact, there is no tangent line at all; $f'(q)$ is not defined. (Recall a similar situation with the graph of $f(x) = |x|$ at $x = 0$.) When $f'(x)$ does not exist for some interior point x in the domain of f, we say that f has a **singular point** at x. We shall call an extremum that occurs at a singular point a **singular extremum.** The points that are either stationary or singular we call collectively the **critical points** of f.

The remaining two extrema are at the **endpoints** of the domain (remember that we do allow relative extrema at endpoints). As we see in the figure, they are (almost) always either relative maxima or relative minima.

We bring all the above information together in Figure 8:

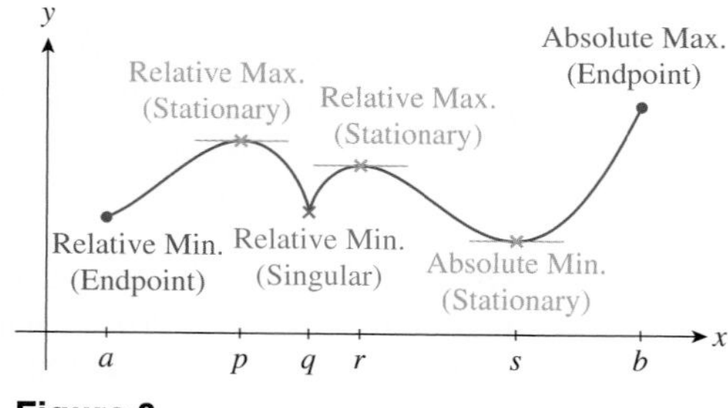

Figure 8

Q: *Are there any other types of relative extrema?*

A: No; relative extrema of a function always occur at critical points or endpoints. (A rigorous proof is beyond the scope of this book.)*

* Here is an outline of the argument. Suppose f has a maximum, say, at $x = a$, at some interior point of its domain. Then either f is differentiable there, or it is not. If it is not, then we have a singular point. If f is differentiable at $x = a$, then consider the slope of the secant line through the points where $x = a$ and $x = a + h$ for small positive h. Because f has a maximum at $x = a$, it is falling (or level) to the right of $x = a$, and so the slope of this secant line must be ≤ 0. Thus, we must have $f'(a) \leq 0$ in the limit as $h \to 0$. On the other hand, if h is small and *negative,* then the corresponding secant line must have slope ≥ 0 because f is also falling (or level) as we move left from $x = a$, and so $f'(a) \geq 0$. Because $f'(a)$ is both ≥ 0 and ≤ 0, it must be zero, and so we have a stationary point at $x = a$.

Locating Candidates for Extrema

If f is a real-valued function, then its extrema occur among the following types of points:

1. **Stationary Points:** f has a stationary point at x if x is in the interior of the domain and $f'(x) = 0$. To locate stationary points, set $f'(x) = 0$ and solve for x.
2. **Singular Points:** f has a singular point at x if x is in the interior of the domain and $f'(x)$ is not defined. To locate singular points, find values of x where $f'(x)$ is *not* defined, but $f(x)$ *is* defined.
3. **Endpoints:** These are the endpoints, if any, of the domain. Recall that closed intervals contain endpoints, but open intervals do not. If the domain of f is an open interval or the whole real line, then there are no endpoints.

Once we have a candidate for an extremum of f, we find the corresponding point (x, y) on the graph of f using $y = f(x)$.

Quick Examples

1. **Stationary Points:** Let $f(x) = x^3 - 12x$. Then to locate the stationary points, set $f'(x) = 0$ and solve for x. This gives $3x^2 - 12 = 0$, so f has stationary points at $x = \pm 2$. The corresponding points on the graph are $(-2, f(-2)) = (-2, 16)$ and $(2, f(2)) = (2, -16)$.
2. **Singular Points:** Let $f(x) = 3(x-1)^{1/3}$. Then $f'(x) = (x-1)^{-2/3} = 1/(x-1)^{2/3}$. $f'(1)$ is not defined, although $f(1)$ *is* defined. Thus, the (only) singular point occurs at $x = 1$. The corresponding point on the graph is $(1, f(1)) = (1, 0)$.
3. **Endpoints:** Let $f(x) = 1/x$, with domain $(-\infty, 0) \cup [1, +\infty)$. Then the only endpoint in the domain of f occurs at $x = 1$. The corresponding point on the graph is $(1, 1)$. The natural domain of $1/x$, on the other hand, has no endpoints.

Remember, though, that the three types of points we identify above are only *candidates* for extrema. It is quite possible, as we shall see, to have a stationary point or a singular point that is neither a maximum nor a minimum. (It is also possible for an endpoint to be neither a maximum nor a minimum, but only in functions whose graphs are rather bizarre—see Exercise 65 associated with this section.)

Now let's look at some examples of finding maxima and minima. In all of these examples, we will use the following procedure: First, we find the derivative, which we examine to find the stationary points and singular points. Next, we make a table listing the x-coordinates of the critical points and endpoints, together with their y-coordinates. We use this table to make a rough sketch of the graph. From the table and rough sketch, we usually have enough data to be able to say where the extreme points are and what kind they are.

EXAMPLE 1 Maxima and Minima

Find the relative and absolute maxima and minima of

$$f(x) = x^2 - 2x$$

on the interval [0, 4].

Solution We first calculate $f'(x) = 2x - 2$. We use this derivative to locate the critical points (stationary and singular points).

Stationary Points To locate the stationary points, we solve the equation $f'(x) = 0$, or

$$2x - 2 = 0,$$

getting $x = 1$. The domain of the function is [0, 4], so $x = 1$ is in the interior of the domain. Thus, the only candidate for a stationary relative extremum occurs when $x = 1$.

Singular Points We look for interior points where the derivative is not defined. However, the derivative is $2x - 2$, which is defined for every x. Thus, there are no singular points and hence no candidates for singular relative extrema.

Endpoints The domain is [0, 4], so the endpoints occur when $x = 0$ and $x = 4$.

We record these values of x in a table, together with the corresponding y-coordinates (values of f):

x	0	1	4
$f(x) = x^2 - 2x$	0	−1	8

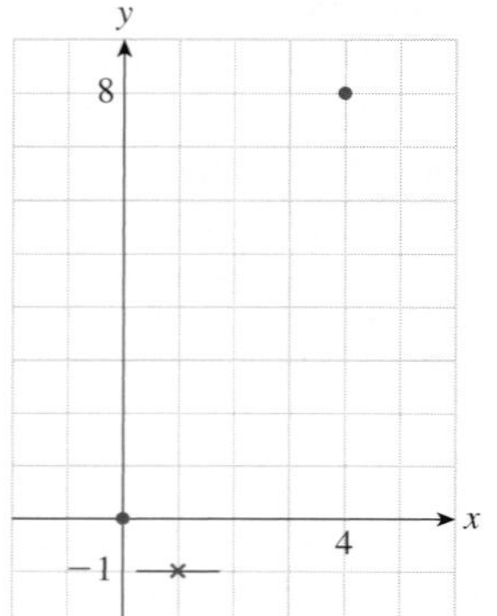

Figure 9

This gives us three points on the graph, (0, 0), (1, −1), and (4, 8), which we plot in Figure 9. We remind ourselves that the point (1, −1) is a stationary point of the graph by drawing in a part of the horizontal tangent line. Connecting these points must give us a graph something like that in Figure 10.

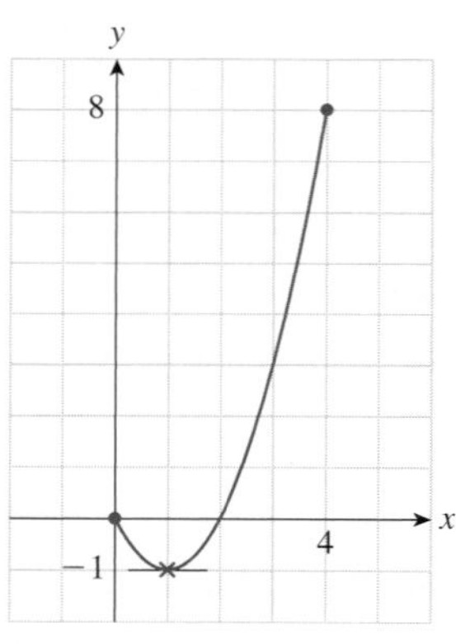

Figure 10

From Figure 10 we can see that f has the following extrema:

x	$y = x^2 - 2x$	*Classification*
0	0	Relative maximum (endpoint)
1	−1	Absolute minimum (stationary point)
4	8	Absolute maximum (endpoint)

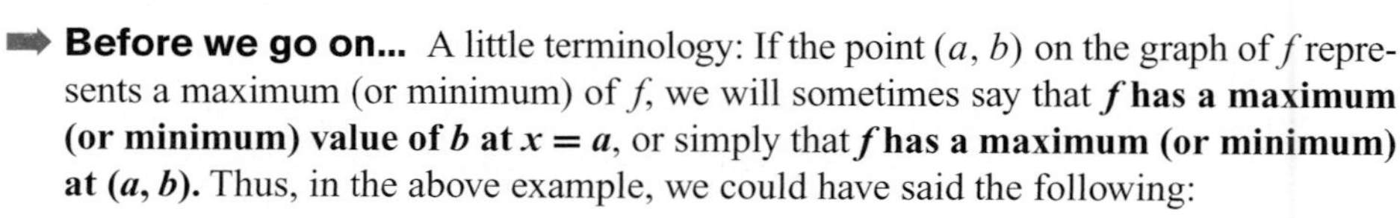

➡ **Before we go on...** A little terminology: If the point (a, b) on the graph of f represents a maximum (or minimum) of f, we will sometimes say that ***f* has a maximum (or minimum) value of *b* at *x* = *a***, or simply that ***f* has a maximum (or minimum) at (*a*, *b*)**. Thus, in the above example, we could have said the following:

- "f has a relative maximum value of 0 at $x = 0$," or "f has a relative maximum at (0, 0)."
- "f has an absolute minimum value of −1 at $x = 1$," or "f has an absolute minimum at (1, −1)."
- "f has an absolute maximum value of 8 at $x = 4$," or "f has an absolute maximum at (4, 8)."

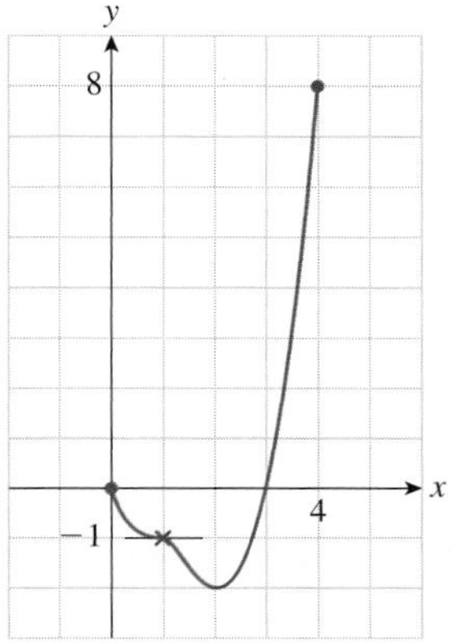

Figure 11

Q: *How can we be sure that the graph in Example 1 doesn't look like Figure 11?*

A: If it did, there would be another critical point somewhere between $x = 1$ and $x = 4$. But we already know that there aren't any other critical points. The table we made listed all of the possible extrema; there can be no more. ■

First Derivative Test

The **first derivative test*** gives another, very systematic, way of checking whether a critical point is a maximum or minimum. To motivate the first derivative test, consider again the critical point $x = 1$ in Example 1. If we look at some values of $f'(x)$ to the left and right of the critical point, we obtain the information shown in the following table:

* Why "first" derivative test? To distinguish it from a test based on the **second derivative** of a function, which we shall discuss in Section 5.3.

	Point to the Left	Critical Point	Point to the Right
x	0.5	1	2
$f'(x) = 2x - 2$	−1	0	2
Direction of Graph	↘	→	↗

At $x = 0.5$ (to the left of the critical point) we see that $f'(0.5) = -1 < 0$, so the graph has negative slope and f is decreasing. We note this with the downward pointing arrow. At $x = 2$ (to the right of the critical point), we find $f'(2) = 2 > 0$, so the graph has positive slope and f is increasing. In fact, because $f'(x) = 0$ only at $x = 1$, we know that $f'(x) < 0$ for all x in (0, 1), and we can say that f is decreasing on the interval (0, 1). Similarly, f is increasing on (1, 4).

So, starting at $x = 0$, the graph of f goes down until we reach $x = 1$ and then it goes back up, telling us that $x = 1$ must be a minimum. Notice how the minimum is suggested by the arrows to the left and right.

First Derivative Test for Extrema

Suppose that c is a critical point of the continuous function f, and that its derivative is defined for x close to, and on both sides of, $x = c$. Then, determine the sign of the derivative to the left and right of $x = c$.

1. If $f'(x)$ is positive to the left of $x = c$ and negative to the right, then f has a maximum at $x = c$.
2. If $f'(x)$ is negative to the left of $x = c$ and positive to the right, then f has a minimum at $x = c$.
3. If $f'(x)$ has the same sign on both sides of $x = c$, then f has neither a maximum nor a minimum at $x = c$.

Quick Examples

1. In Example 1 on previous page, we saw that $f(x) = x^2 - 2x$ has a critical point at $x = 1$ with $f'(x)$ negative to the left of $x = 1$ and positive to the right (see the table). Therefore, f has a minimum at $x = 1$.

2. Here is a graph showing a function f with a singular point at $x = 1$:

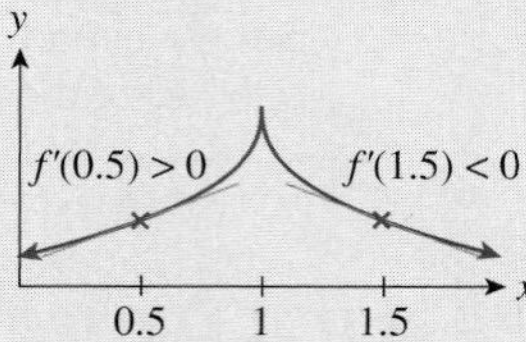

The graph gives us the information shown in the table:

	Point to the Left	Critical Point	Point to the Right
x	0.5	1	1.5
$f'(x)$	+	Undefined	−
Direction of Graph	↗		↘

Since $f'(x)$ is positive to the left of $x = 1$ and negative to the right, we see that f has a maximum at $x = 1$. (Notice again how this is suggested by the direction of the arrows.)

EXAMPLE 2 Unbounded Interval

Find all extrema of $f(x) = 3x^4 - 4x^3$ on $[-1, \infty)$.

Solution We first calculate $f'(x) = 12x^3 - 12x^2$.

Stationary points We solve the equation $f'(x) = 0$, which is

$$12x^3 - 12x^2 = 0 \text{ or}$$
$$12x^2(x - 1) = 0.$$

There are two solutions, $x = 0$ and $x = 1$, and both are in the domain. These are our candidates for the x-coordinates of stationary extrema.

Singular points There are no points where $f'(x)$ is not defined, so there are no singular points.

Endpoints The domain is $[-1, \infty)$, so there is one endpoint, at $x = -1$.

We record these points in a table with the corresponding y-coordinates:

x	−1	0	1
$f(x) = 3x^4 - 4x^3$	7	0	−1

We will illustrate three methods we can use to determine which are minima, which are maxima, and which are neither:

1. Plot these points and sketch the graph by hand.

2. Use the first derivative test.

3. Use technology to help us.

Use the method you find most convenient.

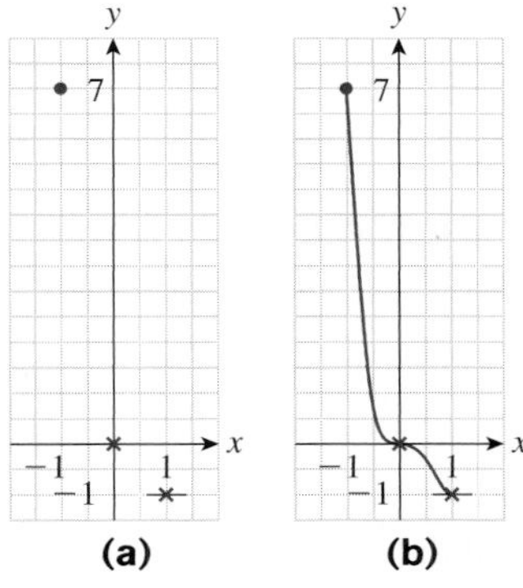

Figure 12

Using a Hand Plot: If we plot these points by hand, we obtain Figure 12(a), which suggests Figure 12(b).

We can't be sure what happens to the right of $x = 1$. Does the curve go up, or does it go down? To find out, let's plot a "test point" to the right of $x = 1$. Choosing $x = 2$, we obtain $y = 3(2)^4 - 4(2)^3 = 16$, so (2, 16) is another point on the graph. Thus, it must turn upward to the right of $x = 1$, as shown in Figure 13.

From the graph, we find that f has the following extrema:

A relative (endpoint) maximum at $(-1, 7)$

An absolute (stationary) minimum at $(1, -1)$

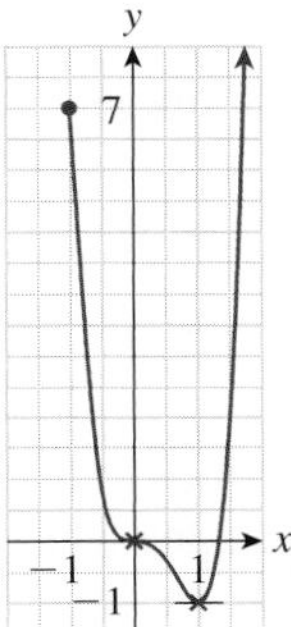

Figure 13

Using the First Derivative Test: List the critical and endpoints in a table, and add additional points as necessary so that each critical point has a noncritical point on either side. Then compute the derivative at each of these points, and draw an arrow to indicate the direction of the graph.

	Endpoint		Critical Point		Critical Point	
x	−1	−0.5	0	0.5	1	2
$f'(x) = 12x^3 - 12x^2$	−24	−1.5	0	−1.5	0	48
Direction of Graph		↘	→	↘	→	↗

Notice that the arrows now suggest the shape of the curve in Figure 13. The first derivative test tells us that the function has a relative maximum at $x = -1$, neither a maximum nor a minimum at $x = 0$, and a relative minimum at $x = 1$. Deciding which of these extrema are absolute and which are relative requires us to compute y-coordinates and plot the corresponding points on the graph by hand, as we did in the first method.

using Technology

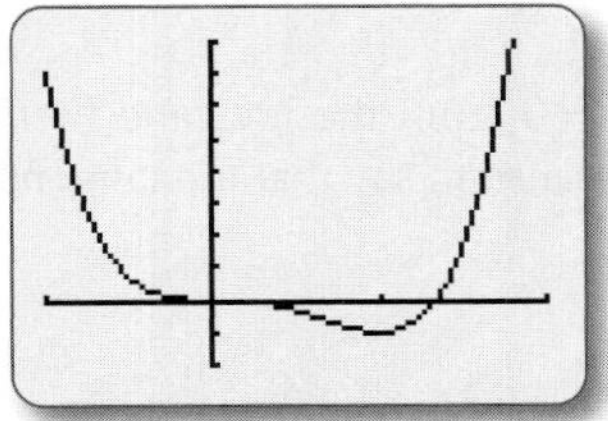
Figure 14

If we use technology to show the graph, we should choose the viewing window so that it contains the three interesting points we found: $x = -1$, $x = 0$, and $x = 1$. Again, we can't be sure yet what happens to the right of $x = 1$; does the graph go up or down from that point? If we set the viewing window to an interval of $[-1, 2]$ for x and $[-2, 8]$ for y, we will leave enough room to the right of $x = 1$ and below $y = -1$ to see what the graph will do. The result will be something like Figure 14.

Now we can tell what happens to the right of $x = 1$: the function increases. We know that it cannot later decrease again because if it did, there would have to be another critical point where it turns around, and we found that there are no other critical points. ■

➡ **Before we go on...** Notice that the stationary point at $x = 0$ in Example 2 is neither a relative maximum nor a relative minimum. It is simply a place where the graph of f flattens out for a moment before it continues to fall. Notice also that f has no absolute maximum because $f(x)$ increases without bound as x gets large. ■

EXAMPLE 3 Singular Point

Find all extrema of $f(t) = t^{2/3}$ on $[-1, 1]$.

Solution First, $f'(t) = \frac{2}{3}t^{-1/3}$.

Stationary points We need to solve

$$\frac{2}{3}t^{-1/3} = 0.$$

We can rewrite this equation without the negative exponent:

$$\frac{2}{3t^{1/3}} = 0.$$

Now, the only way that a fraction can equal 0 is if the numerator is 0, so this fraction can never equal 0. Thus, there are no stationary points.

Singular Points The derivative

$$f'(t) = \frac{2}{3t^{1/3}}$$

is not defined for $t = 0$. However, 0 is in the interior of the domain of f (although f' is not defined at $t = 0$, f itself is). Thus, f has a singular point at $t = 0$.

Endpoints There are two endpoints, -1 and 1.

We now put these three points in a table with the corresponding y-coordinates:

t	-1	0	1
$f(t)$	1	0	1

Using a Hand Plot: The derivative, $f'(t) = 2/(3t^{1/3})$, is not defined at the singular point $t = 0$. To help us sketch the graph, let's use limits to investigate what happens to the derivative as we approach 0 from either side:

$$\lim_{t\to 0^-} f'(t) = \lim_{t\to 0^-} \frac{2}{3t^{1/3}} = -\infty$$

$$\lim_{t\to 0^+} f'(t) = \lim_{t\to 0^+} \frac{2}{3t^{1/3}} = +\infty.$$

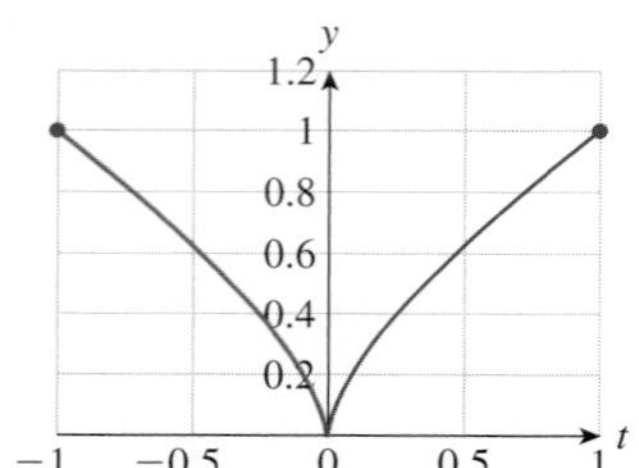

Figure 15

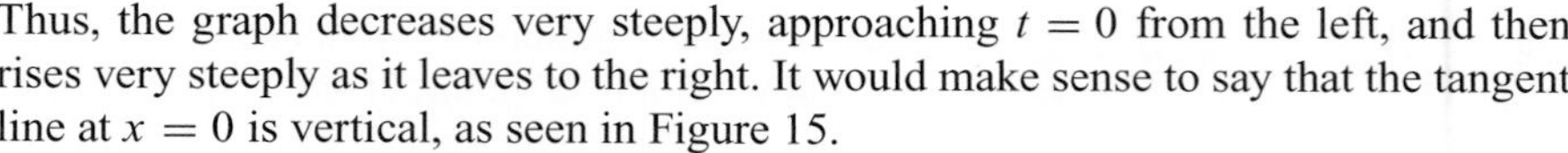

Thus, the graph decreases very steeply, approaching $t = 0$ from the left, and then rises very steeply as it leaves to the right. It would make sense to say that the tangent line at $x = 0$ is vertical, as seen in Figure 15.

From this graph, we find the following extrema for f:

An absolute (endpoint) maximum at $(-1, 1)$

An absolute (singular) minimum at $(0, 0)$

An absolute (endpoint) maximum at $(1, 1)$.

Notice that the absolute maximum value of f is achieved at two values of t: $t = -1$ and $t = 1$.

First Derivative Test: Here is the corresponding table for the first derivative test.

t	-0.5	0	0.5
$f'(t) = \dfrac{2}{3t^{1/3}}$	$-\dfrac{2}{3(0.5)^{1/3}}$	Undefined	$\dfrac{2}{3(0.5)^{1/3}}$
Direction of Graph	↘	↕	↗

(We drew a vertical arrow at $t = 0$ to indicate a vertical tangent.) Again, notice how the arrows suggest the shape of the curve in Figure 15, and the first derivative test confirms that we have a minimum at $x = 0$.

using Technology

Because there is only one critical point, at $t = 0$, it is clear from this table that f must decrease from $t = -1$ to $t = 0$ and then increase from $t = 0$ to $t = 1$. To graph f using technology, choose a viewing window with an interval of $[-1, 1]$ for t and $[0, 1]$ for y. The result will be something like Figure 15.* ■

* Many graphing calculators will give you only the right-hand half of the graph shown in Figure 15 because fractional powers of negative numbers are not, in general, real numbers. To obtain the whole curve, enter the formula as `Y=(x^2)^(1/3)`, a fractional power of the non-negative function x^2.

In Examples 1 and 3, we could have found the absolute maxima and minima without doing any graphing. In Example 1, after finding the critical points and endpoints, we created the following table:

x	0	1	4
$f(x)$	0	−1	8

From this table we can see that f must decrease from its value of 0 at $x = 0$ to -1 at $x = 1$, and then increase to 8 at $x = 4$. The value of 8 must be the largest value it takes on, and the value of -1 must be the smallest, on the interval $[0, 4]$. Similarly, in Example 3 we created the following table:

t	−1	0	1
$f(t)$	1	0	1

From this table we can see that the largest value of f on the interval $[-1, 1]$ is 1 and the smallest value is 0. We are taking advantage of the following fact, the proof of which uses some deep and beautiful mathematics (alas, beyond the scope of this book):

Extreme Value Theorem

If f is *continuous* on a *closed interval* $[a, b]$, then it will have an absolute maximum and an absolute minimum value on that interval. Each absolute extremum must occur at either an endpoint or a critical point. Therefore, the absolute maximum is the largest value in a table of the values of f at the endpoints and critical points, and the absolute minimum is the smallest value.

Quick Example

The function $f(x) = 3x - x^3$ on the interval $[0, 2]$ has one critical point at $x = 1$. The values of f at the critical point and the endpoints of the interval are given in the following table:

	Endpoint	**Critical point**	**Endpoint**
x	0	1	2
$f(x)$	0	2	−2

From this table we can say that the absolute maximum value of f on $[0, 2]$ is 2, which occurs at $x = 1$, and the absolute minimum value of f is -2, which occurs at $x = 2$.

As we can see in Example 2 and the following examples, if the domain is not a closed interval, then f may not have an absolute maximum and minimum, and a table of values as above is of little help in determining whether it does.

EXAMPLE 4 Domain Not a Closed Interval

Find all extrema of $f(x) = x + \frac{1}{x}$.

Solution Because no domain is specified, we take the domain to be as large as possible. The function is not defined at $x = 0$ but is at all other points, so we take its domain to be $(-\infty, 0) \cup (0, +\infty)$. We calculate

$$f'(x) = 1 - \frac{1}{x^2}.$$

Stationary Points Setting $f'(x) = 0$, we solve

$$1 - \frac{1}{x^2} = 0$$

to find $x = \pm 1$. Calculating the corresponding values of f, we get the two stationary points $(1, 2)$ and $(-1, -2)$.

Singular Points The only value of x for which $f'(x)$ is not defined is $x = 0$, but then f is not defined there either, so there are no singular points in the domain.

Endpoints The domain, $(-\infty, 0) \cup (0, +\infty)$, has no endpoints.

From this scant information, it is hard to tell what f does. If we are sketching the graph by hand, or using the first derivative test, we will need to plot additional "test points" to the left and right of the stationary points $x = \pm 1$.

 **using Technology**

For the technology approach, let's choose a viewing window with an interval of $[-3, 3]$ for x and $[-4, 4]$ for y, which should leave plenty of room to see how f behaves near the stationary points. The result is something like Figure 16.

From this graph we can see that f has:

A relative (stationary) maximum at $(-1, -2)$

A relative (stationary) minimum at $(1, 2)$

Curiously, the relative maximum is lower than the relative minimum! Notice also that, because of the break in the graph at $x = 0$, the graph did not need to rise to get from $(-1, -2)$ to $(1, 2)$. ■

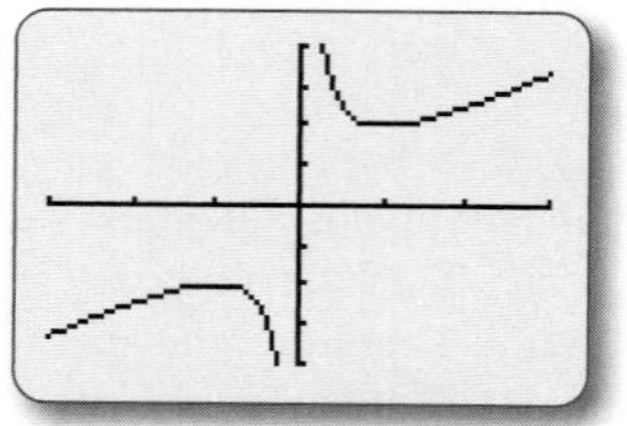

Figure 16

So far we have been solving the equation $f'(x) = 0$ to obtain our candidates for stationary extrema. However, it is often not easy—or even possible—to solve equations analytically. In the next example, we show a way around this problem by using graphing technology.

EXAMPLE 5 T Finding Approximate Extrema Using Technology

Graph the function $f(x) = (x - 1)^{2/3} - \frac{x^2}{2}$ with domain $[-2, +\infty)$. Also graph its derivative and hence locate and classify all extrema of f, with coordinates accurate to two decimal places.

Solution In Example 4 of Section 3.5, we saw how to draw the graphs of f and f' using technology. Note that the technology formula to use for the graph of f is

```
((x-1)^2)^(1/3)-0.5*x^2
```

instead of

```
(x-1)^(2/3)-0.5*x^2
```

(Why?)

Figure 17 shows the resulting graphs of f and f'.

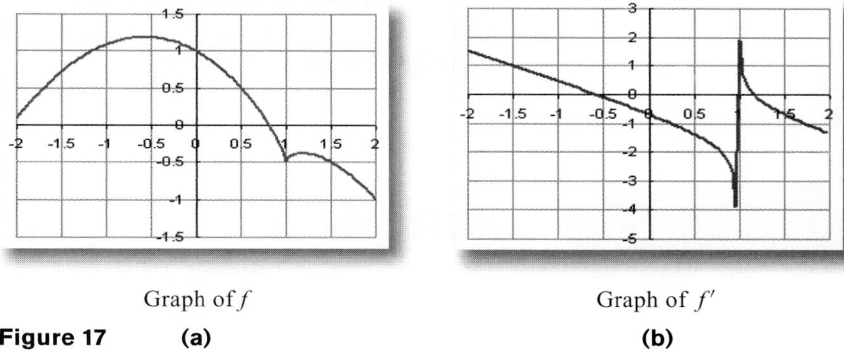

Graph of f Graph of f'

Figure 17 **(a)** **(b)**

If we extend Xmax beyond $x = 2$, we find that the graph continues downward, apparently without any further interesting behavior.

Stationary Points The graph of f shows two stationary points, both maxima, at around $x = -0.6$ and $x = 1.2$. Notice that the graph of f' is zero at these points. Moreover, it is easier to locate these values accurately on the graph of f' because it is easier to pinpoint where a graph crosses the x-axis than to locate a stationary point. Zooming in to the stationary point at $x \approx -0.6$ results in Figure 18.

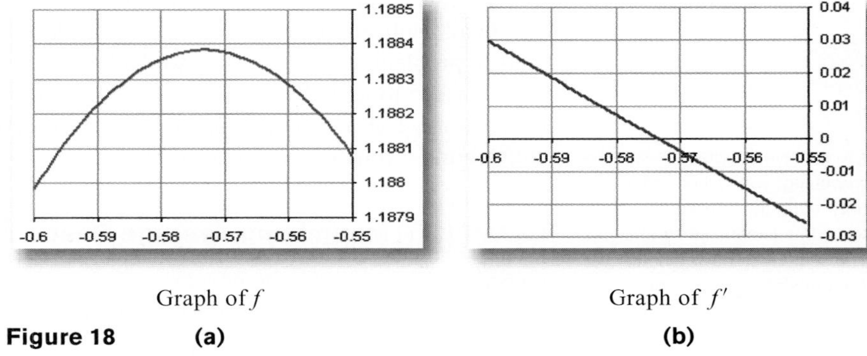

Graph of f Graph of f'

Figure 18 **(a)** **(b)**

From the graph of f, we can see that the stationary point is somewhere between -0.58 and -0.57. The graph of f' shows more clearly that the zero of f', hence the stationary point of f lies somewhat closer to -0.57 than to -0.58. Thus, the stationary point occurs at $x \approx -0.57$, rounded to two decimal places.

In a similar way, we find the second stationary point at $x \approx 1.18$.

Singular Points Going back to Figure 17, we notice what appears to be a cusp (singular point) at the relative minimum around $x = 1$, and this is confirmed by a glance at the graph of f', which seems to take a sudden jump at that value. Zooming in closer suggests that the singular point occurs at exactly $x = 1$. In fact, we can calculate

$$f'(x) = \frac{2}{3(x-1)^{1/3}} - x.$$

From this formula we see clearly that $f'(x)$ is defined everywhere except at $x = 1$.

Endpoints The only endpoint in the domain is $x = -2$, which gives a relative minimum.

Thus, we have found the following approximate extrema for f:

A relative (endpoint) minimum at $(-2, 0.08)$
An absolute (stationary) maximum at $(-0.57, 1.19)$
A relative (singular) minimum at $(1, -0.5)$
A relative (stationary) maximum at $(1.18, -0.38)$.

5.1 EXERCISES

Access end-of-section exercises online at **www.webassign.net**

5.2 Applications of Maxima and Minima

In many applications we would like to find the largest or smallest possible value of some quantity—for instance, the greatest possible profit or the lowest cost. We call this the *optimal* (best) value. In this section we consider several such examples and use calculus to find the optimal value in each.

In all applications the first step is to translate a written description into a mathematical problem. In the problems we look at in this section, there are *unknowns* that we are asked to find, there is an expression involving those unknowns that must be made as large or as small as possible—the **objective function**—and there may be **constraints**—equations or inequalities relating the variables.*

* If you have studied linear programming, you will notice a similarity here, but unlike the situation in linear programming, neither the objective function nor the constraints need be linear.

EXAMPLE 1 Minimizing Average Cost

Gymnast Clothing manufactures expensive hockey jerseys for sale to college bookstores in runs of up to 500. Its cost (in dollars) for a run of x hockey jerseys is

$$C(x) = 2{,}000 + 10x + 0.2x^2.$$

How many jerseys should Gymnast produce per run in order to minimize average cost?†

† Why don't we seek to minimize total cost? The answer would be uninteresting; to minimize total cost, we would make *no* jerseys at all. Minimizing the average cost is a more practical objective.

Solution Here is the procedure we will follow to solve problems like this.

1. ***Identify the unknown(s).*** There is one unknown: x, the number of hockey jerseys Gymnast should produce per run. (We know this because the question is, How many jerseys . . . ?)

2. ***Identify the objective function.*** The objective function is the quantity that must be made as small (in this case) as possible. In this example it is the average cost, which is given by

$$\bar{C}(x) = \frac{C(x)}{x} = \frac{2{,}000 + 10x + 0.2x^2}{x}$$

$$= \frac{2{,}000}{x} + 10 + 0.2x \text{ dollars/jersey.}$$

3. ***Identify the constraints (if any).*** At most 500 jerseys can be manufactured in a run. Also, $\bar{C}(0)$ is not defined. Thus, x is constrained by

$$0 < x \le 500.$$

Put another way, the domain of the objective function $\bar{C}(x)$ is (0, 500].

4. ***State and solve the resulting optimization problem.*** Our optimization problem is:

$$\text{Minimize } \bar{C}(x) = \frac{2,000}{x} + 10 + 0.2x \qquad \text{Objective function}$$

$$\text{subject to } 0 < x \le 500. \qquad \text{Constraint}$$

We now solve this problem as in Section 5.1. We first calculate

$$\bar{C}'(x) = -\frac{2,000}{x^2} + 0.2.$$

We solve $\bar{C}'(x) = 0$ to find $x = \pm 100$. We reject $x = -100$ because -100 is not in the domain of $\bar{C}$ (and makes no sense), so we have one stationary point, at $x = 100$. There, the average cost is $\bar{C}(100) = \$50$ per jersey.

The only point at which the formula for $\bar{C}'$ is not defined is $x = 0$, but that is not in the domain of $\bar{C}$, so we have no singular points. We have one endpoint in the domain, at $x = 500$. There, the average cost is $\bar{C}(500) = \$114$.

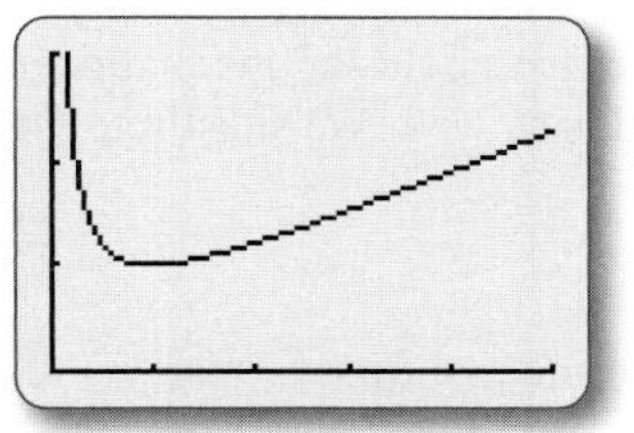

Figure 19

using Technology

Let's plot $\bar{C}$ in a viewing window with the intervals [0, 500] for x and [0, 150] for y, which will show the whole domain and the two interesting points we've found so far. The result is Figure 19.

From the graph of $\bar{C}$, we can see that the stationary point at $x = 100$ gives the absolute minimum. We can therefore say that Gymnast Clothing should produce 100 jerseys per run, for a lowest possible average cost of \$50 per jersey. ■

EXAMPLE 2 Maximizing Area

Slim wants to build a rectangular enclosure for his pet rabbit, Killer, against the side of his house, as shown in Figure 20. He has bought 100 feet of fencing. What are the dimensions of the largest area that he can enclose?

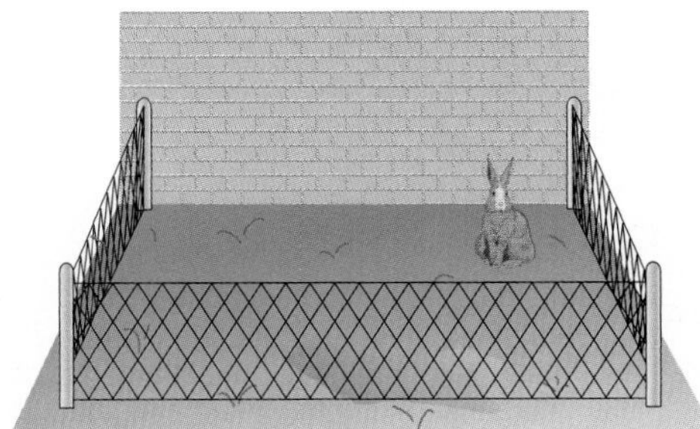

Figure 20

Solution

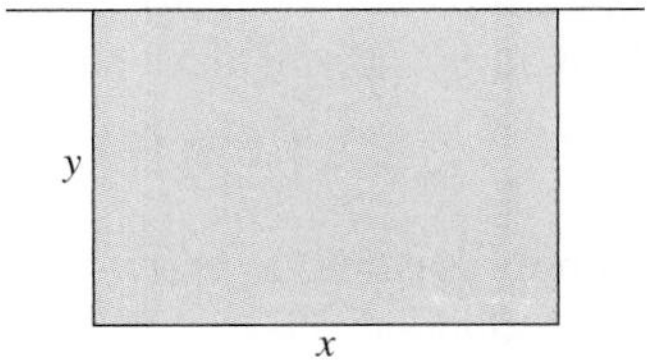

1. ***Identify the unknown(s).*** To identify the unknown(s), we look at the question: What are the *dimensions* of the largest area he can enclose? Thus, the unknowns are the dimensions of the fence. We call these x and y, as shown in Figure 21.

Figure 21

2. ***Identify the objective function.*** We look for what it is that we are trying to maximize (or minimize). The phrase "largest area" tells us that our object is to *maximize the area*, which is the product of length and width, so our objective function is

$$A = xy, \text{ where } A \text{ is the area of the enclosure.}$$

3. ***Identify the constraints (if any).*** What stops Slim from making the area as large as he wants? He has only 100 feet of fencing to work with. Looking again at Figure 21, we see that the sum of the lengths of the three sides must equal 100, so

$$x + 2y = 100.$$

One more point: Because x and y represent the lengths of the sides of the enclosure, neither can be a negative number.

4. ***State and solve the resulting optimization problem.*** Our mathematical problem is:

Maximize $A = xy$ — Objective function

subject to $x + 2y = 100$, $x \geq 0$, and $y \geq 0$. — Constraints

We know how to find maxima and minima of a function of one variable, but A appears to depend on two variables. We can remedy this by using a constraint to express one variable in terms of the other. Let's take the constraint $x + 2y = 100$ and solve for x in terms of y:

$$x = 100 - 2y.$$

Substituting into the objective function gives

$$A = xy = (100 - 2y)y = 100y - 2y^2$$

and we have eliminated x from the objective function. What about the inequalities? One says that $x \geq 0$, but we want to eliminate x from this as well. We substitute for x again, getting

$$100 - 2y \geq 0.$$

Solving this inequality for y gives $y \leq 50$. The second inequality says that $y \geq 0$. Now, we can restate our problem with x eliminated:

Maximize $A(y) = 100y - 2y^2$ subject to $0 \leq y \leq 50$.

We now proceed with our usual method of solving such problems. We calculate $A'(y) = 100 - 4y$. Solving $100 - 4y = 0$, we get one stationary point at $y = 25$. There, $A(25) = 1{,}250$. There are no points at which $A'(y)$ is not defined, so there are no singular points. We have two endpoints, at $y = 0$ and $y = 50$. The corresponding areas are $A(0) = 0$ and $A(50) = 0$. We record the three points we found in a table:

y	0	25	50
A(y)	0	1,250	0

It's clear now how A must behave: It increases from 0 at $y = 0$ to 1,250 at $y = 25$ and then decreases back to 0 at $y = 50$. Thus, the largest possible value of A is 1,250 square feet, which occurs when $y = 25$. To completely answer the question that was asked, we need to know the corresponding value of x. We have $x = 100 - 2y$, so $x = 50$ when $y = 25$. Thus, Slim should build his enclosure 50 feet across and 25 feet deep (with the "missing" 50-foot side being formed by part of the house).

Before we go on... Notice that the problem in Example 2 came down to finding the absolute maximum value of A on the closed and bounded interval [0, 50]. As we noted in the preceding section, the table of values of A at its critical points and the endpoints of the interval gives us enough information to find the absolute maximum. ■

Let's stop for a moment and summarize the steps we've taken in these two examples.

Solving an Optimization Problem

1. **Identify the unknown(s), possibly with the aid of a diagram.** These are usually the quantities asked for in the problem.
2. **Identify the objective function.** This is the quantity you are asked to maximize or minimize. You should name it explicitly, as in "Let $S =$ surface area."
3. **Identify the constraint(s).** These can be equations relating variables or inequalities expressing limitations on the values of variables.
4. **State the optimization problem.** This will have the form "Maximize [minimize] the objective function subject to the constraint(s)."
5. **Eliminate extra variables.** If the objective function depends on several variables, solve the constraint equations to express all variables in terms of one particular variable. Substitute these expressions into the objective function to rewrite it as a function of a single variable. In short, if there is only one constraint equation:

   Solve the constraint for one of the unknowns and substitute into the objective.

   Also substitute the expressions into any inequality constraints to help determine the domain of the objective function.
6. **Find the absolute maximum (or minimum) of the objective function.** Use the techniques of the preceding section.

Now for some further examples.

EXAMPLE 3 Maximizing Revenue

Cozy Carriage Company builds baby strollers. Using market research, the company estimates that if it sets the price of a stroller at p dollars, then it can sell $q = 300{,}000 - 10p^2$ strollers per year.* What price will bring in the greatest annual revenue?

* This equation is, of course, the demand equation for the baby strollers. However, coming up with a suitable demand equation in real life is hard, to say the least. In this regard, the very entertaining and also insightful article, *Camels and Rubber Duckies* by Joel Spolsky at www.joelonsoftware.com/articles/CamelsandRubberDuckies.html is a must-read.

Solution The question we are asked identifies our main unknown, the price p. However, there is another quantity that we do not know, q, the number of strollers the company will sell per year. The question also identifies the objective function, revenue, which is

$$R = pq.$$

Including the equality constraint given to us, that $q = 300{,}000 - 10p^2$, and the "reality" inequality constraints $p \geq 0$ and $q \geq 0$, we can write our problem as

$$\text{Maximize } R = pq \text{ subject to } q = 300{,}000 - 10p^2,\ p \geq 0, \text{ and } q \geq 0.$$

We are given q in terms of p, so let's substitute to eliminate q:

$$R = pq = p(300{,}000 - 10p^2) = 300{,}000p - 10p^3.$$

Substituting in the inequality $q \geq 0$, we get

$$300{,}000 - 10p^2 \geq 0.$$

Thus, $p^2 \leq 30{,}000$, which gives $-100\sqrt{3} \leq p \leq 100\sqrt{3}$. When we combine this with $p \geq 0$, we get the following restatement of our problem:

$$\text{Maximize } R(p) = 300{,}000p - 10p^3 \text{ such that } 0 \leq p \leq 100\sqrt{3}.$$

We solve this problem in much the same way we did the preceding one. We calculate $R'(p) = 300{,}000 - 30p^2$. Setting $300{,}000 - 30p^2 = 0$, we find one stationary point at $p = 100$. There are no singular points and we have the endpoints $p = 0$ and $p = 100\sqrt{3}$. Putting these points in a table and computing the corresponding values of R, we get the following:

p	0	100	$100\sqrt{3}$
$R(p)$	0	20,000,000	0

Thus, Cozy Carriage should price its strollers at \$100 each, which will bring in the largest possible revenue of \$20,000,000.

Figure 22

EXAMPLE 4 Optimizing Resources

The Metal Can Company has an order to make cylindrical cans with a volume of 250 cubic centimeters. What should be the dimensions of the cans in order to use the least amount of metal in their production?

Solution We are asked to find the dimensions of the cans. It is traditional to take as the dimensions of a cylinder the height h and the radius of the base r, as in Figure 22.

We are also asked to minimize the amount of metal used in the can, which is the area of the surface of the cylinder. We can look up the formula or figure it out ourselves: Imagine removing the circular top and bottom and then cutting vertically and flattening out the hollow cylinder to get a rectangle, as shown in Figure 23.

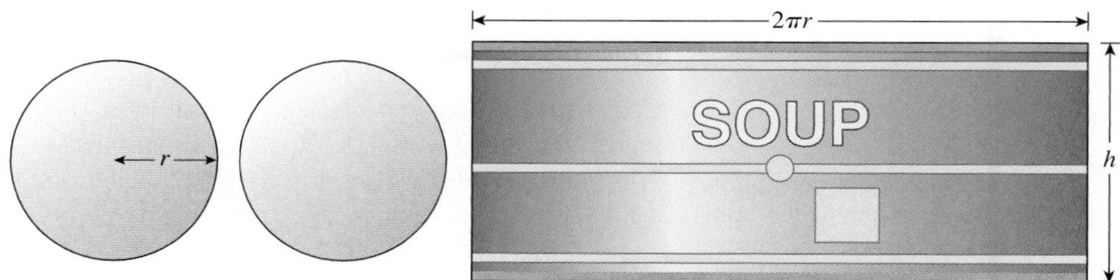

Figure 23

Our objective function is the (total) surface area S of the can. The area of each disc is πr^2, while the area of the rectangular piece is $2\pi rh$. Thus, our objective function is

$$S = 2\pi r^2 + 2\pi rh.$$

As usual, there is a constraint: The volume must be exactly 250 cubic centimeters. The formula for the volume of a cylinder is $V = \pi r^2 h$, so

$$\pi r^2 h = 250.$$

It is easiest to solve this constraint for h in terms of r:

$$h = \frac{250}{\pi r^2}.$$

Substituting in the objective function, we get

$$S = 2\pi r^2 + 2\pi r\frac{250}{\pi r^2} = 2\pi r^2 + \frac{500}{r}.$$

Now r cannot be negative or 0, but it can become very large (a very wide but very short can could have the right volume). We therefore take the domain of $S(r)$ to be $(0, +\infty)$, so our mathematical problem is as follows:

$$\text{Minimize } S(r) = 2\pi r^2 + \frac{500}{r} \text{ subject to } r > 0.$$

Now we calculate

$$S'(r) = 4\pi r - \frac{500}{r^2}.$$

To find stationary points, we set this equal to 0 and solve:

$$\begin{aligned} 4\pi r - \frac{500}{r^2} &= 0 \\ 4\pi r &= \frac{500}{r^2} \\ 4\pi r^3 &= 500 \\ r^3 &= \frac{125}{\pi}. \end{aligned}$$

So

$$r = \sqrt[3]{\frac{125}{\pi}} = \frac{5}{\sqrt[3]{\pi}} \approx 3.41.$$

The corresponding surface area is approximately $S(3.41) \approx 220$. There are no singular points or endpoints in the domain.

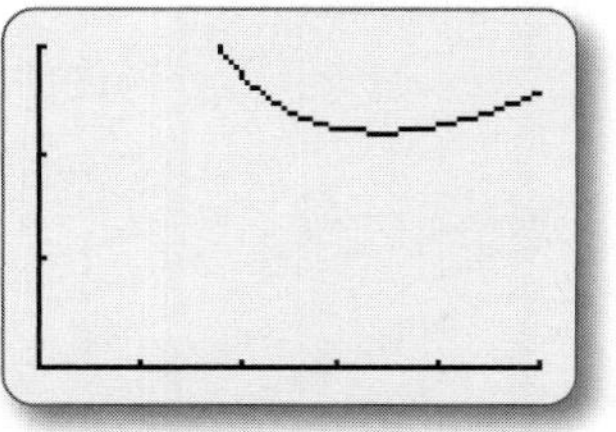

Figure 24

using Technology

To see how S behaves near the one stationary point, let's graph it in a viewing window with interval [0, 5] for r and [0, 300] for S. The result is Figure 24.

From the graph we can clearly see that the smallest surface area occurs at the stationary point at $r \approx 3.41$. The height of the can will be

$$h = \frac{250}{\pi r^2} \approx 6.83.$$

Thus, the can that uses the least amount of metal has a height of approximately 6.83 centimeters and a radius of approximately 3.41 centimeters. Such a can will use approximately 220 square centimeters of metal.

Before we go on... We obtained the value of r in Example 4 by solving the equation

$$4\pi r = \frac{500}{r^2}.$$

This time, let us do things differently: Divide both sides by 4π to obtain

$$r = \frac{500}{4\pi r^2} = \frac{125}{\pi r^2}$$

and compare what we got with the expression for h:

$$h = \frac{250}{\pi r^2},$$

which we see is exactly twice the expression for r. Put another way, the height is exactly equal to the diameter so that the can looks square when viewed from the side. Have you ever seen cans with that shape? Why do you think most cans do not have this shape? ■

EXAMPLE 5 Allocation of Labor

The Gym Sock Company manufactures cotton athletic socks. Production is partially automated through the use of robots. Daily operating costs amount to \$50 per laborer and \$30 per robot. The number of pairs of socks the company can manufacture in a day is given by a Cobb-Douglas* production formula

* Cobb-Douglas production formulas were discussed in Section 4.6.

$$q = 50n^{0.6}r^{0.4},$$

where q is the number of pairs of socks that can be manufactured by n laborers and r robots. Assuming that the company wishes to produce 1,000 pairs of socks per day at a minimum cost, how many laborers and how many robots should it use?

Solution The unknowns are the number of laborers n and the number of robots r. The objective is to minimize the daily cost:

$$C = 50n + 30r.$$

The constraints are given by the daily quota

$$1{,}000 = 50n^{0.6}r^{0.4}$$

and the fact that n and r are nonnegative. We solve the constraint equation for one of the variables; let's solve for n:

$$n^{0.6} = \frac{1{,}000}{50r^{0.4}} = \frac{20}{r^{0.4}}.$$

Taking the $1/0.6$ power of both sides gives

$$n = \left(\frac{20}{r^{0.4}}\right)^{1/0.6} = \frac{20^{1/0.6}}{r^{0.4/0.6}} = \frac{20^{5/3}}{r^{2/3}} \approx \frac{147.36}{r^{2/3}}.$$

Substituting in the objective equation gives us the cost as a function of r:

$$C(r) \approx 50\left(\frac{147.36}{r^{2/3}}\right) + 30r$$
$$= 7{,}368r^{-2/3} + 30r.$$

The only remaining constraint on r is that $r > 0$. To find the minimum value of $C(r)$, we first take the derivative:

$$C'(r) \approx -4{,}912r^{-5/3} + 30.$$

Setting this equal to zero, we solve for r:

$$r^{-5/3} \approx 0.006107$$
$$r \approx (0.006107)^{-3/5} \approx 21.3.$$

The corresponding cost is $C(21.3) \approx \$1{,}600$. There are no singular points or endpoints in the domain of C.

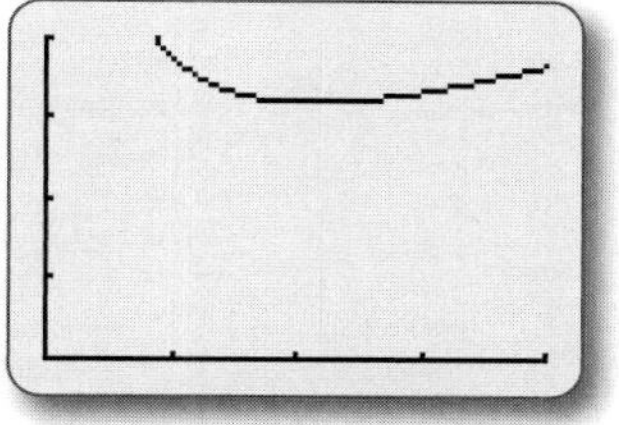

Figure 25

using Technology

To see how C behaves near its stationary point, let's draw its graph in a viewing window with an interval of [0, 40] for r and [0, 2,000] for C. The result is Figure 25.

From the graph we can see that C does have its minimum at the stationary point. The corresponding value of n is

$$n \approx \frac{147.36}{r^{2/3}} \approx 19.2.$$

At this point, our solution appears to be this: Use (approximately) 19.2 laborers and (approximately) 21.3 robots to meet the manufacturing quota at a minimum cost. However, we are not interested in fractions of robots or people, so we need to find integer solutions for n and r. If we round these numbers, we get the solution $(n, r) = (19, 21)$. However, a quick calculation shows that

$$q = 50(19)^{0.6}(21)^{0.4} \approx 989 \text{ pairs of socks,}$$

which fails to meet the quota of 1,000. Thus, we need to round at least one of the quantities n and r *upward* in order to meet the quota. The three possibilities, with corresponding values of q and C, are as follows:

$$(n, r) = (20, 21), \text{ with } q \approx 1{,}020 \text{ and } C = \$1{,}630$$
$$(n, r) = (19, 22), \text{ with } q \approx 1{,}007 \text{ and } C = \$1{,}610$$
$$(n, r) = (20, 22), \text{ with } q \approx 1{,}039 \text{ and } C = \$1{,}660.$$

Of these, the solution that meets the quota at a minimum cost is $(n, r) = (19, 22)$. Thus, the Gym Sock Co. should use 19 laborers and 22 robots, at a cost of $50 \times 19 + 30 \times 22 = \$1{,}610$, to manufacture $50 \times 19^{0.6} \times 22^{0.4} \approx 1{,}007$ pairs of socks.

FAQs

Constraints and Objectives

Q: *How do I know whether or not there are constraints in an applied optimization problem?*

A: There are usually at least *inequality* constraints; the variables usually represent real quantities, such as length or number of items, and so cannot be negative, leading to constraints like $x \geq 0$ (or $0 \leq x \leq 100$ in the event that there is an upper limit). *Equation* constraints usually arise when there is more than one unknown in the objective, and dictate how one unknown is related to others; as in, say "the length is twice the width," or "the demand is 8 divided by the price" (a demand equation).

Q: *How do I know what to use as the objective, and what to use as the constraint(s)?*

A: To identify the objective, look for a phrase such as "find the maximum (or minimum) value of." The amount you are trying to maximize or minimize is the objective. For example,

- . . . *at the least cost*. . . The objective function is the equation for cost, $C = \ldots.$
- . . . *the greatest area*. . . The objective function is the equation for area, $A = \ldots.$

To determine the constraint *inequalities*, ask yourself what limitations are placed on the unknown variables as above—are they nonnegative? are there upper limits? To identify the constraint *equations*, look for sentences that dictate restrictions in the form of relationships between the variables, as in the answer to the first question above.

5.2 EXERCISES

Access end-of-section exercises online at **www.webassign.net** 

5.3 Higher Order Derivatives: Acceleration and Concavity

The **second derivative** is simply the derivative of the derivative function. To explain why we would be interested in such a thing, we start by discussing one of its interpretations.

Acceleration

Recall that if $s(t)$ represents the position of a car at time t, then its velocity is given by the derivative: $v(t) = s'(t)$. But one rarely drives a car at a constant speed; the velocity itself may be changing. The rate at which the velocity is changing is the **acceleration**. Because the derivative measures the rate of change, acceleration is the derivative of velocity: $a(t) = v'(t)$. Because v is the derivative of s, we can express the acceleration in terms of s:

$$a(t) = v'(t) = (s')'(t) = s''(t).$$

That is, a is the derivative of the derivative of s; in other words, the second derivative of s, which we write as s''. (In this context you will often hear the derivative s' referred to as the **first derivative**.)

Second Derivative, Acceleration

If a function f has a derivative that is in turn differentiable, then its **second derivative** is the derivative of the derivative of f, written as f''. If $f''(a)$ exists, we say that f is **twice differentiable at** $x = a$.

Quick Examples

1. If $f(x) = x^3 - x$, then $f'(x) = 3x^2 - 1$, so $f''(x) = 6x$ and $f''(-2) = -12$.
2. If $f(x) = 3x + 1$, then $f'(x) = 3$, so $f''(x) = 0$.
3. If $f(x) = e^x$, then $f'(x) = e^x$, so $f''(x) = e^x$ as well.

The **acceleration** of a moving object is the derivative of its velocity—that is, the second derivative of the position function.

Quick Example

If t is time in hours and the position of a car at time t is $s(t) = t^3 + 2t^2$ miles, then the car's velocity is $v(t) = s'(t) = 3t^2 + 4t$ miles per hour and its acceleration is $a(t) = s''(t) = v'(t) = 6t + 4$ miles per hour per hour.

Differential Notation for the Second Derivative

We have written the second derivative of $f(x)$ as $f''(x)$. We could also use differential notation:

$$f''(x) = \frac{d^2f}{dx^2}.$$

This notation comes from writing the second derivative as the derivative of the derivative in differential notation:

$$f''(x) = \frac{d}{dx}\left[\frac{df}{dx}\right] = \frac{d^2f}{dx^2}.$$

Similarly, if $y = f(x)$, we write $f''(x)$ as $\frac{d}{dx}\left[\frac{dy}{dx}\right] = \frac{d^2y}{dx^2}$. For example, if $y = x^3$, then $\frac{d^2y}{dx^2} = 6x$.

An important example of acceleration is the acceleration due to gravity.

EXAMPLE 1 Acceleration Due to Gravity

According to the laws of physics, the height of an object near the surface of the earth falling in a vacuum from an initial rest position 100 feet above the ground under the influence of gravity is approximately

$$s(t) = 100 - 16t^2 \text{ feet}$$

in t seconds. Find its acceleration.

Solution The velocity of the object is

$$v(t) = s'(t) = -32t \text{ ft/sec}. \qquad \text{Differential notation: } v = \frac{ds}{dt} = -32t \text{ ft/sec}$$

The reason for the negative sign is that the height of the object is decreasing with time, so its velocity is negative. Hence, the acceleration is

$$a(t) = s''(t) = -32 \text{ ft/sec}^2. \qquad \text{Differential notation: } a = \frac{d^2s}{dt^2} = -32 \text{ ft/sec}^2$$

(We write ft/sec^2 as an abbreviation for feet/second/second—that is, feet per second per second. It is often read "feet per second squared.") Thus, the *downward* velocity is increasing by 32 ft/sec every second. We say that 32 ft/sec^2 is the **acceleration due to gravity**. In the absence of air resistance, all falling bodies near the surface of the earth, no matter what their weight, will fall with this acceleration.*

Before we go on... In very careful experiments using balls rolling down inclined planes, Galileo made one of his most important discoveries—that the acceleration due to gravity is constant and does not depend on the weight or composition of the object falling.† A famous, though probably apocryphal, story has him dropping cannonballs of different weights off the Leaning Tower of Pisa to prove his point.§ ■

* On other planets the acceleration due to gravity is different. For example, on Jupiter, it is about three times as large as on Earth.

† An interesting aside: Galileo's experiments depended on getting extremely accurate timings. Because the timepieces of his day were very inaccurate, he used the most accurate time measurement he could: He sang and used the beat as his stopwatch.

§ Here is a true story: The point was made again during the Apollo 15 mission to the moon (July 1971) when astronaut David R. Scott dropped a feather and a hammer from the same height. The moon has no atmosphere, so the two hit the surface of the moon simultaneously.

EXAMPLE 2 Acceleration of Sales

For the first 15 months after the introduction of a new video game, the total sales can be modeled by the curve

$$S(t) = 20e^{0.4t} \text{ units sold,}$$

where t is the time in months since the game was introduced. After about 25 months total sales follow more closely the curve

$$S(t) = 100{,}000 - 20e^{17-0.4t}.$$

How fast are total sales accelerating after 10 months? How fast are they accelerating after 30 months? What do these numbers mean?

Solution By acceleration we mean the rate of change of the rate of change, which is the second derivative. During the first 15 months, the first derivative of sales is

$$\frac{dS}{dt} = 8e^{0.4t}$$

and so the second derivative is

$$\frac{d^2S}{dt^2} = 3.2e^{0.4t}.$$

Thus, after 10 months the acceleration of sales is

$$\left.\frac{d^2S}{dt^2}\right|_{t=10} = 3.2e^4 \approx 175 \text{ units/month/month, or units/month}^2.$$

We can also compute total sales

$$S(10) = 20e^4 \approx 1{,}092 \text{ units}$$

and the rate of change of sales

$$\left.\frac{dS}{dt}\right|_{t=10} = 8e^4 \approx 437 \text{ units/month.}$$

What do these numbers mean? By the end of the tenth month, a total of 1,092 video games have been sold. At that time the game is selling at the rate of 437 units per month. This rate of sales is increasing by 175 units per month per month. More games will be sold each month than the month before.

Analysis of the sales after 30 months is done similarly, using the formula

$$S(t) = 100{,}000 - 20e^{17-0.4t}.$$

The derivative is

$$\frac{dS}{dt} = 8e^{17-0.4t}$$

and the second derivative is

$$\frac{d^2S}{dt^2} = -3.2e^{17-0.4t}.$$

After 30 months,

$$S(30) = 100{,}000 - 20e^{17-12} \approx 97{,}032 \text{ units}$$

$$\left.\frac{dS}{dt}\right|_{t=30} = 8e^{17-12} \approx 1{,}187 \text{ units/month}$$

$$\left.\frac{d^2S}{dt^2}\right|_{t=30} = -3.2e^{17-12} \approx -475 \text{ units/month}^2.$$

By the end of the thirtieth month, 97,032 video games have been sold, the game is selling at a rate of 1,187 units per month, and the rate of sales is *decreasing* by 475 units per month. Fewer games are sold each month than the month before.

Geometric Interpretation of Second Derivative: Concavity

The first derivative of f tells us where the graph of f is rising [where $f'(x) > 0$] and where it is falling [where $f'(x) < 0$]. The second derivative tells in what direction the graph of f *curves* or *bends*. Consider the graphs in Figures 26 and 27.

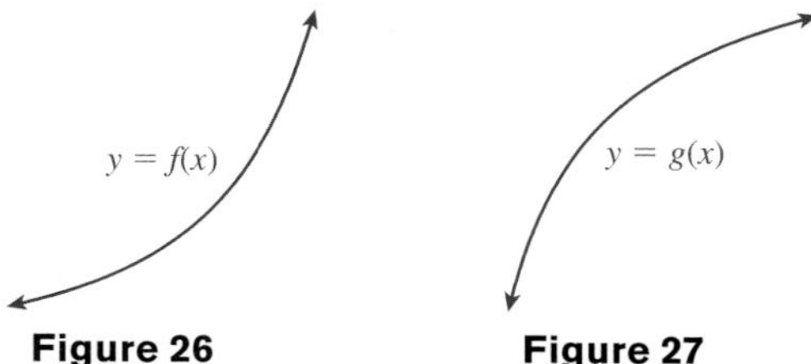

Figure 26 **Figure 27**

Think of a car driving from left to right along each of the roads shown in the two figures. A car driving along the graph of f in Figure 26 will turn to the left (upward); a car driving along the graph of g in Figure 27 will turn to the right (downward). We say that the graph of f is **concave up** and the graph of g is **concave down**. Now think about the derivatives of f and g. The derivative $f'(x)$ starts small but *increases* as the graph gets steeper. Because $f'(x)$ is increasing, its derivative $f''(x)$ must be positive. On the other hand, $g'(x)$ *decreases* as we go to the right. Because $g'(x)$ is decreasing, its derivative $g''(x)$ must be negative. Summarizing, we have the following.

Concavity and the Second Derivative

A curve is **concave up** if its slope is increasing, in which case the second derivative is positive. A curve is **concave down** if its slope is decreasing, in which case the second derivative is negative. A point in the domain of f where the graph of f changes concavity, from concave up to concave down or vice versa, is called a **point of inflection**. At a point of inflection, the second derivative is either zero or undefined.

Locating Points of Inflection

To locate possible points of inflection, list points where $f''(x) = 0$ and also interior points where $f''(x)$ is not defined.

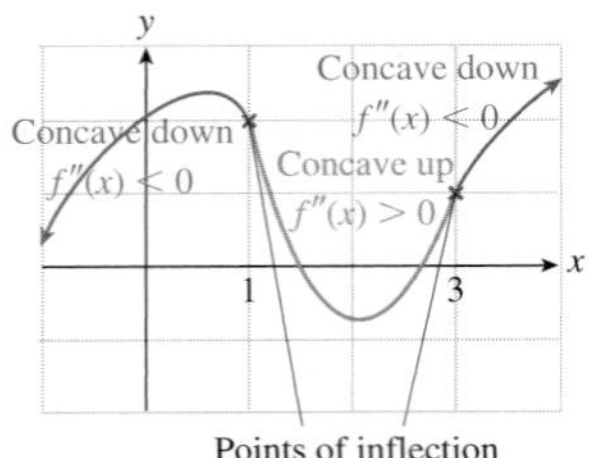

Figure 28

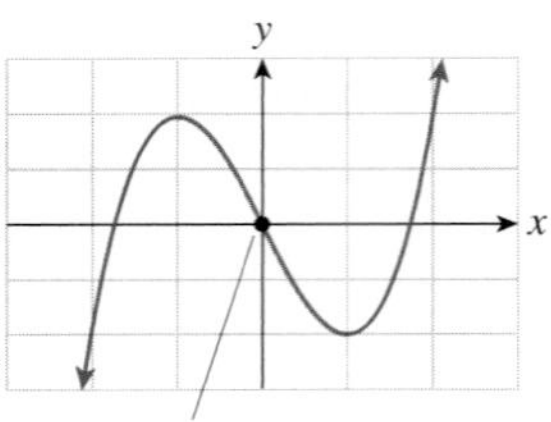

Figure 29

Quick Examples

1. The graph of the function f shown in Figure 28 is concave up when $1 < x < 3$, so $f''(x) > 0$ for $1 < x < 3$. It is concave down when $x < 1$ and $x > 3$, so $f''(x) < 0$ when $x < 1$ and $x > 3$. It has points of inflection at $x = 1$ and $x = 3$.
2. Consider $f(x) = x^3 - 3x$, whose graph is shown in Figure 29. $f''(x) = 6x$ is negative when $x < 0$ and positive when $x > 0$. The graph of f is concave down when $x < 0$ and concave up when $x > 0$. f has a point of inflection at $x = 0$, where the second derivative is 0.

The following example shows one of the reasons it's useful to look at concavity.

EXAMPLE 3 Inflation

Figure 30 shows the value of the U.S. Consumer Price Index (CPI) from January 2010 through April 2011.[1]

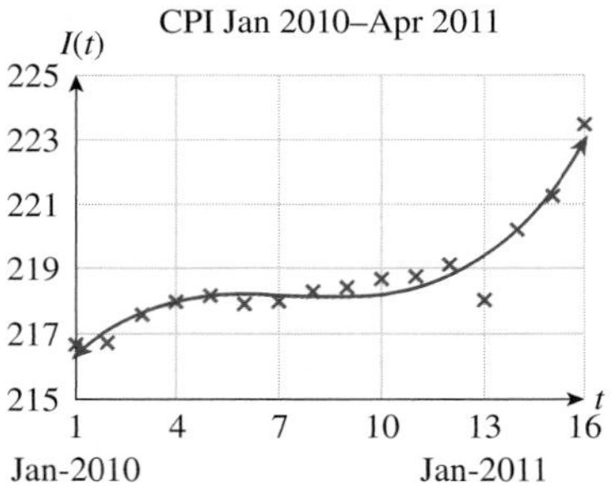

Figure 30

The approximating curve shown on the figure is given by

$$I(t) = 0.0081t^3 - 0.18t^2 + 1.3t + 215 \qquad (1 \le t \le 16),$$

where t is time in months ($t = 1$ represents January 2010). When the CPI is increasing, the U.S. economy is **experiencing inflation**. In terms of the model, this means that the derivative is positive: $I'(t) > 0$. Notice that $I'(t) > 0$ for most of the period shown (the graph is sloping upward), so the U.S. economy experienced inflation for most of $1 \le t \le 16$. We could measure **inflation** by the first derivative $I'(t)$ of the CPI, but we traditionally measure it as a ratio:

$$\text{Inflation rate} = \frac{I'(t)}{I(t)}, \qquad \text{Relative rate of change of the CPI}$$

expressed as a percentage per unit time (per month in this case).

a. Use the model to estimate the inflation rate in March 2010.

b. Was inflation slowing or speeding up in March 2010?

c. When was inflation slowing? When was inflation speeding up? When was inflation slowest?

[1]The CPI is compiled by the Bureau of Labor Statistics and is based upon a 1982 value of 100. For instance, a CPI of 200 means the CPI has doubled since 1982. Source: InflationData.com (www.inflationdata.com).

Solution

a. We need to compute $I'(t)$:

$$I'(t) = 0.0243t^2 - 0.36t + 1.3.$$

Thus, the inflation rate in March 2010 was given by

$$\text{Inflation rate} = \frac{I'(3)}{I(3)} = \frac{0.0243(3)^2 - 0.36(3) + 1.3}{0.0081(3)^3 - 0.18(3)^2 + 1.3(3) + 215}$$

$$= \frac{0.4387}{217.4987} \approx 0.0020,$$

or 0.20% per month.*

b. We say that inflation is "slowing" when the CPI is decelerating ($I''(t) < 0$; the index rises at a slower rate or falls at a faster rate†). Similarly, inflation is "speeding up" when the CPI is accelerating ($I''(t) > 0$; the index rises at a faster rate or falls at a slower rate). From the formula for $I'(t)$, the second derivative is

$$I''(t) = 0.0486t - 0.36$$
$$I''(3) = 0.0486(3) - 0.36 = -0.2142.$$

Because this quantity is negative, we conclude that inflation was slowing down in March 2010.

c. When inflation is slowing, $I''(t)$ is negative, so the graph of the CPI is concave down. When inflation is speeding up, it is concave up. At the point at which it switches, there is a point of inflection (Figure 31).

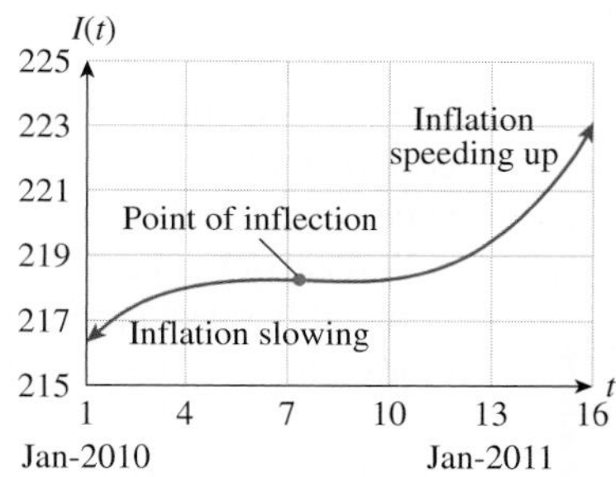

Figure 31

The point of inflection occurs when $I''(t) = 0$; that is,

$$0.0486t - 0.36 = 0$$
$$t = \frac{0.36}{0.0486} \approx 7.4.$$

Thus, inflation was slowing when $t < 7.4$ (that is, until around the middle of July), and speeding up when $t > 7.4$ (after that time). Inflation was slowest at the point when it stopped slowing down and began to speed up, $t \approx 7.4$ (in fact there was slight deflation at that particular point as $I'(7.4)$ is negative); notice that the graph has the least slope at that point.

* The 0.20% monthly inflation rate corresponds to a $12 \times 0.20 =$ 2.40% annual inflation rate. This result could be obtained directly by changing the units of the t-axis from months to years and then redoing the calculation.

† When the CPI is falling, the inflation rate is negative and we experience *deflation*.

EXAMPLE 4 The Point of Diminishing Returns

After the introduction of a new video game, the total worldwide sales are modeled by the curve

$$S(t) = \frac{1}{1 + 50e^{-0.2t}} \text{ million units sold,}$$

where t is the time in months since the game was introduced (compare Example 2). The graphs of $S(t)$, $S'(t)$, and $S''(t)$ are shown in Figure 32. Where is the graph of S concave up, and where is it concave down? Where are any points of inflection? What does this all mean?

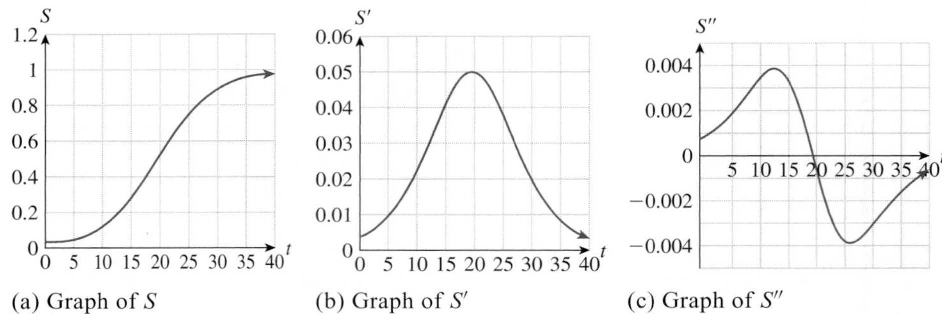

(a) Graph of S (b) Graph of S' (c) Graph of S''

Figure 32

using Technology

You can use a TI-83/84 Plus, a downloadable Excel sheet at the Website, or the Function Evaluator and Grapher at the Website to graph the second derivative of the function in Example 4:

TI-83/84 Plus

```
Y1=1/(1+50*e^(-0.2X))
Y2=nDeriv(Y1,X,X)
Y3=nDeriv(Y2,X,X)
```

Website

www.WanerMath.com

In the Function Evaluator and Grapher, enter the functions as shown.

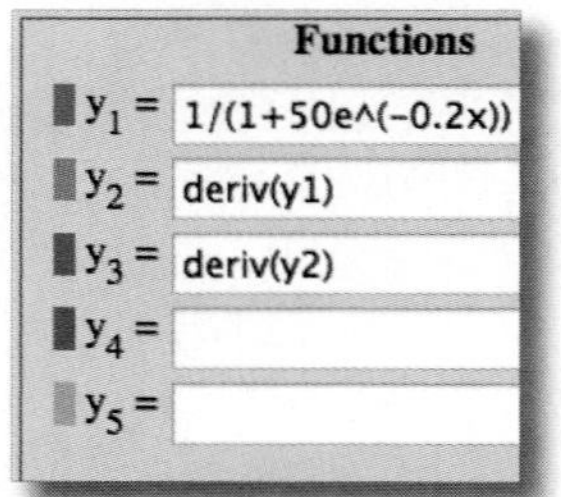

(Use Ymin = 0, yMax = 0.1 for a nice view of S' and Ymin = −0.01 and yMax = 0.01 for S''.)

For an Excel utility, try:

On Line Utilities
↓
Excel First and Second Derivative Graphing Utility

(This utility needs macros, so ensure they are enabled.)

Solution Look at the graph of S. We see that the graph of S is concave up in the early months and then becomes concave down later. The point of inflection, where the concavity changes, is somewhere between 15 and 25 months.

Now look at the graph of S''. This graph crosses the t-axis very close to $t = 20$, is positive before that point, and negative after that point. Because positive values of S'' indicate S is concave up and negative values concave down, we conclude that the graph of S is concave up for about the first 20 months; that is, for $0 < t < 20$ and concave down for $20 < t < 40$. The concavity switches at the point of inflection, which occurs at about $t = 20$ (when $S''(t) = 0$; a more accurate answer is $t \approx 19.56$).

What does this all mean? Look at the graph of S', which shows sales per unit time, or monthly sales. From this graph we see that monthly sales are increasing for $t < 20$: more units are being sold each month than the month before. Monthly sales reach a peak of 0.05 million = 50,000 games per month at the point of inflection $t = 20$ and then begin to drop off. Thus, the point of inflection occurs at the time when monthly sales stop increasing and start to fall off; that is, the time when monthly sales peak. The point of inflection is sometimes called the **point of diminishing returns**. Although the total sales figure continues to rise (see the graph of S: game units continue to be sold), the *rate* at which units are sold starts to drop. (See Figure 33.)

Figure 33

The Second Derivative Test for Relative Extrema

The second derivative often gives us a way of knowing whether or not a stationary point is a relative extremum. Figure 34 shows a graph with two stationary points: a relative maximum at $x = a$ and a relative minimum at $x = b$.

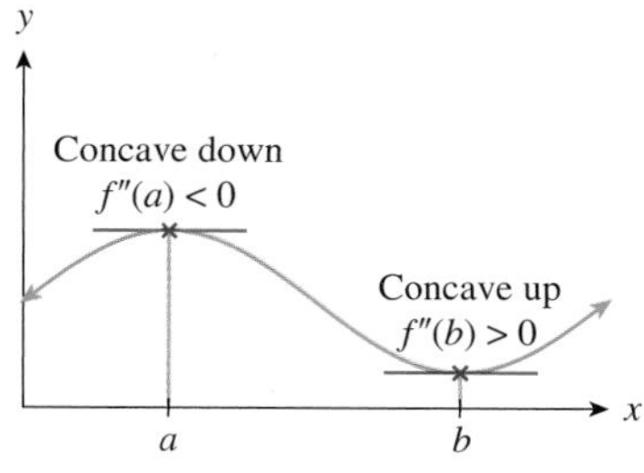

Figure 34

Notice that the curve is *concave down* at the relative maximum ($x = a$), so that $f''(a) < 0$, and *concave up* at the relative minimum ($x = b$), so that $f''(b) > 0$. This suggests the following (compare the First Derivative Test in Section 5.1).

Second Derivative Test for Relative Extrema

Suppose that the function f has a stationary point at $x = c$, and that $f''(c)$ exists. Determine the sign of $f''(c)$.

1. If $f''(c) > 0$, then f has a relative minimum at $x = c$.

2. If $f''(c) < 0$, then f has a relative maximum at $x = c$.

If $f''(c) = 0$, then the test is inconclusive and you need to use one of the methods of Section 5.1 (such as the first derivative test) to determine whether or not f has a relative extremum at $x = c$.

Quick Examples

1. $f(x) = x^2 - 2x$ has $f'(x) = 2x - 2$ and hence a stationary point at $x = 1$. $f''(x) = 2$, and so $f''(1) = 2$, which is positive, so f has a relative minimum at $x = 1$.

2. Let $f(x) = x^3 - 3x^2 - 9x$. Then

$f'(x) = 3x^2 - 6x - 9 = 3(x + 1)(x - 3)$

Stationary points at $x = -1, x = 3$

$f''(x) = 6x - 6$

$f''(-1) = -12$, so there is a relative maximum at $x = -1$

$f''(3) = 12$, so there is a relative minimum at $x = 3$.

3. $f(x) = x^4$ has $f'(x) = 4x^3$ and hence a stationary point at $x = 0$. $f''(x) = 12x^2$ and so $f''(0) = 0$, telling us that the second derivative test is inconclusive. However, we can see from the graph of f or the first derivative test that f has a minimum at $x = 0$.

Higher Order Derivatives

There is no reason to stop at the second derivative; we could once again take the *derivative* of the second derivative to obtain the **third derivative**, f''', and we could take the derivative once again to obtain the **fourth derivative**, written $f^{(4)}$, and then continue to obtain $f^{(5)}$, $f^{(6)}$, and so on (assuming we get a differentiable function at each stage).

Higher Order Derivatives

We define

$$f'''(x) = \frac{d}{dx}[f''(x)]$$

$$f^{(4)}(x) = \frac{d}{dx}[f'''(x)]$$

$$f^{(5)}(x) = \frac{d}{dx}[f^{(4)}(x)],$$

and so on, assuming all these derivatives exist.

Different Notations

$$f'(x),\ f''(x),\ f'''(x),\ f^{(4)}(x), \ldots,\ f^{(n)}(x), \ldots$$

$$\frac{df}{dx}, \frac{d^2f}{dx^2}, \frac{d^3f}{dx^3}, \frac{d^4f}{dx^4}, \ldots, \frac{d^nf}{dx^n}, \ldots$$

$$\frac{dy}{dx}, \frac{d^2y}{dx^2}, \frac{d^3y}{dx^3}, \frac{d^4y}{dx^4}, \ldots, \frac{d^ny}{dx^n}, \ldots \qquad \text{When } y = f(x)$$

$$y, y', y'', y''', y^{(4)}, \ldots, y^{(n)}, \ldots \qquad \text{When } y = f(x)$$

Quick Examples

1. If $f(x) = x^3 - x$, then $f'(x) = 3x^2 - 1$, $f''(x) = 6x$, $f'''(x) = 6$, $f^{(4)}(x) = f^{(5)}(x) = \cdots = 0$.
2. If $f(x) = e^x$, then $f'(x) = e^x$, $f''(x) = e^x$, $f'''(x) = e^x$, $f^{(4)}(x) = f^{(5)}(x) = \cdots = e^x$.

Q: *We know that the second derivative can be interpreted as acceleration. How do we interpret the third derivative; and the fourth, fifth, and so on?*

A: Think of a car traveling down the road (with position $s(t)$ at time t) in such a way that its acceleration $\frac{d^2s}{dt^2}$ is changing with time (for instance, the driver may be slowly increasing pressure on the accelerator, causing the car to accelerate at a greater and greater rate). Then $\frac{d^3s}{dt^3}$ is the rate of change of acceleration. $\frac{d^4s}{dt^4}$ would then be the *acceleration* of the acceleration, and so on.

Q: *How are these higher order derivatives reflected in the graph of a function f?*

A: Because the concavity is measured by f'', its derivative f''' tells us the rate of change of concavity. Similarly, $f^{(4)}$ would tell us the *acceleration* of concavity, and so on. These properties are very subtle and hard to discern by simply looking at the curve; the higher the order, the more subtle the property. There is a remarkable theorem by Taylor* that tells us that, for a large class of functions (including polynomial, exponential, logarithmic, and trigonometric functions) the values of all orders of derivative $f(a)$, $f'(a)$, $f''(a)$, $f'''(a)$, and so on at the single point $x = a$ are enough to describe the entire graph (even at points very far from $x = a$)! In other words, the smallest piece of a graph near any point a contains sufficient information to "clone" the entire graph!

* Brook Taylor (1685–1731) was an English mathematician.

FAQs

Interpreting Points of Inflection and Using the Second Derivative Test

Q: *It says in Example 4 that monthly sales reach a maximum at the point of inflection (second derivative is zero), but the second derivative test says that, for a maximum, the second derivative must be negative. What is going on here?*

A: What is a maximum in Example 4 is the *rate of change of* sales: which is measured in sales per unit time (monthly sales in the example). In other words, it is the *derivative* of the total sales function that is a maximum, so we located the maximum by setting its derivative (which is the *second* derivative of total sales) equal to zero. In general: To find relative (stationary) extrema of the *original* function, set $f'(x)$ equal to zero and solve for x as usual. The second derivative test can then be used to test the stationary point obtained. To find relative (stationary) extrema of the *rate of change of f*, set $f''(x) = 0$ and solve for x.

Q: *I used the second derivative test and it was inconclusive. That means that there is neither a relative maximum nor a relative minimum at $x = a$, right?*

A: Wrong. If (as is often the case) the second derivative is zero at a stationary point, all it means is that the second derivative test itself cannot determine whether the given point is a relative maximum, minimum, or neither. For instance, $f(x) = x^4$ has a stationary minimum at $x = 0$, but the second derivative test is inconclusive. In such cases, one should use another test (such as the first derivative test) to decide if the point is a relative maximum, minimum, or neither.

5.3 EXERCISES

Access end-of-section exercises online at **www.webassign.net** ENHANCED WebAssign

5.4 Analyzing Graphs

Mathematical curves are beautiful—their subtle form can be imitated by only the best of artists—and calculus gives us the tools we need to probe their secrets. While it is easy to use graphing technology to draw a graph, we must use calculus to understand what we are seeing. Following is a list of some of the most interesting features of the graph of a function.

Features of a Graph

1. ***The x- and y-intercepts:*** If $y = f(x)$, find the x-intercept(s) by setting $y = 0$ and solving for x; find the y-intercept by setting $x = 0$ and solving for y:

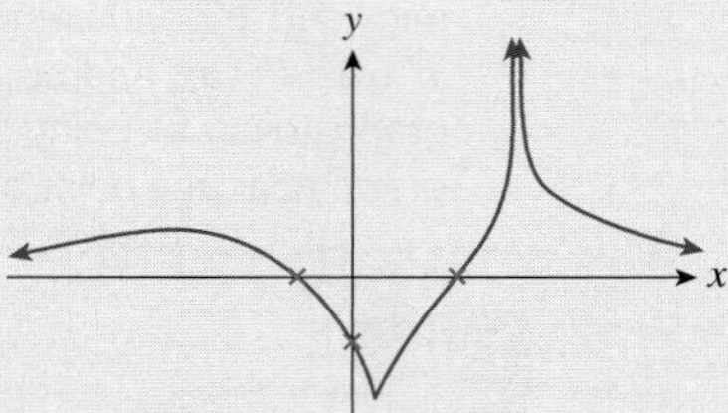

2. ***Extrema:*** Use the techniques of Section 5.1 to locate the maxima and minima:

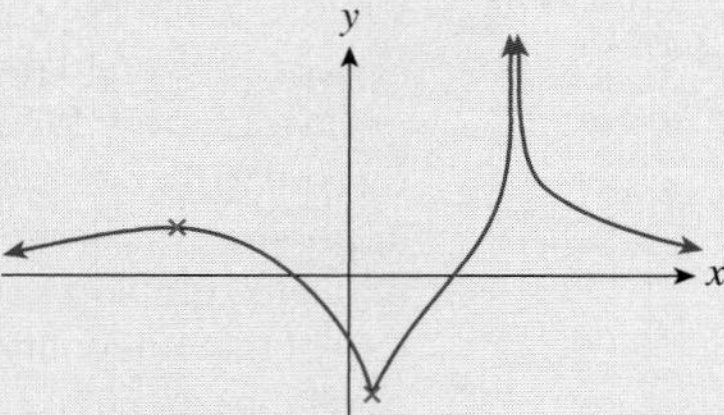

3. ***Points of inflection:*** Use the techniques of Section 5.2 to locate the points of inflection:

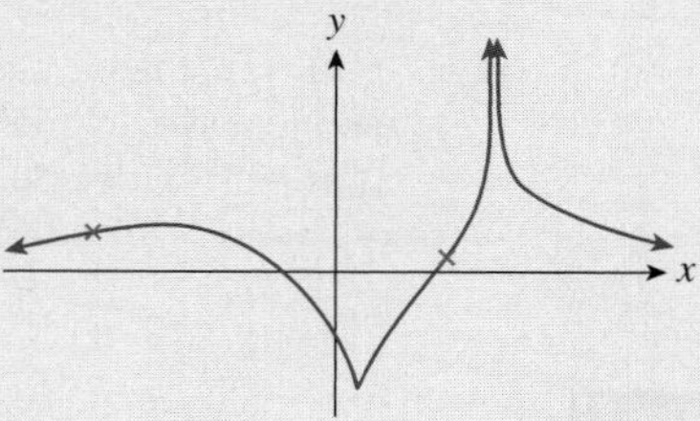

4. ***Behavior near points where the function is not defined:*** If $f(x)$ is not defined at $x = a$, consider $\lim_{x\to a^-} f(x)$ and $\lim_{x\to a^+} f(x)$ to see how the graph of f behaves as x approaches a:

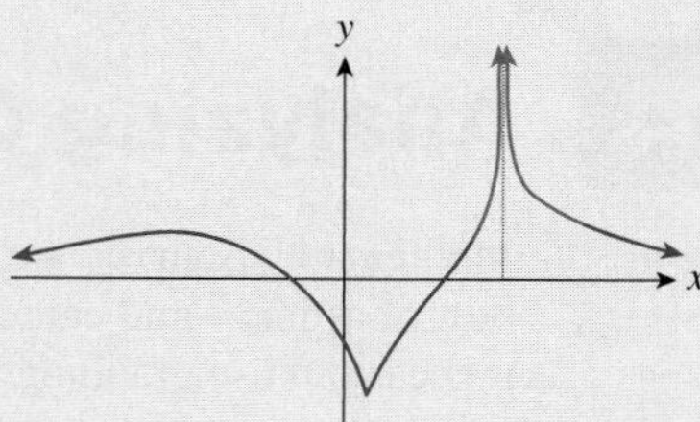

5. ***Behavior at infinity:*** Consider $\lim_{x\to-\infty} f(x)$ and $\lim_{x\to+\infty} f(x)$ if appropriate, to see how the graph of f behaves far to the left and right:

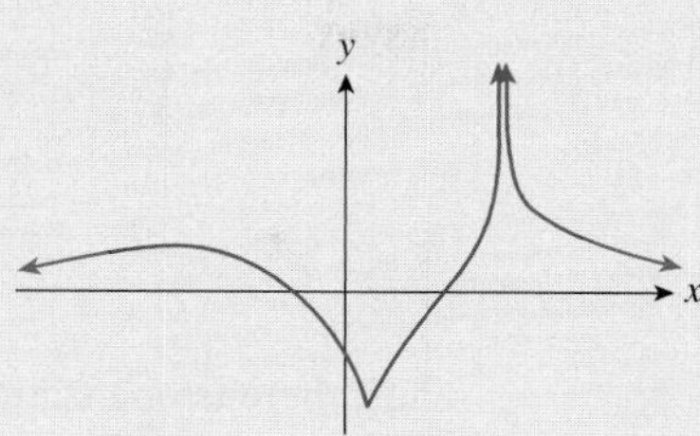

Note It is sometimes difficult or impossible to solve all of the equations that come up in Steps 1, 2, and 3 of the previous analysis. As a consequence, we might not be able to say exactly where the x-intercept, extrema, or points of inflection are. When this happens, we will use graphing technology to assist us in determining accurate numerical approximations. ■

EXAMPLE 1 Analyzing a Graph

Analyze the graph of $f(x) = \dfrac{1}{x} - \dfrac{1}{x^2}$.

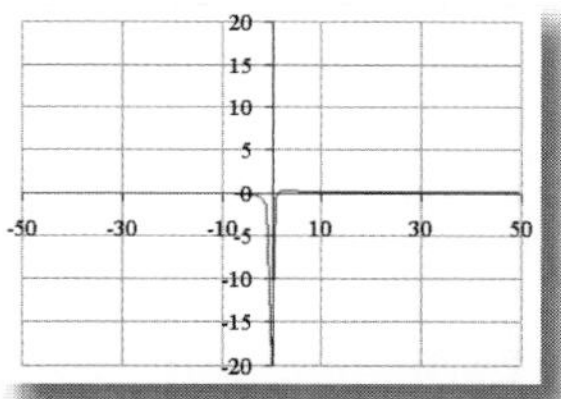

$-50 \le x \le 50, -20 \le y \le 20$

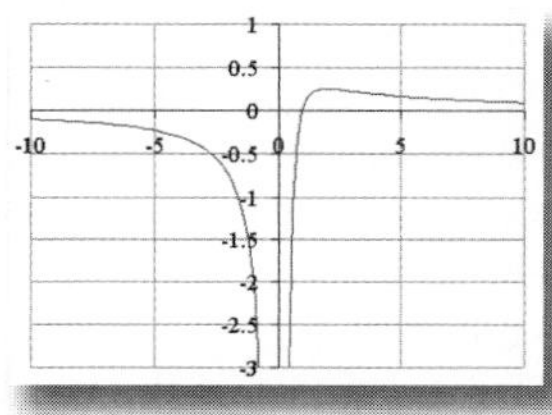

$-10 \le x \le 10, -3 \le y \le 1$

Figure 35

Solution The graph, as drawn using graphing technology, is shown in Figure 35, using two different viewing windows. (Note that $x = 0$ is not in the domain of f.) The second window in Figure 35 seems to show the features of the graph better than the first. Does the second viewing window include *all* the interesting features of the graph? Or are there perhaps some interesting features to the right of $x = 10$ or to the left of $x = -10$? Also, where exactly do features like maxima, minima, and points of inflection occur? In our five-step process of analyzing the interesting features of the graph, we will be able to sketch the curve by hand, and also answer these questions.

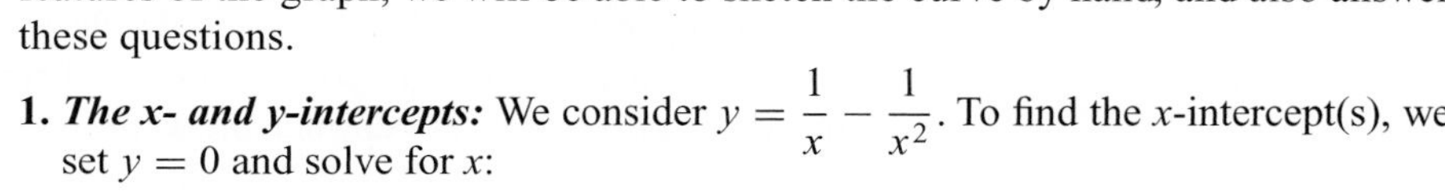

1. ***The x- and y-intercepts:*** We consider $y = \dfrac{1}{x} - \dfrac{1}{x^2}$. To find the x-intercept(s), we set $y = 0$ and solve for x:

$$0 = \frac{1}{x} - \frac{1}{x^2}$$

$$\frac{1}{x} = \frac{1}{x^2}.$$

Multiplying both sides by x^2 (we know that x cannot be zero, so we are not multiplying both sides by 0) gives

$$x = 1.$$

Thus, there is one x-intercept (which we can see in Figure 35) at $x = 1$.

For the y-intercept, we would substitute $x = 0$ and solve for y. However, we cannot substitute $x = 0$; because $f(0)$ is not defined, the graph does not meet the y-axis.

We add features to our freehand sketch as we go. Figure 36 shows what we have so far.

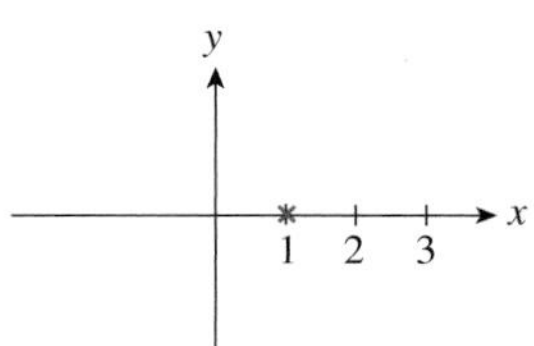

Figure 36

2. ***Relative extrema:*** We calculate $f'(x) = -\frac{1}{x^2} + \frac{2}{x^3}$. To find any stationary points, we set the derivative equal to 0 and solve for x:

$$-\frac{1}{x^2} + \frac{2}{x^3} = 0$$
$$\frac{1}{x^2} = \frac{2}{x^3}$$
$$x = 2.$$

Thus, there is one stationary point, at $x = 2$. We can use a test point to the right to determine that this stationary point is a relative maximum:

x	1 (Intercept)	2	3 (Test point)
$y = \frac{1}{x} - \frac{1}{x^2}$	0	$\frac{1}{4}$	$\frac{2}{9}$

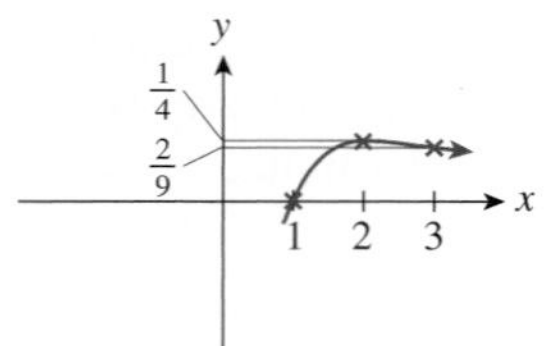

Figure 37

The only possible singular point is at $x = 0$ because $f'(0)$ is not defined. However, $f(0)$ is not defined either, so there are no singular points. Figure 37 shows our graph so far.

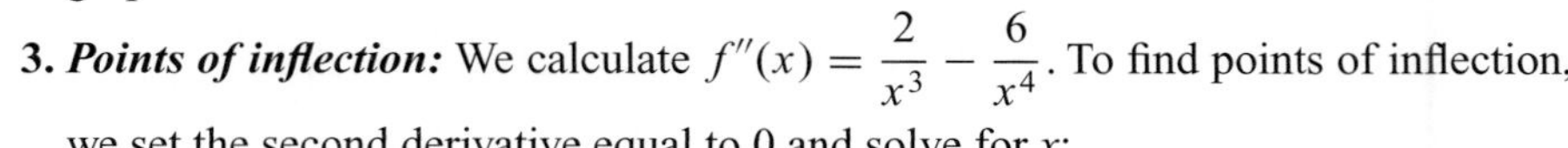

3. ***Points of inflection:*** We calculate $f''(x) = \frac{2}{x^3} - \frac{6}{x^4}$. To find points of inflection, we set the second derivative equal to 0 and solve for x:

$$\frac{2}{x^3} - \frac{6}{x^4} = 0$$
$$\frac{2}{x^3} = \frac{6}{x^4}$$
$$2x = 6$$
$$x = 3.$$

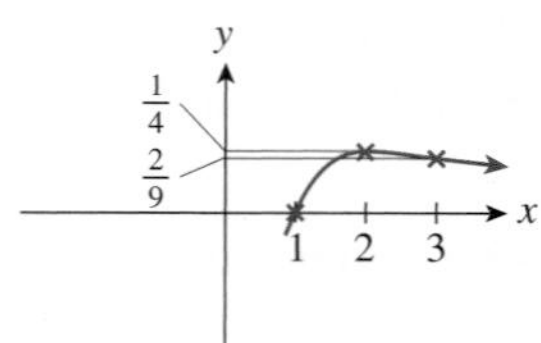

Figure 38

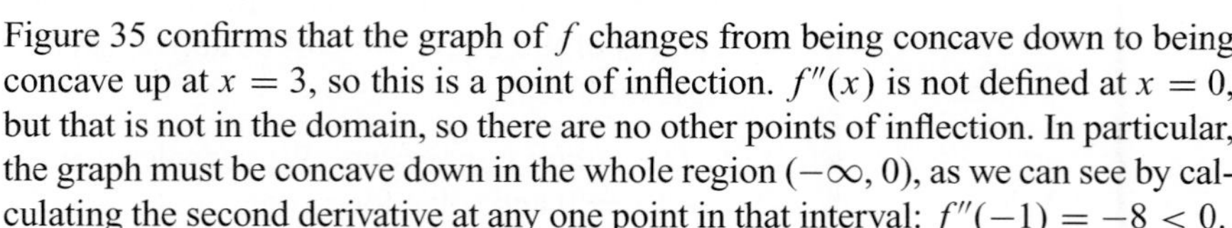

Figure 35 confirms that the graph of f changes from being concave down to being concave up at $x = 3$, so this is a point of inflection. $f''(x)$ is not defined at $x = 0$, but that is not in the domain, so there are no other points of inflection. In particular, the graph must be concave down in the whole region $(-\infty, 0)$, as we can see by calculating the second derivative at any one point in that interval: $f''(-1) = -8 < 0$.

Figure 38 shows our graph so far (we extended the curve near $x = 3$ to suggest a point of inflection at $x = 3$).

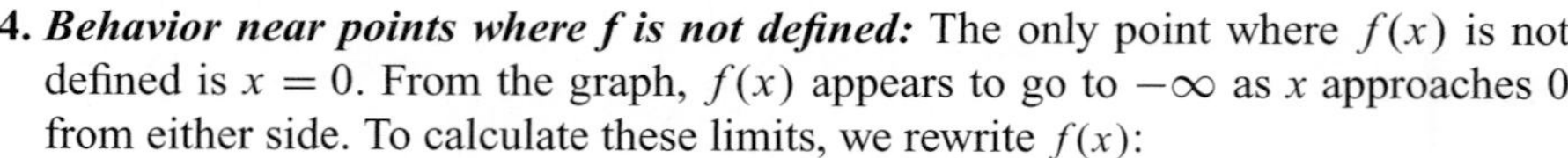

4. ***Behavior near points where f is not defined:*** The only point where $f(x)$ is not defined is $x = 0$. From the graph, $f(x)$ appears to go to $-\infty$ as x approaches 0 from either side. To calculate these limits, we rewrite $f(x)$:

$$f(x) = \frac{1}{x} - \frac{1}{x^2} = \frac{x-1}{x^2}.$$

Now, if x is close to 0 (on either side), the numerator $x - 1$ is close to -1 and the denominator is a very small but positive number. The quotient is therefore a negative number of very large magnitude. Therefore,

$$\lim_{x \to 0^-} f(x) = -\infty$$

and

$$\lim_{x \to 0^+} f(x) = -\infty.$$

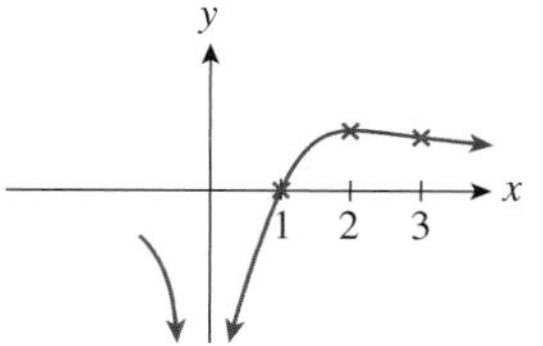

Figure 39

From these limits, we see the following:

(1) Immediately to the *left* of $x = 0$, the graph plunges down toward $-\infty$.

(2) Immediately to the *right* of $x = 0$, the graph also plunges down toward $-\infty$.

Figure 39 shows our graph with these features added. We say that f has a **vertical asymptote** at $x = 0$, meaning that the points on the graph of f get closer and closer to points on a vertical line (the y-axis in this case) further and further from the origin.

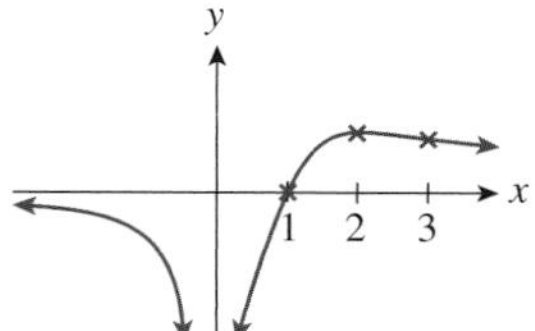

Figure 40

5. ***Behavior at infinity:*** Both $1/x$ and $1/x^2$ go to 0 as x goes to $-\infty$ or $+\infty$; that is,

$$\lim_{x \to -\infty} f(x) = 0$$

and

$$\lim_{x \to +\infty} f(x) = 0.$$

Thus, on the extreme left and right of our picture, the height of the curve levels off toward zero. Figure 40 shows the completed freehand sketch of the graph.

We say that f has a **horizontal asymptote** at $y = 0$. (Notice another thing: We haven't plotted a single point to the left of the y-axis, and yet we have a pretty good idea of what the curve looks like there! Compare the technology-drawn curve in Figure 35.)

In summary, there is one x-intercept at $x = 1$; there is one relative maximum (which, we can now see, is also an absolute maximum) at $x = 2$; there is one point of inflection at $x = 3$, where the graph changes from being concave down to concave up. There is a vertical asymptote at $x = 0$, on both sides of which the graph goes down toward $-\infty$, and a horizontal asymptote at $y = 0$.

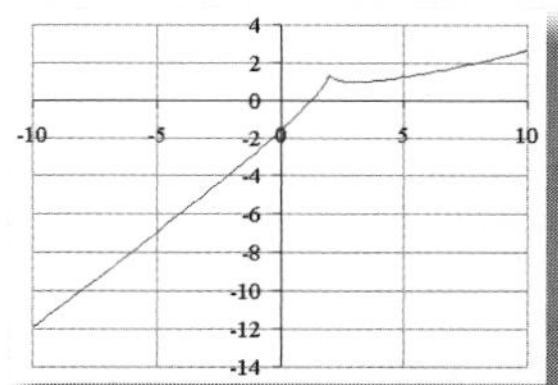

Technology:
`2*x/3-((x-2)^2)^(1/3)`

Figure 41

EXAMPLE 2 Analyzing a Graph

Analyze the graph of $f(x) = \dfrac{2x}{3} - (x - 2)^{2/3}$.

Solution Figure 41 shows a technology-generated version of the graph. Note that in the technology formulation $(x - 2)^{2/3}$ is written as $[(x - 2)^2]^{1/3}$ to avoid problems with some graphing calculators and Excel.

Let us now re-create this graph by hand, and in the process identify the features we see in Figure 41.

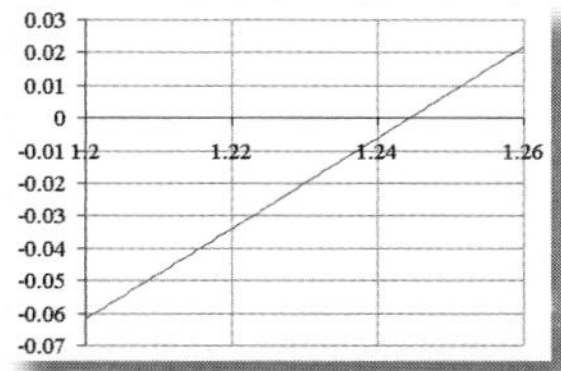

Figure 42

1. ***The x- and y-intercepts:*** We consider $y = \dfrac{2x}{3} - (x - 2)^{2/3}$. For the y-intercept, we set $x = 0$ and solve for y:

$$y = \frac{2(0)}{3} - (0 - 2)^{2/3} = -2^{2/3} \approx -1.59.$$

To find the x-intercept(s), we set $y = 0$ and solve for x. However, if we attempt this, we will find ourselves with a cubic equation that is hard to solve. (Try it!) Following the advice in the note on page 309, we use graphing technology to locate the x-intercept we see in Figure 41 by zooming in (Figure 42). From Figure 42, we find $x \approx 1.24$. We shall see in the discussion to follow that there can be no other x-intercepts.

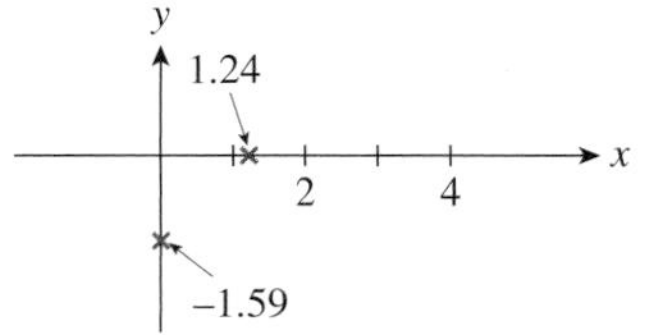

Figure 43

Figure 43 shows our freehand sketch so far.

2. ***Relative extrema:*** We calculate

$$f'(x) = \frac{2}{3} - \frac{2}{3}(x-2)^{-1/3}$$
$$= \frac{2}{3} - \frac{2}{3(x-2)^{1/3}}.$$

To find any stationary points, we set the derivative equal to 0 and solve for x:

$$\frac{2}{3} - \frac{2}{3(x-2)^{1/3}} = 0$$
$$(x-2)^{1/3} = 1$$
$$x - 2 = 1^3 = 1$$
$$x = 3.$$

To check for singular points, look for points where $f(x)$ is defined and $f'(x)$ is not defined. The only such point is $x = 2$: $f'(x)$ is not defined at $x = 2$, whereas $f(x)$ is defined there, so we have a singular point at $x = 2$.

x	2 (Singular point)	3 (Stationary point)	4 (Test point)
$y = \frac{2x}{3} - (x-2)^{2/3}$	$\frac{4}{3}$	1	1.079

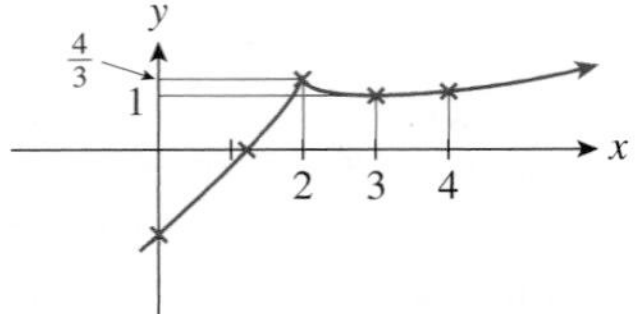

Figure 44

Figure 44 shows our graph so far.

We see that there is a singular relative maximum at (2, 4/3) (we will confirm that the graph eventually gets higher on the right) and a stationary relative minimum at $x = 3$.

3. ***Points of inflection:*** We calculate

$$f''(x) = \frac{2}{9(x-2)^{4/3}}.$$

To find points of inflection, we set the second derivative equal to 0 and solve for x. But the equation

$$0 = \frac{2}{9(x-2)^{4/3}}$$

has no solution for x, so there are no points of inflection on the graph.

4. ***Behavior near points where f is not defined:*** Because $f(x)$ is defined everywhere, there are no such points to consider. In particular, there are no vertical asymptotes.

5. ***Behavior at infinity:*** We estimate the following limits numerically:

$$\lim_{x \to -\infty} \left[\frac{2x}{3} - (x-2)^{2/3} \right] = -\infty$$

and

$$\lim_{x \to +\infty} \left[\frac{2x}{3} - (x-2)^{2/3} \right] = +\infty.$$

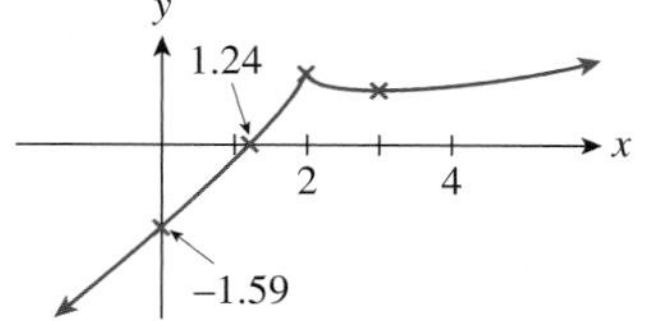

Figure 45

Thus, on the extreme left the curve goes down toward $-\infty$, and on the extreme right the curve rises toward $+\infty$. In particular, there are no horizontal asymptotes. (There can also be no other x-intercepts.)

Figure 45 shows the completed graph.

5.4 EXERCISES

Access end-of-section exercises online at **www.webassign.net**

5.5 Related Rates

We start by recalling some basic facts about the rate of change of a quantity:

Rate of Change of Q

If Q is a quantity changing over time t, then the derivative dQ/dt is the rate at which Q changes over time.

Quick Examples

1. If A is the area of an expanding circle, then dA/dt is the rate at which the area is increasing.
2. *Words:* The radius r of a sphere is currently 3 cm and increasing at a rate of 2 cm/sec.

   Symbols: $r = 3$ cm and $dr/dt = 2$ cm/sec.

In this section we are concerned with what are called **related rates** problems. In such a problem we have two (sometimes more) related quantities, we know the rate at which one is changing, and we wish to find the rate at which another is changing. A typical example is the following.

EXAMPLE 1 The Expanding Circle

The radius of a circle is increasing at a rate of 10 cm/sec. How fast is the area increasing at the instant when the radius has reached 5 cm?

Solution We have two related quantities: the radius of the circle, r, and its area, A. The first sentence of the problem tells us that r is increasing at a certain rate. When we see a sentence referring to speed or change, it is very helpful to rephrase the sentence using the phrase "the rate of change of." Here, we can say

The rate of change of r is 10 cm/sec.

Because the rate of change is the derivative, we can rewrite this sentence as the equation

$$\frac{dr}{dt} = 10.$$

Similarly, the second sentence of the problem asks how A is changing. We can rewrite that question:

What is the rate of change of A when the radius is 5 cm?

Using mathematical notation, the question is:

What is $\frac{dA}{dt}$ *when* $r = 5$?

Thus, knowing one rate of change, dr/dt, we wish to find a related rate of change, dA/dt. To find exactly how these derivatives are related, we need the equation relating the variables, which is

$$A = \pi r^2.$$

To find the relationship between the derivatives, we take the derivative of both sides of this equation *with respect to t*. On the left we get dA/dt. On the right we need to remember that r is a function of t and use the chain rule. We get

$$\frac{dA}{dt} = 2\pi r \frac{dr}{dt}.$$

Now we substitute the given values $r = 5$ and $dr/dt = 10$. This gives

$$\left.\frac{dA}{dt}\right|_{r=5} = 2\pi(5)(10) = 100\pi \approx 314 \text{ cm}^2/\text{sec}.$$

Thus, the area is increasing at the rate of 314 cm^2/sec when the radius is 5 cm.

We can organize our work as follows:

Solving a Related Rates Problem

A. The Problem

1. List the related, changing quantities.
2. Restate the problem in terms of rates of change. Rewrite the problem using mathematical notation for the changing quantities and their derivatives.

B. The Relationship

1. Draw a diagram, if appropriate, showing the changing quantities.
2. Find an equation or equations relating the changing quantities.
3. Take the derivative with respect to time of the equation(s) relating the quantities to get the **derived equation(s)**, which relate the rates of change of the quantities.

C. The Solution

1. Substitute into the derived equation(s) the given values of the quantities and their derivatives.
2. Solve for the derivative required.

We can illustrate the procedure with the "ladder problem" found in almost every calculus textbook.

EXAMPLE 2 The Falling Ladder

Jane is at the top of a 5-foot ladder when it starts to slide down the wall at a rate of 3 feet per minute. Jack is standing on the ground behind her. How fast is the base of the ladder moving when it hits him if Jane is 4 feet from the ground at that instant?

Solution The first sentence talks about (the top of) the ladder sliding down the wall. Thus, one of the changing quantities is the height of the top of the ladder. The question asked refers to the motion of the base of the ladder, so another changing quantity is the distance of the base of the ladder from the wall. Let's record these variables and follow the outline above to obtain the solution.

A. The Problem

1. The changing quantities are

$h =$ height of the top of the ladder
$b =$ distance of the base of the ladder from the wall

2. We rephrase the problem in words, using the phrase "rate of change":

The rate of change of the height of the top of the ladder is -3 *feet per minute. What is the rate of change of the distance of the base from the wall when the top of the ladder is* 4 *feet from the ground?*

We can now rewrite the problem mathematically:

$$\frac{dh}{dt} = -3. \text{ Find } \frac{db}{dt} \text{ when } h = 4.$$

Figure 46

B. The Relationship

1. Figure 46 shows the ladder and the variables h and b. Notice that we put in the figure the fixed length, 5, of the ladder, but any changing quantities, like h and b, we leave as variables. We shall not use any specific values for h or b until the very end.

2. From the figure, we can see that h and b are related by the Pythagorean theorem:

$$h^2 + b^2 = 25.$$

3. Taking the derivative with respect to time of the equation above gives us the derived equation:

$$2h\frac{dh}{dt} + 2b\frac{db}{dt} = 0.$$

C. The Solution

1. We substitute the known values $dh/dt = -3$ and $h = 4$ into the derived equation:

$$2(4)(-3) + 2b\frac{db}{dt} = 0.$$

We would like to solve for db/dt, but first we need the value of b, which we can determine from the equation $h^2 + b^2 = 25$, using the value $h = 4$:

$$\begin{aligned} 16 + b^2 &= 25 \\ b^2 &= 9 \\ b &= 3. \end{aligned}$$

Substituting into the derived equation, we get

$$-24 + 2(3)\frac{db}{dt} = 0.$$

2. Solving for db/dt gives

$$\frac{db}{dt} = \frac{24}{6} = 4.$$

Thus, the base of the ladder is sliding away from the wall at 4 ft/min when it hits Jack.

EXAMPLE 3 Average Cost

The cost to manufacture x cellphones in a day is

$$C(x) = 10{,}000 + 20x + \frac{x^2}{10{,}000} \text{ dollars.}$$

The daily production level is currently $x = 5{,}000$ cellphones and is increasing at a rate of 100 units per day. How fast is the average cost changing?

Solution

A. The Problem

1. The changing quantities are the production level x and the average cost, $\bar{C}$.

2. We rephrase the problem as follows:

> *The daily production level is $x = 5{,}000$ units and the rate of change of x is 100 units/day. What is the rate of change of the average cost, $\bar{C}$?*

In mathematical notation,

$$x = 5{,}000 \text{ and } \frac{dx}{dt} = 100. \text{ Find } \frac{d\bar{C}}{dt}.$$

B. The Relationship

1. In this example the changing quantities cannot easily be depicted geometrically.

2. We are given a formula for the *total* cost. We get the *average* cost by dividing the total cost by x:

$$\bar{C} = \frac{C}{x}.$$

So,

$$\bar{C} = \frac{10{,}000}{x} + 20 + \frac{x}{10{,}000}.$$

3. Taking derivatives with respect to t of both sides, we get the derived equation:

$$\frac{d\bar{C}}{dt} = \left(-\frac{10{,}000}{x^2} + \frac{1}{10{,}000}\right)\frac{dx}{dt}.$$

C. The Solution

Substituting the values from part A into the derived equation, we get

$$\frac{d\bar{C}}{dt} = \left(-\frac{10{,}000}{5{,}000^2} + \frac{1}{10{,}000}\right)100$$

$$= -0.03 \text{ dollars/day.}$$

Thus, the average cost is decreasing by 3¢ per day.

The scenario in the following example is similar to Example 5 in Section 5.2.

EXAMPLE 4 Allocation of Labor

The Gym Sock Company manufactures cotton athletic socks. Production is partially automated through the use of robots. The number of pairs of socks the company can manufacture in a day is given by a Cobb-Douglas production formula:

$$q = 50n^{0.6}r^{0.4},$$

where q is the number of pairs of socks that can be manufactured by n laborers and r robots. The company currently produces 1,000 pairs of socks each day and employs 20 laborers. It is bringing one new robot on line every month. At what rate are laborers being laid off, assuming that the number of socks produced remains constant?

Solution

A. The Problem

1. The changing quantities are the number of laborers n and the number of robots r.

2. $\dfrac{dr}{dt} = 1$. Find $\dfrac{dn}{dt}$ when $n = 20$.

B. The Relationship

1. No diagram is appropriate here.

2. The equation relating the changing quantities:

$$1{,}000 = 50n^{0.6}r^{0.4} \quad \text{Productivity is constant at 1,000 pairs of socks each day.}$$

or

$$20 = n^{0.6}r^{0.4}.$$

3. The derived equation is

$$\begin{aligned} 0 &= 0.6n^{-0.4}\left(\frac{dn}{dt}\right)r^{0.4} + 0.4n^{0.6}r^{-0.6}\left(\frac{dr}{dt}\right) \\ &= 0.6\left(\frac{r}{n}\right)^{0.4}\left(\frac{dn}{dt}\right) + 0.4\left(\frac{n}{r}\right)^{0.6}\left(\frac{dr}{dt}\right). \end{aligned}$$

We solve this equation for dn/dt because we shall want to find dn/dt below and because the equation becomes simpler when we do this:

$$0.6\left(\frac{r}{n}\right)^{0.4}\left(\frac{dn}{dt}\right) = -0.4\left(\frac{n}{r}\right)^{0.6}\left(\frac{dr}{dt}\right)$$

$$\begin{aligned} \frac{dn}{dt} &= -\frac{0.4}{0.6}\left(\frac{n}{r}\right)^{0.6}\left(\frac{n}{r}\right)^{0.4}\left(\frac{dr}{dt}\right) \\ &= -\frac{2}{3}\left(\frac{n}{r}\right)\left(\frac{dr}{dt}\right). \end{aligned}$$

C. The Solution

Substituting the numbers in A into the last equation in B, we get

$$\frac{dn}{dt} = -\frac{2}{3}\left(\frac{20}{r}\right)(1).$$

We need to compute r by substituting the known value of n in the original formula:

$$20 = n^{0.6}r^{0.4}$$
$$20 = 20^{0.6}r^{0.4}$$
$$r^{0.4} = \frac{20}{20^{0.6}} = 20^{0.4}$$
$$r = 20.$$

Thus,

$$\frac{dn}{dt} = -\frac{2}{3}\left(\frac{20}{20}\right)(1) = -\frac{2}{3} \text{ laborers per month.}$$

The company is laying off laborers at a rate of $2/3$ per month, or two every three months.

We can interpret this result as saying that, at the current level of production and number of laborers, one robot is as productive as $2/3$ of a laborer, or 3 robots are as productive as 2 laborers.

5.5 EXERCISES

Access end-of-section exercises online at **www.webassign.net**

5.6 Elasticity

You manufacture an extremely popular brand of sneakers and want to know what will happen if you increase the selling price. Common sense tells you that demand will drop as you raise the price. But will the drop in demand be enough to cause your revenue to fall? Or will it be small enough that your revenue will rise because of the higher selling price? For example, if you raise the price by 1%, you might suffer only a 0.5% loss in sales. In this case, the loss in sales will be more than offset by the increase in price and your revenue will rise. In such a case, we say that the demand is **inelastic**, because it is not very sensitive to the increase in price. On the other hand, if your 1% price increase results in a 2% drop in demand, then raising the price will cause a drop in revenues. We then say that the demand is **elastic** because it reacts strongly to a price change.

We can use calculus to measure the response of demand to price changes if we have a demand equation for the item we are selling.* We need to know the *percentage drop in demand per percentage increase in price.* This ratio is called the **elasticity of demand**, or **price elasticity of demand**, and is usually denoted by E. Let's derive a formula for E in terms of the demand equation.

Assume that we have a demand equation

$$q = f(p),$$

where q stands for the number of items we would sell (per week, per month, or what have you) if we set the price per item at p. Now suppose we increase the price p by a very small amount, Δp. Then our percentage increase in price is $(\Delta p/p) \times 100\%$. This increase in p will presumably result in a decrease in the demand q. Let's denote

* Coming up with a good demand equation is not always easy. We saw in Chapter 1 that it is possible to find a linear demand equation if we know the sales figures at two different prices. However, such an equation is only a first approximation. To come up with a more accurate demand equation, we might need to gather data corresponding to sales at several different prices and use curve-fitting techniques like regression. Another approach would be an analytic one, based on mathematical modeling techniques that an economist might use.

That said, we refer you again to *Camels and Rubber Duckies* by Joel Spolsky at www.joelonsoftware.com/articles/CamelsandRubberDuckies.html just in case you think there is nothing more to demand curves.

this corresponding decrease in q by $-\Delta q$ (we use the minus sign because, by convention, Δq stands for the *increase* in demand). Thus, the percentage decrease in demand is $(-\Delta q/q) \times 100\%$.

Now E is the ratio

$$E = \frac{\text{Percentage decrease in demand}}{\text{Percentage increase in price}}$$

so

$$E = \frac{-\dfrac{\Delta q}{q} \times 100\%}{\dfrac{\Delta p}{p} \times 100\%}.$$

Canceling the 100%s and reorganizing, we get

$$E = -\frac{\Delta q}{\Delta p} \cdot \frac{p}{q}.$$

Q: *What small change in price will we use for Δp?*

A: It should probably be pretty small. If, say, we increased the price of sneakers to \$1 million per pair, the sales would likely drop to zero. But knowing this tells us nothing about how the market would respond to a modest increase in price. In fact, we'll do the usual thing we do in calculus and let Δp approach 0.

In the expression for E, if we let Δp go to 0, then the ratio $\Delta q/\Delta p$ goes to the derivative dq/dp. This gives us our final and most useful definition of the elasticity.

Price Elasticity of Demand

The **price elasticity of demand E** is the percentage rate of decrease of demand per percentage increase in price. E is given by the formula

$$E = -\frac{dq}{dp} \cdot \frac{p}{q}.$$

We say that the demand is **elastic** if $E > 1$, is **inelastic** if $E < 1$, and has **unit elasticity** if $E = 1$.

Quick Example

Suppose that the demand equation is $q = 20{,}000 - 2p$, where p is the price in dollars. Then

$$E = -(-2)\frac{p}{20{,}000 - 2p} = \frac{p}{10{,}000 - p}.$$

If $p = \$2{,}000$, then $E = 1/4$, and demand is inelastic at this price.

If $p = \$8{,}000$, then $E = 4$, and demand is elastic at this price.

If $p = \$5{,}000$, then $E = 1$, and the demand has unit elasticity at this price.

We are generally interested in the price that maximizes revenue and, in ordinary cases, the price that maximizes revenue must give unit elasticity. One way of seeing this is as follows:* If the demand is inelastic (which ordinarily occurs at a low unit price), then raising the price by a small percentage—1% say—results in a smaller percentage drop in demand. For example, in the Quick Example on the previous page, if $p = \$2{,}000$, then the demand would drop by only $\frac{1}{4}\%$ for every 1% increase in price. To see the effect on revenue, we use the fact† that, for small changes in price,

* For another—more rigorous—argument, see Exercise 29 associated with this section.

† See, for example, Exercise 53 associated with Section 5.5.

$$\begin{aligned}\text{Percentage change in revenue} &\approx \text{Percentage change in price} \\ &\quad + \text{Percentage change in demand} \\ &= 1 + \left(-\frac{1}{4}\right) = \frac{3}{4}\%.\end{aligned}$$

Thus, the revenue will increase by about 3/4%. Put another way:

If the demand is inelastic, raising the price increases revenue.

On the other hand, if the price is elastic (which ordinarily occurs at a high unit price), then increasing the price slightly will lower the revenue, so:

If the demand is elastic, lowering the price increases revenue.

The price that results in the largest revenue must therefore be at unit elasticity.

EXAMPLE 1 Price Elasticity of Demand: Dolls

Suppose that the demand equation for *Bobby Dolls* is given by $q = 216 - p^2$, where p is the price per doll in dollars and q is the number of dolls sold per week.

a. Compute the price elasticity of demand when $p = \$5$ and $p = \$10$, and interpret the results.

b. Find the range of prices for which the demand is elastic and the range for which the demand is inelastic.

c. Find the price at which the weekly revenue is maximized. What is the maximum weekly revenue?

using Technology

See the Technology Guides at the end of the chapter to find out how to automate computations like those in part (a) of Example 1 using a graphing calculator or Excel. Here is an outline:

TI-83/84 Plus

```
Y1=216-X^2
Y2=-nDeriv(Y1,X,X)*X/Y1
```

[2ND] [TABLE] Enter $x = 5$

[More details on page 331.]

Spreadsheet

Enter values of p: 4.9, 4.91, . . . , 5.0, 5.01, . . . , 5.1 in A5–A25.

In B5 enter `216-A5^2` and copy down to B25.

In C5 enter `=(A6-A5)/A5` and paste the formula in C5–D24.

In E5 enter =–D5/C5 and copy down to E24. This column contains the values of E for the values of p in column A.

[More details on page 331.]

Solution

a. The price elasticity of demand is

$$E = -\frac{dq}{dp} \cdot \frac{p}{q}.$$

Taking the derivative and substituting for q gives

$$E = 2p \cdot \frac{p}{216 - p^2} = \frac{2p^2}{216 - p^2}.$$

When $p = \$5$,

$$E = \frac{2(5)^2}{216 - 5^2} = \frac{50}{191} \approx 0.26.$$

Thus, when the price is set at \$5, the demand is dropping at a rate of 0.26% per 1% increase in the price. Because $E < 1$, the demand is inelastic at this price, so raising the price will increase revenue.

When $p = \$10$,

$$E = \frac{2(10)^2}{216 - 10^2} = \frac{200}{116} \approx 1.72.$$

Thus, when the price is set at \$10, the demand is dropping at a rate of 1.72% per 1% increase in the price. Because $E > 1$, demand is elastic at this price, so raising the price will decrease revenue; lowering the price will increase revenue.

b. and **c.** We answer part (c) first. Setting $E = 1$, we get

$$\frac{2p^2}{216 - p^2} = 1$$

$$p^2 = 72.$$

Thus, we conclude that the maximum revenue occurs when $p = \sqrt{72} \approx \$8.49$. We can now answer part (b): The demand is elastic when $p > \$8.49$ (the price is too high), and the demand is inelastic when $p < \$8.49$ (the price is too low). Finally, we calculate the maximum weekly revenue, which equals the revenue corresponding to the price of \$8.49:

$$R = qp = (216 - p^2)p = (216 - 72)\sqrt{72} = 144\sqrt{72} \approx \$1{,}222.$$

The concept of elasticity can be applied in other situations. In the following example we consider *income* elasticity of demand—the percentage increase in demand for a particular item per percentage increase in personal income.

EXAMPLE 2 Income Elasticity of Demand: Porsches

You are the sales director at *Suburban Porsche* and have noticed that demand for Porsches depends on income according to

$$q = 0.005e^{-0.05x^2+x} \qquad (1 \le x \le 10).$$

Here, x is the income of a potential customer in hundreds of thousands of dollars and q is the probability that the person will actually purchase a Porsche.* The **income elasticity of demand** is

$$E = \frac{dq}{dx}\frac{x}{q}.$$

* In other words, q is the fraction of visitors to your showroom having income x who actually purchase a Porsche.

Compute and interpret E for $x = 2$ and 9.

Solution

Q: *Why is there no negative sign in the formula?*

A: Because we anticipate that the demand will increase as income increases, the ratio

$$\frac{\text{Percentage increase in demand}}{\text{Percentage increase in income}}$$

will be positive, so there is no need to introduce a negative sign.

Turning to the calculation, since $q = 0.005e^{-0.05x^2+x}$,

$$\frac{dq}{dx} = 0.005e^{-0.05x^2+x}(-0.1x + 1)$$

and so

$$\begin{aligned} E &= \frac{dq}{dx}\frac{x}{q} \\ &= 0.005e^{-0.05x^2+x}(-0.1x+1)\frac{x}{0.005e^{-0.05x^2+x}} \\ &= x(-0.1x+1). \end{aligned}$$

When $x = 2$, $E = 2[-0.1(2) + 1)] = 1.6$. Thus, at an income level of \$200,000, the probability that a customer will purchase a Porsche increases at a rate of 1.6% per 1% increase in income.

When $x = 9$, $E = 9[-0.1(9) + 1)] = 0.9$. Thus, at an income level of \$900,000, the probability that a customer will purchase a Porsche increases at a rate of 0.9% per 1% increase in income.

5.6 EXERCISES

Access end-of-section exercises online at **www.webassign.net**

CHAPTER 5 REVIEW

KEY CONCEPTS

Website www.WanerMath.com

Go to the Website at www.WanerMath.com to find a comprehensive and interactive Web-based summary of Chapter 5.

5.1 Maxima and Minima

Relative maximum, relative minimum *p. 279*

Absolute maximum, absolute minimum *p. 279*

Stationary points, singular points, endpoints *p. 281*

Finding and classifying maxima and minima *p. 282*

First derivative test for relative extrema *p. 283*

Extreme value theorem *p. 287*

Using technology to locate approximate extrema *p. 288*

5.2 Applications of Maxima and Minima

Minimizing average cost *p. 290*

Maximizing area *p. 291*

Steps in solving optimization problems *p. 293*

Maximizing revenue *p. 293*

Optimizing resources *p. 294*

Allocation of labor *p. 296*

5.3 Higher Order Derivatives: Acceleration and Concavity

The second derivative of a function f is the derivative of the derivative of f, written as f'' *p. 298*

The acceleration of a moving object is the second derivative of the position function *p. 298*

Acceleration due to gravity *p. 299*

Acceleration of sales *p. 300*

Concave up, concave down, point of inflection *p. 301*

Locating points of inflection *p. 301*

Application to inflation *p. 302*

Second derivative test for relative extrema *p. 305*

Higher order derivatives *p. 306*

5.4 Analyzing Graphs

Features of a graph: x- and y-intercepts, relative extrema, points of inflection; behavior near points where the function is not defined, behavior at infinity *pp. 308–309*

Analyzing a graph *p. 309*

5.5 Related Rates

If Q is a quantity changing over time t, then the derivative dQ/dt is the rate at which Q changes over time *p. 313*

The expanding circle *p. 313*

Steps in solving related rates problems *p. 314*

The falling ladder *p. 314*

Average cost *p. 316*

Allocation of labor *p. 317*

5.6 Elasticity

Price elasticity of demand $E = -\dfrac{dq}{dp} \cdot \dfrac{p}{q}$; demand is elastic if $E > 1$, inelastic if $E < 1$, has unit elasticity if $E = 1$ *p. 319*

Computing and interpreting elasticity, and maximizing revenue *p. 320*

Using technology to compute elasticity *p. 320*

Income elasticity of demand *p. 321*

REVIEW EXERCISES

In Exercises 1–8, find all the relative and absolute extrema of the given functions on the given domain (if supplied) or on the largest possible domain (if no domain is supplied).

1. $f(x) = 2x^3 - 6x + 1$ on $[-2, +\infty)$

2. $f(x) = x^3 - x^2 - x - 1$ on $(-\infty, \infty)$

3. $g(x) = x^4 - 4x$ on $[-1, 1]$

4. $f(x) = \dfrac{x+1}{(x-1)^2}$ for $-2 \le x \le 2, x \ne 1$

5. $g(x) = (x-1)^{2/3}$ **6.** $g(x) = x^2 + \ln x$ on $(0, +\infty)$

7. $h(x) = \dfrac{1}{x} + \dfrac{1}{x^2}$ **8.** $h(x) = e^{x^2} + 1$

In Exercises 9–12, the graph of the function f or its derivative is given. Find the approximate x-coordinates of all relative extrema and points of inflection of the original function f (if any).

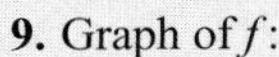

9. Graph of f:

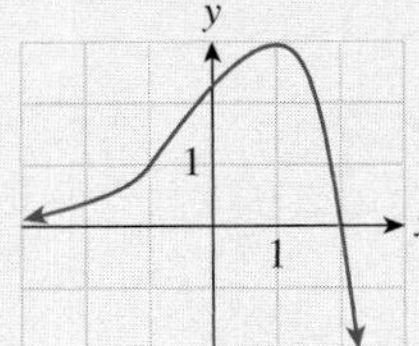

10. Graph of f:

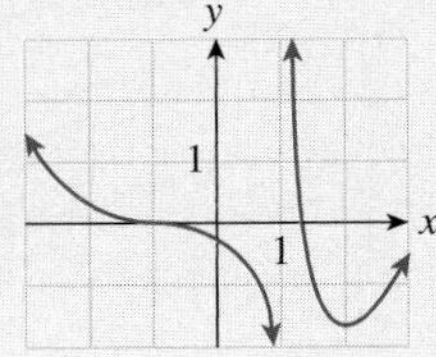

11. Graph of f':

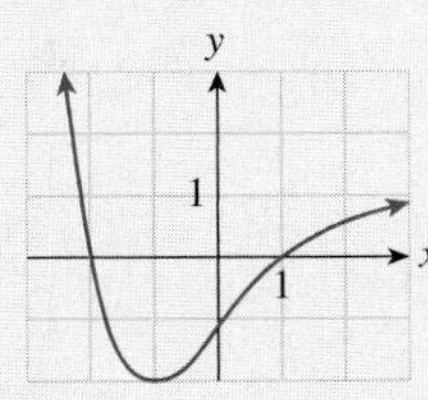

12. Graph of f':

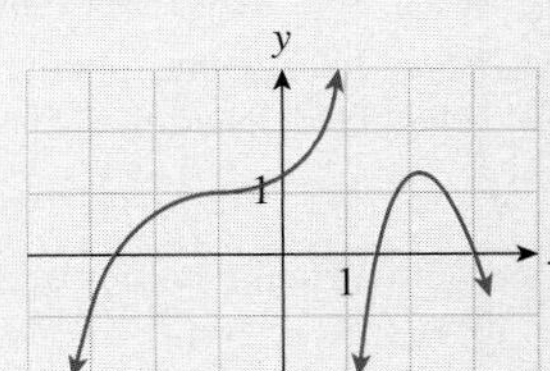

In Exercises 13 and 14, the graph of the second derivative of a function f is given. Find the approximate x-coordinates of all points of inflection of the original function f (if any).

13. Graph of f''

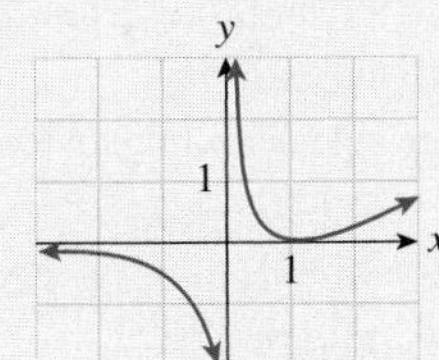

14. Graph of f''

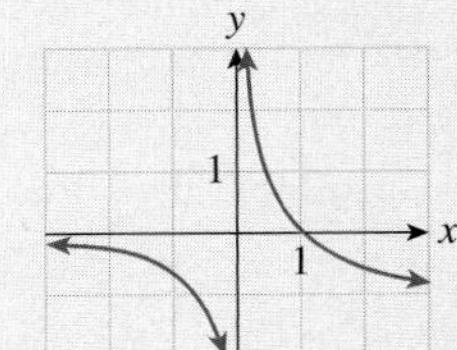

*In Exercises 15 and 16, the position s of a point (in meters) is given as a function of time t (in seconds). Find **(a)** its acceleration as a function of t and **(b)** its acceleration at the specified time.*

15. $s = \frac{2}{3t^2} - \frac{1}{t}; t = 1$ **16.** $s = \frac{4}{t^2} - \frac{3t}{4}; t = 2$

In Exercises 17–22, sketch the graph of the given function, indicating all relative and absolute extrema and points of inflection. Find the coordinates of these points exactly, where possible. Also indicate any horizontal and vertical asymptotes.

17. $f(x) = x^3 - 12x$ on $[-2, +\infty)$

18. $g(x) = x^4 - 4x$ on $[-1, 1]$

19. $f(x) = \frac{x^2 - 3}{x^3}$

20. $f(x) = (x - 1)^{2/3} + \frac{2x}{3}$

21. $g(x) = (x - 3)\sqrt{x}$

22. $g(x) = (x + 3)\sqrt{x}$

APPLICATIONS: OHaganBooks.com

23. ***Revenue*** Demand for the latest best-seller at OHaganBooks .com, *A River Burns through It*, is given by

$$q = -p^2 + 33p + 9 \qquad (18 \le p \le 28)$$

copies sold per week when the price is p dollars. What price should the company charge to obtain the largest revenue?

24. ***Revenue*** Demand for *The Secret Loves of John O*, a romance novel by Margó Dufón that flopped after two weeks on the market, is given by

$$q = -2p^2 + 5p + 6 \qquad (0 \le p \le 3.3)$$

copies sold per week when the price is p dollars. What price should OHaganBooks charge to obtain the largest revenue?

25. ***Profit*** Taking into account storage and shipping, it costs OHaganBooks.com

$$C = 9q + 100$$

dollars to sell q copies of *A River Burns through It* in a week (see Exercise 23).

a. If demand is as in Exercise 23, express the weekly profit earned by OHaganBooks.com from the sale of *A River Burns through It* as a function of unit price p.

b. What price should the company charge to get the largest weekly profit? What is the maximum possible weekly profit?

c. Compare your answer in part (b) with the price the company should charge to obtain the largest revenue (Exercise 23). Explain any difference.

26. ***Profit*** Taking into account storage and shipping, it costs OHaganBooks.com

$$C = 3q$$

dollars to sell q copies of Margó Dufón's *The Secret Loves of John O* in a week (see Exercise 24).

a. If demand is as in Exercise 24, express the weekly profit earned by OHaganBooks.com from the sale of *The Secret Loves of John O* as a function of unit price p.

b. What price should the company charge to get the largest weekly profit? What is the maximum possible weekly profit?

c. Compare your answer in part (b) with the price the company should charge to obtain the largest revenue (Exercise 24). Explain any difference.

27. ***Box Design*** The sales department at OHaganBooks.com, which has decided to send chocolate lobsters to each of its customers, is trying to design a shipping box with a square base. It has a roll of cardboard 36 inches wide from which to make the boxes. Each box will be obtained by cutting out corners from a rectangle of cardboard as shown in the following diagram:

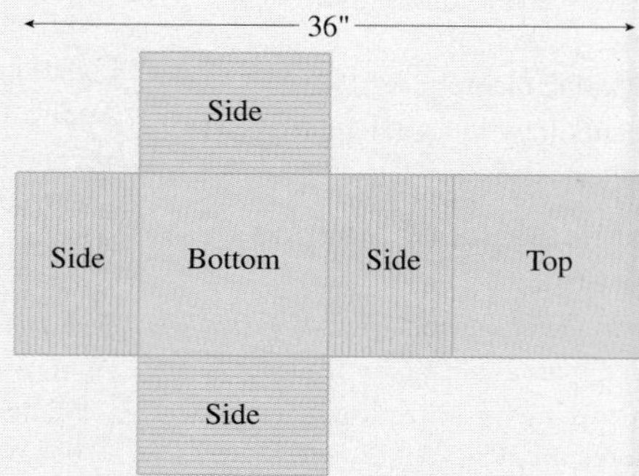

(Notice that the top and bottom of each box will be square, but the sides will not necessarily be square.) What are the dimensions of the boxes with the largest volume that can be made in this way? What is the maximum volume?

28. ***Box Redesign*** The sales department at OHaganBooks.com was not pleased with the result of the box design in the preceding exercise; the resulting box was too large for the chocolate lobsters, so, following a suggestion by a math major student intern, the department decided to redesign the boxes to meet the following specifications: As in Exercise 27, each box would be obtained by cutting out corners from a rectangle of cardboard as shown in the following diagram:

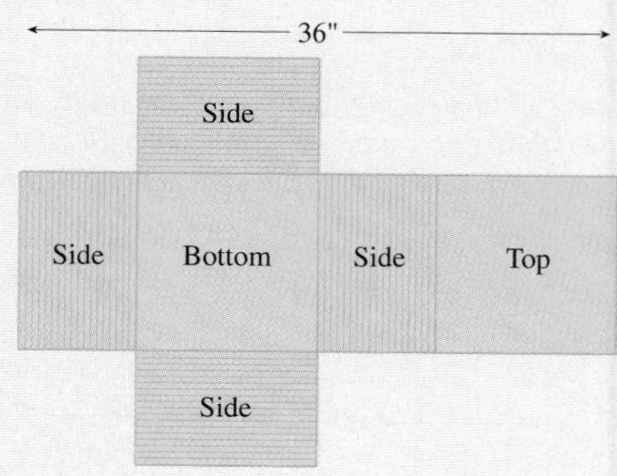

(Notice that the top and bottom of each box would be square, but not necessarily the sides.) The dimensions would be such that the total surface area of the sides plus the bottom of the box would be as large as possible. What are the dimensions of the boxes with the largest area that can be made in this way? How does this box compare with that obtained in Exercise 27?

29. ***Elasticity of Demand*** (Compare Exercise 23.) Demand for the latest best-seller at OHaganBooks.com, *A River Burns through It*, is given by

$$q = -p^2 + 33p + 9 \qquad (18 \le p \le 28)$$

copies sold per week when the price is p dollars.

a. Find the price elasticity of demand as a function of p.
b. Find the elasticity of demand for this book at a price of \$20 and at a price of \$25. (Round your answers to two decimal places.) Interpret the answers.
c. What price should the company charge to obtain the largest revenue?

30. ***Elasticity of Demand*** (Compare Exercise 24.) Demand for *The Secret Loves of John O*, a romance novel by Margó Dufón that flopped after two weeks on the market, is given by

$$q = -2p^2 + 5p + 6 \qquad (0 \le p \le 3.3)$$

copies sold per week when the price is p dollars.

a. Find the price elasticity of demand as a function of p.
b. Find the elasticity of demand for this book at a price of \$2 and at a price of \$3. (Round your answers to two decimal places.) Interpret the answers.
c. What price should the company charge to obtain the largest revenue?

31. ***Elasticity of Demand*** Last year OHaganBooks.com experimented with an online subscriber service, Red On Line (ROL), for its electronic book service. The consumer demand for ROL was modeled by the equation

$$q = 1{,}000e^{-p^2+p},$$

where p was the monthly access charge and q is the number of subscribers.

a. Obtain a formula for the price elasticity of demand, E, for ROL services.
b. Compute the elasticity of demand if the monthly access charge is set at \$2 per month. Interpret the result.
c. How much should the company have charged in order to obtain the maximum monthly revenue? What would this revenue have been?

32. ***Elasticity of Demand*** JungleBooks.com (one of OHaganBooks' main competitors) responded with its own online subscriber service, Better On Line (BOL), for its electronic book service. The consumer demand for BOL was modeled by the equation

$$q = 2{,}000e^{-3p^2+2p},$$

where p was the monthly access charge and q is the number of subscribers.

a. Obtain a formula for the price elasticity of demand, E, for BOL services.
b. Compute the elasticity of demand if the monthly access charge is set at \$2 per month. Interpret the result.
c. How much should the company have charged in order to obtain the maximum monthly revenue? What would this revenue have been?

33. ***Sales*** OHaganBooks.com modeled its weekly sales over a period of time with the function

$$s(t) = 6{,}053 + \frac{4{,}474}{1 + e^{-0.55(t-4.8)}},$$

where t is the time in weeks. Following are the graphs of s, s', and s'':

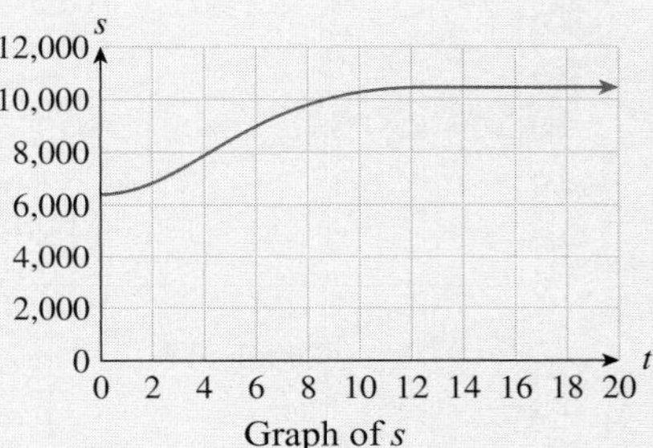

Graph of s

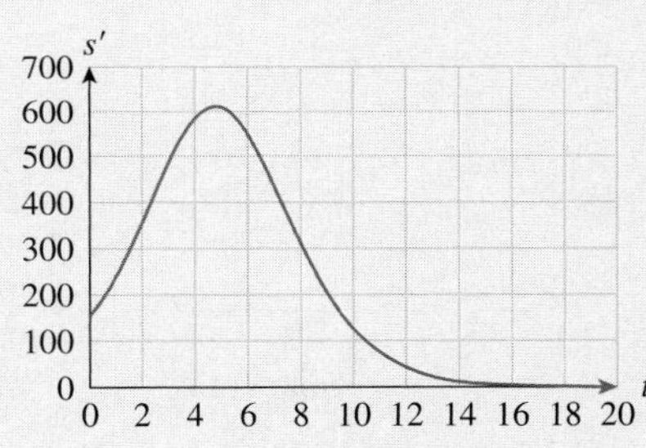

Graph of s'

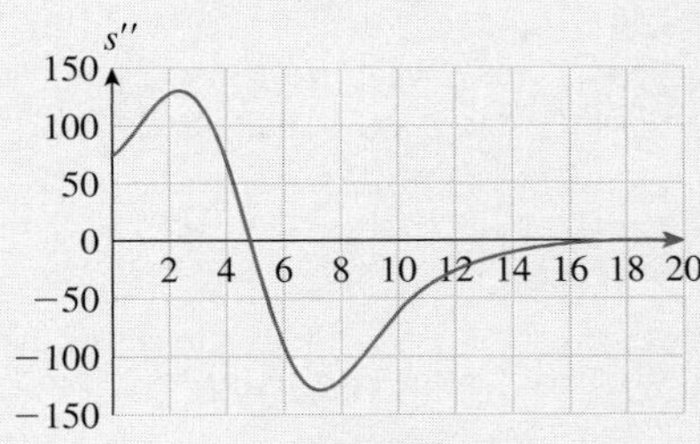

Graph of s''

a. Estimate when, to the nearest week, the weekly sales were growing fastest.
b. To what features on the graphs of s, s', and s'' does your answer to part (a) correspond?
c. The graph of s has a horizontal asymptote. What is the approximate value (s-coordinate) of this asymptote, and what is its significance in terms of weekly sales at OHaganBooks.com?

d. The graph of s' has a horizontal asymptote. What is the value (s'-coordinate) of this asymptote, and what is its significance in terms of weekly sales at OHaganBooks.com?

34. ***Sales*** The quarterly sales of OHagan *oPods* (OHaganBooks' answer to the *iPod*; a portable audio book unit with an incidental music feature) from the fourth quarter of 2009 can be roughly approximated by the function

$$N(t) = \frac{1{,}100}{1 + 9(1.8)^{-t}} \text{ oPods } (t \geq 0),$$

where t is time in quarters since the fourth quarter of 2009. Following are the graphs of N, N', and N'':

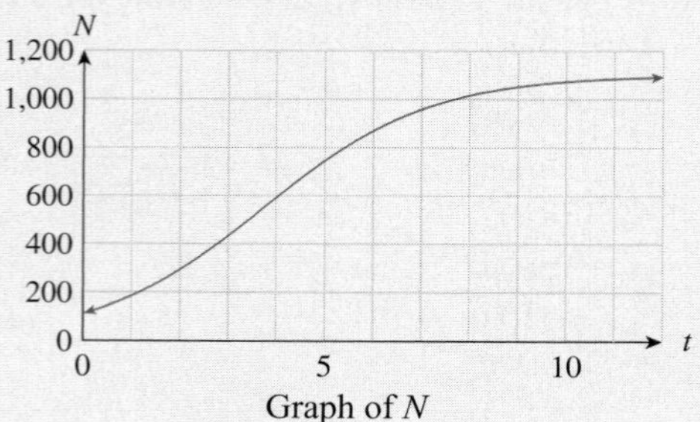

Graph of N

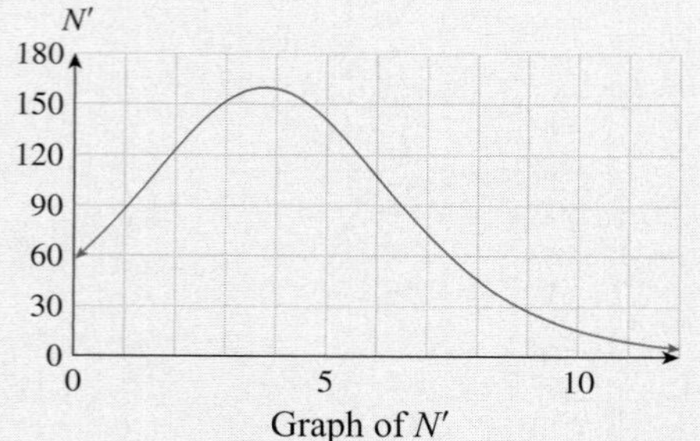

Graph of N'

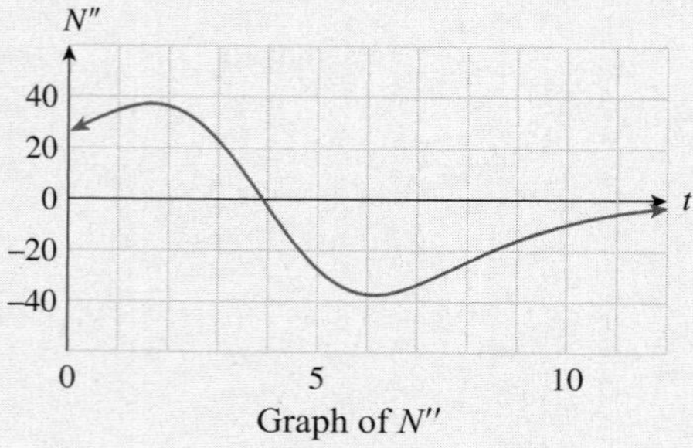

Graph of N''

a. Estimate when, to the nearest quarter, the quarterly sales were growing fastest.

b. To what features on the graphs of N, N', and N'' does your answer to part (a) correspond?

c. The graph of N has a horizontal asymptote. What is the approximate value (N-coordinate) of this asymptote, and what is its significance in terms of quarterly sales of *oPods*?

d. The graph of N' has a horizontal asymptote. What is the value (N'-coordinate) of this asymptote, and what is its significance in terms of quarterly sales of *oPods*?

35. ***Chance Encounter*** Marjory Duffin is walking north towards the corner entrance of OHaganBooks.com company headquarters at 5 ft/sec, while John O'Hagan is walking west toward the same entrance, also at 5 ft/sec. How fast is their distance apart decreasing when

a. each of them is 2 ft from the corner?

b. each of them is 1 ft from the corner?

c. each of them is h ft from the corner?

d. they collide on the corner?

36. ***Company Logos*** OHaganBooks.com's Web site has an animated graphic with its name in a rectangle whose height and width change; on either side of the rectangle are semicircles, as in the figure, whose diameters are the same as the height of the rectangle.

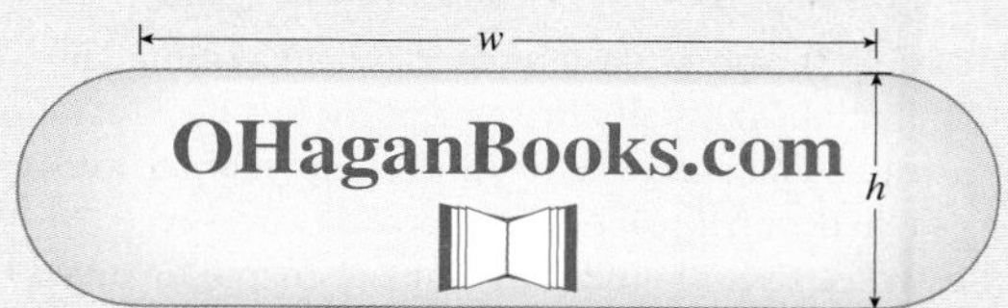

For reasons too complicated to explain, the designer wanted the combined area of the rectangle and semicircles to remain constant. At one point during the animation, the width of the rectangle is 1 inch, growing at a rate of 0.5 inches per second, while the height is 3 inches. How fast is the height changing?

Case Study

Production Lot Size Management

SERDAR/Alamy

Your publishing company, *Knockem Dead Paperbacks, Inc.*, is about to release its next best-seller, *Henrietta's Heaving Heart* by Celestine A. Lafleur. The company expects to sell 100,000 books each month in the next year. You have been given the job of scheduling print runs to meet the anticipated demand and minimize total costs to the company. Each print run has a setup cost of \$5,000, each book costs \$1 to produce, and monthly storage costs for books awaiting shipment average 1¢ per book. What will you do?

If you decide to print all 1,200,000 books (the total demand for the year, 100,000 books per month for 12 months) in a single run at the start of the year and sales run as predicted, then the number of books in stock would begin at 1,200,000 and decrease to zero by the end of the year, as shown in Figure 47.

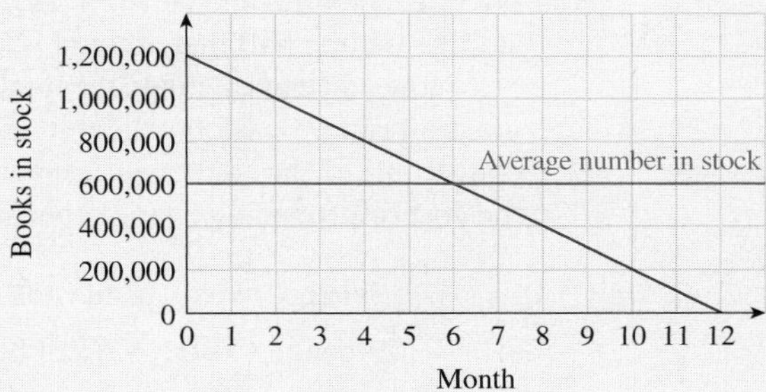

Figure 47

On average, you would be storing 600,000 books for 12 months at 1¢ per book, giving a total storage cost of $600{,}000 \times 12 \times .01 = \$72{,}000$. The setup cost for the single print run would be \$5,000. When you add to these the total cost of producing 1,200,000 books at \$1 per book, your total cost would be \$1,277,000.

If, on the other hand, you decide to cut down on storage costs by printing the book in two runs of 600,000 each, you would get the picture shown in Figure 48.

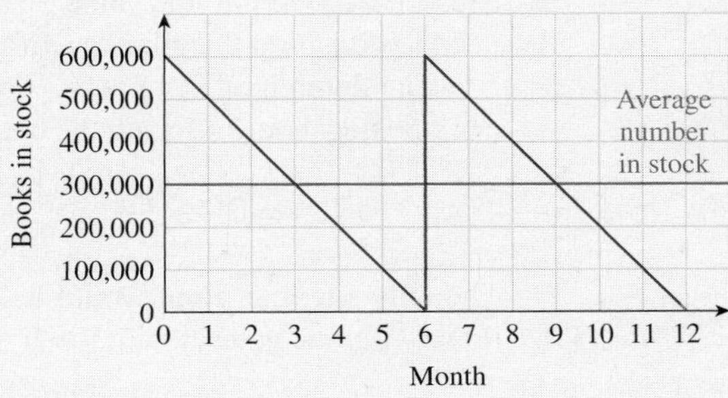

Figure 48

Now, the storage cost would be cut in half because on average there would be only 300,000 books in stock. Thus, the total storage cost would be \$36,000, and the setup cost would double to \$10,000 (because there would now be two runs). The production costs would be the same: 1,200,000 books @ \$1 per book. The total cost would therefore be reduced to \$1,246,000, a savings of \$31,000 compared to your first scenario.

"Aha!" you say to yourself, after doing these calculations. "Why not drastically cut costs by setting up a run every month?" You calculate that the setup costs alone would be $12 \times \$5{,}000 = \$60{,}000$, which is already more than the setup plus storage costs for two runs, so a run every month will cost too much. Perhaps, then, you should investigate three runs, four runs, and so on, until you find the lowest cost. This strikes you as too laborious a process, especially considering that you will have to do it all over again when planning for Lafleur's sequel, *Lorenzo's Lost Love,* due to be released next year. Realizing that this is an optimization problem, you decide to use some calculus to help you come up with a *formula* that you can use for all future plans. So you get to work.

Instead of working with the number 1,200,000, you use the letter N so that you can be as flexible as possible. (What if *Lorenzo's Lost Love* sells more copies?) Thus, you have a total of N books to be produced for the year. You now calculate the total cost of using x print runs per year. Because you are to produce a total of N books in x print runs, you will have to produce N/x books in each print run. N/x is called the **lot size**. As you can see from the diagrams above, the average number of books in storage will be half that amount, $N/(2x)$.

Now you can calculate the total cost for a year. Write P for the setup cost of a single print run ($P = \$5{,}000$ in your case) and c for the *annual* cost of storing a book (to convert all of the time measurements to years; $c = \$0.12$ here). Finally, write b for the cost of producing a single book ($b = \$1$ here). The costs break down as follows.

Setup Costs: x print runs @ P dollars per run:	Px
Storage Costs: $N/(2x)$ books stored @ c dollars per year:	$cN/(2x)$
Production Costs: N books @ b dollars per book:	Nb
Total Cost:	$Px + \dfrac{cN}{2x} + Nb$

Remember that P, N, c, and b are all constants and x is the only variable. Thus, your cost function is

$$C(x) = Px + \frac{cN}{2x} + Nb$$

and you need to find the value of x that will minimize $C(x)$. But that's easy! All you need to do is find the relative extrema and select the absolute minimum (if any).

The domain of $C(x)$ is $(0, +\infty)$ because there is an x in the denominator and x can't be negative. To locate the extrema, you start by locating the critical points:

$$C'(x) = P - \frac{cN}{2x^2}.$$

The only singular point would be at $x = 0$, but 0 is not in the domain. To find stationary points, you set $C'(x) = 0$ and solve for x:

$$P - \frac{cN}{2x^2} = 0$$

$$2x^2 = \frac{cN}{P}$$

so

$$x = \sqrt{\frac{cN}{2P}}.$$

There is only one stationary point, and there are no singular points or endpoints. To graph the function you will need to put in numbers for the various constants. Substituting $N = 1{,}200{,}000$, $P = 5{,}000$, $c = 0.12$, and $b = 1$, you get

$$C(x) = 5{,}000x + \frac{72{,}000}{x} + 1{,}200{,}000$$

with the stationary point at

$$x = \sqrt{\frac{(0.12)(1{,}200{,}000)}{2(5000)}} \approx 3.79.$$

The total cost at the stationary point is

$$C(3.79) \approx 1{,}237{,}900.$$

You now graph $C(x)$ in a window that includes the stationary point, say, $0 \le x \le 12$ and $1{,}100{,}000 \le C \le 1{,}500{,}000$, getting Figure 49.

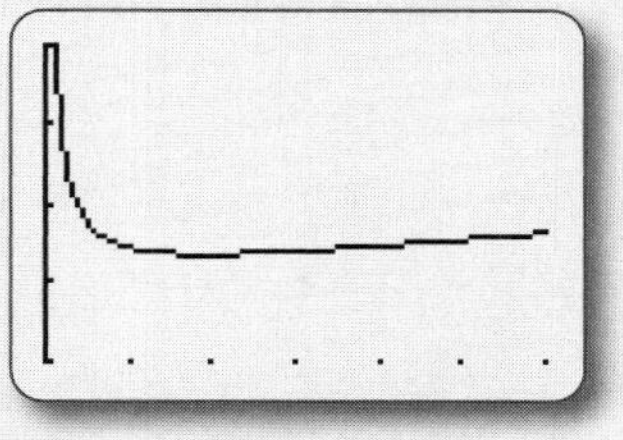

Figure 49

From the graph, you can see that the stationary point is an absolute minimum. In the graph it appears that the graph is always concave up, which also tells you that your stationary point is a minimum. You can check the concavity by computing the second derivative:

$$C''(x) = \frac{cN}{x^3} > 0.$$

The second derivative is always positive because c, N, and x are all positive numbers, so indeed the graph is always concave up. Now you also know that it works regardless of the particular values of the constants.

So now you are practically done! You know that the absolute minimum cost occurs when you have $x \approx 3.79$ print runs per year. Don't be disappointed that the answer is not a whole number; whole number solutions are rarely found in real scenarios. What the answer (and the graph) do indicate is that either three or four print runs per year will cost the least money. If you take $x = 3$, you get a total cost of

$$C(3) = \$1{,}239{,}000.$$

If you take $x = 4$, you get a total cost of

$$C(4) = \$1{,}238{,}000.$$

So, four print runs per year will allow you to minimize your total costs.

EXERCISES

1. *Lorenzo's Lost Love* will sell 2,000,000 copies in a year. The remaining costs are the same. How many print runs should you use now?
2. In general, what happens to the number of runs that minimizes cost if both the setup cost and the total number of books are doubled?
3. In general, what happens to the number of runs that minimizes cost if the setup cost increases by a factor of 4?
4. Assuming that the total number of copies and storage costs are as originally stated, find the setup cost that would result in a single print run.
5. Assuming that the total number of copies and setup cost are as originally stated, find the storage cost that would result in a print run each month.

6. In Figure 48 we assumed that all the books in each run were manufactured in a very short time; otherwise the figure might have looked more like the following graph, which shows the inventory, assuming a slower rate of production.

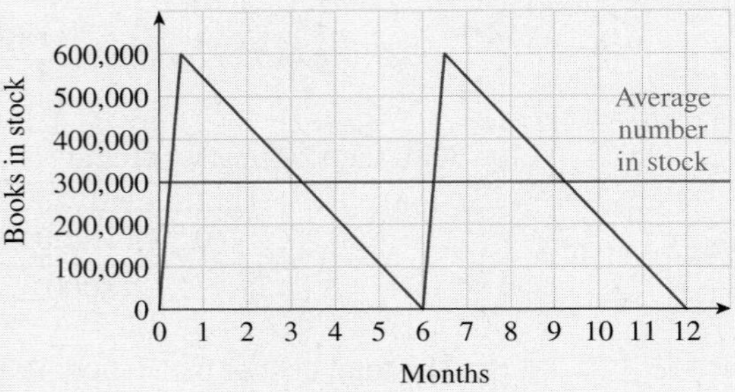

How would this affect the answer?

7. Referring to the general situation discussed in the text, find the cost as a function of the total number of books produced, assuming that the number of runs is chosen to minimize total cost. Also find the average cost per book.

8. Let $\bar{C}$ be the average cost function found in the preceding exercise. Calculate $\lim_{N\to+\infty} \bar{C}(N)$ and interpret the result.

TI-83/84 Plus Technology Guide

Section 5.6

Example 1(a) (page 320) Suppose that the demand equation for *Bobby Dolls* is given by $q = 216 - p^2$, where p is the price per doll in dollars and q is the number of dolls sold per week. Compute the price elasticity of demand when $p = \$5$ and $p = \$10$, and interpret the results.

Solution with Technology

The TI-83/84 Plus function `nDeriv` can be used to compute approximations of the elasticity E at various prices.

1. Set

```
Y1= 216-X^2                        Demand equation
Y2= -nDeriv(Y1,X,X)* X/Y1          Formula for E
```

2. Use the table feature to list the values of elasticity for a range of prices. For part (a) we chose values of `X` close to 5:

X	Y1	Y2
4.7	193.91	.22784
4.8	192.96	.23881
4.9	191.99	.25012
5	191	[illegible]
5.1	189.99	.2738
5.2	188.96	.2862
5.3	187.91	.29897

Y2=.261780104712

SPREADSHEET Technology Guide

Section 5.6

Example 1(a) (page 320) Suppose that the demand equation for *Bobby Dolls* is given by $q = 216 - p^2$, where p is the price per doll in dollars and q is the number of dolls sold per week. Compute the price elasticity of demand when $p = \$5$ and $p = \$10$, and interpret the results.

Solution with Technology

To approximate E in a spreadsheet, we can use the following approximation of E.

$$E \approx \frac{\text{Percentage decrease in demand}}{\text{Percentage increase in price}} \approx -\frac{\left(\dfrac{\Delta q}{q}\right)}{\left(\dfrac{\Delta p}{p}\right)}$$

The smaller Δp is, the better the approximation. Let's use $\Delta p = 1¢$, or 0.01 (which is small compared with the typical prices we consider—around \$5 to \$10).

1. We start by setting up our worksheet to list a range of prices, in increments of Δp, on either side of a price in which we are interested, such as $p_0 = \$5$:

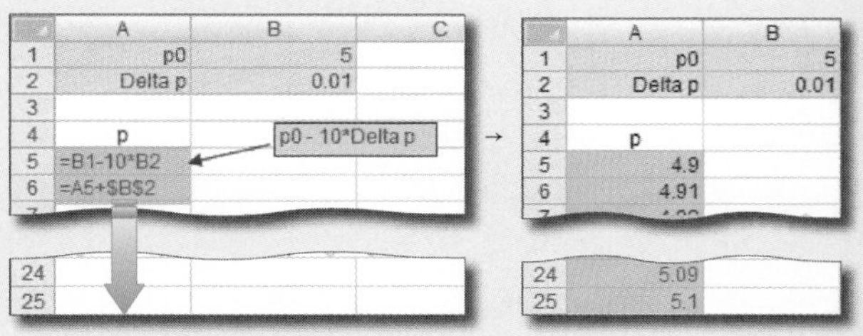

We start in cell A5 with the formula for $p_0 - 10\Delta p$ and then successively add Δp going down column A. You will find that the value $p_0 = 5$ appears midway down the list.

2. Next, we compute the corresponding values for the demand q in column B.

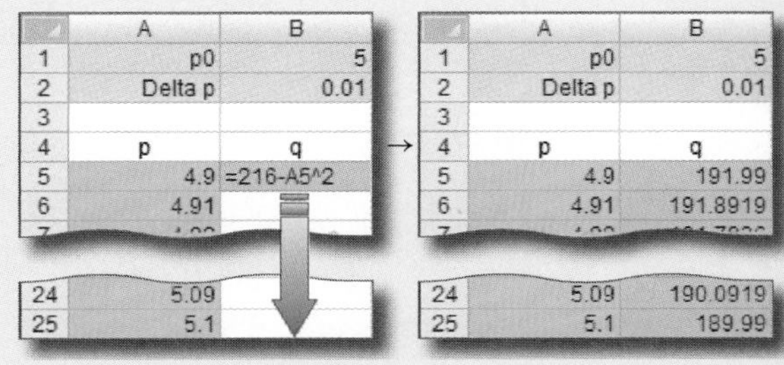

3. We add two new columns for the percentage changes in p and q. The formula shown in cell C5 is copied down columns C and D, to Row 24. (Why not row 25?)

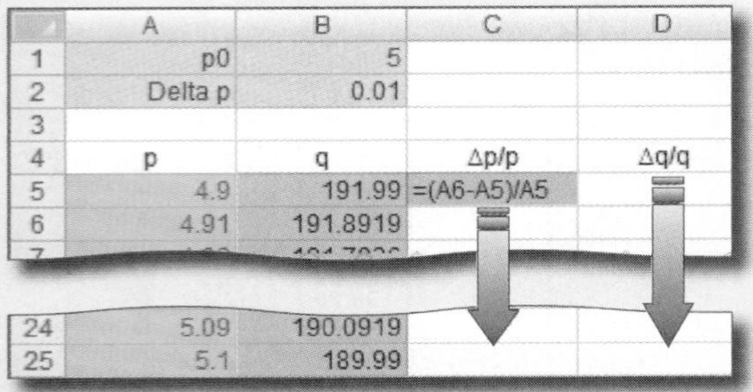

	A	B	C	D
1	p0	5		
2	Delta p	0.01		
3				
4	p	q	Δp/p	Δq/q
5	4.9	191.99	=(A6-A5)/A5	
6	4.91	191.8919		
24	5.09	190.0919		
25	5.1	189.99		

4. The elasticity can now be computed in column E as shown:

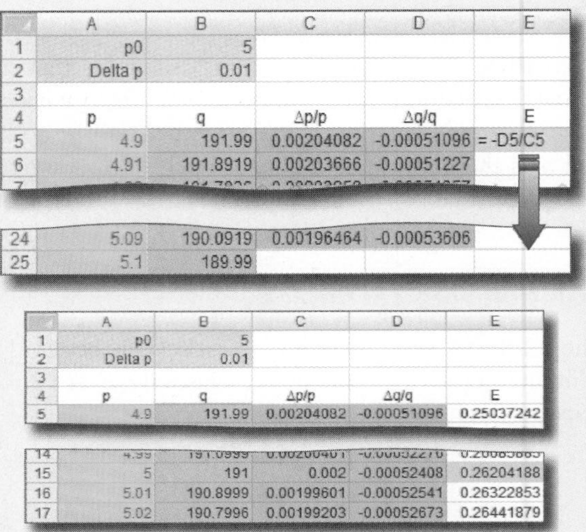

	A	B	C	D	E
1	p0	5			
2	Delta p	0.01			
3					
4	p	q	Δp/p	Δq/q	E
5	4.9	191.99	0.00204082	-0.00051096	= -D5/C5
6	4.91	191.8919	0.00203666	-0.00051227	
24	5.09	190.0919	0.00196464	-0.00053606	
25	5.1	189.99			

	A	B	C	D	E
1	p0	5			
2	Delta p	0.01			
3					
4	p	q	Δp/p	Δq/q	E
5	4.9	191.99	0.00204082	-0.00051096	0.25037242
15	5	191	0.002	-0.00052408	0.26204188
16	5.01	190.8999	0.00199601	-0.00052541	0.26322853
17	5.02	190.7996	0.00199203	-0.00052673	0.26441879

6

The Integral

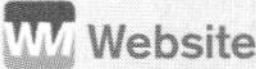
Website

www.WanerMath.com

At the Website you will find:

- Section-by-section tutorials, including game tutorials with randomized quizzes
- A detailed chapter summary
- A true/false quiz
- Additional review exercises
- A numerical integration utility with Riemann sum grapher
- Graphing calculator programs for numerical integration
- Graphers, Excel tutorials, and other resources
- The following extra topic:

  Numerical Integration

Case Study Spending on Housing Construction

It is March 2007, and *Time* magazine, in its latest edition, is asking "Will the Housing Bubble Burst in 2007?" You are a summer intern at *Ronald Ramp Real Estate Development, Inc.,* which is considering a major luxury condominium development in Tampa, Florida, nicknamed the "Ramp Towers Tampa." You have been asked to find formulas for monthly spending on housing construction in the United States and for the average spent per month starting one year ago. You have data about percentage spending changes. **How will you model the trend and estimate the total?**

Bill Varie/Flirt/Corbis

Introduction

Roughly speaking, calculus is divided into two parts: **differential calculus** (the calculus of derivatives) and **integral calculus**, which is the subject of this chapter and the next. Integral calculus is concerned with problems that are in some sense the reverse of the problems seen in differential calculus. For example, where differential calculus shows how to compute the rate of change of a quantity, integral calculus shows how to find the quantity if we know its rate of change. This idea is made precise in the **Fundamental Theorem of Calculus**. Integral calculus and the Fundamental Theorem of Calculus allow us to solve many problems in economics, physics, and geometry, including one of the oldest problems in mathematics—computing areas of regions with curved boundaries.

6.1 The Indefinite Integral

Suppose that we knew the marginal cost to manufacture an item and we wanted to reconstruct the cost function. We would have to *reverse* the process of differentiation, to go from the derivative (the marginal cost function) back to the original function (the total cost). We'll first discuss how to do that and then look at some applications.

Here is an example: If the derivative of $F(x)$ is $4x^3$, what was $F(x)$? We recognize $4x^3$ as the derivative of x^4. So, we might have $F(x) = x^4$. However, $F(x) = x^4 + 7$ works just as well. In fact, $F(x) = x^4 + C$ works for any number C. Thus, there are *infinitely many* possible answers to this question.

In fact, we will see shortly that the formula $F(x) = x^4 + C$ covers *all* possible answers to the question. Let's give a name to what we are doing.

Antiderivative

An **antiderivative** of a function f is a function F such that $F' = f$.

Quick Examples

1. An antiderivative of $4x^3$ is x^4. Because the derivative of x^4 is $4x^3$
2. Another antiderivative of $4x^3$ is $x^4 + 7$. Because the derivative of $x^4 + 7$ is $4x^3$
3. An antiderivative of $2x$ is $x^2 + 12$. Because the derivative of $x^2 + 12$ is $2x$

Thus,

If the derivative of A(x) is B(x), then an antiderivative of B(x) is A(x).

We call the set of *all* antiderivatives of a function the **indefinite integral** of the function.

Indefinite Integral

$$\int f(x)\,dx$$

is read "the **indefinite integral** of $f(x)$ with respect to x" and stands for the set of all antiderivatives of f. Thus, $\int f(x)\,dx$ is a *collection of functions*; it is not a single function or a number. The function f that is being **integrated** is called the **integrand**, and the variable x is called the **variable of integration**.

Quick Examples

1. $\int 4x^3\,dx = x^4 + C$ Every possible antiderivative of $4x^3$ has the form $x^4 + C$.
2. $\int 2x\,dx = x^2 + C$ Every possible antiderivative of $2x$ has the form $x^2 + C$.

The **constant of integration** C reminds us that we can add any constant and get a different antiderivative.

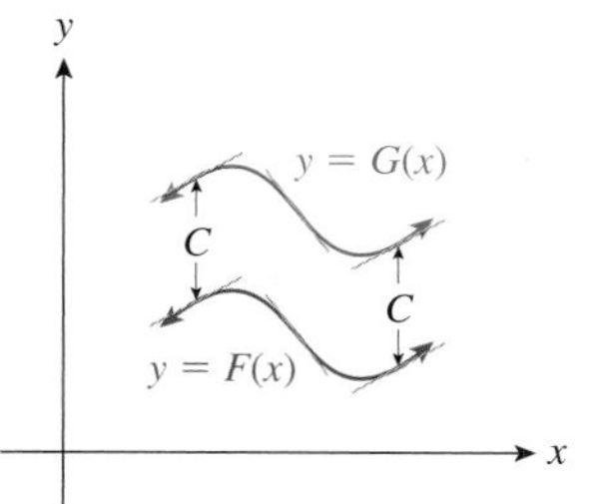

Figure 1

Q: *If $F(x)$ is one antiderivative of $f(x)$, why must all other antiderivatives have the form $F(x) + C$?*

A: Suppose $F(x)$ and $G(x)$ are both antiderivatives of $f(x)$, so that $F'(x) = G'(x)$. Consider what this means by looking at Figure 1. If $F'(x) = G'(x)$ for all x, then F and G have the *same slope* at each value of x. This means that their graphs must be *parallel* and hence remain exactly the same vertical distance apart. But that is the same as saying that the functions differ by a constant—that is, that $G(x) = F(x) + C$ for some constant C.*

* This argument can be turned into a more rigorous proof—that is, a proof that does not rely on geometric concepts such as "parallel graphs." We should also say that the result (and our geometric argument as well!) requires that the domain of F and G be a single (possibly infinite) open interval.

EXAMPLE 1 Indefinite Integral

Check that

a. $\int x\,dx = \frac{x^2}{2} + C$ **b.** $\int x^2\,dx = \frac{x^3}{3} + C$ **c.** $\int x^{-1}\,dx = \ln|x| + C$

Solution We check each equation by taking the derivative of its right-hand side and checking whether it equals the integrand on the left:

a. $\frac{d}{dx}\left(\frac{x^2}{2} + C\right) = \frac{2x}{2} + 0 = x$ ✔

b. $\frac{d}{dx}\left(\frac{x^3}{3} + C\right) = \frac{3x^2}{3} + 0 = x^2$ ✔

c. $\frac{d}{dx}(\ln|x| + C) = \frac{1}{x} + 0 = x^{-1}.$ ✔

Because the derivative of the right-hand side is the integrand in each case, we can conclude that the given statements are all valid.†

† We are glossing over a subtlety in part (c): The constant of integration C can be different for $x < 0$ and $x > 0$ because the graph breaks at $x = 0$. (See the comment at the end of the last marginal note.) In general, our understanding will be that the constant of integration may be different on disconnected intervals of the domain.

➡ **Before we go on...** Example 1 gives us a very useful technique to check our answer every time we calculate an integral:

Take the derivative of the answer and check that it equals the integrand. ■

Now, we would like to make the process of finding indefinite integrals (antiderivatives) more mechanical. For example, it would be nice to have a power rule for indefinite integrals similar to the one we already have for derivatives. Example 1 already suggests such a rule for us:

Power Rule for the Indefinite Integral

$$\int x^n\,dx = \frac{x^{n+1}}{n+1} + C$$ This holds only if $n \neq -1$.*

$$\int x^{-1}\,dx = \ln|x| + C$$ For the special case $n = -1$

Equivalent Form of Second Formula: $\int \frac{1}{x}\,dx = \ln|x| + C$ Because $x^{-1} = \frac{1}{x}$

In Words For n other than -1, to find the integral of x^n, add 1 to the exponent, and then divide by the new exponent. When $n = -1$, the answer is the natural logarithm of the absolute value of x.

* Note that the right-hand side of the formula makes no sense if $n = -1$ because it has $n + 1$ in the denominator.

Quick Examples

1. $\int x^{55}\,dx = \frac{x^{56}}{56} + C$

2. $\int \frac{1}{x^{55}}\,dx = \int x^{-55}\,dx$ Exponent form

$= \frac{x^{-54}}{-54} + C$ When we add 1 to -55, we get -54, *not* -56.

$= -\frac{1}{54x^{54}} + C$

3. $\int 1\,dx = x + C$ Because $1 = x^0$. This is an important special case.

4. $\int \sqrt{x}\,dx = \int x^{1/2}dx$ Exponent form

$= \frac{x^{3/2}}{3/2} + C$

$= \frac{2x^{3/2}}{3} + C$

Notes

1. The integral $\int 1\,dx$ is commonly written as $\int dx$. Similarly, the integral $\int \frac{1}{x^{55}}\,dx$ may be written as $\int \frac{dx}{x^{55}}$.

2. We can easily check the power rule formula by taking the derivative of the right-hand side:

$$\frac{d}{dx}\left(\frac{x^{n+1}}{n+1} + C\right) = \frac{(n+1)x^n}{n+1} = x^n \quad ✔$$

3. Because the derivative of $\ln x$ is also $1/x$, you might be tempted to write $\int x^{-1}dx = \ln x + C$. But $\ln x$, being defined only for positive x, does not have the same domain as $1/x$, whereas $\ln|x|$ does. So, we must use $\ln|x| + C$ instead. ■

Following are more indefinite integrals that come from formulas for differentiation we have encountered before:

Indefinite Integral of e^x, b^x, and $|x|$

$$\int e^x\,dx = e^x + C \qquad \text{Because } \frac{d}{dx}(e^x) = e^x$$

If b is any positive number other than 1, then

$$\int b^x\,dx = \frac{b^x}{\ln b} + C \qquad \text{Because } \frac{d}{dx}\left(\frac{b^x}{\ln b}\right) = \frac{b^x \ln b}{\ln b} = b^x$$

$$\int |x|\,dx = \frac{x|x|}{2} + C. \qquad \text{Because } \frac{d}{dx}\left(\frac{x|x|}{2}\right) = |x| \text{ (Check this yourself!)}$$

Quick Example

$$\int 2^x\,dx = \frac{2^x}{\ln 2} + C$$

For more complicated functions, like $2x^3 + 6x^5 - 1$, we need the following rules for integrating sums, differences, and constant multiples.

Sums, Differences, and Constant Multiples

Sum and Difference Rules

$$\int [f(x) \pm g(x)]\,dx = \int f(x)\,dx \pm \int g(x)\,dx$$

In Words: The integral of a sum is the sum of the integrals, and the integral of a difference is the difference of the integrals.

Constant Multiple Rule

$$\int kf(x)\,dx = k\int f(x)\,dx \qquad (k \text{ constant})$$

In Words: The integral of a constant times a function is the constant times the integral of the function. (In other words, the constant "goes along for the ride.")

Quick Examples

Sum Rule: $\int (x^3 + 1)\,dx = \int x^3\,dx + \int 1\,dx = \frac{x^4}{4} + x + C$

$f(x) = x^3$; $g(x) = 1$

Constant Multiple Rule: $\int 5x^3\,dx = 5\int x^3\,dx = 5\frac{x^4}{4} + C$

$k = 5$; $f(x) = x^3$

Constant Multiple Rule: $\int 4\,dx = 4\int 1\,dx = 4x + C$

$k = 4$; $f(x) = 1$

Constant Multiple Rule: $\int 4e^x\,dx = 4\int e^x\,dx = 4e^x + C$

$k = 4$; $f(x) = e^x$

Proof of the Sum Rule

We saw above that if two functions have the same derivative, they differ by a (possibly zero) constant. Look at the rule for sums:

$$\int [f(x) + g(x)]\,dx = \int f(x)\,dx + \int g(x)\,dx$$

If we take the derivative of the left-hand side with respect to x, we get the integrand, $f(x) + g(x)$. If we take the derivative of the right-hand side, we get

$$\frac{d}{dx}\left[\int f(x)\,dx + \int g(x)\,dx\right] = \frac{d}{dx}\left[\int f(x)\,dx\right] + \frac{d}{dx}\left[\int g(x)\,dx\right]$$

Derivative of a sum = Sum of derivatives.

$$= f(x) + g(x)$$

Because the left- and right-hand sides have the same derivative, they differ by a constant. But, because both expressions are indefinite integrals, adding a constant does not affect their value, so they are the same as indefinite integrals.

Notice that a key step in the proof was the fact that the derivative of a sum is the sum of the derivatives.

A similar proof works for the difference and constant multiple rules.

EXAMPLE 2 Using the Sum and Difference Rules

Find the integrals.

a. $\int (x^3 + x^5 - 1)\,dx$ **b.** $\int \left(x^{2.1} + \frac{1}{x^{1.1}} + \frac{1}{x} + e^x\right) dx$ **c.** $\int (e^x + 3^x - |x|)\,dx$

Solution

a. $\int (x^3 + x^5 - 1)\,dx = \int x^3\,dx + \int x^5\,dx - \int 1\,dx$ Sum/difference rule

$= \frac{x^4}{4} + \frac{x^6}{6} - x + C$ Power rule

b. $\int \left(x^{2.1} + \frac{1}{x^{1.1}} + \frac{1}{x} + e^x\right) dx$

$$= \int (x^{2.1} + x^{-1.1} + x^{-1} + e^x)\, dx \qquad \text{Exponent form}$$

$$= \int x^{2.1}\, dx + \int x^{-1.1}\, dx + \int x^{-1}\, dx + \int e^x\, dx \qquad \text{Sum rule}$$

$$= \frac{x^{3.1}}{3.1} + \frac{x^{-0.1}}{-0.1} + \ln|x| + e^x + C \qquad \text{Power rule and exponential rule}$$

$$= \frac{x^{3.1}}{3.1} - \frac{10}{x^{0.1}} + \ln|x| + e^x + C$$

c. $\int (e^x + 3^x - |x|)\, dx = \int e^x\, dx + \int 3^x\, dx - \int |x|\, dx$ Sum/difference rule

$$= e^x + \frac{3^x}{\ln 3} - \frac{x|x|}{2} + C \qquad \text{Rules for powers, exponentials, and absolute value}$$

➡ **Before we go on...** You should check each of the answers in Example 2 by differentiating.

Q: *Why is there only a single arbitrary constant C in each of the answers?*

A: We could have written the answer to part (a) as

$$\frac{x^4}{4} + D + \frac{x^6}{6} + E - x + F$$

where D, E, and F are all arbitrary constants. Now suppose, for example, we set $D = 1$, $E = -2$, and $F = 6$. Then the particular antiderivative we get is $x^4/4 + x^6/6 - x + 5$, which has the form $x^4/4 + x^6/6 - x + C$. Thus, we could have chosen the single constant C to be 5 and obtained the same answer. In other words, the answer $x^4/4 + x^6/6 - x + C$ is just as general as the answer $x^4/4 + D + x^6/6 + E - x + F$, but simpler.

■

In practice we do not explicitly write the integral of a sum as a sum of integrals but just "integrate term by term," much as we learned to differentiate term by term.

EXAMPLE 3 Combining the Rules

Find the integrals.

a. $\int (10x^4 + 2x^2 - 3e^x)\, dx$ **b.** $\int \left(\frac{2}{x^{0.1}} + \frac{x^{0.1}}{2} - \frac{3}{4x}\right) dx$

c. $\int (3e^x - 2(1.2^x) + 5|x|)\, dx$

Solution

a. We need to integrate separately each of the terms $10x^4$, $2x^2$, and $3e^x$. To integrate $10x^4$ we use the rules for constant multiples and powers:

$$\int 10x^4\,dx = 10\int x^4\,dx = 10\frac{x^5}{5} + C = 2x^5 + C.$$

The other two terms are similar. We get

$$\int (10x^4 + 2x^2 - 3e^x)\,dx = 10\frac{x^5}{5} + 2\frac{x^3}{3} - 3e^x + C = 2x^5 + \frac{2}{3}x^3 - 3e^x + C.$$

b. We first convert to exponent form and then integrate term by term:

$$\int \left(\frac{2}{x^{0.1}} + \frac{x^{0.1}}{2} - \frac{3}{4x}\right)dx = \int \left(2x^{-0.1} + \frac{1}{2}x^{0.1} - \frac{3}{4}x^{-1}\right)dx \quad \text{Exponent form}$$

$$= 2\frac{x^{0.9}}{0.9} + \frac{1}{2}\frac{x^{1.1}}{1.1} - \frac{3}{4}\ln|x| + C \quad \text{Integrate term by term.}$$

$$= \frac{20x^{0.9}}{9} + \frac{x^{1.1}}{2.2} - \frac{3}{4}\ln|x| + C. \quad \text{Back to rational form}$$

c. $\displaystyle\int (3e^x - 2(1.2^x) + 5|x|)\,dx = 3e^x - 2\frac{1.2^x}{\ln 1.2} + 5\frac{x|x|}{2} + C$

EXAMPLE 4 Different Variable Name

Find $\displaystyle\int \left(\frac{1}{u} + \frac{1}{u^2}\right)du.$

Solution This integral may look a little strange because we are using the letter u instead of x, but there is really nothing special about x. Using u as the variable of integration, we get

$$\int \left(\frac{1}{u} + \frac{1}{u^2}\right)du = \int (u^{-1} + u^{-2})\,du \quad \text{Exponent form.}$$

$$= \ln|u| + \frac{u^{-1}}{-1} + C \quad \text{Integrate term by term.}$$

$$= \ln|u| - \frac{1}{u} + C. \quad \text{Simplify the result.}$$

➡ **Before we go on...** When we compute an indefinite integral, we want the independent variable in the answer to be the same as the variable of integration. Thus, if the integral in Example 4 had been written in terms of x rather than u, we would have written

$$\int \left(\frac{1}{x} + \frac{1}{x^2}\right)dx = \ln|x| - \frac{1}{x} + C.$$

■

APPLICATIONS

EXAMPLE 5 Finding Cost from Marginal Cost

The marginal cost to produce baseball caps at a production level of x caps is $4 - 0.001x$ dollars per cap, and the cost of producing 100 caps is $500. Find the cost function.

Solution We are asked to find the cost function $C(x)$, given that the *marginal* cost function is $4 - 0.001x$. Recalling that the marginal cost function is the derivative of the cost function, we can write

$$C'(x) = 4 - 0.001x$$

and must find $C(x)$. Now $C(x)$ must be an antiderivative of $C'(x)$, so

$$\begin{aligned} C(x) &= \int (4 - 0.001x)\,dx \\ &= 4x - 0.001\frac{x^2}{2} + K \qquad K \text{ is the constant of integration.}^* \\ &= 4x - 0.0005x^2 + K. \end{aligned}$$

* We used K and not C for the constant of integration because we are using C for cost.

Now, unless we have a value for K, we don't really know what the cost function is. However, there is another piece of information we have ignored: The cost of producing 100 baseball caps is $500. In symbols

$$C(100) = 500.$$

Substituting in our formula for $C(x)$, we have

$$\begin{aligned} C(100) &= 4(100) - 0.0005(100)^2 + K \\ 500 &= 395 + K \\ K &= 105. \end{aligned}$$

Now that we know what K is, we can write down the cost function:

$$C(x) = 4x - 0.0005x^2 + 105.$$

➡ **Before we go on...** Let us consider the significance of the constant term 105 in Example 5. If we substitute $x = 0$ into the cost function, we get

$$C(0) = 4(0) - 0.0005(0)^2 + 105 = 105.$$

Thus, $105 is the cost of producing zero items; in other words, it is the **fixed cost**. ■

EXAMPLE 6 Total Sales from Annual Sales

By the start of 2008, Apple had sold a total of about 3.5 million iPhones. From the start of 2008 through the end of 2010, sales of iPhones were approximately

$$s(t) = 4.5t^2 + 5.8t + 6.5 \text{ million iPhones per year} \qquad (0 \le t \le 3),$$

where t is time in years since the start of 2008.[1]

[1] Source for data: Apple quarterly press releases (www.apple.com/investor/).

a. Find an expression for the total sales of iPhones up to time t.

b. Use the answer to part (a) to estimate the total sales of iPhones by the end of 2010. (The actual figure was 89.7 million.)

Solution

a. Let $S(t)$ be the total sales of iPhones, in millions, up to time t, where t is measured in years since the start of 2008, so we know that $S(0) = 3.5$. We are also given an expression for the number of iPhones sold per year. This function is the *derivative* of $S(t)$:

$$S'(t) = 4.5t^2 + 5.8t + 6.5.$$

Thus, the desired total sales function must be an antiderivative of $S'(t)$:

$$\begin{aligned} S(t) &= \int (4.5t^2 + 5.8t + 6.5)\,dt \\ &= \frac{4.5t^3}{3} + 5.8\frac{t^2}{2} + 6.5t + C \\ &= 1.5t^3 + 2.9t^2 + 6.5t + C. \end{aligned}$$

To calculate the value of the constant C, we can, as in the preceding example, use the known value of S: $S(0) = 3.5$.

$$S(0) = 1.5(0)^3 + 2.9(0)^2 + 6.5(0) + C = 3.5,$$

so

$$C = 3.5.$$

We can now write down the total sales function:

$$S(t) = 1.5t^3 + 2.9t^2 + 6.5t + 3.5 \text{ million iPhones.}$$

b. Because the end of 2010 (or the start of 2011) corresponds to $t = 3$, we calculate total sales as

$$S(3) = 1.5(3)^3 + 2.9(3)^2 + 6.5(3) + 3.5 = 89.6 \text{ million iPhones.}$$

Motion in a Straight Line

An important application of the indefinite integral is to the study of motion. The application of calculus to problems about motion is an example of the intertwining of mathematics and physics. We begin by bringing together some facts, scattered through the last several chapters, that have to do with an object moving in a straight line, and then restating them in terms of antiderivatives.

Position, Velocity, and Acceleration: Derivative Form

If $s = s(t)$ is the **position** of an object at time t, then its **velocity** is given by the derivative

$$v = \frac{ds}{dt}.$$

In Words: Velocity is the derivative of position.

The **acceleration** of an object is given by the derivative

$$a = \frac{dv}{dt}.$$

In Words: Acceleration is the derivative of velocity.

Position, Velocity, and Acceleration: Integral Form

$$s(t) = \int v(t)\,dt \qquad \text{Because } v = \frac{ds}{dt}$$

$$v(t) = \int a(t)\,dt \qquad \text{Because } a = \frac{dv}{dt}$$

Quick Examples

1. If the velocity of a particle moving in a straight line is given by $v(t) = 4t + 1$, then its position after t seconds is given by $s(t) = \int v(t)\,dt = \int (4t + 1)\,dt = 2t^2 + t + C$.
2. If sales are accelerating at 2 golf balls/day^2, then the rate of change of sales ("velocity of sales") is $v(t) = \int a(t)\,dt = \int 2\,dt = 2t + C$ golf balls/day.
3. If the rate of change of sales is $v(t) = 2t + 5$ golf balls per day, then the total sales are $s(t) = \int v(t)\,dt = \int (2t + 5)\,dt = t^2 + 5t + C$ golf balls sold through time t.

EXAMPLE 7 Motion in a Straight Line

a. The velocity of a particle moving along a straight line is given by $v(t) = 4t + 1$ m/sec. Given that the particle is at position $s = 2$ meters at time $t = 1$, find an expression for s in terms of t.

b. For a freely falling body experiencing no air resistance and zero initial velocity, find an expression for the velocity v in terms of t. [Note: On Earth, a freely falling body experiencing no air resistance accelerates downward at approximately 9.8 meters per second per second, or 9.8 m/sec^2 (or 32 ft/sec^2).]

Solution

a. As we saw in the Quick Example above, the position of the particle after t seconds is given by

$$s(t) = \int v(t)\,dt$$

$$= \int (4t + 1)\,dt = 2t^2 + t + C.$$

But what is the value of C? Now, we are told that the particle is at position $s = 2$ at time $t = 1$. In other words, $s(1) = 2$. Substituting this into the expression for $s(t)$ gives

$$2 = 2(1)^2 + 1 + C$$

so

$$C = -1.$$

Hence the position after t seconds is given by

$$s(t) = 2t^2 + t - 1 \text{ meters.}$$

b. Let's measure heights above the ground as positive, so that a rising object has positive velocity and the acceleration due to gravity is negative. (It causes the upward velocity to decrease in value.) Thus, the acceleration of the object is given by

$$a(t) = -9.8 \text{ m/sec}^2.$$

We wish to know the velocity, which is an antiderivative of acceleration, so we compute

$$v(t) = \int a(t)\, dt = \int (-9.8)\, dt = -9.8t + C.$$

To find the value of C, we use the given information that at time $t = 0$ the velocity is 0: $v(0) = 0$. Substituting this into the expression for $v(t)$ gives

$$0 = -9.8(0) + C$$

so

$$C = 0.$$

Hence, the velocity after t seconds is given by

$$v(t) = -9.8t \text{ m/sec.}$$

EXAMPLE 8 Vertical Motion under Gravity

You are standing on the edge of a cliff and toss a stone upward at a speed of $v_0 = 30$ feet per second (v_0 is called the *initial velocity*).

a. Find the stone's velocity as a function of time. How fast and in what direction is it going after 5 seconds? (Neglect the effects of air resistance.)

b. Find the position of the stone as a function of time. Where will it be after 5 seconds?

c. When and where will the stone reach its zenith, its highest point?

Solution

a. This is similar to Example 7(b): Measuring height above the ground as positive, the acceleration of the stone is given by $a(t) = -32$ ft/s^2, and so

$$v(t) = \int (-32)\, dt = -32t + C.$$

To obtain C, we use the fact that you tossed the stone upward at 30 ft/s; that is, when $t = 0$, $v = 30$, or $v(0) = 30$. Thus,

$$30 = v(0) = -32(0) + C.$$

So, $C = 30$ and the formula for velocity is

$$v(t) = -32t + 30 \text{ ft/sec.} \qquad v(t) = -32t + v_0$$

In particular, after 5 seconds the velocity will be

$$v(5) = -32(5) + 30 = -130 \text{ ft/sec.}$$

After 5 seconds the stone is *falling* with a speed of 130 ft/sec.

b. We wish to know the position, but position is an antiderivative of velocity. Thus,

$$s(t) = \int v(t)\, dt = \int (-32t + 30)\, dt = -16t^2 + 30t + C.$$

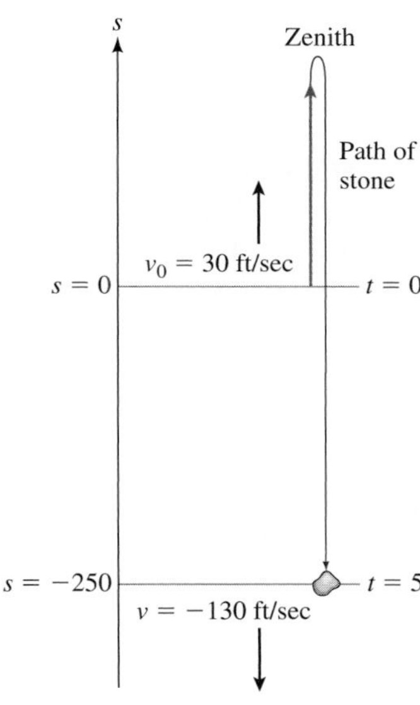

Figure 2

Now to find C, we need to know the initial position $s(0)$. We are not told this, so let's measure heights so that the initial position is zero. Then

$$0 = s(0) = C$$

and

$$s(t) = -16t^2 + 30t \text{ ft.}$$

$s(t) = -16t^2 + v_0t + s_0$
s_0 = initial position

In particular, after 5 seconds the stone has a height of

$$s(5) = -16(5)^2 + 30(5) = -250 \text{ ft.}$$

In other words, the stone is now 250 ft *below* where it was when you first threw it, as shown in Figure 2.

c. The stone reaches its zenith when its height $s(t)$ is at its maximum value, which occurs when $v(t) = s'(t)$ is zero. So we solve

$$v(t) = -32t + 30 = 0$$

getting $t = 30/32 = 15/16 = 0.9375$ sec. This is the time when the stone reaches its zenith. The height of the stone at that time is

$$s(15/16) = -16(15/16)^2 + 30(15/16) = 14.0625 \text{ ft.}$$

➡ **Before we go on...** Here again are the formulas we obtained in Example 8, together with their metric equivalents:

Vertical Motion under Gravity: Velocity and Position

If we ignore air resistance, the vertical velocity and position of an object moving under gravity are given by

	British Units	Metric Units
Velocity:	$v(t) = -32t + v_0$ ft/sec	$v(t) = -9.8t + v_0$ m/sec
Position:	$s(t) = -16t^2 + v_0t + s_0$ ft	$s(t) = -4.9t^2 + v_0t + s_0$ m

v_0 = initial velocity = velocity at time 0
s_0 = initial position = position at time 0

Quick Example

If a ball is thrown down at 2 ft/sec from a height of 200 ft, then its velocity and position after t seconds are $v(t) = -32t - 2$ ft/sec and $s(t) = -16t^2 - 2t + 200$ ft. ■

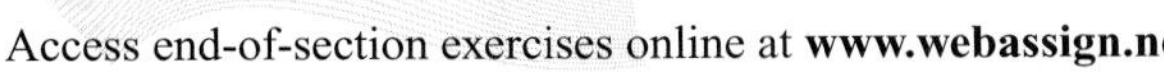

6.1 EXERCISES

Access end-of-section exercises online at **www.webassign.net**

6.2 Substitution

The chain rule for derivatives gives us an extremely useful technique for finding antiderivatives. This technique is called **change of variables** or **substitution**.

Recall that to differentiate a function like $(x^2+1)^6$, we first think of the function as $g(u)$, where $u = x^2 + 1$ and $g(u) = u^6$. We then compute the derivative, using the chain rule, as

$$\frac{d}{dx}g(u) = g'(u)\frac{du}{dx}.$$

Any rule for derivatives can be turned into a technique for finding antiderivatives by writing it in integral form. The integral form of the above formula is

$$\int g'(u)\frac{du}{dx}\,dx = g(u) + C.$$

But, if we write $g(u) + C = \int g'(u)\,du$, we get the following interesting equation:

$$\int g'(u)\frac{du}{dx}\,dx = \int g'(u)\,du.$$

This equation is the one usually called the *change of variables formula.* We can turn it into a more useful integration technique as follows. Let $f = g'(u)(du/dx)$. We can rewrite the above change of variables formula using f:

$$\int f\,dx = \int \left(\frac{f}{du/dx}\right)du.$$

In essence, we are making the formal substitution

$$dx = \frac{1}{du/dx}\,du.$$

Here's the technique:

Substitution Rule

If u is a function of x, then we can use the following formula to evaluate an integral:

$$\int f\,dx = \int \left(\frac{f}{du/dx}\right)du.$$

Rather than use the formula directly, we use the following step-by-step procedure:

1. Write u as a function of x.
2. Take the derivative du/dx and solve for the quantity dx in terms of du.
3. Use the expression you obtain in step 2 to substitute for dx in the given integral and substitute u for its defining expression.

Now let's see how this procedure works in practice.

EXAMPLE 1 Substitution

Find $\int 4x(x^2+1)^6\,dx$.

Solution To use substitution we need to choose an expression to be u. There is no hard and fast rule, but here is one hint that often works:

Take u to be an expression that is being raised to a power.

In this case, let's set $u = x^2 + 1$. Continuing the procedure above, we place the calculations for Step 2 in a box.

$u = x^2 + 1$	Write u as a function of x.
$\frac{du}{dx} = 2x$	Take the derivative of u with respect to x.
$dx = \frac{1}{2x}\,du$	Solve for dx: $dx = \frac{1}{du/dx}\,du$.

Now we *substitute u for its defining expression and substitute for dx* in the original integral:

$$\int 4x(x^2+1)^6\,dx = \int 4xu^6\frac{1}{2x}\,du \qquad \text{Substitute* for } u \text{ and } dx.$$

$$= \int 2u^6\,du. \qquad \text{Cancel the } x\text{s and simplify.}$$

* This step is equivalent to using the formula stated in the Substitution Rule box. If it should bother you that the integral contains both x and u, note that x is now a function of u.

We have boiled the given integral down to the much simpler integral $\int 2u^6\,du$, and we can now write down the solution:

$$2\frac{u^7}{7} + C = \frac{2(x^2+1)^7}{7} + C. \qquad \text{Substitute } (x^2+1) \text{ for } u \text{ in the answer.}$$

Before we go on... There are two points to notice in Example 1. First, before we can actually integrate with respect to u, *we must eliminate all x's from the integrand*. If we cannot, we may have chosen the wrong expression for u. Second, after integrating, we must substitute back to obtain an expression involving x.

It is easy to check our answer. We differentiate:

$$\frac{d}{dx}\left[\frac{2(x^2+1)^7}{7}\right] = \frac{2(7)(x^2+1)^6(2x)}{7} = 4x(x^2+1)^6. \quad ✔$$

Notice how we used the chain rule to check the result obtained by substitution. ■

When we use substitution, the first step is always to decide what to take as u. Again, there are no set rules, but we see some common cases in the examples.

EXAMPLE 2 More Substitution

Evaluate the following:

a. $\int x^2(x^3+1)^2\,dx$ **b.** $\int 3xe^{x^2}\,dx$ **c.** $\int \frac{1}{2x+5}\,dx$ **d.** $\int \left(\frac{1}{2x+5}+4x^2+1\right)dx$

Solution

a. As we said in Example 1, it often works to take u to be an expression that is being raised to a power. We usually also want to see the derivative of u as a factor in the integrand so that we can cancel terms involving x. In this case, x^3+1 is being raised to a power, so let's set $u=x^3+1$. Its derivative is $3x^2$; in the integrand we see x^2, which is missing the factor 3, but missing or incorrect constant factors are not a problem.

$u = x^3 + 1$	Write u as a function of x.
$\frac{du}{dx} = 3x^2$	Take the derivative of u with respect to x.
$dx = \frac{1}{3x^2}\,du$	Solve for dx: $dx = \frac{1}{du/dx}\,du$.

$$\int x^2(x^3+1)^2\,dx = \int x^2u^2\frac{1}{3x^2}du \qquad \text{Substitute for } u \text{ and } dx.$$

$$= \int \frac{1}{3}u^2\,du \qquad \text{Cancel the terms with } x.$$

$$= \frac{1}{9}u^3 + C \qquad \text{Take the antiderivative.}$$

$$= \frac{1}{9}(x^3+1)^3 + C \qquad \text{Substitute for } u \text{ in the answer.}$$

b. When we have an exponential with an expression in the exponent, it often works to substitute u for that expression. In this case, let's set $u=x^2$. (Notice again that we see a constant multiple of its derivative $2x$ as a factor in the integrand—a good sign.)

$$u = x^2$$

$$\frac{du}{dx} = 2x$$

$$dx = \frac{1}{2x}\,du$$

Substituting into the integral, we have

$$\int 3xe^{x^2}dx = \int 3xe^u\frac{1}{2x}\,du = \int \frac{3}{2}e^u du$$

$$= \frac{3}{2}e^u + C = \frac{3}{2}e^{x^2} + C.$$

c. We begin by rewriting the integrand as a power:

$$\int \frac{1}{2x+5}\,dx = \int (2x+5)^{-1}\,dx.$$

Now we take our earlier advice and set u equal to the expression that is being raised to a power:

$$u = 2x + 5$$

$$\frac{du}{dx} = 2$$

$$dx = \frac{1}{2}\,du$$

Substituting into the integral, we have

$$\int \frac{1}{2x+5}\,dx = \int \frac{1}{2}u^{-1}\,du = \frac{1}{2}\ln|u| + C$$

$$= \frac{1}{2}\ln|2x+5| + C.$$

d. Here, the substitution $u = 2x + 5$ works for the first part of the integrand, $1/(2x+5)$, but not for the rest of it, so we break up the integral:

$$\int \left(\frac{1}{2x+5} + 4x^2 + 1\right) dx = \int \frac{1}{2x+5}\,dx + \int (4x^2+1)\,dx.$$

For the first integral we can use the substitution $u = 2x + 5$ [which we did in part (a)], and for the second, no substitution is necessary:

$$\int \frac{1}{2x+5}\,dx + \int (4x^2+1)\,dx = \frac{1}{2}\ln|2x+5| + \frac{4x^3}{3} + x + C.$$

EXAMPLE 3 Choosing u

Evaluate $\int (x+3)\sqrt{x^2+6x}\,dx$.

Solution There are two parenthetical expressions. Notice, however, that the derivative of the expression (x^2+6x) is $2x+6$, which is twice the term $(x+3)$ in front of the radical. Recall that we would like the derivative of u to appear as a factor. Thus, let's take $u = x^2 + 6x$.

$$u = x^2 + 6x$$

$$\frac{du}{dx} = 2x + 6 = 2(x+3)$$

$$dx = \frac{1}{2(x+3)}\,du$$

Substituting into the integral, we have

$$\int (x+3)\sqrt{x^2+6x}\,dx$$

$$= \int (x+3)\sqrt{u}\left(\frac{1}{2(x+3)}\right) du$$

$$= \int \frac{1}{2}\sqrt{u}\,du = \frac{1}{2}\int u^{1/2}du$$

$$= \frac{1}{2}\frac{2}{3}u^{3/2} + C = \frac{1}{3}(x^2+6x)^{3/2} + C.$$

Some cases require a little more work.

EXAMPLE 4 When the *x* Terms Do Not Cancel

Evaluate $\int \frac{2x}{(x-5)^2}\,dx$.

Solution We first rewrite

$$\int \frac{2x}{(x-5)^2}\,dx = \int 2x(x-5)^{-2}\,dx.$$

This suggests that we should set $u = x - 5$.

$$u = x - 5$$
$$\frac{du}{dx} = 1$$
$$dx = du$$

Substituting, we have

$$\int \frac{2x}{(x-5)^2}\,dx = \int 2xu^{-2}\,du.$$

Now, there is nothing in the integrand to cancel the x that appears. If, as here, there is still an x in the integrand after substituting, we go back to the expression for u, solve for x, and substitute the expression we obtain for x in the integrand. So, we take $u = x - 5$ and solve for $x = u + 5$. Substituting, we get

$$\begin{aligned}\int 2xu^{-2}\,du &= \int 2(u+5)u^{-2}\,du \\ &= 2\int (u^{-1} + 5u^{-2})\,du \\ &= 2\ln|u| - \frac{10}{u} + C \\ &= 2\ln|x-5| - \frac{10}{x-5} + C.\end{aligned}$$

EXAMPLE 5 Application: Bottled Water for Pets

Annual sales of bottled spring water for pets can be modeled by the logistic function

$$s(t) = \frac{3{,}000e^{0.5t}}{3 + e^{0.5t}} \text{ million gallons per year} \qquad (0 \le t \le 12),$$

where t is time in years since the start of 2000.[2]

a. Find an expression for the total amount of bottled spring water for pets sold since the start of 2000.

b. How much bottled spring water for pets was sold from the start of 2005 to the start of 2008?

Solution

a. If we write the total amount of pet spring water sold since the start of 2000 as $S(t)$, then the information we are given says that

$$S'(t) = s(t) = \frac{3{,}000e^{0.5t}}{3 + e^{0.5t}}.$$

[2]Based on data through 2008 and the recovery in 2010 of the general bottled water market to pre-recession levels. Sources: "*Liquid Assets: America's Expensive Love Affair with Bottled Water*" Daniel Gross, April 26, 2011 (finance.yahoo.com), Beverage Marketing Corporation (www.beveragemarketing.com).

Thus,

$$S(t) = \int \frac{3{,}000e^{0.5t}}{3 + e^{0.5t}}\,dt$$

is the function we are after. To integrate the expression, take u to be the denominator of the integrand:

$$u = 3 + e^{0.5t}$$
$$\frac{du}{dt} = 0.5e^{0.5t}$$
$$dt = \frac{1}{0.5e^{0.5t}}\,du$$

$$\begin{aligned} S(t) &= \int \frac{3{,}000e^{0.5t}}{3 + e^{0.5t}}\,dt \\ &= \int \frac{3{,}000e^{0.5t}}{u} \cdot \frac{1}{0.5e^{0.5t}}\,du \\ &= \frac{3{,}000}{0.5} \int \frac{1}{u}\,du \\ &= 6{,}000 \ln|u| + C = 6{,}000 \ln(3 + e^{0.5t}) + C. \end{aligned}$$

(Why could we drop the absolute value in the last step?)

Now what is C? Because $S(t)$ represents the total amount of bottled spring water for pets sold *since time* $t = 0$, we have $S(0) = 0$ (because that is when we started counting). Thus,

$$\begin{aligned} 0 &= 6{,}000 \ln\big(3 + e^{0.5(0)}\big) + C \\ &= 6{,}000 \ln 4 + C \\ C &= -6{,}000 \ln 4 \approx -8{,}318. \end{aligned}$$

Therefore, the total sales from the start of 2000 is approximately

$$S(t) = 6{,}000 \ln(3 + e^{0.5t}) - 8{,}318 \text{ million gallons.}$$

b. The period from the start of 2005 to the start of 2008 is represented by the interval [5, 8]. From part (a):

$$\begin{aligned} &\text{Sales through the start of 2005} = S(5) \\ &\quad = 6{,}000 \ln\big(3 + e^{0.5(5)}\big) - 8{,}318 \approx 8{,}003 \text{ million gallons.} \end{aligned}$$

$$\begin{aligned} &\text{Sales through the start of 2008} = S(8) \\ &\quad = 6{,}000 \ln\big(3 + e^{0.5(8)}\big) - 8{,}318 \approx 16{,}003 \text{ million gallons.} \end{aligned}$$

Therefore, sales over the period were about $16{,}003 - 8{,}003 = 8{,}000$ million gallons.

➡ **Before we go on...** You might wonder why we are writing a logistic function in the form we used in Example 5 rather than in one of the "standard" forms $\frac{N}{1 + Ab^{-t}}$ or $\frac{N}{1 + Ae^{-kt}}$. Our only reason for doing this is to make the substitution work. To convert from the second "standard" form to the form we used in the example, multiply top and bottom by e^{kt}. (See Exercises 81 and 82 associated with Section 6.4 for further discussion.) ■

Shortcuts

The following rule allows us to simply write down the antiderivative in cases where we would otherwise need the substitution $u = ax + b$, as in Example 2 (a and b are constants with $a \neq 0$):

Shortcut Rule: Integrals of Expressions Involving $(ax + b)$

If $\int f(x)\,dx = F(x) + C$ and a and b are constants, with $a \neq 0$, then*

$$\int f(ax+b)\,dx = \frac{1}{a}F(ax+b) + C.$$

Quick Example

(Also see the examples in the table below.)

Because $\int x^4\,dx = \frac{x^5}{5} + C$, it follows that

$$\int (3x-1)^4\,dx = \frac{1}{3}\frac{(3x-1)^5}{5} + C = \frac{(3x-1)^5}{15} + C.$$

* The shortcut rule can be justified by making the substitution $u = ax + b$ in the general case as stated.

Below are some instances of the shortcut rule with additional examples (their individual derivations using the substitution $u = ax + b$ will appear in the exercises associated with this section):

Shortcut Rule	***Example***
$\int (ax+b)^n\,dx = \frac{1}{a}\frac{(ax+b)^{n+1}}{n+1} + C$ (if $n \neq -1$)	$\int (3x-1)^2\,dx = \frac{(3x-1)^3}{3(3)} + C = \frac{(3x-1)^3}{9} + C$
$\int (ax+b)^{-1}\,dx = \frac{1}{a}\ln\lvert ax+b\rvert + C$	$\int (3-2x)^{-1}\,dx = \frac{1}{(-2)}\ln\lvert 3-2x\rvert + C = -\frac{1}{2}\ln\lvert 3-2x\rvert + C$
$\int e^{ax+b}\,dx = \frac{1}{a}e^{ax+b} + C$	$\int e^{-x+4}\,dx = \frac{1}{(-1)}e^{-x+4} + C = -e^{-x+4} + C$
$\int c^{ax+b}\,dx = \frac{1}{a\ln c}c^{ax+b} + C$	$\int 2^{-3x+4}\,dx = \frac{1}{(-3\ln 2)}2^{-3x+4} + C = -\frac{1}{3\ln 2}2^{-3x+4} + C$
$\int \lvert ax+b\rvert\,dx = \frac{1}{2a}(ax+b)\lvert ax+b\rvert + C$	$\int \lvert 2x-1\rvert\,dx = \frac{1}{4}(2x-1)\lvert 2x-1\rvert + C$

The following more general version of the shortcut rule was suggested by a student.[†] We have marked it as optional so that you could skip it on a first reading, but we strongly suggest you come back to it afterward, as the rule will allow you to

† Mike Fuschetto, who was a Hofstra business calculus student in Spring 2010.

easily write down the answers in most of the exercises associated with this section, as well as in almost all the examples of this section.

(Optional) Mike's Shortcut Rule: Integrals of More General Expressions

If $\int f(x)\,dx = F(x) + C$ and g and u are any differentiable functions of x, then

$$\int g \cdot f(u)\,dx = \frac{g}{u'} \cdot F(u) + C \text{ provided that } \frac{g}{u'} \text{ is constant.}$$

Quick Example

(Also see the examples in the table below.)

$$\int x^4\,dx = \frac{x^5}{5} + C, \quad \text{so}$$

$$\int 5x^2(x^3-1)^4\,dx = \frac{5x^2}{3x^2} \cdot \frac{(x^3-1)^5}{5} + C$$

$$= \frac{5}{3} \cdot \frac{(x^3-1)^5}{5} + C = \frac{(x^3-1)^5}{3} + C.$$

Caution

If g/u' is not constant, then Mike's rule does not apply; for instance, the following calculation is *wrong*:

$$\int x(2x-1)^2 dx = \frac{x}{2} \cdot \frac{(2x-1)^3}{3} + C \quad \text{✗ } \textit{WRONG!} \text{ because } \frac{x}{2} \text{ is not constant.}$$

Here are some instances of Mike's rule with additional examples.

Shortcut Rule	*Example*
$\int g \cdot u^n\,dx = \frac{g}{u'}\frac{u^{n+1}}{n+1} + C$ (if $n \neq -1$)	$\int 3x(x^2-1)^3\,dx = \frac{3x}{2x}\frac{(x^2-1)^4}{4} + C = \frac{3(x^2-1)^4}{8} + C$
$\int g \cdot u^{-1}\,dx = \frac{g}{u'}\ln\|u\| + C$	$\int e^x(3-2e^x)^{-1}\,dx = \frac{e^x}{-2e^x}\ln\|3-2e^x\| + C = -\frac{1}{2}\ln\|3-2e^x\| + C$
$\int g \cdot e^u dx = \frac{g}{u'}e^u + C$	$\int x^2e^{-x^3+4}\,dx = \frac{x^2}{-3x^2}e^{-x^3+4} + C = -\frac{1}{3}e^{-x^3+4} + C$
$\int g \cdot c^u dx = \frac{g}{u'\ln c}c^u + C$	$\int x^3 2^{x^4-1}\,dx = \frac{x^3}{4x^3\ln 2}2^{x^4-1} + C = -\frac{1}{4\ln 2}2^{x^4-1} + C$
$\int g \cdot \|u\|\,dx = \frac{g}{2u'}u\|u\| + C$	$\int x\|x^2-1\|\,dx = \frac{x}{4x}(x^2-1)\|x^2-1\| + C = \frac{1}{4}(x^2-1)\|x^2-1\| + C$

FAQs

When to Use Substitution and What to Use for *u*

Q: *If I am asked to calculate an antiderivative, how do I know when to use a substitution and when* not *to use one?*

A: Do *not* use substitution when integrating sums, differences, and/or constant multiples of powers of x and exponential functions, such as $2x^3 - \frac{4}{x^2} + \frac{1}{2x} + 3^x + \frac{2^x}{3}$.

To recognize when you should try a substitution, pretend that you are *differentiating* the given expression instead of integrating it. If differentiating the expression would require use of the chain rule, then integrating that expression may well require a substitution, as in, say, $x(3x^2 - 4)^3$ or $(x + 1)e^{x^2+2x-1}$. (In the first we have a *quantity* cubed, and in the second we have e raised to a *quantity*.)

Q: *If an integral seems to call for a substitution, what should I use for u?*

A: There are no set rules for deciding what to use for u, but the preceding examples show some common patterns:

- If you see a linear expression raised to a power, try setting u equal to that linear expression. For example, in $(3x - 2)^{-3}$, set $u = 3x - 2$. (Alternatively, try using the shortcuts above.)
- If you see a constant raised to a linear expression, try setting u equal to that linear expression. For example, in $3^{(2x+1)}$, set $u = 2x + 1$. (Alternatively, try a shortcut.)
- If you see an expression raised to a power multiplied by the derivative of that expression (or a constant multiple of the derivative), try setting u equal to that expression. For example, in $x^2(3x^3 - 4)^{-1}$, set $u = 3x^3 - 4$.
- If you see a constant raised to an expression, multiplied by the derivative of that expression (or a constant multiple of its derivative), try setting u equal to that expression. For example, in $5(x + 1)e^{x^2+2x-1}$, set $u = x^2 + 2x - 1$.
- If you see an expression in the denominator and its derivative (or a constant multiple of its derivative) in the numerator, try setting u equal to that expression. For example, in $\frac{2^{3x}}{3 - 2^{3x}}$, set $u = 3 - 2^{3x}$.

Persistence often pays off: If a certain substitution does not work, try another approach or a different substitution.

6.2 EXERCISES

Access end-of-section exercises online at **www.webassign.net**

ENHANCED WebAssign

6.3 The Definite Integral: Numerical and Graphical Viewpoints

In Sections 6.1 and 6.2, we discussed the indefinite integral. There is an older, related concept called the **definite integral**. Let's introduce this new idea with an example. (We'll drop hints now and then about how the two types of integral are related. In Section 6.4 we discuss the exact relationship, which is one of the most important results in calculus.)

In Section 6.1, we used antiderivatives to answer questions of the form "Given the marginal cost, compute the total cost." (See Example 5 in Section 6.1.) In this section we approach such questions more directly, and we will forget about antiderivatives for now.

EXAMPLE 1 Oil Spill

Your deep ocean oil rig has suffered a catastrophic failure, and oil is leaking from the ocean floor wellhead at a rate of

$$v(t) = 0.08t^2 - 4t + 60 \text{ thousand barrels per day } (0 \le t \le 20),$$

where t is time in days since the failure.[3] Use a numerical calculation to estimate the total volume of oil released during the first 20 days.

Solution The graph of $v(t)$ is shown in Figure 3.

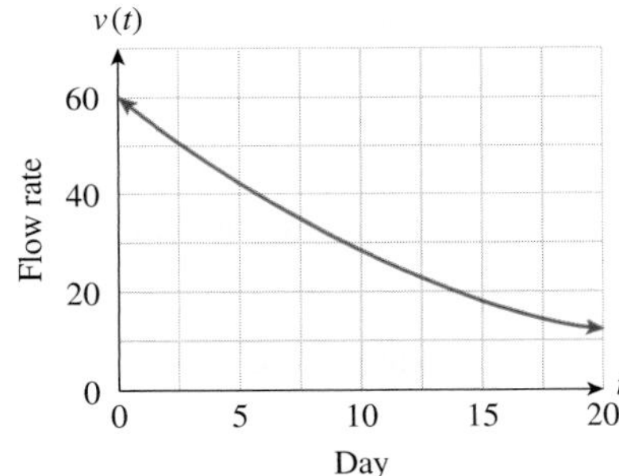

Figure 3

Let's start with a very crude estimate of the total volume of oil released, using the graph as a guide. The rate of change of this total volume at the beginning of the time period is $v(0) = 60$ thousand barrels per day. If this rate were to remain constant for the entire 20-day period, the total volume of oil released would be

$$\begin{aligned}\text{Total volume} &= \text{Volume per day} \times \text{Number of days} = 60 \times 20 \\ &= 1{,}200 \text{ thousand barrels.}\end{aligned}$$

[3]The model is consistent with the order of magnitude of the BP Deepwater Horizon oil spill of April 20–June 15, 2010, when the rate of flow of oil was estimated by the Federal Emergency Management Agency's Flow Rate Technical Group to be between 35,000 and 60,000 barrels per day (one barrel of oil is equivalent to about 0.16 cubic meters). Source: www.doi.gov/deepwaterhorizon.

Figure 4 shows how we can represent this calculation on the graph of $v(t)$.

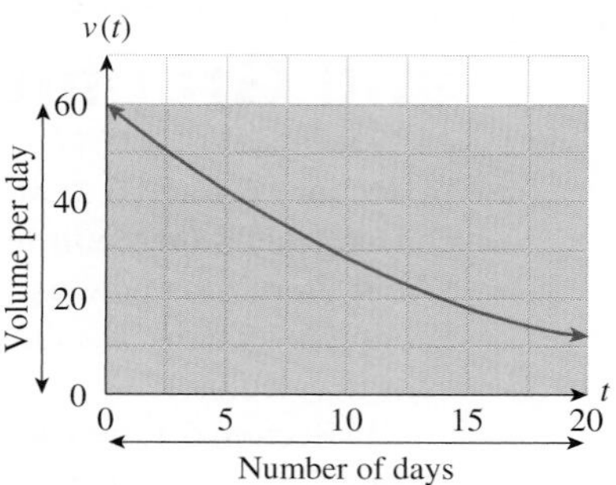

Figure 4

The volume per day based on $v(0) = 60$ is represented by the y-coordinate of the graph at its left edge, while the number of days is represented by the width of the interval [0, 20] on the x-axis. Therefore, computing the area of the shaded rectangle in the figure gives the same calculation:

$$\begin{aligned}\text{Area of rectangle} &= \text{Volume per day} \times \text{Number of days}\\ &= 60 \times 20 = 1{,}200 = \text{Total volume.}\end{aligned}$$

But, as we see in the graph, the flow rate does not remain constant, but goes down quite significantly over the course of the 20-day interval. We can obtain a somewhat more accurate estimate of the total volume by re-estimating the volume using 10-day periods—that is, by dividing the interval [0, 20] into two equal intervals, or subdivisions. We estimate the volume over each 10-day period using the flow rate at the beginning of that period.

$$\begin{aligned}\text{Volume in first period} &= \text{Volume per day} \times \text{Number of days}\\ &= v(0) \times 10 = 60 \times 10 = 600 \text{ thousand barrels}\\ \text{Volume in second period} &= \text{Volume per day} \times \text{Number of days}\\ &= v(10) \times 10 = 28 \times 10 = 280 \text{ thousand barrels}\end{aligned}$$

Adding these volumes gives us the more accurate estimate

$$v(0) \times 10 + v(10) \times 10 = 880 \text{ thousand barrels.}$$ Calculation using 2 subdivisions

In Figure 5 we see that we are computing the combined area of two rectangles, each of whose heights is determined by the height of the graph at its left edge:

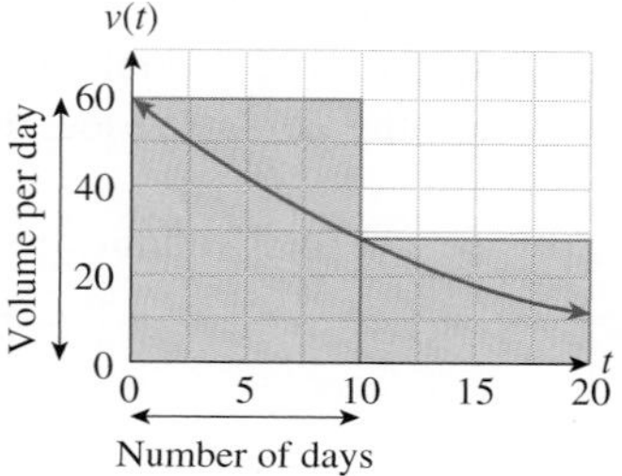

The areas of the rectangles are estimates of the volumes for successive 10-day periods.

Figure 5

$$\text{Area of first rectangle} = \text{Volume per day} \times \text{Number of days}$$
$$= v(0) \times 10 = 60 \times 10 = 600 = \text{Volume for first 10 days}$$
$$\text{Area of second rectangle} = \text{Volume per day} \times \text{Number of days} = v(10) \times 10$$
$$= 28 \times 10 = 280 = \text{Volume for second 10 days.}$$

We can get an even better estimate of the volume by using four divisions of [0, 20] instead of two:

$$v(0) \times 5 + v(5) \times 5 + v(10) \times 5 + v(15) \times 5$$ Calculation using 4 subdivisions
$$= 300 + 210 + 140 + 90 = 740 \text{ thousand barrels.}$$

As we see in Figure 6, we have now computed the combined area of *four* rectangles, each of whose heights is again determined by the height of the graph at its left edge.

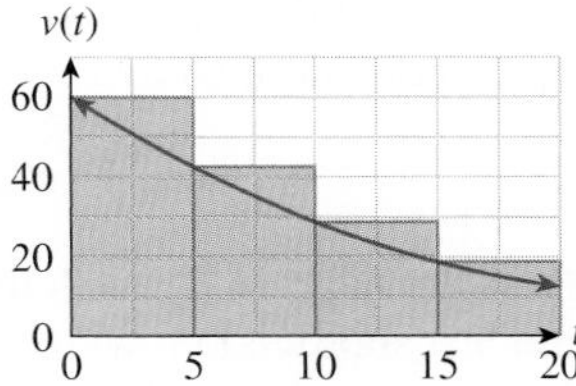

Estimated Volume Using 4 Subdivisions
The areas of the rectangles are estimates of the volumes for successive 5-day periods.

Figure 6

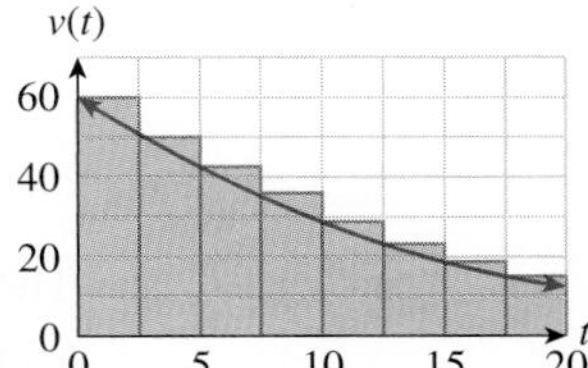

Estimated Volume Using 8 Subdivisions
The areas of the rectangles are estimates of the volumes for successive 2.5-day periods.

Figure 7

Notice how the volume seems to be decreasing as we use more subdivisions. More importantly, total volume seems to be getting closer to the area under the graph. Figure 7 illustrates the calculation for 8 equal subdivisions. The approximate total volume using 8 subdivisions is the total area of the shaded region in Figure 7:

$$v(0) \times 2.5 + v(2.5) \times 2.5 + v(5) \times 2.5 + \cdots + v(17.5) \times 2.5$$
$$= 675 \text{ thousand barrels.}$$ Calculation using 8 subdivisions

Looking at Figure 7, we still get the impression that we are overestimating the volume. If we want to be *really* accurate in our estimation of the volume, we should really be calculating the volume *continuously* every few hours or, better yet, minute by minute, as illustrated in Figure 8.

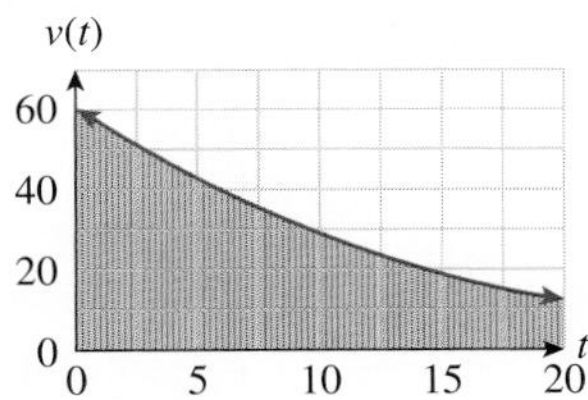

Every Six Hours (80 subdivisions)
Volume ≈ 619.35 thousand barrels

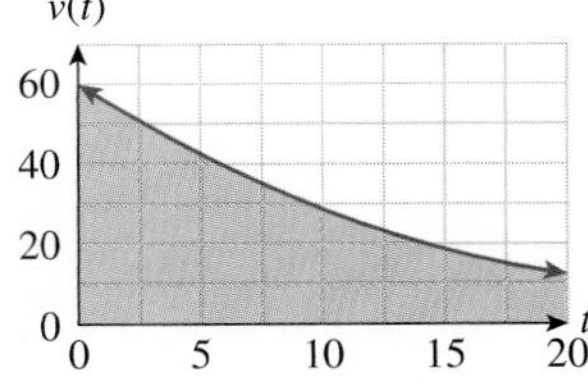

Every Minute (28,800 subdivisions)
Volume ≈ 613.35 thousand barrels

Figure 8

using Technology

Website
www.WanerMath.com
At the Website, select the Online Utilities tab, and choose the Numerical Integration Utility and Grapher. There, enter the formula

```
0.08x^2-4x+60
```

for $f(x)$, enter 0 and 20 for the lower and upper limits, and press "Left Sum" to compute and see a Riemann sum. You can then adjust the number of subdivisions to see the effect on the graph and on the calculation, and thus replicate the information in Figure 8.

Figure 8 strongly suggests that the more accurately we estimate the total volume, the closer the answer gets to the exact area under the portion of the graph of $v(t)$ with $0 \le t \le 20$, and leads us to the conclusion that the *exact* total volume is the exact area under the rate of change of volume curve for $0 \le t \le 20$. In other words, we have made the following remarkable discovery:

Total volume is the area under the rate of change of volume curve!

➡ **Before we go on...** The 80-subdivision calculation in Example 1 is tedious to do by hand, and no one in his or her right mind would even *attempt* to do the minute-by-minute calculation by hand! Below we discuss ways of doing these calculations with the aid of technology. ■

The type of calculation done in Example 1 is useful in many applications. Let's look at the general case and give the result a name.

In general, we have a function f (such as the function v in the example), and we consider an interval $[a, b]$ of possible values of the independent variable x. We subdivide the interval $[a, b]$ into some number of segments of equal length. Write n for the number of segments, or **subdivisions**.

Next, we label the endpoints of these subdivisions x_0 for a, x_1 for the end of the first subdivision, x_2 for the end of the second subdivision, and so on until we get to x_n, the end of the nth subdivision, so that $x_n = b$. Thus,

$$a = x_0 < x_1 < \cdots < x_n = b.$$

The first subdivision is the interval $[x_0, x_1]$, the second subdivision is $[x_1, x_2]$, and so on until we get to the last subdivision, which is $[x_{n-1}, x_n]$. We are dividing the interval $[a, b]$ into n subdivisions of equal length, so each segment has length $(b - a)/n$. We write Δx for $(b - a)/n$ (Figure 9).

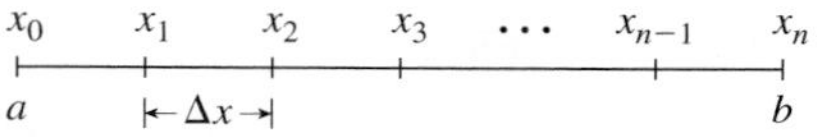

x_0 x_1 x_2 x_3 . . . x_{n-1} x_n

a $\vdash \Delta x \dashv$ b

Figure 9

Having established this notation, we can write the calculation that we want to do as follows: For each subdivision $[x_{k-1}, x_k]$, compute $f(x_{k-1})$, the value of the function f at the left endpoint. Multiply this value by the length of the interval, which is Δx. Then add together all n of these products to get the number

$$f(x_0)\Delta x + f(x_1)\Delta x + \cdots + f(x_{n-1})\Delta x.$$

This sum is called a **(left) Riemann* sum** for f. In Example 1 we computed several different Riemann sums. Here is the computation for $n = 4$ we used in the oil spill example (see Figure 10):

* After Georg Friedrich Bernhard Riemann (1826–1866).

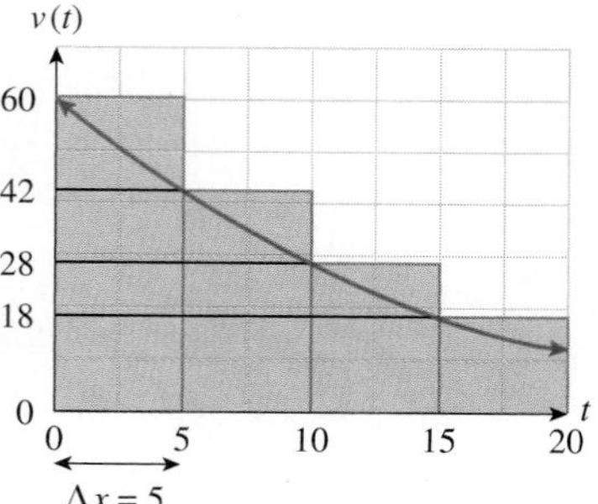

Figure 10

$$\begin{aligned}\text{Left Riemann sum} &= f(x_0)\Delta x + f(x_1)\Delta x + \cdots + f(x_{n-1})\Delta x \\ &= f(0)(5) + f(5)(5) + f(10)(5) + f(15)(5) \\ &= 60(5) + 42(5) + 28(5) + 18(5) = 740.\end{aligned}$$

Because sums are often used in mathematics, mathematicians have developed a shorthand notation for them. We write

$$f(x_0)\Delta x + f(x_1)\Delta x + \cdots + f(x_{n-1})\Delta x \text{ as } \sum_{k=0}^{n-1} f(x_k)\Delta x.$$

The symbol $\sum$ is the Greek letter sigma and stands for **summation**. The letter k here is called the index of summation, and we can think of it as counting off the segments. We read the notation as "the sum from $k = 0$ to $n - 1$ of the quantities $f(x_k)\Delta x$." Think of it as a set of instructions:

Set $k = 0$, and calculate $f(x_0)\Delta x$. — $f(0)(5)$ in the above calculation

Set $k = 1$, and calculate $f(x_1)\Delta x$. — $f(5)(5)$ in the above calculation

. . .

Set $k = n - 1$, and calculate $f(x_{n-1})\Delta x$. — $f(15)(5)$ in the above calculation

Then sum all the quantities so calculated.

Riemann Sum

If f is a continuous function, the **left Riemann sum** with n equal subdivisions for f over the interval $[a, b]$ is defined to be

$$\begin{aligned}\text{Left Riemann sum} &= \sum_{k=0}^{n-1} f(x_k)\Delta x \\ &= f(x_0)\Delta x + f(x_1)\Delta x + \cdots + f(x_{n-1})\Delta x \\ &= [f(x_0) + f(x_1) + \cdots + f(x_{n-1})]\Delta x,\end{aligned}$$

where $a = x_0 < x_1 < \cdots < x_n = b$ are the endpoints of the subdivisions, and $\Delta x = (b - a)/n$.

Interpretation of the Riemann Sum

If f is the rate of change of a quantity F (that is, $f = F'$), then the Riemann sum of f approximates the total change of F from $x = a$ to $x = b$. The approximation improves as the number of subdivisions increases toward infinity.

Quick Examples

1. If $f(t)$ is the rate of change in the number of bats in a belfry and $[a, b] = [2, 3]$, then the Riemann sum approximates the total change in the number of bats in the belfry from time $t = 2$ to time $t = 3$.
2. If $c(x)$ is the marginal cost of producing the xth item and $[a, b] = [10, 20]$, then the Riemann sum approximates the cost of producing items 11 through 20.

Visualizing a Left Riemann Sum (Non-negative Function)

Graphically, we can represent a left Riemann sum of a non-negative function as an approximation of the area under a curve:

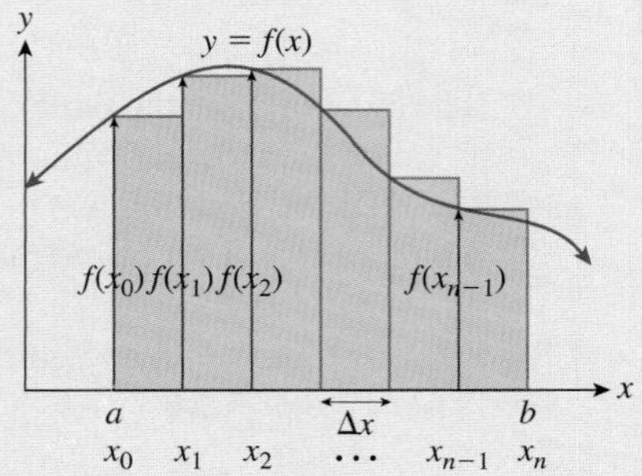

Riemann sum = Shaded area = Area of first rectangle + Area of second rectangle + · · · + Area of nth rectangle $= f(x_0)\Delta x + f(x_1)\Delta x + f(x_2)\Delta x + \cdots + f(x_{n-1})\Delta x$.

Quick Example

In Example 1 we computed several Riemann sums, including these:

$n = 1$: Riemann sum $= v(0)\Delta t = 60 \times 20 = 1{,}200$

$n = 2$: Riemann sum $= [v(t_0) + v(t_1)]\Delta t$

$= [v(0) + v(10)](10) = 880$

$n = 4$: Riemann sum $= [v(t_0) + v(t_1) + v(t_2) + v(t_3)]\Delta t$

$= [v(0) + v(5) + v(10) + v(15)](5) = 740$

$n = 8$: Riemann sum $= [v(t_0) + v(t_1) + \cdots + v(t_7)]\Delta t$

$= [v(0) + v(2.5) + \cdots + v(17.5)](2.5) = 675$

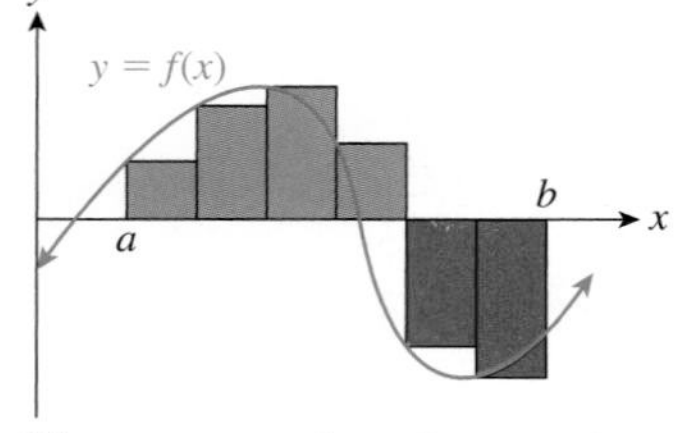

Riemann sum = Area above x-axis − Area below x-axis

Figure 11

Note To visualize the Riemann sum of a function that is negative, look again at the formula $f(x_0)\Delta x + f(x_1)\Delta x + f(x_2)\Delta x + \cdots + f(x_{n-1})\Delta x$ for the Riemann sum. Each term $f(x_k)\Delta x_k$ represents the area of one rectangle in the figure above. So, the areas of the rectangles with negative values of $f(x_k)$ are automatically counted as negative. They appear as red rectangles in Figure 11. ■

EXAMPLE 2 Computing a Riemann Sum from a Formula

Compute the left Riemann sum for $f(x) = x^2 + 1$ over the interval $[-1, 1]$, using $n = 5$ subdivisions.

Solution Because the interval is $[a, b] = [-1, 1]$ and $n = 5$, we have

$$\Delta x = \frac{b-a}{n} = \frac{1-(-1)}{5} = 0.4. \qquad \text{Width of subdivisions}$$

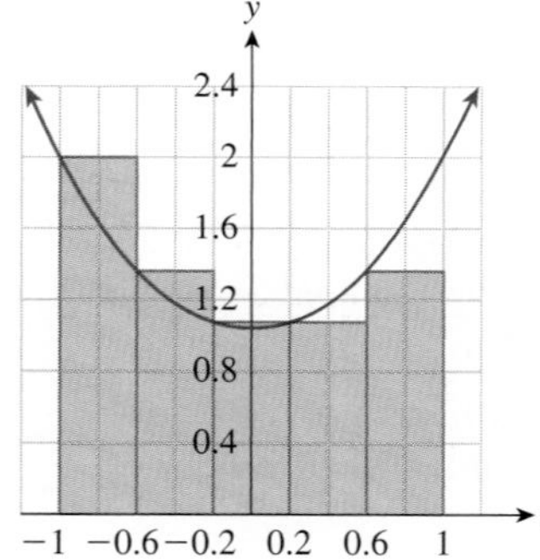

Figure 12

Thus, the subdivisions of $[-1, 1]$ are determined by

$$-1 < -0.6 < -0.2 < 0.2 < 0.6 < 1. \qquad \text{Start with } -1 \text{ and keep adding } \Delta x = 0.4.$$

Figure 12 shows the graph with a representation of the Riemann sum.

The Riemann sum we want is

$$[f(x_0) + f(x_1) + \cdots + f(x_4)]\Delta x$$
$$= [f(-1) + f(-0.6) + f(-0.2) + f(0.2) + f(0.6)]0.4.$$

We can conveniently organize this calculation in a table as follows:

x	-1	-0.6	-0.2	0.2	0.6	**Total**
$f(x) = x^2 + 1$	2	1.36	1.04	1.04	1.36	6.8

The Riemann sum is therefore

$$6.8\Delta x = 6.8 \times 0.4 = 2.72.$$

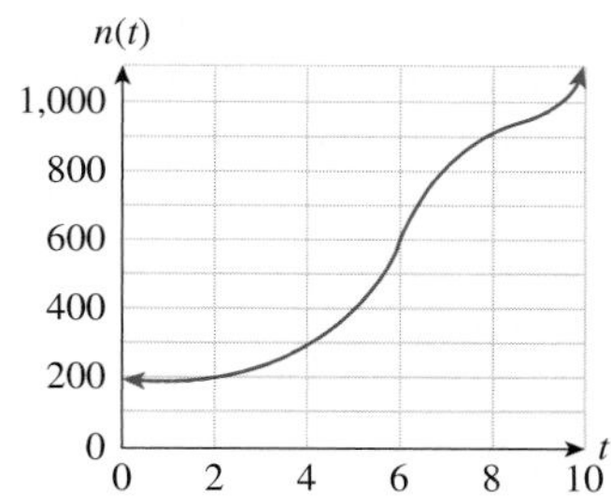

Figure 13

EXAMPLE 3 Computing a Riemann Sum from a Graph

Figure 13 shows the approximate annual production $n(t)$ of engineering and technology Ph.D. graduates in Mexico during the period 2000–2010.[4] Use a left Riemann sum with four subdivisions to estimate the total number of Ph.D. graduates from 2002 to 2010.

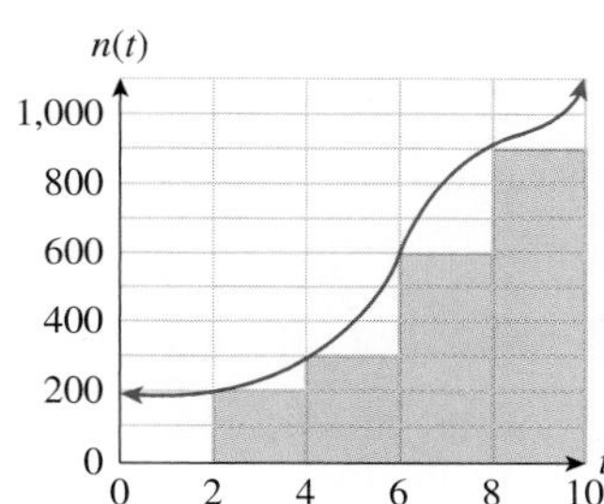

Figure 14

Solution Let us represent the total number of (engineering and technology) Ph.D. graduates up to time t (measured in years since 2000) by $N(t)$. The total number of Ph.D. graduates from 2002 to 2010 is then the total change in $N(t)$ over the interval [2, 10]. In view of the above discussion, we can approximate the total change in $N(t)$ using a Riemann sum of its rate of change $n(t)$. Because $n = 4$ subdivisions are specified, the width of each subdivision is

$$\Delta t = \frac{b-a}{n} = \frac{10-2}{4} = 2.$$

We can therefore represent the left Riemann sum by the shaded area shown in Figure 14.

From the graph,

$$\text{Left sum} = n(2)\Delta t + n(4)\Delta t + n(6)\Delta t + n(8)\Delta t$$
$$= (200)(2) + (300)(2) + (600)(2) + (900)(2) = 4{,}000.$$

So, we estimate that there was a total of about 4,000 engineering and technology Ph.D. graduates during the given period.

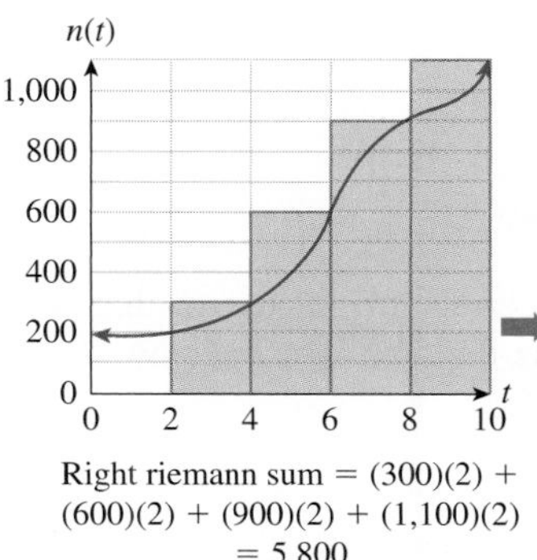

Right riemann sum = (300)(2) + (600)(2) + (900)(2) + (1,100)(2) = 5,800

Height of each ractangle is determined by height of graph at right edge.

Figure 15

Before we go on... A glance at Figure 14 tells us that we have significantly underestimated the actual number of graduates, as the actual area under the curve is considerably larger. Figure 15 shows a **right Riemann sum** and gives us a much larger estimate. For continuous functions, the difference between these two types of Riemann sums approaches zero as the number of subdivisions approaches infinity (see below) so we will focus primarily on only one type: left Riemann sums. ■

As in Example 1, we're most interested in what happens to the Riemann sum when we let n get very large. When f is continuous,* its Riemann sums will always approach a limit as n goes to infinity. (This is not meant to be obvious. Proofs may be found in advanced calculus texts.) We give the limit a name.

* This applies to some other functions as well, including "piecewise continuous" functions discussed in the next section.

[4] Source for data: Instituto Nacional de Estadística y Geografía (www.inegi.org.mx).

The Definite Integral

If f is a continuous function, the **definite integral of f from a to b** is defined to be the limit of the Riemann sums as the number of subdivisions approaches infinity:

$$\int_a^b f(x)\,dx = \lim_{n\to\infty} \sum_{k=0}^{n-1} f(x_k)\,\Delta x.$$

In Words: The integral, from a to b, of $f(x)\,dx$ equals the limit, as $n \to \infty$, of the Riemann Sum with a partition of n subdivisions.

The function f is called the **integrand**, the numbers a and b are the **limits of integration**, and the variable x is the **variable of integration**. A Riemann sum with a large number of subdivisions may be used to approximate the definite integral.

Interpretation of the Definite Integral

If f is the rate of change of a quantity F (that is, $f = F'$), then $\int_a^b f(x)\,dx$ is the (exact) total change of F from $x = a$ to $x = b$.

Quick Examples

1. If $f(t)$ is the rate of change in the number of bats in a belfry and $[a, b] = [2, 3]$, then $\int_2^3 f(t)\,dt$ is the total change in the number of bats in the belfry from time $t = 2$ to time $t = 3$.
2. If, at time t hours, you are selling wall posters at a rate of $s(t)$ posters per hour, then

$$\text{Total number of posters sold from hour 3 to hour 5} = \int_3^5 s(t)\,dt.$$

Visualizing the Definite Integral

Non-negative Functions: If $f(x) \ge 0$ for all x in $[a, b]$, then $\int_a^b f(x)\,dx$ is the area under the graph of f over the interval $[a, b]$, as shaded in the figure.

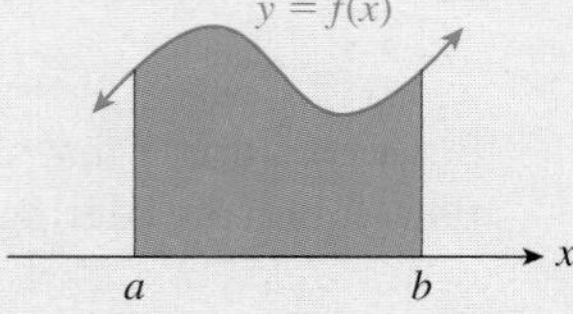

General Functions: $\int_a^b f(x)\,dx$ is the area between $x = a$ and $x = b$ that is above the x-axis and below the graph of f, minus the area that is below the x-axis and above the graph of f:

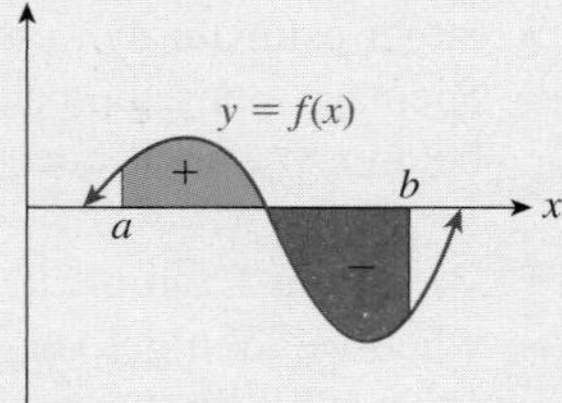

$$\int_a^b f(x)\,dx = \text{Area above } x\text{-axis} - \text{Area below } x\text{-axis}$$

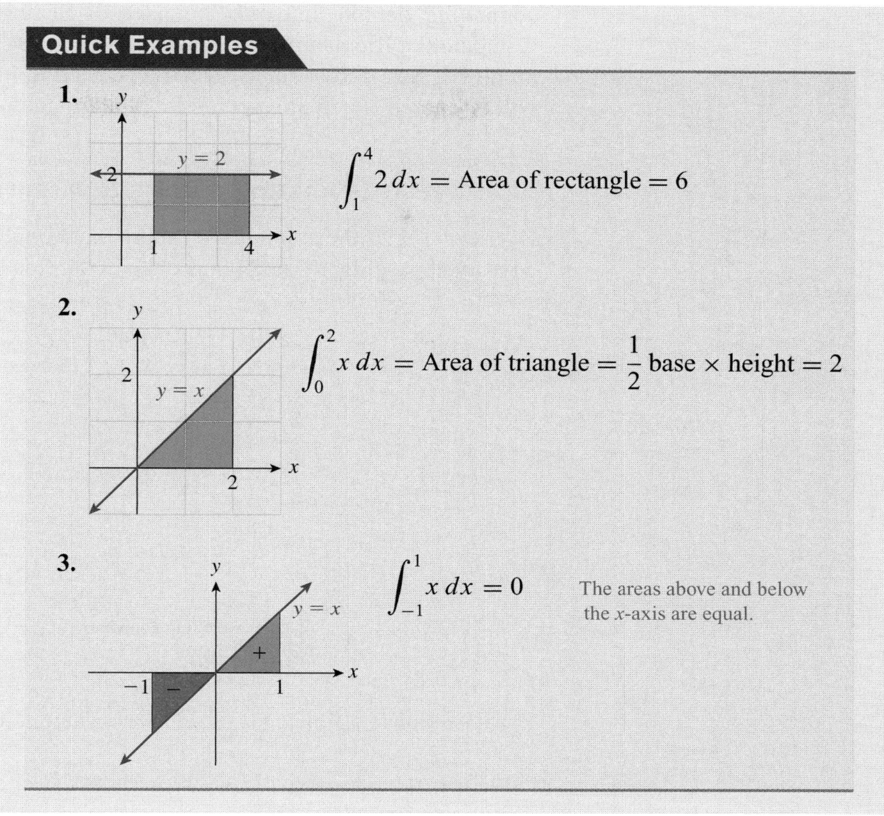

Quick Examples
1.
y
y = 2
2
1
4
x
$\int_1^4 2\,dx = \text{Area of rectangle} = 6$
2.
y
2
y = x
2
x
$\int_0^2 x\,dx = \text{Area of triangle} = \frac{1}{2}\text{base} \times \text{height} = 2$
3.
y
y = x
+
−1
−
1
x
$\int_{-1}^1 x\,dx = 0$
The areas above and below the x-axis are equal.

Notes

1. Remember that $\int_a^b f(x)\,dx$ stands for a number that depends on f, a, and b. The variable of integration x that appears is called a **dummy variable** because it has no effect on the answer. In other words,

$$\int_a^b f(x)\,dx = \int_a^b f(t)\,dt. \qquad \text{x or t is just a name we give the variable.}$$

2. The notation for the definite integral (due to Leibniz) comes from the notation for the Riemann sum. The integral sign $\int$ is an elongated S, the Roman equivalent of the Greek $\sum$. The d in dx is the lowercase Roman equivalent of the Greek Δ.
3. The definition above is adequate for continuous functions, but more complicated definitions are needed to handle other functions. For example, we broke the interval $[a, b]$ into n subdivisions of equal length, but other definitions allow a **partition** of the interval into subdivisions of possibly unequal lengths. We have evaluated f at the left endpoint of each subdivision, but we could equally well have used the right endpoint or any other point in the subdivision. All of these variations lead to the same answer when f is continuous.
4. The similarity between the notations for the definite integral and the indefinite integral is no mistake. We will discuss the exact connection in the next section. ■

Computing Definite Integrals

In some cases, we can compute the definite integral directly from the graph (see the quick examples above and the next example below). In general, the only method of computing definite integrals we have discussed so far is numerical estimation: Compute the Riemann sums for larger and larger values of n and then estimate the number it seems to be approaching as we did in Example 1. (In the next section we will discuss an algebraic method for computing them.)

EXAMPLE 4 Estimating a Definite Integral from a Graph

Figure 16 shows the graph of the (approximate) rate $f'(t)$, at which the United States consumed gasoline from 2000 through 2008. (t is time in years since 2000.)[5]

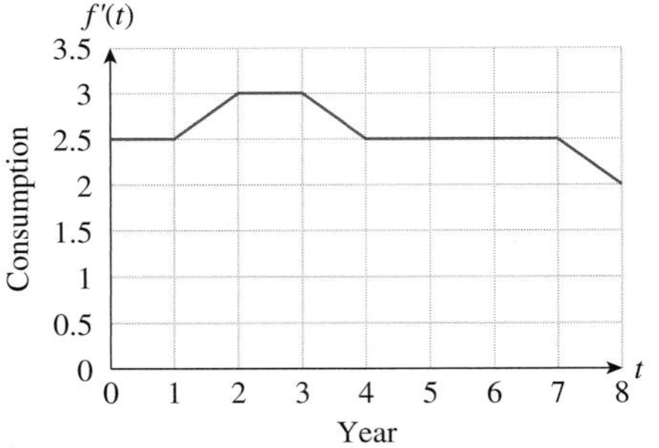

$f'(t)$ = Gasoline consumption (100 million gals per year)

Figure 16

Use the graph to estimate the total U.S. consumption of gasoline over the period shown.

Solution The derivative $f'(t)$ represents the rate of change of the total U.S. consumption of gasoline, and so the total U.S. consumption of gasoline over the given period [0, 8] is given by the definite integral:

$$\text{Total U.S. consumption of gasoline} = \text{Total change in } f(t) = \int_0^8 f'(t)\,dt$$

and is given by the area under the graph (Figure 17).

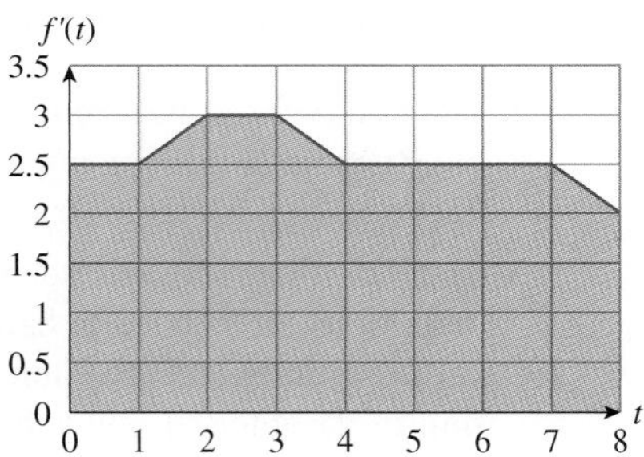

Figure 17

One way to determine the area is to count the number of filled rectangles as defined by the grid. Each rectangle has an area of $1 \times 0.5 = 0.5$ units (and the half-rectangles determined by diagonal portions of the graph have half that area). Counting rectangles, we find a total of 41.5 complete rectangles, so

$$\text{Total area} = 20.75.$$

[5]Source: Energy Information Administration (Department of Energy). (www.eia.doe.gov).

Because $f'(t)$ is in 100 million gallons per year, we conclude that the total U.S. consumption of gasoline over the given period was about 2,075 million gallons, or 2.075 billion gallons.

While counting rectangles might seem easy, it becomes awkward in cases involving large numbers of rectangles or partial rectangles whose area is not easy to determine. In a case like this, in which the graph consists of straight lines, rather than counting rectangles, we can get the area by averaging the left and right Riemann sums whose subdivisions are determined by the grid:

$$\text{Left sum} = (2.5 + 2.5 + 3 + 3 + 2.5 + 2.5 + 2.5 + 2.5)(1) = 21$$
$$\text{Right sum} = (2.5 + 3 + 3 + 2.5 + 2.5 + 2.5 + 2.5 + 2)(1) = 20.5$$
$$\text{Average} = \frac{21 + 20.5}{2} = 20.75.$$

To see why this works, look at the single interval [1, 2]. The left sum contributes $2.5 \times 1 = 2.5$ and the right sum contributes $3 \times 1 = 3$. The exact area is their average, 2.75 (Figure 18).

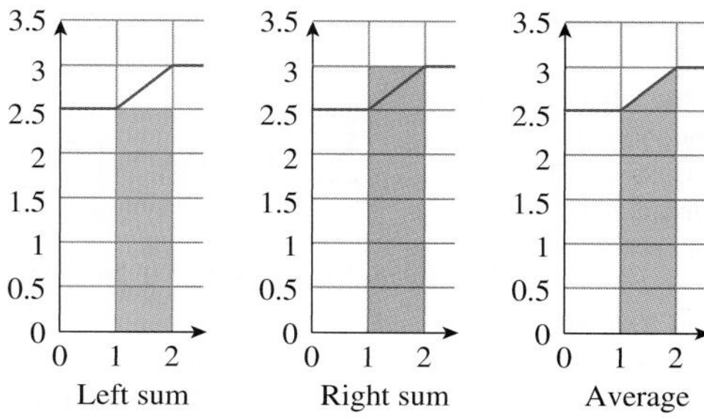

Figure 18

The average of the left and right Riemann sums is frequently a better estimate of the definite integral than either is alone.

Before we go on... It is important to check that the units we are using in Example 4 match up correctly: t is given in *years* and $f'(t)$ is given in 100 million gallons per *year*. The integral is then given in

$$\text{Years} \times \frac{\text{100 million gallons}}{\text{Year}} = \text{100 million gallons.}$$

If we had specified $f'(t)$ in, say, 100 million gallons per *day* but t in years, then we would have needed to convert either t or $f'(t)$ so that the units of time match. ■

The next example illustrates the use of technology in estimating definite integrals using Riemann sums.

EXAMPLE 5 Using Technology to Approximate the Definite Integral

Use technology to estimate the area under the graph of $f(x) = 1 - x^2$ over the interval [0, 1] using $n = 100$, $n = 200$, and $n = 500$ subdivisions.

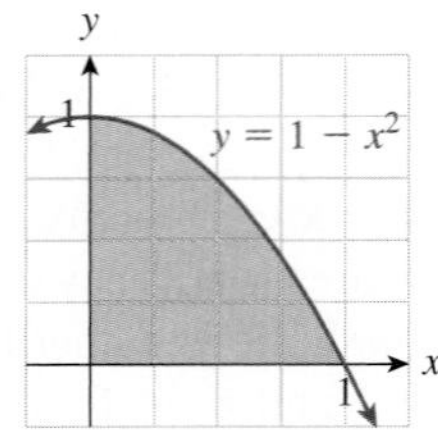

Figure 19

Solution We need to estimate the area under the parabola shown in Figure 19.

From the discussion above,

$$\text{Area} = \int_0^1 (1 - x^2)\, dx.$$

The Riemann sum with $n = 100$ has $\Delta x = (b - a)/n = (1 - 0)/100 = 0.01$ and is given by

$$\sum_{k=0}^{99} f(x_k)\Delta x = [f(0) + f(0.01) + \cdots + f(0.99)](0.01).$$

Similarly, the Riemann sum with $n = 200$ has $\Delta x = (b - a)/n = (1 - 0)/200 = 0.005$ and is given by

$$\sum_{k=0}^{199} f(x_k)\Delta x = [f(0) + f(0.005) + \cdots + f(0.995)](0.005).$$

For $n = 500$, $x = (b - a)/n = (1 - 0)/500 = 0.002$ and the Riemann sum is

$$\sum_{k=0}^{499} f(x_k)\Delta x = [f(0) + f(0.002) + \cdots + f(0.998)](0.002).$$

using Technology

See the Technology Guides at the end of the chapter to find out how to compute these sums on a TI-83/84 Plus and a spreadsheet.

Website

www.WanerMath.com

At the Website, select the Online Utilities tab, and choose the Numerical Integration Utility and Grapher. This utility computes left and right Riemann sums, as well as a good numerical approximation of the integral. There is also a downloadable Excel spreadsheet that computes and graphs Riemann sums (Riemann Sum Grapher).

Using technology to evaluate these Riemann sums, we find:

$$n = 100: \sum_{k=0}^{99} f(x_k)\Delta x = 0.67165$$

$$n = 200: \sum_{k=0}^{199} f(x_k)\Delta x = 0.6691625$$

$$n = 500: \sum_{k=0}^{499} f(x_k)\Delta x = 0.667666$$

so we estimate that the area under the curve is about 0.667. (The exact answer is 2/3, as we will be able to verify using the techniques in the next section.)

EXAMPLE 6 Motion

A fast car has velocity $v(t) = 6t^2 + 10t$ ft/sec after t seconds (as measured by a radar gun). Use several values of n to find the distance covered by the car from time $t = 3$ seconds to time $t = 4$ seconds.

Solution Because the velocity $v(t)$ is rate of change of position, the total change in position over the interval [3, 4] is

$$\text{Distance covered} = \text{Total change in position} = \int_3^4 v(t)\, dt = \int_3^4 (6t^2 + 10t)\, dt.$$

Website
www.WanerMath.com
At the Website you will find the following optional online interactive section: Internet Topic: Numerical Integration.

As in Examples 1 and 5, we can subdivide the one-second interval [3, 4] into smaller and smaller pieces to get more and more accurate approximations of the integral. By computing Riemann sums for various values of n, we get the following results.

$$n = 10: \sum_{k=0}^{9} v(t_k)\Delta t = 106.41 \qquad n = 100: \sum_{k=0}^{99} v(t_k)\Delta t \approx 108.740$$

$$n = 1{,}000: \sum_{k=0}^{999} v(t_k)\Delta t \approx 108.974 \qquad n = 10{,}000: \sum_{k=0}^{9999} v(t_k)\Delta t \approx 108.997$$

These calculations suggest that the total distance covered by the car, the value of the definite integral, is approximately 109 feet.

➡ **Before we go on...** Do Example 6 using antiderivatives instead of Riemann sums, as in Section 6.1. Do you notice a relationship between antiderivatives and definite integrals? This will be explored in the next section. ■

6.3 EXERCISES

Access end-of-section exercises online at **www.webassign.net**

6.4 The Definite Integral: Algebraic Viewpoint and the Fundamental Theorem of Calculus

In Section 6.3 we saw that the definite integral of the marginal cost function gives the total cost. However, in Section 6.1 we used antiderivatives to recover the cost function from the marginal cost function, so we *could* use antiderivatives to compute total cost. The following example, based on Example 5 in Section 6.1, compares these two approaches.

EXAMPLE 1 Finding Cost from Marginal Cost

The marginal cost of producing baseball caps at a production level of x caps is $4 - 0.001x$ dollars per cap. Find the total change of cost if production is increased from 100 to 200 caps.

Solution

Method 1: Using an Antiderivative (based on Example 5 in Section 6.1): Let $C(x)$ be the cost function. Because the marginal cost function is the derivative of the cost function, we have $C'(x) = 4 - 0.001x$ and so

$$\begin{aligned} C(x) &= \int (4 - 0.001x)\,dx \\ &= 4x - 0.001\frac{x^2}{2} + K \qquad \text{K is the constant of integration.} \\ &= 4x - 0.0005x^2 + K. \end{aligned}$$

Although we do not know what to use for the value of the constant K, we can say:

$$\begin{aligned} \text{Cost at production level of 100 caps} &= C(100) \\ &= 4(100) - 0.0005(100)^2 + K \\ &= \$395 + K \\ \text{Cost at production level of 200 caps} &= C(200) \\ &= 4(200) - 0.0005(200)^2 + K \\ &= \$780 + K. \end{aligned}$$

Therefore,

$$\begin{aligned} \text{Total change in cost} &= C(200) - C(100) \\ &= (\$780 + K) - (\$395 + K) = \$385. \end{aligned}$$

Notice how the constant of integration simply canceled out! So, we could choose any value for K that we wanted (such as $K = 0$) and still come out with the correct total change. Put another way, we could use *any antiderivative* of $C'(x)$, such as

$$F(x) = 4x - 0.0005x^2$$

$F(x)$ is *any* antiderivative of $C'(x)$ whereas $C(x)$ is the actual cost function.

or

$$F(x) = 4x - 0.0005x^2 + 4,$$

compute $F(200) - F(100)$, and obtain the total change, \$385.

Summarizing this method: To compute the total change of $C(x)$ over the interval [100, 200], use any antiderivative $F(x)$ of $C'(x)$, and compute $F(200) - F(100)$.

Method 2: Using a Definite Integral (based on the interpretation of the definite integral as total change discussed in Section 6.3): Because the marginal cost $C'(x)$ is the rate of change of the total cost function $C(x)$, the total change in $C(x)$ over the interval [100, 200] is given by

$$\begin{aligned} \text{Total change in cost} &= \text{Area under the marginal cost function curve} \\ &= \int_{100}^{200} C'(x)\,dx \\ &= \int_{100}^{200} (4 - 0.001x)\,dx \qquad \text{See Figure 20.} \\ &= \$385. \qquad \text{Using geometry or Riemann sums} \end{aligned}$$

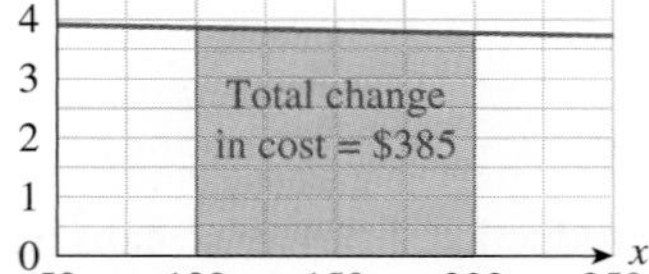

Figure 20

Putting these two methods together gives us the following surprising result:

$$\int_{100}^{200} C'(x)\,dx = F(200) - F(100),$$

where $F(x)$ is any antiderivative of $C'(x)$.

Now, there is nothing special in Example 1 about the specific function $C'(x)$ or the choice of endpoints of integration. So if we replace $C'(x)$ by a general continuous function $f(x)$, we can write

$$\int_a^b f(x)\,dx = F(b) - F(a),$$

where $F(x)$ is any antiderivative of $f(x)$. This result is known as the **Fundamental Theorem of Calculus**.

The Fundamental Theorem of Calculus (FTC)

Let f be a continuous function defined on the interval $[a, b]$ and let F be *any* antiderivative of f defined on $[a, b]$. Then

$$\int_a^b f(x)\,dx = F(b) - F(a).$$

Moreover, an antiderivative of f is guaranteed to exist.

In Words: Every continuous function has an antiderivative. To compute the definite integral of $f(x)$ over $[a, b]$, first find an antiderivative $F(x)$, then evaluate it at $x = b$, evaluate it at $x = a$, and subtract the two answers.

Quick Example

Because $F(x) = x^2$ is an antiderivative of $f(x) = 2x$,

$$\int_0^1 2x\,dx = F(1) - F(0) = 1^2 - 0^2 = 1.$$

Note The Fundamental Theorem of Calculus actually applies to some other functions besides the continuous ones. The function f is **piecewise continuous** on $[a, b]$ if it is defined and continuous at all but finitely many points in the interval, and at each point where the function is not defined or is discontinuous, the left and right limits of f exist and are finite. (See Figure 21.)

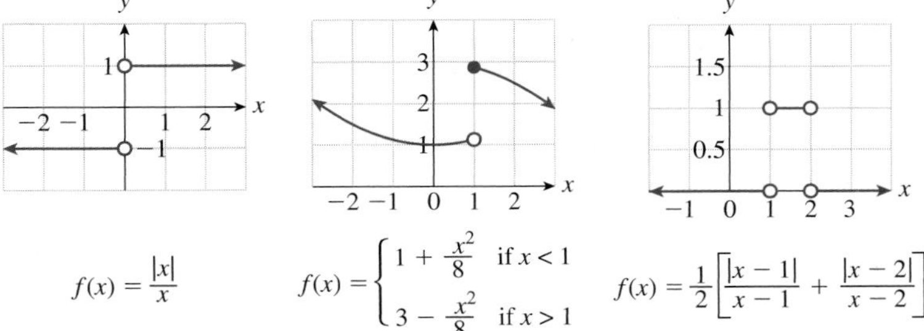

Figure 21

The FTC also applies to any piecewise continuous function f, as long as we specify that the antiderivative F that we choose be continuous. To be precise, to say that F is an antiderivative of f means here that $F'(x) = f(x)$ except at the points at which f is discontinuous or not defined, where $F'(x)$ may not exist. For example, if f is the step function $f(x) = |x|/x$, shown on the left in Figure 21, then we can use $F(x) = |x|$. (Note that F is continuous, and $F'(x) = |x|/x$ except when $x = 0$.) ■

EXAMPLE 2 Using the FTC to Calculate a Definite Integral

Calculate $\int_0^1 (1 - x^2)\,dx$.

Solution To use the FTC, we need to find an antiderivative of $1 - x^2$. But we know that

$$\int (1 - x^2)\,dx = x - \frac{x^3}{3} + C.$$

We need only one antiderivative, so let's take $F(x) = x - x^3/3$. The FTC tells us that

$$\int_0^1 (1 - x^2)\,dx = F(1) - F(0) = \left(1 - \frac{1}{3}\right) - (0) = \frac{2}{3}$$

which is the value we estimated in Section 6.3.

➡ **Before we go on...** A useful piece of notation is often used here. We write*

$$[F(x)]_a^b = F(b) - F(a).$$

* There seem to be several notations in use, actually. Another common notation is $F(x)\Big|_a^b$.

Thus, we can rewrite the computation in Example 2 more simply as follows:

$$\int_0^1 (1 - x^2)\,dx = \left[x - \frac{x^3}{3}\right]_0^1$$

Substitute $x = 1$. Substitute $x = 0$.

$$= \left(1 - \frac{1}{3}\right) - \left(0 - \frac{0}{3}\right)$$

$$= \left(1 - \frac{1}{3}\right) - (0) = \frac{2}{3}.$$

■

EXAMPLE 3 More Use of the FTC

Compute the following definite integrals.

a. $\int_0^1 (2x^3 + 10x + 1)\,dx$ **b.** $\int_1^5 \left(\frac{1}{x^2} + \frac{1}{x}\right)dx$

Solution

a. $$\int_0^1 (2x^3 + 10x + 1)\,dx = \left[\frac{1}{2}x^4 + 5x^2 + x\right]_0^1$$

Substitute $x = 1$. Substitute $x = 0$.

$$= \left(\frac{1}{2} + 5 + 1\right) - \left(\frac{1}{2}(0) + 5(0) + 0\right)$$

$$= \left(\frac{1}{2} + 5 + 1\right) - (0) = \frac{13}{2}$$

b. $\displaystyle\int_1^5 \left(\frac{1}{x^2}+\frac{1}{x}\right) dx = \int_1^5 (x^{-2}+x^{-1})\,dx$

$$= \left[-x^{-1} + \ln|x|\right]_1^5$$

Substitute $x = 5$. Substitute $x = 1$.

$$= \left(-\frac{1}{5} + \ln 5\right) - (-1 + \ln 1)$$

$$= \frac{4}{5} + \ln 5$$

When calculating a definite integral, we may have to use substitution to find the necessary antiderivative. We could substitute, evaluate the indefinite integral with respect to u, express the answer in terms of x, and then evaluate at the limits of integration. However, there is a shortcut, as we shall see in the next example.

EXAMPLE 4 Using the FTC with Substitution

Evaluate $\displaystyle\int_1^2 (2x-1)e^{2x^2-2x}\,dx$.

Solution The shortcut we promised is to put *everything* in terms of u, including the limits of integration.

$$u = 2x^2 - 2x$$

$$\frac{du}{dx} = 4x - 2$$

$$dx = \frac{1}{4x-2}\,du$$

When $x = 1$, $u = 0$. Substitute $x = 1$ in the formula for u.

When $x = 2$, $u = 4$. Substitute $x = 2$ in the formula for u.

We get the value $u = 0$, for example, by substituting $x = 1$ in the equation $u = 2x^2 - 2x$. We can now rewrite the integral.

$$\int_1^2 (2x-1)e^{2x^2-2x}\,dx = \int_0^4 (2x-1)e^u \frac{1}{4x-2}\,du$$

$$= \int_0^4 \frac{1}{2}e^u\,du$$

$$= \left[\frac{1}{2}e^u\right]_0^4 = \frac{1}{2}e^4 - \frac{1}{2}$$

➡ **Before we go on...** The alternative, longer calculation in Example 4 is first to calculate the indefinite integral:

$$\int (2x-1)e^{2x^2-2x}\,dx = \int \frac{1}{2}e^u\,du$$

$$= \frac{1}{2}e^u + C = \frac{1}{2}e^{2x^2-2x} + C.$$

Then we can say that

$$\int_1^2 (2x-1)e^{2x^2-2x}\,dx = \left[\frac{1}{2}e^{2x^2-2x}\right]_1^2 = \frac{1}{2}e^4 - \frac{1}{2}.$$ ■

APPLICATIONS

EXAMPLE 5 Oil Spill

In Section 6.3 we considered the following example: Your deep ocean oil rig has suffered a catastrophic failure, and oil is leaking from the ocean floor wellhead at a rate of

$$v(t) = 0.08t^2 - 4t + 60 \text{ thousand barrels per day } (0 \le t \le 20),$$

where t is time in days since the failure. Compute the total volume of oil released during the first 20 days.

using Technology

TI-83/84 Plus

Home Screen:

`fnInt(0.08x^2-4x+60,x,0,20)`

(`fnInt` is `MATH` → `9`)

Website

www.WanerMath.com

At the Website, select the Online Utilities tab, and choose the Numerical Integration Utility and Grapher. There, enter the formula

`0.08x^2-4x+60`

for $f(x)$, enter 0 and 20 for the lower and upper limits, and press "Integral".

Solution We calculate

$$\text{Total volume} = \int_0^{20} (0.08t^2 - 4t + 60)\,dt = \left[0.08\frac{t^3}{3} - 2t^2 + 60t\right]_0^{20}$$

$$= \left[0.08\frac{20^3}{3} - 2(20)^2 + 60(20)\right] - [0.08(0) - 2(0) + 60(0)]$$

$$= \frac{640}{3} - 800 + 1{,}200 \approx 613.3 \text{ thousand barrels.}$$

EXAMPLE 6 Computing Area

Find the total area of the region enclosed by the graph of $y = xe^{x^2}$, the x-axis, and the vertical lines $x = -1$ and $x = 1$.

Solution The region whose area we want is shown in Figure 22. Notice the symmetry of the graph. Also, half the region we are interested in is above the x-axis, while the other half is below. If we calculated the integral $\int_{-1}^{1} xe^{x^2}\,dx$, the result would be

$$\text{Area above } x\text{-axis} - \text{Area below } x\text{-axis} = 0,$$

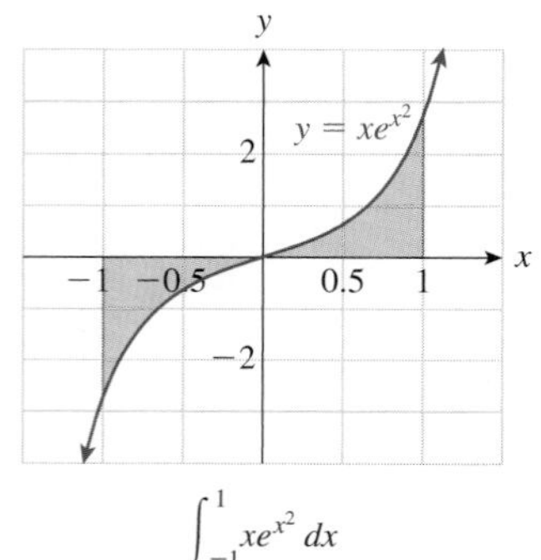

$\int_{-1}^{1} xe^{x^2}\,dx$

Figure 22

which does not give us the total area. To prevent the area below the x-axis from being combined with the area above the axis, we do the calculation in two parts, as illustrated in Figure 23.

(In Figure 23 we broke the integral at $x = 0$ because that is where the graph crosses the x-axis.) These integrals can be calculated using the substitution $u = x^2$:

$$\int_{-1}^{0} xe^{x^2}\,dx = \frac{1}{2}\left[e^{x^2}\right]_{-1}^{0} = \frac{1}{2}(1-e) \approx -0.85914$$ Why is it negative?

$$\int_{0}^{1} xe^{x^2}\,dx = \frac{1}{2}\left[e^{x^2}\right]_{0}^{1} = \frac{1}{2}(e-1) \approx 0.85914$$

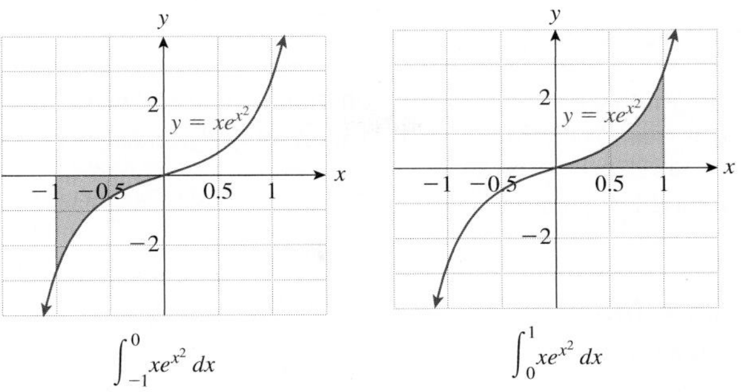

Figure 23

To obtain the total area, we should add the *absolute* values of these answers because we don't wish to count any area as negative. Thus,

$$\text{Total area} \approx 0.85914 + 0.85914 = 1.71828.$$

6.4 EXERCISES

Access end-of-section exercises online at **www.webassign.net**

CHAPTER 6 REVIEW

KEY CONCEPTS

WW Website www.WanerMath.com
Go to the Website at www.WanerMath.com to find a comprehensive and interactive Web-based summary of Chapter 6.

6.1 The Indefinite Integral

An antiderivative of a function f is a function F such that $F' = f$. *p. 334*

Indefinite integral $\int f(x)\,dx$ *p. 334*

Power rule for the indefinite integral:

$$\int x^n\,dx = \frac{x^{n+1}}{n+1} + C \quad (\text{if } n \neq -1) \quad p.\ 336$$

$$\int x^{-1}\,dx = \ln|x| + C \quad p.\ 336$$

Indefinite integral of e^x and b^x:

$$\int e^x\,dx = e^x + C$$

$$\int b^x\,dx = \frac{b^x}{\ln b} + C \quad p.\ 337$$

Indefinite integral of $|x|$:

$$\int |x|\,dx = \frac{x|x|}{2} + C \quad p.\ 337$$

Sums, differences, and constant multiples:

$$\int [f(x) \pm g(x)]\,dx = \int f(x)\,dx \pm \int g(x)\,dx$$

$$\int kf(x)\,dx = k\int f(x)\,dx \quad (k \text{ constant}) \quad p.\ 337$$

Combining the rules *p. 339*

Position, velocity, and acceleration:

$$v = \frac{ds}{dt} \qquad s(t) = \int v(t)\,dt$$

$$a = \frac{dv}{dt} \qquad v(t) = \int a(t)\,dt$$

pp. 342–343

Motion in a straight line *p. 343*

Vertical motion under gravity *p. 345*

6.2 Substitution

Substitution rule:

$$\int f\,dx = \int \left(\frac{f}{du/dx}\right) du \quad p.\ 346$$

Using the substitution rule *p. 347*

Shortcuts: integrals of expressions involving $(ax + b)$:

$$\int (ax+b)^n\,dx = \frac{(ax+b)^{n+1}}{a(n+1)} + C \quad (\text{if } n \neq -1)$$

$$\int (ax+b)^{-1}\,dx = \frac{1}{a}\ln|ax+b| + C$$

$$\int e^{ax+b}\,dx = \frac{1}{a}e^{ax+b} + C$$

$$\int c^{ax+b}\,dx = \frac{1}{a\ln c}c^{ax+b} + C$$

$$\int |ax+b|\,dx = \frac{1}{2a}(ax+b)|ax+b| + C \quad p.\ 352$$

Mike's shortcut rule:

If $\int f(x)\,dx = F(x) + C$ and g and u are differentiable functions of x, then $\int g \cdot f(u)\,dx = \frac{g}{u'} \cdot F(u) + C$ provided that $\frac{g}{u'}$ is constant. *p. 353*

6.3 The Definite Integral: Numerical and Graphical Viewpoints

Left Riemann sum:

$$\sum_{k=0}^{n-1} f(x_k)\Delta x = [f(x_0) + f(x_1) + \cdots + f(x_{n-1})]\Delta x \quad p.\ 359$$

Computing the Riemann sum from a formula *p. 360*

Computing the Riemann sum from a graph *p. 361*

Definite integral of f from a to b:

$$\int_a^b f(x)\,dx = \lim_{n\to\infty} \sum_{k=0}^{n-1} f(x_k)\Delta x. \quad p.\ 362$$

Estimating the definite integral from a graph *p. 364*

Estimating the definite integral using technology *p. 365*

Application to motion in a straight line *p. 366*

6.4 The Definite Integral: Algebraic Viewpoint and the Fundamental Theorem of Calculus

Computing total cost from marginal cost *p. 367*

The Fundamental Theorem of Calculus (FTC) *p. 369*

Using the FTC to compute definite integrals *p. 370*

Using the FTC with substitution *p. 371*

Computing area *p. 372*

REVIEW EXERCISES

Evaluate the indefinite integrals in Exercises 1–18.

1. $\int (x^2 - 10x + 2)\,dx$

2. $\int (e^x + \sqrt{x})\,dx$

3. $\int \left(\frac{4x^2}{5} - \frac{4}{5x^2}\right) dx$

4. $\int \left(\frac{3x}{5} - \frac{3}{5x}\right) dx$

5. $\int (2x)^{-1}\,dx$

6. $\int (-2x + 2)^{-2}\,dx$

7. $\int e^{-2x+11}\,dx$

8. $\int \frac{dx}{(4x-3)^2}$

9. $\int x(x^2+1)^{1.3}\,dx$

10. $\int x(x^2+4)^{10}\,dx$

11. $\int \frac{4x}{(x^2-7)}\,dx$

12. $\int \frac{x}{(3x^2-1)^{0.4}}\,dx$

13. $\int (x^3 - 1)\sqrt{x^4 - 4x + 1}\,dx$

14. $\int \frac{x^2+1}{(x^3+3x+2)^2}\,dx$

15. $\int (-xe^{x^2/2})dx$

16. $\int xe^{-x^2/2}\,dx$

17. $\int \frac{x+1}{x+2}\,dx$

18. $\int x\sqrt{x-1}\,dx$

In Exercises 19 and 20, use the given graph to estimate the left Riemann sum for the given interval with the stated number of subdivisions.

19. $[0, 3]$, $n = 6$

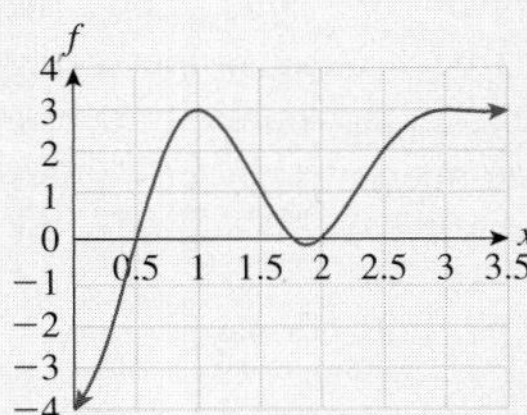

20. $[1, 3]$, $n = 4$

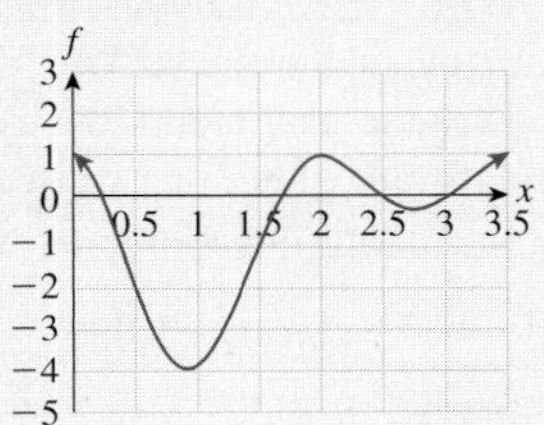

Calculate the left Riemann sums for the given functions over the given interval in Exercises 21–24, using the given values of n. (When rounding, round answers to four decimal places.)

21. $f(x) = x^2 + 1$ over $[-1, 1]$, $n = 4$

22. $f(x) = (x-1)(x-2) - 2$ over $[0, 4]$, $n = 4$

23. $f(x) = x(x^2 - 1)$ over $[0, 1]$, $n = 5$

24. $f(x) = \frac{x-1}{x-2}$ over $[0, 1.5]$, $n = 3$

T *In Exercises 25 and 26, use technology to approximate the given definite integrals using left Riemann sums with n = 10, 100, and 1,000. (Round answers to four decimal places.)*

25. $\int_0^1 e^{-x^2}\,dx$

26. $\int_1^3 x^{-x}\,dx$

In Exercises 27 and 28, the graph of the derivative $f'(x)$ of $f(x)$ is shown. Compute the total change of $f(x)$ over the given interval.

27. $[-1, 2]$

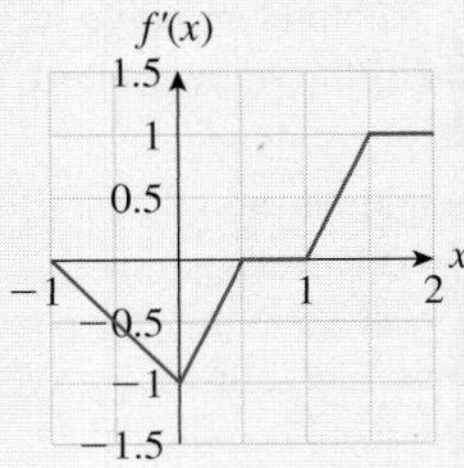

28. $[0, 2]$

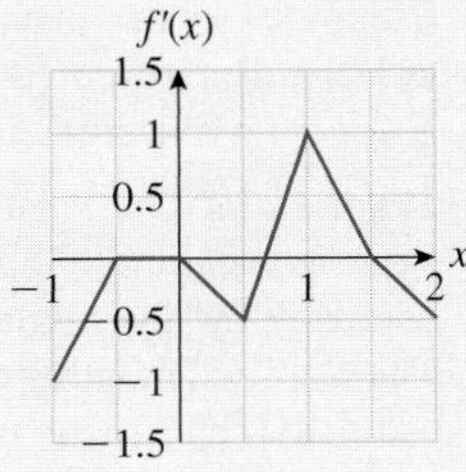

Evaluate the definite integrals in Exercises 29–38, using the Fundamental Theorem of Calculus.

29. $\int_{-1}^{1} (x - x^3 + |x|)\,dx$

30. $\int_0^9 (x + \sqrt{x})\,dx$

31. $\int_{-1}^{1} \frac{3}{(2x-5)^2}\,dx$

32. $\int_0^9 \frac{1}{x+1}\,dx$

33. $\int_0^{50} e^{-0.02x-1}\,dx$

34. $\int_{-20}^{0} 3e^{2.2x}\,dx$

35. $\int_0^2 x^2\sqrt{x^3+1}\,dx$

36. $\int_0^2 \frac{x^2}{\sqrt{x^3+1}}\,dx$

37. $\int_0^{\ln 2} \frac{e^{-2x}}{1+4e^{-2x}}\,dx$

38. $\int_0^{\ln 3} e^{2x}(1-3e^{2x})^2\,dx$

In Exercises 39–42, find the areas of the specified regions. (Do not count area below the x-axis as negative.)

39. The area bounded by $y = 4 - x^2$, the x-axis, and the lines $x = -2$ and $x = 2$

40. The area bounded by $y = 4 - x^2$, the x-axis, and the lines $x = 0$ and $x = 5$

41. The area bounded by $y = xe^{-x^2}$, the x-axis, and the lines $x = 0$ and $x = 5$

42. The area bounded by $y = |2x|$, the x-axis, and the lines $x = -1$ and $x = 1$

APPLICATIONS: OHaganBooks.com

43. ***Sales*** The rate of net sales (sales minus returns) of *The Secret Loves of John O*, a romance novel by Margó Dufón, can be approximated by

$$n(t) = 196 + t^2 - 0.16t^5 \text{ copies per week}$$

t weeks since its release.

a. Find the total net sales N as a function of time t.

b. How many books are still held by customers after 6 weeks? (Round your answer to the nearest book.)

44. ***Demand*** If OHaganBooks.com were to give away its latest bestseller, *A River Burns through It*, the demand q would be 100,000 books. The marginal demand (dq/dp) for the book is $-20p$ at a price of p dollars.

a. What is the demand function for this book?

b. At what price does demand drop to zero?

45. ***Motion under Gravity*** Billy-Sean O'Hagan's friend Juan (Billy-Sean is John O'Hagan's son, currently a senior in college) says he can throw a baseball vertically upward at 100 feet per second. Assuming Juan's claim is true,

a. Where would the baseball be at time t seconds?

b. How high would the ball go?

c. When would it return to Juan's hand?

46. ***Motion under Gravity*** An overworked employee at OHaganBooks.com goes to the top of the company's 100-foot-tall headquarters building and flings a book up into the air at a speed of 60 feet per second.

a. When will the book hit the ground 100 feet below? (Neglect air resistance.)

b. How fast will it be traveling when it hits the ground?

c. How high will the book go?

47. *Sales* Sales at the OHaganBooks.com Web site of *Larry Potter and the Riemann Sum* fluctuated rather wildly in the first 5 months of last year as the following graph shows:

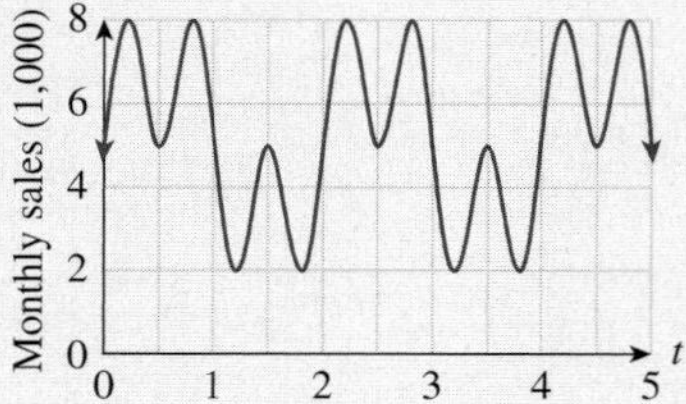

Puzzled by the graph, CEO John O'Hagan asks Jimmy Duffin[6] to estimate the total sales over the entire 5-month period shown. Jimmy decides to use a left Riemann sum with 10 partitions to estimate the total sales. What does he find?

48. *Sales* The following graph shows the approximate rate of change $s(t)$ of the total value, in thousands of dollars, of Spanish books sold online at OHaganBooks.com (t is the number of months since January 1):

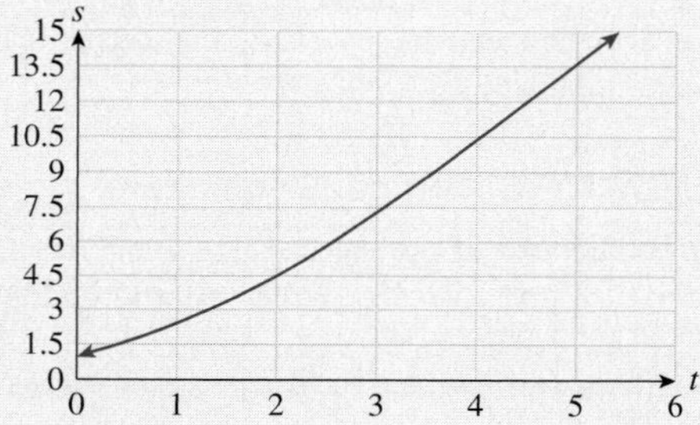

Use the graph to estimate the total value of Spanish books sold from March 1 through June 1. (Use a left Riemann sum with three subdivisions.)

49. *Promotions* Unlike sales of *Larry Potter and the Riemann Sum*, sales at OHaganBooks.com of the special leather-bound gift editions of *Calculus for Vampires* have been suffering lately, as shown in the following graph (negative sales indicate returns by dissatisfied customers; t is time in months since January 1 of this year):

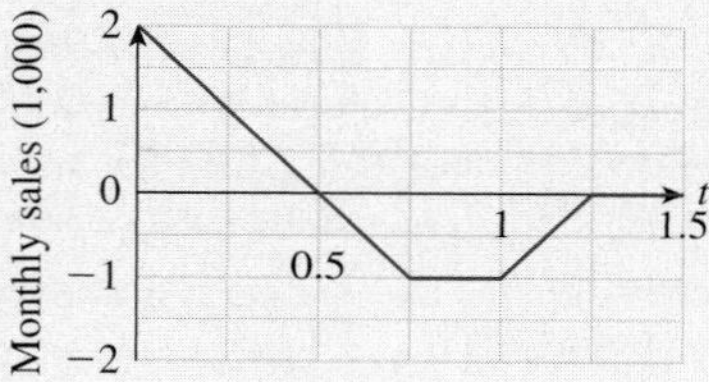

Use the graph to compute the total (net) sales over the period shown.

50. *Sales* Even worse than with the leather-bound *Calculus for Vampires*, sales of *Real Estate for Werewolves* have been dismal, as shown in the following graph (negative sales indicate returns by dissatisfied customers; t is time in months since January 1 of this year):

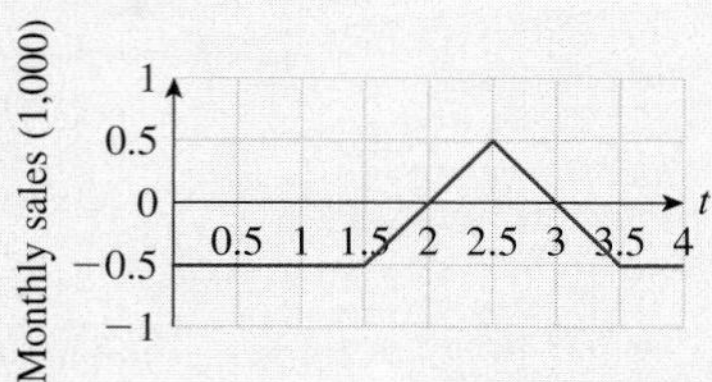

Use the graph to compute the total (net) sales over the period shown.

51. *Web Site Activity* The number of "hits" on the OHaganBooks.com Web site has been steadily increasing over the past month in response to recent publicity over a software glitch that caused the company to pay customers for buying books online. The activity can be modeled by

$$n(t) = 1{,}000t - 10t^2 + t^3 \text{ hits per day,}$$

where t is time in days since news about the software glitch was first publicized on GrungeReport.com. Use a left Riemann sum with five partitions to estimate the total number of hits during the first 10 days of the period.

52. *Web Site Crashes* The latest DoorsXL servers OHaganBooks.com has been using for its Web site have been crashing with increasing frequency lately. One of the student summer interns has estimated the number of crashes to be

$$q(t) = 0.05t^2 + 0.4t + 9 \text{ crashes per week} \quad (0 \le t \le 10),$$

where t is the number of weeks since the DoorsXL system was first installed. Use a Riemann sum with five partitions to estimate the total number of crashes from the start of week 5 to the start of week 10. (Round your answer to the nearest crash.)

53. *Student Intern Costs* The marginal monthly cost of maintaining a group of summer student interns at OHaganBooks .com is calculated to be

$$c(x) = \frac{1{,}000(x+3)^2}{(8+(x+3)^3)^{3/2}} \text{ thousand dollars per additional student.}$$

Compute, to the nearest \$100, the total monthly cost if O'HaganBooks increases the size of the student intern program from five students to seven students.

54. *Legal Costs* The legal team maintained by OHaganBooks .com to handle the numerous lawsuits brought against the company by disgruntled clients may have to be expanded. The marginal monthly cost to maintain a team of x lawyers is estimated (by a method too complicated to explain) as

$$c(x) = (x-2)^2[8-(x-2)^3]^{3/2} \text{ thousand dollars per additional lawyer.}$$

Compute, to the nearest \$1,000, the total monthly cost if O'HaganBooks goes ahead with a proposal to increase the size of the legal team from two to four.

[6] Marjory Duffin's nephew, currently at OHaganBooks.com on a summer internship.

55. ***Projected Sales*** When OHaganBooks.com was about to go online, it estimated that its weekly sales would begin at about 6,400 books per week, with sales increasing at such a rate that weekly sales would double about every 2 weeks. If these estimates had been correct, how many books would the company have sold in the first 5 weeks? (Round your answer to the nearest 1,000 books.)

56. ***Projected Sales*** Once OHaganBooks.com actually went online, its weekly sales began at about 7,500 books per week, with weekly sales doubling every 3 weeks. How many books did the company actually sell in the first 5 weeks? (Round your answer to the nearest 1,000 books.)

57. ***Actual Sales*** OHaganBooks.com modeled its revised weekly sales over a period of time after it went online with the function

$$s(t) = 6{,}053 + \frac{4{,}474e^{0.55t}}{e^{0.55t} + 14.01},$$

where t is the time in weeks after it went online. According to this model, how many books did it actually sell in the first 5 weeks?

58. ***Computer Usage*** A consultant recently hired by OHaganBooks .com estimates total weekly computer usage as

$$w(t) = 620 + \frac{900e^{0.25t}}{3 + e^{0.25t}} \text{ hours} \quad (0 \le t \le 20),$$

where t is time in weeks since January 1 of this year. Use the model to estimate the total computer usage during the first 14 weeks of the year.

Spending on Housing Construction

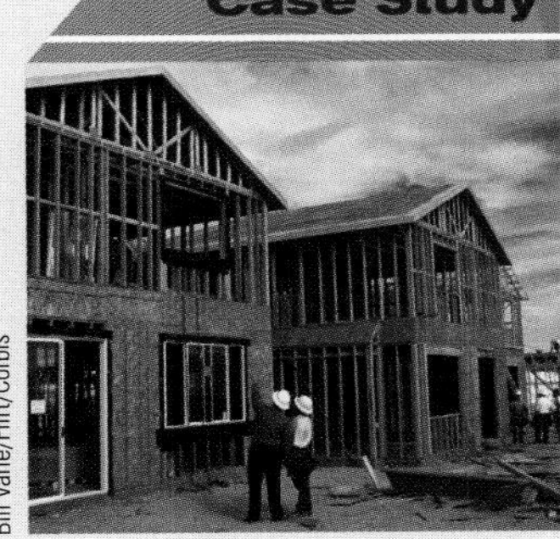

Bill Varie/Flirt/Corbis

It is March 2007, and *Time* magazine, in its latest edition, is asking "Will the Housing Bubble Burst in 2007?"[7] You are a summer intern at *Ronald Ramp Real Estate Development, Inc.,* which is considering a major luxury condominium development in Tampa, Florida nicknamed the "Ramp Towers Tampa."

Yesterday, you received the following memo from your supervisor:

> **DATE:** March 15, 2007
> **TO:** SW@EnormousStateU.edu
> **FROM:** SC@RonaldRampDev.com (Junior VP Development)
> **CC:** SGLombardoVP@RonaldRampDev.com (S. G. Lombardo, Senior VP Development)
> **SUBJECT:** Residential Construction Trends. Urgent!
>
> Help! There is a management meeting in two hours and Michelle Homestead, who, as you know, is spearheading the latest Ramp Towers Tampa feasibility study, must report to "the Ronald" by tomorrow and has asked me to immediately produce some mathematical formulas to (1) model the trend in residential construction spending since January 2006, when it was \$618.7 billion, and (2) estimate the average spent per month on residential construction over a specified period of time. All I have on hand so far is data giving the month-over-month percentage changes (attached). Do you have any ideas?

[7] *Time,* February, 2007 (www.time.com/time/business/article/0,8599,1592751,00.html).

ATTACHMENT*

Month	% Change	Month	% Change
1	1.16	16	−1.59
2	1.17	17	−1.62
3	1.33	18	−1.58
4	0.67	19	−1.67
5	0.42	20	−1.77
6	−0.04	21	−1.92
7	−0.24	22	−2.06
8	−0.44	23	−2.19
9	−0.70	24	−2.45
10	−0.94	25	−2.50
11	−1.31	26	−2.65
12	−1.34	27	−2.85
13	−1.49	28	−2.65
14	−1.75	29	−2.66
15	−1.74	30	−2.66

*Based on 12-month moving average; Source for data: U.S. Census Bureau: Manufacturing, Mining and Construction Statistics, Data 360 (www.data360.org/dataset.aspx?Data_Set_Id=3627).

Getting to work, you decide that the first thing to do is fit these data to a mathematical curve that you can use to project future changes in construction spending. You graph the data to get a sense of what mathematical models might be appropriate (Figure 24).

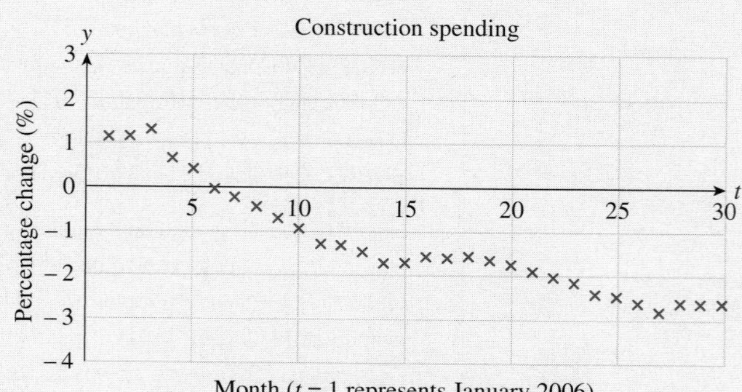

Figure 24

The graph suggests a decreasing trend, possibly concave up. You recall that there are a variety of curves that can behave this way, one of the simplest being

$$y = at^c + b \quad (t > 0),$$

where a, b, and c are constants.

You convert all the percentages to decimals, giving the following table of data:

t	y	t	y
1	0.0116	16	−0.0159
2	0.0117	17	−0.0162
3	0.0133	18	−0.0158
4	0.0067	19	−0.0167
5	0.0042	20	−0.0177
6	−0.0004	21	−0.0192
7	−0.0024	22	−0.0206
8	−0.0044	23	−0.0219
9	−0.0070	24	−0.0245
10	−0.0094	25	−0.0250
11	−0.0131	26	−0.0265
12	−0.0134	27	−0.0285
13	−0.0149	28	−0.0265
14	−0.0175	29	−0.0266
15	−0.0174	30	−0.0266

* To do this, you can use Excel's Solver or the Website function evaluator and grapher with model `$1*x^$2+$3`.

You then find the values of a, b, and c that best fit the given data:*

$$a = -0.0200974,\ b = 0.0365316,\ c = 0.345051.$$

These values give you the following model for construction spending (with figures rounded to five significant digits):

$$y = -0.020097t^{0.34505} + 0.036532.$$

Figure 25 shows the graph of y superimposed on the data.

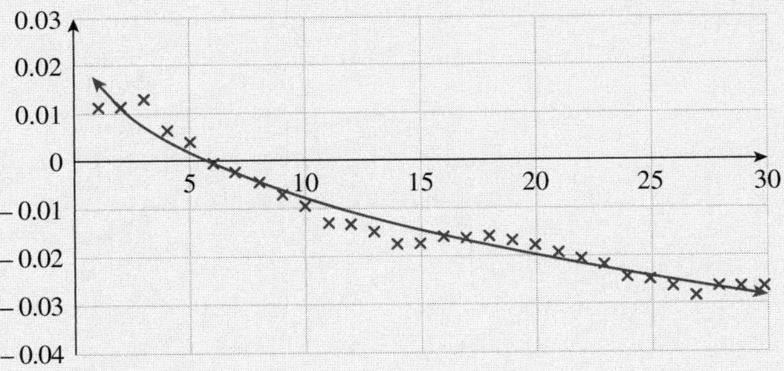

Figure 25

Now that you have a model for the month-over-month change in construction spending, you must use it to find the actual spending on construction. First, you realize that the model gives the *fractional rate of increase* of construction spending (because it is specified as a percentage, or fraction, of the total spending). In other words, if $p(t)$ represents the construction cost in month t, then

$$y = \frac{dp/dt}{p} = \frac{d}{dt}(\ln p). \qquad \text{By the chain rule for derivatives}$$

You find an equation for actual monthly construction cost at time t by solving for p:

$$\begin{aligned}\ln p &= \int y\,dt \\ &= \int (at^c + b)\,dt \\ &= \frac{at^{c+1}}{c+1} + bt + K \\ &= dt^{c+1} + bt + K,\end{aligned}$$

where

$$d = \frac{a}{c+1} = \frac{-0.020097}{0.34505+1} \approx -0.014941,$$

b and c are as above, and K is the constant of integration. So

$$p(t) = e^{dt^{c+1}+bt+K}.$$

To compute K, you substitute the initial data from the memo: $p(1) = 618.7$. Thus,

$$618.7 = e^{d+b+K} = e^{-0.014941+0.036532+K} = e^{0.021591+K}.$$

Thus,

$$\ln(618.7) = 0.021591 + K,$$

which gives

$$K = \ln(618.7) - 0.021591 \approx 6.4060 \text{ (to five significant digits).}$$

Now you can write down the following formula for the monthly spending on residential construction as a function of t, the number of months since the beginning of 2006:

$$p(t) = e^{dt^{c+1}+bt+K} = e^{-0.014941t^{0.34505+1}+0.036532t+6.4060}.$$

What remains is the calculation of the average spent per month over a specified period $[r, s]$. Since p is the rate of change of the total spent, the total spent on housing construction over this period is

$$P = \int_r^s p(t)\,dt$$

and so the average spent per month is

$$\bar{P} = \frac{1}{s-r}\int_r^s p(t)\,dt.$$

$\frac{1}{\text{Number of months}} \times \text{Total spent}$

Substituting the formula for $p(t)$ gives

$$\bar{P} = \frac{1}{s-r}\int_r^s e^{-0.014941t^{0.34505+1}+0.036532t+6.4060}\,dt.$$

You cannot find an explicit antiderivative for the integrand, so you decide that the only way to compute it is numerically. You send the following memo to SC.

DATE: March 15, 2007

TO: SC@RonaldRampDev.com (Junior VP Development)

FROM: SW@EnormousStateU.edu

CC: SGLombardoVP@RonaldRampDev.com (S. G. Lombardo, Senior VP Development)

SUBJECT: The formula you wanted

Spending in the U.S. on housing construction in the tth month of 2006 can be modeled by

$$p(t) = e^{-0.014941t^{0.34505+1}+0.036532t+6.4060} \text{ million dollars.}$$

Further, the average spent per month from month r to month s (since the start of January 2006) can be computed as

$$\bar{P} = \frac{1}{s-r}\int_r^s e^{-0.014941t^{0.34505+1}+0.036532t+6.4060}\,dt$$

To calculate it easily (and impress the subcommittee members), I suggest you have a graphing calculator on hand and enter the following on your graphing calculator (watch the parentheses!):

```
Y1=1/(S-R)*fnInt(e^(-0.014941T^(0.34505+1)+0.036532T+6.4060),T,R,S)
```

Then suppose, for example, you need to estimate the average for the period March 1, 2006 ($t = 3$) to February 1, 2007 ($t = 14$). All you do is enter

```
3→R
14→S
Y1
```

and your calculator will give you the result: The average spending was $628 million per month.

Good luck with the meeting!

EXERCISES

1. Use the actual January 2006 spending figure of $618.7 million and the percentage changes in the table to compute the actual spending in February, March, and April of that year. Also use the model of monthly spending to estimate those figures, and compare the predicted values with the actual figures. Is it unacceptable that the April figures agree to only one significant digit? Explain.
2. Use the model developed above to estimate the average monthly spending on residential construction over the 12-month period beginning June 1, 2006. (Round your answer to the nearest $1 million.)
3. What (if any) advantages are there to using a model for residential construction spending when the actual residential construction spending figures are available?
4. The formulas for $p(t)$ and $\bar{P}$ were based on the January 2006 spending figure of $618.7 million. Change the models to allow for a possibly revised January 2006 spending figure of $\$p_0$ million.
5. If we had used quadratic regression to model the construction spending data, we would have obtained

$$y = 0.00005t^2 - 0.0028t + 0.0158.$$

(See the graph.)

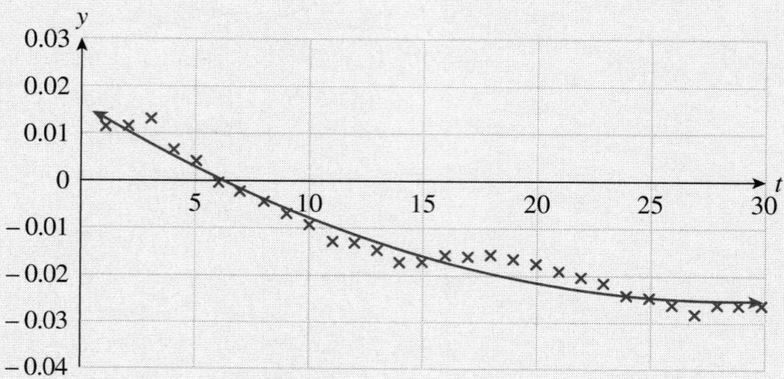

Use this formula and the given January 2006 spending figure to obtain corresponding models for $p(t)$ and $\bar{P}$.

6. Compare the model in the text with the quadratic model in Exercise 5 in terms of both short- and long-term predictions; in particular, when does the quadratic model predict construction spending will have reached its biggest monthly decrease? Are either of these models realistic in the near term? in the long term?

TI-83/84 Plus Technology Guide

Section 6.3

Example 5 (page 365) Estimate the area under the graph of $f(x) = 1 - x^2$ over the interval [0, 1] using $n = 100$, $n = 200$, and $n = 500$ partitions.

Solution with Technology

There are several ways to compute Riemann sums with a graphing calculator. We illustrate one method. For $n = 100$, we need to compute the sum

$$\sum_{k=0}^{99} f(x_k)\Delta x = [f(0) + f(0.01) + \cdots + f(0.99)](0.01).$$

See discussion in Example 5.

Thus, we first need to calculate the numbers $f(0)$, $f(0.01)$, and so on, and add them up. The TI-83/84 Plus has a built-in `sum` function (available in the LIST MATH menu), which, like the SUM function in a spreadsheet, sums the entries in a list.

1. To generate a list that contains the numbers we want to add together, use the `seq` function (available in the LIST OPS menu). If we enter

```
seq(1-X^2,X,0,0.99,0.01)
```

seq: [2ND] [LIST] OPS [5]

the calculator will calculate a list by evaluating `1-X^2` for values of `X` from 0 to 0.99 in steps of 0.01.

2. To take the sum of all these numbers, we wrap the `seq` function in a call to `sum`:

```
sum(seq(1-X^2,X,0,0.99,0.01))
```

sum: [2ND] [LIST] MATH [5]

This gives the sum

$$f(0) + f(0.01) + \cdots + f(0.99) = 67.165.$$

3. To obtain the Riemann sum, we need to multiply this sum by $\Delta x = 0.01$, and we obtain the estimate of $67.165 \times 0.01 = 0.67165$ for the Riemann sum:

```
sum(seq(1-X^2,X,
0,0.99,0.01))
           67.165
```

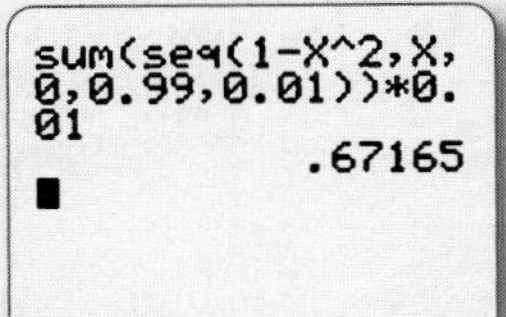

We obtain the other Riemann sums similarly, as shown here:

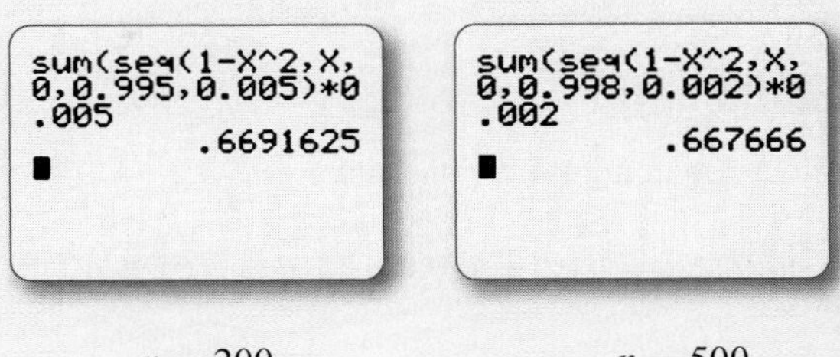

$n = 200$ $\qquad$ $n = 500$

One disadvantage of this method is that the TI-83/84 Plus can generate and sum a list of at most 999 entries. The LEFTSUM program below calculates left Riemann sums for any n. The TI-83/84 Plus also has a built-in function `fnInt`, which finds a very accurate approximation of a definite integral, using a more sophisticated technique than the one we are discussing here.

The LEFTSUM program for the TI-83/84 Plus

The following program calculates (left) Riemann sums for any n. The latest version of this program (and others) is available at the Website.

PROGRAM: LEFTSUM	
`:Input "LEFT ENDPOINT? ",A`	Prompts for the left endpoint a
`:Input "RIGHT ENDPOINT? ",B`	Prompts for the right endpoint b
`:Input "N? ",N`	Prompts for the number of rectangles
`:(B-A)/N→D`	D is $\Delta x = (b - a)/n$.
`:Ø→L`	L will eventually be the left sum.
`:A→X`	X is the current x-coordinate.
`:For(I,1,N)`	Start of a loop—recall the sigma notation.
`:L+Y1→L`	Add $f(x_{i-1})$ to L.
`:A+I*D→X`	Uses formula $x_i = a + i\Delta x$
`:End`	End of loop
`:L*D→L`	Multiply by Δx.
`:Disp "LEFT SUM IS ",L`	
`:Stop`	

SPREADSHEET Technology Guide

Section 6.3

Example 5 (page 365) Estimate the area under the graph of $f(x) = 1 - x^2$ over the interval [0, 1] using $n = 100$, $n = 200$, and $n = 500$ partitions.

Solution with Technology

We need to compute various sums.

$$\sum_{k=0}^{99} f(x_k)\Delta x = [f(0) + f(0.01) + \cdots + f(0.99)](0.01)$$

See discussion in Example 5.

$$\sum_{k=0}^{199} f(x_k)\Delta x = [f(0) + f(0.005) + \cdots + f(0.995)](0.005)$$

$$\sum_{k=0}^{499} f(x_k)\Delta x = [f(0) + f(0.002) + \cdots + f(0.998)](0.002).$$

Here is how you can compute them all on the same spreadsheet.

1. Enter the values for the endpoints a and b, the number of subdivisions n, and the formula $\Delta x = (b - a)/n$:

	A	B	C	D
1	x	f(x)	a	0
2			b	1
3			n	100
4			Delta x	=(D2-D1)/D3

2. Next, we compute all the x-values we might need in column A. Because the largest value of n that we will be using is 500, we will need a total of 501 values of x. Note that the value in each cell below A3 is obtained from the one above by adding Δx.

	A	B	C	D
1	x	f(x)	a	0
2	=D1		b	1
3	=A2+$D$4		n	100
4			Delta x	0.01
501				
502				

	A	B	C	D
1	x	f(x)	a	0
2	0		b	1
3	0.01		n	100
4	0.02		Delta x	0.01
501	4.99			
502	5			

(The fact that the values of x currently go too far will be corrected in the next step.)

3. We need to calculate the numbers $f(0)$, $f(0.01)$, and so on, but only those for which the corresponding x-value is less than b. To do this, we use a logical formula like we did with piecewise-defined functions in Chapter 1:

	A	B	C	D
1	x	f(x)	a	0
2	0	=(1-A2^2)*(A2<$D$2)	b	1
3	0.01		n	100
4	0.02		Delta x	0.01
501	4.99			
502	5			

When the value of x is b or above, the function will evaluate to zero, because we do not want to count it.

4. Finally, we compute the Riemann sum by adding up everything in column B and multiplying by Δx:

	A	B	C	D
1	x	f(x)	a	0
2	0	1	b	1
3	0.01	0.9999	n	100
4	0.02	0.9996	Delta x	0.01
5	0.03	0.9991	Left Sum	=SUM(B:B)*D4
6	0.04	0.9984		

↓

	A	B	C	D
1	x	f(x)	a	0
2	0	1	b	1
3	0.01	0.9999	n	100
4	0.02	0.9996	Delta x	0.01
5	0.03	0.9991	Left Sum	0.67165
6	0.04	0.9984		

Now it is easy to obtain the sums for $n = 200$ and $n = 500$: Simply change the value of n in cell D3:

	A	B	C	D
1	x	f(x)	a	0
2	0	1	b	1
3	0.005	0.999975	n	200
4	0.01	0.9999	Delta x	0.005
5	0.015	0.999775	Left Sum	0.6691625
6	0.02	0.9996		

	A	B	C	D
1	x	f(x)	a	0
2	0	1	b	1
3	0.002	0.999996	n	500
4	0.004	0.999984	Delta x	0.002
5	0.006	0.999964	Left Sum	0.667666
6	0.008	0.999936		

7

Further Integration Techniques and Applications of the Integral

Website

www.WanerMath.com

At the Website you will find:

- A detailed chapter summary
- A true/false quiz
- Additional review exercises
- A numerical integration utility
- Graphing calculator programs for numerical integration
- Graphers, Excel tutorials, and other resources

Case Study Estimating Tax Revenues

You have just been hired by the incoming administration to coordinate national tax policy, and the so-called experts on your staff can't seem to agree on which of two tax proposals will result in more revenue for the government. The data you have are the two income tax proposals (graphs of tax vs. income) and the distribution of incomes in the country. **How do you use this information to decide which tax policy will result in more revenue?**

Dmitriy Shironosov/Shutterstock.com

Introduction

In the preceding chapter, we learned how to compute many integrals and saw some of the applications of the integral. In this chapter, we look at some further techniques for computing integrals and then at more applications of the integral. We also see how to extend the definition of the definite integral to include integrals over infinite intervals, and show how such integrals can be used for long-term forecasting. Finally, we introduce the beautiful theory of differential equations and some of its numerous applications.

7.1 Integration by Parts

Integration by parts is an integration technique that comes from the product rule for derivatives. The tabular method we present here has been around for some time and makes integration by parts quite simple, particularly in problems where it has to be used several times.*

* The version of the tabular method we use was developed and taught to us by Dan Rosen at Hofstra University.

We start with a little notation to simplify things while we introduce integration by parts. (We use this notation only in the next few pages.) If u is a function, denote its derivative by $D(u)$ and an antiderivative by $I(u)$. Thus, for example, if $u = 2x^2$, then

$$D(u) = 4x$$

and

$$I(u) = \frac{2x^3}{3}.$$

[If we wished, we could instead take $I(u) = \dfrac{2x^3}{3} + 46$, but we usually opt to take the simplest antiderivative.]

Integration by Parts

If u and v are continuous functions of x, and u has a continuous derivative, then

$$\int u \cdot v\, dx = u \cdot I(v) - \int D(u) I(v)\, dx.$$

Quick Example

(Discussed more fully in Example 1 on the next page)

$$\int x \cdot e^x dx = xI(e^x) - \int D(x)I(e^x)\, dx$$

$$= xe^x - \int 1 \cdot e^x\, dx \qquad I(e^x) = e^x;\ D(x) = 1$$

$$= xe^x - e^x + C. \qquad \int e^x\, dx = e^x + C$$

As the Quick Example shows, although we could not immediately integrate $u \cdot v = x \cdot e^x$, we could easily integrate $D(u)I(v) = 1 \cdot e^x = e^x$.

Derivation of Integration-by-Parts Formula

As we mentioned, the integration-by-parts formula comes from the product rule for derivatives. We apply the product rule to the function $uI(v)$

$$\begin{aligned} D[u \cdot I(v)] &= D(u)I(v) + uD(I(v)) \\ &= D(u)I(v) + uv \end{aligned}$$

because $D(I(v))$ is the derivative of an antiderivative of v, which is v. Integrating both sides gives

$$u \cdot I(v) = \int D(u)I(v)\,dx + \int uv\,dx.$$

A simple rearrangement of the terms now gives us the integration-by-parts formula.

The integration-by-parts formula is easiest to use via the tabular method illustrated in the following example, where we repeat the calculation we did in the Quick Example on the previous page.

EXAMPLE 1 Integration by Parts: Tabular Method

Calculate $\int xe^x\,dx$.

Solution First, the reason we *need* to use integration by parts to evaluate this integral is that none of the other techniques of integration that we've talked about up to now will help us. Furthermore, we cannot simply find antiderivatives of x and e^x and multiply them together. [You should check that $(x^2/2)e^x$ is *not* an antiderivative of xe^x.] However, as we saw above, this integral can be found by integration by parts. We want to find the integral of the *product* of x and e^x. We must make a decision: Which function will play the role of u and which will play the role of v in the integration-by-parts formula? Because the derivative of x is just 1, differentiating makes it simpler, so we try letting x be u and letting e^x be v. We need to calculate $D(u)$ and $I(v)$, which we record in the following table.

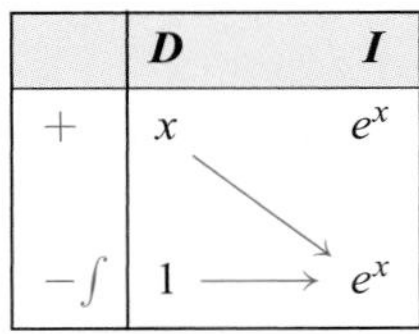

	D	I
$+$	x	e^x
$-\int$	1	e^x

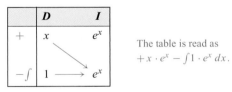

The table is read as $+x \cdot e^x - \int 1 \cdot e^x\,dx$.

Below x in the D column, we put $D(x) = 1$; below e^x in the I column, we put $I(e^x) = e^x$. The arrow at an angle connecting x and $I(e^x)$ reminds us that the product $xI(e^x)$ will appear in the answer; the plus sign on the left of the table reminds us that it is $+xI(e^x)$ that appears. The integral sign and the horizontal arrow connecting $D(x)$ and $I(e^x)$ remind us that the *integral* of the product $D(x)I(e^x)$ also appears in the answer; the minus sign on the left reminds us that we need to subtract this integral. Combining these two contributions, we get

$$\int xe^x\,dx = xe^x - \int e^x\,dx.$$

The integral that appears on the right is much easier than the one we began with, so we can complete the problem:

$$\int xe^x\,dx = xe^x - \int e^x\,dx = xe^x - e^x + C.$$

➡ **Before we go on...** In Example 1, what if we had made the opposite decision and put e^x in the D column and x in the I column? Then we would have had the following table:

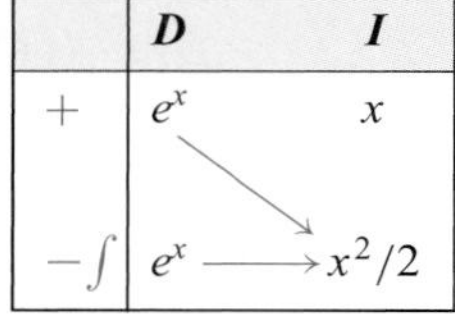

	D	I
$+$	e^x	x
$-\int$	e^x	$x^2/2$

This gives

$$\int xe^x\,dx = \frac{x^2}{2}e^x - \int \frac{x^2}{2}e^x\,dx.$$

The integral on the right is harder than the one we started with, not easier! How do we know beforehand which way to go? We don't. We have to be willing to do a little trial and error: We try it one way, and if it doesn't make things simpler, we try it another way. *Remember, though, that the function we put in the I column must be one that we can integrate.* ■

EXAMPLE 2 Repeated Integration by Parts

Calculate $\int x^2e^{-x}\,dx$.

Solution Again, we have a product—the integrand is the product of x^2 and e^{-x}. Because differentiating x^2 makes it simpler, we put it in the D column and get the following table:

	D	I
$+$	x^2	e^{-x}
$-\int$	$2x$	$-e^{-x}$

This table gives us

$$\int x^2e^{-x}\,dx = x^2(-e^{-x}) - \int 2x(-e^{-x})\,dx.$$

The last integral is simpler than the one we started with, but it still involves a product. It's a good candidate for another integration by parts. The table we would use would start with $2x$ in the D column and $-e^{-x}$ in the I column, which is exactly what we see in the last row of the table we've already made. Therefore, we *continue the process*, elongating the table above:

	D	I
$+$	x^2	e^{-x}
$-$	$2x$	$-e^{-x}$
$+\int$	2	e^{-x}

(Notice how the signs on the left alternate. Here's why: To compute $-\int 2x(-e^{-x})\,dx$ we use the negative of the following table:

	D	*I*
+	$2x$	$-e^{-x}$
$-\int$	2	e^{-x}

so we reverse all the signs.)

Now, we still have to compute an integral (the integral of the product of the functions in the bottom row) to complete the computation. But why stop here? Let's continue the process one more step:

	D	*I*
+	x^2	e^{-x}
−	$2x$	$-e^{-x}$
+	2	e^{-x}
$-\int$	0	$-e^{-x}$

In the bottom line we see that all that is left to integrate is $0(-e^{-x}) = 0$. Because the indefinite integral of 0 is C, we can read the answer from the table as

$$\begin{aligned}\int x^2e^{-x}\,dx &= x^2(-e^{-x}) - 2x(e^{-x}) + 2(-e^{-x}) + C\\ &= -x^2e^{-x} - 2xe^{-x} - 2e^{-x} + C\\ &= -e^{-x}(x^2 + 2x + 2) + C.\end{aligned}$$

In Example 2 we saw a technique that we can summarize as follows:

Integrating a Polynomial Times a Function

If one of the factors in the integrand is a polynomial and the other factor is a function that can be integrated repeatedly, put the polynomial in the D column and keep differentiating until you get zero. Then complete the I column to the same depth, and read off the answer.

For practice, redo Example 1 using this technique.

It is not always the case that the integrand is a polynomial times something easy to integrate, so we can't always expect to end up with a zero in the D column. In that case we hope that at some point we will be able to integrate the product of the functions in the last row. Here are some examples.

EXAMPLE 3 Polynomial Times a Logarithm

Calculate: **a.** $\int x \ln x\, dx$ **b.** $\int (x^2 - x) \ln x\, dx$ **c.** $\int \ln x\, dx$

Solution

a. This is a product and therefore a good candidate for integration by parts. Our first impulse is to differentiate x, but that would mean integrating $\ln x$, and we do not (yet) know how to do that. So we try it the other way around and hope for the best.

	D	*I*
$+$	$\ln x$	x
$-\int$	$\frac{1}{x}$	$\frac{x^2}{2}$

Why did we stop? If we continued the table, both columns would get more complicated. However, if we stop here we get

$$\begin{aligned}\int x \ln x\, dx &= (\ln x)\left(\frac{x^2}{2}\right) - \int \left(\frac{1}{x}\right)\left(\frac{x^2}{2}\right) dx \\ &= \frac{x^2}{2}\ln x - \frac{1}{2}\int x\, dx \\ &= \frac{x^2}{2}\ln x - \frac{x^2}{4} + C.\end{aligned}$$

b. We can use the same technique we used in part (a) to integrate any polynomial times the logarithm of x:

	D	*I*
$+$	$\ln x$	$x^2 - x$
$-\int$	$\frac{1}{x}$	$\frac{x^3}{3} - \frac{x^2}{2}$

$$\begin{aligned}\int (x^2 - x) \ln x\, dx &= (\ln x)\left(\frac{x^3}{3} - \frac{x^2}{2}\right) - \int \left(\frac{1}{x}\right)\left(\frac{x^3}{3} - \frac{x^2}{2}\right) dx \\ &= \left(\frac{x^3}{3} - \frac{x^2}{2}\right)\ln x - \int \left(\frac{x^2}{3} - \frac{x}{2}\right) dx \\ &= \left(\frac{x^3}{3} - \frac{x^2}{2}\right)\ln x - \frac{x^3}{9} + \frac{x^2}{4} + C.\end{aligned}$$

c. The integrand $\ln x$ is not a product. We can, however, *make* it into a product by thinking of it as $1 \cdot \ln x$. Because this is a polynomial times $\ln x$, we proceed as in parts (a) and (b):

	D	*I*
$+$	$\ln x$	1
$-\int$	$1/x$	x

We notice that the product of $1/x$ and x is just 1, which we know how to integrate, so we can stop here:

$$\begin{aligned}\int \ln x\, dx &= x \ln x - \int \left(\frac{1}{x}\right) x\, dx \\ &= x \ln x - \int 1\, dx \\ &= x \ln x - x + C.\end{aligned}$$

FAQs

Whether to Use Integration by Parts, and What Goes in the *D* and *I* Columns

Q: *Will integration by parts always work to integrate a product?*

A: No. Although integration by parts often works for products in which one factor is a polynomial, it will almost *never* work in the examples of products we saw when discussing substitution in Section 6.2. For example, although integration by parts can be used to compute $\int(x^2 - x)e^{2x-1}\,dx$ (put $x^2 - x$ in the *D* column and e^{2x-1} in the *I* column), it *cannot* be used to compute $\int(2x - 1)e^{x^2-x}\,dx$ (put $u = x^2 - x$). Recognizing when to use integration by parts is best learned by experience.

Q: *When using integration by parts, which expression goes in the D column, and which in the I column?*

A: Although there is no general rule, the following guidelines are useful:

- To integrate a product in which one factor is a polynomial and the other can be integrated several times, put the polynomial in the *D* column and the other factor in the *I* column. Then differentiate the polynomial until you get zero.
- If one of the factors is a polynomial but the other factor cannot be integrated easily, put the polynomial in the *I* column and the other factor in the *D* column. Stop when the product of the functions in the bottom row can be integrated.
- If neither factor is a polynomial, put the factor that seems easier to integrate in the *I* column and the other factor in the *D* column. Again, stop the table as soon as the product of the functions in the bottom row can be integrated.
- If your method doesn't work, try switching the functions in the *D* and *I* columns or try breaking the integrand into a product in a different way. If none of this works, maybe integration by parts isn't the technique to use on this problem.

7.1 EXERCISES

Access end-of-section exercises online at **www.webassign.net** ENHANCED WebAssign

7.2 Area between Two Curves and Applications

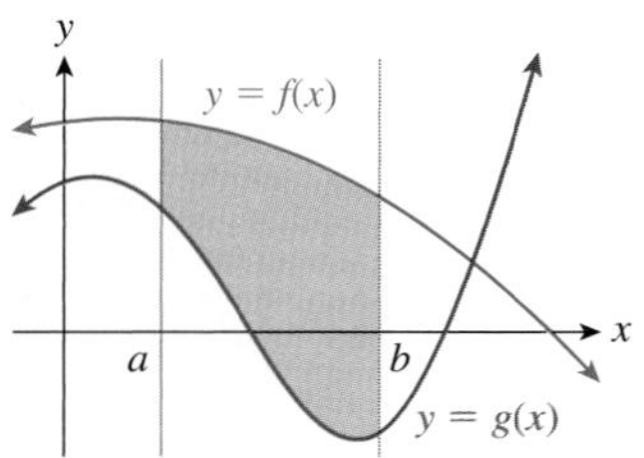

Figure 1

As we saw in the preceding chapter, we can use the definite integral to calculate the area between the graph of a function and the x-axis. With only a little more work, we can use it to calculate the area between two graphs. Figure 1 shows the graphs of two functions, $f(x)$ and $g(x)$, with $f(x) \geq g(x)$ for every x in the interval $[a, b]$.

To find the shaded area between the graphs of the two functions, we use the following formula.

Area between Two Graphs

If $f(x) \geq g(x)$ for all x in $[a, b]$ (so that the graph of f does not move below that of g), then the area of the region between the graphs of f and g and between $x = a$ and $x = b$ is given by

$$A = \int_a^b [f(x) - g(x)]\, dx. \quad \text{Integral of (Top – Bottom)}$$

Caution If the graphs of f and g cross in the interval, the above formula does not hold; for instance, if $f(x) = x$ and $g(x) = -x$, then the total area shown in the figure is 2 square units, whereas $\int_{-1}^{1} [f(x) - g(x)]\, dx = 0$.

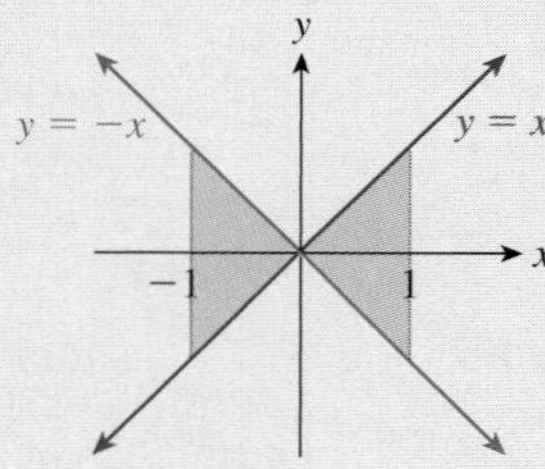

Let's look at an example and then discuss why the formula works.

EXAMPLE 1 The Area between Two Curves

Find the areas of the following regions:

a. Between $f(x) = -x^2 - 3x + 4$ and $g(x) = x^2 - 3x - 4$ and between $x = -1$ and $x = 1$

b. Between $f(x) = |x|$ and $g(x) = -|x - 1|$ over $[-1, 2]$

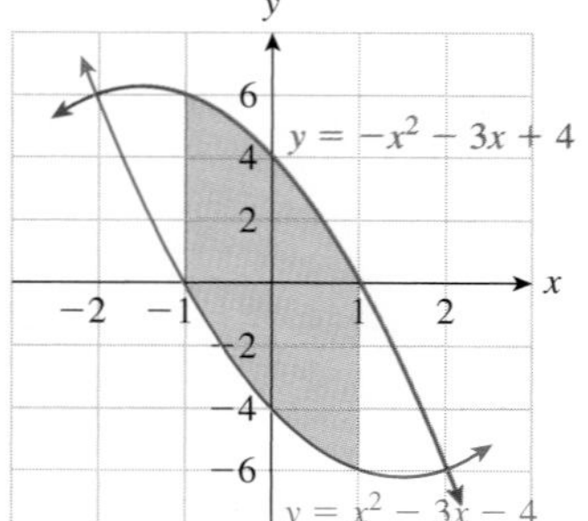

Figure 2

Solution

a. The area in question is shown in Figure 2. Because the graph of f lies above the graph of g in the interval $[-1, 1]$, we have $f(x) \geq g(x)$ for all x in $[-1, 1]$. Therefore, we can use the formula given above and calculate the area as follows:

$$A = \int_{-1}^{1} [f(x) - g(x)]\, dx$$
$$= \int_{-1}^{1} [(-x^2 - 3x + 4) - (x^2 - 3x - 4)]\, dx$$
$$= \int_{-1}^{1} (8 - 2x^2)\, dx$$
$$= \left[8x - \frac{2}{3}x^3\right]_{-1}^{1}$$
$$= \frac{44}{3}.$$

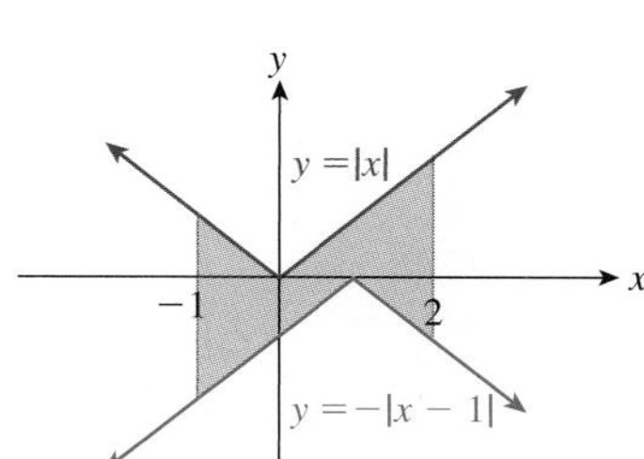

Figure 3

b. The given area (see Figure 3) can be broken up into triangles and rectangles, but we already know a formula for the antiderivative of $|ax + b|$ for constants a and b, so we can use calculus instead:

$$A = \int_{-1}^{2} [f(x) - g(x)]\, dx$$
$$= \int_{-1}^{2} [|x| - (-|x - 1|)]\, dx$$
$$= \int_{-1}^{2} [|x| + |x - 1|]\, dx$$
$$= \frac{1}{2}[x|x| + (x - 1)|x - 1|]_{-1}^{2} \qquad \int |ax + b|\, dx = \frac{1}{2a}(ax + b)|ax + b| + C$$
$$= \frac{1}{2}[(4 + 1) - (-1 - 4)]$$
$$= \frac{1}{2}(10) = 5.$$

Q: *Why does the formula for the area between two curves work?*

A: Let's go back once again to the general case illustrated in Figure 1, where we were given two functions f and g with $f(x) \geq g(x)$ for every x in the interval $[a, b]$. To avoid complicating the argument by the fact that the graph of g, or f, or both may dip below the x-axis in the interval $[a, b]$ (as occurs in Figure 1 and also in Example 1), we shift both graphs vertically upward by adding a big enough constant M to lift them both above the x-axis in the interval $[a, b]$, as shown in Figure 4.

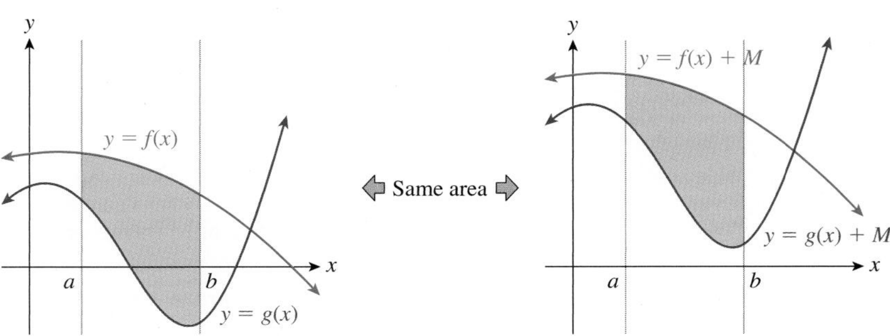

Figure 4

As the figure illustrates, the area of the region between the graphs is not affected, so we will calculate the area of the region shown on the right of Figure 4. That calculation is shown in Figure 5.

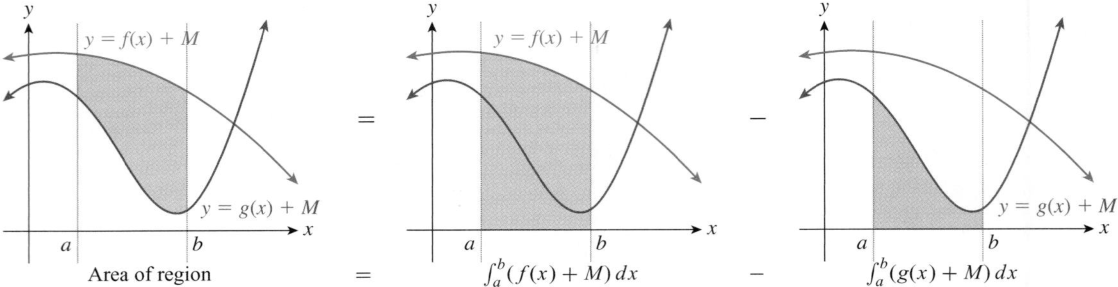

Figure 5

From the figure, the area we want is

$$\int_a^b (f(x) + M)\,dx - \int_a^b (g(x) + M)\,dx = \int_a^b [(f(x) + M) - (g(x) + M)]\,dx$$

$$= \int_a^b [f(x) - g(x)]\,dx,$$

which is the formula we gave originally.

So far, we've been assuming that $f(x) \geq g(x)$, so that the graph of f never dips below the graph of g and so the graphs cannot cross (although they can touch). Example 2 shows how we compute the area between graphs that *do* cross.

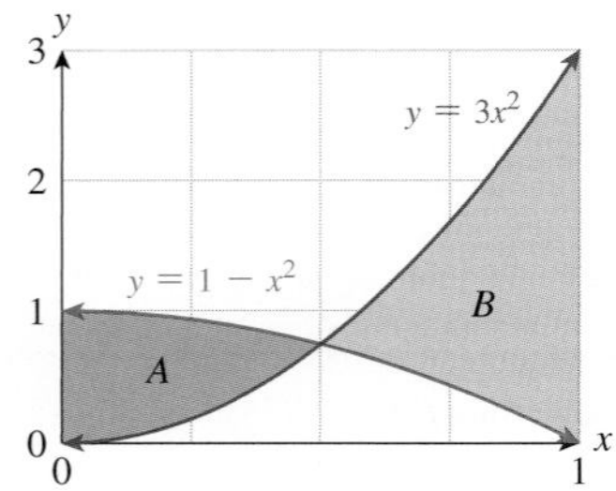

Figure 6

EXAMPLE 2 Regions Enclosed by Crossing Curves

Find the area of the region between $y = 3x^2$ and $y = 1 - x^2$ and between $x = 0$ and $x = 1$.

Solution The area we wish to calculate is shown in Figure 6. From the figure, we can see that neither graph lies above the other over the whole interval. To get around this, we break the area into the two pieces on either side of the point at which the graphs cross and then compute each area separately. To do this, we need to know exactly where that crossing point is. The crossing point is where $3x^2 = 1 - x^2$, so we solve for x:

$$\begin{aligned} 3x^2 &= 1 - x^2 \\ 4x^2 &= 1 \\ x^2 &= \frac{1}{4} \\ x &= \pm\frac{1}{2}. \end{aligned}$$

Because we are interested only in the interval [0, 1], the crossing point we're interested in is at $x = 1/2$.

Now, to compute the areas A and B, we need to know which graph is on top in each of these areas. We can see that from the figure, but what if the functions were more complicated and we could not easily draw the graphs? To be sure, we can test the values of the two functions at some point in each region. But we really need not worry. If we make the wrong choice for the top function, the integral will yield the negative of the area (why?), so we can simply take the absolute value of the integral to get the area of the region in question. For this example, we have

$$\begin{aligned} A &= \int_0^{1/2} [(1 - x^2) - 3x^2]\,dx = \int_0^{1/2} (1 - 4x^2)\,dx \\ &= \left[x - \frac{4x^3}{3}\right]_0^{1/2} \\ &= \left(\frac{1}{2} - \frac{1}{6}\right) - (0 - 0) = \frac{1}{3} \end{aligned}$$

and

$$\begin{aligned} B &= \int_{1/2}^{1} [3x^2 - (1 - x^2)]\,dx = \int_{1/2}^{1} (4x^2 - 1)\,dx \\ &= \left[\frac{4x^3}{3} - x\right]_{1/2}^{1} \\ &= \left(\frac{4}{3} - 1\right) - \left(\frac{1}{6} - \frac{1}{2}\right) = \frac{2}{3}. \end{aligned}$$

This gives a total area of $A + B = \dfrac{1}{3} + \dfrac{2}{3} = 1$.

Before we go on... What would have happened in Example 2 if we had not broken the area into two pieces but had just calculated the integral of the difference of the two functions? We would have calculated

$$\int_0^1 [(1 - x^2) - 3x^2]\,dx = \int_0^1 [1 - 4x^2]\,dx = \left[x - \frac{4x^3}{3}\right]_0^1 = -\frac{1}{3},$$

which is not even close to the right answer. What this integral calculated was actually $A - B$ rather than $A + B$. Why? ■

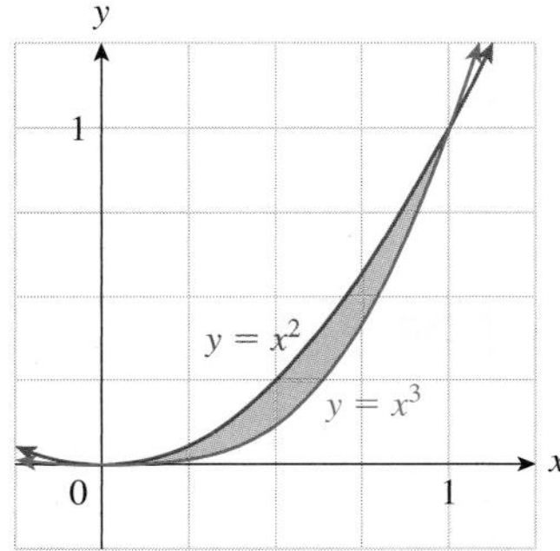

Figure 7

EXAMPLE 3 The Area Enclosed by Two Curves

Find the area enclosed by $y = x^2$ and $y = x^3$.

Solution This example has a new wrinkle: We are not told what interval to use for x. However, if we look at the graph in Figure 7, we see that the question can have only one meaning.

We are being asked to find the area of the shaded sliver, which is the only region that is actually *enclosed* by the two graphs. This sliver is bounded on either side by the two points where the graphs cross, so our first task is to find those points. They are the points where $x^2 = x^3$, so we solve for x:

$$\begin{aligned} x^2 &= x^3 \\ x^3 - x^2 &= 0 \\ x^2(x - 1) &= 0 \\ x = 0 \quad &\text{or} \quad x = 1 \end{aligned}$$

Thus, we must integrate over the interval [0, 1]. Although we see from the diagram (or by substituting $x = 1/2$) that the graph of $y = x^2$ is above that of $y = x^3$, if we didn't notice that we might calculate

$$\int_0^1 (x^3 - x^2)\, dx = \left[\frac{x^4}{4} - \frac{x^3}{3}\right]_0^1 = -\frac{1}{12}.$$

This tells us that the required area is $1/12$ square units and also that we had our integral reversed. Had we calculated $\int_0^1 (x^2 - x^3)\, dx$ instead, we would have found the correct answer, $1/12$.

We can summarize the procedure we used in the preceding two examples.

Finding the Area between the Graphs of $f(x)$ and $g(x)$

1. Find all points of intersection by solving $f(x) = g(x)$ for x. This either determines the interval over which you will integrate or breaks up a given interval into regions between the intersection points.
2. Determine the area of each region you found by integrating the difference of the larger and the smaller function. (If you accidentally take the smaller minus the larger, the integral will give the negative of the area, so just take the absolute value.)
3. Add together the areas you found in step 2 to get the total area.

Q: *Is there any quick and easy method to find the area between two graphs without having to find all points of intersection? What if it is hard or impossible to find out where the curves intersect?*

A: We can use technology to give the approximate area between two graphs. First recall that, if $f(x) \geq g(x)$ for all x in $[a, b]$, then the area between their graphs over $[a, b]$ is given by $\int_a^b [f(x) - g(x)]\, dx$, whereas if $g(x) \geq f(x)$, the area is given by $\int_a^b [g(x) - f(x)]\, dx$. Notice that both expressions are equal to

$$\int_a^b |f(x) - g(x)|\, dx$$

telling us that we can use this same formula in both cases.

T Area between Two Graphs: Approximation Using Technology

The area of the region between the graphs of f and g and between $x = a$ and $x = b$ is given by

$$A = \int_a^b |f(x) - g(x)|\, dx.$$

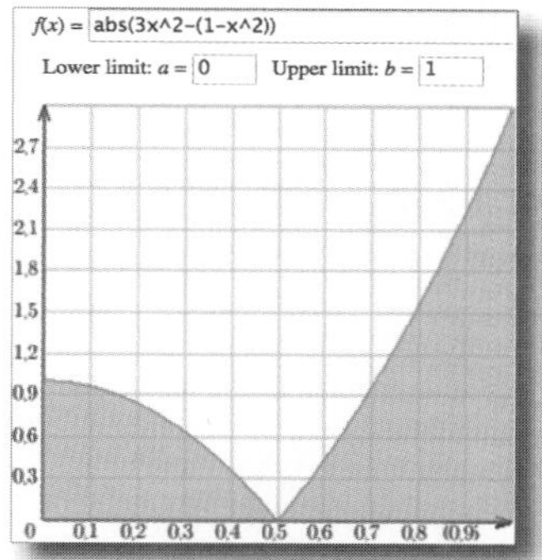

Figure 8

Quick Example

To approximate the area of the region between $y = 3x^2$ and $y = 1 - x^2$ and between $x = 0$ and $x = 1$ we calculated in Example 2, use technology to compute

$$\int_0^1 |3x^2 - (1 - x^2)|\, dx = 1.$$

TI-83/84 Plus: Enter `fnInt(abs(3x^2-(1-x^2)),X,0,1)`

Website: Online Utilities → Numerical Integration Utility
Enter `abs(3x^2-(1-x^2))` for $f(x)$ and 0 and 1 for the lower and upper limits, and press "Integral". (See Figure 8.)

7.2 EXERCISES

Access end-of-section exercises online at **www.webassign.net** ENHANCED WebAssign

7.3 Averages and Moving Averages

Averages

To find the average of, say, 20 numbers, we simply add them up and divide by 20. More generally, if we want to find the **average**, or **mean**, of the n numbers $y_1, y_2, y_3, \ldots y_n$, we add them up and divide by n. We write this average as $\bar{y}$ ("y-bar").

Average, or Mean, of a Collection of Values

$$\bar{y} = \frac{y_1 + y_2 + \cdots + y_n}{n}$$

Quick Example

The average of $\{0, 2, -1, 5\}$ is $\bar{y} = \dfrac{0 + 2 - 1 + 5}{4} = \dfrac{6}{4} = 1.5.$

But, we also use the word *average* in other senses. For example, we speak of the average speed of a car during a trip.

EXAMPLE 1 Average Speed

Over the course of 2 hours, my speed varied from 50 miles per hour to 60 miles per hour, following the function $v(t) = 50 + 2.5t^2$, $0 \leq t \leq 2$. What was my average speed over those 2 hours?

Solution Recall that average speed is simply the total distance traveled divided by the time it took. Recall, also, that we can find the distance traveled by integrating the speed:

$$\begin{aligned}\text{Distance traveled} &= \int_0^2 v(t)\,dt \\ &= \int_0^2 (50 + 2.5t^2)\,dt \\ &= \left[50t + \frac{2.5}{3}t^3\right]_0^2 \\ &= 100 + \frac{20}{3} \\ &\approx 106.67 \text{ miles.}\end{aligned}$$

It took 2 hours to travel this distance, so the average speed was

$$\text{Average speed} \approx \frac{106.67}{2} \approx 53.3 \text{ mph.}$$

In general, if we travel with velocity $v(t)$ from time $t = a$ to time $t = b$, we will travel a distance of $\int_a^b v(t)\,dt$ in time $b - a$, which gives an average velocity of

$$\text{Average velocity} = \frac{1}{b-a}\int_a^b v(t)\,dt.$$

Thinking of this calculation as finding the average value of the velocity function, we generalize and make the following definition.

Average Value of a Function

The **average**, or **mean**, of a function $f(x)$ on an interval $[a, b]$ is

$$\bar{f} = \frac{1}{b-a}\int_a^b f(x)\,dx.$$

Quick Example

The average of $f(x) = x$ on $[1, 5]$ is

$$\begin{aligned}\bar{f} &= \frac{1}{b-a}\int_a^b f(x)\,dx \\ &= \frac{1}{5-1}\int_1^5 x\,dx \\ &= \frac{1}{4}\left[\frac{x^2}{2}\right]_1^5 = \frac{1}{4}\left(\frac{25}{2} - \frac{1}{2}\right) = 3.\end{aligned}$$

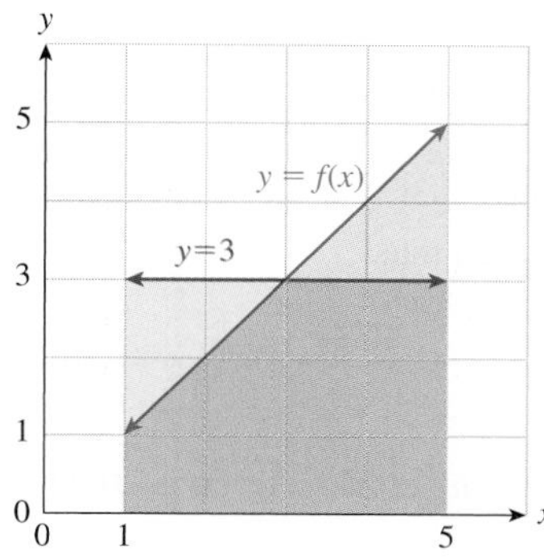

Figure 9

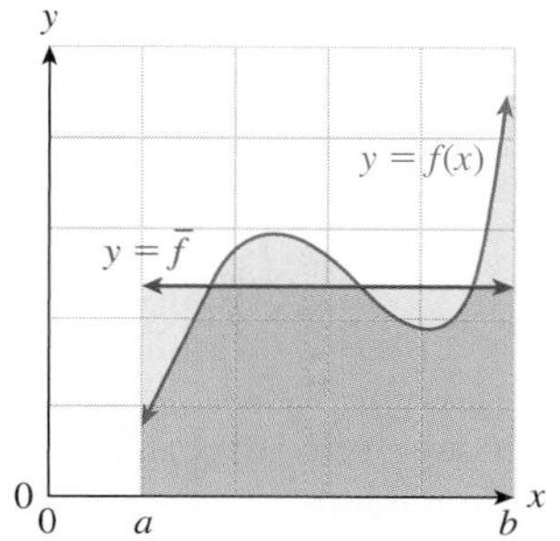

Figure 10

Interpreting the Average of a Function Geometrically

The average of a function has a geometric interpretation. Referring to the Quick Example on previous page, we can compare the graph of $y = f(x)$ with the graph of $y = 3$, both over the interval [1, 5] (Figure 9). We can find the area under the graph of $f(x) = x$ by geometry or by calculus; it is 12. The area in the rectangle under $y = 3$ is also 12.

In general, the average $\bar{f}$ of a positive function over the interval $[a, b]$ gives the height of the rectangle over the interval $[a, b]$ that has the same area as the area under the graph of $f(x)$ as illustrated in Figure 10. The equality of these areas follows from the equation

$$(b - a)\bar{f} = \int_a^b f(x)\,dx.$$

EXAMPLE 2 Average Balance

A savings account at the People's Credit Union pays 3% interest, compounded continuously, and at the end of the year you get a bonus of 1% of the average balance in the account during the year. If you deposit \$10,000 at the beginning of the year, how much interest and how large a bonus will you get?

Solution We can use the continuous compound interest formula to calculate the amount of money you have in the account at time t:

$$A(t) = 10{,}000e^{0.03t},$$

where t is measured in years. At the end of 1 year, the account will have

$$A(1) = \$10{,}304.55$$

so you will have earned \$304.55 interest. To compute the bonus, we need to find the average amount in the account, which is the average of $A(t)$ over the interval [0, 1]. Thus,

$$\begin{aligned}\bar{A} &= \frac{1}{b-a}\int_a^b A(t)\,dt \\ &= \frac{1}{1-0}\int_0^1 10{,}000e^{0.03t}\,dt = \frac{10{,}000}{0.03}\left[e^{0.03t}\right]_0^1 \\ &\approx \$10{,}151.51.\end{aligned}$$

The bonus is 1% of this, or \$101.52.

➡ **Before we go on...** The 1% bonus in Example 2 was one third of the total interest. Why did this happen? What fraction of the total interest would the bonus be if the interest rate was 4%, 5%, or 10%? ■

Moving Averages

Suppose you follow the performance of a company's stock by recording the daily closing prices. The graph of these prices may seem jagged or "jittery" due to random day-to-day fluctuations. To see any trends, you would like a way to "smooth out" these data. The **moving average** is one common way to do that.

EXAMPLE 3 Stock Prices

The following table shows Colossal Conglomerate's closing stock prices for 20 consecutive trading days.

Day	1	2	3	4	5	6	7	8	9	10
Price	20	22	21	24	24	23	25	26	20	24
Day	11	12	13	14	15	16	17	18	19	20
Price	26	26	25	27	28	27	29	27	25	24

Plot these prices and the 5-day moving average.

Solution The 5-day moving average is the average of each day's price together with the prices of the preceding 4 days. We can compute the 5-day moving averages starting on the fifth day. We get these numbers:

Day	1	2	3	4	5	6	7	8	9	10
Moving Average					22.2	22.8	23.4	24.4	23.6	23.6
Day	11	12	13	14	15	16	17	18	19	20
Moving Average	24.2	24.4	24.2	25.6	26.4	26.6	27.2	27.6	27.2	26.4

The closing stock prices and moving averages are plotted in Figure 11.

Figure 11

As you can see, the moving average is less volatile than the closing price. Because the moving average incorporates the stock's performance over 5 days at a time, a single day's fluctuation is smoothed out. Look at day 9 in particular. The moving average also tends to lag behind the actual performance because it takes past performance into account. Look at the downturns at days 6 and 18 in particular.

using Technology

See the Technology Guides at the end of the chapter to find out how to tabulate and graph moving averages using a graphing calculator or a spreadsheet. Outline:

TI-83/84 Plus

STAT_EDIT; days in L_1, prices in L_2.

Home screen:

```
seq((L2(X)+L2(X-1)
+L2(X-2)+L2(X-3)
+L2(X-4))/5,X,5,20)
→L3
```

[More details on page 431.]

Spreadsheet

Day data in A2–A21

Price data in B2–B21

Enter `=AVERAGE(B2:B6)` in C6, copy down to C21.

Graph the data in columns A–C.

[More details on page 432.]

The period of 5 days for a moving average, as used in Example 3, is arbitrary. Using a longer period of time would smooth the data more but increase the lag. For data used as economic indicators, such as housing prices or retail sales, it is common to compute the four-quarter moving average to smooth out seasonal variations.

It is also sometimes useful to compute moving averages of continuous functions. We may want to do this if we use a mathematical model of a large collection of data. Also, some physical systems have the effect of converting an input function (an electrical signal, for example) into its moving average. By an ***n*-unit moving average** of a function $f(x)$ we mean the function $\bar{f}$ for which $\bar{f}(x)$ is the average of the value of $f(x)$ on $[x-n, x]$. Using the formula for the average of a function, we get the following formula.

n-Unit Moving Average of a Function

The n-unit moving average of a function f is

$$\bar{f}(x) = \frac{1}{n}\int_{x-n}^{x} f(t)\,dt.$$

Quick Example

The 2-unit moving average of $f(x) = x^2$ is

$$\bar{f}(x) = \frac{1}{2}\int_{x-2}^{x} t^2\,dt = \frac{1}{6}\left[t^3\right]_{x-2}^{x} = x^2 - 2x + \frac{4}{3}.$$

The graphs of $f(x)$ and $\bar{f}(x)$ are shown in Figure 12.

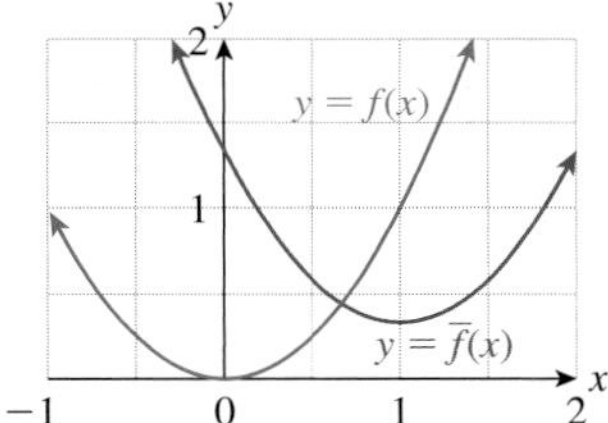

Figure 12

EXAMPLE 4 Moving Averages: Sawtooth and Step Functions

Graph the following functions, and then compute and graph their 1-unit moving averages.

$$f(x) = |x| - |x-1| + |x-2| \qquad \text{Sawtooth}$$

$$g(x) = \frac{1}{2}\left[1 + \frac{|x-1|}{x-1}\right] \qquad \text{Unit step at } x = 1$$

Solution The graphs of f and g are shown in Figure 13. (Notice that the step function is not defined at $x = 1$. Most graphers will show the step function as an actual step by connecting the points (1, 0) and (1, 1) with a vertical line.)

The 1-step moving averages are:

$$\bar{f}(x) = \int_{x-1}^{x} f(t)\,dt = \int_{x-1}^{x} [|t| - |t-1| + |t-2|]\,dt$$

$$= \frac{1}{2}[t|t| - (t-1)|t-1| + (t-2)|t-2|]_{x-1}^{x}$$

$$= \frac{1}{2}([x|x| - (x-1)|x-1| + (x-2)|x-2|]$$
$$- [(x-1)|x-1| - (x-2)|x-2| + (x-3)|x-3|])$$

$$= \frac{1}{2}[x|x| - 2(x-1)|x-1| + 2(x-2)|x-2| - (x-3)|x-3|].$$

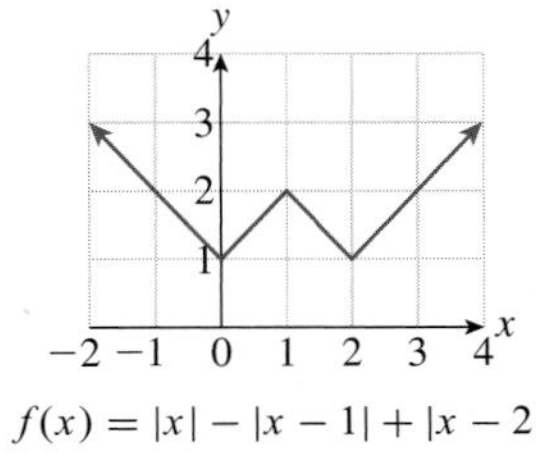

$f(x) = |x| - |x-1| + |x-2|$

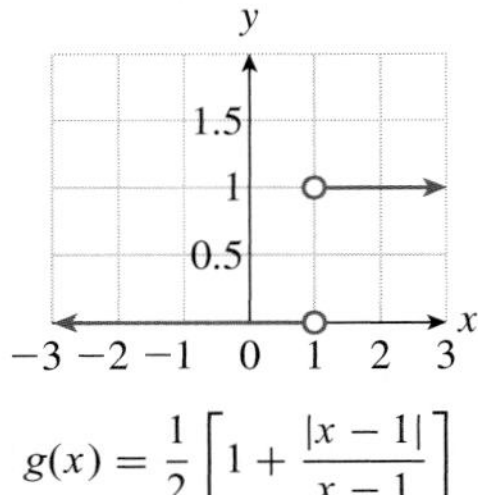

$g(x) = \frac{1}{2}\left[1 + \frac{|x-1|}{x-1}\right]$

Figure 13

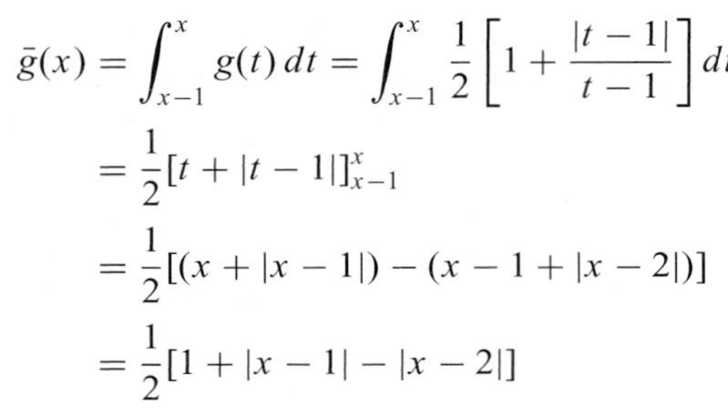

$$\bar{g}(x) = \int_{x-1}^{x} g(t)\,dt = \int_{x-1}^{x} \frac{1}{2}\left[1 + \frac{|t-1|}{t-1}\right] dt$$

$$= \frac{1}{2}[t + |t-1|]_{x-1}^{x}$$

$$= \frac{1}{2}[(x + |x-1|) - (x - 1 + |x-2|)]$$

$$= \frac{1}{2}[1 + |x-1| - |x-2|]$$

using Technology

TI-83/84 Plus

We can graph the moving average of f in Example 4 on a TI-83/84 Plus as follows:

```
Y1=abs(X)-abs(X-1)
+abs(X-2)
Y2=fnInt(Y1(T),T,
X-1,X)
ZOOM 0
```

For g, change Y_1 to

```
Y1=.5*(1+abs(X-1)/(X-1))
```

The graphs of $\bar{f}$ and $\bar{g}$ are shown in Figure 14.

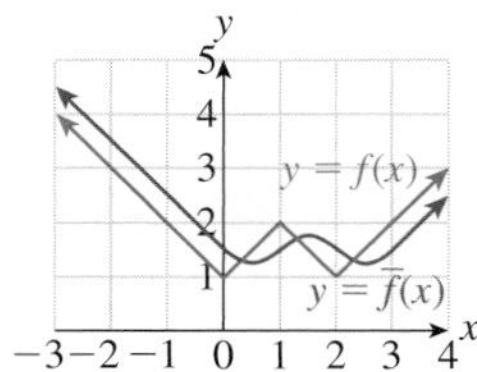

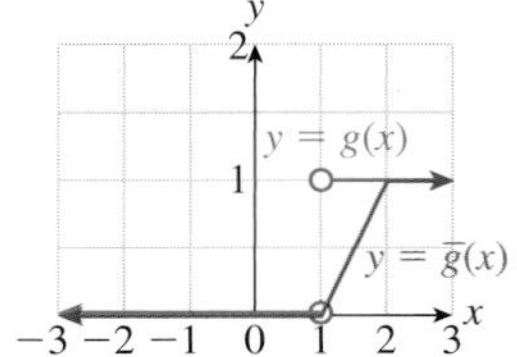

Figure 14

Notice how the graph of $\bar{f}$ smooths out the zig-zags of the sawtooth function.

Before we go on... Figure 15 shows the 2-unit moving average of f in Example 4,

$$\bar{f}(x) = \frac{1}{2}\int_{x-2}^{x} f(t)\,dt$$

$$= \frac{1}{4}(x|x| - (x-1)|x-1| + (x-3)|x-3| - (x-4)|x-4|).$$

Figure 15

Notice how the 2-point moving average has completely eliminated the zig-zags, illustrating how moving averages can be used to remove seasonal fluctuations in real-life situations. ■

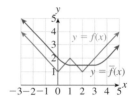

7.3 EXERCISES

Access end-of-section exercises online at **www.webassign.net** ENHANCED WebAssign

7.4 Applications to Business and Economics: Consumers' and Producers' Surplus and Continuous Income Streams

Figure 16

Consumers' Surplus

Consider a general demand curve presented, as is traditional in economics, as $p = D(q)$, where p is unit price and q is demand measured, say, in annual sales (Figure 16). Thus, $D(q)$ is the price at which the demand will be q units per year. The price p_0 shown on the graph is the highest price that customers are willing to pay.

Suppose, for example, that the graph in Figure 16 is the demand curve for a particular new model of computer. When the computer first comes out and supplies are low (q is small), "early adopters" will be willing to pay a high price. This is the part of the graph on the left, near the p-axis. As supplies increase and the price drops, more consumers will be willing to pay and more computers will be sold. We can ask the following question: How much are consumers willing to spend for the first $\bar{q}$ units?

Consumers' Willingness to Spend

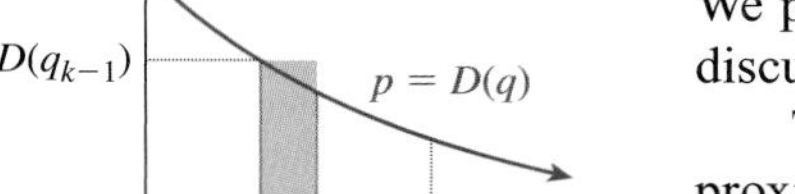

Figure 17

We can approximate consumers' willingness to spend on the first $\bar{q}$ units as follows. We partition the interval $[0, \bar{q}]$ into n subintervals of equal length, as we did when discussing Riemann sums. Figure 17 shows a typical subinterval, $[q_{k-1}, q_k]$.

The price consumers are willing to pay for each of units q_{k-1} through q_k is approximately $D(q_{k-1})$, so the total that consumers are willing to spend for these units is approximately $D(q_{k-1})(q_k - q_{k-1}) = D(q_{k-1})\Delta q$, the area of the shaded region in Figure 17. Thus, the total amount that consumers are willing to spend for items 0 through $\bar{q}$ is

$$W \approx D(q_0)\Delta q + D(q_1)\Delta q + \cdots + D(q_{n-1})\Delta q = \sum_{k=0}^{n-1} D(q_k)\Delta q,$$

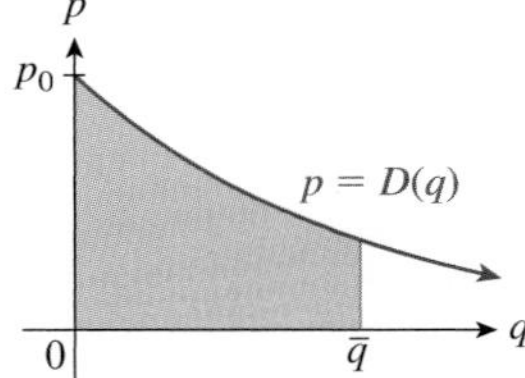

Figure 18

which is a Riemann sum. The approximation becomes better the larger n becomes, and in the limit the Riemann sums converge to the integral

$$W = \int_0^{\bar{q}} D(q)\, dq.$$

This quantity, the area shaded in Figure 18, is the total consumers' willingness to spend to buy the first $\bar{q}$ units.

Consumers' Expenditure

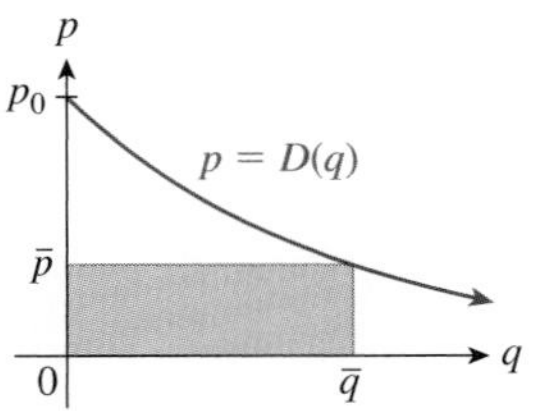

Figure 19

Now suppose that the manufacturer simply sets the price at $\bar{p}$, with a corresponding demand of $\bar{q}$, so $D(\bar{q}) = \bar{p}$. Then the amount that consumers will actually spend to buy these $\bar{q}$ is $\bar{p}\,\bar{q}$, the product of the unit price and the quantity sold. This is the area of the rectangle shown in Figure 19. Notice that we can write $\bar{p}\bar{q} = \int_0^{\bar{q}} \bar{p}\, dq$, as suggested by the figure.

The difference between what consumers are willing to pay and what they actually pay is money in their pockets and is called the **consumers' surplus**.

Consumers' Surplus

If demand for an item is given by $p = D(q)$, the selling price is $\bar{p}$, and $\bar{q}$ is the corresponding demand [so that $D(\bar{q}) = \bar{p}$], then the **consumers' surplus** is the difference between willingness to spend and actual expenditure:*

$$CS = \int_0^{\bar{q}} D(q)\, dq - \bar{p}\bar{q} = \int_0^{\bar{q}} (D(q) - \bar{p})\, dq.$$

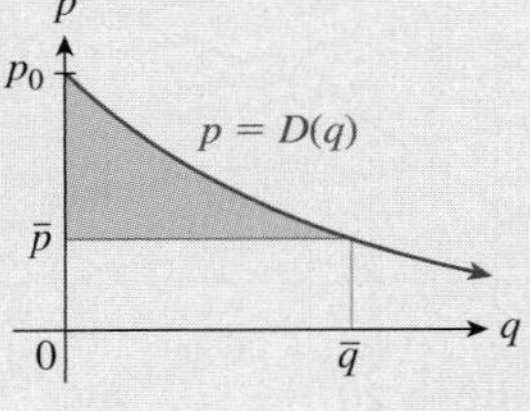

Graphically, it is the area between the graphs of $p = D(q)$ and $p = \bar{p}$, as shown in the figure.

* Multiletter variables like CS used here may be unusual in a math textbook but are traditional in the math of finance. In particular, the notations PV and FV used later in this section are almost universally used in finance textbooks, calculators (such as the TI-83/84 Plus), and such places as study guides for the finance portion of the Society of Actuaries exams.

EXAMPLE 1 Consumers' Surplus

Your used-CD store has an exponential demand equation of the form

$$p = 15e^{-0.01q}$$

where q represents daily sales of used CDs and p is the price you charge per CD. Calculate the daily consumers' surplus if you sell your used CDs at \$5 each.

Solution We are given $D(q) = 15e^{-0.01q}$ and $\bar{p} = 5$. We also need $\bar{q}$. By definition,

$$D(\bar{q}) = \bar{p}$$

or $\quad 15e^{-0.01\bar{q}} = 5,$

which we must solve for $\bar{q}$:

$$e^{-0.01\bar{q}} = \frac{1}{3}$$

$$-0.01\bar{q} = \ln\left(\frac{1}{3}\right) = -\ln 3$$

$$\bar{q} = \frac{\ln 3}{0.01} \approx 109.8612.$$

We now have

$$\begin{aligned} CS &= \int_0^{\bar{q}} (D(q) - \bar{p})\, dq \\ &= \int_0^{109.8612} (15e^{-0.01q} - 5)\, dq \\ &= \left[\frac{15}{-0.01}e^{-0.01q} - 5q\right]_0^{109.8612} \\ &\approx (-500 - 549.31) - (-1{,}500 - 0) \\ &= \$450.69 \text{ per day.} \end{aligned}$$

Producers' Surplus

We can also calculate extra income earned by producers. Consider a supply equation of the form $p = S(q)$, where $S(q)$ is the price at which a supplier is willing to supply q items (per time period). Because a producer is generally willing to supply more units at a higher price per unit, a supply curve usually has a positive slope, as shown in Figure 20. The price p_0 is the lowest price that a producer is willing to charge.

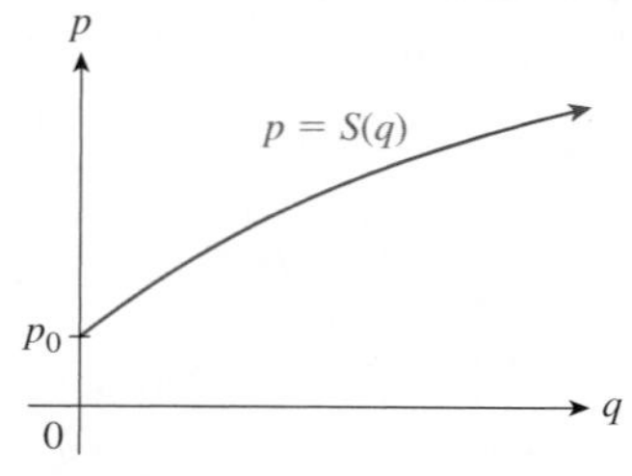

Figure 20

Arguing as before, we see that the minimum amount of money producers are willing to receive in exchange for $\bar{q}$ items is $\int_0^{\bar{q}} S(q)\, dq$. On the other hand, if the producers charge $\bar{p}$ per item for $\bar{q}$ items, their actual revenue is $\bar{p}\bar{q} = \int_0^{\bar{q}} \bar{p}\, dq$.

The difference between the producers' actual revenue and the minimum they would have been willing to receive is the **producers' surplus**.

Producers' Surplus

The **producers' surplus** is the extra amount earned by producers who were willing to charge less than the selling price of $\bar{p}$ per unit and is given by

$$PS = \int_0^{\bar{q}} [\bar{p} - S(q)]\,dq,$$

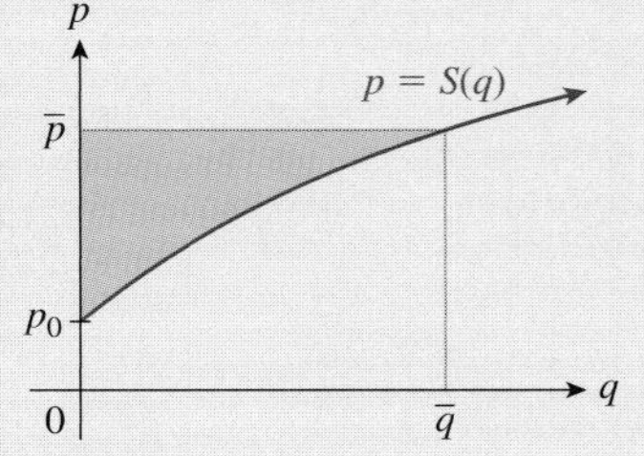

where $S(\bar{q}) = \bar{p}$. Graphically, it is the area of the region between the graphs of $p = \bar{p}$ and $p = S(q)$ for $0 \le q \le \bar{q}$, as in the figure.

EXAMPLE 2 Producers' Surplus

My tie-dye T-shirt enterprise has grown to the extent that I am now able to produce T-shirts in bulk, and several campus groups have begun placing orders. I have informed one group that I am prepared to supply $20\sqrt{p-4}$ T-shirts at a price of p dollars per shirt. What is my total surplus if I sell T-shirts to the group at \$8 each?

Solution We need to calculate the producers' surplus when $\bar{p} = 8$. The supply equation is

$$q = 20\sqrt{p-4}$$

but in order to use the formula for producers' surplus, we need to express p as a function of q. First, we square both sides to remove the radical sign:

$$q^2 = 400(p-4)$$

so

$$p - 4 = \frac{q^2}{400}$$

giving

$$p = S(q) = \frac{q^2}{400} + 4.$$

We now need the value of $\bar{q}$ corresponding to $\bar{p} = 8$. Substituting $p = 8$ in the original equation gives

$$\bar{q} = 20\sqrt{8-4} = 20\sqrt{4} = 40.$$

Thus,

$$\begin{aligned} PS &= \int_0^{\bar{q}} (\bar{p} - S(q))\,dq \\ &= \int_0^{40} \left[8 - \left(\frac{q^2}{400} + 4\right)\right] dq \\ &= \int_0^{40} \left(4 - \frac{q^2}{400}\right) dq \\ &= \left[4q - \frac{q^3}{1{,}200}\right]_0^{40} \approx \$106.67. \end{aligned}$$

Thus, I earn a surplus of \$106.67 if I sell T-shirts to the group at \$8 each.

EXAMPLE 3 Equilibrium

To continue the preceding example: A representative informs me that the campus group is prepared to order only $\sqrt{200(16-p)}$ T-shirts at p dollars each. I would like to produce as many T-shirts for them as possible but avoid being left with unsold T-shirts. Given the supply curve from the preceding example, what price should I charge per T-shirt, and what are the consumers' and producers' surpluses at that price?

Solution The price that guarantees neither a shortage nor a surplus of T-shirts is the **equilibrium price**, the price where supply equals demand. We have

$$\text{Supply:} \quad q = 20\sqrt{p-4}.$$

$$\text{Demand:} \quad q = \sqrt{200(16-p)}.$$

Equating these gives

$$20\sqrt{p-4} = \sqrt{200(16-p)}$$

$$400(p-4) = 200(16-p),$$

$$400p - 1{,}600 = 3{,}200 - 200p$$

$$600p = 4{,}800$$

$$p = \$8 \text{ per T-shirt.}$$

We therefore take $\bar{p} = 8$ (which happens to be the price we used in the preceding example). We get the corresponding value for q by substituting $p = 8$ into either the demand or supply equation:

$$\bar{q} = 20\sqrt{8-4} = 40.$$

Thus, $\bar{p} = 8$ and $\bar{q} = 40$.

We must now calculate the consumers' surplus and the producers' surplus. We calculated the producers' surplus for $\bar{p} = 8$ in the preceding example:

$$PS = \$106.67.$$

For the consumers' surplus, we must first express p as a function of q for the demand equation. Thus, we solve the demand equation for p as we did for the supply equation and we obtain

$$\text{Demand:} \quad D(q) = 16 - \frac{q^2}{200}.$$

Therefore,

$$\begin{aligned} CS &= \int_0^{\bar{q}} (D(q) - \bar{p})\, dq \\ &= \int_0^{40} \left[\left(16 - \frac{q^2}{200}\right) - 8\right] dq \\ &= \int_0^{40} \left(8 - \frac{q^2}{200}\right) dq \\ &= \left[8q - \frac{q^3}{600}\right]_0^{40} \approx \$213.33. \end{aligned}$$

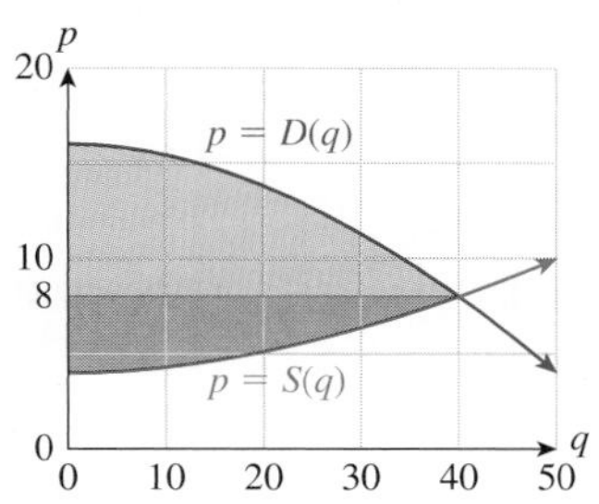

Figure 21

➡ **Before we go on...** Figure 21 shows both the consumers' surplus (top portion) and the producers' surplus (bottom portion) from Example 3. Because extra money in people's pockets is a good thing, the total of the consumers' and the producers' surpluses is called the **total social gain**. In this case it is

$$\text{Social gain} = CS + PS = 213.33 + 106.67 = \$320.00.$$

As you can see from the figure, the total social gain is also the area between two curves and equals

$$\int_0^{40} (D(q) - S(q))\,dq.$$

■

Continuous Income Streams

For purposes of calculation, it is often convenient to assume that a company with a high sales volume receives money continuously. In such a case, we have a function $R(t)$ that represents the rate at which money is being received by the company at time t.

Jim West/PhotoEdit

EXAMPLE 4 Continuous Income

An ice cream store's business peaks in late summer; the store's summer revenue is approximated by

$$R(t) = 300 + 4.5t - 0.05t^2 \text{ dollars per day} \quad (0 \le t \le 92),$$

where t is measured in days after June 1. What is its total revenue for the months of June, July, and August?

Solution Let's approximate the total revenue by breaking up the interval [0, 92] representing the three months into n subintervals $[t_{k-1}, t_k]$, each with length Δt. In the interval $[t_{k-1}, t_k]$ the store receives money at a rate of approximately $R(t_{k-1})$ dollars per day for Δt days, so it will receive a total of $R(t_{k-1})\Delta t$ dollars. Over the whole summer, then, the store will receive approximately

$$R(t_0)\Delta t + R(t_1)\Delta t + \cdots + R(t_{n-1})\Delta t \text{ dollars.}$$

As we let n become large to better approximate the total revenue, this Riemann sum approaches the integral

$$\text{Total revenue} = \int_0^{92} R(t)\,dt.$$

Substituting the function we were given, we get

$$\begin{aligned}\text{Total revenue} &= \int_0^{92} (300 + 4.5t - 0.05t^2)\,dt \\ &= \left[300t + 2.25t^2 - \frac{0.05}{3}t^3\right]_0^{92} \\ &\approx \$33{,}666.\end{aligned}$$

Before we go on... We could approach the calculation in Example 4 another way: $R(t) = S'(t)$, where $S(t)$ is the total revenue earned up to day t. By the Fundamental Theorem of Calculus,

$$\text{Total revenue} = S(92) - S(0) = \int_0^{92} R(t)\,dt.$$

We did the calculation using Riemann sums mainly as practice for the next example. ■

Generalizing Example 4, we can say the following:

Total Value of a Continuous Income Stream

If the rate of receipt of income is $R(t)$ dollars per unit of time, then the total income received from time $t = a$ to $t = b$ is

$$\text{Total value} = TV = \int_a^b R(t)\,dt.$$

EXAMPLE 5 Future Value

Suppose the ice cream store in the preceding example deposits its receipts in an account paying 5% interest per year compounded continuously. How much money will it have in its account at the end of August?

Solution Now we have to take into account not only the revenue but also the interest it earns in the account. Again, we break the interval [0, 92] into n subintervals. During the interval $[t_{k-1}, t_k]$, approximately $R(t_{k-1})\Delta t$ dollars are deposited in the account. That money will earn interest until the end of August, a period of $92 - t_{k-1}$ days, or $(92 - t_{k-1})/365$ years. The formula for continuous compounding tells us that by the end of August, those $R(t_{k-1})\Delta t$ dollars will have turned into

$$R(t_{k-1})\Delta t e^{0.05(92-t_{k-1})/365} = R(t_{k-1})e^{0.05(92-t_{k-1})/365}\Delta t \text{ dollars.}$$

(Recall that 5% is the *annual* interest rate.) Adding up the contributions from each subinterval, we see that the total in the account at the end of August will be approximately

$$R(t_0)e^{0.05(92-t_0)/365}\Delta t + R(t_1)e^{0.05(92-t_1)/365}\Delta t + \cdots + R(t_{n-1})e^{0.05(92-t_{n-1})/365}\Delta t.$$

This is a Riemann sum; as n gets large the sum approaches the integral

$$\text{Future value} = FV = \int_0^{92} R(t)e^{0.05(92-t)/365}\,dt.$$

Substituting $R(t) = 300 + 4.5t - 0.05t^2$, we obtain

$$FV = \int_0^{92} (300 + 4.5t - 0.05t^2)e^{0.05(92-t)/365}\,dt$$

$$\approx \$33{,}880. \qquad \text{Using technology or integration by parts}$$

➡ **Before we go on...** The interest earned in the account in Example 5 was fairly small. (Compare this answer to that in Example 4.) Not only was the money in the account for only three months, but much of it was put in the account toward the end of that period, so had very little time to earn interest. ■

Generalizing again, we have the following:

Future Value of a Continuous Income Stream

If the rate of receipt of income from time $t = a$ to $t = b$ is $R(t)$ dollars per unit of time and the income is deposited as it is received in an account paying interest at rate r per unit of time, compounded continuously, then the amount of money in the account at time $t = b$ is

$$\text{Future value} = FV = \int_a^b R(t)e^{r(b-t)}\,dt.$$

EXAMPLE 6 Present Value

You are thinking of buying the ice cream store discussed in the preceding two examples. What is its income stream worth to you on June 1? Assume that you have access to the same account paying 5% per year compounded continuously.

Solution The value of the income stream on June 1 is the amount of money that, if deposited June 1, would give you the same future value as the income stream will. If we let PV denote this "present value," its value after 92 days will be

$$PVe^{0.05\times 92/365}.$$

We equate this with the future value of the income stream to get

$$PVe^{0.05\times 92/365} = \int_0^{92} R(t)e^{0.05(92-t)/365}\,dt$$

so

$$PV = \int_0^{92} R(t)e^{-0.05t/365}\,dt.$$

Substituting the formula for $R(t)$ and integrating using technology or integration by parts, we get

$$PV \approx \$33{,}455.$$

The general formula is the following:

Present Value of a Continuous Income Stream

If the rate of receipt of income from time $t = a$ to $t = b$ is $R(t)$ dollars per unit of time and the income is deposited as it is received in an account paying interest at rate r per unit of time, compounded continuously, then the value of the income stream at time $t = a$ is

$$\text{Present value} = PV = \int_a^b R(t)e^{r(a-t)}\,dt.$$

We can derive this formula from the relation

$$FV = PVe^{r(b-a)}$$

because the present value is the amount that would have to be deposited at time $t = a$ to give a future value of FV at time $t = b$.

Note These formulas are more general than we've said. They still work when $R(t) < 0$ if we interpret negative values as money flowing *out* rather than in. That is, we can use these formulas for income we receive or for payments that we make, or for situations where we sometimes receive money and sometimes pay it out. These formulas can also be used for flows of quantities other than money. For example, if we use an exponential model for population growth and we let $R(t)$ represent the rate of immigration $[R(t) > 0]$ or emigration $[R(t) < 0]$, then the future value formula gives the future population. ■

7.4 EXERCISES

Access end-of-section exercises online at **www.webassign.net**

7.5 Improper Integrals and Applications

All the definite integrals we have seen so far have had the form $\int_a^b f(x)\,dx$, with a and b finite and $f(x)$ piecewise continuous on the closed interval $[a, b]$. If we relax one or both of these requirements somewhat, we obtain what are called **improper integrals**. There are various types of improper integrals.

Integrals in Which a Limit of Integration Is Infinite

Integrals in which one or more limits of integration are infinite can be written as

$$\int_a^{+\infty} f(x)\,dx, \quad \int_{-\infty}^{b} f(x)\,dx, \quad \text{or} \quad \int_{-\infty}^{+\infty} f(x)\,dx.$$

Let's concentrate for a moment on the first form, $\int_a^{+\infty} f(x)\,dx$. What does the $+\infty$ mean here? As it often does, it means that we are to take a limit as something gets large. Specifically, it means the limit as the upper bound of integration gets large.

Improper Integral with an Infinite Limit of Integration

We define

$$\int_a^{+\infty} f(x)\,dx = \lim_{M\to+\infty} \int_a^{M} f(x)\,dx,$$

provided the limit exists. If the limit exists, we say that $\int_a^{+\infty} f(x)\,dx$ **converges**. Otherwise, we say that $\int_a^{+\infty} f(x)\,dx$ **diverges**. Similarly, we define

$$\int_{-\infty}^{b} f(x)\,dx = \lim_{M\to-\infty}\int_M^b f(x)\,dx,$$

provided the limit exists. Finally, we define

$$\int_{-\infty}^{+\infty} f(x)\,dx = \int_{-\infty}^{a} f(x)\,dx + \int_a^{+\infty} f(x)\,dx$$

for some convenient a, provided *both* integrals on the right converge.

Quick Examples

1. $$\int_1^{+\infty}\frac{dx}{x^2} = \lim_{M\to+\infty}\int_1^M\frac{dx}{x^2} = \lim_{M\to+\infty}\left[-\frac{1}{x}\right]_1^M = \lim_{M\to+\infty}\left(-\frac{1}{M}+1\right) = 1$$
   Converges

2. $$\int_1^{+\infty}\frac{dx}{x} = \lim_{M\to+\infty}\int_1^M\frac{dx}{x} = \lim_{M\to+\infty}[\ln|x|]_1^M = \lim_{M\to+\infty}(\ln M - \ln 1) = +\infty$$
   Diverges

3. $$\int_{-\infty}^{-1}\frac{dx}{x^2} = \lim_{M\to-\infty}\int_M^{-1}\frac{dx}{x^2} = \lim_{M\to-\infty}\left[-\frac{1}{x}\right]_M^{-1} = \lim_{M\to-\infty}\left(1+\frac{1}{M}\right) = 1$$
   Converges

4. $$\begin{aligned}\int_{-\infty}^{+\infty} e^{-x}\,dx &= \int_{-\infty}^{0} e^{-x}\,dx + \int_0^{+\infty} e^{-x}\,dx\\ &= \lim_{M\to-\infty}\int_M^0 e^{-x}\,dx + \lim_{M\to+\infty}\int_0^M e^{-x}\,dx\\ &= \lim_{M\to-\infty} -[e^{-x}]_M^0 + \lim_{M\to+\infty} -[e^{-x}]_0^M\\ &= \lim_{M\to-\infty}(e^{-M}-1) + \lim_{M\to+\infty}(1-e^{-M})\\ &= +\infty + 1\end{aligned}$$
   Diverges

5. $$\begin{aligned}\int_{-\infty}^{+\infty} xe^{-x^2}\,dx &= \int_{-\infty}^{0} xe^{-x^2}\,dx + \int_0^{+\infty} xe^{-x^2}\,dx\\ &= \lim_{M\to-\infty}\int_M^0 xe^{-x^2}\,dx + \lim_{M\to+\infty}\int_0^M xe^{-x^2}\,dx\\ &= \lim_{M\to-\infty}\left[-\frac{1}{2}e^{-x^2}\right]_M^0 + \lim_{M\to+\infty}\left[-\frac{1}{2}e^{-x^2}\right]_0^M\\ &= \lim_{M\to-\infty}\left(-\frac{1}{2}+\frac{1}{2}e^{-M^2}\right) + \lim_{M\to+\infty}\left(-\frac{1}{2}e^{-M^2}+\frac{1}{2}\right)\\ &= -\frac{1}{2}+\frac{1}{2} = 0\end{aligned}$$
   Converges

Q: *We learned that the integral can be interpreted as the area under the curve. Is this still true for improper integrals?*

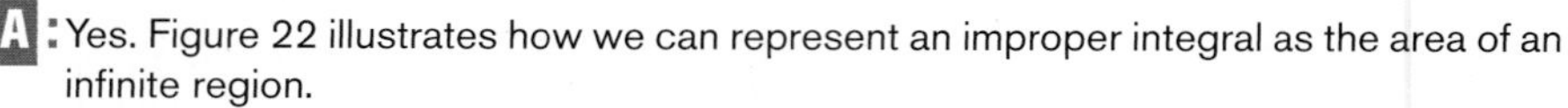

A: Yes. Figure 22 illustrates how we can represent an improper integral as the area of an infinite region.

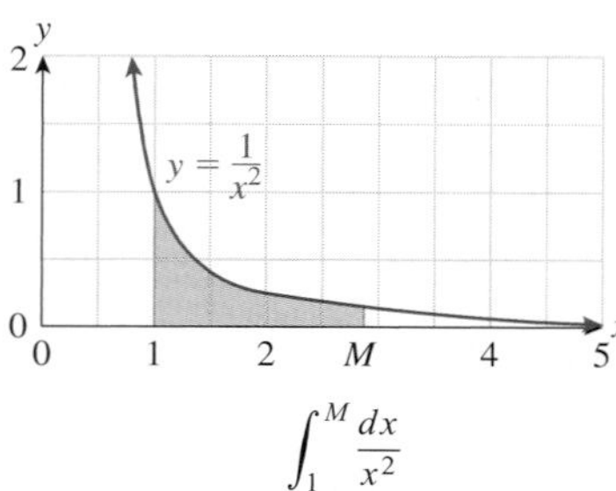

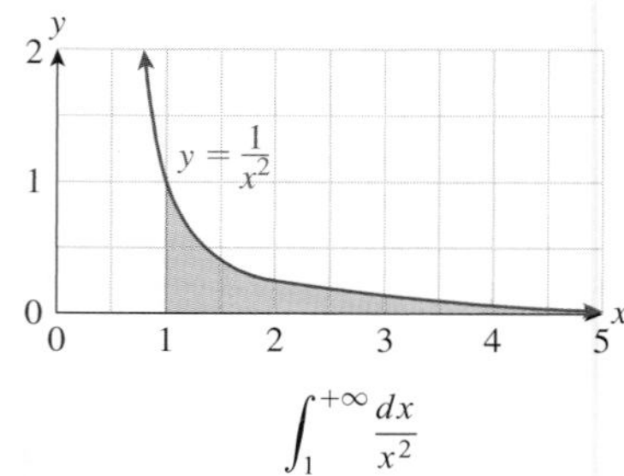

Figure 22

On the left we see the area represented by $\int_1^M dx/x^2$. As M gets larger, the integral approaches $\int_1^{+\infty} dx/x^2$. In the picture, think of M being moved farther and farther along the x-axis in the direction of increasing x, resulting in the region shown on the right.

Q: *Wait! We calculated $\int_1^{+\infty} dx/x^2 = 1$. Does this mean that the infinitely long area in Figure 22 has an area of only 1 square unit?*

A: That is exactly what it means. If you had enough paint to cover 1 square unit, you would never run out of paint while painting the region in Figure 22. This is one of the places where mathematics seems to contradict common sense. But common sense is notoriously unreliable when dealing with infinities.

Andy Crawford/Dorling Kindersley/ Getty Images

EXAMPLE 1 Future Sales of CDs

By 2009, music downloads were making serious inroads into the sales of physical CDs. Approximately 290 million CD albums were sold in 2009 and sales declined by about 23% per year the following year.[1] Suppose that this rate of decrease were to continue indefinitely and continuously. How many CD albums, total, would be sold from 2009 on?

Solution Recall that the total sales between two dates can be computed as the definite integral of the rate of sales. So, if we wanted the sales between 2009 and a time far in the future, we would compute $\int_0^M s(t)\,dt$ with a large M, where $s(t)$ is the annual sales t years after 2009. Because we want to know the *total* number of CD albums sold from 2009 on, we let $M \to +\infty$; that is, we compute $\int_0^{+\infty} s(t)\,dt$.

Because sales of CD albums are decreasing by 23% per year, we can model $s(t)$ by

$$s(t) = 290(0.77)^t \text{ million CD albums per year,}$$

where t is the number of years since 2009.

[1]Source: *2010 Year-End Shipment Statistics*, Recording Industry Association of America (www.riaa.com).

using Technology

You can estimate the integral in Example 1 with technology by computing $\int_0^M 290(0.77)^t\,dt$ for $M = 10, 100, 1000, \ldots$. You will find that the resulting values appear to converge to about 1,110. (Stop when the effect of further increases of M has no effect at this level of accuracy.)

TI-83/84 Plus

```
Y1=290*0.77^X
```
Home screen:
```
fnInt(Y1,X,0,10)
fnInt(Y1,X,0,100)
fnInt(Y1,X,0,1000)
```

Website

www.WanerMath.org:

On-Line Utilities

→ Numerical Integration Utility and Grapher

Enter
```
290*0.77^X
```
for $f(x)$. Enter 0 and 10 for the lower and upper limits and press "Integral" for the most accurate estimate of the integral.

Repeat with the upper limit set to 100, 1000, and higher.

$$
\begin{aligned}
\text{Total sales from 2009 on} &= \int_0^{+\infty} 290(0.77)^t\,dt \\
&= \lim_{M\to+\infty} \int_0^M 290(0.77)^t\,dt \\
&= \frac{290}{\ln 0.77} \lim_{M\to+\infty} [(0.77)^t]_0^M \\
&= \frac{290}{\ln 0.77} \lim_{M\to+\infty} (0.77^M - 0.77^0) \\
&= \frac{290}{\ln 0.77}(-1) \\
&\approx 1{,}110 \text{ million CD albums}
\end{aligned}
$$

Integrals in Which the Integrand Becomes Infinite

We can sometimes compute integrals $\int_a^b f(x)\,dx$ in which $f(x)$ becomes infinite. As we'll see in Example 4, the Fundamental Theorem of Calculus does not work for such integrals. The first case to consider is when $f(x)$ approaches $\pm\infty$ at either a or b.

EXAMPLE 2 Integrand Infinite at One Endpoint

Calculate $\displaystyle\int_0^1 \frac{1}{\sqrt{x}}\,dx$.

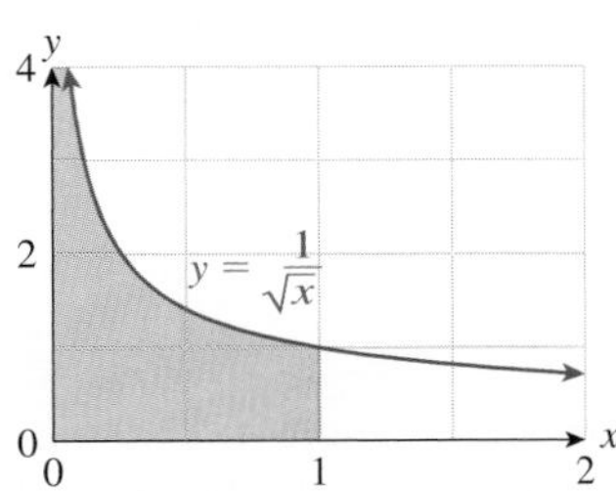

Figure 23

Solution Notice that the integrand approaches $+\infty$ as x approaches 0 from the right and is not defined at 0. This makes the integral an improper integral. Figure 23 shows the region whose area we are trying to calculate; it extends infinitely vertically rather than horizontally.

Now, if $0 < r < 1$, the integral $\int_r^1 (1/\sqrt{x})\,dx$ is a proper integral because we avoid the bad behavior at 0. This integral gives the area shown in Figure 24. If we let r approach 0 from the right, the area in Figure 24 will approach the area in Figure 23. So, we calculate

$$
\begin{aligned}
\int_0^1 \frac{1}{\sqrt{x}}\,dx &= \lim_{r\to 0^+} \int_r^1 \frac{1}{\sqrt{x}}\,dx \\
&= \lim_{r\to 0^+} [2\sqrt{x}]_r^1 \\
&= \lim_{r\to 0^+} (2 - 2\sqrt{r}) \\
&= 2.
\end{aligned}
$$

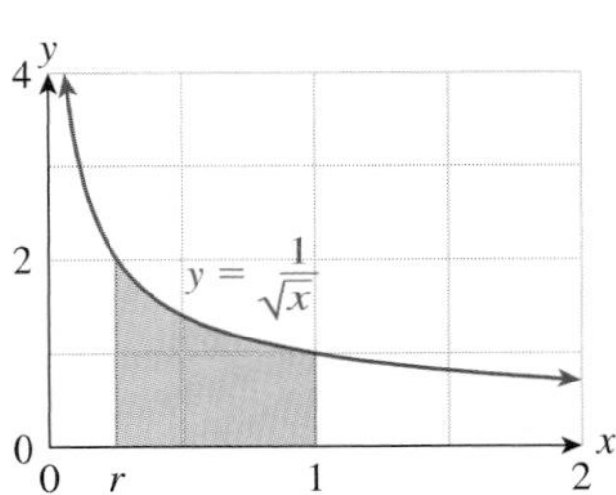

Figure 24

Thus, we again have an infinitely long region with finite area.

Generalizing, we make the following definition.

Improper Integral in Which the Integrand Becomes Infinite

If $f(x)$ is defined for all x with $a < x \leq b$ but approaches $\pm\infty$ as x approaches a, we define

$$\int_a^b f(x)\,dx = \lim_{r \to a^+} \int_r^b f(x)\,dx$$

provided the limit exists. Similarly, if $f(x)$ is defined for all x with $a \leq x < b$ but approaches $\pm\infty$ as x approaches b, we define

$$\int_a^b f(x)\,dx = \lim_{r \to b^-} \int_a^r f(x)\,dx$$

provided the limit exists. In either case, if the limit exists, we say that $\int_a^b f(x)\,dx$ **converges**. Otherwise, we say that $\int_a^b f(x)\,dx$ **diverges**.

Note We saw in Chapter 6 that the Fundamental Theorem of Calculus applies to piecewise continuous functions as well as continuous ones. Examples are $f(x) = |x|/x$ and $(x^2 - 1)/(x - 1)$. The integrals of such functions are not improper, and we can use the Fundamental Theorem of Calculus to evaluate such integrals in the usual way. ■

EXAMPLE 3 Testing for Convergence

Does $\displaystyle\int_{-1}^{3} \frac{x}{x^2 - 9}\,dx$ converge? If so, to what?

Solution We first check to see where, if anywhere, the integrand approaches $\pm\infty$. That will happen where the denominator becomes 0, so we solve $x^2 - 9 = 0$.

$$x^2 - 9 = 0$$
$$x^2 = 9$$
$$x = \pm 3$$

The solution $x = -3$ is outside of the range of integration, so we ignore it. The solution $x = 3$ is, however, the right endpoint of the range of integration, so the integral is improper. We need to investigate the following limit:

$$\int_{-1}^{3} \frac{x}{x^2 - 9}\,dx = \lim_{r \to 3^-} \int_{-1}^{r} \frac{x}{x^2 - 9}\,dx.$$

Now, to calculate the integral we use a substitution:

$$u = x^2 - 9$$
$$\frac{du}{dx} = 2x$$
$$dx = \frac{1}{2x}\,du$$

when $x = r$, $u = r^2 - 9$

when $x = -1$, $u = (-1)^2 - 9 = -8$

Thus,

$$\begin{aligned}\int_{-1}^{r} \frac{x}{x^2-9}\,dx &= \int_{-8}^{r^2-9} \frac{1}{2u}\,du \\ &= \frac{1}{2}[\ln|u|]_{-8}^{r^2-9} \\ &= \frac{1}{2}(\ln|r^2-9| - \ln 8).\end{aligned}$$

Now we take the limit:

$$\begin{aligned}\int_{-1}^{3} \frac{x}{x^2-9}\,dx &= \lim_{r\to 3^-} \int_{-1}^{r} \frac{x}{x^2-9}\,dx \\ &= \lim_{r\to 3^-} \frac{1}{2}\,(\ln|r^2-9| - \ln 8) \\ &= -\infty\end{aligned}$$

because, as $r \to 3$, $r^2-9 \to 0$, and so $\ln|r^2-9| \to -\infty$. Thus, this integral diverges.

EXAMPLE 4 Integrand Infinite between the Endpoints

Does $\displaystyle\int_{-2}^{3} \frac{1}{x^2}\,dx$ converge? If so, to what?

Solution Again we check to see if there are any points at which the integrand approaches $\pm\infty$. There is such a point, at $x = 0$. This is between the endpoints of the range of integration. To deal with this we break the integral into two integrals:

$$\int_{-2}^{3} \frac{1}{x^2}\,dx = \int_{-2}^{0} \frac{1}{x^2}\,dx + \int_{0}^{3} \frac{1}{x^2}\,dx.$$

Each integral on the right is an improper integral with the integrand approaching $\pm\infty$ at an endpoint. If both of the integrals on the right converge, we take the sum as the value of the integral on the left. So now we compute

$$\begin{aligned}\int_{-2}^{0} \frac{1}{x^2}\,dx &= \lim_{r\to 0^-} \int_{-2}^{r} \frac{1}{x^2}\,dx \\ &= \lim_{r\to 0^-} \left[-\frac{1}{x}\right]_{-2}^{r} \\ &= \lim_{r\to 0^-} \left(-\frac{1}{r} - \frac{1}{2}\right),\end{aligned}$$

which diverges to $+\infty$. There is no need now to check $\int_0^3 (1/x^2)\,dx$; because one of the two pieces of the integral diverges, we simply say that $\int_{-2}^{3}(1/x^2)\,dx$ diverges.

➡ **Before we go on...** What if we had been sloppy in Example 4 and had not checked first whether the integrand approached $\pm\infty$ somewhere? Then we probably would have applied the Fundamental Theorem of Calculus and done the following:

$$\int_{-2}^{3} \frac{1}{x^2}\,dx = \left(-\frac{1}{x}\right)_{-2}^{3} = \left(-\frac{1}{3} - \frac{1}{2}\right) = -\frac{5}{6}. \qquad ✗ \textit{ WRONG!}$$

Notice that the answer this "calculation" gives is patently ridiculous. Because $1/x^2 > 0$ for all x for which it is defined, any definite integral of $1/x^2$ must give a positive answer. *Moral:* Always check to see whether the integrand blows up anywhere in the range of integration. If it does, the FTC does not apply, and we must use the methods of this example. ■

We end with an example of what to do if an integral is improper for more than one reason.

EXAMPLE 5 An Integral Improper in Two Ways

Does $\displaystyle\int_0^{+\infty} \frac{1}{\sqrt{x}}\,dx$ converge? If so, to what?

Solution This integral is improper for two reasons. First, the range of integration is infinite. Second, the integrand blows up at the endpoint 0. In order to separate these two problems, we break up the integral at some convenient point:

$$\int_0^{+\infty} \frac{1}{\sqrt{x}}\,dx = \int_0^{1} \frac{1}{\sqrt{x}}\,dx + \int_1^{+\infty} \frac{1}{\sqrt{x}}\,dx$$

We chose to break the integral at 1. Any positive number would have sufficed, but 1 is generally easier to use in calculations.

The first piece, $\int_0^1 (1/\sqrt{x})\,dx$, we discussed in Example 2; it converges to 2. For the second piece we have

$$\begin{aligned}\int_1^{+\infty} \frac{1}{\sqrt{x}}\,dx &= \lim_{M\to+\infty} \int_1^{M} \frac{1}{\sqrt{x}}\,dx \\ &= \lim_{M\to+\infty} [2\sqrt{x}]_1^M \\ &= \lim_{M\to+\infty} (2\sqrt{M} - 2),\end{aligned}$$

which diverges to $+\infty$. Because the second piece of the integral diverges, we conclude that $\int_0^{+\infty} (1/\sqrt{x})\,dx$ diverges.

7.5 EXERCISES

Access end-of-section exercises online at **www.webassign.net** ENHANCED WebAssign

7.6 Differential Equations and Applications

A **differential equation** is an equation that involves a derivative of an unknown function. A **first-order differential equation** involves only the first derivative of the unknown function. A **second-order differential equation** involves the second derivative of the unknown function (and possibly the first derivative). Higher order differential equations are defined similarly. In this book, we will deal only with first-order differential equations.

To **solve** a differential equation means to find the unknown function. Many of the laws of science and other fields describe how things change. When expressed mathematically, these laws take the form of equations involving derivatives—that is, differential equations. The field of differential equations is a large and very active area of study in mathematics, and we shall see only a small part of it in this section.

EXAMPLE 1 Motion

A dragster accelerates from a stop so that its speed t seconds after starting is $40t$ ft/sec. How far will the car go in 8 seconds?

Solution We wish to find the car's position function $s(t)$. We are told about its speed, which is ds/dt. Precisely, we are told that

$$\frac{ds}{dt} = 40t.$$

This is the differential equation we have to solve to find $s(t)$. But we already know how to solve this kind of differential equation; we integrate:

$$s(t) = \int 40t\, dt = 20t^2 + C.$$

We now have the **general solution** to the differential equation. By letting C take on different values, we get all the possible solutions. We can specify the one **particular solution** that gives the answer to our problem by imposing the **initial condition** that $s(0) = 0$. Substituting into $s(t) = 20t^2 + C$, we get

$$0 = s(0) = 20(0)^2 + C = C$$

so $C = 0$ and $s(t) = 20t^2$. To answer the question, the car travels $20(8)^2 = 1{,}280$ feet in 8 seconds.

We did not have to work hard to solve the differential equation in Example 1. In fact, any differential equation of the form $dy/dx = f(x)$ can (in theory) be solved by integrating. (Whether we can actually carry out the integration is another matter!)

Simple Differential Equations

A **simple** differential equation has the form

$$\frac{dy}{dx} = f(x).$$

Its general solution is

$$y = \int f(x)\, dx.$$

Quick Example

The differential equation

$$\frac{dy}{dx} = 2x^2 - 4x^3$$

is simple and has general solution

$$y = \int f(x)\, dx = \frac{2x^3}{3} - x^4 + C.$$

Not all differential equations are simple, as the next example shows.

EXAMPLE 2 Separable Differential Equation

Consider the differential equation $\dfrac{dy}{dx} = \dfrac{x}{y^2}$.

a. Find the general solution.

b. Find the particular solution that satisfies the initial condition $y(0) = 2$.

Solution

a. This is not a simple differential equation because the right-hand side is a function of both x and y. We cannot solve this equation by just integrating; the solution to this problem is to "separate" the variables.

Step 1: *Separate the variables algebraically.* We rewrite the equation as

$$y^2\, dy = x\, dx.$$

Step 2: *Integrate both sides.*

$$\int y^2\, dy = \int x\, dx$$

giving

$$\frac{y^3}{3} = \frac{x^2}{2} + C$$

Step 3: *Solve for the dependent variable.* We solve for y:

$$y^3 = \frac{3}{2}x^2 + 3C = \frac{3}{2}x^2 + D$$

(Rewriting $3C$ as D, an equally arbitrary constant), so

$$y = \left(\frac{3}{2}x^2 + D\right)^{1/3}.$$

This is the general solution of the differential equation.

b. We now need to find the value for D that will give us the solution satisfying the condition $y(0) = 2$. Substituting 0 for x and 2 for y in the general solution, we get

$$2 = \left(\frac{3}{2}(0)^2 + D\right)^{1/3} = D^{1/3}$$

so

$$D = 2^3 = 8.$$

Thus, the particular solution we are looking for is

$$y = \left(\frac{3}{2}x^2 + 8\right)^{1/3}.$$

➡ **Before we go on...** We can check the general solution in Example 2 by calculating both sides of the differential equation and comparing.

$$\frac{dy}{dx} = \frac{d}{dx}\left(\frac{3}{2}x^2 + D\right)^{1/3} = x\left(\frac{3}{2}x^2 + D\right)^{-2/3}$$

$$\frac{x}{y^2} = \frac{x}{\left(\frac{3}{2}x^2 + 8\right)^{2/3}} = x\left(\frac{3}{2}x^2 + D\right)^{-2/3} \quad ✓$$

■

Q: *In Example 2, we wrote $y^2\,dy$ and $x\,dx$. What do they mean?*

A: Although it is possible to give meaning to these symbols, for us they are just a notational convenience. We could have done the following instead:

$$y^2\frac{dy}{dx} = x.$$

Now we integrate both sides with respect to x.

$$\int y^2\frac{dy}{dx}\,dx = \int x\,dx$$

We can use substitution to rewrite the left-hand side:

$$\int y^2\frac{dy}{dx}\,dx = \int y^2\,dy,$$

which brings us back to the equation

$$\int y^2\,dy = \int x\,dx.$$

We were able to separate the variables in the preceding example because the right-hand side, x/y^2, was a *product* of a function of x and a function of y—namely,

$$\frac{x}{y^2} = x\left(\frac{1}{y^2}\right).$$

In general, we can say the following:

Separable Differential Equation

A **separable** differential equation has the form

$$\frac{dy}{dx} = f(x)g(y).$$

We solve a separable differential equation by separating the xs and the ys algebraically, writing

$$\frac{1}{g(y)}\,dy = f(x)\,dx$$

and then integrating:

$$\int \frac{1}{g(y)}\,dy = \int f(x)\,dx.$$

EXAMPLE 3 Rising Medical Costs

Spending on Medicare from 2010 to 2021 was projected to rise continuously at an instantaneous rate of 5.6% per year.[2] Find a formula for Medicare spending y as a function of time t in years since 2010.

Solution When we say that Medicare spending y was going up continuously at an instantaneous rate of 5.6% per year, we mean that

the instantaneous rate of increase of y was 5.6% of its value

or
$$\frac{dy}{dt} = 0.056y.$$

This is a separable differential equation. Separating the variables gives

$$\frac{1}{y}dy = 0.056\,dt.$$

Integrating both sides, we get

$$\int \frac{1}{y}\,dy = \int 0.056\,dt$$

so
$$\ln y = 0.056t + C.$$

(We should write $\ln|y|$, but we know that the medical costs are positive.) We now solve for y.

$$y = e^{0.056t+C} = e^C e^{0.056t} = Ae^{0.056t},$$

where A is a positive constant. This is the formula we used before for continuous percentage growth.

---

[2] Spending is in constant 2010 dollars. Source for projected data: Congressional Budget Office, *March 2011 Medicare Baseline* (www.cbo.gov).

Before we go on... To determine A in Example 3 we need to know, for example, Medicare spending at time $t = 0$ (the initial condition). The source cited gives Medicare spending as \$525.6 billion in 2010. Substituting $t = 0$ in the equation on the previous page gives

$$525.6 = Ae^0 = A.$$

Thus, projected Medicare spending is

$$y = 525.6e^{0.056t} \text{ billion dollars}$$

t years after 2010. ■

EXAMPLE 4 Newton's Law of Cooling

Newton's Law of Cooling states that a hot object cools at a rate proportional to the difference between its temperature and the temperature of the surrounding environment (the **ambient temperature**). If a hot cup of coffee, at 170°F, is left to sit in a room at 70°F, how will the temperature of the coffee change over time?

Solution We let $H(t)$ denote the temperature of the coffee at time t. Newton's Law of Cooling tells us that $H(t)$ *decreases* at a rate proportional to the difference between $H(t)$ and 70°F, the ambient temperature. In other words,

$$\frac{dH}{dt} = -k(H - 70),$$

where k is some positive constant.* Note that $H \geq 70$: The coffee will never cool to less than the ambient temperature.

* When we say that a quantity Q is *proportional* to a quantity R, we mean that $Q = kR$ for some constant k. The constant k is referred to as the **constant of proportionality**.

The variables here are H and t, which we can separate as follows:

$$\frac{dH}{H - 70} = -k\,dt.$$

Integrating, we get

$$\int \frac{dH}{H - 70} = \int (-k)\,dt$$

so $\quad \ln(H - 70) = -kt + C.$

(Note that $H - 70$ is positive, so we don't need absolute values.) We now solve for H:

$$\begin{aligned} H - 70 &= e^{-kt+C} \\ &= e^C e^{-kt} \\ &= Ae^{-kt} \end{aligned}$$

so

$$H(t) = 70 + Ae^{-kt},$$

where A is some positive constant. We can determine the constant A using the initial condition $H(0) = 170$:

$$170 = 70 + Ae^0 = 70 + A$$

so $\quad A = 100.$

Therefore,

$$H(t) = 70 + 100e^{-kt}.$$

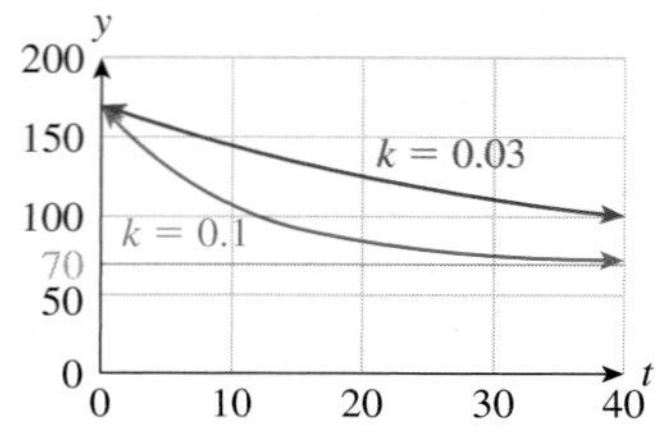

Figure 25

Q: *But what is k?*

A: The constant k determines the rate of cooling. Its value depends on the units of time we are using, on the substance cooling—in this case the coffee—and on its container. Because k depends so heavily on the particular circumstances, it's usually easiest to determine it experimentally. Figure 25 shows two possible graphs, one with $k = 0.1$ and the other with $k = 0.03$ ($k \approx 0.03$ would be reasonable for a cup of coffee in a polystyrene container with t measured in minutes).

In any case, we can see from the graph or the formula for $H(t)$ that the temperature of the coffee will approach the ambient temperature exponentially.

➡ **Before we go on...** Notice that the calculation in Example 4 shows that the temperature of an object cooling according to Newton's Law is given in general by

$$H(t) = T_a + (T_0 - T_a)e^{-kt},$$

where T_a is the ambient temperature (70° in the example) and T_0 is the initial temperature (170° in the example). The formula also holds if the ambient temperature is higher than the initial temperature ("Newton's Law of Heating"). ■

7.6 EXERCISES

Access end-of-section exercises online at **www.webassign.net**

CHAPTER 7 REVIEW

KEY CONCEPTS

WW Website www.WanerMath.com
Go to the Website at www.WanerMath.com to find a comprehensive and interactive Web-based summary of Chapter 7.

7.1 Integration by Parts

Integration-by-parts formula:

$$\int u \cdot v\,dx = u \cdot I(v) - \int D(u)I(v)\,dx$$

p. 386

Tabular method for integration by parts *p. 387*

Integrating a polynomial times a logarithm *p. 390*

7.2 Area between Two Curves and Applications

If $f(x) \geq g(x)$ for all x in $[a, b]$, then the area of the region between the graphs of f and g and between $x = a$ and $x = b$ is given by

$$A = \int_a^b [f(x) - g(x)]\,dx \quad p.\ 392$$

Regions enclosed by crossing curves *p. 394*

Area enclosed by two curves *p. 395*

General instructions for finding the area between the graphs of $f(x)$ and $g(x)$ *p. 396*

Approximating the area between two curves using technology:

$$A = \int_a^b |f(x) - g(x)|\,dx \quad p.\ 396$$

7.3 Averages and Moving Averages

Average, or mean, of a collection of values

$$\bar{y} = \frac{y_1 + y_2 + \cdots + y_n}{n} \quad p.\ 397$$

The *average*, or *mean*, of a function $f(x)$ on an interval $[a, b]$ is

$$\bar{f} = \frac{1}{b-a}\int_a^b f(x)\,dx. \quad p.\ 398$$

Average balance *p. 399*

Computing the moving average of a set of data *p. 400*

n-Unit moving average of a function:

$$\bar{f}(x) = \frac{1}{n}\int_{x-n}^{x} f(t)\,dt \quad p.\ 401$$

Computing moving averages of sawtooth and step functions *p. 401*

7.4 Applications to Business and Economics: Consumers' and Producers' Surplus and Continuous Income Streams

Consumers' surplus:

$$CS = \int_0^{\bar{q}} [D(q) - \bar{p}]\,dq \quad p.\ 403$$

Producers' surplus:

$$PS = \int_0^{\bar{q}} [\bar{p} - S(q)]\,dq \quad p.\ 405$$

Equilibrium price *p. 406*

Social gain $= CS + PS$ *p. 407*

Total value of a continuous income stream: $TV = \int_a^b R(t)\,dt$ *p. 408*

Future value of a continuous income stream: $FV = \int_a^b R(t)e^{r(b-t)}\,dt$ *p. 409*

Present value of a continuous income stream: $PV = \int_a^b R(t)e^{r(a-t)}\,dt$ *p. 409*

7.5 Improper Integrals and Applications

Improper integral with an infinite limit of integration:

$$\int_a^{+\infty} f(x)\,dx, \int_{-\infty}^{b} f(x)\,dx, \int_{-\infty}^{+\infty} f(x)\,dx \quad p.\ 410$$

Improper integral in which the integrand becomes infinite *p. 414*

Testing for convergence *p. 414*

Integrand infinite between the endpoints *p. 415*

Integral improper in two ways *p. 416*

7.6 Differential Equations and Applications

Simple differential equations:

$$\frac{dy}{dx} = f(x) \quad p.\ 417$$

Separable differential equations:

$$\frac{dy}{dx} = f(x)g(y) \quad p.\ 420$$

Newton's Law of Cooling *p. 421*

REVIEW EXERCISES

Evaluate the integrals in Exercises 1–10.

1. $\int (x^2 + 2)e^x\,dx$

2. $\int (x^2 - x)e^{-3x+1}\,dx$

3. $\int x^2 \ln(2x)\,dx$

4. $\int \log_5 x\,dx$

5. $\int 2x|2x + 1|\,dx$

6. $\int 3x|-x + 5|\,dx$

7. $\int 5x\frac{|-x + 3|}{-x + 3}\,dx$

8. $\int 2x\frac{|3x + 1|}{3x + 1}\,dx$

9. $\int_{-2}^{2} (x^3 + 1)e^{-x}\,dx$

10. $\int_1^e x^2 \ln x\,dx$

In Exercises 11–14, find the areas of the given regions.

11. Between $y = x^3$ and $y = 1 - x^3$ for x in $[0, 1]$

12. Between $y = e^x$ and $y = e^{-x}$ for x in $[0, 2]$

13. Enclosed by $y = 1 - x^2$ and $y = x^2$

14. Between $y = x$ and $y = xe^{-x}$ for x in $[0, 2]$

In Exercises 15–18, find the average value of the given function over the indicated interval.

15. $f(x) = x^3 - 1$ over $[-2, 2]$

16. $f(x) = \frac{x}{x^2 + 1}$ over $[0, 1]$

17. $f(x) = x^2e^x$ over $[0, 1]$

18. $f(x) = (x + 1)\ln x$ over $[1, 2e]$

In Exercises 19–22, find the 2-unit moving averages of the given function.

19. $f(x) = 3x + 1$

20. $f(x) = 6x^2 + 12$

21. $f(x) = x^{4/3}$

22. $f(x) = \ln x$

In Exercises 23 and 24, calculate the consumers' surplus at the indicated unit price $\bar{p}$ for the given demand equation.

23. $p = 50 - \frac{1}{2}q;\ \bar{p} = 10$

24. $p = 10 - q^{1/2};\ \bar{p} = 4$

In Exercises 25 and 26, calculate the producers' surplus at the indicated unit price $\bar{p}$ for the given supply equation.

25. $p = 50 + \frac{1}{2}q;\ \bar{p} = 100$

26. $p = 10 + q^{1/2};\ \bar{p} = 40$

In Exercises 27–32, decide whether the given integral converges. If the integral converges, compute its value.

27. $\int_1^{\infty} \frac{1}{x^5}\,dx$

28. $\int_0^1 \frac{1}{x^5}\,dx$

29. $\int_{-1}^{1} \frac{x}{(x^2-1)^{5/3}}\,dx$

30. $\int_0^2 \frac{x}{(x^2-1)^{1/3}}\,dx$

31. $\int_0^{+\infty} 2xe^{-x^2}\,dx$

32. $\int_0^{+\infty} x^2e^{-6x^3}\,dx$

Solve the differential equations in Exercises 33–36.

33. $\frac{dy}{dx} = x^2y^2$

34. $\frac{dy}{dx} = xy + 2x$

35. $xy\frac{dy}{dx} = 1;\ y(1) = 1$

36. $y(x^2+1)\frac{dy}{dx} = xy^2;\ y(0) = 2$

APPLICATIONS: OHaganBooks.com

37. ***Spending on Stationery*** Alarmed by the volume of pointless memos and reports being copied and circulated by management at OHaganBooks.com, John O'Hagan ordered a 5-month audit of paper usage at the company. He found that management consumed paper at a rate of

$$q(t) = 45t + 200 \text{ thousand sheets per month} \quad (0 \le t \le 5)$$

(t is the time in months since the audit began). During the same period, the price of paper was escalating; the company was charged approximately

$$p(t) = 9e^{0.09t} \text{ dollars per thousand sheets.}$$

Use an integral to estimate, to the nearest hundred dollars, the total spent on paper for management during the given period.

38. ***Spending on Shipping*** During the past 10 months, OHaganBooks.com shipped orders at a rate of about

$$q(t) = 25t + 3{,}200 \text{ packages per month} \quad (0 \le t \le 10)$$

(t is the time in months since the beginning of the year). During the same period, the cost of shipping a package averaged approximately

$$p(t) = 4e^{0.04t} \text{ dollars per package.}$$

Use an integral to estimate, to the nearest thousand dollars, the total spent on shipping orders during the given period.

39. ***Education Costs*** Billy-Sean O'Hagan, having graduated *summa cum laude* from college, has been accepted by the doctoral program in biophysics at Oxford. John O'Hagan estimates that the total cost (minus scholarships) he will need to pay is \$2,000 per month, but that this cost will escalate at a continuous compounding rate of 1% per month.

a. What, to the nearest dollar, is the average monthly cost over the course of two years?

b. Find the four-month moving average of the monthly cost.

40. ***Investments*** OHaganBooks.com keeps its cash reserves in a hedge fund paying 6% compounded continuously. It starts a year with \$1 million in reserves and does not withdraw or deposit any money.

a. What is the average amount it will have in the fund over the course of two years?

b. Find the one-month moving average of the amount it has in the fund.

41. ***Consumers' and Producers' Surplus*** Currently, the hottest selling item at OHaganBooks.com is *Mensa for Dummies*[3] with a demand curve of $q = 20{,}000(28 - p)^{1/3}$ books per week, and a supply curve of $q = 40{,}000(p - 19)^{1/3}$ books per week.

a. Find the equilibrium price and demand.

b. Find the consumers' and producers' surpluses at the equilibrium price.

42. ***Consumers' and Producers' Surplus*** OHaganBooks.com is about to start selling a new coffee table book, *Computer Designs of the Late Twentieth Century*. It estimates the demand curve to be $q = 1{,}000\sqrt{200 - 2p}$, and its willingness to order books from the publisher is given by the supply curve $q = 1{,}000\sqrt{10p - 400}$.

a. Find the equilibrium price and demand.

b. Find the consumers' and producers' surpluses at the equilibrium price.

43. ***Revenue*** Sales of the bestseller *A River Burns through It* are dropping at OHaganBooks.com. To try to bolster sales, the company is dropping the price of the book, now \$40, at a rate of \$2 per week. As a result, this week OHaganBooks.com will sell 5,000 copies, and it estimates that sales will fall continuously at a rate of 10% per week. How much revenue will it earn on sales of this book over the next 8 weeks?

44. ***Foreign Investments*** Panicked by the performance of the U.S. stock market, Marjory Duffin is investing her 401(k) money in a Russian hedge fund at a rate of approximately

$$q(t) = 1.7t^2 - 0.5t + 8 \text{ thousand shares per month,}$$

[3] The actual title is: *Let Us Just Have A Ball! Mensa for Dummies*, by Wendu Mekbib, Silhouette Publishing Corporation.

where t is time in months since the stock market began to plummet. At the time she started making the investments, the hedge fund was selling for $1 per share, but subsequently declined in value at a continuous rate of 5% per month. What was the total amount of money Marjory Duffin invested after one year? (Answer to the nearest $1,000.)

45. ***Investments*** OHaganBooks.com CEO John O'Hagan has started a gift account for the Marjory Duffin Foundation. The account pays 6% compounded continuously and is initially empty. OHaganBooks.com deposits money continuously into it, starting at the rate of $100,000 per month and increasing continuously by $10,000 per month.

a. How much money will the company have in the account at the end of two years?

b. How much of the amount you found in part (a) was principal deposited and how much was interest earned? (Round answers to the nearest $1,000.)

46. ***Savings*** John O'Hagan had been saving money for Billy-Sean's education since he had been a wee lad. O'Hagan began depositing money at the rate of $1,000 per month and increased his deposits continuously by $50 per month. If the account earned 5% compounded continuously and O'Hagan continued these deposits for 15 years,

a. How much money did he accumulate?

b. How much was money deposited and how much was interest?

47. ***Acquisitions*** The Megabucks Corporation is considering buying OHaganBooks.com. It estimates OHaganBooks.com's revenue stream at $50 million per year, growing continuously at a 10% rate. Assuming interest rates of 6%, how much is OHaganBooks.com's revenue for the next year worth now?

48. ***More Acquisitions*** OHaganBooks.com is thinking of buying JungleBooks and would like to recoup its investment after three years. The estimated net profit for JungleBooks is $40 million per year, growing linearly by $5 million per year. Assuming interest rates of 4%, how much should OHaganBooks.com pay for JungleBooks?

49. ***Incompetence*** OHaganBooks.com is shopping around for a new bank. A junior executive at one bank offers it the following interesting deal: The bank will pay OHaganBooks.com interest continuously at a rate numerically equal to 0.01% of the square of the amount of money it has in the account at any time. By considering what would happen if $10,000 was deposited in such an account, explain why the junior executive was fired shortly after this offer was made.

50. ***Shrewd Bankers*** The new junior officer at the bank (who replaced the one fired in the preceding exercise) offers OHaganBook.com the following deal for the $800,000 they plan to deposit: While the amount in the account is less than $1 million, the bank will pay interest continuously at a rate equal to 10% of the difference between $1 million and the amount of money in the account. When it rises over $1 million, the bank will pay interest of 20%. Why should OHaganBooks.com not take this offer?

Case Study

Estimating Tax Revenues

Dmitriy Shironosov/Shutterstock.com

You have just been hired by the incoming administration of your country as chief consultant for national tax policy, and you have been getting conflicting advice from the finance experts on your staff. Several of them have come up with plausible suggestions for new tax structures, and your job is to choose the plan that results in more revenue for the government.

Before you can evaluate their plans, you realize that it is essential to know your country's income distribution—that is, how many people earn how much money.* You might think that the most useful way of specifying income distribution would be to use a function that gives the exact number $f(x)$ of people who earn a given salary x. This would necessarily be a discrete function—it makes sense only if x happens to be a whole number of cents. There is, after all, no one earning a salary of exactly $22,000.142567! Furthermore, this function would behave rather erratically, because there are, for example, probably many more people making a salary of exactly $30,000 than exactly $30,000.01. Given these problems, it is far more convenient to start with the function defined by

$$N(x) = \textit{the total number of people earning between } 0 \textit{ and } x \textit{ dollars.}$$

* To simplify our discussion, we are assuming that (1) all tax revenues are based on earned income and that (2) everyone in the population we consider earns some income.

Actually, you would want a "smoothed" version of this function. The graph of $N(x)$ might look like the one shown in Figure 26.

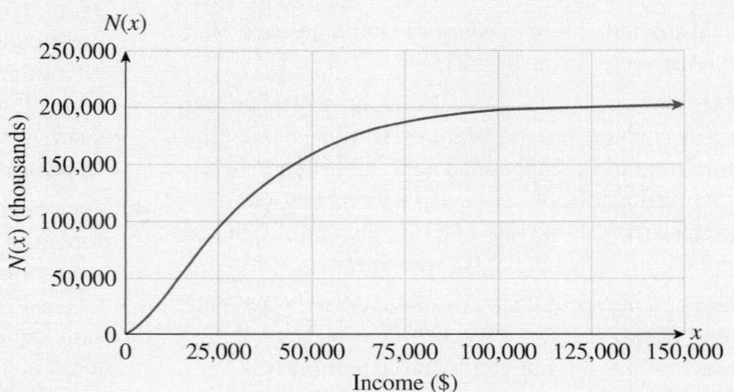

Figure 26

If we take the *derivative* of $N(x)$, we get an income distribution function. Its graph might look like the one shown in Figure 27.

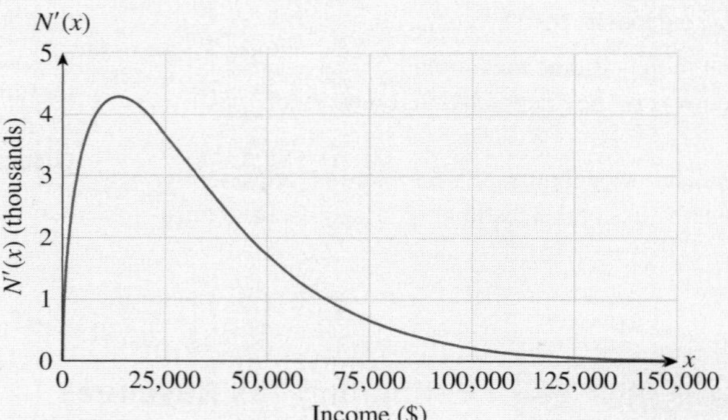

Figure 27

Because the derivative measures the rate of change, its value at x is the additional number of taxpayers per \$1 increase in salary. Thus, the fact that $N'(25{,}000) \approx 3{,}700$ tells us that approximately 3,700 people are earning a salary of between \$25,000 and \$25,001. In other words, N' shows the distribution of incomes among the population—hence, the name "distribution function."*

You thus send a memo to your experts requesting the income distribution function for the nation. After much collection of data, they tell you that the income distribution function is

$$N'(x) = 14x^{0.672}e^{-x/20{,}400}.$$

This is in fact the function whose graph is shown in Figure 27 and is an example of a **gamma distribution.**† (You might find it odd that you weren't given the original function N, but it will turn out that you don't need it. How would you compute it?)

Given this income distribution, your financial experts have come up with the two possible tax policies illustrated in Figures 28 and 29.

* A very similar idea is used in probability. See the optional chapter "Calculus Applied to Probability and Statistics" on the Website.

† Gamma distributions are often good models for income distributions. The one used in the text is the authors' approximation of the income distribution in the United States in 2009. Source for data: U.S. Census Bureau, Current Population Survey, 2010 Annual Social and Economic Supplement. (www.census.gov).

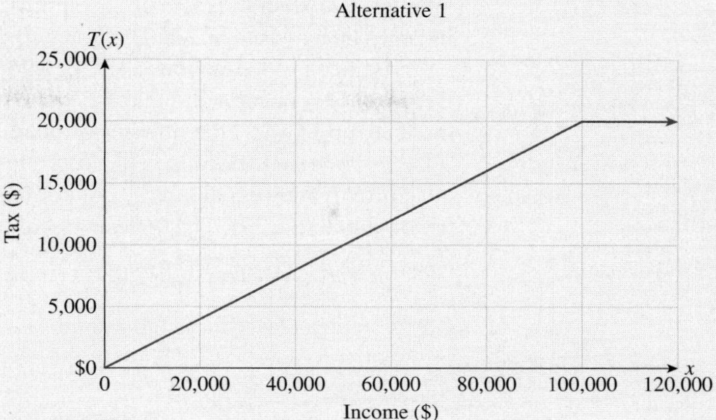

Figure 28

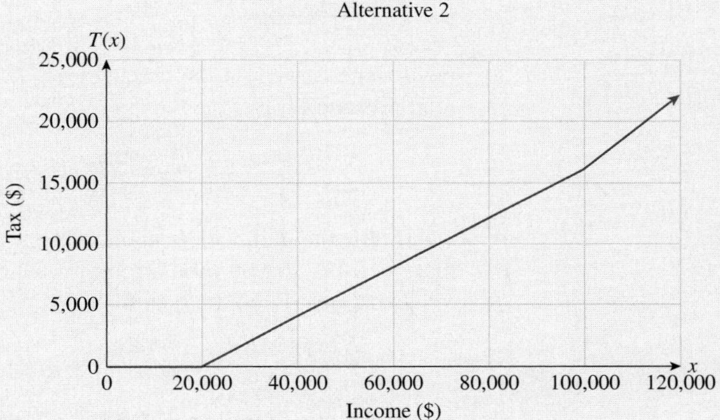

Figure 29

In the first alternative, all taxpayers pay 20% of their income in taxes, except that no one pays more than \$20,000 in taxes. In the second alternative, there are three tax brackets, described by the following table:

Income	*Marginal tax rate*
\$0–20,000	0%
\$20,000–100,000	20%
Above \$100,000	30%

Now you must determine which alternative will generate more annual tax revenue.

Each of Figures 28 and 29 is the graph of a function, T. Rather than using the formulas for these particular functions, you begin by working with the general situation. You have an income distribution function N' and a tax function T, both functions of annual income. You need to find a formula for total tax revenues. First you decide to use a cutoff so that you need to work only with incomes in some finite interval $[0, M]$; you might use, for example, $M = \$10$ million. (Later you will let M approach $+\infty$.) Next, you subdivide the interval $[0, M]$ into a large number of intervals of small width, Δx. If $[x_{k-1}, x_k]$ is a typical such interval, you wish to calculate the approximate tax revenue from people whose total incomes lie between x_{k-1} and x_k. You will then sum over k to get the total revenue.

You need to know how many people are making incomes between x_{k-1} and x_k. Because $N(x_k)$ people are making incomes *up to* x_k and $N(x_{k-1})$ people are making incomes up to x_{k-1}, the number of people making incomes between x_{k-1} and x_k is $N(x_k) - N(x_{k-1})$. Because x_k is very close to x_{k-1}, the incomes of these people are all approximately equal to x_{k-1} dollars, so each of these taxpayers is paying an annual tax of about $T(x_{k-1})$. This gives a tax revenue of

$$[N(x_k) - N(x_{k-1})]T(x_{k-1}).$$

Now you do a clever thing. You write $x_k - x_{k-1} = \Delta x$ and replace $N(x_k) - N(x_{k-1})$ by

$$\frac{N(x_k) - N(x_{k-1})}{\Delta x}\Delta x.$$

This gives you a tax revenue of about

$$\frac{N(x_k) - N(x_{k-1})}{\Delta x}T(x_{k-1})\Delta x$$

from wage-earners in the bracket $[x_{k-1}, x_k]$. Summing over k gives an approximate total revenue of

$$\sum_{k=1}^{n}\frac{N(x_k) - N(x_{k-1})}{\Delta x}T(x_{k-1})\Delta x,$$

where n is the number of subintervals. The larger n is, the more accurate your estimate will be, so you take the limit of the sum as $n \to \infty$. When you do this, two things happen. First, the quantity

$$\frac{N(x_k) - N(x_{k-1})}{\Delta x}$$

approaches the derivative, $N'(x_{k-1})$. Second, the sum, which you recognize as a Riemann sum, approaches the integral

$$\int_0^M N'(x)T(x)\,dx.$$

You now take the limit as $M \to +\infty$ to get

$$\text{Total tax revenue} = \int_0^{+\infty} N'(x)T(x)\,dx.$$

This improper integral is fine in theory, but the actual calculation will have to be done numerically, so you stick with the upper limit of \$10 million for now. You will have to check that it is reasonable at the end. (Notice that, by the graph of N', it appears that extremely few, if any, people earn that much.) Now you already have a formula for $N'(x)$, but you still need to write formulas for the tax functions $T(x)$ for both alternatives.

Alternative 1 The graph in Figure 28 rises linearly from 0 to 20,000 as x ranges from 0 to 100,000, and then stays constant at 20,000. The slope of the first part is $20{,}000/100{,}000 = 0.2$. The taxation function is therefore

$$T_1(x) = \begin{cases} 0.2x & \text{if } 0 \le x < 100{,}000 \\ 20{,}000 & \text{if } x \ge 100{,}000 \end{cases}.$$

For use of technology, it's convenient to express this in closed form using absolute values:*

$$T_1(x) = 0.2x + \frac{1}{2}\left(1 + \frac{|x - 100{,}000|}{x - 100{,}000}\right)(20{,}000 - 0.2x).$$

The total revenue generated by this tax scheme is, therefore,

$$R_1 = \int_0^{10{,}000{,}000} (14x^{0.672}e^{-x/20{,}400}) \times \left[0.2x + \frac{1}{2}\left(1 + \frac{|x - 100{,}000|}{x - 100{,}000}\right)(20{,}000 - 0.2x)\right] dx.$$

You decide not to attempt this by hand! You use numerical integration software to obtain a grand total of $R_1 = \$1{,}360{,}990{,}000{,}000$, or \$1.36099 trillion (rounded to six significant digits).

(If you use the Numerical Integration utility on the Website, enter

```
14x^(0.672)*exp(-x/20400)*(0.2x+0.5*(20000-0.2x)*
        (1+abs(x-100000)/(x-100000)))
```

for $f(x)$, and 0 and 10000000 for a and b respectively, and press "Integral".)

* To see how to obtain the formula, consult the introduction to Exercises 67–68 in Section 7.1.

Alternative 2 The graph in Figure 29 rises with a slope of 0.2 from 0 to 16,000 as x ranges from 20,000 to 100,000, then rises from that point on with a slope of 0.3. (This is why we say that the *marginal* tax rates are 20% and 30%, respectively.) The taxation function is therefore

$$T_2(x) = \begin{cases} 0 & \text{if } 0 \le x < 20{,}000 \\ 0.2(x - 20{,}000) & \text{if } 20{,}000 \le x < 100{,}000. \\ 16{,}000 + 0.3(x - 100{,}000) & \text{if } x \ge 100{,}000 \end{cases}$$

Again, you express this in closed form using absolute values:

$$\begin{aligned} T_2(x) &= [0.2(x - 20{,}000)]\frac{1}{2}\left(\frac{|x - 20{,}000|}{x - 20{,}000} - \frac{|x - 100{,}000|}{x - 100{,}000}\right) \\ &\quad + [16{,}000 + 0.3(x - 100{,}000)]\frac{1}{2}\left(1 + \frac{|x - 100{,}000|}{x - 100{,}000}\right) \\ &= 0.1(x - 20{,}000)\left(\frac{|x - 20{,}000|}{x - 20{,}000} - \frac{|x - 100{,}000|}{x - 100{,}000}\right) \\ &\quad + [8{,}000 + 0.15(x - 100{,}000)]\left(1 + \frac{|x - 100{,}000|}{x - 100{,}000}\right). \end{aligned}$$

Values of x between 0 and 20,000 do not contribute to the integral, so

$$R_2 = \int_{20{,}000}^{10{,}000{,}000} 14x^{0.672}e^{-x/20{,}400}\,T_2(x)\,dx$$

with $T_2(x)$ as above. Numerical integration software gives $R_2 = \$0.713465$ trillion—considerably less than Alternative 1. Thus, even though Alternative 2 taxes the wealthy more heavily, it yields less total revenue.

Now what about the cutoff at \$10 million annual income? If you try either integral again with an upper limit of \$100 million, you will see no change in either result to six significant digits. There simply are not enough taxpayers earning an income above \$10,000,000 to make a difference. You conclude that your answers are sufficiently accurate and that the first alternative provides more tax revenue.

EXERCISES

In Exercises 1–4, calculate the total tax revenue for a country with the given income distribution and tax policies (all currency in dollars).

1. T $N'(x) = 100x^{0.466}e^{-x/23,000}$; 25% tax on all income

2. T $N'(x) = 100x^{0.4}e^{-x/30,000}$; 45% tax on all income

3. T $N'(x) = 100x^{0.466}e^{-x/23,000}$; tax brackets as in the following tax table:

Income	*Marginal Tax Rate*
\$0–30,000	0%
\$30,000–250,000	10%
Above \$250,000	80%

4. T $N'(x) = 100x^{0.4}e^{-x/30,000}$; no tax on any income below \$250,000, 100% marginal tax rate on any income above \$250,000

5. Let $N'(x)$ be an income distribution function.

a. If $0 \le a < b$, what does $\int_a^b N'(x)\,dx$ represent? HINT [Use the Fundamental Theorem of Calculus.]
b. What does $\int_0^{+\infty} N'(x)\,dx$ represent?

6. Let $N'(x)$ be an income distribution function. What does $\int_0^{+\infty} xN'(x)\,dx$ represent? HINT [Argue as in the text.]

7. Let $P(x)$ be the number of people earning more than x dollars.

a. What is $N(x) + P(x)$?
b. Show that $P'(x) = -N'(x)$.
c. Use integration by parts to show that, if $T(0) = 0$, then the total tax revenue is

$$\int_0^{+\infty} P(x)T'(x)\,dx.$$

[Note: You may assume that $T'(x)$ is continuous, but the result is still true if we assume only that $T(x)$ is continuous and piecewise continuously differentiable.]

8. Income tax functions T are most often described, as in the text, by tax brackets and marginal tax rates.

a. If one tax bracket is $a < x \le b$, show that $\int_a^b P(x)\,dx$ is the total income earned in the country that falls into that bracket (P as in the preceding exercise).
b. Use (a) to explain directly why $\int_0^{+\infty} P(x)T'(x)\,dx$ gives the total tax revenue in the case where T is described by tax brackets and constant marginal tax rates in each bracket.

TI-83/84 Plus Technology Guide

Section 7.3

Example 3 (page 400) The following table shows Colossal Conglomerate's closing stock prices for 20 consecutive trading days:

Day	1	2	3	4	5	6	7	8	9	10
Price	20	22	21	24	24	23	25	26	20	24
Day	11	12	13	14	15	16	17	18	19	20
Price	26	26	25	27	28	27	29	27	25	24

Plot these prices and the 5-day moving average.

Solution with Technology

Here is how to automate this calculation on a TI-83/84 Plus.

1. Use

```
seq(X,X,1,20)→L1
```

2ND STAT → OPS → 5
STO 2ND STAT → L1

to enter the sequence of numbers 1 through 20 into the list L_1, representing the trading days.

2. Using the list editor accessible through the STAT menu, enter the daily stock prices in list L_2.
3. Calculate the list of 5-day moving averages by using the following command:

```
seq((L2(X)+L2(X-1)+L2(X-2)+L2(X-3)
  +L2(X-4))/5,X,5,20)→L3
```

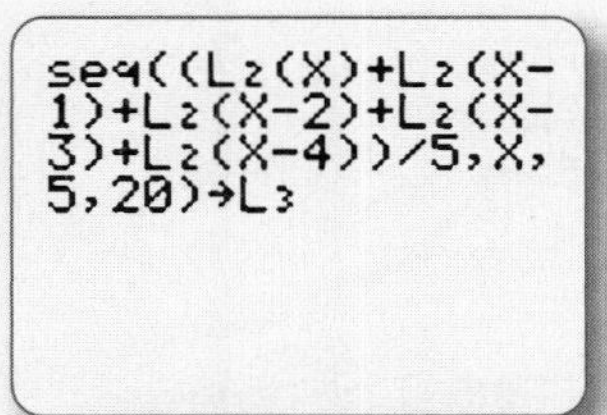

This has the effect of putting the moving averages into elements 1 through 15 of list L_3.

4. If you wish to plot the moving average on the same graph as the daily prices, you will want the averages in L_3 to match up with the prices in L_2. One way to do this is to put four more entries at the beginning of L_3—say, copies of the first four entries of L_2. The following command accomplishes this:

```
augment(seq(L2(X),X,1,4),L3)→L3
```

2ND STAT → OPS → 9

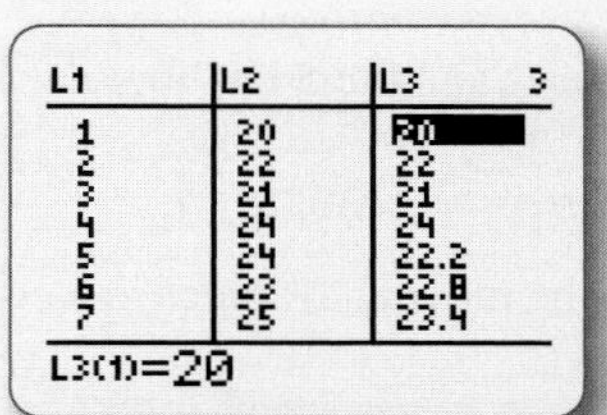

L1	L2	L3
1	20	20
2	22	22
3	21	21
4	24	24
5	24	22.2
6	23	22.8
7	25	23.4

L3(1)=20

5. You can now graph the prices and moving averages by creating an xyLine scatter plot through the STAT PLOT menu, with L_1 being the Xlist and L_2 being the Ylist for Plot1, and L_1 being the Xlist and L_3 the Ylist for Plot2:

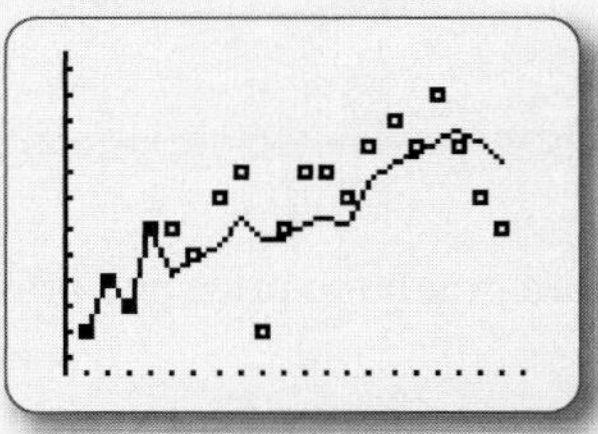

SPREADSHEET Technology Guide

Section 7.3

Example 3 (page 400) The following table shows Colossal Conglomerate's closing stock prices for 20 consecutive trading days:

Day	1	2	3	4	5	6	7	8	9	10
Price	20	22	21	24	24	23	25	26	20	24
Day	11	12	13	14	15	16	17	18	19	20
Price	26	26	25	27	28	27	29	27	25	24

Plot these prices and the 5-day moving average.

Solution with Technology

1. Compute the moving averages in a column next to the daily prices, as shown here:

	A	B	C
1	Day	Price	Moving Average
2	1	20	
3	2	22	
4	3	21	
5	4	24	
6	5	24	=AVERAGE(B2:B6)
7	6	23	
21	20	24	

→

	A	B	C
1	Day	Price	Moving Average
2	1	20	
3	2	22	
4	3	21	
5	4	24	
6	5	24	22.2
7	6	23	22.8
21	20	24	26.4

2. You can then graph the price and moving average using a scatter plot.

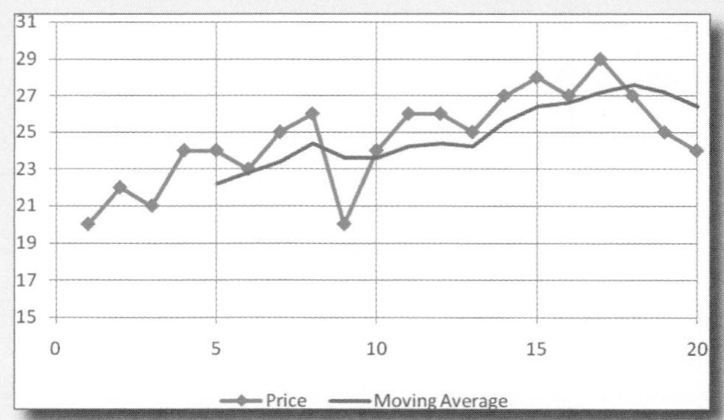

8

Functions of Several Variables

Website

www.WanerMath.com

At the Website you will find:

- A detailed chapter summary
- A true/false quiz
- A surface grapher
- An Excel surface grapher
- A linear multiple regression utility
- The following optional extra sections:

  Maxima and Minima: Boundaries and the Extreme Value Theorem

  The Chain Rule for Functions of Several Variables

Case Study Modeling College Population

College Malls, Inc. is planning to build a national chain of shopping malls in college neighborhoods. The company is planning to lease only to stores that target the specific age demographics of the national college student population. To decide which age brackets to target, the company has asked you, a paid consultant, for an analysis of the college population by student age, and of its trends over time. **How can you analyze the relevant data?**

david pearson/Alamy

Introduction

We have studied functions of a single variable extensively. But not every useful function is a function of only one variable. In fact, most are not. For example, if you operate an online bookstore in competition with Amazon.com, BN.com, and BooksAMillion.com, your sales may depend on those of your competitors. Your company's daily revenue might be modeled by a function such as

$$R(x, y, z) = 10{,}000 - 0.01x - 0.02y - 0.01z + 0.00001yz,$$

where x, y, and z are the online daily revenues of Amazon.com, BN.com, and BooksAMillion.com, respectively. Here, R is a function of three variables because it *depends on* x, y, and z. As we shall see, the techniques of calculus extend readily to such functions. Among the applications we shall look at is optimization: finding, where possible, the maximum or minimum of a function of two or more variables.

8.1 Functions of Several Variables from the Numerical, Algebraic, and Graphical Viewpoints

Numerical and Algebraic Viewpoints

Recall that a function of one variable is a rule for manufacturing a new number $f(x)$ from a single independent variable x. A function of two or more variables is similar, but the new number now depends on more than one independent variable.

Function of Several Variables

A **real-valued function**, f, **of** $x, y, z, \ldots$ is a rule for manufacturing a new number, written $f(x, y, z, \ldots)$, from the values of a sequence of independent variables $(x, y, z, \ldots)$. The function f is called a **real-valued function of two variables** if there are two independent variables, a **real-valued function of three variables** if there are three independent variables, and so on.

Quick Examples

1. $f(x, y) = x - y$ — Function of two variables

   $f(1, 2) = 1 - 2 = -1$ — Substitute 1 for x and 2 for y.

   $f(2, -1) = 2 - (-1) = 3$ — Substitute 2 for x and -1 for y.

   $f(y, x) = y - x$ — Substitute y for x and x for y.

2. $g(x, y) = x^2 + y^2$ — Function of two variables

   $g(-1, 3) = (-1)^2 + 3^2 = 10$ — Substitute -1 for x and 3 for y.

3. $h(x, y, z) = x + y + xz$ — Function of three variables

   $h(2, 2, -2) = 2 + 2 + 2(-2) = 0$ — Substitute 2 for x, 2 for y, and -2 for z.

Note It is often convenient to use $x_1, x_2, x_3, \ldots$ for the independent variables, so that, for instance, the third example above would be $h(x_1, x_2, x_3) = x_1 + x_2 + x_1x_3$. ■

Figure 1 illustrates the concept of a function of two variables: In goes a pair of numbers and out comes a single number.

$(x, y) \longrightarrow \boxed{g} \longrightarrow x^2 + y^2 \qquad (2, -1) \longrightarrow \boxed{g} \longrightarrow 5$

Figure 1

As with functions of one variable, functions of several variables can be represented numerically (using a table of values), algebraically (using a formula as in the above examples), and sometimes graphically (using a graph).

Let's now look at a number of examples of interesting functions of several variables.

Roy Mehta/The Image Bank/Getty Images

using Technology

See the Technology Guides at the end of the chapter to see how you can use a TI-83/84 Plus and a spreadsheet to display various values of $C(x, y)$ in Example 1. Here is an outline:

TI-83/84 Plus

```
Y1=10000+20X+40Y
```

To evaluate $C(10, 30)$:

```
10 → X
30 → Y
Y1
```

[More details on page 480.]

Spreadsheet

x-values down column A starting in A2

y-values down column B starting in B2

```
=10000+20*A2+40*B2
```

in C2; copy down column C. [More details on page 480.]

EXAMPLE 1 Cost Function

You own a company that makes two models of speakers: the Ultra Mini and the Big Stack. Your total monthly cost (in dollars) to make x Ultra Minis and y Big Stacks is given by

$$C(x, y) = 10{,}000 + 20x + 40y.$$

What is the significance of each term in this formula?

Solution The terms have meanings similar to those we saw for linear cost functions of a single variable. Let us look at the terms one at a time.

Constant Term Consider the monthly cost of making no speakers at all ($x = y = 0$). We find

$$C(0, 0) = 10{,}000. \qquad \text{Cost of making no speakers is \$10,000.}$$

Thus, the constant term 10,000 is the **fixed cost**, the amount you have to pay each month even if you make no speakers.

Coefficients of x and y Suppose you make a certain number of Ultra Minis and Big Stacks one month and the next month you increase production by one Ultra Mini. The costs are

$$\begin{aligned} C(x, y) &= 10{,}000 + 20x + 40y && \text{First month} \\ C(x + 1, y) &= 10{,}000 + 20(x + 1) + 40y && \text{Second month} \\ &= 10{,}000 + 20x + 20 + 40y \\ &= C(x, y) + 20. \end{aligned}$$

Thus, each Ultra Mini adds \$20 to the total cost. We say that \$20 is the **marginal cost** of each Ultra Mini. Similarly, because of the term $40y$, each Big Stack adds \$40 to the total cost. The marginal cost of each Big Stack is \$40.

This cost function is an example of a *linear function of two variables*. The coefficients of x and y play roles similar to that of the slope of a line. In particular, they give the rates of change of the function as each variable increases while the other stays constant (think about it). We shall say more about linear functions below.

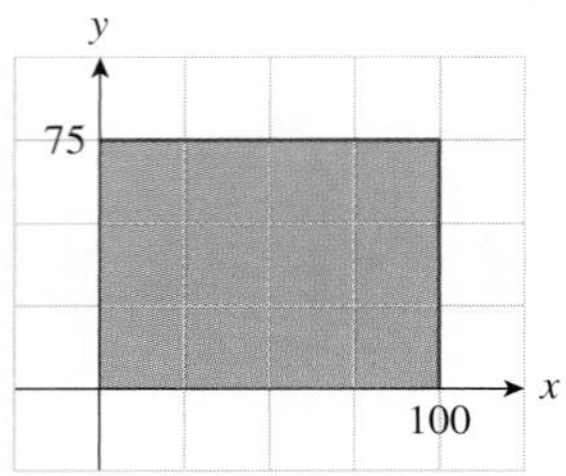

Figure 2

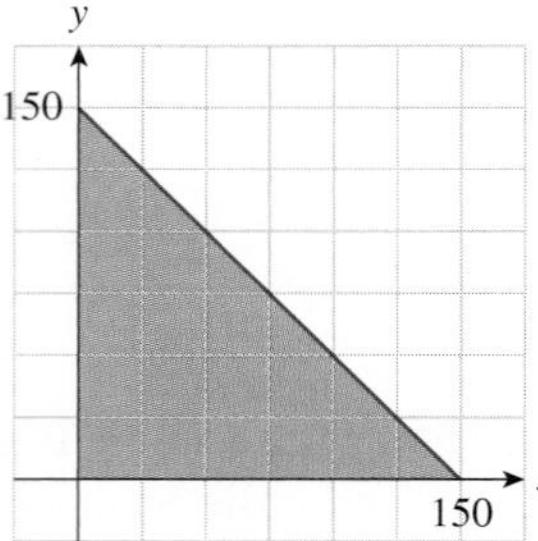

Figure 3

Before we go on... In Example 1 which values of x and y may we substitute into $C(x, y)$? Certainly we must have $x \geq 0$ and $y \geq 0$ because it makes no sense to speak of manufacturing a negative number of speakers. Also, there is certainly some upper bound to the number of speakers that can be made in a month. The bound might take one of several forms. The number of each model may be bounded—say $x \leq 100$ and $y \leq 75$. The inequalities $0 \leq x \leq 100$ and $0 \leq y \leq 75$ describe the region in the plane shaded in Figure 2.

Another possibility is that the *total* number of speakers is bounded—say, $x + y \leq 150$. This, together with $x \geq 0$ and $y \geq 0$, describes the region shaded in Figure 3.

In either case, the region shown represents the pairs (x, y) for which $C(x, y)$ is defined. Just as with a function of one variable, we call this region the **domain** of the function. As before, when the domain is not given explicitly, we agree to take the largest domain possible. ■

EXAMPLE 2 Faculty Salaries

David Katz came up with the following function for the salary of a professor with 10 years of teaching experience in a large university.

$$S(x, y, z) = 13{,}005 + 230x + 18y + 102z$$

Here, S is the salary in 1969–1970 in dollars per year, x is the number of books the professor has published, y is the number of articles published, and z is the number of "excellent" articles published.[1] What salary do you expect that a professor with 10 years' experience earned in 1969–1970 if she published 2 books, 20 articles, and 3 "excellent" articles?

Solution All we need to do is calculate

$$\begin{aligned} S(2, 20, 3) &= 13{,}005 + 230(2) + 18(20) + 102(3) \\ &= \$14{,}131. \end{aligned}$$

Before we go on... In Example 1, we gave a linear function of two variables. In Example 2 we have a linear function of three variables. Katz came up with his model by surveying a large number of faculty members and then finding the linear function "best" fitting the data. Such models are called **multiple linear regression** models. In the Case Study at the end of this chapter, we shall see a spreadsheet method of finding the coefficients of a multiple regression model from a set of observed data.

What does this model say about the value of a single book or a single article? If a book takes 15 times as long to write as an article, how would you recommend that a professor spend her writing time? ■

Here are two simple kinds of functions of several variables.

Linear Function

A function f of n variables is **linear** if f has the property that

$$f(x_1, x_2, \ldots, x_n) = a_0 + a_1x_1 + \cdots + a_nx_n \qquad (a_0, a_1, a_2, \ldots, a_n \text{ constants}).$$

[1]David A. Katz, "Faculty Salaries, Promotions and Productivity at a Large University," *American Economic Review*, June 1973, pp. 469–477. Prof. Katz's equation actually included other variables, such as the number of dissertations supervised; our equation assumes that all of these are zero.

Quick Examples

1. $f(x, y) = 3x - 5y$ — Linear function of x and y
2. $C(x, y) = 10{,}000 + 20x + 40y$ — Example 1
3. $S(x_1, x_2, x_3) = 13{,}005 + 230x_1 + 18x_2 + 102x_3$ — Example 2

Interaction Function

If we add to a linear function one or more terms of the form bx_ix_j (b a nonzero constant and $i \neq j$), we get a **second-order interaction function**.

Quick Examples

1. $C(x, y) = 10{,}000 + 20x + 40y + 0.1xy$
2. $R(x_1, x_2, x_3) = 10{,}000 - 0.01x_1 - 0.02x_2 - 0.01x_3 + 0.00001x_2x_3$

So far, we have been specifying functions of several variables **algebraically**—by using algebraic formulas. If you have ever studied statistics, you are probably familiar with statistical tables. These tables may also be viewed as representing functions **numerically**, as the next example shows.

EXAMPLE 3 Function Represented Numerically: Body Mass Index

The following table lists some values of the "body mass index," which gives a measure of the massiveness of your body, taking height into account.* The variable w represents your weight in pounds, and h represents your height in inches. An individual with a body mass index of 25 or above is generally considered overweight.

* It is interesting that weight-lifting competitions are usually based on weight, rather than body mass index. As a consequence, taller people are at a significant disadvantage in these competitions because they must compete with shorter, stockier people of the same weight. (An extremely thin, very tall person can weigh as much as a muscular short person, although his body mass index would be significantly lower.)

$w \rightarrow$; $h \downarrow$

h \ w	130	140	150	160	170	180	190	200	210
60	25.2	27.1	29.1	31.0	32.9	34.9	36.8	38.8	40.7
61	24.4	26.2	28.1	30.0	31.9	33.7	35.6	37.5	39.4
62	23.6	25.4	27.2	29.0	30.8	32.7	34.5	36.3	38.1
63	22.8	24.6	26.4	28.1	29.9	31.6	33.4	35.1	36.9
64	22.1	23.8	25.5	27.2	28.9	30.7	32.4	34.1	35.8
65	21.5	23.1	24.8	26.4	28.1	29.7	31.4	33.0	34.7
66	20.8	22.4	24.0	25.6	27.2	28.8	30.4	32.0	33.6
67	20.2	21.8	23.3	24.9	26.4	28.0	29.5	31.1	32.6
68	19.6	21.1	22.6	24.1	25.6	27.2	28.7	30.2	31.7
69	19.0	20.5	22.0	23.4	24.9	26.4	27.8	29.3	30.8
70	18.5	19.9	21.4	22.8	24.2	25.6	27.0	28.5	29.9
71	18.0	19.4	20.8	22.1	23.5	24.9	26.3	27.7	29.1
72	17.5	18.8	20.2	21.5	22.9	24.2	25.6	26.9	28.3
73	17.0	18.3	19.6	20.9	22.3	23.6	24.9	26.2	27.5
74	16.6	17.8	19.1	20.4	21.7	22.9	24.2	25.5	26.7
75	16.1	17.4	18.6	19.8	21.1	22.3	23.6	24.8	26.0
76	15.7	16.9	18.1	19.3	20.5	21.7	22.9	24.2	25.4

using Technology

See the Technology Guides at the end of the chapter to see how to use a spreadsheet to create the table in Example 3. Here is an outline:

Spreadsheet

w-values 130 to 210 in B1–J1

h-values 60 to 76 in A2–A18

```
=0.45*B$1/
(0.0254*$A2)^2
```

in B2; copy down and across through J18. [More details on page 480.]

As the table shows, the value of the body mass index depends on two quantities: w and h. Let us write $M(w, h)$ for the body mass index function. What are $M(140, 62)$ and $M(210, 63)$?

Solution We can read the answers from the table:

$$M(140, 62) = 25.4 \qquad w = 140 \text{ lb}, h = 62 \text{ in}$$

and

$$M(210, 63) = 36.9. \qquad w = 210 \text{ lb}, h = 63 \text{ in}$$

The function $M(w, h)$ is actually given by the formula

$$M(w, h) = \frac{0.45w}{(0.0254h)^2}.$$

[The factor 0.45 converts the weight to kilograms, and 0.0254 converts the height to meters. If w is in kilograms and h is in meters, the formula is simpler: $M(w, h) = w/h^2$.]

Geometric Viewpoint: Three-Dimensional Space and the Graph of a Function of Two Variables

Just as functions of a single variable have graphs, so do functions of two or more variables. Recall that the graph of $f(x)$ consists of all points $(x, f(x))$ in the xy-plane. By analogy, we would like to say that the graph of a function of *two* variables, $f(x, y)$, consists of all points of the form $(x, y, f(x, y))$. Thus, we need three axes: the x-, y-, and z-axes. In other words, our graph will live in **three-dimensional space**, or **3-space**.*

Just as we had two mutually perpendicular axes in two-dimensional space (the xy-plane; see Figure 4(a)), so we have three mutually perpendicular axes in three-dimensional space (Figure 4(b)).

* If we were dealing instead with a function of *three* variables, then we would need to go to *four-dimensional* space. Here we run into visualization problems (to say the least!) so we won't discuss the graphs of functions of three or more variables in this text.

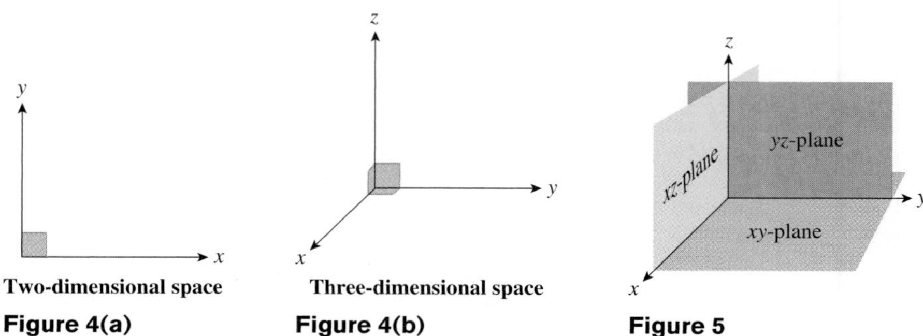

Two-dimensional space

Figure 4(a)

Three-dimensional space

Figure 4(b)

Figure 5

In both 2-space and 3-space, the axis labeled with the last letter goes up. Thus, the z-direction is the "up" direction in 3-space, rather than the y-direction.

Three important planes are associated with these axes: the xy-plane, the yz-plane, and the xz-plane. These planes are shown in Figure 5. Any two of these planes intersect in one of the axes (for example, the xy- and xz-planes intersect in the x-axis) and all three meet at the origin. Notice that the xy-plane consists of all points with z-coordinate zero, the xz-plane consists of all points with $y = 0$, and the yz-plane consists of all points with $x = 0$.

In 3-space, each point has *three* coordinates, as you might expect: the x-coordinate, the y-coordinate, and the z-coordinate. To see how this works, look at the following examples.

EXAMPLE 4 Plotting Points in Three Dimensions

Locate the points $P(1, 2, 3)$, $Q(-1, 2, 3)$, $R(1, -1, 0)$, and $S(1, 2, -2)$ in 3-space.

Solution To locate P, the procedure is similar to the one we used in 2-space: Start at the origin, proceed 1 unit in the x direction, then proceed 2 units in the y direction, and finally, proceed 3 units in the z direction. We wind up at the point P shown in Figures 6(a) and 6(b).

Here is another, extremely useful way of thinking about the location of P. First, look at the x- and y-coordinates, obtaining the point $(1, 2)$ in the xy-plane. The point we want is then 3 units vertically above the point $(1, 2)$ because the z-coordinate of a point is just its height above the xy-plane. This strategy is shown in Figure 6(c).

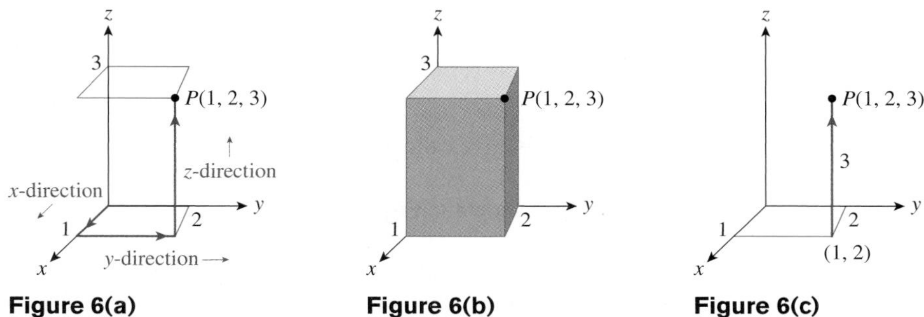

Figure 6(a) **Figure 6(b)** **Figure 6(c)**

Plotting the points Q, R, and S is similar, using the convention that negative coordinates correspond to moves back, left, or down. (See Figure 7.)

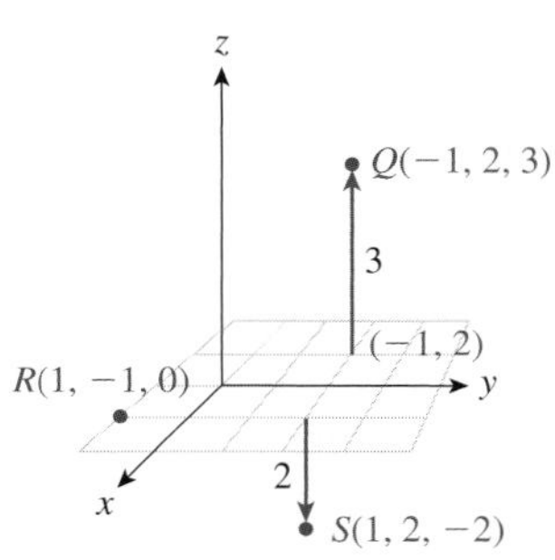

Figure 7

Our next task is to describe the graph of a function $f(x, y)$ of two variables.

Graph of a Function of Two Variables

The **graph of the function f of two variables** is the set of all points $(x, y, f(x, y))$ in three-dimensional space, where we restrict the values of (x, y) to lie in the domain of f. In other words, the graph is the set of all the points (x, y, z) with $z = f(x, y)$.

Note For *every* point (x, y) in the domain of f, the z-coordinate of the corresponding point on the graph is given by evaluating the function at (x, y). Thus, there will be a point of the graph on the vertical line through *every* point in the domain of f, so that the graph is usually a *surface* of some sort (see the figure).

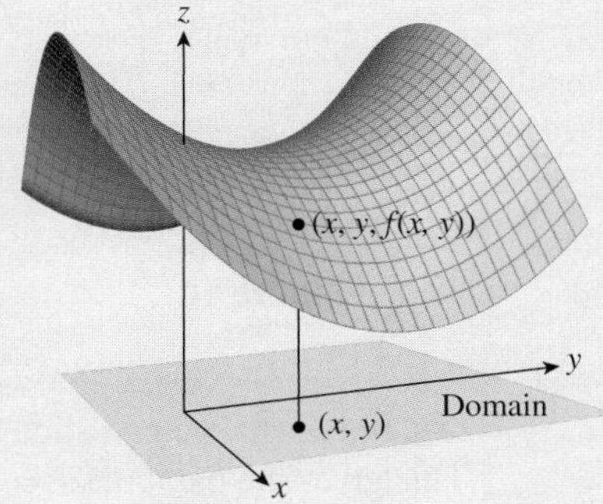

EXAMPLE 5 Graph of a Function of Two Variables

Describe the graph of $f(x, y) = x^2 + y^2$.

Solution Your first thought might be to make a table of values. You could choose some values for x and y and then, for each such pair, calculate $z = x^2 + y^2$. For example, you might get the following table:

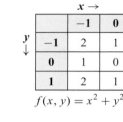

$y \downarrow$ \ $x \rightarrow$	**−1**	**0**	**1**
−1	2	1	2
0	1	0	1
1	2	1	2

$f(x, y) = x^2 + y^2$

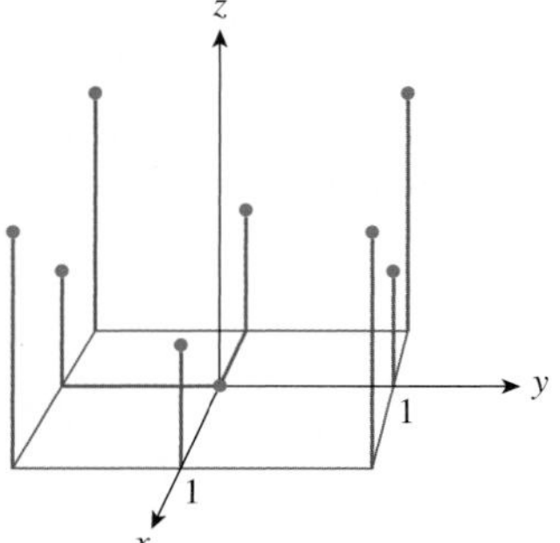

Figure 8

This gives the following nine points on the graph of f: $(-1, -1, 2)$, $(-1, 0, 1)$, $(-1, 1, 2)$, $(0, -1, 1)$, $(0, 0, 0)$, $(0, 1, 1)$, $(1, -1, 2)$, $(1, 0, 1)$, and $(1, 1, 2)$. These points are shown in Figure 8.

The points on the xy-plane we chose for our table are the grid points in the xy-plane, and the corresponding points on the graph are marked with solid dots. The problem is that this small number of points hardly tells us what the surface looks like, and even if we plotted more points, it is not clear that we would get anything more than a mass of dots on the page.

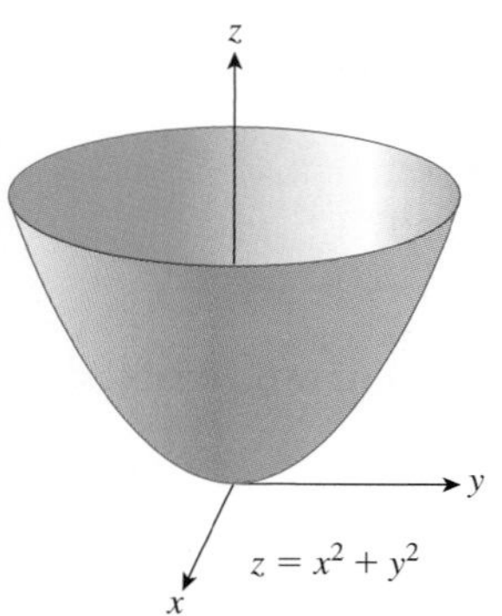

Figure 9

What can we do? There are several alternatives. One place to start is to use technology to draw the graph. (See the technology note on the next page.) We then obtain something like Figure 9. This particular surface is called a **paraboloid**.

If we slice vertically through this surface along the yz-plane, we get the picture in Figure 10. The shape of the front edge, where we cut, is a parabola. To see why, note that the yz-plane is the set of points where $x = 0$. To get the intersection of $x = 0$ and $z = x^2 + y^2$, we substitute $x = 0$ in the second equation, getting $z = y^2$. This is the equation of a parabola in the yz-plane.

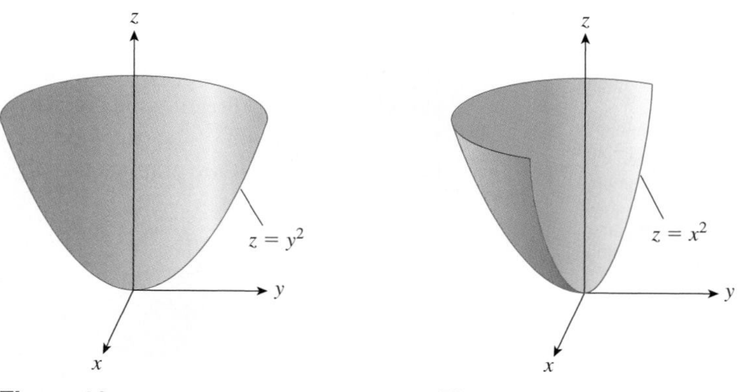

Figure 10 **Figure 11**

Similarly, we can slice through the surface with the xz-plane by setting $y = 0$. This gives the parabola $z = x^2$ in the xz-plane (Figure 11).

We can also look at horizontal slices through the surface, that is, slices by planes parallel to the xy-plane. These are given by setting $z = c$ for various numbers c. For example, if we set $z = 1$, we will see only the points with height 1. Substituting in the equation $z = x^2 + y^2$ gives the equation

$$1 = x^2 + y^2,$$

which is the equation of a circle of radius 1.* If we set $z = 4$, we get the equation of a circle of radius 2:

$$4 = x^2 + y^2.$$

In general, if we slice through the surface at height $z = c$, we get a circle (of radius $\sqrt{c}$). Figure 12 shows several of these circles.

* See Section 0.7 for a discussion of equations of circles.

using Technology

We can use technology to obtain the graph of the function in Example 5:

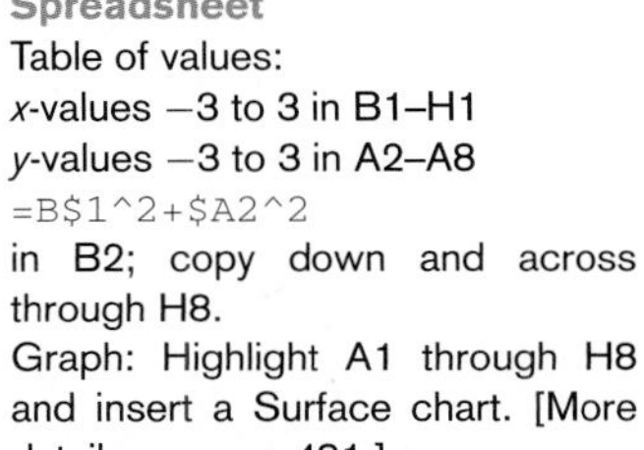

Spreadsheet

Table of values:
x-values −3 to 3 in B1–H1
y-values −3 to 3 in A2–A8
`=B$1^2+$A2^2`
in B2; copy down and across through H8.
Graph: Highlight A1 through H8 and insert a Surface chart. [More details on page 481.]

Website

www.WanerMath.com
Online Utilities
→ Surface Graphing Utility
Enter `x^2+y^2` for $f(x, y)$
Set xMin = −3, xMax = 3, yMin = −3, yMax = 3
Press "Graph".

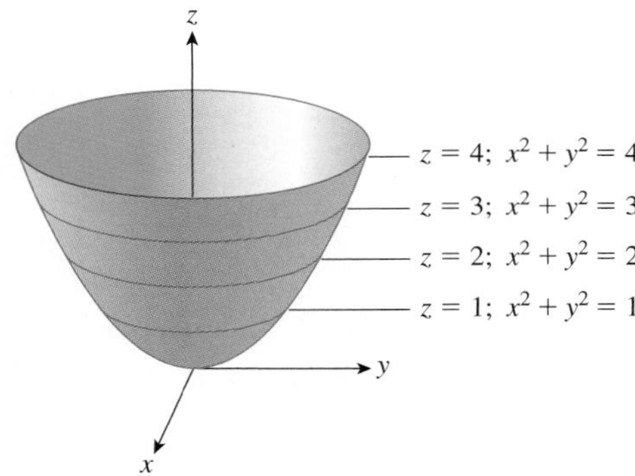

Figure 12

Looking at these circular slices, we see that this surface is the one we get by taking the parabola $z = x^2$ and spinning it around the z-axis. This is an example of what is known as a **surface of revolution**.

➡ **Before we go on...** The graph of any function of the form $f(x, y) = Ax^2 + By^2 + Cxy + Dx + Ey + F$ ($A, B, \dots, F$ constants), with $4AB - C^2$ positive, can be shown to be a paraboloid of the same general shape as that in Example 5 if A and B are positive, or upside-down if A and B are negative. If $A \neq B$, the horizontal slices will be ellipses rather than circles.

Notice that each horizontal slice through the surface in Example 5 was obtained by putting $z = constant$. This gave us an equation in x and y that described a curve. These curves are called the **level curves** of the surface $z = f(x, y)$ (see the discussion on the next page). In Example 5, the equations are of the form $x^2 + y^2 = c$ (c constant), and so the level curves are circles. Figure 13 shows the level curves for $c = 0$, 1, 2, 3, and 4.

The level curves give a contour map or topographical map of the surface. Each curve shows all of the points on the surface at a particular height c. You can use this contour map to visualize the shape of the surface. Imagine moving the contour at $c = 1$ to a height of 1 unit above the xy-plane, the contour at $c = 2$ to a height of 2 units above the xy-plane, and so on. You will end up with something like Figure 12. ■

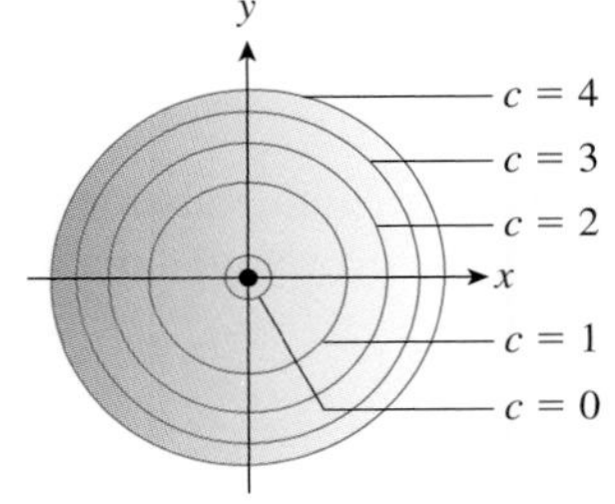

Level curves of the paraboloid $z = x^2 + y^2$

Figure 13

The following summary includes the techniques we have just used plus some additional ones:

Analyzing the Graph of a Function of Two Variables

If possible, use technology to render the graph $z = f(x, y)$ of a given function f of two variables. You can analyze its graph as follows:

Step 1 Obtain the ***x*-, *y*-, and *z*-intercepts** (the places where the surface crosses the coordinate axes).

***x*-Intercept(s):** Set $y = 0$ and $z = 0$ and solve for x.

***y*-Intercept(s):** Set $x = 0$ and $z = 0$ and solve for y.

***z*-Intercept:** Set $x = 0$ and $y = 0$ and compute z.

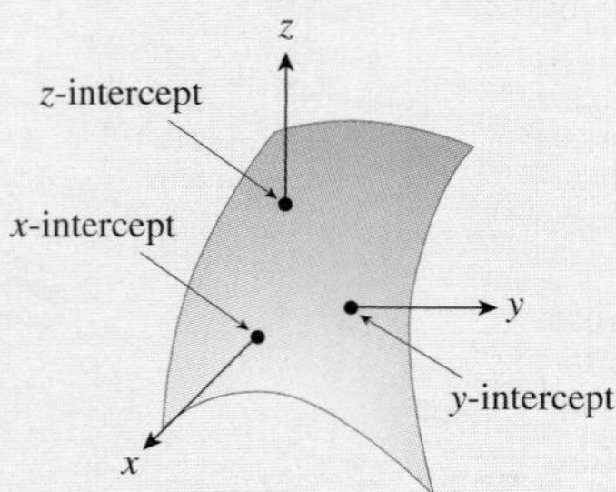

Step 2 Slice the surface along planes parallel to the xy-, yz-, and xz-planes.

z* = *constant Set $z =$ *constant* and analyze the resulting curves. These are the curves resulting from horizontal slices, and are called the **level curves** (see below).

x* = *constant Set $x =$ *constant* and analyze the resulting curves. These are the curves resulting from slices parallel to the yz-plane.

y* = *constant Set $y =$ *constant* and analyze the resulting curves. These are the curves resulting from slices parallel to the xz-plane.

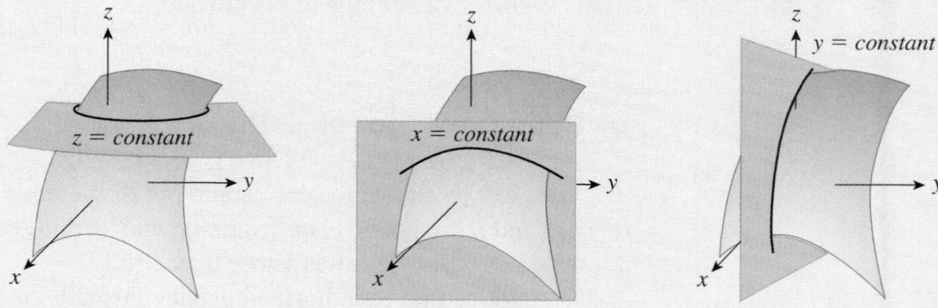

Level Curves

The **level curves** of a function f of two variables are the curves with equations of the form $f(x, y) = c$, where c is constant. These are the curves obtained from the graph of f by slicing it horizontally as above.

Quick Examples

1. Figure 13 on previous page shows some level curves of $f(x, y) = x^2 + y^2$. The ones shown have equations $f(x, y) = 0, 1, 2, 3$, and 4.

2. Let $f(x, y) = y - x^2 + 4$. Its level curves have the form $y - x^2 + 4 = c$ (c constant). If we solve this equation for y, we see that $y = x^2 + c - 4$, the equation of a parabola with its vertex on the y-axis at the point $c - 4$. The following figure shows a portion of the graph of f and some of its level curves.

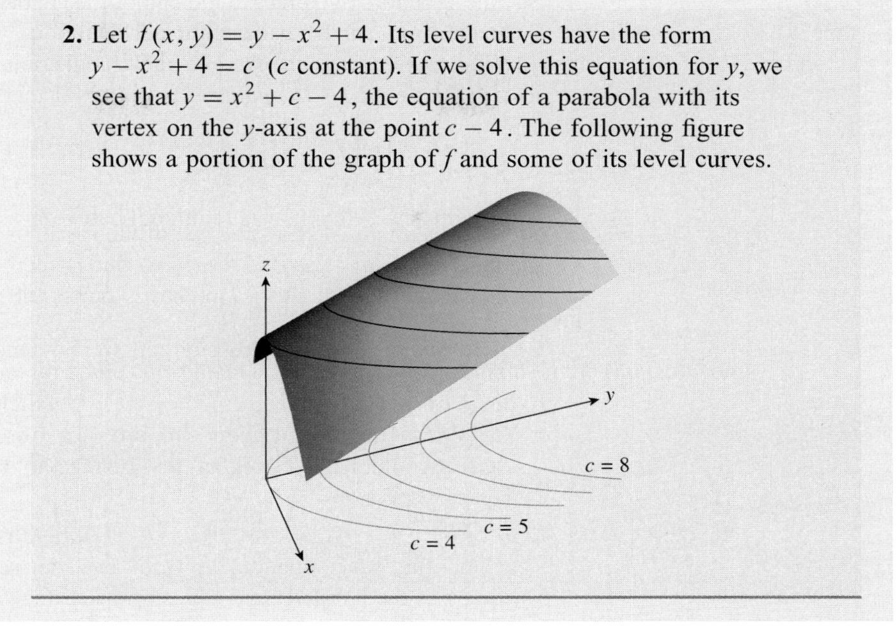

Spreadsheets often have built-in features to render surfaces such as the paraboloid in Example 5. In the following example, we use Excel to graph another surface and then analyze it as above.

EXAMPLE 6 **Analyzing a Surface**

Describe the graph of $f(x, y) = x^2 - y^2$.

Solution First we obtain a picture of the graph using technology. Figure 14 shows two graphs obtained using resources at the Website.

Chapter 8 → Math Tools for Chapter 8 → Surface Graphing Utility

Chapter 8 → Math Tools for Chapter 8 → Excel Surface Graphing Utility

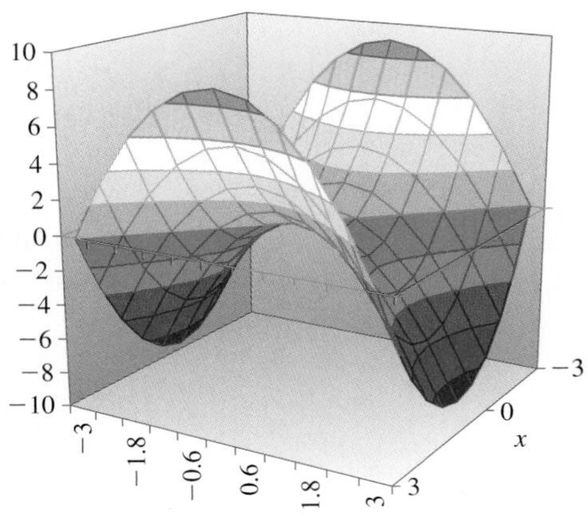

Figure 14

See the Technology Guides at the end of the chapter to find out how to obtain a similar graph from scratch using a spreadsheet.

The graph shows an example of a "saddle point" at the origin. (We return to this idea in Section 8.3.) To analyze the graph for the features shown in the box above, replace $f(x, y)$ by z to obtain

$$z = x^2 - y^2.$$

Step 1: *Intercepts* Setting any two of the variables x, y, and z equal to zero results in the third also being zero, so the x-, y-, and z-intercepts are all 0. In other words, the surface touches all three axes in exactly one point, the origin.

Step 2: *Slices* Slices in various directions show more interesting features.

Slice by $x = c$ This gives $z = c^2 - y^2$, which is the equation of a parabola that opens downward. You can see two of these slices ($c = -3, c = 3$) as the front and back edges of the surface in Figure 14. (More are shown in Figure 15(a).)

Slice by $y = c$ This gives $z = x^2 - c^2$, which is the equation of a parabola once again—this time, opening upward. You can see two of these slices ($c = -3$, $c = 3$) as the left and right edges of the surface in Figure 14. (More are shown in Figure 15(b).)

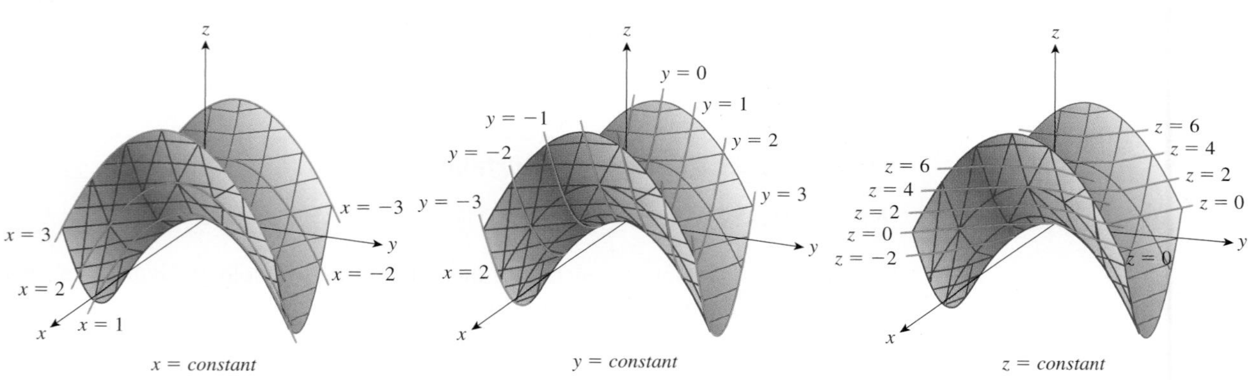

Figure 15(a) **Figure 15(b)** **Figure 15(c)**

Level Curves: Slice by $z = c$ This gives $x^2 - y^2 = c$, which is a hyperbola. The level curves for various values of c are visible in Figure 14 as the horizontal slices. (See Figure 15(c).) The case $c = 0$ is interesting: The equation $x^2 - y^2 = 0$ can be rewritten as $x = \pm y$ (why?), which represents two lines at right angles to each other.

To obtain really beautiful renderings of surfaces, you could use one of the commercial computer algebra software packages, such as Mathematica® or Maple®, or, if you use a Mac, the built-in grapher (grapher.app located in the Utilities folder).

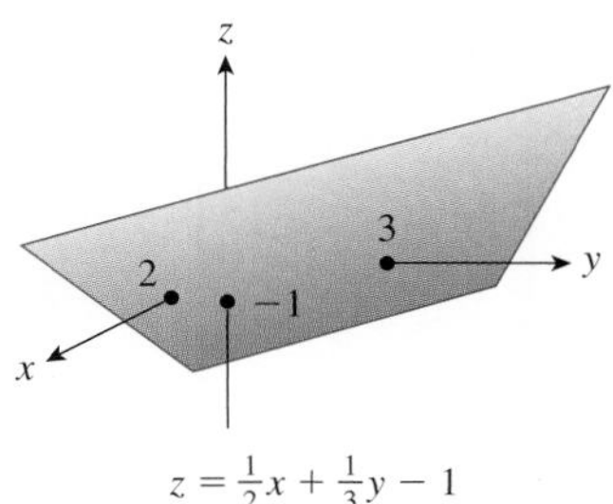

$z = \frac{1}{2}x + \frac{1}{3}y - 1$

Figure 16

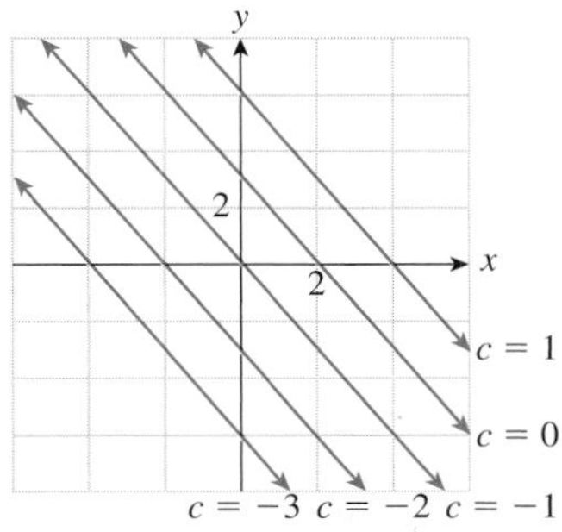

Level curves: $3x + 2y = 12$

Figure 17

* Think about what happens when the function is constant.

EXAMPLE 7 Graph and Level Curves of a Linear Function

Describe the graph of $g(x, y) = \dfrac{1}{2}x + \dfrac{1}{3}y - 1$.

Solution Notice first that g is a linear function of x and y. Figure 16 shows a portion of the graph, which is a plane.

We can get a good idea of what plane this is by looking at the x-, y-, and z-intercepts.

x-intercept Set $y = z = 0$, which gives $x = 2$.

y-intercept Set $x = z = 0$, which gives $y = 3$.

z-intercept Set $x = y = 0$, which gives $z = -1$.

Three points are enough to define a plane, so we can say that the plane is the one passing through the three points (2, 0, 0), (0, 3, 0), and (0, 0, −1). It can be shown that the graph of every linear function of two variables is a plane.

Level curves: Set $g(x, y) = c$ to obtain $\frac{1}{2}x + \frac{1}{3}y - 1 = c$, or $\frac{1}{2}x + \frac{1}{3}y = c + 1$. We can rewrite this equation as $3x + 2y = 6(c + 1)$, which is the equation of a straight line. Choosing different values of c gives us a family of parallel lines as shown in Figure 17. (For example, the line corresponding to $c = 1$ has equation $3x + 2y = 6(1 + 1) = 12$.) In general, the set of level curves of every non-constant linear function is a set of parallel straight lines.*

EXAMPLE 8 Using Level Curves

A certain function f of two variables has level curves $f(x, y) = c$ for $c = -2, -1, 0,$ 1, and 2, as shown in Figure 18. (Each grid square is 1×1.)

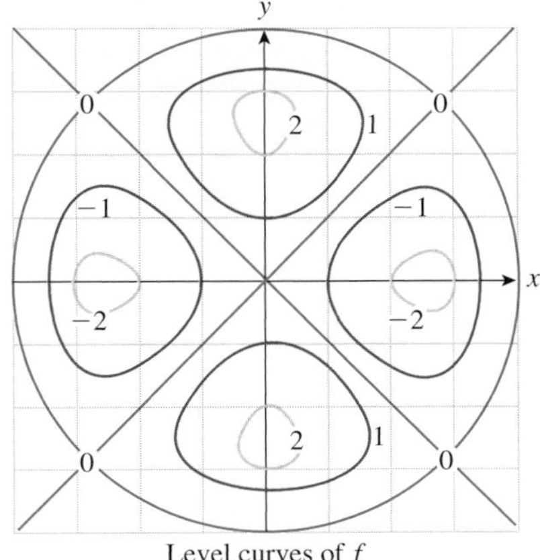

Level curves of f

Figure 18

Estimate the following: $f(1, 1)$, $f(1.5, -2)$, $f(1.5, 0)$, and $f(1, 2)$.

Solution The point (1, 1) appears to lie exactly on the red level curve $c = 0$, so $f(1, 1) \approx 0$. Similarly, the point (1.5, −2) appears to lie exactly on the blue level curve $c = 1$, so $f(1.5, -2) \approx 1$. The point (1.5, 0) appears to lie midway between the level curves $c = -1$ and $c = -2$, so we estimate $f(1.5, 2) \approx -1.5$. Finally, the point (1, 2) lies between the level curves $c = 1$ and $c = 2$, but closer to $c = 1$, so we can estimate $f(1, 2)$ at around 1.3.

8.1 EXERCISES

Access end-of-section exercises online at **www.webassign.net** ENHANCED WebAssign

8.2 Partial Derivatives

Recall that if f is a function of x, then the derivative df/dx measures how fast f changes as x increases. If f is a function of two or more variables, we can ask how fast f changes as each variable increases while the others remain fixed. These rates of change are called the "partial derivatives of f," and they measure how each variable contributes to the change in f. Here is a more precise definition.

Partial Derivatives

The **partial derivative of f with respect to x** is the derivative of f with respect to x, when all other variables are treated as constant. Similarly, the **partial derivative of f with respect to y** is the derivative of f with respect to y, with all other variables treated as constant, and so on for other variables. The partial derivatives are written as $\frac{\partial f}{\partial x}, \frac{\partial f}{\partial y}$, and so on. The symbol ∂ is used (instead of d) to remind us that there is more than one variable and that we are holding the other variables fixed.

Quick Examples

1. Let $f(x, y) = x^2 + y^2$.

$$\frac{\partial f}{\partial x} = 2x + 0 = 2x$$ Because y^2 is treated as a constant

$$\frac{\partial f}{\partial y} = 0 + 2y = 2y$$ Because x^2 is treated as a constant

2. Let $z = x^2 + xy$.

$$\frac{\partial z}{\partial x} = 2x + y$$ $\frac{\partial}{\partial x}(xy) = \frac{\partial}{\partial x}(x \cdot \text{constant}) = \text{constant} = y$

$$\frac{\partial z}{\partial y} = 0 + x$$ $\frac{\partial}{\partial y}(xy) = \frac{\partial}{\partial y}(\text{constant} \cdot y) = \text{constant} = x$

3. Let $f(x, y) = x^2y + y^2x - xy + y$.

$$\frac{\partial f}{\partial x} = 2xy + y^2 - y$$ y is treated as a constant.

$$\frac{\partial f}{\partial y} = x^2 + 2xy - x + 1$$ x is treated as a constant.

Interpretation

$\frac{\partial f}{\partial x}$ is the rate at which f changes as x changes, for a fixed (constant) y.

$\frac{\partial f}{\partial y}$ is the rate at which f changes as y changes, for a fixed (constant) x.

EXAMPLE 1 Marginal Cost: Linear Model

We return to Example 1 from Section 8.1. Suppose that you own a company that makes two models of speakers, the Ultra Mini and the Big Stack. Your total monthly cost (in dollars) to make x Ultra Minis and y Big Stacks is given by

$$C(x, y) = 10{,}000 + 20x + 40y.$$

What is the significance of $\frac{\partial C}{\partial x}$ and of $\frac{\partial C}{\partial y}$?

Solution First we compute these partial derivatives:

$$\frac{\partial C}{\partial x} = 20$$

$$\frac{\partial C}{\partial y} = 40.$$

We interpret the results as follows: $\frac{\partial C}{\partial x} = 20$ means that the cost is increasing at a rate of \$20 per additional Ultra Mini (if production of Big Stacks is held constant); $\frac{\partial C}{\partial y} = 40$ means that the cost is increasing at a rate of \$40 per additional Big Stack (if production of Ultra Minis is held constant). In other words, these are the **marginal costs** of each model of speaker.

➡ **Before we go on...** How much does the cost rise if you increase x by Δx and y by Δy? In Example 1, the change in cost is given by

$$\Delta C = 20\Delta x + 40\Delta y = \frac{\partial C}{\partial x}\Delta x + \frac{\partial C}{\partial y}\Delta y.$$

This suggests the **chain rule for several variables**. Part of this rule says that if x and y are both functions of t, then C is a function of t through them, and the rate of change of C with respect to t can be calculated as

$$\frac{dC}{dt} = \frac{\partial C}{\partial x} \cdot \frac{dx}{dt} + \frac{\partial C}{\partial y} \cdot \frac{dy}{dt}.$$

See the optional section on the chain rule for several variables for further discussion and applications of this interesting result. ■

EXAMPLE 2 Marginal Cost: Interaction Model

Another possibility for the cost function in the preceding example is an interaction model

$$C(x, y) = 10{,}000 + 20x + 40y + 0.1xy.$$

a. *Now* what are the marginal costs of the two models of speakers?

b. What is the marginal cost of manufacturing Big Stacks at a production level of 100 Ultra Minis and 50 Big Stacks per month?

Solution

a. We compute the partial derivatives:

$$\frac{\partial C}{\partial x} = 20 + 0.1y$$

$$\frac{\partial C}{\partial y} = 40 + 0.1x.$$

Thus, the marginal cost of manufacturing Ultra Minis increases by \$0.1 or 10¢ for each Big Stack that is manufactured. Similarly, the marginal cost of manufacturing Big Stacks increases by 10¢ for each Ultra Mini that is manufactured.

b. From part (a), the marginal cost of manufacturing Big Stacks is

$$\frac{\partial C}{\partial y} = 40 + 0.1x.$$

At a production level of 100 Ultra Minis and 50 Big Stacks per month, we have $x = 100$ and $y = 50$. Thus, the marginal cost of manufacturing Big Stacks at these production levels is

$$\left.\frac{\partial C}{\partial y}\right|_{(100,50)} = 40 + 0.1(100) = \$50 \text{ per Big Stack.}$$

Partial derivatives of functions of three variables are obtained in the same way as those for functions of two variables, as the following example shows.

EXAMPLE 3 Function of Three Variables

Calculate $\frac{\partial f}{\partial x}$, $\frac{\partial f}{\partial y}$, and $\frac{\partial f}{\partial z}$ if $f(x, y, z) = xy^2z^3 - xy$.

Solution Although we now have three variables, the calculation remains the same: $\partial f/\partial x$ is the derivative of f with respect to x, with *both* other variables, y and z, held constant:

$$\frac{\partial f}{\partial x} = y^2z^3 - y.$$

Similarly, $\partial f/\partial y$ is the derivative of f with respect to y, with both x and z held constant:

$$\frac{\partial f}{\partial y} = 2xyz^3 - x.$$

Finally, to find $\partial f/\partial z$, we hold both x and y constant and take the derivative with respect to z.

$$\frac{\partial f}{\partial z} = 3xy^2z^2.$$

Note The procedure for finding a partial derivative is the same for any number of variables: To get the partial derivative with respect to any one variable, we treat all the others as constants. ■

Geometric Interpretation of Partial Derivatives

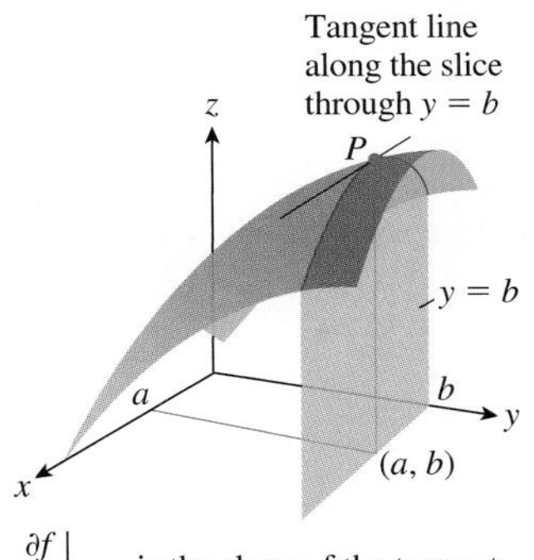

$\left.\frac{\partial f}{\partial x}\right|_{(a,b)}$ is the slope of the tangent line at the point $P(a, b, f(a, b))$ along the slice through $y = b$.

Figure 19

Recall that if f is a function of one variable x, then the derivative df/dx gives the slopes of the tangent lines to its graph. Now, suppose that f is a function of x and y. By definition, $\partial f/\partial x$ is the derivative of the function of x we get by holding y fixed. If we evaluate this derivative at the point (a, b), we are holding y fixed at the value b, taking the ordinary derivative of the resulting function of x, and evaluating this at $x = a$. Now, holding y fixed at b amounts to slicing through the graph of f along the plane $y = b$, resulting in a curve. Thus, the partial derivative is the slope of the tangent line to this curve at the point where $x = a$ and $y = b$, along the plane $y = b$ (Figure 19). This fits with our interpretation of $\partial f/\partial x$ as the rate of increase of f with increasing x when y is held fixed at b.

The other partial derivative, $\partial f/\partial y|_{(a,b)}$, is, similarly, the slope of the tangent line at the same point $P(a, b, f(a, b))$ but along the slice by the plane $x = a$. You should draw the corresponding picture for this on your own.

EXAMPLE 4 Marginal Cost

Referring to the interactive cost function $C(x, y) = 10{,}000 + 20x + 40y + 0.1xy$ in Example 2, we can identify the marginal costs $\partial C/\partial x$ and $\partial C/\partial y$ of manufacturing Ultra Minis and Big Stacks at a production level of 100 Ultra Minis and 50 Big Stacks per month as the slopes of the tangent lines to the two slices by $y = 50$ and $x = 100$ at the point on the graph where $(x, y) = (100, 50)$ as seen in Figure 20.

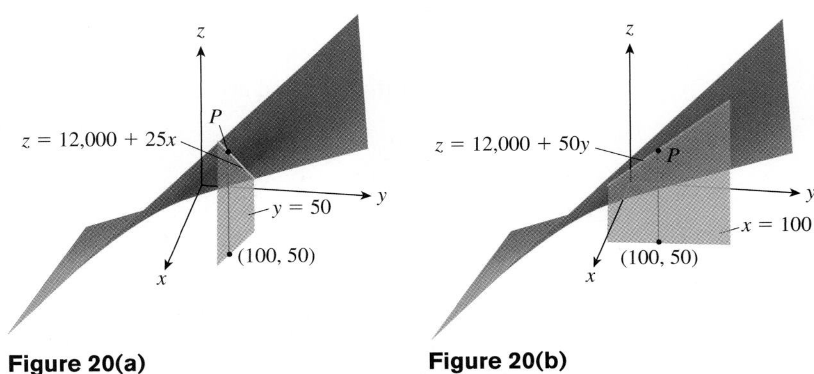

Figure 20(a) **Figure 20(b)**

Figure 20(a) shows the slice at $y = 50$ through the point $P = (100, 50, C(100, 50)) = (100, 50, 14{,}500)$. The equation of that slice is given by substituting $y = 50$ in the cost equation:

$$C(x, 50) = 10{,}000 + 20x + 40(50) + 0.1x(50) = 12{,}000 + 25x. \quad \text{A line of slope 25}$$

Because the slice is already a line, it coincides with the tangent line through P as depicted in Figure 19. This slope is equal to $\partial C/\partial x|_{(100,50)}$:

$$\frac{\partial C}{\partial x} = 20 + 0.1y \quad \text{See Example 2.}$$

so $$\left.\frac{\partial C}{\partial x}\right|_{(100,50)} = 20 + 0.1(50) = 25.$$

Similarly, Figure 20(b) shows the slice at $x = 100$ through the same point P. The equation of that slice is given by substituting $x = 100$ in the cost equation:

$$C(100, y) = 10{,}000 + 20(100) + 40y + 0.1(100)y = 12{,}000 + 50y. \quad \text{A line of slope 50}$$

This slope is equal to $\partial C/\partial y|_{(100,50)} = 50$ as we calculated in Example 2.

Second-Order Partial Derivatives

Just as for functions of a single variable, we can calculate second derivatives. Suppose, for example, that we have a function of x and y, say, $f(x, y) = x^2 - x^2y^2$. We know that

$$\frac{\partial f}{\partial x} = 2x - 2xy^2.$$

If we take the partial derivative with respect to x once again, we obtain

$$\frac{\partial}{\partial x}\left(\frac{\partial f}{\partial x}\right) = 2 - 2y^2. \qquad \text{Take } \frac{\partial}{\partial x} \text{ of } \frac{\partial f}{\partial x}.$$

(The symbol $\partial/\partial x$ means "the partial derivative with respect to x," just as d/dx stands for "the derivative with respect to x.") This is called the **second-order partial derivative** and is written $\dfrac{\partial^2 f}{\partial x^2}$. We get the following derivatives similarly:

$$\frac{\partial f}{\partial y} = -2x^2y$$

$$\frac{\partial^2 f}{\partial y^2} = -2x^2. \qquad \text{Take } \frac{\partial}{\partial y} \text{ of } \frac{\partial f}{\partial y}.$$

Now what if we instead take the partial derivative with respect to y of $\partial f/\partial x$?

$$\frac{\partial^2 f}{\partial y \partial x} = \frac{\partial}{\partial y}\left(\frac{\partial f}{\partial x}\right) \qquad \text{Take } \frac{\partial}{\partial y} \text{ of } \frac{\partial f}{\partial x}.$$

$$= \frac{\partial}{\partial y}[2x - 2xy^2] = -4xy$$

Here, $\dfrac{\partial^2 f}{\partial y \partial x}$ means "first take the partial derivative with respect to x and then with respect to y" and is called a **mixed partial derivative**. If we differentiate in the opposite order, we get

$$\frac{\partial^2 f}{\partial x \partial y} = \frac{\partial}{\partial x}\left(\frac{\partial f}{\partial y}\right) = \frac{\partial}{\partial x}[-2x^2y] = -4xy,$$

the same expression as $\dfrac{\partial^2 f}{\partial y \partial x}$. This is no coincidence: The mixed partial derivatives $\dfrac{\partial^2 f}{\partial x \partial y}$ and $\dfrac{\partial^2 f}{\partial y \partial x}$ are always the same as long as the first partial derivatives are both differentiable functions of x and y and the mixed partial derivatives are continuous. Because all the functions we shall use are of this type, we can take the derivatives in any order we like when calculating mixed derivatives.

Here is another notation for partial derivatives that is especially convenient for second-order partial derivatives:

$$f_x \text{ means } \frac{\partial f}{\partial x}$$

$$f_y \text{ means } \frac{\partial f}{\partial y}$$

$$f_{xy} \text{ means } (f_x)_y = \frac{\partial^2 f}{\partial y \partial x} \quad \text{(Note the order in which the derivatives are taken.)}$$

$$f_{yx} \text{ means } (f_y)_x = \frac{\partial^2 f}{\partial x \partial y}.$$

8.2 EXERCISES

Access end-of-section exercises online at **www.webassign.net** ENHANCED WebAssign

8.3 Maxima and Minima

In Chapter 5, on applications of the derivative, we saw how to locate relative extrema of a function of a single variable. In this section we extend our methods to functions of two variables. Similar techniques work for functions of three or more variables.

Figure 21 shows a portion of the graph of the function

$$f(x, y) = 2(x^2 + y^2) - (x^4 + y^4) + 1.$$

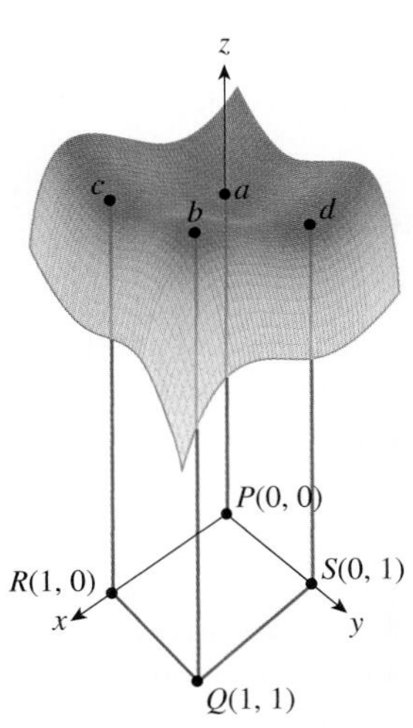

Figure 21

The graph in Figure 21 resembles a "flying carpet," and several interesting points, marked a, b, c, and d, are shown.

1. The point a has coordinates $(0, 0, f(0, 0))$, is directly above the origin $(0, 0)$, and is the lowest point in its vicinity; water would puddle there. We say that f has a **relative minimum** at $(0, 0)$ because $f(0, 0)$ is smaller than $f(x, y)$ for any (x, y) near $(0, 0)$.
2. Similarly, the point b is higher than any point in its vicinity. Thus, we say that f has a **relative maximum** at $(1, 1)$.
3. The points c and d represent a new phenomenon and are called **saddle points**. They are neither relative maxima nor relative minima but seem to be a little of both.

To see more clearly what features a saddle point has, look at Figure 22, which shows a portion of the graph near the point d.

If we slice through the graph along $y = 1$, we get a curve on which d is the *lowest* point. Thus, d looks like a relative minimum along this slice. On the other hand, if we slice through the graph along $x = 0$, we get another curve, on which d is the *highest*

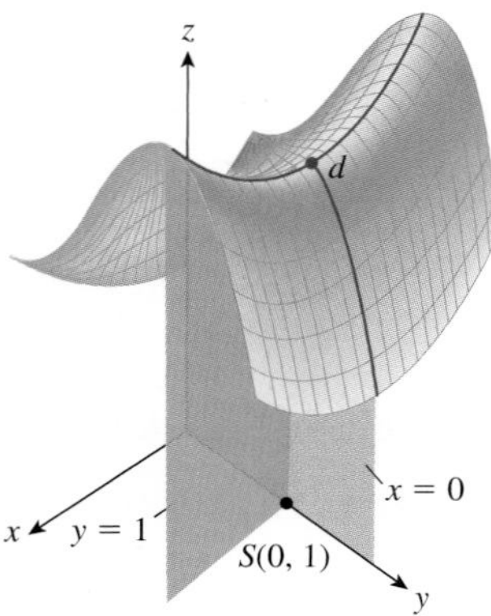

Figure 22

point, so d looks like a relative maximum along this slice. This kind of behavior characterizes a saddle point: f has a **saddle point** at (r, s) if f has a relative minimum at (r, s) along some slice through that point and a relative maximum along another slice through that point. If you look at the other saddle point, c, in Figure 21, you see the same characteristics.

While numerical information can help us locate the approximate positions of relative extrema and saddle points, calculus permits us to locate these points accurately, as we did for functions of a single variable. Look once again at Figure 21, and notice the following:

- The points P, Q, R, and S are all in the **interior** of the domain of f; that is, none lie on the boundary of the domain. Said another way, we can move some distance in any direction from any of these points without leaving the domain of f.
- The tangent lines along the slices through these points parallel to the x- and y-axes are *horizontal*. Thus, the partial derivatives $\partial f/\partial x$ and $\partial f/\partial y$ are zero when evaluated at any of the points P, Q, R, and S. This gives us a way of locating candidates for relative extrema and saddle points.

The following summary generalizes and also expands on some of what we have just said:

Relative and Absolute Maxima and Minima

The function f of n variables has a **relative maximum** at $(x_1, x_2, \ldots, x_n) = (r_1, r_2, \ldots, r_n)$ if $f(r_1, r_2, \ldots, r_n) \geq f(x_1, x_2, \ldots, x_n)$ for every point $(x_1, x_2, \ldots, x_n)$ near* $(r_1, r_2, \ldots, r_n)$ in the domain of f. We say that f has an **absolute maximum** at $(r_1, r_2, \ldots, r_n)$ if $f(r_1, r_2, \ldots, r_n) \geq f(x_1, x_2, \ldots, x_n)$ for every point $(x_1, x_2, \ldots, x_n)$ in the domain of f. The terms **relative minimum** and **absolute minimum** are defined in a similar way. Note that, as with functions of a single variable, absolute extrema are special kinds of relative extrema.

* For $(x_1, x_2, \ldots, x_n)$ to be near $(r_1, r_2, \ldots, r_n)$ we mean that x_1 is in some open interval centered at r_1, x_2 is in some open interval centered at r_2, and so on.

Locating Candidates for Extrema and Saddle Points in the Interior of the Domain of *f*

- Set $\dfrac{\partial f}{\partial x_1} = 0, \dfrac{\partial f}{\partial x_2} = 0, \ldots, \dfrac{\partial f}{\partial x_n} = 0$, simultaneously, and solve for $x_1, x_2, \ldots, x_n$.
- Check that the resulting points $(x_1, x_2, \ldots, x_n)$ are in the interior of the domain of f.

Points at which all the partial derivatives of f are zero are called **critical points**. The critical points are the only candidates for extrema and saddle points in the interior of the domain of f, assuming that its partial derivatives are defined at every point.†

† One can use the techniques of the next section to find extrema on the *boundary* of the domain of a function; for a complete discussion, see the optional extra section: *Maxima and Minima: Boundaries and the Extreme Value Theorem.* (We shall not consider the analogs of the singular points.)

Quick Examples

In each of the following Quick Examples, the domain is the whole Cartesian plane, and the partial derivatives are defined at every point, so the critical points give us the only candidates for extrema and saddle points:

1. Let $f(x, y) = x^3 + (y - 1)^2$. Then $\dfrac{\partial f}{\partial x} = 3x^2$ and $\dfrac{\partial f}{\partial y} = 2(y - 1)$.

Thus, we solve the system

$$3x^2 = 0 \quad \text{and} \quad 2(y - 1) = 0.$$

The first equation gives $x = 0$, and the second gives $y = 1$. Thus, the only critical point is (0, 1).

2. Let $f(x, y) = x^2 - 4xy + 8y$. Then $\dfrac{\partial f}{\partial x} = 2x - 4y$ and $\dfrac{\partial f}{\partial y} = -4x + 8$. Thus, we solve

$$2x - 4y = 0 \quad \text{and} \quad -4x + 8 = 0.$$

The second equation gives $x = 2$, and the first then gives $y = 1$. Thus, the only critical point is (2, 1).

3. Let $f(x, y) = e^{-(x^2+y^2)}$. Taking partial derivatives and setting them equal to zero gives

$$-2xe^{-(x^2+y^2)} = 0 \qquad \text{We set } \frac{\partial f}{\partial x} = 0.$$

$$-2ye^{-(x^2+y^2)} = 0. \qquad \text{We set } \frac{\partial f}{\partial y} = 0.$$

The first equation implies that $x = 0$,* and the second implies that $y = 0$. Thus, the only critical point is (0, 0).

* Recall that if a product of two numbers is zero, then one or the other must be zero. In this case the number $e^{-(x^2+y^2)}$ can't be zero (because e^u is never zero), which gives the result claimed.

In the remainder of this section we will be interested in locating all critical points of a given function and then classifying each one as a relative maximum, minimum, saddle point, or none of these. Whether or not any relative extrema we find are in fact absolute is a subject we discuss in the next section.†

† In some of the applications in the exercises you will, however, need to consider whether the extrema you find are absolute.

EXAMPLE 1 Locating and Classifying Critical Points

Locate all critical points of $f(x, y) = x^2y - x^2 - 2y^2$. Graph the function to classify the critical points as relative maxima, minima, saddle points, or none of these.

Solution The partial derivatives are

$$f_x = 2xy - 2x = 2x(y - 1)$$
$$f_y = x^2 - 4y.$$

Setting these equal to zero gives

$$x = 0 \text{ or } y = 1$$
$$x^2 = 4y.$$

We get a solution by choosing either $x = 0$ or $y = 1$ and substituting into $x^2 = 4y$.

***Case 1: x* = 0** Substituting into $x^2 = 4y$ gives $0 = 4y$ and hence $y = 0$. Thus, the critical point for this case is $(x, y) = (0, 0)$.

***Case 2: y* = 1** Substituting into $x^2 = 4y$ gives $x^2 = 4$ and hence $x = \pm 2$. Thus, we get two critical points for this case: (2, 1) and (−2, 1).

We now have three critical points altogether: (0, 0), (2, 1), and (−2, 1). Because the domain of f is the whole Cartesian plane and the partial derivatives are defined at every point, these critical points are the only candidates for relative extrema and saddle points. We get the corresponding points on the graph by substituting for x and y in the equation for f to get the z-coordinates. The points are (0, 0, 0), (2, 1, −2), and (−2, 1, −2).

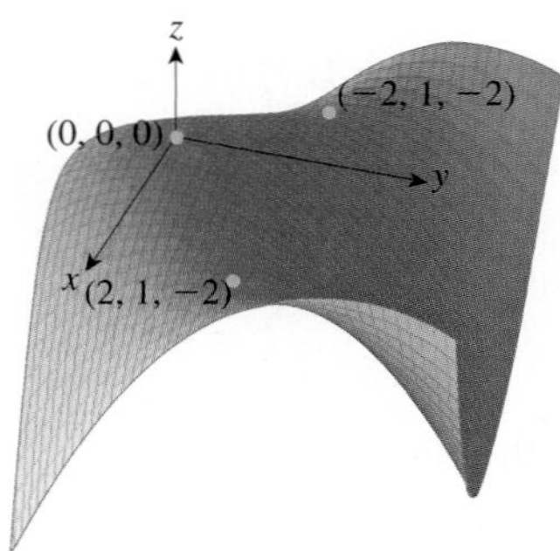

Figure 23

T ***Classifying the Critical Points Graphically*** To classify the critical points graphically, we look at the graph of f shown in Figure 23.

Examining the graph carefully, we see that the point (0, 0, 0) is a relative maximum. As for the other two critical points, are they saddle points or are they relative maxima? They are relative maxima along the y-direction, but they are relative minima along the lines $y = \pm x$ (see the top edge of the picture, which shows a dip at (−2, 1, 2)) and so they are saddle points. If you don't believe this, we will get more evidence following and in a later example.

T ***Classifying the Critical Points Numerically*** We can use a tabular representation of the function to classify the critical points numerically. The following tabular representation of the function can be obtained using a spreadsheet. (See the Spreadsheet Technology Guide discussion of Section 8.1 Example 3 at the end of the chapter for information on using a spreadsheet to generate such a table.)

$x \rightarrow$, $y \downarrow$

	−3	**−2**	**−1**	**0**	**1**	**2**	**3**
−3	−54	−34	−22	−18	−22	−34	−54
−2	−35	−20	−11	−8	−11	−20	−35
−1	−20	−10	−4	−2	−4	−10	−20
0	−9	−4	−1	0	−1	−4	−9
1	−2	−2	−2	−2	−2	−2	−2
2	1	−4	−7	−8	−7	−4	1
3	0	−10	−16	−18	−16	−10	0

The shaded and colored cells show rectangular neighborhoods of the three critical points (0, 0), (2, 1), and (−2, 1). (Notice that they overlap.) The values of f at the critical points are at the centers of these rectangles. Looking at the gray neighborhood of $(x, y) = (0, 0)$, we see that $f(0, 0) = 0$ is the largest value of f in the shaded cells, suggesting that f has a maximum at (0, 0). The shaded neighborhood of (2, 1) on the right shows $f(2, 1) = -2$ as the maximum along some slices (e.g., the vertical slice), and a minimum along the diagonal slice from top left to bottom right. This is what results in a saddle point on the graph. The point (−2, 1) is similar, and thus f also has a saddle point at (−2, 1).

Q: *Is there an algebraic way of deciding whether a given point is a relative maximum, relative minimum, or saddle point?*

A: There is a "second derivative test" for functions of two variables, stated as follows.

Second Derivative Test for Functions of Two Variables

Suppose (a, b) is a critical point in the interior of the domain of the function f of two variables. Let H be the quantity

$$H = f_{xx}(a, b) f_{yy}(a, b) - [f_{xy}(a, b)]^2.$$ *H is called the Hessian.*

Then, if H is *positive*,

- f has a relative minimum at (a, b) if $f_{xx}(a, b) > 0$.
- f has a relative maximum at (a, b) if $f_{xx}(a, b) < 0$.

If H is *negative*,

- f has a saddle point at (a, b).

If $H = 0$, the test tells us nothing, so we need to look at the graph or a numerical table to see what is going on.

Quick Examples

1. Let $f(x, y) = x^2 - y^2$. Then

$$f_x = 2x \quad \text{and} \quad f_y = -2y,$$

which gives $(0, 0)$ as the only critical point. Also,

$$f_{xx} = 2,\ f_{xy} = 0, \quad \text{and} \quad f_{yy} = -2,$$ Note that these are constant.

which gives $H = (2)(-2) - 0^2 = -4$. Because H is negative, we have a saddle point at $(0, 0)$.

2. Let $f(x, y) = x^2 + 2y^2 + 2xy + 4x$. Then

$$f_x = 2x + 2y + 4 \quad \text{and} \quad f_y = 2x + 4y.$$

Setting these equal to zero gives a system of two linear equations in two unknowns:

$$x + y = -2$$
$$x + 2y = 0.$$

This system has solution $(-4, 2)$, so this is our only critical point. The second partial derivatives are $f_{xx} = 2$, $f_{xy} = 2$, and $f_{yy} = 4$, so $H = (2)(4) - 2^2 = 4$. Because $H > 0$ and $f_{xx} > 0$, we have a relative minimum at $(-4, 2)$.

Note There is a second derivative test for functions of three or more variables, but it is considerably more complicated. We stick with functions of two variables for the most part in this book. The justification of the second derivative test is beyond the scope of this book. ■

EXAMPLE 2 Using the Second Derivative Test

Use the second derivative test to analyze the function $f(x, y) = x^2y - x^2 - 2y^2$ discussed in Example 1, and confirm the results we got there.

Solution We saw in Example 1 that the first-order derivatives are

$$f_x = 2xy - 2x = 2x(y - 1)$$
$$f_y = x^2 - 4y$$

and the critical points are $(0, 0)$, $(2, 1)$, and $(-2, 1)$. We also need the second derivatives:

$$f_{xx} = 2y - 2$$
$$f_{xy} = 2x$$
$$f_{yy} = -4.$$

The point (0, 0): $f_{xx}(0, 0) = -2$, $f_{xy}(0, 0) = 0$, $f_{yy}(0, 0) = -4$, so $H = 8$. Because $H > 0$ and $f_{xx}(0, 0) < 0$, the second derivative test tells us that f has a relative maximum at (0, 0).

The point (2, 1): $f_{xx}(2, 1) = 0$, $f_{xy}(2, 1) = 4$ and $f_{yy}(2, 1) = -4$, so $H = -16$. Because $H < 0$, we know that f has a saddle point at (2, 1).

The point (−2, 1): $f_{xx}(-2, 1) = 0$, $f_{xy}(-2, 1) = -4$ and $f_{yy}(-2, 1) = -4$, so once again $H = -16$, and f has a saddle point at (−2, 1).

Deriving the Formulas for Linear Regression

Back in Section 1.4, we presented the following set of formulas for the **regression** or **best-fit** line associated with a given set of data points $(x_1, y_1), (x_2, y_2), \ldots, (x_n, y_n)$.

Regression Line

The line that best fits the n data points $(x_1, y_1), (x_2, y_2), \ldots, (x_n, y_n)$ has the form

$$y = mx + b,$$

where

$$m = \frac{n(\sum xy) - (\sum x)(\sum y)}{n(\sum x^2) - (\sum x)^2}$$

$$b = \frac{\sum y - m(\sum x)}{n}$$

n = number of data points.

Derivation of the Regression Line Formulas

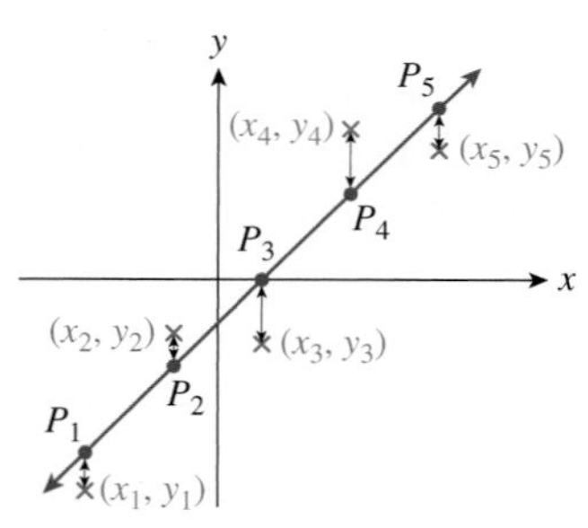

Figure 24

Recall that the regression line is defined to be the line that minimizes the sum of the squares of the **residuals**, measured by the vertical distances shown in Figure 24, which shows a regression line associated with $n = 5$ data points. In the figure, the points $P_1, \ldots, P_n$ on the regression line have coordinates $(x_1, mx_1 + b), (x_2, mx_2 + b), \ldots, (x_n, mx_n + b)$. The residuals are the quantities $y_{\text{Observed}} - y_{\text{Predicted}}$:

$$y_1 - (mx_1 + b), y_2 - (mx_2 + b), \ldots, y_n - (mx_n + b).$$

The sum of the squares of the residuals is therefore

$$S(m, b) = [y_1 - (mx_1 + b)]^2 + [y_2 - (mx_2 + b)]^2 + \cdots + [y_n - (mx_n + b)]^2$$

and this is the quantity we must minimize by choosing m and b. Because we reason that there is a line that minimizes this quantity, there must be a relative minimum at that point. We shall see in a moment that the function S has at most one critical point, which must therefore be the desired absolute minimum. To obtain the critical points of S, we set the partial derivatives equal to zero and solve:

$$S_m = 0: \quad -2x_1[y_1 - (mx_1 + b)] - \cdots - 2x_n[y_n - (mx_n + b)] = 0$$

$$S_b = 0: \quad -2[y_1 - (mx_1 + b)] - \cdots - 2[y_n - (mx_n + b)] = 0.$$

Dividing by −2 and gathering terms allows us to rewrite the equations as

$$\begin{aligned} m(x_1^2+\cdots+x_n^2)+b(x_1+\cdots+x_n) &= x_1y_1+\cdots+x_ny_n \\ m(x_1+\cdots+x_n)+nb &= y_1+\cdots+y_n. \end{aligned}$$

We can rewrite these equations more neatly using $\sum$-notation:

$$\begin{aligned} m\left(\sum x^2\right)+b\left(\sum x\right) &= \sum xy \\ m\left(\sum x\right)+nb &= \sum y. \end{aligned}$$

This is a system of two linear equations in the two unknowns m and b. It may or may not have a unique solution. When there is a unique solution, we can conclude that the best-fit line is given by solving these two equations for m and b. Alternatively, there is a general formula for the solution of any system of two equations in two unknowns, and if we apply this formula to our two equations, we get the regression formulas above.

8.3 EXERCISES

Access end-of-section exercises online at **www.webassign.net** ENHANCED WebAssign

8.4 Constrained Maxima and Minima and Applications

So far we have looked only at the relative extrema of functions with no constraints. However, in Section 5.2 we saw examples in which we needed to find the maximum or minimum of an objective function subject to one or more constraints on the independent variables. For instance, consider the following problem:

$$\text{Minimize } S = xy + 2xz + 2yz \quad \text{subject to } xyz = 4 \text{ with } x > 0,\ y > 0,\ z > 0.$$

One strategy for solving such problems is essentially the same as the strategy we used earlier: Solve the constraint equation for one of the variables, substitute into the objective function, and then optimize the resulting function using the methods of the preceding section. We will call this the *substitution method.** An alternative method, called the *method of Lagrange multipliers*, is useful when it is difficult or impossible to solve the constraint equation for one of the variables, and even when it is possible to do so.

* Although often the method of choice, the substitution method is not infallible (see Exercises 19 and 20 associated with this section).

Substitution Method

EXAMPLE 1 Using Substitution

Minimize $S = xy + 2xz + 2yz$ subject to $xyz = 4$ with $x > 0$, $y > 0$, $z > 0$.

Solution As suggested in the above discussion, we proceed as follows:

Solve the constraint equation for one of the variables and then substitute in the objective function. The constraint equation is $xyz = 4$. Solving for z gives

$$z = \frac{4}{xy}.$$

The objective function is $S = xy + 2xz + 2yz$, so substituting $z = 4/xy$ gives

$$\begin{aligned} S &= xy + 2x\frac{4}{xy} + 2y\frac{4}{xy} \\ &= xy + \frac{8}{y} + \frac{8}{x}. \end{aligned}$$

Minimize the resulting function of two variables. We use the method in Section 8.4 to find the minimum of $S = xy + \frac{8}{y} + \frac{8}{x}$ for $x > 0$ and $y > 0$. We look for critical points:

$$S_x = y - \frac{8}{x^2} \qquad S_y = x - \frac{8}{y^2}$$

$$S_{xx} = \frac{16}{x^3} \qquad S_{xy} = 1 \qquad S_{yy} = \frac{16}{y^3}.$$

We now equate the first partial derivatives to zero:

$$y = \frac{8}{x^2} \quad \text{and} \quad x = \frac{8}{y^2}.$$

To solve for x and y, we substitute the first of these equations in the second, getting

$$\begin{aligned} x &= \frac{x^4}{8} \\ x^4 - 8x &= 0 \\ x(x^3 - 8) &= 0. \end{aligned}$$

The two solutions are $x = 0$, which we reject because x cannot be zero, and $x = 2$. Substituting $x = 2$ in $y = 8/x^2$ gives $y = 2$ also. Thus, the only critical point is $(2, 2)$. To apply the second derivative test, we compute

$$S_{xx}(2, 2) = 2 \qquad S_{xy}(2, 2) = 1 \qquad S_{yy}(2, 2) = 2$$

and find that $H = 3 > 0$ and $S_{xx}(2, 2) > 0$, so we have a relative minimum at $(2, 2)$.

The corresponding value of z is given by the constraint equation:

$$z = \frac{4}{xy} = \frac{4}{4} = 1.$$

The corresponding value of the objective function is

$$S = xy + \frac{8}{y} + \frac{8}{x} = 4 + \frac{8}{2} + \frac{8}{2} = 12.$$

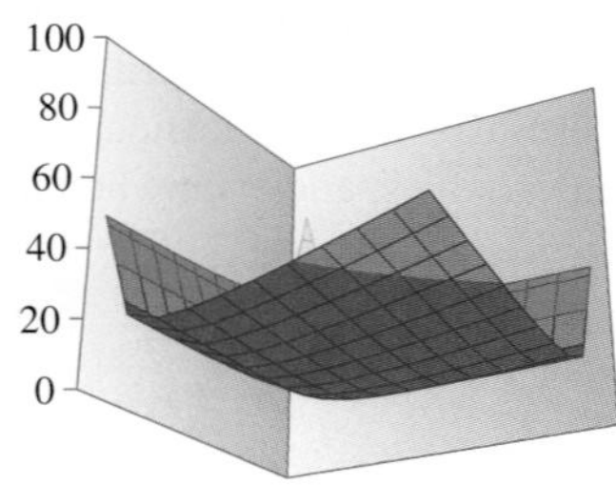

Graph of $S = xy + \frac{8}{y} + \frac{8}{x}$
$(0.2 \le x \le 5, 0.2 \le y \le 5)$

Figure 25

Figure 25 shows a portion of the graph of $S = xy + \frac{8}{y} + \frac{8}{x}$ for positive x and y (drawn using the Excel Surface Grapher in the Chapter 8 utilities at the Website) and suggests that there is a single absolute minimum, which must be at our only candidate point $(2, 2)$.

We conclude that the minimum of S is 12 and occurs at $(2, 2, 1)$.

The Method of Lagrange Multipliers

As we mentioned above, the method of Lagrange multipliers has the advantage that it can be used in constrained optimization problems when it is difficult or impossible to solve a constraint equation for one of the variables. We restrict attention to the case of a single constraint equation, although the method also generalizes to any number of constraint equations.

Locating Relative Extrema Using the Method of Lagrange Multipliers

To locate the candidates for relative extrema of a function $f(x, y, \ldots)$ subject to the constraint $g(x, y, \ldots) = 0$:

1. Construct the **Lagrangian function**

$$L(x, y, \ldots) = f(x, y, \ldots) - \lambda g(x, y, \ldots)$$

   where λ is a new unknown called a **Lagrange multiplier.**

2. The candidates for the relative extrema occur at the critical points of $L(x, y, \ldots)$. To find them, set all the partial derivatives of $L(x, y, \ldots)$ equal to zero and solve the resulting system, together with the constraint equation $g(x, y, \ldots) = 0$, for the unknowns $x, y, \ldots$ and λ.

The points $(x, y, \ldots)$ that occur in solutions are then the candidates for the relative extrema of f subject to $g = 0$.

Although the justification for the method of Lagrange multipliers is beyond the scope of this text (a derivation can be found in many vector calculus textbooks), we will demonstrate by example how it is used.

EXAMPLE 2 Using Lagrange Multipliers

Use the method of Lagrange multipliers to find the maximum value of $f(x, y) = 2xy$ subject to $x^2 + 4y^2 = 32$.

Solution We start by rewriting the problem with the constraint in the form $g(x, y) = 0$:

$$\text{Maximize } f(x, y) = 2xy \text{ subject to } x^2 + 4y^2 - 32 = 0.$$

Here, $g(x, y) = x^2 + 4y^2 - 32$, and the Lagrangian function is

$$\begin{aligned} L(x, y) &= f(x, y) - \lambda g(x, y) \\ &= 2xy - \lambda(x^2 + 4y^2 - 32). \end{aligned}$$

The system of equations we need to solve is thus

$$\begin{aligned} L_x = 0: &\quad 2y - 2\lambda x = 0 \\ L_y = 0: &\quad 2x - 8\lambda y = 0 \\ g = 0: &\quad x^2 + 4y^2 - 32 = 0. \end{aligned}$$

It is often convenient to solve such a system by first solving one of the equations for λ and then substituting in the remaining equations. Thus, we start by solving the first equation to obtain

$$\lambda = \frac{y}{x}.$$

(A word of caution: Because we divided by x, we made the implicit assumption that $x \neq 0$, so before continuing we should check what happens if $x = 0$. But if $x = 0$, then the first equation, $2y = 2\lambda x$, tells us that $y = 0$ as well, and this contradicts the third equation: $x^2 + 4y^2 - 32 = 0$. Thus, we can rule out the possibility that $x = 0$.) Substituting the value of λ in the second equation gives

$$2x - 8\left(\frac{y}{x}\right)y = 0 \quad \text{or} \quad x^2 = 4y^2.$$

We can now substitute $x^2 = 4y^2$ in the constraint equation, obtaining

$$\begin{aligned} 4y^2 + 4y^2 - 32 &= 0 \\ 8y^2 &= 32 \\ y &= \pm 2. \end{aligned}$$

We now substitute back to obtain

$$x^2 = 4y^2 = 16,$$

or

$$x = \pm 4.$$

We don't need the value of λ, so we won't solve for it. Thus, the candidates for relative extrema are given by $x = \pm 4$ and $y = \pm 2$; that is, the four points $(-4, -2)$, $(-4, 2)$, $(4, -2)$, and $(4, 2)$. Recall that we are seeking the values of x and y that give the maximum value for $f(x, y) = 2xy$. Because we now have only four points to choose from, we compare the values of f at these four points and conclude that the maximum value of f occurs when $(x, y) = (-4, -2)$ or $(4, 2)$.

Something is suspicious in Example 2. We didn't check to see whether these candidates were relative extrema to begin with, let alone absolute extrema! How do we justify this omission? One of the difficulties with using the method of Lagrange multipliers is that it does not provide us with a test analogous to the second derivative test for functions of several variables. However, if you grant that the function in question does have an absolute maximum, then we require no test, because one of the candidates must give this maximum.

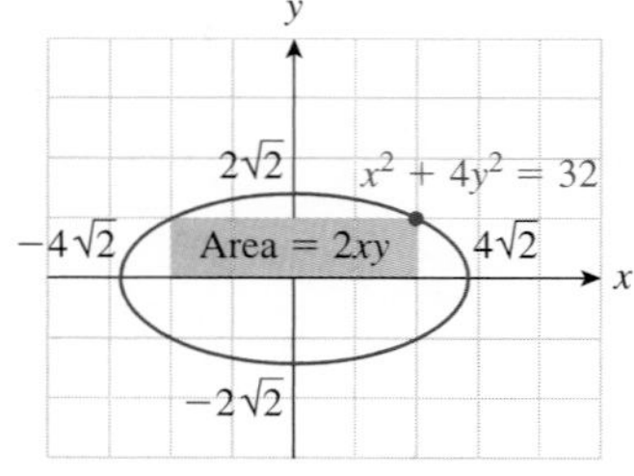

Figure 26

Q: *But how do we know that the given function has an absolute maximum?*

A: The best way to see this is by giving a geometric interpretation. The constraint $x^2 + 4y^2 = 32$ tells us that the point (x, y) must lie on the ellipse shown in Figure 26. The function $f(x, y) = 2xy$ gives the area of the rectangle shaded in the figure. There must be a *largest* such rectangle, because the area varies continuously from 0 when (x, y) is on the x-axis, to positive when (x, y) is in the first quadrant, to 0 again when (x, y) is on the y-axis, so f must have an absolute maximum for at least one pair of coordinates (x, y).

We now show how to use Lagrange multipliers to solve the minimization problem in Example 1:

EXAMPLE 3 Using Lagrange Multipliers: Function of Three Variables

Use the method of Lagrange multipliers to find the minimum value of $S = xy + 2xz + 2yz$ subject to $xyz = 4$ with $x > 0, y > 0, z > 0$.

Solution We start by rewriting the problem in standard form:

$$\text{Maximize } f(x, y, z) = xy + 2xz + 2yz$$
$$\text{subject to } xyz - 4 = 0 \text{ (with } x > 0, y > 0, z > 0).$$

Here, $g(x, y, z) = xyz - 4$, and the Lagrangian function is

$$\begin{aligned} L(x, y, z) &= f(x, y, z) - \lambda g(x, y, z) \\ &= xy + 2xz + 2yz - \lambda(xyz - 4). \end{aligned}$$

The system of equations we need to solve is thus

$$\begin{aligned} L_x = 0: &\quad y + 2z - \lambda yz = 0 \\ L_y = 0: &\quad x + 2z - \lambda xz = 0 \\ L_z = 0: &\quad 2x + 2y - \lambda xy = 0 \\ g = 0: &\quad xyz - 4 = 0. \end{aligned}$$

As in the last example, we solve one of the equations for λ and substitute in the others. The first equation gives

$$\lambda = \frac{1}{z} + \frac{2}{y}.$$

Substituting this into the second equation gives

$$x + 2z = x + \frac{2xz}{y}$$

or $\quad 2 = \dfrac{2x}{y},$ Subtract x from both sides and then divide by z.

giving $\quad y = x.$

Substituting the expression for λ into the third equation gives

$$2x + 2y = \frac{xy}{z} + 2x$$

or $\quad 2 = \dfrac{x}{z},$ Subtract $2x$ from both sides and then divide by y.

giving $\quad z = \dfrac{x}{2}.$

Now we have both y and z in terms of x. We substitute these values in the last (constraint) equation:

$$\begin{aligned} x(x)\left(\frac{x}{2}\right) - 4 &= 0 \\ x^3 &= 8 \\ x &= 2. \end{aligned}$$

Thus, $y = x = 2$, and $z = \frac{x}{2} = 1$. Therefore, the only critical point occurs at (2, 2, 1), as we found in Example 1, and the corresponding value of S is

$$S = xy + 2xz + 2yz = (2)(2) + 2(2)(1) + 2(2)(1) = 12.$$

➡ **Before we go on...** Again, the method of Lagrange multipliers does not tell us whether the critical point in Example 3 is a maximum, minimum, or neither. However, if you grant that the function in question does have an absolute minimum, then the values we found must give this minimum value. ■

APPLICATIONS

EXAMPLE 4 Minimizing Area

Find the dimensions of an open-top rectangular box that has a volume of 4 cubic feet and the smallest possible surface area.

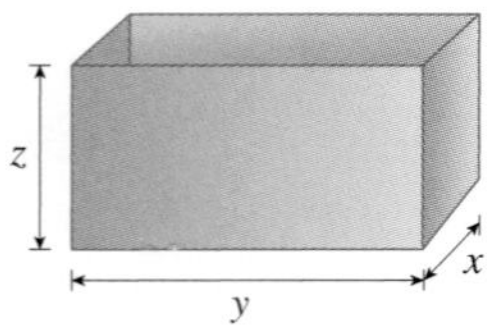

Figure 27

Solution Our first task is to rephrase this request as a mathematical optimization problem. Figure 27 shows a picture of the box with dimensions x, y, and z. We want to minimize the total surface area, which is given by

$$S = xy + 2xz + 2yz. \qquad \text{Base + Sides + Front and Back}$$

This is our objective function. We can't simply choose x, y, and z to all be zero, however, because the enclosed volume must be 4 cubic feet. So,

$$xyz = 4. \qquad \text{Constraint}$$

This is our constraint equation. Other unstated constraints are $x > 0$, $y > 0$, and $z > 0$, because the dimensions of the box must be positive. We now restate the problem as follows:

$$\text{Minimize } S = xy + 2xz + 2yz \quad \text{subject to } xyz = 4,\ x > 0,\ y > 0,\ z > 0.$$

But this is exactly the problem in Examples 1 and 3, and has a solution $x = 2$, $y = 2$, $z = 1$, $S = 12$. Thus, the required dimensions of the box are

$$x = 2 \text{ ft},\ y = 2 \text{ ft},\ z = 1 \text{ ft},$$

requiring a total surface area of 12 ft^2.

Q: *In Example 1 we checked that we had a relative minimum at $(x, y) = (2, 2)$ and we were persuaded graphically that this was probably an absolute minimum. Can we be sure that this relative minimum is an absolute minimum?*

A: Yes. There must be a least surface area among all boxes that hold 4 cubic feet. (Why?) Because this would give a relative minimum of S and because the only possible relative minimum of S occurs at (2, 2), this is the absolute minimum.

EXAMPLE 5 Maximizing productivity

An electric motor manufacturer uses workers and robots on its assembly line and has a Cobb-Douglas productivity function* of the form

$$P(x, y) = 10x^{0.2}y^{0.8} \text{ motors manufactured per day,}$$

where x is the number of assembly-line workers and y is the number of robots. Daily operating costs amount to \$100 per worker and \$16 per robot. How many workers and robots should be used to maximize productivity if the manufacturer has a daily budget of \$4,000?

* Cobb-Douglas production formulas were discussed in Section 4.6.

Solution Our objective function is the productivity $P(x, y)$, and the constraint is

$$100x + 16y = 4{,}000.$$

So, the optimization problem is:

$$\text{Maximize } P(x, y) = 10x^{0.2}y^{0.8} \text{ subject to } 100x + 16y = 4{,}000 \ (x \geq 0, y \geq 0).$$

Here, $g(x, y) = 100x + 16y - 4{,}000$, and the Lagrangian function is

$$\begin{aligned} L(x, y) &= P(x, y) - \lambda g(x, y) \\ &= 10x^{0.2}y^{0.8} - \lambda(100x + 16y - 4{,}000). \end{aligned}$$

The system of equations we need to solve is thus

$$\begin{aligned} L_x &= 0: \quad && 2x^{-0.8}y^{0.8} - 100\lambda = 0 \\ L_y &= 0: && 8x^{0.2}y^{-0.2} - 16\lambda = 0 \\ g &= 0: && 100x + 16y = 4{,}000. \end{aligned}$$

We can rewrite the first two equations as:

$$2\left(\frac{y}{x}\right)^{0.8} = 100\lambda \qquad 8\left(\frac{x}{y}\right)^{0.2} = 16\lambda.$$

Dividing the first by the second to eliminate λ gives

$$\frac{1}{4}\left(\frac{y}{x}\right)^{0.8}\left(\frac{y}{x}\right)^{0.2} = \frac{100}{16}$$

that is,
$$\frac{1}{4}\frac{y}{x} = \frac{25}{4},$$

giving
$$y = 25x.$$

Substituting this result into the constraint equation gives

$$\begin{aligned} 100x + 16(25x) &= 4{,}000 \\ 500x &= 4{,}000 \end{aligned}$$

so
$$x = 8 \text{ workers}, \quad y = 25x = 200 \text{ robots}$$

for a productivity of

$$P(8{,}200) = 10(8)^{0.2}(200)^{0.8} \approx 1{,}051 \text{ motors manufactured per day.}$$

FAQ

When to Use Lagrange Multipliers

Q: *When can I use the method of Lagrange multipliers? When should I use it?*

A: We have discussed the method only when there is a single equality constraint. There is a generalization, which we have not discussed, that works when there are more equality constraints (we need to introduce one multiplier for each constraint). So, if you have a problem with more than one equality constraint, or with any inequality constraints, you must use the substitution method. On the other hand, if you have one equality constraint, and it would be difficult to solve it for one of the variables, then you should use Lagrange multipliers.

8.4 EXERCISES

Access end-of-section exercises online at **www.webassign.net**

8.5 Double Integrals and Applications

When discussing functions of one variable, we computed the area under a graph by integration. The analog for the graph of a function of two variables is the *volume V* under the graph, as in Figure 28. Think of the region R in the xy-plane as the "shadow" under the portion of the surface $z = f(x, y)$ shown.

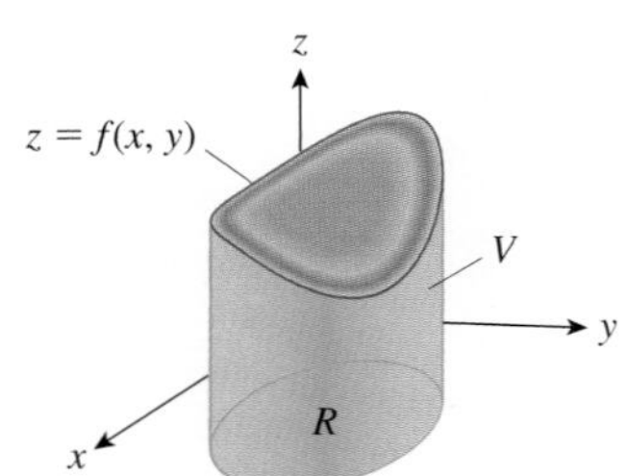

Figure 28

By analogy with the definite integral of a function of one variable, we make the following definition:

Geometric Definition of the Double Integral

The **double integral of $f(x, y)$ over the region R in the xy-plane** is defined as

$$\begin{aligned}&(\text{Volume } above \text{ the region } R \text{ and under the graph of } f)\\ &\quad - (\text{Volume } below \text{ the region } R \text{ and above the graph of } f).\end{aligned}$$

We denote the double integral of $f(x, y)$ over the region R by $\iint_R f(x, y)\,dx\,dy$.

Quick Example

Take $f(x, y) = 2$ and take R to be the rectangle $0 \le x \le 1, 0 \le y \le 1$. Then the graph of f is a flat horizontal surface $z = 2$, and

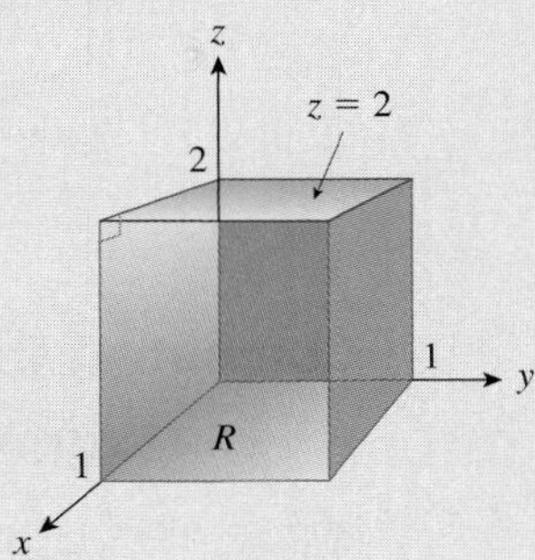

$$\iint_R f(x, y)\, dx\, dy = \text{Volume of box}$$
$$= \text{Width} \times \text{Length} \times \text{Height} = 1 \times 1 \times 2 = 2.$$

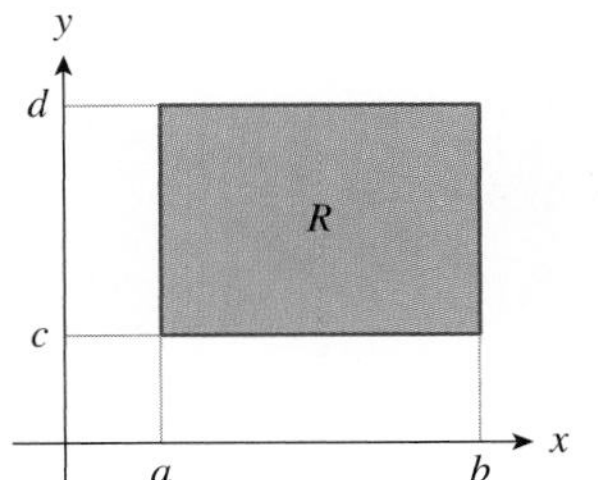

Figure 29

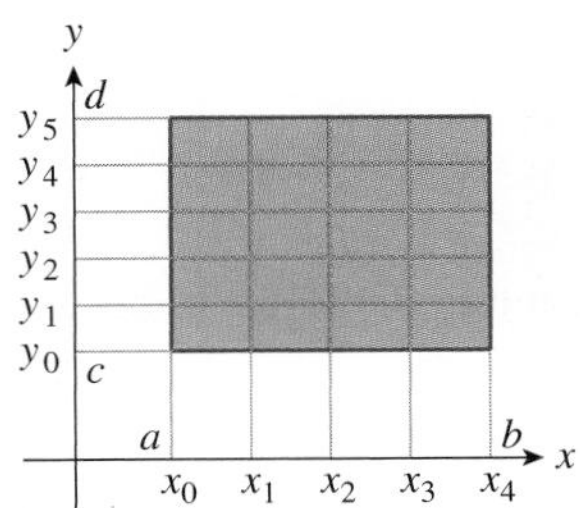

Figure 30

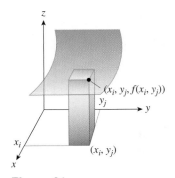

Figure 31

As we saw in the case of the definite integral of a function of one variable, we also desire *numerical* and *algebraic* definitions for two reasons: (1) to make the mathematical definition more precise, so as not to rely on the notion of "volume," and (2) for direct computation of the integral using technology or analytical tools.

We start with the simplest case, when the region R is a rectangle $a \le x \le b$ and $c \le y \le d$. (See Figure 29.) To compute the volume over R, we mimic what we did to find the area under the graph of a function of one variable. We break up the interval $[a, b]$ into m intervals all of width $\Delta x = (b - a)/m$, and we break up $[c, d]$ into n intervals all of width $\Delta y = (d - c)/n$. Figure 30 shows an example with $m = 4$ and $n = 5$.

This gives us mn rectangles defined by $x_{i-1} \le x \le x_i$ and $y_{j-1} \le y \le y_j$. Over one of these rectangles, f is approximately equal to its value at one corner—say $f(x_i, y_j)$. The volume under f over this small rectangle is then approximately the volume of the rectangular brick (size exaggerated) shown in Figure 31. This brick has height $f(x_i, y_j)$, and its base is Δx by Δy. Its volume is therefore $f(x_i, y_j)\Delta x\,\Delta y$. Adding together the volumes of all of the bricks over the small rectangles in R, we get

$$\iint_R f(x, y)\, dx\, dy \approx \sum_{j=1}^{n}\sum_{i=1}^{m} f(x_i, y_j)\Delta x\, \Delta y.$$

This double sum is called a **double Riemann sum**. We define the double integral to be the limit of the Riemann sums as m and n go to infinity.

Algebraic Definition of the Double Integral

$$\iint_R f(x, y)\, dx\, dy = \lim_{n\to\infty}\lim_{m\to\infty}\sum_{j=1}^{n}\sum_{i=1}^{m} f(x_i, y_j)\Delta x\, \Delta y$$

Note This definition is adequate (the limit exists) when f is continuous. More elaborate definitions are needed for general functions. ■

This definition also gives us a clue about how to compute a double integral. The innermost sum is $\sum_{i=1}^{m} f(x_i, y_j)\Delta x$, which is a Riemann sum for $\int_a^b f(x, y_j)\, dx$. The innermost limit is therefore

$$\lim_{m\to\infty} \sum_{i=1}^{m} f(x_i, y_j)\Delta x = \int_a^b f(x, y_j)\, dx.$$

The outermost limit is then also a Riemann sum, and we get the following way of calculating double integrals:

Computing the Double Integral over a Rectangle

If R is the rectangle $a \le x \le b$ and $c \le y \le d$, then

$$\iint_R f(x, y)\, dx\, dy = \int_c^d \left(\int_a^b f(x, y)\, dx\right) dy = \int_a^b \left(\int_c^d f(x, y)\, dy\right) dx.$$

The second formula comes from switching the order of summation in the double sum.

Quick Example

If R is the rectangle $1 \le x \le 2$ and $1 \le y \le 3$, then

$$\iint_R 1\, dx\, dy = \int_1^3 \left(\int_1^2 1\, dx\right) dy$$

$$= \int_1^3 [x]_{x=1}^2\, dy \qquad \text{Evaluate the inner integral.}$$

$$= \int_1^3 1\, dy \qquad [x]_{x=1}^2 = 2 - 1 = 1.$$

$$= [y]_{y=1}^3 = 3 - 1 = 2.$$

The Quick Example used a constant function for the integrand. Here is an example in which the integrand is not constant.

EXAMPLE 1 Double Integral over a Rectangle

Let R be the rectangle $0 \le x \le 1$ and $0 \le y \le 2$. Compute $\iint_R xy\, dx\, dy$. This integral gives the volume of the part of the boxed region under the surface $z = xy$ shown in Figure 32.

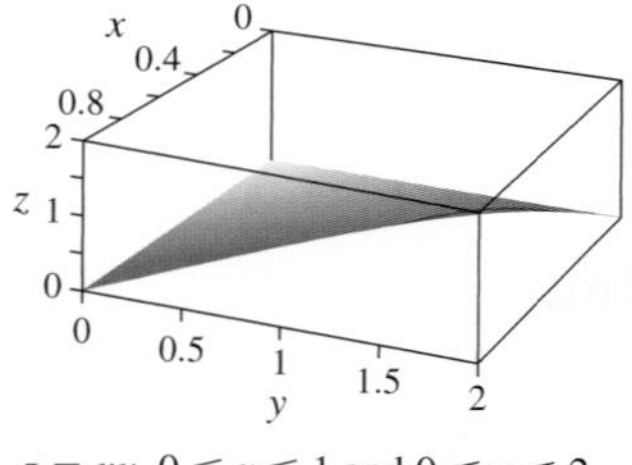

$z = xy$, $0 \le x \le 1$ and $0 \le y \le 2$

Figure 32

Solution

$$\iint_R xy\, dx\, dy = \int_0^2 \int_0^1 xy\, dx\, dy$$

(We usually drop the parentheses around the inner integral like this.) As in the Quick Example, we compute this **iterated integral** from the inside out. First we compute

$$\int_0^1 xy\, dx.$$

To do this computation, we do as we did when finding partial derivatives: We treat y as a constant. This gives

$$\int_0^1 xy\,dx = \left[\frac{x^2}{2}\cdot y\right]_{x=0}^1 = \frac{1}{2}y - 0 = \frac{y}{2}.$$

We can now calculate the outer integral.

$$\int_0^2\int_0^1 xy\,dx\,dy = \int_0^2 \frac{y}{2}\,dy = \left[\frac{y^2}{4}\right]_0^2 = 1$$

➡ **Before we go on...** We could also reverse the order of integration in Example 1:

$$\int_0^1\int_0^2 xy\,dy\,dx = \int_0^1 \left[x\cdot\frac{y^2}{2}\right]_{y=0}^2 = \int_0^1 2x\,dx = \left[x^2\right]_0^1 = 1.$$ ■

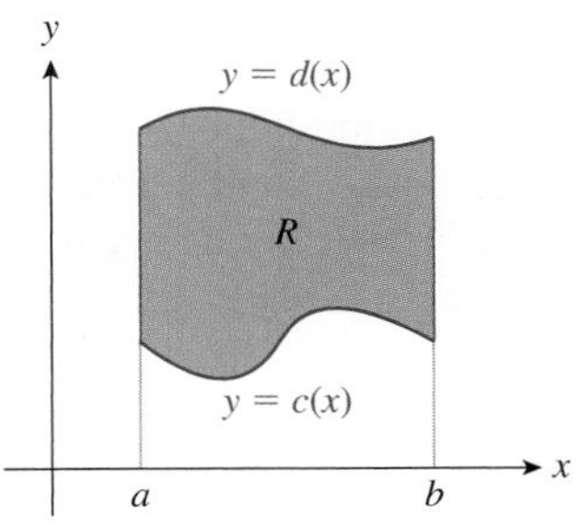

Figure 33

Often we need to integrate over regions R that are not rectangular. There are two cases that come up. The first is a region like the one shown in Figure 33. In this region, the bottom and top sides are defined by functions $y = c(x)$ and $y = d(x)$, respectively, so that the whole region can be described by the inequalities $a \le x \le b$ and $c(x) \le y \le d(x)$. To evaluate a double integral over such a region, we have the following formula:

Computing the Double Integral over a Nonrectangular Region

If R is the region $a \le x \le b$ and $c(x) \le y \le d(x)$ (Figure 33), then we integrate over R according to the following equation:

$$\iint_R f(x, y)\,dx\,dy = \int_a^b\int_{c(x)}^{d(x)} f(x, y)\,dy\,dx.$$

EXAMPLE 2 Double Integral over a Nonrectangular Region

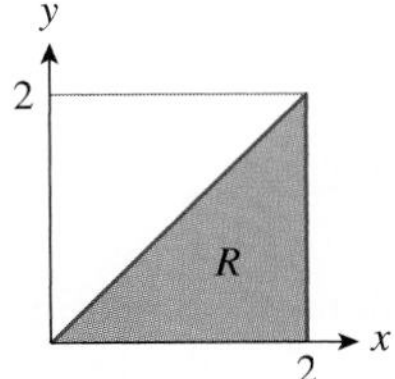

Figure 34

R is the triangle shown in Figure 34. Compute $\iint_R x\,dx\,dy$.

Solution R is the region described by $0 \le x \le 2$, $0 \le y \le x$. We have

$$\begin{aligned}\iint_R x\,dx\,dy &= \int_0^2\int_0^x x\,dy\,dx\\ &= \int_0^2 [xy]_{y=0}^x\,dx\\ &= \int_0^2 x^2\,dx\\ &= \left[\frac{x^3}{3}\right]_0^2 = \frac{8}{3}.\end{aligned}$$

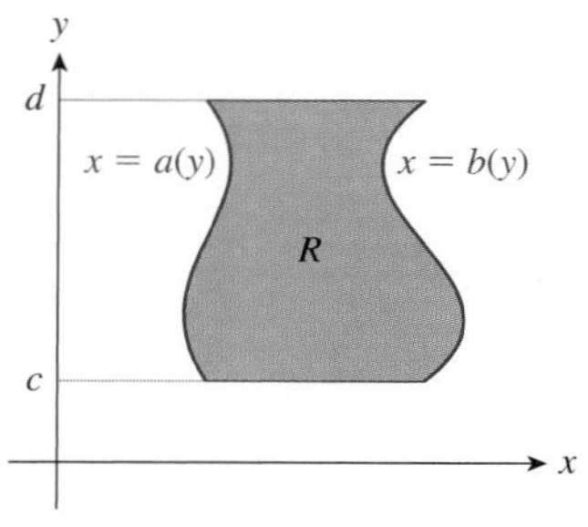

Figure 35

The second type of region is shown in Figure 35. This is the region described by $c \le y \le d$ and $a(y) \le x \le b(y)$. To evaluate a double integral over such a region, we have the following formula:

Double Integral over a Nonrectangular Region (continued)

If R is the region $c \le y \le d$ and $a(y) \le x \le b(y)$ (Figure 35), then we integrate over R according to the following equation:

$$\iint_R f(x, y)\, dx\, dy = \int_c^d \int_{a(y)}^{b(y)} f(x, y)\, dx\, dy.$$

EXAMPLE 3 Double Integral over a Nonrectangular Region

Redo Example 2, integrating in the opposite order.

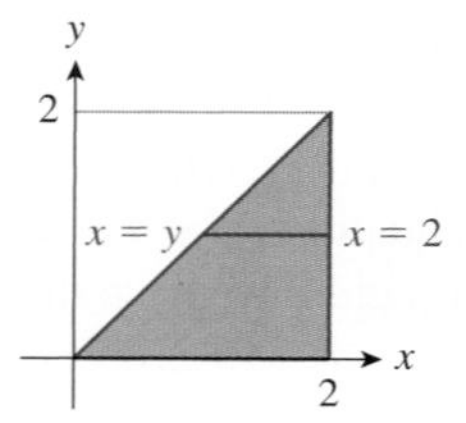

Figure 36

Solution We can integrate in the opposite order if we can describe the region in Figure 34 in the way shown in Figure 35. In fact, it is the region $0 \le y \le 2$ and $y \le x \le 2$. To see this, we draw a horizontal line through the region, as in Figure 36. The line extends from $x = y$ on the left to $x = 2$ on the right, so $y \le x \le 2$. The possible heights for such a line are $0 \le y \le 2$. We can now compute the integral:

$$\begin{aligned}
\iint_R x\, dx\, dy &= \int_0^2 \int_y^2 x\, dx\, dy \\
&= \int_0^2 \left[\frac{x^2}{2}\right]_{x=y}^{2} dy \\
&= \int_0^2 \left(2 - \frac{y^2}{2}\right) dy \\
&= \left[2y - \frac{y^3}{6}\right]_0^2 = \frac{8}{3}.
\end{aligned}$$

Note Many regions can be described in two different ways, as we saw in Examples 2 and 3. Sometimes one description will be much easier to work with than the other, so it pays to consider both. ■

APPLICATIONS

There are many applications of double integrals besides finding volumes. For example, we can use them to find *averages*. Remember that the average of $f(x)$ on $[a, b]$ is given by $\int_a^b f(x)\, dx$ divided by $(b - a)$, the length of the interval.

Average of a Function of Two Variables

The average of $f(x, y)$ on the region R is

$$\bar{f} = \frac{1}{A} \iint_R f(x, y)\, dx\, dy.$$

Here, A is the area of R. We can compute the area A geometrically, or by using the techniques from the chapter on applications of the integral, or by computing

$$A = \iint_R 1\, dx\, dy.$$

Quick Example

The average value of $f(x, y) = xy$ on the rectangle given by $0 \le x \le 1$ and $0 \le y \le 2$ is

$$\bar{f} = \frac{1}{2}\iint_R xy\, dx\, dy \qquad \text{The area of the rectangle is 2.}$$

$$= \frac{1}{2}\int_0^2\int_0^1 xy\, dx\, dy$$

$$= \frac{1}{2}\cdot 1 = \frac{1}{2}. \qquad \text{We calculated the integral in Example 1.}$$

EXAMPLE 4 Average Revenue

Your company is planning to price its new line of subcompact cars at between \$10,000 and \$15,000. The marketing department reports that if the company prices the cars at p dollars per car, the demand will be between $q = 20{,}000 - p$ and $q = 25{,}000 - p$ cars sold in the first year. What is the average of all the possible revenues your company could expect in the first year?

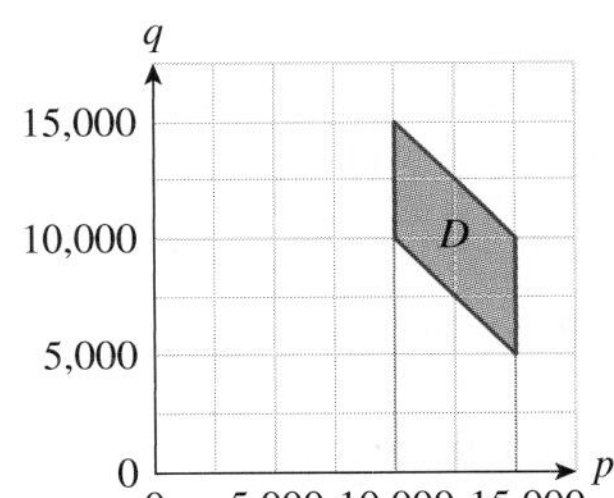

Figure 37

Solution Revenue is given by $R = pq$ as usual, and we are told that

$$10{,}000 \le p \le 15{,}000$$

and $$20{,}000 - p \le q \le 25{,}000 - p.$$

This domain D of prices and demands is shown in Figure 37.

To average the revenue R over the domain D, we need to compute the area A of D. Using either calculus or geometry, we get $A = 25{,}000{,}000$. We then need to integrate R over D:

$$\iint_D pq\, dp\, dq = \int_{10{,}000}^{15{,}000}\int_{20{,}000-p}^{25{,}000-p} pq\, dq\, dp$$

$$= \int_{10{,}000}^{15{,}000}\left[\frac{pq^2}{2}\right]_{q=20{,}000-p}^{25{,}000-p} dp$$

$$= \frac{1}{2}\int_{10{,}000}^{15{,}000}[p(25{,}000-p)^2 - p(20{,}000-p)^2]\, dp$$

$$= \frac{1}{2}\int_{10{,}000}^{15{,}000}[225{,}000{,}000p - 10{,}000p^2]\, dp$$

$$\approx 3{,}072{,}900{,}000{,}000{,}000.$$

The average of all the possible revenues your company could expect in the first year is therefore

$$\bar{R} = \frac{3{,}072{,}900{,}000{,}000{,}000}{25{,}000{,}000} \approx \$122{,}900{,}000.$$

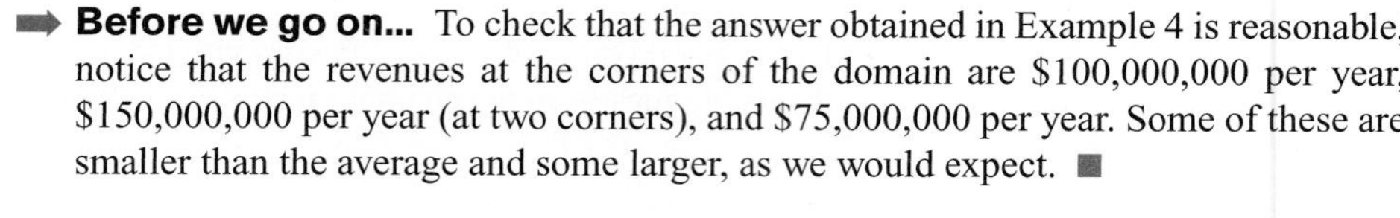

Before we go on... To check that the answer obtained in Example 4 is reasonable, notice that the revenues at the corners of the domain are \$100,000,000 per year, \$150,000,000 per year (at two corners), and \$75,000,000 per year. Some of these are smaller than the average and some larger, as we would expect. ■

Another useful application of the double integral comes about when we consider density. For example, suppose that $P(x, y)$ represents the population density (in people per square mile, say) in the city of Houston, shown in Figure 38.

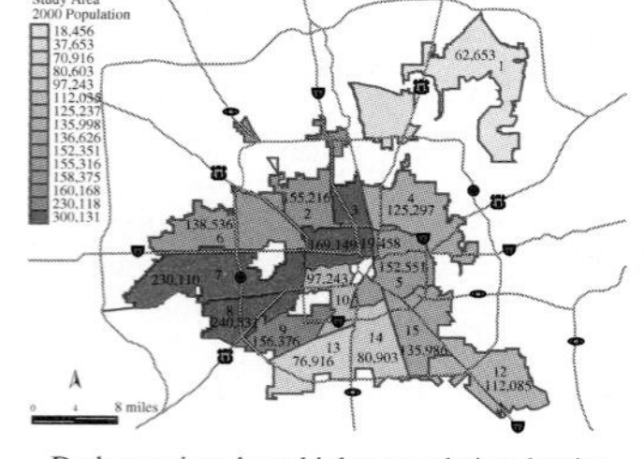

Darker regions have higher population density

Figure 38

If we break the city up into small rectangles (for example, city blocks), then the population in the small rectangle $x_{i-1} \le x \le x_i$ and $y_{j-1} \le y \le y_j$ is approximately $P(x_i, y_j)\Delta x \Delta y$. Adding up all of these population estimates, we get

$$\text{Total population} \approx \sum_{j=1}^{n}\sum_{i=1}^{m} P(x_i, y_j)\,\Delta x\,\Delta y.$$

Because this is a double Riemann sum, when we take the limit as m and n go to infinity, we get the following calculation of the population of the city:

$$\text{Total population} = \iint_{\text{City}} P(x, y)\,dx\,dy.$$

EXAMPLE 5 Population

Squaresville is a city in the shape of a square 5 miles on a side. The population density at a distance of x miles east and y miles north of the southwest corner is $P(x, y) = x^2 + y^2$ thousand people per square mile. Find the total population of Squaresville.

Squaresville

Figure 39

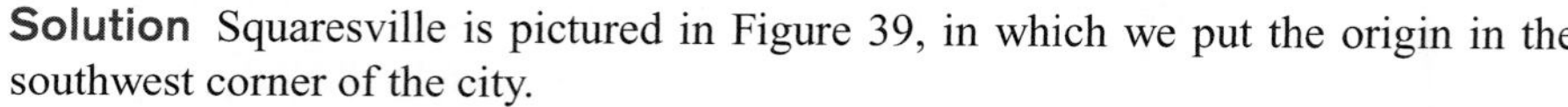

Solution Squaresville is pictured in Figure 39, in which we put the origin in the southwest corner of the city.

To compute the total population, we integrate the population density over the city S.

$$\begin{aligned}
\text{Total population} &= \iint_{\text{Squaresville}} P(x, y)\,dx\,dy \\
&= \int_0^5\int_0^5 (x^2 + y^2)\,dx\,dy \\
&= \int_0^5 \left[\frac{x^3}{3} + xy^2\right]_{x=0}^{5} dy \\
&= \int_0^5 \left[\frac{125}{3} + 5y^2\right] dy \\
&= \frac{1{,}250}{3} \approx 417 \text{ thousand people}
\end{aligned}$$

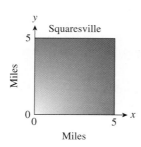

Before we go on... Note that the average population density is the total population divided by the area of the city, which is about 17,000 people per square mile. Compare this calculation with the calculations of averages in the previous two examples. ■

8.5 EXERCISES

Access end-of-section exercises online at **www.webassign.net** ENHANCED WebAssign

CHAPTER 8 REVIEW

KEY CONCEPTS

Website www.WanerMath.com

Go to the Website at www.WanerMath.com to find a comprehensive and interactive Web-based summary of Chapter 8.

8.1 Functions of Several Variables from the Numerical, Algebraic, and Graphical Viewpoints

8.2 Partial Derivatives

8.3 Maxima and Minima

8.4 Constrained Maxima and Minima and Applications

8.5 Double Integrals and Applications

REVIEW EXERCISES

1. Let $f(x, y, z) = \dfrac{x}{y + xz} + x^2 y$. Evaluate $f(0, 1, 1)$, $f(2, 1, 1)$, $f(-1, 1, -1)$, $f(z, z, z)$, and $f(x + h, y + k, z + l)$.

2. Let $g(x, y, z) = xy(x + y - z) + x^2$. Evaluate $g(0, 0, 0)$, $g(1, 0, 0)$, $g(0, 1, 0)$, $g(x, x, x)$, and $g(x, y + k, z)$.

3. Let $f(x, y, z) = 2.72 - 0.32x - 3.21y + 12.5z$. Complete the following: f___ by___ units for every 1 unit of increase in x, and ___ by ___ units for every unit of increase in z.

4. Let $g(x, y, z) = 2.16x + 11y - 1.53z + 31.4$. Complete the following: g___ by___ units for every 1 unit of increase in y, and ___ by ___ units for every unit of increase in z.

In Exercises 5–6 complete the given table for values for $h(x, y) = 2x^2 + xy - x$.

5.

$y \downarrow$ \ $x \rightarrow$	−1	0	1
−1			
0			
1			

6.

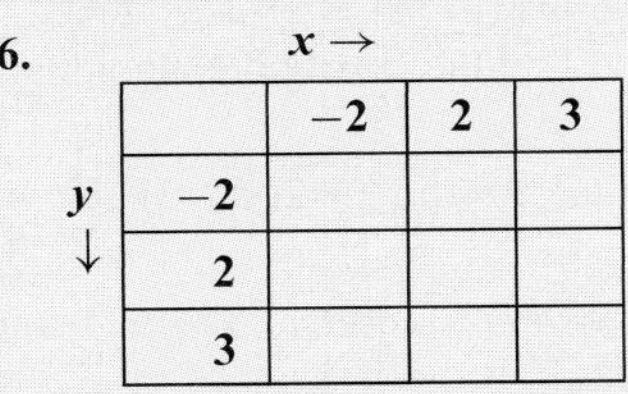

$y \downarrow$ \ $x \rightarrow$	−2	2	3
−2			
2			
3			

7. Give a formula for a (single) function f with the property that $f(x, y) = -f(y, x)$ and $f(1, -1) = 3$.

8. Let $f(x, y) = x^2 + (y + 1)^2$. Show that $f(y, x) = f(x + 1, y - 1)$.

Sketch the graphs of the functions in Exercises 9–14.

9. $r(x, y) = x + y$

10. $r(x, y) = x - y$

11. $t(x, y) = x^2 + 2y^2$. Show cross sections at $x = 0$ and $z = 1$.

12. $t(x, y) = \frac{1}{2}x^2 + y^2$. Show cross sections at $x = 0$ and $z = 1$.

13. $f(x, y) = -2\sqrt{x^2 + y^2}$. Show cross sections at $z = -4$ and $y = 1$.

14. $f(x, y) = 2 + 2\sqrt{x^2 + y^2}$. Show cross sections at $z = 4$ and $y = 1$.

In Exercises 15–20, compute the partial derivatives shown for the given function.

15. $f(x, y) = x^2 + xy$; find f_x, f_y, and f_{yy}

16. $f(x, y) = \frac{6}{xy} + \frac{xy}{6}$; find f_x, f_y, and f_{yy}

17. $f(x, y) = 4x + 5y - 6xy$; find $f_{xx}(1, 0) - f_{xx}(3, 2)$

18. $f(x, y) = e^{xy} + e^{3x^2 - y^2}$; find $\frac{\partial f}{\partial x}$ and $\frac{\partial^2 f}{\partial x \partial y}$

19. $f(x, y, z) = \frac{x}{x^2 + y^2 + z^2}$; find $\frac{\partial f}{\partial x}, \frac{\partial f}{\partial y}, \frac{\partial f}{\partial z}$, and $\left.\frac{\partial f}{\partial x}\right|_{(0,1,0)}$.

20. $f(x, y, z) = x^2 + y^2 + z^2 + xyz$; find $f_{xx} + f_{yy} + f_{zz}$

In Exercises 21–26, locate and classify all critical points.

21. $f(x, y) = (x - 1)^2 + (2y - 3)^2$

22. $g(x, y) = (x - 1)^2 - 3y^2 + 9$

23. $k(x, y) = x^2 y - x^2 - y^2$

24. $j(x, y) = xy + x^2$

25. $h(x, y) = e^{xy}$

26. $f(x, y) = \ln(x^2 + y^2) - (x^2 + y^2)$

In Exercises 27–30, solve the given constrained optimization problem by using substitution to eliminate a variable. (Do not use Lagrange multipliers.)

27. Find the largest value of xyz subject to $x + y + z = 1$ with $x > 0, y > 0, z > 0$. Also find the corresponding point(s) (x, y, z).

28. Find the minimum value of $f(x, y, z) = x^2 + y^2 + z^2 - 1$ subject to $x = y + z$. Also find the corresponding point(s) (x, y, z).

29. Find the point on the surface $z = \sqrt{x^2 + 2(y - 3)^2}$ closest to the origin.

30. Minimize $S = xy + x^2 z^2 + 4yz$ subject to $xyz = 1$ with $x > 0, y > 0, z > 0$.

In Exercises 31–34, use Lagrange multipliers to solve the given optimization problem.

31. Find the minimum value of $f(x, y) = x^2 + y^2$ subject to $xy = 2$. Also find the corresponding point(s) (x, y).

32. The problem in Exercise 28.

33. The problem in Exercise 29.

34. The problem in Exercise 30.

In Exercises 35–40, compute the given quantities.

35. $\int_0^1 \int_0^2 (2xy)\, dx\, dy$

36. $\int_1^2 \int_0^1 xye^{x+y}\, dx\, dy$

37. $\int_0^2 \int_0^{2x} \frac{1}{x^2 + 1}\, dy\, dx$

38. The average value of xye^{x+y} over the rectangle $0 \le x \le 1$, $1 \le y \le 2$

39. $\iint_R (x^2 - y^2)\, dx\, dy$, where R is the region shown in the figure

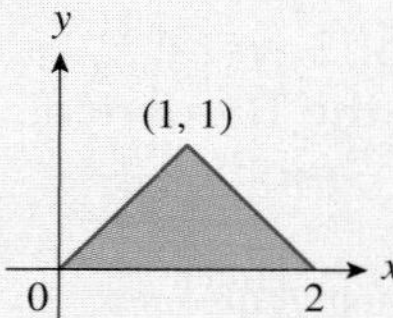

40. The volume under the graph of $z = 1 - y$ over the region in the xy plane between the parabola $y = 1 - x^2$ and the x-axis

APPLICATIONS: OHaganBooks.com

41. ***Web Site Traffic*** OHaganBooks.com has two principal competitors: JungleBooks.com and FarmerBooks.com. Current Web site traffic at OHaganBooks.com is estimated at 5,000 hits per day. This number is predicted to decrease by 0.8 for every new customer of JungleBooks.com and by 0.6 for every new customer of FarmerBooks.com.

   a. Use this information to model the daily Website traffic at OHaganBooks.com as a linear function of the new customers of its two competitors.

   b. According to the model, if Junglebooks.com gets 100 new customers and OHaganBooks.com traffic drops to 4,770 hits per day, how many new customers has FarmerBooks.com obtained?

   c. The model in part (a) did not take into account the growth of the total online consumer base. OHaganBooks.com expects to get approximately one additional hit per day for every 10,000 new Internet shoppers. Modify your model in part (a) so as to include this information using a new independent variable.

   d. How many new Internet shoppers would it take to offset the effects on traffic at OHaganBooks.com of 100 new customers at each of its competitor sites?

42. ***Productivity*** Billy-Sean O'Hagan is writing up his Ph.D. thesis in biophysics but finds that his productivity is affected by the temperature and the number of text messages he receives per hour. On a brisk winter's day when the temperature is 0°C and there are no text messages, Billy-Sean can produce 15 pages of his thesis. His productivity goes down by 0.3 pages per degree Celsius increase in the temperature and by 1.2 pages for each additional text message per hour.

   a. Use this information to model Billy-Sean's productivity p as a function of the temperature and the hourly rate of text messages.

   b. The other day the temperature was 20°C and Billy-Sean managed to produce only three pages of his thesis. What was the hourly rate of incoming text messages?

c. Billy-Sean reasons that each cup of coffee he drinks per hour can counter the effect on his productivity of two text messages per hour. Modify the model in part (a) to take consumption of coffee into account.

d. What would the domain of your function look like to ensure that p is never negative?

43. ***Internet Advertising*** To increase business at OHaganBooks.com, you have purchased banner ads at well-known Internet portals and have advertised on television. The following interaction model shows the average number h of hits per day as a function of monthly expenditures x on banner ads and y on television advertising (x and y are in dollars).

$$h(x, y) = 1{,}800 + 0.05x + 0.08y + 0.00003xy$$

a. Based on your model, how much traffic can you anticipate if you spend \$2,000 per month for banner ads and \$3,000 per month on television advertising?

b. Evaluate $\frac{\partial h}{\partial y}$, specify its units of measurement, and indicate whether it increases or decreases with increasing x.

c. How much should the company spend on banner ads to obtain 1 hit per day for each \$5 spent per month on television advertising?

44. ***Company Retreats*** Their companies having recently been bailed out by the government at taxpayer expense, Marjory Duffin and John O'Hagan are planning a joint winter business retreat in Cancun, but they are not sure how many sales reps to take along. The following interaction model shows the estimated cost C to their companies (in dollars) as a function of the number of sales reps x and the length of time t in days.

$$C(x, t) = 20{,}000 - 100x + 600t + 300xt$$

a. Based on the model, how much would it cost to take five sales reps along for a 10-day retreat?

b. Evaluate $\frac{\partial C}{\partial t}$, specify its units of measurement, and indicate whether it increases or decreases with increasing x.

c. How many reps should they take along if they wish to limit the marginal daily cost to \$1,200?

45. ***Internet Advertising*** Refer to the model in Exercise 43. One or more of the following statements is correct. Identify which one(s).

(A) If nothing is spent on television advertising, one more dollar spent per month in banner ads will buy approximately 0.05 hits per day at OHaganBooks.com.

(B) If nothing is spent on television advertising, one more hit per day at OHaganBooks.com will cost the company about 5¢ per month in banner ads.

(C) If nothing is spent on banner ads, one more hit per day at OHaganBooks.com will cost the company about 5¢ per month in banner ads.

(D) If nothing is spent on banner ads, one more dollar spent per month in banner ads will buy approximately 0.05 hits per day at OHaganBooks.com.

(E) Hits at OHaganBooks.com cost approximately 5¢ per month spent on banner ads, and this cost increases at a rate of 0.003¢ per month, per hit.

46. ***Company Retreats*** Refer to the model in Exercise 44. One or more of the following statements is correct. Identify which one(s).

(A) If the retreat lasts for 10 days, the daily cost per sales rep is \$400.

(B) If the retreat lasts for 10 days, each additional day will cost the company \$2,900.

(C) If the retreat lasts for 10 days, each additional sales rep will cost the company \$800.

(D) If the retreat lasts for 10 days, the daily cost per sales rep is \$2,900.

(E) If the retreat lasts for 10 days, each additional sales rep will cost the company \$2,900.

47. ***Productivity*** The holiday season is now at its peak and OHaganBooks.com has been understaffed and swamped with orders. The current backlog (orders unshipped for two or more days) has grown to a staggering 50,000, and new orders are coming in at a rate of 5,000 per day. Research based on productivity data at OHaganBooks.com results in the following model:

$$P(x, y) = 1{,}000x^{0.9}y^{0.1} \text{ additional orders filled per day,}$$

where x is the number of additional personnel hired and y is the daily budget (excluding salaries) allocated to eliminating the backlog.

a. How many additional orders will be filled per day if the company hires 10 additional employees and budgets an additional \$1,000 per day? (Round the answer to the nearest 100.)

b. In addition to the daily budget, extra staffing costs the company \$150 per day for every new staff member hired. In order to fill at least 15,000 additional orders per day at a minimum total daily cost, how many new staff members should the company hire? (Use the method of Lagrange multipliers.)

48. ***Productivity*** The holiday season has now ended, and orders at OHaganBooks.com have plummeted, leaving staff members in the shipping department with little to do besides spend their time on their Facebook pages, so the company is considering laying off a number of personnel and slashing the shipping budget. Research based on productivity data at OHaganBooks.com results in the following model:

$$C(x, y) = 1{,}000x^{0.8}y^{0.2} \text{ fewer orders filled per day,}$$

where x is the number of personnel laid off and y is the cut in the shipping budget (excluding salaries).

a. How many fewer orders will be filled per day if the company lays off 15 additional employees and cuts the budget by an additional \$2,000 per day? (Round the answer to the nearest 100.)

b. In addition to the cut in the shipping budget, the layoffs will save the company \$200 per day for every new staff member laid off. The company needs to meet a target of 20,000 fewer orders per day but, for tax reasons, it must minimize the total resulting savings. How many new staff members should the company lay off? (Use the method of Lagrange multipliers.)

49. ***Profit*** If OHaganBooks.com sells x paperback books and y hardcover books per week, it will make an average weekly profit of

$$P(x, y) = 3x + 10y \text{ dollars.}$$

If it sells between 1,200 and 1,500 paperback books and between 1,800 and 2,000 hardcover books per week, what is the average of all its possible weekly profits?

50. ***Cost*** It costs Duffin House

$$C(x, y) = x^2 + 2y \text{ dollars}$$

to produce x coffee table art books and y paperback books per week. If it produces between 100 and 120 art books and between 800 and 1,000 paperbacks per week, what is the average of all its possible weekly costs?

Case Study — Modeling College Population

david pearson/Alamy

College Malls, Inc. is planning to build a national chain of shopping malls in college neighborhoods. However, malls in general have been experiencing large numbers of store closings due to, among other things, misjudgments of the shopper demographics. As a result, the company is planning to lease only to stores that target the specific age demographics of the national college student population.

As a marketing consultant to College Malls, you will be providing the company with a report that addresses the following specific issues:

- A quick way of estimating the number of students of any specified age and in any particular year, and the effect of increasing age on the college population
- The ages that correspond to relatively high and low college populations
- How fast the 20-year old and 25-year old student populations are increasing
- Some near-term projections of the student population trend

You decide that a good place to start would be with a visit to the Census Bureau's Web site at www.census.gov. After some time battling with search engines, all you can find is some data on college enrollment for three age brackets for the period 1980–2009, as shown in the following table:[2]

College Enrollment (Thousands)

Year	1980	1985	1990	1995	2000	2001	2002	2003	2004	2005	2006	2007	2008	2009
18–24	7,229	7,537	7,964	8,541	9,451	9,629	10,033	10,365	10,611	10,834	10,587	11,161	11,466	12,072
25–34	2,703	3,063	3,161	3,349	3,207	3,422	3,401	3,494	3,690	3,600	3,658	3,838	4,013	6,141
35–44	700	963	1,344	1,548	1,454	1,557	1,678	1,526	1,615	1,657	1,548	1,520	1,672	1,848

The data are inadequate for several reasons: The data are given only for certain years, and in age brackets rather than year-by-year; nor is it obvious as to how you would project the figures. However, you notice that the table is actually a numerical representation of a function of two variables: year and age. Since the age brackets are of different sizes, you "normalize" the data by dividing each figure by the number of years represented in the corresponding age bracket; for instance, you divide the 1980 figure for the first age group by 7 in order to obtain the average enrollment for each year of age in that group. You then rewrite the resulting table representing the years by values of t and each age bracket by the (rounded) age x at its center (enrollment values are rounded):

[2]Source: Census Bureau (www.census.gov/population/www/socdemo/school.html).

$x \downarrow$ \ $t \rightarrow$	0	5	10	15	20	21	22	23	24	25	26	27	28	29
21	1,033	1,077	1,138	1,220	1,350	1,376	1,433	1,481	1,516	1,548	1,512	1,594	1,638	1,725
30	270	306	316	335	321	342	340	349	369	360	366	384	401	614
40	70	96	134	155	145	156	168	153	162	166	155	152	167	185

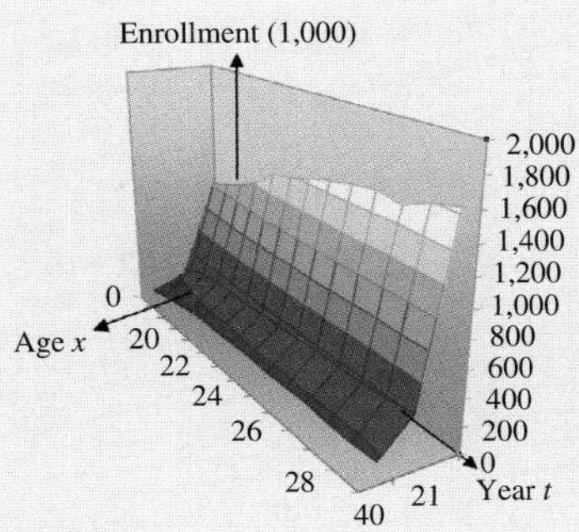

Figure 40

In order to see a visual representation of what the data are saying, you use Excel to graph the data as a surface (Figure 40). It is important to notice that Excel does not scale the t-axis as you would expect: It uses one subdivision for each year shown in the chart, and the result is an uneven scaling of the t-axis. Despite this drawback, you do see two trends after looking at views of the graph from various angles. First, enrollment of 21-year olds (the back edge of the graph) seems to be increasing faster than enrollment of other age groups. Second, the enrollment for all ages seem to be increasing approximately linearly with time, although at different rates for different age groups; for instance, the front and rear edges rise more-or-less linearly, but do not seem to be parallel.

At this point you realize that a mathematical model of these data would be useful; not only would it "smooth out the bumps" but it would give you a way to estimate enrollment N at each specific age, project the enrollments, and thereby complete the project for College Malls. Although technology can give you a regression model for data such as this, it is up to you to decide on the form of the model. It is in choosing an appropriate model that your analysis of the graph comes in handy. Because N should vary linearly with time t for each value of x, you would like

$$N = mt + k$$

for each value of x. Also, because there are three values of x for every value of time, you try a quadratic model for N as a function of x:

$$N = a + bx + cx^2.$$

Putting these together, you get the following candidate model:

$$N(t, x) = a_1 + a_2t + a_3x + a_4x^2,$$

where a_1, a_2, a_3, and a_4 are constants. However, for each specific age $x = k$, you get

$$N(t, k) = a_1 + a_2t + a_3k + a_4k^2 = \text{Constant} + a_2t$$

with the same slope a_2 for every choice of the age k, contrary to your observation that enrollment for different age groups is rising at different rates, so you will need a more elaborate model. You recall from your applied calculus course that interaction functions give a way to model the effect of one variable on the rate of change of another, so, as an experiment, you try adding interaction terms to your model:

$$\text{Model 1: } N(t, x) = a_1 + a_2t + a_3x + a_4x^2 + a_5xt \qquad \text{Second-order model}$$

$$\text{Model 2: } N(t, x) = a_1 + a_2t + a_3x + a_4x^2 + a_5xt + a_6x^2t. \qquad \text{Third-order model}$$

(Model 1 is referred to as a second-order model because it contains no products of more than two independent variables, whereas Model 2 contains the third-order term $x^2t = x \cdot x \cdot t$.) If you study these two models for specific values k of x you get:

$$\text{Model 1: } N = \text{Constant} + (a_2 + a_5k)t \qquad \text{Slope depends linearly on age.}$$

$$\text{Model 2: } N = \text{Constant} + (a_2 + a_5k + a_6k^2)t. \qquad \text{Slope depends quadratically on age.}$$

◇	A	B	C
1	*N*	*t*	*x*
2	1033	0	21
3	1077	5	21
4	1138	10	21
5	1220	15	21
6	1350	20	21
7	1376	21	21
8	1433	22	21
9	1481	23	21
10	1516	24	21
11	1548	25	21
12	1512	26	21
13	1594	27	21
14	1638	28	21
15	1725	29	21
16	270	0	30
17	306	5	30
18	316	10	30
19	335	15	30
20	321	20	30
21	342	21	30
22	340	22	30
23	349	23	30
24	369	24	30
25	360	25	30
26	366	26	30
27	384	27	30
28	401	28	30
29	614	29	30
30	70	0	40
31	96	5	40
32	134	10	40
33	155	15	40
34	145	20	40
35	156	21	40
36	168	22	40
37	153	23	40
38	162	24	40
39	166	25	40
40	155	26	40
41	152	27	40
42	167	28	40
43	185	29	40

Figure 41

This is encouraging: Both models show different slopes for different ages. Model 1 would predict that the slope either increases with increasing age (a_5 positive) or decreases with increasing age (a_5 negative). However, the graph suggests that the slope is larger for both younger and older students, but smaller for students of intermediate age, contrary to what Model 1 predicts, so you decide to go with the more flexible Model 2, which permits the slope to decrease and then increase with increasing age, which is exactly what you observe on the graph, and so you decide to go ahead with Model 2.

You decide to use Excel to generate your model. However, the data as shown in the table are not in a form Excel can use for regression; the data need to be organized into columns; Column A for the dependent variable N and Columns B–C for the independent variables, as shown in Figure 41.

You then add columns for the higher order terms x^2, xt, and x^2t as shown below:

◇	A	B	C	D	E	F
1	*N*	*t*	*x*	*x*^2	*x***t*	*x*^2**t*
2	1033	0	21	=C2^2	=C2*B2	=C2^2*B2
3	1077	5	21			
4	1138	10	21			
5	1220	15	21			
6	1350	20	21			
7	1376	21	21			
8	1433	22	21			
9	1481	23	21			

Figure 42

◇	A	B	C	D	E	F
1	*N*	*t*	*x*	*x*^2	*x***t*	*x*^2**t*
2	1033	0	21	441	0	0
3	1077	5	21	441	105	2205
4	1138	10	21	441	210	4410
5	1220	15	21	441	315	6615
6	1350	20	21	441	420	8820
7	1376	21	21	441	441	9261
8	1433	22	21	441	462	9702
9	1481	23	21	441	483	10143

Figure 43

Next, highlight a vacant 5×6 block (the block A46:F50 say), type the formula `=LINEST(A2:A43,B2:F43,,TRUE)`, and press Ctrl+Shift+Enter (not just Enter!). You will see a table of statistics like the following:

43	185	29	40	1600	1160	46400
44						
45						
46	=LINEST(A2:A43,B2:F43,,TRUE)					
47						
48						
49						
50						

Figure 44

43	185	29	40	1600	1160	46400
44						
45						
46	0.088570354	-6.46546922	3.21423521	-241.273271	120.009278	4594.62978
47	0.02096866	1.28767845	0.44955302	27.6069027	18.6958148	400.824866
48	0.99337478	49.506512	#N/A	#N/A	#N/A	#N/A
49	1079.5571	36	#N/A	#N/A	#N/A	#N/A
50	13229404.1	88232.2104	#N/A	#N/A	#N/A	#N/A

Figure 45

The desired constants $a_1, a_2, a_3, a_4, a_5, a_6$ appear in the first row of the data, but in *reverse order*. Thus, if we round to 5 significant digits, we have

$$a_1 = 4{,}594.6 \quad a_2 = 120.01 \quad a_3 = -241.27$$
$$a_4 = 3.2142 \quad a_5 = -6.4655 \quad a_6 = 0.088570,$$

which gives our regression model:

$$N(t, x) = 4{,}594.6 + 120.01t - 241.27x + 3.2142x^2 - 6.4655xt + 0.088570x^2t.$$

Fine, you say to yourself, now you have the model, but how good a fit is it to the data? That is where rest of the data shown in the output comes in: In the second row are the "standard errors" corresponding to the corresponding coefficients. Notice that each of the standard errors is small compared with the magnitude of the coefficient above it; for instance, 0.021 is only around 1/4 of the magnitude of $a_6 \approx 0.088$ and indicates that the dependence of N on x^2t is statistically significant. (What we do not want to see are standard errors of magnitudes comparable to the coefficients, as those could indicate the wrong choice of independent variables.) The third figure in the left column, 0.99337478, is R^2, where R generalizes the coefficient of correlation discussed in the section on regression in Chapter 1: The closer R is to 1, the better the fit. We can interpret R^2 as indicating that approximately 99.3% of the variation in college enrollment is explained by the regression model, indicating an excellent fit. The figure 1,079.5571 beneath R^2 is called the "F-statistic." The higher the F-statistic (typically, anything above 4 or so would be considered "high"), the more confident we can be that N does depend on the independent variables we are using.*

* We are being deliberately vague about the exact meaning of these statistics, which are discussed fully in many applied statistics texts.

As comforting as these statistics are, nothing can be quite as persuasive as a graph. You turn to the graphing software of your choice and notice that the graph of the model appears to be a faithful representation of the data. (See Figure 46.)

Now you get to work, using the model to address the questions posed by College Malls.

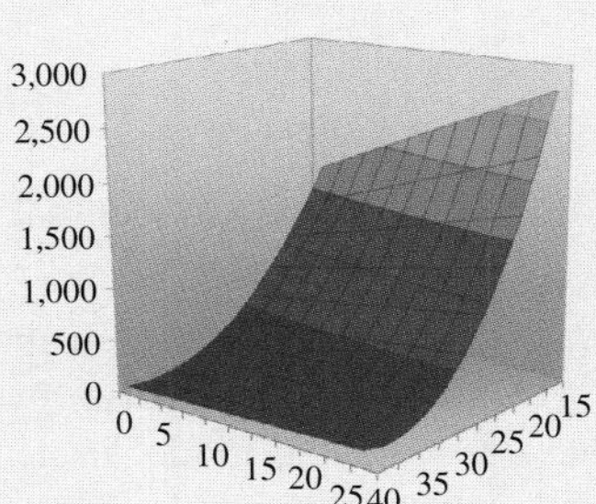

Figure 46

1. *A quick way of estimating the number of students of any specified age and in any particular year, and the effect of increasing age on the college population.* You already have a quantitative relationship in the form of the regression model. As for the second part of the question, the rate of change of college enrollment with respect to age is given by the partial derivative:

$$\frac{\partial N}{\partial x} = -241.27 + 6.4284x - 6.4655t + 0.17714xt \quad \text{thousand students per additional year of age.}$$

Thus, for example, with $x = 20$ in 2004 ($t = 24$), we have

$$\frac{\partial N}{\partial x} = -241.27 + 6.4284(20) - 6.4655(24) + 0.17714(20)(24)$$
$$\approx -183 \text{ thousand students per additional year of age,}$$

so there were about 183,000 fewer students of age 21 than age 20 in 2004.

On the other hand, when $x = 38$ in the same year, we have

$$\frac{\partial N}{\partial x} = -241.27 + 6.4284(38) - 6.4655(24) + 0.17714(38)(24)$$
$$\approx 9.4 \text{ thousand students per additional year of age,}$$

so there were about 9,400 more students of age 39 than age 38 that year.

2. *The ages that correspond to relatively high and low college populations.* Although a glance at the graph shows you that there are no relative maxima, holding t constant (that is, on any given year) gives a parabola along the corresponding slice and hence a minimum somewhere along the slice.

$$\frac{\partial N}{\partial x} = 0$$

when $$-241.27 + 6.4284x - 6.4655t + 0.17714xt = 0,$$

which gives $x = \dfrac{241.27 + 6.4655t}{6.4284 + 0.17714t}$ years of age

For instance, in 2010 ($t = 30$; we are extrapolating the model slightly), the age at which there were fewest students (in the given range) is 37 years of age. The relative maxima for each slice occur at the front and back edges of the surface, meaning that there are relatively more students of the lowest and highest ages represented. The absolute maximum for each slice occurs, as expected, at the lowest age. In short, a mall catering to college students in 2010 should have focused mostly on freshman age students, least on 37-year-olds, and somewhat more on people around age 40.

3. *How fast the 20-year-old and 25-year-old student populations are increasing.* The rate of change of student population with respect to time is

$$\frac{\partial N}{\partial t} = 120.01 - 6.4655x + 0.088570x^2 \text{ thousand students per year.}$$

For the two age groups in question, we obtain

$$x = 20\text{: } 120.01 - 6.4655(20) + 0.088570(20)^2 \approx 26.1 \text{ thousand students/year}$$

$$x = 25\text{: } 120.01 - 6.4655(25) + 0.088570(25)^2 \approx 13.7 \text{ thousand students/year.}$$

(Note that these rates of change are independent of time as we chose a model that is linear in time.)

4. *Some near-term projections of the student population trend.* As we have seen throughout the book, extrapolation can be a risky venture; however, near-term extrapolation from a good model can be reasonable. You enter the model in an Excel spreadsheet to obtain the following predicted college enrollments (in thousands) for the years 2010–2015:

$t \rightarrow$

$x \downarrow$	**30**	**31**	**32**	**33**	**34**	**35**
21	1,644	1,668	1,691	1,714	1,737	1,761
30	422	428	434	439	445	451
40	180	183	186	189	192	195

EXERCISES

1. Use a spreadsheet to obtain Model 1:

$$N(t, x) = a_1 + a_2t + a_3x + a_4x^2 + a_5xt.$$

Compare the fit of this model with that of the quadratic model above. Comment on the result.

2. Obtain Model 2 using only the data through 2005, and also obtain the projections for 2010–2015 using the resulting model. Compare the projections with those based on the more complete set of data in the text.
3. Compute and interpret $\left.\frac{\partial N}{\partial t}\right|_{(10,18)}$ and $\left.\frac{\partial^2 N}{\partial t \partial x}\right|_{(10,18)}$ for the model in the text. What are their units of measurement?
4. Notice that the derivatives in the preceding exercise do not depend on time. What additional polynomial term(s) would make both $\partial N/\partial t$ and $\partial^2 N/\partial t \partial x$ depend on time? (Write down the entire model.) Of what order is your model?
5. Test the model you constructed in the preceding question by inspecting the standard errors associated with the additional coefficients.

TECHNOLOGY GUIDE

TI-83/84 Plus Technology Guide

Section 8.1

Example 1 (page 435) You own a company that makes two models of speakers: the Ultra Mini and the Big Stack. Your total monthly cost (in dollars) to make x Ultra Minis and y Big Stacks is given by

$$C(x, y) = 10{,}000 + 20x + 40y.$$

Compute several values of this function.

Solution with Technology

You can have a TI-83/84 Plus compute $C(x, y)$ numerically as follows:

1. In the "Y=" screen, enter

```
Y1=10000+20X+40Y
```

2. To evaluate, say, $C(10, 30)$ (the cost to make 10 Ultra Minis and 30 Big Stacks), enter

```
10→X
30→Y
Y1
```

```
10→X
                10
30→Y
                30
Y1
             11400
```

and the calculator will evaluate the function and give the answer $C(10, 30) = 11{,}400$.

This procedure is too laborious if you want to calculate $f(x, y)$ for a large number of different values of x and y.

SPREADSHEET Technology Guide

Section 8.1

Example 1 (page 435) You own a company that makes two models of speakers: the Ultra Mini and the Big Stack. Your total monthly cost (in dollars) to make x Ultra Minis and y Big Stacks is given by

$$C(x, y) = 10{,}000 + 20x + 40y.$$

Compute several values of this function.

Solution with Technology

Spreadsheets handle functions of several variables easily. The following setup shows how a table of values of C can be created, using values of x and y you enter:

	A	B	C
1	x	y	C(x, y)
2	10	30	=10000+20*A2+40*B2
3	20	30	
4	15	0	
5	0	30	
6	30	30	

↓

	A	B	C
1	x	y	C(x, y)
2	10	30	11400
3	20	30	11600
4	15	0	10300
5	0	30	11200
6	30	30	11800

A disadvantage of this layout is that it's not easy to enter values of x and y systematically in two columns. Can you find a way to remedy this? (See Example 3 for one method.)

Example 3 (page 437) Use technology to create a table of values of the body mass index

$$M(w, h) = \frac{0.45w}{(0.0254h)^2}.$$

Solution with Technology

We can use this formula to recreate a table in a spreadsheet, as follows:

	A	B	C	D
1		130	140	150
2	60	=0.45*B$1/(0.0254*$A2)^2		
3	61			
4	62			
5	63			
6	64			
7	65			
8	66			
9	67			

In the formula in cell B2 we have used B$1 instead of B1 for the w-coordinate because we want all references to w to use the same row (1). Similarly, we want all references to h to refer to the same column (A), so we used $A2 instead of A2.

We copy the formula in cell B2 to all of the red shaded area to obtain the desired table:

	A	B	C	D	
1		130	140	150	
2	60	25.18755038	27.12505425	29.06255813	3
3	61	24.36849808	26.24299793	28.11749778	29.9
4	62	23.58875685	25.40327661	27.21779637	29.0
5	63	22.84585068	24.60322381	26.36059694	28.1
6	64	22.13749545	23.84037971	25.54326398	27.2
7	65	21.46158138	23.11247226	24.76336314	26.4
8	66	20.81615733	22.41740021	24.01864308	25.6
9	67	20.19941665	21.75321793	23.30701921	24.8

Example 5 (page 440) Obtain the graph of $f(x, y) = x^2 + y^2$.

Solution with Technology

1. Set up a table showing a range of values of x and y and the corresponding values of the function (see Example 3):

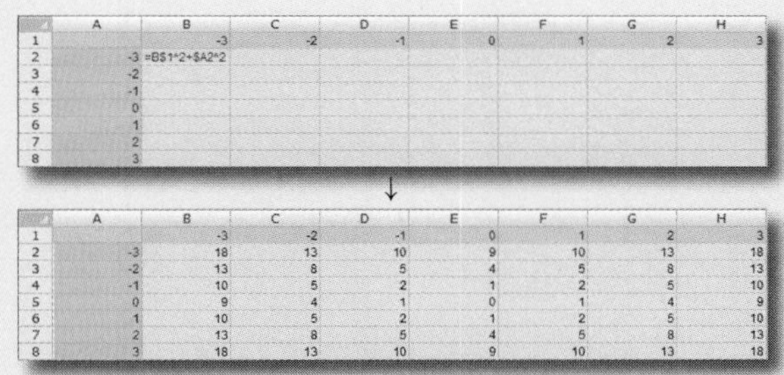

	A	B	C	D	E	F	G	H
1		-3	-2	-1	0	1	2	3
2	-3	=B1^2+A2^2						
3	-2							
4	-1							
5	0							
6	1							
7	2							
8	3							

↓

	A	B	C	D	E	F	G	H
1		-3	-2	-1	0	1	2	3
2	-3	18	13	10	9	10	13	18
3	-2	13	8	5	4	5	8	13
4	-1	10	5	2	1	2	5	10
5	0	9	4	1	0	1	4	9
6	1	10	5	2	1	2	5	10
7	2	13	8	5	4	5	8	13
8	3	18	13	10	9	10	13	18

2. Select the cells with the values (B2: H8) and insert a chart, with the "Surface" option selected and "Series in Columns" selected as the data option, to obtain a graph like the following:

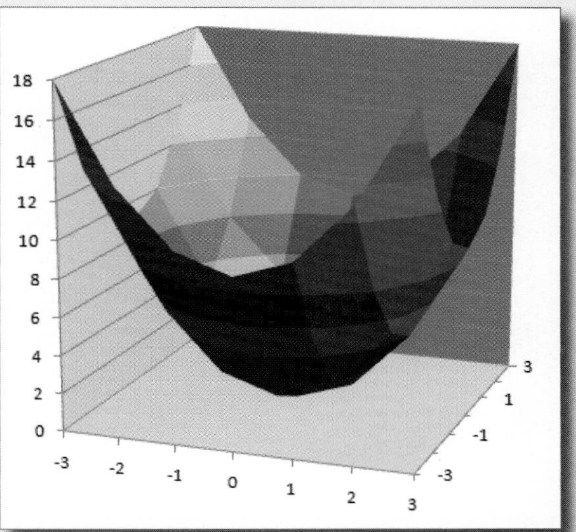

TECHNOLOGY GUIDE

9

Trigonometric Models

Website

www.WanerMath.com

At the Website you will find:

- A detailed chapter summary
- A true/false quiz
- Graphers, Excel tutorials, and other resources

Case Study Predicting Airline Empty Seat Volume

You are a consultant to the Department of Transportation's Special Task Force on Air Traffic Congestion and have been asked to model the volume of empty seats on U.S. airline flights, to make short-term projections of this volume, and to give a formula that estimates the accumulated volume over a specified period of time. You have data from the Bureau of Transportation Statistics on the number of seats and passengers each month starting with January 2002. **How will you analyze these data to prepare your report?**

Lawrence Manning/Flirt/Corbis

Introduction

Cyclical behavior is common in the business world: There are seasonal fluctuations in the demand for surfing equipment, swim wear, snow shovels, and many other items. The nonlinear functions we have studied up to now cannot model this kind of behavior. To model cyclical behavior, we need the **trigonometric** functions.

In the first section, we study the basic trigonometric functions—especially the **sine** and **cosine** functions from which all the trigonometric functions are built—and see how to model various kinds of periodic behavior using these functions. The rest of the chapter is devoted to the calculus of the trigonometric functions—their derivatives and integrals—and to its numerous applications.

9.1 Trigonometric Functions, Models, and Regression

The Sine Function

Figure 1 shows the approximate average daily high temperatures in New York's Central Park.[1] If we draw the graph for several years, we get the repeating pattern shown in Figure 2, where the x-coordinate represents time in years, with $x = 0$ corresponding to August 1, and where the y-coordinate represents the temperature in degrees F. This is an example of **cyclical** or **periodic** behavior.

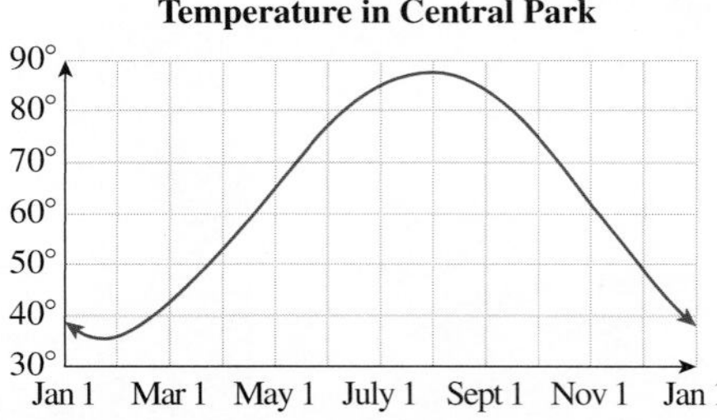

Figure 1

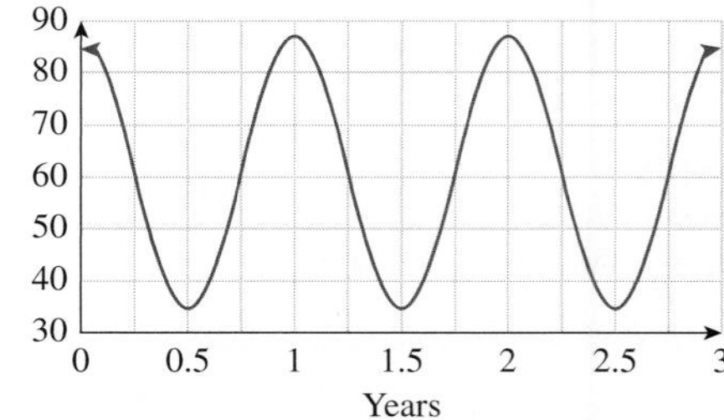

Figure 2

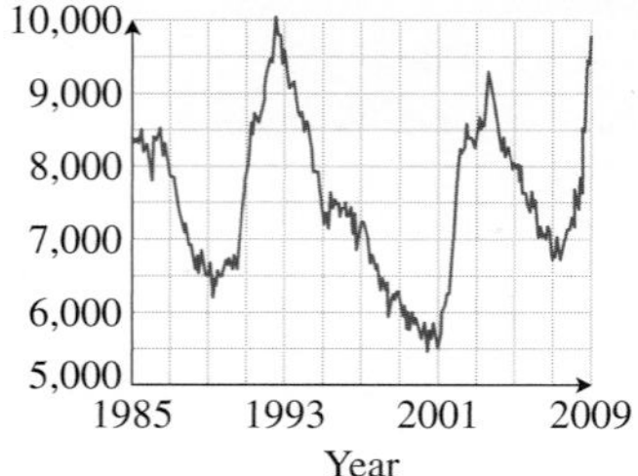

Source: Bureau of Labor Statistics, December 2008 (www.data.bls.gov).

Figure 3

Cyclical behavior is also common in the business world. The graph in Figure 3 suggests cyclical behavior in the U.S. unemployment level.

From a mathematical point of view, the simplest models of cyclical behavior are the **sine** and **cosine** functions. An easy way to describe these functions is as follows. Imagine a bicycle wheel whose radius is one unit, with a marker attached to the rim of the rear wheel, as shown in Figure 4.

[1] Source: National Weather Service/*New York Times*, January 7, 1996, p. 36.

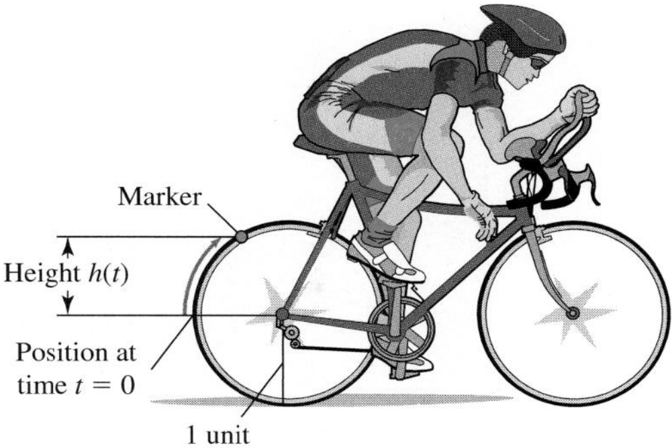

Figure 4

Now, we can measure the height $h(t)$ of the marker above the center of the wheel. As the wheel rotates, $h(t)$ fluctuates between -1 and $+1$. Suppose that, at time $t = 0$, the marker was at height zero as shown in the diagram, so $h(0) = 0$. Because the wheel has a radius of 1 unit, its circumference (the distance all around) is 2π, where $\pi = 3.14159265\ldots$. If the cyclist happens to be moving at a speed of 1 unit per second, it will take the bicycle wheel 2π seconds to make one complete revolution. During the time interval $[0, 2\pi]$, the marker will first rise to a maximum height of $+1$, drop to a low point of -1, and then return to the starting position of 0 at $t = 2\pi$. This function $h(t)$ is called the **sine function**, denoted by $\sin(t)$. Figure 5 shows its graph.

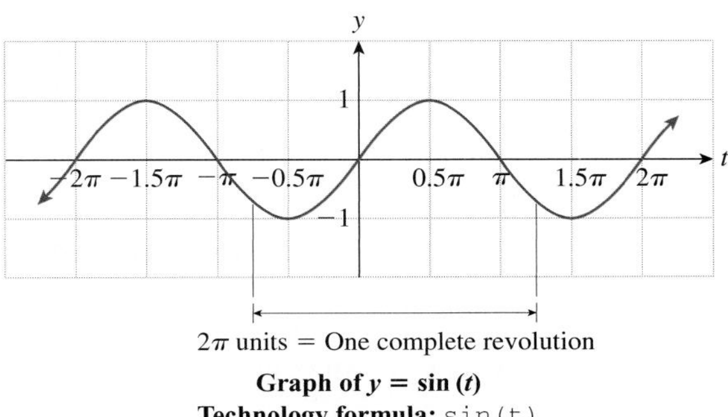

Graph of $y = \sin(t)$
Technology formula: `sin(t)`

Figure 5

Sine Function

"Bicycle Wheel" Definition

If a wheel of radius 1 unit rolls forward at a speed of 1 unit per second, then $\sin(t)$ is the height after t seconds of a marker on the rim of the wheel, starting in the position shown in Figure 4.

Geometric Definition

The **sine** of a real number t is the y-coordinate (height) of the point P in the following diagram, where $|t|$ is the length of the arc shown.

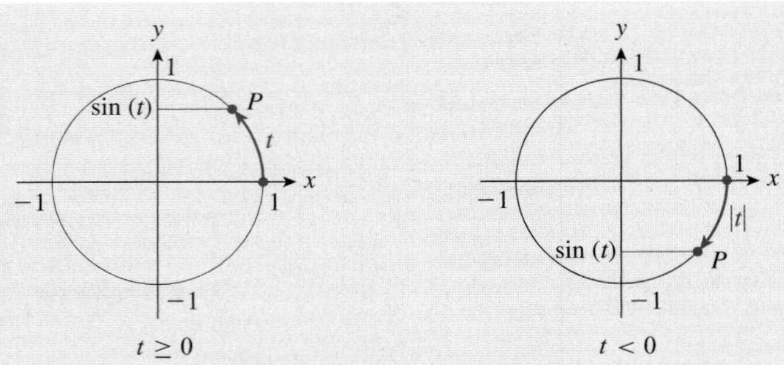

$\sin(t) = y$-coordinate of the point P

Quick Examples

From the graph, we see that

1. $\sin(\pi) = 0$ — Graphing Calculator: `sin(π)` Spreadsheet: `sin(PI())`

2. $\sin\left(\frac{\pi}{2}\right) = 1$ — Graphing Calculator: `sin(π/2)` Spreadsheet: `sin(PI()/2)`

3. $\sin\left(\frac{3\pi}{2}\right) = -1.$ — Graphing Calculator: `sin(3π/2)` Spreadsheet: `sin(3*PI()/2)`

Note We shall often write "sin x" without the parentheses to mean $\sin(x)$; for instance, we may write $\sin(\pi)$ above as $\sin \pi$. Remember, however, that this does *not* mean we are "multiplying" sin by π (which makes no sense). *Always read* $\sin x$ as "*the sine **of** x*." ■

EXAMPLE 1 Some Trigonometric Functions

Use technology to plot the following pairs of graphs on the same set of axes.

a. $f(x) = \sin x$; $g(x) = 2\sin x$

b. $f(x) = \sin x$; $g(x) = \sin(x + 1)$

c. $f(x) = \sin x$; $g(x) = \sin(2x)$

Solution

a. (Important note: If you are using a calculator, make sure it is set to *radian mode*, not degree mode.) We enter these functions as `sin(x)` and `2*sin(x)`, respectively. We use the range $-2\pi \le x \le 2\pi$ (approximately $-6.28 \le x \le 6.28$) for x suggested by the graph in Figure 5, but with larger range of y-coordinates (why?): $-3 \le y \le 3$. The graphs are shown in Figure 6. Here, $f(x) = \sin x$ is shown in red, and $g(x) = 2\sin x$ in blue. Notice that multiplication by 2 has doubled the **amplitude**, or *distance it oscillates up and down*. Where the original sine curve oscillates between −1 and 1, the new curve oscillates between −2 and 2. In general:

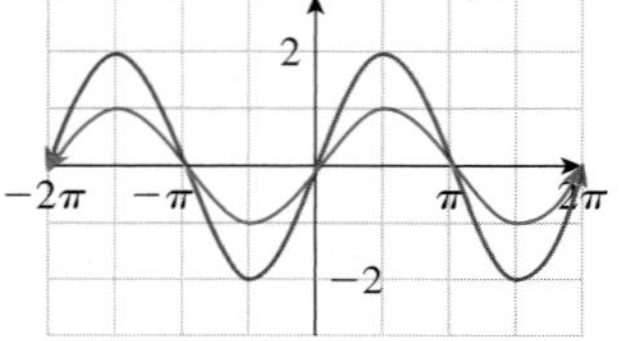

Figure 6

The graph of $A\sin(x)$ *has amplitude* A.

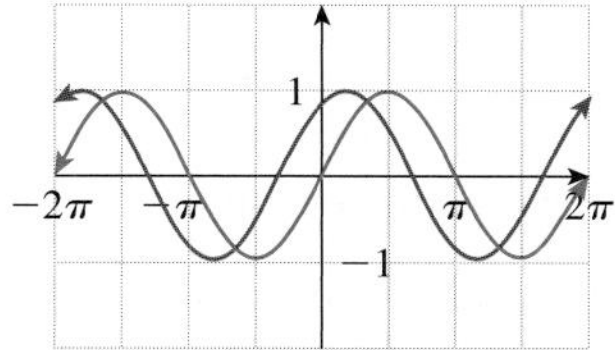

Figure 7

b. We enter these functions as `sin(x)` and `sin(x+1)`, respectively, and we get Figure 7. Once again, $f(x) = \sin x$ is shown in red and $g(x) = \sin(x+1)$ is in blue. The addition of 1 to the argument has shifted the graph to the left by 1 unit. In general:

Replacing x by x + c shifts the graph to the left c units.

(How would we shift the graph to the *right* 1 unit?)

c. We enter these functions as `sin(x)` and `sin(2*x)`, respectively, and get the graph in Figure 8. The graph of $\sin(2x)$ oscillates twice as fast as the graph of $\sin x$. In other words, the graph of $\sin(2x)$ makes two complete cycles on the interval $[0, 2\pi]$ whereas the graph of $\sin x$ completes only one cycle. In general:

Replacing x by bx multiplies the rate of oscillation by b.

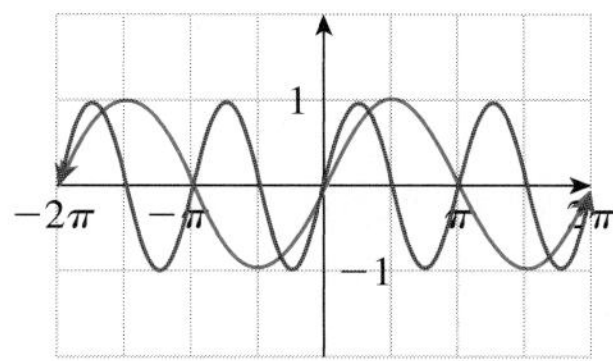

Figure 8

We can combine the operations in Example 1, and a vertical shift as well, to obtain the following.

Website
www.WanerMath.com
At the Website you will find the following optional online interactive section:

Internet Topic: New Functions from Old: Scaled and Shifted Functions

The General Sine Function

The general sine function is

$$f(x) = A\sin[\omega(x - \alpha)] + C.$$

Its graph is shown here.

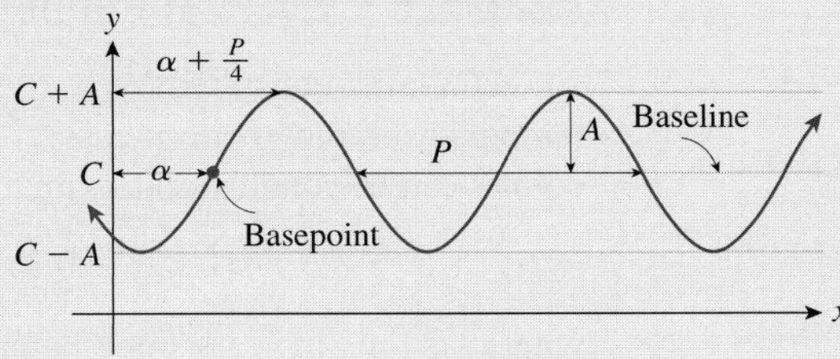

- A is the **amplitude** (the height of each peak above the baseline).
- C is the **vertical offset** (height of the baseline).
- P is the **period** or **wavelength** (the length of each cycle) and is related to ω by

$$P = 2\pi/\omega \quad \text{or} \quad \omega = 2\pi/P.$$

- ω is the **angular frequency** (the number of cycles in every interval of length 2π).
- α is the **phase shift**.

EXAMPLE 2 Electrical Current

The typical voltage V supplied by an electrical outlet in the United States is a sinusoidal function that oscillates between −165 volts and +165 volts with a frequency of 60 cycles per second. Find an equation for the voltage as a function of time t.

Solution What we are looking for is a function of the form

$$V(t) = A\sin[\omega(t-\alpha)] + C.$$

Referring to the figure on the preceding page, we can determine the constants.

Amplitude A and Vertical Offset C: Because the voltage oscillates between -165 volts and $+165$ volts, we see that $A = 165$ and $C = 0$.

Period P: The electric current completes 60 cycles in one second, so the length of time it takes to complete one cycle is $1/60$ second. Thus, the period is $P = 1/60$.

Angular Frequency ω: This is given by the formula

$$\omega = \frac{2\pi}{P} = 2\pi(60) = 120\pi.$$

Phase Shift α: The phase shift α tells us when the curve first crosses the t-axis as it ascends. As we are free to specify what time $t = 0$ represents, let us say that the curve crosses 0 when $t = 0$, so $\alpha = 0$.

Thus, the equation for the voltage at time t is

$$\begin{aligned} V(t) &= A\ \sin[\omega(t-\alpha)] + C \\ &= 165\sin(120\pi t) \end{aligned}$$

where t is time in seconds.

EXAMPLE 3 Cyclical Employment Patterns

An economist consulted by your employment agency indicates that the demand for temporary employment (measured in thousands of job applications per week) in your county can be roughly approximated by the function

$$d = 4.3\sin(0.82t - 0.3) + 7.3,$$

where t is time in years since January 2000. Calculate the amplitude, the vertical offset, the phase shift, the angular frequency, and the period, and interpret the results.

Solution To calculate these constants, we write

$$\begin{aligned} d = A\sin[\omega(t-\alpha)] + C &= A\sin[\omega t - \omega\alpha] + C \\ &= 4.3\sin(0.82t - 0.3) + 7.3 \end{aligned}$$

and we see right away that $A = 4.3$ (the amplitude), $C = 7.3$ (vertical offset), and $\omega = 0.82$ (angular frequency). We also have

$$\omega\alpha = 0.3$$

so that $$\alpha = \frac{0.3}{\omega} = \frac{0.3}{0.82} \approx 0.37$$

(rounding to two significant digits; notice that all the constants are given to two digits). Finally, we get the period using the formula

$$P = \frac{2\pi}{\omega} = \frac{2\pi}{0.82} \approx 7.7.$$

We can interpret these numbers as follows: The demand for temporary employment fluctuates in cycles of 7.7 years about a baseline of 7,300 job applications per week. Every cycle, the demand peaks at 11,600 applications per week (4,300 above the baseline) and dips to a low of 3,000. In May 2000 ($t = 0.37$) the demand for employment was at the baseline level and rising.

Note The generalized sine function in Example 3 was given in the form

$$f(x) = A \sin(\omega t + d) + C$$

for some constant d. Every generalized sine function can be written in this form:

$$A \sin[\omega(t - \alpha)] + C = A \sin(\omega t - \omega\alpha) + C. \qquad d = -\omega\alpha$$

Generalized sine functions are often written in this form. ■

The Cosine Function

Closely related to the sine function is the cosine function, defined as follows. (Refer to the definition of the sine function for comparison.)

Cosine Function

Geometric Definition

The **cosine** of a real number t is the x-coordinate of the point P in the following diagram, in which $|t|$ is the length of the arc shown.

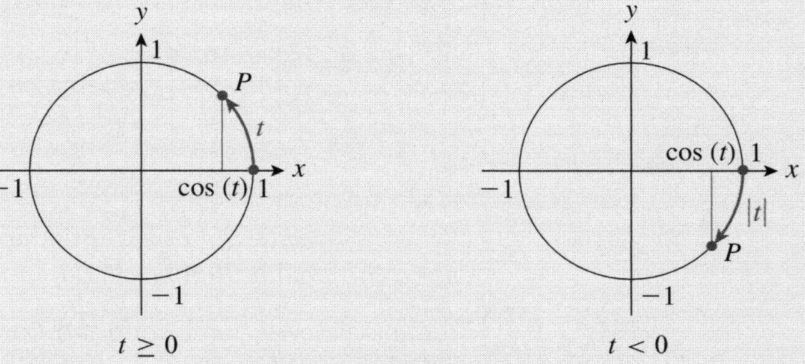

cos (t) = x-coordinate of the point P

Graph of the Cosine Function

The graph of the cosine function is identical to the graph of the sine function, except that it is shifted $\pi/2$ units to the left.

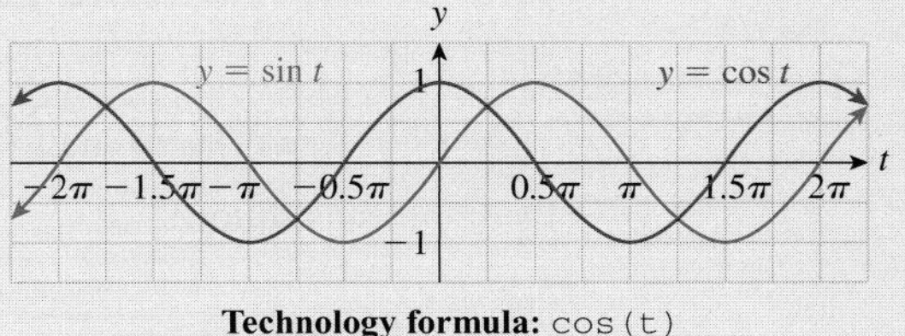

Technology formula: `cos(t)`

Notice that the coordinates of the point P in the diagram above are $(\cos t, \sin t)$ and that the distance from P to the origin is 1 unit. It follows from the Pythagorean theorem that the distance from a point (x, y) to the origin is $\sqrt{x^2 + y^2}$. Thus:

$$\text{Square of the distance from } P \text{ to } (0, 0) = 1$$

$$(\sin t)^2 + (\cos t)^2 = 1.$$

We often write $(\sin t)^2$ as $\sin^2 t$ and similarly for the cosine, so we can rewrite the equation as

$$\sin^2 t + \cos^2 t = 1.$$

This equation is one of the important relationships between the sine and cosine functions.

Fundamental Trigonometric Identities: Relationships between Sine and Cosine

The sine and cosine of a number t are related by

$$\sin^2 t + \cos^2 t = 1.$$

We can obtain the cosine curve by shifting the sine curve to the left a distance of $\pi/2$. [See Example 1(b) for a shifted sine function.] Conversely, we can obtain the sine curve from the cosine curve by shifting it $\pi/2$ units to the right. These facts can be expressed as

$$\cos t = \sin(t + \pi/2)$$
$$\sin t = \cos(t - \pi/2).$$

Alternative Formulation

We can also obtain the cosine curve by first inverting the sine curve horizontally (replace t by $-t$) and then shifting to the *right* a distance of $\pi/2$. This gives us two alternative formulas (which are easier to remember):

$$\cos t = \sin(\pi/2 - t)$$ Cosine is the sine of the complementary angle.
$$\sin t = \cos(\pi/2 - t).$$

Q: *We can rewrite the cosine function in terms of the sine function, so do we really need the cosine function?*

A: Technically, we don't need the cosine function and could get by with only the sine function. On the other hand, it is convenient to have the cosine function because it starts at its highest point rather than at zero. These two functions and their relationship play important roles throughout mathematics.

The General Cosine Function

The general cosine function is

$$f(x) = A\cos[\omega(x - \beta)] + C.$$

Its graph is as follows.

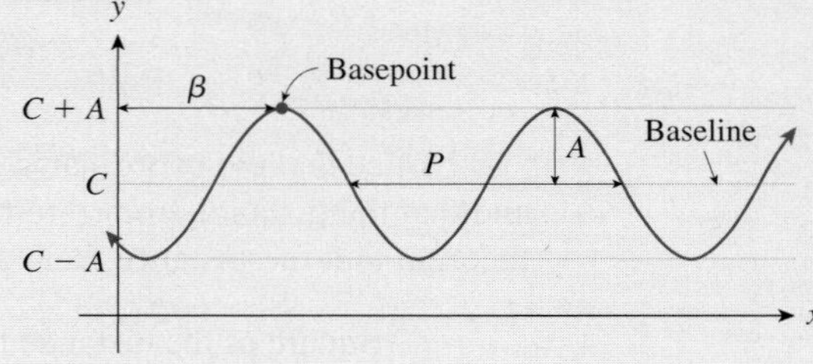

Note that the basepoint of the cosine curve is at the highest point of the curve. All the constants have the same meaning as for the general sine curve:

- A is the **amplitude** (the height of each peak above the baseline).
- C is the **vertical offset** (height of the baseline).
- P is the **period** or **wavelength** (the length of each cycle) and is related to ω by

$$P = 2\pi/\omega \quad \text{or} \quad \omega = 2\pi/P.$$

- ω is the **angular frequency** (the number of cycles in every interval of length 2π).
- β is the phase shift.

Notes

1. We can also describe the above curve as a generalized sine function: Observe by comparing the picture on the preceding page to the one on page 487 that $\beta = \alpha + P/4$. Thus, $\alpha = \beta - P/4$, and the above curve is also

$$f(x) = A\sin[\omega(x - \beta + P/4)] + C.$$

2. As is the case with the generalized sine function, the cosine function above can be written in the form

$$f(x) = A\cos(\omega t + d) + C. \qquad d = -\omega\beta$$

■

EXAMPLE 4 Cash Flows into Stock Funds

The annual cash flow into stock funds (measured as a percentage of total assets) has fluctuated in cycles of approximately 40 years since 1955, when it was at a high point. The highs were roughly +15% of total assets, whereas the lows were roughly −10% of total assets.[2]

a. Model this cash flow with a cosine function of the time t in years, with $t = 0$ representing 1955.

b. Convert the answer in part (a) to a sine function model.

Solution

a. Cosine modeling is similar to sine modeling; we are seeking a function of the form

$$P(t) = A\cos[\omega(t - \beta)] + C.$$

Amplitude A and Vertical Offset C: The cash flow fluctuates between −10% and +15%. We can express this as a fluctuation of $A = 12.5$ about the average $C = 2.5$.

Period P: This is given as $P = 40$.

Angular Frequency ω: We find ω from the formula

$$\omega = \frac{2\pi}{P} = \frac{2\pi}{40} = \frac{\pi}{20} \approx 0.157.$$

[2]Source: Investment Company Institute/*New York Times*, February 2, 1997, p. F8.

Phase Shift β: The basepoint is at the high point of the curve, and we are told that cash flow was at its high point at $t = 0$. Therefore, the basepoint occurs at $t = 0$, and so $\beta = 0$.

Putting the model together gives

$$\begin{aligned} P(t) &= A\cos[\omega(t - \beta)] + C \\ &\approx 12.5\cos(0.157t) + 2.5, \end{aligned}$$

where t is time in years since 1955.

b. To convert between a sine and cosine model, we can use one of the relationships given earlier. Let us use the formula

$$\cos x = \sin(x + \pi/2).$$

Therefore,

$$\begin{aligned} P(t) &\approx 12.5\cos(0.157t) + 2.5 \\ &= 12.5\sin(0.157t + \pi/2) + 2.5. \end{aligned}$$

The Other Trigonometric Functions

The ratios and reciprocals of sine and cosine are given their own names.

Tangent, Cotangent, Secant, Cosecant

Tangent: $\tan x = \dfrac{\sin x}{\cos x}$

Cotangent: $\cot x = \operatorname{cotan} x = \dfrac{\cos x}{\sin x} = \dfrac{1}{\tan x}$

Secant: $\sec x = \dfrac{1}{\cos x}$

Cosecant: $\csc x = \operatorname{cosec} x = \dfrac{1}{\sin x}$

Trigonometric Regression

In the examples so far, we were given enough information to obtain a sine (or cosine) model directly. Often, however, we are given data that only *suggest* a sine curve. In such cases we can use regression to find the best-fit generalized sine (or cosine) curve.

EXAMPLE 5 T Spam

The authors of this book tend to get inundated with spam email. One of us systematically documented the number of spam emails arriving at his email account, and noticed a curious cyclical pattern in the average number of emails arriving each week.[3] Figure 9 shows the daily spam for a 16-week period[4] (each point is a one-week average):

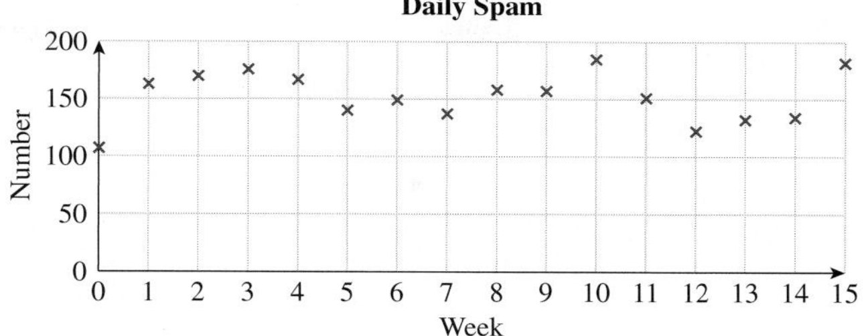

Figure 9

Week	0	1	2	3	4	5	6	7	8	9	10	11	12	13	14	15
Number	107	163	170	176	167	140	149	137	158	157	185	151	122	132	134	182

a. Use technology to find the best-fit sine curve of the form $S(t) = A\sin[\omega(t-\alpha)] + C$.

b. Use your model to estimate the period of the cyclical pattern in spam mail, and also to predict the daily spam average for week 23.

Solution

a. Following are the models obtained by using the TI-83/84 Plus, Excel with Solver, and the Function Evaluator and Grapher on the Website. (See the Technology Guides at the end of the chapter to find out how to obtain these models. For the Website utility, we used the initial guess $A = 50$, $\omega = 1$, $\alpha = 0$, and $C = 150$.)

TI-83/84 Plus: $S(t) \approx 11.6\sin[0.910(t - 1.63)] + 155$

Excel and Website grapher: $S(t) \approx 25.8\sin[0.96(t - 1.22)] + 153$

Q: *Why do the models from the TI-83/84 Plus differ so drastically from the Solver and Website model?*

A: Not all regression algorithms are identical, and it seems that the TI-83/84 Plus's algorithm is not very efficient at finding the best-fit sine curve. Indeed, the value for the sum-of-squares error (SSE) for the TI-83/84 Plus regression curve is around 5,030, whereas it is around 2,148 for the Excel curve, indicating a far better fit.* Notice another thing: The sine curve does not appear to fit the data well in either graph. In general, we can expect better agreement between the different forms of technology for data that follow a sine curve more closely.

* This comparison is actually unfair: The method using Excel's Solver or the Website grapher starts with an initial guess of the coefficients, so the TI-83/84 Plus algorithm, which does not require an initial guess, is starting at a significant disadvantage. An initial guess that is way off can result in Solver or the Website grapher coming up with a very different result! On the other hand, the TI-83/84 Plus algorithm seems problematic and tends to fail (giving an error message) on many sets of data.

b. This model gives a period of approximately

TI-83/84 Plus: $P = \frac{2\pi}{\omega} \approx \frac{2\pi}{0.910} \approx 6.9$ weeks

Excel and Website grapher: $P = \frac{2\pi}{\omega} \approx \frac{2\pi}{0.96} \approx 6.5$ weeks.

So, both models predict a very similar period.

[3]Confirming the notion that academics have little else to do but fritter away their time in pointless pursuits.

[4]Beginning June 6, 2005.

Website
www.WanerMath.com
At the Website you will find the following optional online interactive section, in which you can find further discussion of the graphs of the trigonometric functions and their relationship to right triangles:
Internet Topic: Trigonometric Functions and Calculus
→ The Six Trigonometric Functions.

In week 23, we obtain the following predictions:

TI-83/84 Plus: $S(23) \approx 11.6 \sin[0.910(23 - 1.63)] + 155$
≈ 162 spam emails per day

Excel and Website grapher: $S(23) \approx 25.8 \sin[0.96(23 - 1.22)] + 153$
≈ 176 spam emails per day.

Note The actual figure for week 23 was 213 spam emails per day. The discrepancy illustrates the danger of using regression models to extrapolate. ■

9.1 EXERCISES

Access end-of-section exercises online at **www.webassign.net** ENHANCED WebAssign

9.2 Derivatives of Trigonometric Functions and Applications

We start with the derivatives of the sine and cosine functions.

Theorem Derivatives of the Sine and Cosine Functions

The sine and cosine functions are differentiable with

$$\frac{d}{dx}\sin x = \cos x$$

$$\frac{d}{dx}\cos x = -\sin x \quad \text{Notice the sign change.}$$

Quick Examples

1. $\frac{d}{dx}(x\cos x) = 1 \cdot \cos x + x \cdot (-\sin x)$ Product rule: $x \cos x$ is a product.*
$= \cos x - x \sin x$

2. $\frac{d}{dx}\left(\frac{x^2+x}{\sin x}\right) = \frac{(2x+1)(\sin x) - (x^2+x)(\cos x)}{\sin^2 x}$ Quotient rule

* Apply the calculation thought experiment: If we were to compute $x \cos x$, the last operation we would perform is the multiplication of x and $\cos x$. Hence, $x \cos x$ is a product.

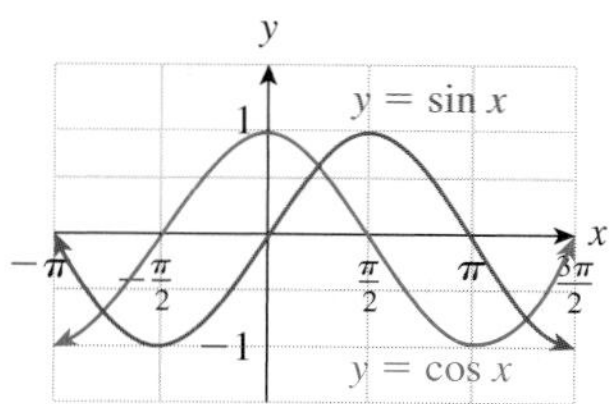

Figure 10(a)

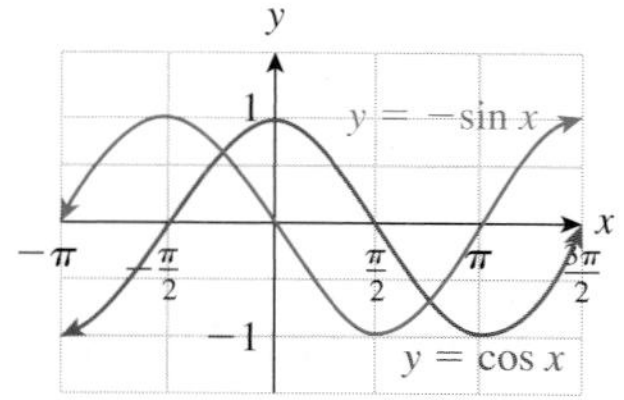

Figure 10(b)

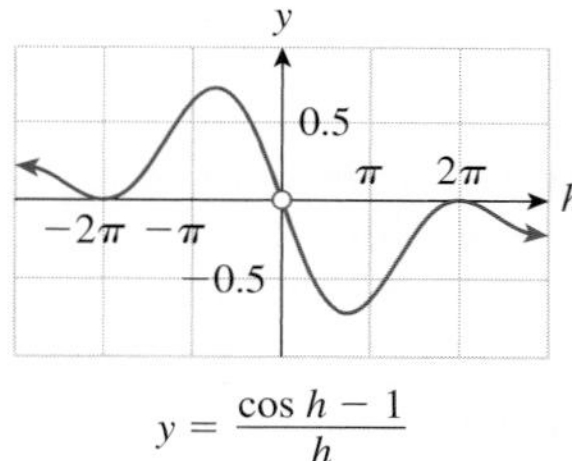

$y = \frac{\cos h - 1}{h}$

Figure 11

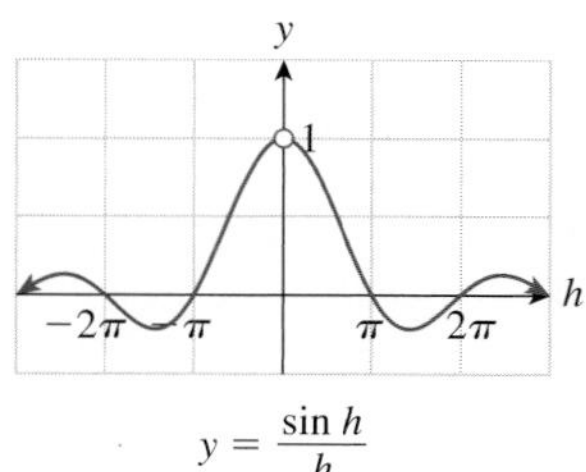

$y = \frac{\sin h}{h}$

Figure 12

Before deriving these formulas, we can see right away that they are plausible by examining Figure 10, which shows the graphs of the sine and cosine functions together with their derivatives. Notice, for instance, that in Figure 10(a) the graph of $\sin x$ is rising most rapidly when $x = 0$, corresponding to the maximum value of its derivative, $\cos x$. When $x = \pi/2$, the graph of $\sin x$ levels off, so that its derivative, $\cos x$, is 0. Another point to notice: Because periodic functions (such as sine and cosine) repeat their behavior, their derivatives must also be periodic.

Derivation of Formulas for Derivatives of the Sine and Cosine Functions

We first calculate the derivative of $\sin x$ from scratch, using the definition of the derivative:

$$\frac{d}{dx} f(x) = \lim_{h \to 0} \frac{f(x+h) - f(x)}{h}.$$

Thus,

$$\frac{d}{dx} \sin x = \lim_{h \to 0} \frac{\sin(x+h) - \sin x}{h}.$$

We now use the addition formula, $\sin(x+y) = \sin x \cos y + \cos x \sin y$:

$$\sin(x+h) = \sin x \cos h + \cos x \sin h.$$

Substituting this expression for $\sin(x+h)$ gives

$$\frac{d}{dx} \sin x = \lim_{h \to 0} \frac{\sin x \cos h + \cos x \sin h - \sin x}{h}$$

Grouping the first and third terms together and factoring out the term $\sin x$ gives

$$\begin{aligned}\frac{d}{dx} \sin x &= \lim_{h \to 0} \frac{\sin x(\cos h - 1) + \cos x \sin h}{h} \\ &= \lim_{h \to 0} \frac{\sin x(\cos h - 1)}{h} + \lim_{h \to 0} \frac{\cos x \sin h}{h} \qquad \text{Limit of a sum} \\ &= \sin x \lim_{h \to 0} \frac{\cos h - 1}{h} + \cos x \lim_{h \to 0} \frac{\sin h}{h}.\end{aligned}$$

and we are left with two limits to evaluate. Calculating these limits analytically requires a little trigonometry.* Alternatively, we can get a good idea of what these two limits are by estimating them numerically or graphically. Figures 11 and 12 show the graphs of $(\cos h - 1)/h$ and $(\sin h)/h$, respectively.

We find that:

$$\lim_{h \to 0} \frac{\cos h - 1}{h} = 0$$

and

$$\lim_{h \to 0} \frac{\sin h}{h} = 1.$$

* You can find these calculations on the Website by following the path:

Website
→ Everything for Calculus
→ Chapter 9
→ Proof of Some Trigonometric Limits

Therefore,

$$\frac{d}{dx}\sin x = (\sin x)(0) + (\cos x)(1) = \cos x.$$

This is the required formula for the derivative of $\sin x$.

Turning to the derivative of the cosine function, we use the identity

$$\cos x = \sin(\pi/2 - x)$$

from Section 9.1. If $y = \cos x = \sin(\pi/2 - x)$, then, using the chain rule, we have

$$\begin{aligned}\frac{dy}{dx} &= \cos(\pi/2 - x)\frac{d}{dx}(\pi/2 - x)\\ &= (-1)\cos(\pi/2 - x)\\ &= -\sin x. \qquad \text{Using the identity } \cos(\pi/2 - x) = \sin x\end{aligned}$$

This is the required formula for the derivative of $\cos x$.

Just as with logarithmic and exponential functions, the chain rule can be used to find more general derivatives.

Derivatives of Sines and Cosines of Functions

Original Rule	*Generalized Rule*	*In Words*
$\frac{d}{dx}\sin x = \cos x$	$\frac{d}{dx}\sin u = \cos u \frac{du}{dx}$	*The derivative of the sine of a quantity is the cosine of that quantity, times the derivative of that quantity.*
$\frac{d}{dx}\cos x = -\sin x$	$\frac{d}{dx}\cos u = -\sin u \frac{du}{dx}$	*The derivative of the cosine of a quantity is negative sine of that quantity, times the derivative of that quantity.*

Quick Examples

1. $\frac{d}{dx}\sin(3x^2 - 1) = \cos(3x^2 - 1)\frac{d}{dx}(3x^2 - 1)$ $\quad u = 3x^2 - 1$ (See margin note*)

$= 6x\cos(3x^2 - 1)$ $\quad$ We placed the $6x$ in front—see Note below.

2. $\frac{d}{dx}\cos(x^3 + x) = -\sin(x^3 + x)\frac{d}{dx}(x^3 + x)$ $\quad u = x^3 + x$

$= -(3x^2 + 1)\sin(x^3 + x)$

* If we were to evaluate $\sin(3x^2 - 1)$, the last operation we would perform is taking the sine of a quantity. Thus, the calculation thought experiment tells us that we are dealing with the *sine of a quantity*, and we use the generalized rule.

Note Avoid writing ambiguous expressions like $\cos(3x^2 - 1)(6x)$. Does this mean

$\cos[(3x^2 - 1)(6x)]$? $\quad$ The cosine of the quantity $(3x^2 - 1)(6x)$

Or does it mean

$[\cos(3x^2 - 1)](6x)$? $\quad$ The product of $\cos(3x^2 - 1)$ and $6x$

To avoid the ambiguity, use parentheses or brackets and write $\cos[(3x^2 - 1)(6x)]$ if you mean the former. If you mean the latter, place the $6x$ in front of the cosine expression and write

$6x\cos(3x^2 - 1)$. $\quad$ The product of $6x$ and $\cos(3x^2 - 1)$ ■

EXAMPLE 1 Derivatives of Trigonometric Functions

Find the derivatives of the following functions.

a. $f(x) = \sin^2 x$ **b.** $g(x) = \sin^2(x^2)$ **c.** $h(x) = e^{-x}\cos(2x)$

Solution

a. Recall that $\sin^2 x = (\sin x)^2$. The calculation thought experiment tells us that $f(x)$ is the square of a quantity.* Therefore, we use the chain rule (or generalized power rule) for differentiating the square of a quantity:

$$\frac{d}{dx}(u^2) = 2u\frac{du}{dx}$$

$$\frac{d}{dx}(\sin x)^2 = 2(\sin x)\frac{d(\sin x)}{dx} \qquad u = \sin x$$

$$= 2\sin x\cos x.$$

Thus, $f'(x) = 2\sin x\cos x$.

* Notice the difference between $\sin^2 x$ and $\sin(x^2)$. The first is the square of $\sin x$, whereas the second is the sine of the quantity x^2.

b. We rewrite the function $g(x) = \sin^2(x^2)$ as $[\sin(x^2)]^2$. Because $g(x)$ is the square of a quantity, we have

$$\frac{d}{dx}\sin^2(x^2) = \frac{d}{dx}[\sin(x^2)]^2 \qquad \text{Rewrite } \sin^2(-) \text{ as } [\sin(-)]^2.$$

$$= 2\sin(x^2)\frac{d[\sin(x^2)]}{dx} \qquad \frac{d}{dx}[u^2] = 2u\frac{du}{dx} \text{ with } u = \sin(x^2)$$

$$= 2\sin(x^2)\cdot\cos(x^2)\cdot 2x. \qquad \frac{d}{dx}\sin u = \cos u\frac{du}{dx} \text{ with } u = x^2$$

Thus, $g'(x) = 4x\sin(x^2)\cos(x^2)$.

c. Because $h(x)$ is the product of e^{-x} and $\cos(2x)$, we use the product rule:

$$h'(x) = (-e^{-x})\cos(2x) + e^{-x}\frac{d}{dx}[\cos(2x)]$$

$$= (-e^{-x})\cos(2x) - e^{-x}\sin(2x)\frac{d}{dx}[2x] \qquad \frac{d}{dx}\cos u = -\sin u\frac{du}{dx}$$

$$= -e^{-x}\cos(2x) - 2e^{-x}\sin(2x)$$

$$= -e^{-x}[\cos(2x) + 2\sin(2x)].$$

Derivatives of Other Trigonometric Functions

Because the remaining trigonometric functions are ratios of sines and cosines, we can use the quotient rule to find their derivatives. For example, we can find the derivative of tan as follows:

$$\frac{d}{dx}\tan x = \frac{d}{dx}\left(\frac{\sin x}{\cos x}\right)$$

$$= \frac{(\cos x)(\cos x) - (\sin x)(-\sin x)}{\cos^2 x}$$

$$= \frac{\cos^2 x + \sin^2 x}{\cos^2 x}$$

$$= \frac{1}{\cos^2 x}$$

$$= \sec^2 x.$$

We ask you to derive the other three derivatives in the exercises. Here is a list of the derivatives of all six trigonometric functions and their chain rule variants.

Derivatives of the Trigonometric Functions

Original Rule	*Generalized Rule*
$\frac{d}{dx}\sin x = \cos x$	$\frac{d}{dx}\sin u = \cos u\frac{du}{dx}$
$\frac{d}{dx}\cos x = -\sin x$	$\frac{d}{dx}\cos u = -\sin u\frac{du}{dx}$
$\frac{d}{dx}\tan x = \sec^2 x$	$\frac{d}{dx}\tan u = \sec^2 u\frac{du}{dx}$
$\frac{d}{dx}\cot x = -\csc^2 x$	$\frac{d}{dx}\cot u = -\csc^2 u\frac{du}{dx}$
$\frac{d}{dx}\sec x = \sec x\tan x$	$\frac{d}{dx}\sec u = \sec u\tan u\frac{du}{dx}$
$\frac{d}{dx}\csc x = -\csc x\cot x$	$\frac{d}{dx}\csc u = -\csc u\cot u\frac{du}{dx}$

Quick Examples

1. $\frac{d}{dx}\tan(x^2-1) = \sec^2(x^2-1)\frac{d(x^2-1)}{dx}$ $\quad u = x^2 - 1$

$= 2x\sec^2(x^2-1)$

2. $\frac{d}{dx}\csc(e^{3x}) = -\csc(e^{3x})\cot(e^{3x})\frac{d(e^{3x})}{dx}$ $\quad u = e^{3x}$

$= -3e^{3x}\csc(e^{3x})\cot(e^{3x})$ $\quad$ The derivative of e^{3x} is $3e^{3x}$.

EXAMPLE 2 Gas Heating Demand

In the preceding section we saw that seasonal fluctuations in temperature suggested a sine function. For instance, we can use the function

$$T = 60 + 25\sin\left[\frac{\pi}{6}(x-4)\right] \qquad T = \text{temperature in °F; } x = \text{months since Jan 1}$$

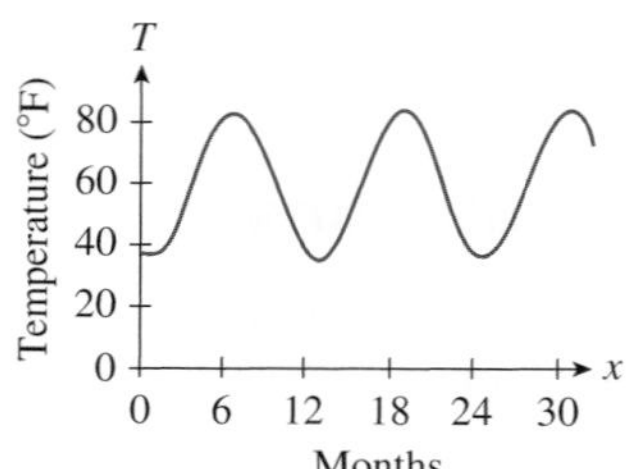

Figure 13

to model a temperature that fluctuates between 35°F on Feb. 1 ($x = 1$) and 85°F on Aug. 1 ($x = 7$). (See Figure 13.)

The demand for gas at a utility company can be expected to fluctuate in a similar way because demand grows with increased heating requirements. A reasonable model might therefore be

$$G = 400 - 100\sin\left[\frac{\pi}{6}(x-4)\right], \qquad \text{Why did we subtract the sine term?}$$

where G is the demand for gas in cubic yards per day. Find and interpret $G'(10)$.

Solution First, we take the derivative of G:

$$G'(x) = -100\cos\left[\frac{\pi}{6}(x-4)\right]\cdot\frac{\pi}{6}$$

$$= -\frac{50\pi}{3}\cos\left[\frac{\pi}{6}(x-4)\right] \text{ cubic yards per day, per month.}$$

Thus,

$$G'(10) = -\frac{50\pi}{3}\cos\left[\frac{\pi}{6}(10-4)\right]$$
$$= -\frac{50\pi}{3}\cos(\pi) = \frac{50\pi}{3}. \qquad \text{Because } \cos\pi = -1$$

The units of $G'(10)$ are cubic yards per day per month, so we interpret the result as follows: On November 1 ($x = 10$) the daily demand for gas is increasing at a rate of $50\pi/3 \approx 52$ cubic yards per day, per month. This is consistent with Figure 13, which shows the temperature decreasing on that date.

9.2 EXERCISES

Access end-of-section exercises online at **www.webassign.net**

9.3 Integrals of Trigonometric Functions and Applications

We saw in Section 6.1 that every calculation of a derivative also gives us a calculation of an antiderivative. For instance, because we know that $\cos x$ is the derivative of $\sin x$, we can say that an antiderivative of $\cos x$ is $\sin x$:

$$\int \cos x\, dx = \sin x + C. \qquad \text{An antiderivative of } \cos x \text{ is } \sin x.$$

The rules for the derivatives of sine, cosine, and tangent give us the following antiderivatives.

Indefinite Integrals of Some Trig Functions

$$\int \cos x\, dx = \sin x + C \qquad \text{Because } \frac{d}{dx}(\sin x) = \cos x$$

$$\int \sin x\, dx = -\cos x + C \qquad \text{Because } \frac{d}{dx}(-\cos x) = \sin x$$

$$\int \sec^2 x\, dx = \tan x + C \qquad \text{Because } \frac{d}{dx}(\tan x) = \sec^2 x$$

Quick Examples

1. $\int(\sin x + \cos x)\, dx = -\cos x + \sin x + C$ Integral of sum = Sum of integrals
2. $\int(4\sin x - \cos x)\, dx = -4\cos x - \sin x + C$ Integral of constant multiple
3. $\int(e^x - \sin x + \cos x)\, dx = e^x + \cos x + \sin x + C$

EXAMPLE 1 Substitution

Evaluate $\int (x+3)\sin(x^2+6x)\,dx$.

Solution There are two parenthetical expressions that we might replace with u. Notice, however, that the derivative of the expression (x^2+6x) is $2x+6$, which is twice the term $(x+3)$ in front of the sine. Recall that we would like the derivative of u to appear as a factor. Thus, let us take $u = x^2+6x$.

$$u = x^2 + 6x$$
$$\frac{du}{dx} = 2x + 6 = 2(x+3)$$
$$dx = \frac{1}{2(x+3)}\,du$$

Substituting into the integral, we get

$$\int (x+3)\sin(x^2+6x)\,dx = \int (x+3)\sin u\left(\frac{1}{2(x+3)}\right)du = \int \frac{1}{2}\sin u\,du$$
$$= -\frac{1}{2}\cos u + C = -\frac{1}{2}\cos(x^2+6x) + C.$$

EXAMPLE 2 Definite Integrals

Compute the following.

a. $\int_0^{\pi} \sin x\,dx$ **b.** $\int_0^{\pi} x\sin(x^2)\,dx$

Solution

a. $\int_0^{\pi} \sin x\,dx = \left[-\cos x\right]_0^{\pi} = (-\cos\pi) - (-\cos 0) = -(-1) - (-1) = 2$

Thus, the area under one "arch" of the sine curve is exactly two square units!

b. $\int_0^{\pi} x\sin(x^2)\,dx = \int_0^{\pi^2} \frac{1}{2}\sin u\,du$ After substituting $u = x^2$

$$= \left[-\frac{1}{2}\cos u\right]_0^{\pi^2}$$
$$= \left[-\frac{1}{2}\cos(\pi^2)\right] - \left[-\frac{1}{2}\cos(0)\right]$$
$$= -\frac{1}{2}\cos(\pi^2) + \frac{1}{2}$$ $\cos(0) = 1$

We can approximate $\frac{1}{2}\cos(\pi^2)$ by a decimal or leave it in the above form, depending on what we want to do with the answer.

Antiderivatives of the Six Trigonometric Functions

The following table gives the indefinite integrals of the six trigonometric functions. (The first two we have already seen.)

Integrals of the Trigonometric Functions

$$\int \sin x\, dx = -\cos x + C$$

$$\int \cos x\, dx = \sin x + C$$

$$\int \tan x\, dx = -\ln|\cos x| + C$$ Shown below.

$$\int \cot x\, dx = \ln|\sin x| + C$$ See the Exercise Set.

$$\int \sec x\, dx = \ln|\sec x + \tan x| + C$$ Shown below.

$$\int \csc x\, dx = -\ln|\csc x + \cot x| + C$$ See the Exercise Set.

Derivations of Formulas for Antiderivatives of Trigonometic Functions

To show that $\int \tan x\, dx = -\ln|\cos x| + C$, we first write $\tan x$ as $\dfrac{\sin x}{\cos x}$ and put $u = \cos x$ in the integral:

$$\begin{aligned}\int \tan x\, dx &= \int \frac{\sin x}{\cos x}\, dx \\ &= -\int \frac{\sin x}{u}\, \frac{du}{\sin x} \\ &= -\int \frac{du}{u} \\ &= -\ln|u| + C \\ &= -\ln|\cos x| + C.\end{aligned}$$

$$u = \cos x$$
$$\frac{du}{dx} = -\sin x$$
$$dx = -\frac{du}{\sin x}$$

To show that $\int \sec x\, dx = \ln|\sec x + \tan x| + C$, first use a little "trick": Write $\sec x$ as $\sec x\left(\dfrac{\sec x + \tan x}{\sec x + \tan x}\right)$ and put u equal to the denominator:

$$\begin{aligned}\int \sec x\, dx &= \int \sec x\left(\frac{\sec x + \tan x}{\sec x + \tan x}\right) dx \\ &= \int \sec x\, \frac{\sec x + \tan x}{u}\, \frac{du}{\sec x(\tan x + \sec x)} \\ &= \int \frac{du}{u} \\ &= \ln|u| + C \\ &= \ln|\sec x + \tan x| + C.\end{aligned}$$

$$u = \sec x + \tan x$$
$$\frac{du}{dx} = \sec x \tan x + \sec^2 x = \sec x\,(\tan x + \sec x)$$
$$dx = \frac{du}{\sec x\,(\tan x + \sec x)}$$

Shortcuts

If a and b are constants with $a \neq 0$, then we have the following formulas. (All of them can be obtained using the substitution $u = ax + b$. They will appear in the exercises.)

Shortcuts: Integrals of Expressions Involving $(ax + b)$

Rule	Quick Example
$\int \sin(ax+b)\,dx = -\frac{1}{a}\cos(ax+b) + C$	$\int \sin(-4x)\,dx = \frac{1}{4}\cos(-4x) + C$
$\int \cos(ax+b)\,dx = \frac{1}{a}\sin(ax+b) + C$	$\int \cos(x+1)\,dx = \sin(x+1) + C$
$\int \tan(ax+b)\,dx = -\frac{1}{a}\ln\lvert\cos(ax+b)\rvert + C$	$\int \tan(-2x)\,dx = \frac{1}{2}\ln\lvert\cos(-2x)\rvert + C$
$\int \cot(ax+b)\,dx = \frac{1}{a}\ln\lvert\sin(ax+b)\rvert + C$	$\int \cot(3x-1)\,dx = \frac{1}{3}\ln\lvert\sin(3x-1)\rvert + C$
$\int \sec(ax+b)\,dx = \frac{1}{a}\ln\lvert\sec(ax+b)+\tan(ax+b)\rvert + C$	$\int \sec(9x)\,dx = \frac{1}{9}\ln\lvert\sec(9x)+\tan(9x)\rvert + C$
$\int \csc(ax+b)\,dx = -\frac{1}{a}\ln\lvert\csc(ax+b)+\cot(ax+b)\rvert + C$	$\int \csc(x+7)\,dx = -\ln\lvert\csc(x+7)+\cot(x+7)\rvert + C$

EXAMPLE 3 Sales

The rate of sales of cypods (one-bedroom living units) in the city-state of Utarek, Mars[5] can be modeled by

$$s(t) = 7.5\cos(\pi t/6) + 87.5 \text{ units per month,}$$

where t is time in months since January 1. How many cypods are sold in a calendar year?

[5] Based on www.Marsnext.com, a now extinct virtual society.

Solution Total sales over one calendar year are given by

$$\begin{aligned}\int_0^{12} s(t)\,dt &= \int_0^{12} [7.5\cos(\pi t/6) + 87.5]\,dt \\ &= \left[7.5\frac{6}{\pi}\sin(\pi t/6) + 87.5t\right]_0^{12} && \text{We used a shortcut on the first term.} \\ &= \left[7.5\frac{6}{\pi}\sin(2\pi) + 87.5(12)\right] - \left[7.5\frac{6}{\pi}\sin(0) + 87.5(0)\right] \\ &= 87.5(12) && \sin(2\pi) = \sin(0) = 0. \\ &= 1{,}050 \text{ cypods.}\end{aligned}$$

➡ **Before we go on...** Would it have made any difference in Example 3 if we had computed total sales over the period [12, 24], [6, 18], or any interval of the form $[a, a + 12]$? ■

Using Integration by Parts with Trig Functions

EXAMPLE 4 Integrating a Polynomial Times Sine or Cosine

Calculate $\int (x^2 + 1)\sin(x + 1)\,dx$.

Solution We use the column method of integration by parts described in Section 7.1. Because differentiating $x^2 + 1$ makes it simpler, we put it in the D column and get the following table:

	D	*I*
+	$x^2 + 1$	$\sin(x + 1)$
−	$2x$	$-\cos(x + 1)$
+	2	$-\sin(x + 1)$
$-\int$	0	$\cos(x + 1)$

[Notice that we used the shortcut formulas to repeatedly integrate $\sin(x + 1)$.] We can now read the answer from the table:

$$\begin{aligned}\int (x^2 + 1)\sin(x + 1)\,dx &= (x^2 + 1)[-\cos(x + 1)] - 2x[-\sin(x + 1)] \\ &\quad + 2[\cos(x + 1)] + C \\ &= (-x^2 - 1 + 2)\cos(x + 1) + 2x\sin(x + 1) + C \\ &= (-x^2 + 1)\cos(x + 1) + 2x\sin(x + 1) + C.\end{aligned}$$

EXAMPLE 5 Integrating an Exponential Times Sine or Cosine

Calculate $\int e^x \sin x\, dx$.

Solution The integrand is the product of e^x and $\sin x$, so we put one in the D column and the other in the I column. For this example, it doesn't matter much which we put where.

	D	I
$+$	$\sin x$	e^x
$-$	$\cos x$	e^x
$+\int$	$-\sin x \longrightarrow$	e^x

It looks like we're just spinning our wheels. Let's stop and see what we have:

$$\int e^x \sin x\, dx = e^x \sin x - e^x \cos x - \int e^x \sin x\, dx.$$

At first glance, it appears that we are back where we started, still having to evaluate $\int e^x \sin x\, dx$. However, if we add this integral to both sides of the equation above, we can solve for it:

$$2\int e^x \sin x\, dx = e^x \sin x - e^x \cos x + C.$$

(Why $+ C$?) So,

$$\int e^x \sin x\, dx = \frac{1}{2}e^x \sin x - \frac{1}{2}e^x \cos x + \frac{C}{2}.$$

Because $C/2$ is just as arbitrary as C, we write C instead of $C/2$, and obtain

$$\int e^x \sin x\, dx = \frac{1}{2}e^x \sin x - \frac{1}{2}e^x \cos x + C.$$

9.3 EXERCISES

Access end-of-section exercises online at **www.webassign.net** ENHANCED WebAssign

CHAPTER 9 REVIEW

KEY CONCEPTS

 Website www.WanerMath.com

Go to the Website at www.WanerMath.com to find a comprehensive and interactive Web-based summary of Chapter 9.

9.1 Trigonometric Functions, Models, and Regression

The **sine** of a real number *p. 485*

Plotting the graphs of functions based on $\sin x$ *p. 486*

The general sine function:

$f(x) = A\sin[\omega(x-\alpha)] + C$

A is the **amplitude**.

C is the **vertical offset** or height of the **baseline**.

ω is the **angular frequency**.

$P = 2\pi/\omega$ is the **period** or **wavelength**.

α is the **phase shift**. *p. 487*

Modeling with the general sine function *p. 488*

The **cosine** of a real number *p. 489*

Fundamental trigonometric identities:

$\sin^2 t + \cos^2 t = 1$

$\cos t = \sin(t + \pi/2)$

$\cos t = \sin(\pi/2 - t)$

$\sin t = \cos(t - \pi/2)$

$\sin t = \cos(\pi/2 - t)$ *p. 490*

The general cosine function:

$f(x) = A\cos[\omega(x-\beta)] + C$ *p. 490*

Modeling with the general cosine function *p. 491*

Other trig functions:

$\tan x = \dfrac{\sin x}{\cos x}$

$\cot x = \text{cotan}\ x = \dfrac{\cos x}{\sin x} = \dfrac{1}{\tan x}$

$\sec x = \dfrac{1}{\cos x}$

$\csc x = \text{cosec}\ x = \dfrac{1}{\sin x}$ *p. 492*

9.2 Derivatives of Trigonometric Functions and Applications

Derivatives of sine and cosine:

$\dfrac{d}{dx}\sin x = \cos x$

$\dfrac{d}{dx}\cos x = -\sin x$ *p. 494*

Some trigonometric limits:

$\lim_{h\to 0} \dfrac{\sin h}{h} = 1$

$\lim_{h\to 0} \dfrac{\cos h - 1}{h} = 0$ *p. 495*

Derivatives of sines and cosines of functions:

$\dfrac{d}{dx}\sin u = \cos u \dfrac{du}{dx}$

$\dfrac{d}{dx}\cos u = -\sin u \dfrac{du}{dx}$ *p. 496*

Derivatives of the other trigonometric functions:

$\dfrac{d}{dx}\tan x = \sec^2 x$

$\dfrac{d}{dx}\cot x = -\csc^2 x$

$\dfrac{d}{dx}\sec x = \sec x \tan x$

$\dfrac{d}{dx}\csc x = -\csc x \cot x$ *p. 498*

9.3 Integrals of Trigonometric Functions and Applications

$\int \cos x\, dx = \sin x + C$

$\int \sin x\, dx = -\cos x + C$

$\int \sec^2 x\, dx = \tan x + C$ *p. 499*

Substitution in integrals involving trig functions *p. 500*

Definite integrals involving trig functions *p. 500*

Antiderivatives of the other trigonometric functions:

$\int \tan x\, dx = -\ln|\cos x| + C$

$\int \cot x\, dx = \ln|\sin x| + C$

$\int \sec x\, dx = \ln|\sec x + \tan x| + C$

$\int \csc x\, dx = -\ln|\csc x + \cot x| + C$

p. 501

Shortcuts: Integrals of expressions involving $(ax+b)$ *p. 502*

Using integration by parts with trig functions *p. 503*

REVIEW EXERCISES

In Exercises 1–4, model the given curve with a sine function. (The scales on the two axes may not be the same.)

1.

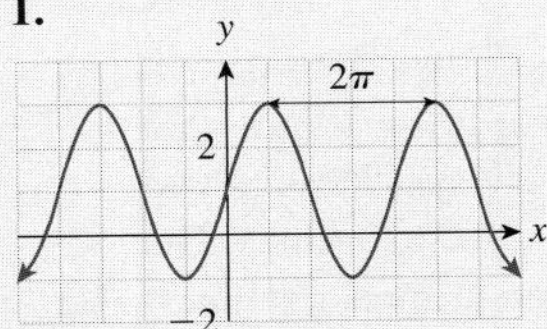

2.

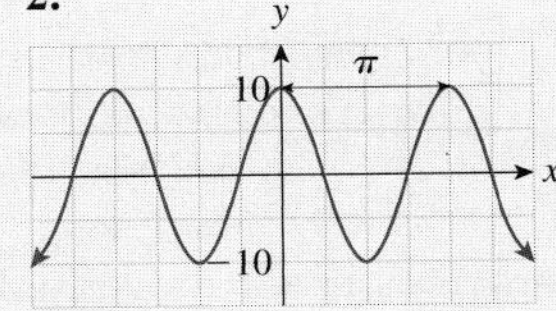

3.

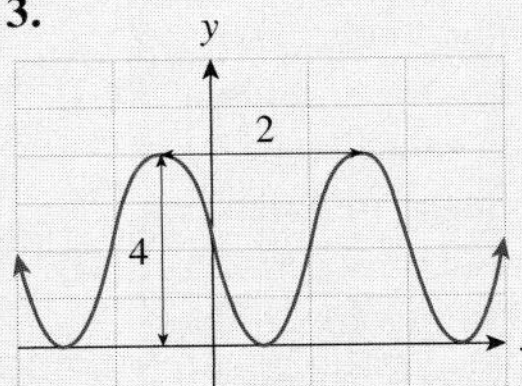

4.

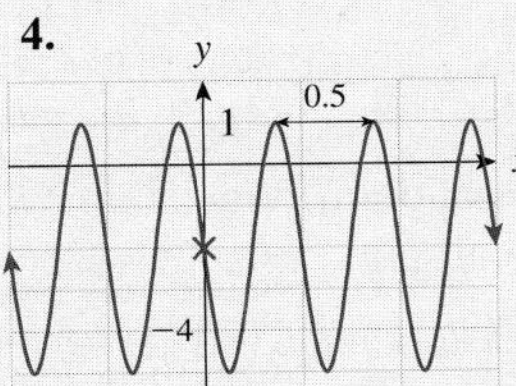

In Exercises 5–8, model the curves in Exercises 1–4 with cosine functions.

5. The curve in Exercise 1 **6.** The curve in Exercise 2

7. The curve in Exercise 3 **8.** The curve in Exercise 4

In Exercises 9–14, find the derivative of the given function.

9. $f(x) = \cos(x^2 - 1)$

10. $f(x) = \sin(x^2 + 1)\cos(x^2 - 1)$

11. $f(x) = \tan(2e^x - 1)$ **12.** $f(x) = \sec\sqrt{x^2 - x}$

13. $f(x) = \sin^2(x^2)$ **14.** $f(x) = \cos^2[1 - \sin(2x)]$

In Exercises 15–22, evaluate the given integral.

15. $\int 4\cos(2x - 1)\, dx$

16. $\int (x-1)\sin(x^2 - 2x + 1)\, dx$

17. $\int 4x \sec^2(2x^2 - 1)\,dx$

18. $\int \dfrac{\cos\left(\dfrac{1}{x}\right)}{x^2 \sin\left(\dfrac{1}{x}\right)}\,dx$

19. $\int x \tan(x^2 + 1)\,dx$

20. $\int_0^{\pi} \cos(x + \pi/2)\,dx$

21. $\int_{\ln(\pi/2)}^{\ln(\pi)} e^x \sin(e^x)\,dx$

22. $\int_{\pi}^{2\pi} \tan(x/6)\,dx$

Use integration by parts to evaluate the integrals in Exercises 23 and 24.

23. $\int x^2 \sin x\,dx$

24. $\int e^x \sin 2x\,dx$

APPLICATIONS: OHaganBooks.com

25. ***Sales*** After several years in the business, OHaganBooks.com noticed that its sales showed seasonal fluctuations, so that weekly sales oscillated in a sine wave from a low of 9,000 books per week to a high of 12,000 books per week, with the high point of the year being three quarters of the way through the year, in October. Model OHaganBooks.com's weekly sales as a generalized sine function of t, the number of weeks into the year.

26. ***Mood Swings*** The shipping personnel at OHaganBooks.com are under considerable pressure to cope with the large volume of orders, and periodic emotional outbursts are commonplace. The human resources department has been logging these outbursts over the course of several years, and has noticed a peak of 50 outbursts a week during the holiday season each December and a low point of 15 per week each June (probably attributable to the mild June weather). Model the weekly number of outbursts as a generalized cosine function of t, the number of months into the year ($t = 1$ represents January).

27. ***Precalculus for Geniuses*** The "For Geniuses" series of books has really been taking off since Duffin House first gained exclusive rights to the series 6 months ago, and revenues from *Precalculus for Geniuses* are expected to follow the curve

$$R(t) = 100{,}000 + 20{,}000e^{-0.05t} \sin\left[\frac{\pi}{6}(t - 2)\right] \text{ dollars} \quad (0 \le t \le 72),$$

where t is time in months from now and $R(t)$ is the monthly revenue. How fast, to the nearest dollar, will the revenue be changing 20 months from now?

28. ***Elvish for Dummies*** The sales department at OHaganBooks.com predicts that the revenue from sales of the latest blockbuster *Elvish for Dummies* will vary in accordance with annual releases of episodes of the movie series "Lord of the Rings Episodes 9–12." It has come up with the following model (which includes the effect of diminishing sales):

$$R(t) = 20{,}000 + 15{,}000e^{-0.12t} \cos\left[\frac{\pi}{6}(t - 4)\right] \text{ dollars} \quad (0 \le t \le 72),$$

where t is time in months from now and $R(t)$ is the monthly revenue. How fast, to the nearest dollar, will the revenue be changing 10 months from now?

29. ***Revenue*** Refer back to Question 27. Use technology or integration by parts to estimate, to the nearest \$100, the total revenue from sales of *Precalculus for Geniuses* over the next 20 months.

30. ***Revenue*** Refer back to Question 28. Use technology or integration by parts to estimate, to the nearest \$100, the total revenue from sales of *Elvish for Dummies* over the next 10 months.

31. ***Mars Missions*** Having completed his doctorate in biophysics, Billy-Sean O'Hagan will be accompanying the first manned mission to Mars. For reasons too complicated to explain (but having to do with the continuation of his doctoral research project and the timing of messages from his fiancée), during the voyage he will be consuming protein at a rate of

$$P(t) = 150 + 50 \sin\left[\frac{\pi}{2}(t - 1)\right] \text{ grams per day}$$

t days into the voyage. Find the total amount of protein he will consume as a function of time t.

32. ***Utilities*** Expenditure for utilities at OHaganBooks.com fluctuated from a high of \$9,500 in October ($t = 0$) to a low of \$8,000 in April ($t = 6$). Construct a sinusoidal model for the monthly expenditure on utilities and use your model to estimate the total annual cost.

Case Study

Predicting Airline Empty Seat Volume

Lawrence Manning/Flirt/Corbis

You are a consultant to the Department of Transportation's Special Task Force on Air Traffic Congestion and have been asked to model the volume of empty seats on U.S. airline flights, to make short-term projections of this volume, and to give a formula that estimates the accumulated volume over a specified period of time. You have data from the Bureau of Transportation Statistics showing, for each month starting January 2002, the number of available seat-miles (the total of the number of seats times the number of miles flown), and also the number of revenue passenger-miles (the total of the number of seats occupied by paying passengers times the number of miles flown), so their difference (if you ignore nonpaying passengers) measures the number of empty seat-miles. (The data can be downloaded at the Website by following Everything for Calculus → Chapter 9 Case Study.)[6]

Month	Empty Seat-Miles (billions)	Month	Empty Seat-Miles (billions)	Month	Empty Seat-Miles (billions)	Month	Empty Seat-Miles (billions)
1	25	21	23	41	20	61	23
2	21	22	22	42	16	62	20
3	18	23	21	43	15	63	17
4	21	24	21	44	18	64	17
5	21	25	26	45	21	65	18
6	18	26	23	46	21	66	13
7	19	27	20	47	19	67	14
8	19	28	19	48	20	68	15
9	25	29	21	49	22	69	21
10	24	30	16	50	19	70	19
11	24	31	16	51	17	71	19
12	21	32	19	52	17	72	21
13	26	33	22	53	17	73	23
14	22	34	21	54	14	74	21
15	22	35	22	55	14	75	17
16	22	36	23	56	18	76	18
17	19	37	24	57	21	77	18
18	16	38	22	58	20	78	15
19	15	39	18	59	19	79	15
20	17	40	20	60	20	80	16

On the graph (Figure 14) you notice two trends: A 12-month cyclical pattern, and also an overall declining trend. This overall trend is often referred to as the *secular trend* and can be seen more clearly using the 12-month moving average (Figure 15).

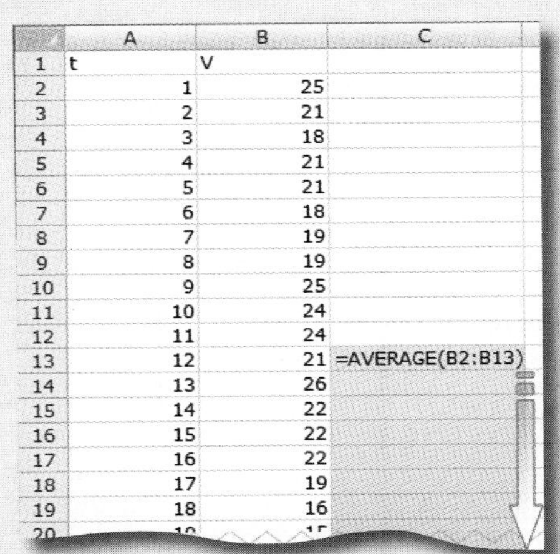

	A	B	C
1	t	V	
2	1	25	
3	2	21	
4	3	18	
5	4	21	
6	5	21	
7	6	18	
8	7	19	
9	8	19	
10	9	25	
11	10	24	
12	11	24	
13	12	21	=AVERAGE(B2:B13)
14	13	26	
15	14	22	
16	15	22	
17	16	22	
18	17	19	
19	18	16	

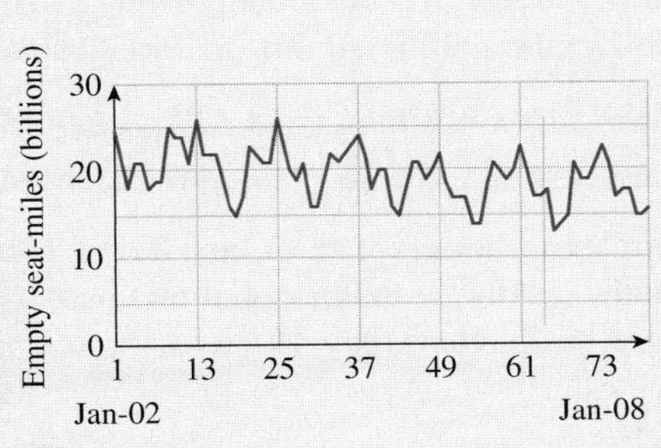

Figure 14

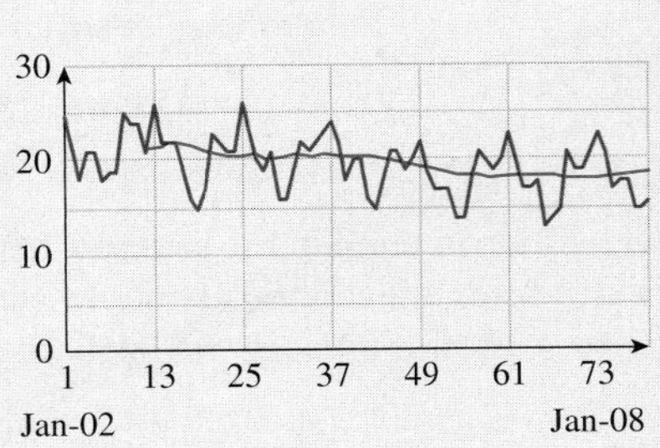

Figure 15

[6]Source: Bureau of Transportation Statistics (www.bts.gov).

You notice that the secular trend appears more-or-less linear. The simplest model of a cyclical term with 12-month period added to a linear secular trend has the form

$$V(t) = \underbrace{A\sin(\pi t/6 + d)}_{\text{Cyclical term}} + \underbrace{Bt + C}_{\text{Secular trend}}. \qquad \omega = 2\pi/12 = \pi/6$$

You are about to use Solver to construct the model when you discover that your copy of Excel has mysteriously lost its Solver, so you wonder whether there is an alternative way to construct the model. After consulting various statistics textbooks, you discover that there is: You can use the addition formula to write

$$\begin{aligned} A\sin(\pi t/6 + d) &= A[\sin(\pi t/6)\cos d + \cos(\pi t/6)\sin d] \\ &= P\sin(\pi t/6) + Q\cos(\pi t/6) \end{aligned}$$

for constants $P = A\cos d$ and $Q = A\sin d$, so instead you could use an equivalent model of the form

$$V(t) = P\sin(\pi t/6) + Q\cos(\pi t/6) + Bt + C.$$

Note that the equations $P = A\cos d$ and $Q = A\sin d$ give

$$P^2 + Q^2 = A^2\cos^2 d + A^2\sin^2 d = A^2(\cos^2 d + \sin^2 d) = A^2,$$

so $\quad A = \sqrt{P^2 + Q^2},$

giving the amplitude in terms of P and Q.

But what has all of this to do with avoiding Solver? The point is that, now, *V is a linear function of the variables* $\sin(\pi t/6)$, $\cos(\pi t/6)$, and t, meaning that you can model y using ordinary linear regression along the lines of the Case Study in Chapter 8: First you rearrange the data so that the V column is first, and add columns to calculate $\sin(\pi t/6)$ and $\cos(\pi t/6)$:

	A	B	C	D
1	V	t	sin(πt/6)	cos(πt/6)
2	25	1	=sin(PI()*B2/6)	=cos(PI()*B2/6)
3	21	2		
4	18	3		
5	21	4		
6	21	5		
7	18	6		
8	19	7		
9	19	8		
10				

↓

	A	B	C	D
1	V	t	sin(πt/6)	cos(πt/6)
2	25	1	0.5	0.866025404
3	21	2	0.866025404	0.5
4	18	3	1	6.12574E-17
5	21	4	0.866025404	-0.5
6	21	5	0.5	-0.866025404
7	18	6	1.22515E-16	-1
8	19	7	-0.5	-0.866025404
9	19	8	-0.866025404	-0.5
10				

Next, in the "Analysis" section of the Data tab, choose "Data analysis." (If this command is not available, you will need to load the Analysis ToolPak add-in.) Choose "Regression" from the list that appears, and in the resulting dialogue box enter the location of the data and where you want to put the results as shown on the left in Figure 16; identify where the dependent and independent variables are (A1–A81 for the Y range, and B1–D81 for the X range), check "Labels", and click "OK".

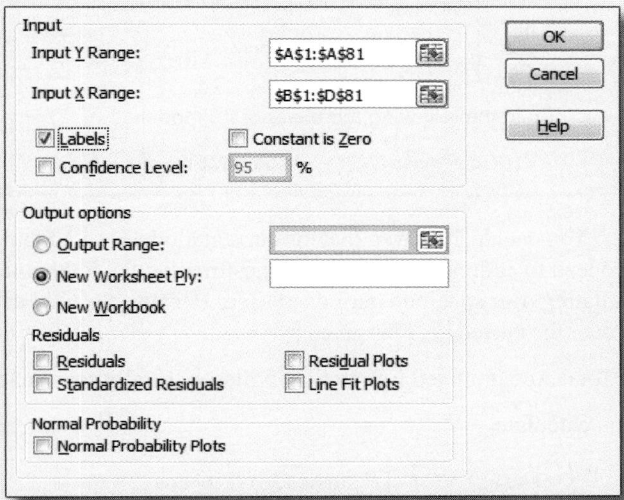

ANOVA	
	df
Regression	3
Residual	76
Total	79
	Coefficients
Intercept	21.6833363
t	-0.051345
sin(πt/6)	0.2044244
cos(πt/6)	2.87103179

Figure 16

A portion of the output is shown above on the right, with the coefficients highlighted. You use the output to write down the regression equation (with coefficients rounded to four significant digits):

$$V(t) = 0.2044 \sin(\pi t/6) + 2.871 \cos(\pi t/6) - 0.05135t + 21.68.$$

Figure 17 shows the original data with the graph of V superimposed.

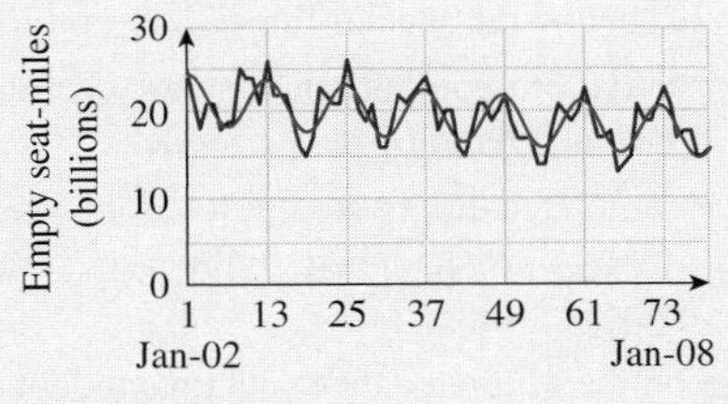

Excel Formula:

```
0.2044*sin(PI()*t/6)+2.871*cos
(PI()*t/6)-0.05135*t+21.68
```

Figure 17

Figure 18

Although the graph does not give a perfect fit, the model captures the behavior quite accurately. Figure 18 shows the 1-year projection of the model.

Q: *How would one alter the model to capture the sharp upward spike every November and the downward spike every June-July?*

A: One could add additional terms, called *seasonal variables:*

$$x_1 = \begin{cases} 1 & \text{if } t = 11, 23, 35, \ldots \\ 0 & \text{if not} \end{cases} \qquad x_1 = 1 \text{ every November}$$

$$x_2 = \begin{cases} 1 & \text{if } t = 6, 7, 18, 19, \ldots \\ 0 & \text{if not} \end{cases} \qquad x_2 = 1 \text{ every June and July}$$

to obtain the following more elaborate model:

$$V(t) = P\sin(\pi t/6) + Q\cos(\pi t/6) + Bt + Cx_1 + Dx_2 + E.$$

You decide, however, that the current model is satisfactory for your purposes, and proceed to address the second task: estimating the total volume of empty seats accumulating over specified periods of time. Because the total accumulation of a function V over the period $[a, b]$ is given by

$$\text{Total Accumulated Empty Seat-Miles} = \text{Total Accumulation of } V = \int_a^b V(t)\,dt,$$

you calculate

$$\int_a^b V(t)\,dt = \int_a^b [P\sin(\pi t/6) + Q\cos(\pi t/6) + Bt + C]\,dt$$

$$= \frac{6P}{\pi}[\cos(\pi a/6) - \cos(\pi b/6)] + \frac{6Q}{\pi}[\sin(\pi b/6) - \sin(\pi a/6)]$$

$$+ \frac{B(b^2 - a^2)}{2} + C(b - a)$$

billion empty seat-miles.

This is a formula which, on plugging in the values of P, Q, B, and C calculated above together with the values for a and b defining the period you're interested in, gives you the total accumulated empty seat-miles over that period.

EXERCISES

1. According to the regression model, the volume of empty seat-miles fluctuates by ____ billion seat-miles below the secular line to ____ billion miles above it.
2. Use the observed data to compute the actual accumulated empty seat-miles for 2007 and compare it to the value predicted by the model.
3. T Graph the accumulated empty seat-miles from January 2002 to month t as a function of t and use your graph to project when, to the nearest month, the accumulated total will pass 1,900 billion empty seat-miles.
4. Use regression on the original data to obtain a model of the form

$$V(t) = P\sin(\pi t/6) + Q\cos(\pi t/6) + Bt + Cx_1 + Dx_2 + E,$$

where $x_1 = \begin{cases} 1 & \text{if } t = 11, 23, 35, \ldots \\ 0 & \text{if not} \end{cases}$ and $x_2 = \begin{cases} 1 & \text{if } t = 6, 7, 18, 19, \ldots \\ 0 & \text{if not} \end{cases}$

as discussed in the text. Graph the resulting model together with the original data. (Round model coefficients to four significant digits.) (Use two additional columns for the independent variables, one showing the values of x_1 and the other x_2.)

5. Redo the model of the text, but using instead available seat-miles (in billions). For the data, go to the Website and follow Everything for Calculus → Chapter 9 Case Study. Note that the data are in thousands of seat-miles, so you would first need to divide by 1,000,000.

TI-83/84 Plus Technology Guide

Section 9.1

Example 5(a) (page 492) The following data show the daily spam for a 16-week period (each figure is a one-week average):

Week	0	1	2	3	4	5	6	7	8	9	10	11	12	13	14	15
Number	107	163	170	176	167	140	149	137	158	157	185	151	122	132	134	182

Use technology to find the best-fit sine curve of the form $S(t) = A \sin[\omega(t - \alpha)] + C$.

Solution with Technology The TI-83/84 Plus has a built-in sine regression utility.

1. As with the other forms of regression discussed in Chapter 2, we start by entering the coordinates of the data points in the lists L_1 and L_2, as shown below on the left:

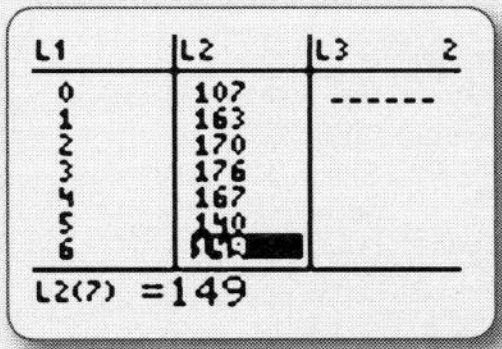

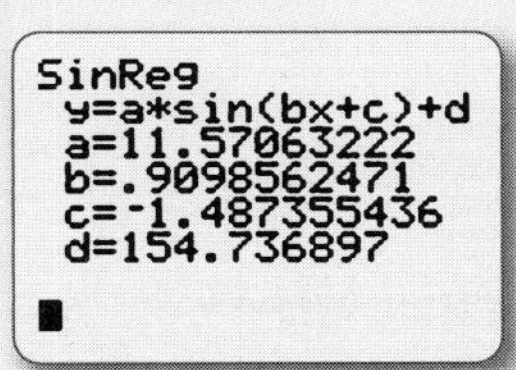

2. Press STAT, select CALC, and choose option #C: `SinReg`.
3. Pressing ENTER gives the sine regression equation in the home screen as seen above right (we have rounded the coefficients):

$$S(t) \approx 11.57 \sin(0.9099t - 1.487) + 154.7.$$

Although this is not exactly in the form we want, we can rewrite it:

$$S(t) \approx 11.57 \sin\left[0.9099\left(t - \frac{1.487}{0.9099}\right)\right] + 154.7$$
$$\approx 11.6 \sin[0.910(t - 1.63)] + 155.$$

4. To graph the points and regression line in the same window, turn Stat Plot on by pressing 2nd STAT PLOT, selecting `1`, and turning `Plot1` on, as shown below on the left:

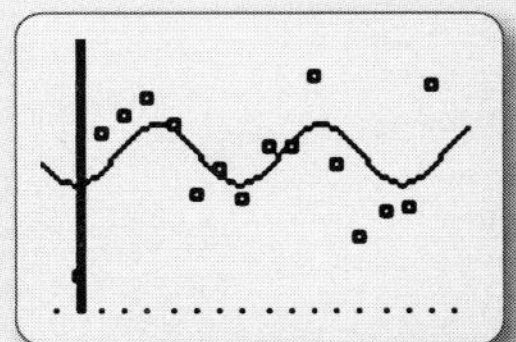

5. Enter the regression equation in the Y= screen by pressing Y=, clearing out whatever function is there, and pressing VARS 5 and selecting `EQ` option #1: `RegEq`.
6. To obtain a convenient window showing all the points and the lines, press ZOOM and choose option #9: `ZoomStat`, and you will obtain the output shown above on the right.

TECHNOLOGY GUIDE

SPREADSHEET Technology Guide

Section 9.1

Example 5(a) (page 492) The following data show the daily spam for a 16-week period (each figure is a one-week average):

Week	0	1	2	3	4	5	6	7	8	9	10	11	12	13	14	15
Number	107	163	170	176	167	140	149	137	158	157	185	151	122	132	134	182

Use technology to find the best-fit sine curve of the form $S(t) = A \sin[\omega(t - \alpha)] + C$.

Solution with Technology We set up our worksheet, shown below, as we did for logistic regression in Section 2.4.

1. For our initial guesses, let us roughly estimate the parameters from the graph. The amplitude is around $A = 30$, and the vertical offset is roughly $C = 150$. The period seems to be around 7 weeks, so let us choose $P = 7$. This gives $\omega = 2\pi/P \approx 0.9$. Finally, let us take $\alpha = 0$ to begin with.

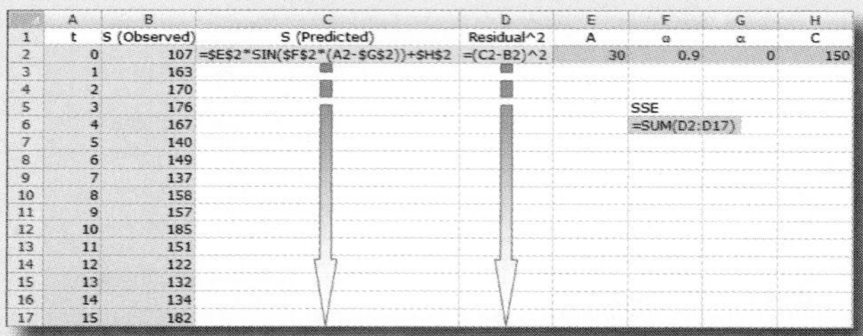

	A	B	C	D	E	F	G	H
1	t	S (Observed)	S (Predicted)	Residual^2	A	ω	α	C
2	0	107	=$E$2*SIN($F$2*(A2-$G$2))+$H$2	=(C2-B2)^2	30	0.9	0	150
3	1	163						
4	2	170						
5	3	176				SSE		
6	4	167				=SUM(D2:D17)		
7	5	140						
8	6	149						
9	7	137						
10	8	158						
11	9	157						
12	10	185						
13	11	151						
14	12	122						
15	13	132						
16	14	134						
17	15	182						

2. We then use Solver to minimize SSE by changing cells E2 through H2 as in Section 2.4 (see the setup below on the left). We obtain the following model (we have rounded the coefficients to three decimal places):

$$S(t) = 25.8 \sin[0.96(t - 1.22)] + 153$$

with SSE $\approx$ 2,148. Plotting the observed and predicted values of S gives the graph shown on the right:

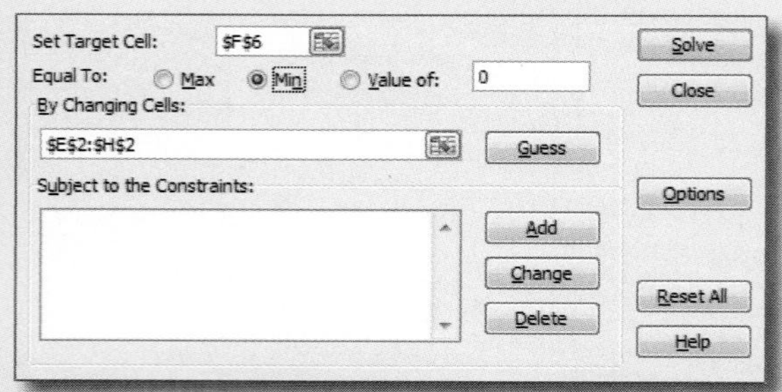

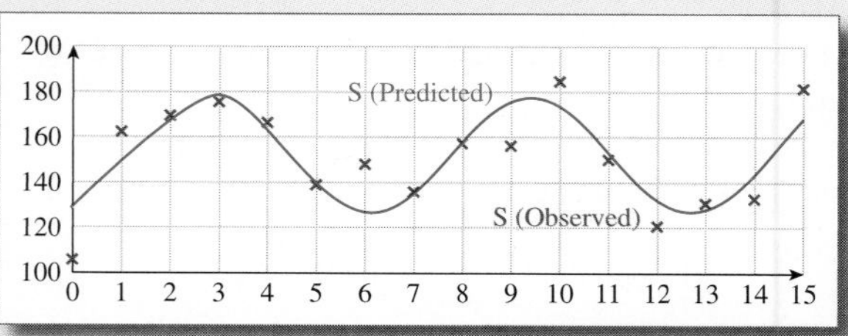

Note that Solver estimated the period for us. However, in many situations, we know what to use for the period beforehand: For example, we expect sales of snow shovels to fluctuate according to an annual cycle. Thus, if we were using regression to fit a regression sine or cosine curve to snow shovel sales data, we would set $P = 1$ year, compute ω, and have Solver estimate only the remaining coefficients: A, C, and α.

Answers to Chapter Review Exercises

Chapter 1

Review

1. a. 1 **b.** −2 **c.** 0 **d.** −1 **3. a.** 1 **b.** 0 **c.** 0 **d.** −1

5.

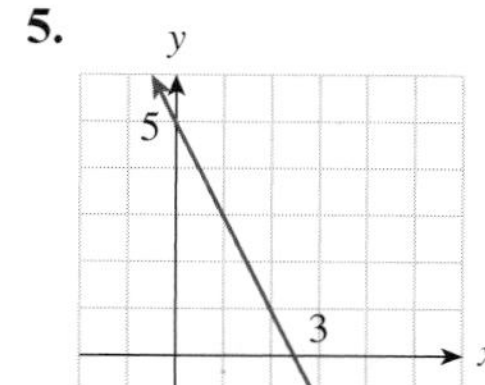

7.

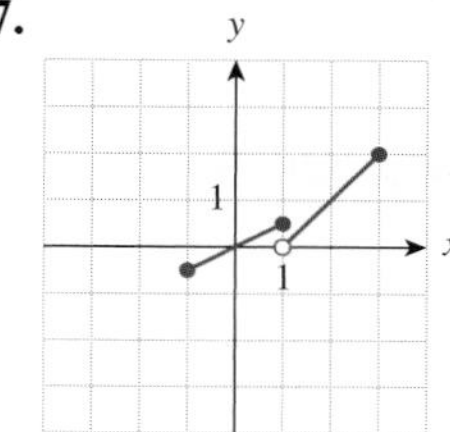

9. Absolute value **11.** Linear **13.** Quadratic
15. $y = -3x + 11$ **17.** $y = 1.25x - 4.25$
19. $y = (1/2)x + 3/2$ **21.** $y = 4x - 12$
23. $y = -x/4 + 1$ **25.** $y = -0.214x + 1.14$, $r \approx -0.33$
27. a. Exponential. Graph:

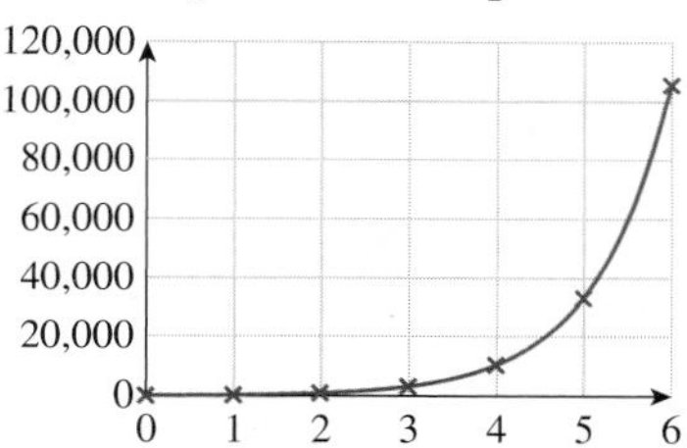

b. The ratios (rounded to 1 decimal place) are:

V(1)/V(0)	V(2)/V(1)	V(3)/V(2)	V(4)/V(3)	V(5)/V(4)	V(6)/V(5)
3	3.3	3.3	3.2	3.2	3.2

They are close to 3.2. **c.** About 343,700 visits/day
29. a. 2.3; 3.5; 6 **b.** For Web site traffic of up to 50,000 visits per day, the number of crashes is increasing by 0.03 per additional thousand visits. **c.** 140,000 **31. a.** (A) **b.** (A) Leveling off (B) Rising (C) Rising; begins to fall after 7 months (D) Rising **33. a.** The number of visits would increase by 30 per day. **b.** No; it would increase at a slower and slower rate and then begin to decrease. **c.** Probably not. This model predicts that Web site popularity will start to decrease as advertising increases beyond \$8,500 per month, and then drop toward zero. **35. a.** $v = 0.05c + 1{,}800$ **b.** 2,150 new visits per day **c.** \$14,000 per month
37. $d = 0.95w + 8$; 86 kg **39. a.** Cost: $C = 5.5x + 500$; Revenue: $R = 9.5x$; Profit $P = 4x - 500$ **b.** More than 125 albums per week **c.** More than 200 albums per week
41. a. $q = -80p + 1{,}060$ **b.** 100 albums per week **c.** \$9.50, for a weekly profit of \$700
43. a. $q = -74p + 1{,}015.5$ **b.** 239 albums per week

Chapter 2

Review

1.

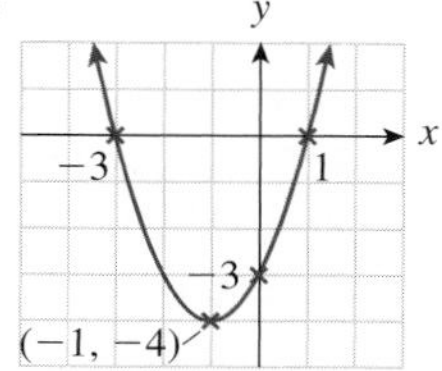

3. f: $f(x) = 5(1/2)^x$, or $5(2^{-x})$

5.

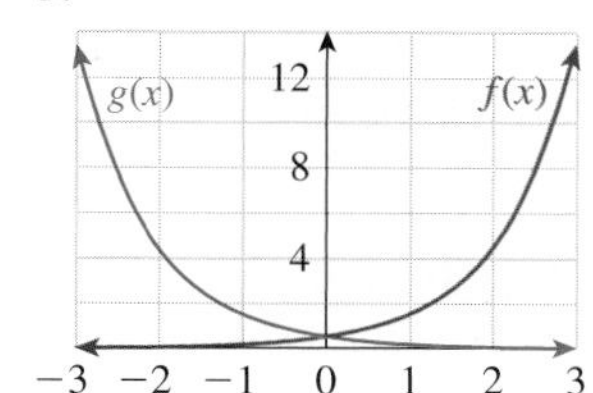

7.

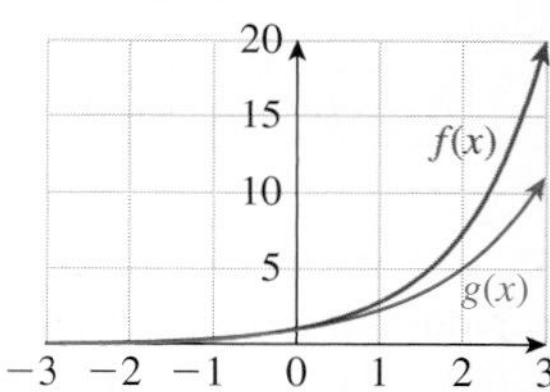

9. \$3,484.85 **11.** \$3,705.48 **13.** \$3,485.50
15. $f(x) = 4.5(9^x)$ **17.** $f(x) = \frac{2}{3}3^x$ **19.** $-\frac{1}{2}\log_3 4$
21. $\frac{1}{3}\log 1.05$

23.

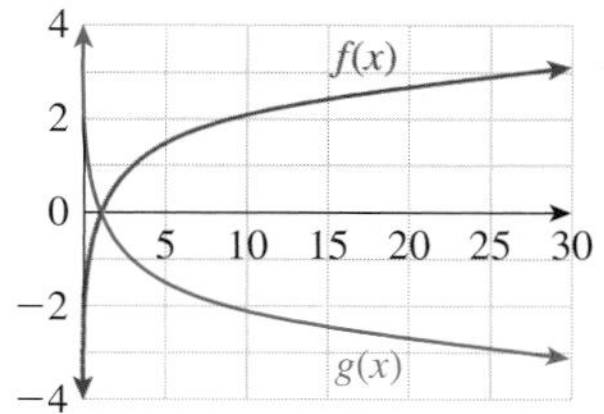

25. $Q = 5e^{-0.00693t}$ **27.** $Q = 2.5e^{0.347t}$
29. 10.2 years **31.** 10.8 years
33. $f(x) = \dfrac{900}{1 + 8(1.5)^{-x}}$
35. $f(x) = \dfrac{20}{1 + 3(0.8)^{-x}}$ **37. a.** \$8,500 per month; an average of approximately 2,100 hits per day

b. \$29,049 per month **c.** The fact that −0.000005, the coefficient of c^2, is negative.
39. a. $R = -60p^2 + 950p$; $p =$ \$7.92 per novel, Monthly revenue = \$3,760.42
b. $P = -60p^2 + 1{,}190p - 4{,}700$; $p =$ \$9.92 per novel, Monthly profit = \$1,200.42 **41. a.** 9.1, 19
b. About 310,000 pounds **43.** 2016
45. 1.12 million pounds **47.** $n(t) = 9.6(0.80^t)$ million pounds of lobster **49.** (C)

Chapter 3

Review

1. 5 **3.** Does not exist **5. a.** −1 **b.** 3 **c.** Does not exist **7.** −4/5 **9.** −1 **11.** −1 **13.** Does not exist
15. 10/7 **17.** Does not exist **19.** $+\infty$ **21.** 0
23. Diverges to $-\infty$ **25.** 0 **27.** 2/5 **29.** 1

31.

h	1	0.01	0.001
Ave. Rate of Change	−0.5	−0.9901	−0.9990

Slope ≈ −1

33.

h	1	0.01	0.001
Avg. Rate of Change	6.3891	2.0201	2.0020

Slope ≈ 2
35. a. P **b.** Q **c.** R **d.** S **37. a.** Q **b.** None **c.** None **d.** None **39. a.** (B) **b.** (B) **c.** (B) **d.** (A) **e.** (C) **41.** $2x + 1$ **43.** $2/x^2$

45.

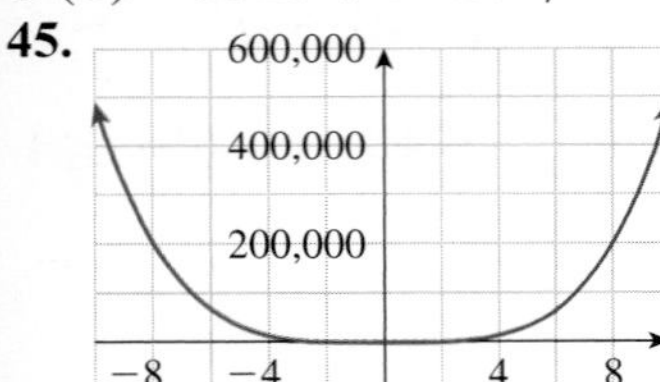

47.

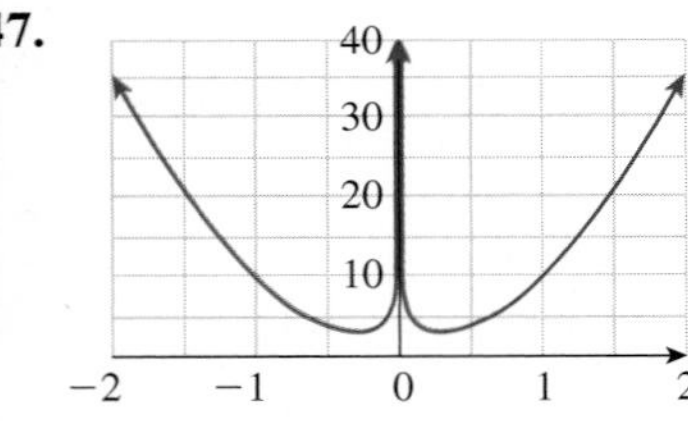

49. a. $P(3) = 25$: O'Hagan purchased the stock at \$25. $\lim_{t\to 3^-} P(t) = 25$: The value of the stock had been approaching \$25 up to the time he bought it. $\lim_{t\to 3^+} P(t) = 10$: The value of the stock dropped to \$10 immediately after he bought it. **b.** Continuous but not differentiable. Interpretation: the stock price changed continuously but suddenly reversed direction (and started to go up) the instant O'Hagan sold it.
51. a. $\lim_{t\to 3} p(t) \approx 40$; $\lim_{t\to +\infty} p(t) = +\infty$. Close to 2007 ($t = 3$), the home price index was about 40. In the long term, the home price index will rise without bound. **b.** 10 (The slope of the linear portion of the curve is 10.) In the long term, the home price index will rise about 10 points per year.
53. a. 500 books per week **b.** [3, 4], [4, 5] **c.** [3, 5]; 650 books per week **55. a.** 3 percentage points per year **b.** 0 percentage points per year **c.** (D)
57. a. $72t + 250$ **b.** 322 books per week
c. 754 books per week.

Chapter 4

Review

1. $50x^4 + 2x^3 - 1$ **3.** $9x^2 + x^{-2/3}$ **5.** $1 - 2/x^3$
7. $-\dfrac{4}{3x^2} + \dfrac{0.2}{x^{1.1}} + \dfrac{1.1x^{0.1}}{3.2}$ **9.** $e^x(x^2 + 2x - 1)$
11. $(-3x|x| + |x|/x - 6x)/(3x^2 + 1)^2$
13. $-4(4x - 1)^{-2}$ **15.** $20x(x^2 - 1)^9$
17. $-0.43(x + 1)^{-1.1}[2 + (x + 1)^{-0.1}]^{3.3}$
19. $e^x(x^2 + 1)^9(x^2 + 20x + 1)$
21. $3^x[(x - 1)\ln 3 - 1]/(x - 1)^2$ **23.** $2xe^{x^2-1}$
25. $2x/(x^2 - 1)$ **27.** $x = 7/6$ **29.** $x = \pm 2$
31. $x = (1 - \ln 2)/2$ **33.** None **35.** $\dfrac{2x - 1}{2y}$
37. $-y/x$ **39.** $\dfrac{(2x - 1)^4(3x + 4)}{(x + 1)(3x - 1)^3} \times \left[\dfrac{8}{2x - 1} + \dfrac{3}{3x + 4} - \dfrac{1}{x + 1} - \dfrac{9}{3x - 1}\right]$
41. $y = -x/4 + 1/2$ **43.** $y = -3ex - 2e$
45. $y = x + 2$ **47. a.** 274 books per week **b.** 636 books per week **c.** The function w begins to decrease more and more rapidly after $t = 14$ Graph:

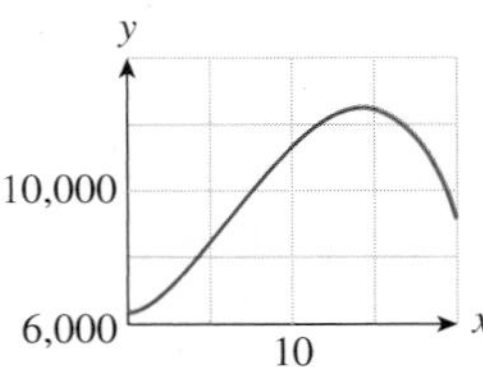

d. Because the data suggest an upward curving parabola, the long-term prediction of sales for a quadratic model would be that sales will increase without bound, in sharp contrast to (c) **49. a.** \$2.88 per book **b.** \$3.715 per book **c.** Approximately −\$0.000104 per book, per additional book sold. **d.** At a sales level of 8,000 books per week, the cost is increasing at a rate of \$2.88 per book (so that the 8,001st book costs approximately \$2.88 to sell), and it costs an average of \$3.715 per book to sell the first 8,000 books.

Moreover, the average cost is decreasing at a rate of \$0.000104 per book, per additional book sold.
51. a. \$3,000 per week (rising) **b.** 300 books per week
53. $R = pq$ gives $R' = p'q + pq'$. Thus, $R'/R = R'/(pq) = (p'q + pq')/pq = p'/p + q'/q$
55. \$110 per year **57. a.** $s'(t) = \dfrac{2{,}475e^{-0.55(t-4.8)}}{(1 + e^{-0.55(t-4.8)})^2}$; 556 books per week **b.** 0; In the long term, the rate of increase of weekly sales slows to zero. **59.** 616.8 hits per day per week. **61. a.** −17.24 copies per \$1. The demand for the gift edition of *The Complete Larry Potter* is dropping at a rate of about 17.24 copies per \$1 increase in the price. **b.** \$138 per dollar is positive, so the price should be raised.

Chapter 5

Review

1. Relative max: (−1, 5), absolute min: (−2, −3) and (1, −3) **3.** Absolute max: (−1, 5), absolute min: (1, −3) **5.** Absolute min: (1, 0) **7.** Absolute min: (−2, −1/4) **9.** Relative max at $x = 1$, point of inflection at $x = -1$ **11.** Relative max at $x = -2$, relative min at $x = 1$, point of inflection at $x = -1$
13. One point of inflection, at $x = 0$
15. a. $a = 4/t^4 - 2/t^3$ m/sec^2 **b.** 2 m/sec^2
17. Relative max: (−2, 16); absolute min: (2, −16); point of inflection: (0, 0); no horizontal or vertical asymptotes

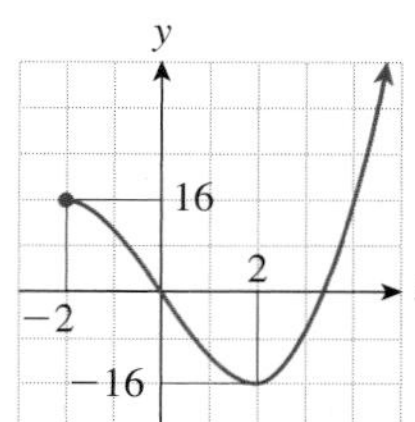

19. Relative min: (−3, −2/9); relative max: (3, 2/9); inflection: $(-3\sqrt{2}, -5\sqrt{2}/36)$, $(3\sqrt{2}, 5\sqrt{2}/36)$; vertical asymptote: $x = 0$; horizontal asymptote: $y = 0$

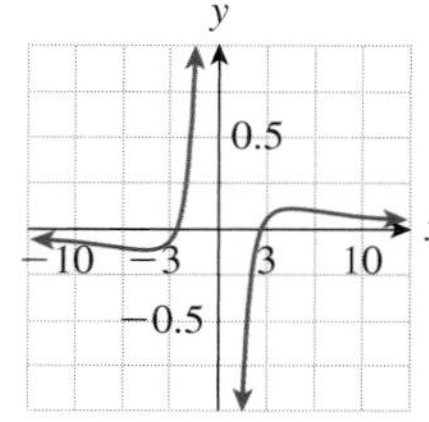

21. Relative max at (0, 0), absolute min at (1, −2), no asymptotes

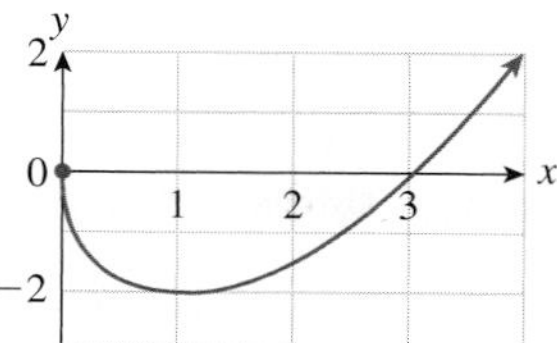

23. \$22.14 per book **25. a.** Profit $= -p^3 + 42p^2 - 288p - 181$ **b.** \$24 per copy; \$3,275 **c.** For maximum revenue, the company should charge \$22.14 per copy. At this price, the cost per book is decreasing with increasing price, while the revenue is not decreasing (its derivative is zero). Thus, the profit is increasing with increasing price, suggesting that the maximum profit will occur at a higher price.
27. 12 in × 12 in × 6 in, for a volume of 864 in^3
29. a. $E = \dfrac{2p^2 - 33p}{-p^2 + 33p + 9}$ **b.** 0.52, 2.03; when the price is \$20, demand is dropping at a rate of 0.52% per 1% increase in the price; when the price is \$25, demand is dropping at a rate of 2.03% per 1% increase in the price. **c.** \$22.14 per book **31. a.** $E = 2p^2 - p$ **b.** 6; the demand is dropping at a rate of 6% per 1% increase in the price. **c.** \$1.00, for a monthly revenue of \$1,000 **33. a.** Week 5 **b.** Point of inflection on the graph of s; maximum on the graph of s', t-intercept in the graph of s''. **c.** 10,500; if weekly sales continue as predicted by the model, they will level off at around 10,500 books per week in the long term. **d.** 0; if weekly sales continue as predicted by the model, the rate of change of sales approaches zero in the long term. **35. a.–d.** $10/\sqrt{2}$ ft/sec

Chapter 6

Review

1. $x^3/3 - 5x^2 + 2x + C$ **3.** $4x^3/15 + 4/(5x) + C$
5. $(1/2)\ln|2x| + C$ or $(1/2)\ln|x| + C$
7. $-e^{-2x+11}/2 + C$ **9.** $(x^2 + 1)^{2.3}/4.6 + C$
11. $2\ln|x^2 - 7| + C$ **13.** $\frac{1}{6}(x^4 - 4x + 1)^{3/2} + C$
15. $-e^{x^2/2} + C$ **17.** $(x + 2) - \ln|x + 2| + C$ or $x - \ln|x + 2| + C$ **19.** 1 **21.** 2.75
23. −0.24 **25.** 0.7778, 0.7500, 0.7471
27. 0 **29.** 1 **31.** 2/7 **33.** $50(e^{-1} - e^{-2})$
35. 52/9 **37.** $[\ln 5 - \ln 2]/8 = \ln(2.5)/8$
39. 32/3 **41.** $(1 - e^{-25})/2$
43. a. $N(t) = 196t + t^2/2 - 0.16t^6/6$ **b.** 4 books
45. a. At a height of $-16t^2 + 100t$ ft **b.** 156.25 ft **c.** After 6.25 seconds **47.** 25,000 copies. **49.** 0 books **51.** 39,200 hits **53.** \$8,200 **55.** About 86,000 books **57.** About 35,800 books

Chapter 7

Review

1. $(x^2 - 2x + 4)e^x + C$ **3.** $(1/3)x^3 \ln 2x - x^3/9 + C$
5. $\frac{1}{2}x(2x+1)|2x+1| - \frac{1}{12}(2x+1)^2|2x+1| + C$
7. $-5x|-x+3| - \frac{5}{2}(-x+3)|-x+3| + C$
9. $-e^2 - 39/e^2$ **11.** $\frac{3}{2 \cdot 2^{1/3}} - \frac{1}{2}$ **13.** $\frac{2\sqrt{2}}{3}$ **15.** -1
17. $e - 2$ **19.** $3x - 2$ **21.** $\frac{3}{14}[x^{7/3} - (x-2)^{7/3}]$
23. \$1,600 **25.** \$2,500 **27.** 1/4 **29.** Diverges **31.** 1
33. $y = -\frac{3}{x^3 + C}$ **35.** $y = \sqrt{2 \ln |x| + 1}$
37. \$18,200 **39. a.** \$2,260
b. $50{,}000e^{0.01t}(1 - e^{-0.04}) \approx 1{,}960.53e^{0.01t}$
41. a. $\bar{p} = 20, \bar{q} = 40{,}000$ **b.** $CS = \$240{,}000$, $PS = \$30{,}000$ **43.** Approximately \$910,000
45. a. \$5,549,000 **b.** Principal: \$5,280,000, interest: \$269,000 **47.** \$51 million **49.** The amount in the account would be given by $y = 10{,}000/(1 - t)$, where t is time in years, so would approach infinity 1 year after the deposit.

Chapter 8

Review

1. 0; 14/3; 1/2; $\frac{1}{1+z} + z^3$; $\frac{x+h}{y+k+(x+h)(z+l)} + (x+h)^2(y+k)$ **3.** Decreases by 0.32 units; increases by 12.5 units **5.** Reading left to right, starting at the top: 4, 0, 0, 3, 0, 1, 2, 0, 2 **7.** Answers may vary; two examples are $f(x, y) = 3(x - y)/2$ and $f(x, y) = 3(x - y)^3/8$.
9.

z = x + y

11.

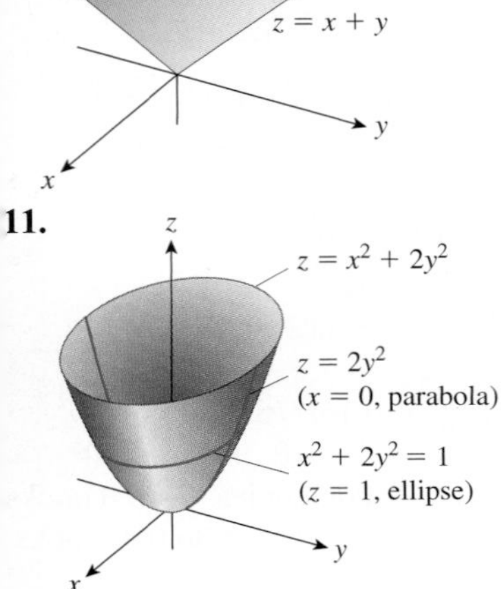

13.

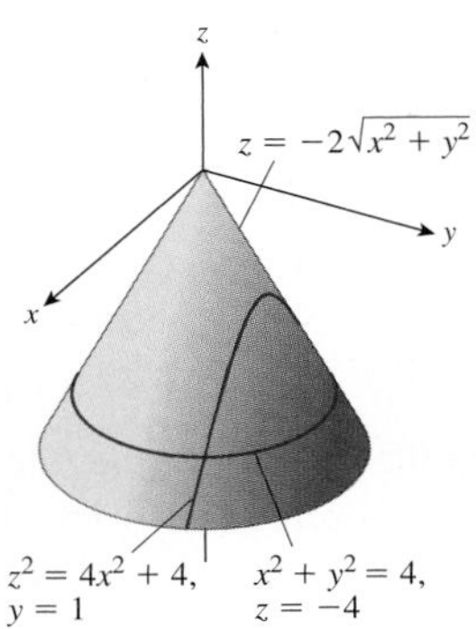

15. $f_x = 2x + y, f_y = x, f_{yy} = 0$ **17.** 0
19. $\frac{\partial f}{\partial x} = \frac{-x^2 + y^2 + z^2}{(x^2 + y^2 + z^2)^2}, \frac{\partial f}{\partial y} = \frac{2xy}{(x^2 + y^2 + z^2)^2}$, $\frac{\partial f}{\partial z} = -\frac{2xz}{(x^2 + y^2 + z^2)^2}, \left.\frac{\partial f}{\partial x}\right|_{(0,1,0)} = 1$
21. Absolute minimum at $(1, 3/2)$ **23.** Maximum at $(0, 0)$, saddle points at $(\pm\sqrt{2}, 1)$ **25.** Saddle point at $(0, 0)$ **27.** 1/27 at $(1/3, 1/3, 1/3)$ **29.** $(0, 2, \sqrt{2})$ **31.** 4; $(\sqrt{2}, \sqrt{2})$ and $(-\sqrt{2}, -\sqrt{2})$ **33.** $(0, 2, \sqrt{2})$ **35.** 2 **37.** ln 5 **39.** 1 **41. a.** $h(x, y) = 5{,}000 - 0.8x - 0.6y$ hits per day (x = number of new customers at JungleBooks.com, y = number of new customers at FarmerBooks.com) **b.** 250 **c.** $h(x, y, z) = 5{,}000 - 0.8x - 0.6y + 0.0001z$ (z = number of new Internet shoppers) **d.** 1.4 million **43. a.** 2,320 hits per day **b.** $0.08 + 0.00003x$ hits (daily) per dollar spent on television advertising per month; increases with increasing x **c.** \$4,000 per month **45.** (A) **47. a.** About 15,800 additional orders per day **b.** 11 **49.** \$23,050

Chapter 9

Review

1. $f(x) = 1 + 2 \sin x$
3. $f(x) = 2 + 2 \sin[\pi(x - 1)] = 2 + 2 \sin[\pi(x + 1)]$
5. $f(x) = 1 + 2 \cos(x - \pi/2)$
7. $f(x) = 2 + 2 \cos[\pi(x + 1/2)] = 2 + 2 \cos[\pi(x - 3/2)]$ **9.** $-2x \sin(x^2 - 1)$
11. $2e^x \sec^2(2e^x - 1)$ **13.** $4x \sin(x^2) \cos(x^2)$
15. $2 \sin(2x - 1) + C$ **17.** $\tan(2x^2 - 1) + C$
19. $-\frac{1}{2} \ln|(\cos(x^2 + 1)| + C$ **21.** 1
23. $-x^2 \cos x + 2x \sin x + 2 \cos x + C$
25. $s(t) = 10{,}500 + 1{,}500 \sin[(2\pi/52)t - \pi] = 10{,}500 + 1{,}500 \sin(0.12083t - 3.14159)$
27. Decreasing at a rate of \$3,852 per month
29. \$2,029,700
31. $150t - \frac{100}{\pi} \cos\left[\frac{\pi}{2}(t - 1)\right]$ grams

Index

Index

Index of Applications

Business and Economics

General Interest

SURVEYING YACHTS AND SMALL CRAFT

PAUL STEVENS

ADLARD COLES NAUTICAL
LONDON

Published by Adlard Coles Nautical
an imprint of Bloomsbury Publishing Plc
50 Bedford Square, London WC1B 3DP
www.adlardcoles.com

First edition published 2010
This reprint published 2018

ISBN 978-1-4081-1403-2

A CIP catalogue record for this book is available from the British Library.

This book is produced using paper that is made from wood grown in managed, sustainable forests. It is natural, renewable and recyclable. The logging and manufacturing processes conform to the environmental regulations of the country of origin.

Typeset in Helvetica Neue 9 pt on 12 pt.

Note: while all reasonable care has been taken in the publication of this book, the publisher takes no responsibility for the use of the methods or products described in the book